Lecture Notes in Artificial Intelligence 5803

Edited by R. Goebel, J. Siekmann, and W. Wahlster

Subseries of Lecture Notes in Computer Science

Bärbel Mertsching Marcus Hund
Zaheer Aziz (Eds.)

KI 2009: Advances in Artificial Intelligence

32nd Annual German Conference on AI
Paderborn, Germany, September 15-18, 2009
Proceedings

Series Editors

Randy Goebel, University of Alberta, Edmonton, Canada
Jörg Siekmann, University of Saarland, Saarbrücken, Germany
Wolfgang Wahlster, DFKI and University of Saarland, Saarbrücken, Germany

Volume Editors

Bärbel Mertsching
Marcus Hund
Zaheer Aziz
University of Paderborn, GET Lab
Pohlweg 47-49, 33098 Paderborn, Germany
E-mail: {mertsching, hund, aziz}@get.uni-paderborn.de

Library of Congress Control Number: 2009934858

CR Subject Classification (1998): I.2, I.2.6, F.1.1, I.5.1, H.5.2

LNCS Sublibrary: SL 7 – Artificial Intelligence

ISSN 0302-9743
ISBN-10 3-642-04616-9 Springer Berlin Heidelberg New York
ISBN-13 978-3-642-04616-2 Springer Berlin Heidelberg New York

springer.com

Printed in Germany

Typesetting: Camera-ready by author, data conversion by Scientific Publishing Services, Chennai, India
Printed on acid-free paper SPIN: 12764280 06/3180 5 4 3 2 1 0

Preface

The 32nd Annual German Conference on Artificial Intelligence, KI 2009 (KI being the German acronym for AI), was held at the University of Paderborn, Germany on September 15–18, 2009, continuing a series of successful events. Starting back in 1975 as a national meeting, the conference now gathers researchers and developers from academic fields and industries worldwide to share their research results covering all aspects of artificial intelligence. This year we received submissions from 23 countries and 4 continents. Besides the international orientation, we made a major effort to include as many branches of AI as possible under the roof of the KI conference. A total of 21 area chairs representing different communities within the field of AI selected further members of the program committee and helped the local organizers to acquire papers. The new approach appealed to the AI community: we had 126 submissions, which constituted an increase of more than 50%, and which resulted in 14 parallel sessions on the following topics

agents and intelligent virtual environments
AI and engineering
automated reasoning
cognition
evolutionary computation Robotics
experience and knowledge management
history and philosophical foundations
knowledge representation and reasoning
machine learning and mining
natural language processing
planning and scheduling
spatial and temporal reasoning
vision and perception

offering cutting edge presentations and discussions with leading experts. Thirty-one percent of the contributions came from outside German-speaking countries.

This volume contains the papers and the posters selected for presentation after a thorough review process (three reviews per paper). Sixty-one percent of the submissions could be accepted for oral presentation. The best papers selected for the Springer Best Paper Award were honored in a special ceremony during the conference dinner.

As in previous years, priority was given to workshops and tutorials, for which an entire day was reserved. To our great pleasure 12 workshops and 5 tutorials could be offered, which constituted a new record of active (and passive) participation at the KI conference. These are covered in a separate volume. The conference website http://ki2009.upb.de provides information and references to their contents.

While the parallel sessions of the main conference and the workshops encouraged the exchange of recent research results and the discussion of trends in the field, 11 invited talks addressed all participants of the conference and helped to bridge the gaps between the different branches of AI research.

In addition to the scientific program of the conference, three further layers were addressed:

- AI and automation
- AI in the current excellence clusters, collaborative research centers and graduate schools, and
- AI at the universities of applied sciences.

This year's major theme was the impact of AI methods and theories on the engineering domain. Nowadays, AI can be found everywhere in this field, in the production process as well as in smart products, although very often the influence of AI research remains hidden. The special focus on AI and automation presented state-of-the-art topics in the main conference as well as in the accompanying exhibition and helped to close the gap between academia and industrial applications.

We are proud that we could host the first AI mashup challenge at a KI conference, enabling all participants of the conference to experience new forms of interactive user participation and vote for the best mashups, a new breed of web-based data integration application that is sprouting up all across the Internet. Further information may be found by following the link: http://ki09.de/.

Last but not least, the conference offered ample opportunity to talk to colleagues and friends. Coffee breaks and an interesting social program provided lots of opportunities for networking and made a conference a community event.

Such a big conference is not possible without the help of many people. First of all we would like to thank all colleagues who submitted contributions in the form of papers, workshops, tutorials or other input. Additionally, we thank all those who helped to organize KI 2009 and who are listed on the following pages. The conference management software ConfTool Pro proved very helpful, and its author, Harald Weinreich, gave valuable support. As the conference benefited so much from their generosity, we wish to thank our partners and sponsors. Furthermore, we are especially grateful to our local colleagues, who did a tremendous job in making the conference possible.

August 2009

Bärbel Mertsching
Marcus Hund
Muhammad Zaheer Aziz

Organization

General Chair

Bärbel Mertsching — University of Paderborn

Workshop Chair

Klaus-Dieter Althoff — University of Hildesheim

Tutorial Chair

Joachim Baumeister — University of Würzburg

Publicity Chair

Andreas Dengel — DFKI, Kaiserslautern

Exhibition and Industrial Relations Chair

Muhammad Zaheer Aziz — University of Paderborn

Local Chair

Marcus Hund — University of Paderborn

Administrative Assistance

Christine Harms — ccHa, Sankt Augustin

Local Arrangements

Irtiza Ali
Dirk Fischer
Astrid Haar
Zubair Kamran
Ludmilla Kleinmann
Tobias Kotthäuser
Frank Schmidtmeier
Mohamed Salah Shafik

Area Chairs

AI and Automation	Hans Kleine Büning University of Paderborn
AI at the Universities of Applied Sciences	Volker Lohweg Ostwestfalen-Lippe University of Applied Sciences
AI in Clusters	Helge Ritter Bielefeld University
Agents & Intelligent Virtual Environments	Ipke Wachsmuth Bielefeld University
AI and Economics	Torsten Eymann University of Bayreuth
AI and Engineering	Hans Kleine Büning University of Paderborn
Automated Reasoning	Ulrich Furbach University of Koblenz
Cognition	Kerstin Schill University of Bremen
Distributed AI	Ingo Timm University of Frankfurt
Evolutionary Computation	Günter Rudolph Technical University of Dortmund
Experience and Knowledge Management	Ralph Bergmann University of Trier
Games and Interactive Entertainment	Christoph Schlieder University of Bamberg
History and Philosophical Foundations	Ruth Hagengruber University of Paderborn
Human-Machine-Interaction	Gerhard Rigoll Technical University of Munich
Intelligent VR/AR	Didier Stricker DFKI, Kaiserslautern
Knowledge Representation & Reasoning	Steffen Hölldobler Technical University of Dresden

Machine Learning & Mining	Stefan Wrobel Fraunhofer IAIS, St. Augustin
Natural Language Processing	Hans Uszkoreit DFKI Saarbrücken
Ontologies & Semantic Web	Rudi Studer / Pascal Hitzler University of Karlsruhe
Planning & Scheduling	Susanne Biundo-Stephan University of Ulm
Robotics	Wolfram Burgard University of Freiburg
Spatial & Temporal Reasoning	Sabine Timpf University of Augsburg
Vision & Perception	Bärbel Mertsching University of Paderborn

Program Committee

Klaus-Dieter Althoff	University of Hildesheim
Elisabeth André	University of Augsburg
Pedro Barahona	Universidade Nova de Lisboa
Itzhak Benenson	Tel-Aviv University
Maren Bennewitz	University of Freiburg
Bettina Berendt	Katholieke Universiteit Leuven
Ralph Bergmann	University of Trier
Hans-Georg Beyer	FH Vorarlberg
Thomas Bittner	New York State University at Buffalo
Susanne Biundo-Stephan	University of Ulm
Daniel Borrajo	Universidad Carlos III de Madrid
Jürgen Branke	University of Warwick
Thorsten Brants	Google Inc., Mountain View
Guido Brunnett	Chemnitz University of Technology
Wolfram Burgard	University of Freiburg
Maria Paola Bonacina	Università degli Studi di Verona
Marc Cavazza	University of Teesside
Amedeo Cesta	ISTC-CNR Rome
Herbert Dawid	Universität Bielefeld
Jim Delgrande	Simon Fraser Univ. Vancouver
Andreas Dengel	DFKI, Kaiserslautern
Rüdiger Dillmann	University of Karlsruhe
Stefan Edelkamp	University of Bremen
Michael Emmerich	LIACS, Leiden University
Martin Ester	Simon Fraser University
Nick Evans	EURECOM
Torsten Eymann	University of Bayreuth

Maria Fasli	University of Essex
Gernot Fink	TU Dortmund
Luciano Floridi	University of Hertfordshire
Enrico Franconi	Free University Bozen-Bolzano
Andrew Frank	Technical University of Vienna
Christian Freksa	University of Bremen
Udo Frese	University of Bremen
Ulrich Furbach	University of Koblenz
Johannes Fürnkranz	TU Darmstadt
Andreas Gerber	X-aitment GmbH, Saarbrücken
Jürgen Giesl	RWTH Aachen
Asunción Gómez Pérez	Universidad Politécnica de Madrid
Horst-Michael Gross	TU Ilmenau
Barbara M. Grüter	Hochschule Bremen University of Applied Sciences
Andreas Günter	University of Hamburg
Claudio Gutierrez	Universidad de Chile, Santiago
Ruth Hagengruber	University of Paderborn
Malte Helmert	University of Freiburg
Joachim Hertzberg	University of Osnabrück
Otthein Herzog	TZI Bremen
Pascal Hitzler	University of Karlsruhe
Eyke Hüllermeyer	Philipps-Universität Marburg
Alejandro Jaimes	Telefonica Research
Thorsten Joachims	Cornell University
Bernhard Jung	TU Bergakademie Freiberg
Michael Kipp	DFKI Saarbrücken
Hans Kleine Büning	University of Paderborn
Franziska Klügl	Örebro University
Stefan Kirn	University of Hohenheim
Koichi Kise	Osaka Prefecture University
Stanislav Klimenko	Moscow Institute of Physics and Technology
Philipp Koehn	University of Edinburgh
Antonio Krüger	University of Münster
Werner Kuhn	University of Münster
Kai-Uwe Kühnberger	University of Osnabrück
Volker Lohweg	Ostwestfalen-Lippe University of Applied Sciences
Peter Loos	DFKI Saarbrücken
Alfredo Martins	LSA - Autonomous Systems Laboratory
Maic Masuch	University of Duisburg-Essen
Peter McBurney	University of Liverpool
Bärbel Mertsching	University of Paderborn
Katharina Morik	TU Dortmund
Wolfgang Nejdl	University of Hanover
Frank Neumann	MPI für Informatik

Invited Talks

Franz Baader	Technical University of Dresden
Oliver Brock	Technische Universität Berlin
Gordon Cheng	ATR Kyoto, Japan
Frank van Harmelen	Vrije Universiteit Amsterdam
Marc Erich Latoschik	University of Bayreuth
Klaus Mainzer	Technical University of Munich
Dana Nau	University of Maryland
Han La Poutré	Centrum Wiskunde & Informatica, Amsterdam
Ulrich Reimer	FHS St. Gallen
Tobias Scheffer	University of Postsdam
Wolfgang Wahlster	DFKI, Saarbrücken

AI 2009 Mashup Challenge

Brigitte Endres-Niggemeyer	University of Applied Sciences, Hanover
Valentin Zacharias	FZI Forschungszentrum Informatik
Pascal Hitzler	Kno.e.sis Center, Wright State University, Dayton, Ohio

Sponsors

A3plus GmbH
Center of Excellence Cognitive Interaction Technology (CITEC)
Deutsches Forschungszentrum für Künstliche Intelligenz GmbH (DFKI)
Hamburger Informatik Technologie-Center e. V. (HITeC)
Institut Industrial IT (inIT)
Linguatec Sprachtechnologien GmbH
O'Reilly Media, Inc.
ScienceDirect
Springer
Semantic Technology Institute (STI) Germany
t3n magazin
University of Paderborn

Partners

advanced system engineering center (asec)
Gesellschaft für Informatik e.V. (GI)
AG Industrielle Bildverarbeitung OWL
InnoZent OWL
OstWestfallenLippe: Leadership durch intelligente Systeme

Table of Contents

Planning and Scheduling

Solving Fully-Observable Non-deterministic Planning Problems via Translation into a General Game 1
Peter Kissmann and Stefan Edelkamp

Planning with h^+ in Theory and Practice 9
Christoph Betz and Malte Helmert

A Framework for Interactive Hybrid Planning 17
Bernd Schattenberg, Julien Bidot, Sascha Geßler, and Susanne Biundo

A Memory-Efficient Search Strategy for Multiobjective Shortest Path Problems 25
L. Mandow and J.L. Pérez de la Cruz

Perfect Hashing for State Spaces in BDD Representation 33
Martin Dietzfelbinger and Stefan Edelkamp

On the Benefit of Fusing DL-Reasoning with HTN-Planning 41
Ronny Hartanto and Joachim Hertzberg

Flexible Timeline-Based Plan Verification 49
A. Cesta, A. Finzi, S. Fratini, A. Orlandini, and E. Tronci

Solving Non-deterministic Planning Problems with Pattern Database Heuristics 57
Pascal Bercher and Robert Mattmüller

An Exploitative Monte-Carlo Poker Agent 65
Immanuel Schweizer, Kamill Panitzek, Sang-Hyeun Park, and Johannes Fürnkranz

Vision and Perception

Interaction of Control and Knowledge in a Structural Recognition System 73
Eckart Michaelsen, Michael Arens, and Leo Doktorski

Attention Speeds Up Visual Information Processing: Selection for Perception or Selection for Action? 81
Katharina Weiß and Ingrid Scharlau

Real-Time Scan-Line Segment Based Stereo Vision for the Estimation of Biologically Motivated Classifier Cells 89
M. Salah E.-N. Shafik and Bärbel Mertsching

Occlusion as a Monocular Depth Cue Derived from Illusory Contour Perception 97
Marcus Hund and Bärbel Mertsching

Fast Hand Detection Using Posture Invariant Constraints 106
Nils Petersen and Didier Stricker

A Novel and Efficient Method to Extract Features and Vector Creation in Iris Recognition System 114
Amir Azizi and Hamid Reza Pourreza

What You See is What You Set – The Position of Moving Objects 123
Heinz-Werner Priess and Ingrid Scharlau

Parameter Evolution: A Design Pattern for Active Vision 128
Michael Müller, Dennis Senftleben, and Josef Pauli

Machine Learning and Data Mining

Clustering Objects from Multiple Collections 136
Vera Hollink, Maarten van Someren, and Viktor de Boer

Generalized Clustering via Kernel Embeddings 144
Stefanie Jegelka, Arthur Gretton, Bernhard Schölkopf, Bharath K. Sriperumbudur, and Ulrike von Luxburg

Context-Based Clustering of Image Search Results 153
Hongqi Wang, Olana Missura, Thomas Gärtner, and Stefan Wrobel

Variational Bayes for Generic Topic Models 161
Gregor Heinrich and Michael Goesele

Evolutionary Computation

Surrogate Constraint Functions for CMA Evolution Strategies 169
Oliver Kramer, André Barthelmes, and Günter Rudolph

Rake Selection: A Novel Evolutionary Multi-Objective Optimization Algorithm 177
Oliver Kramer and Patrick Koch

A Comparison of Neighbourhood Topologies for Staff Scheduling with Particle Swarm Optimisation 185
Maik Günther and Volker Nissen

Controlling a Four Degree of Freedom Arm in 3D Using the XCSF Learning Classifier System 193
Patrick O. Stalph, Martin V. Butz, and Gerulf K.M. Pedersen

An Evolutionary Graph Transformation System as a Modelling Framework for Evolutionary Algorithms 201
Hauke Tönnies

Natural Language Processing

Semi-automatic Creation of Resources for Spoken Dialog Systems 209
Tatjana Scheffler, Roland Roller, and Norbert Reithinger

Correlating Natural Language Parser Performance with Statistical Measures of the Text 217
Yi Zhang and Rui Wang

Comparing Two Approaches for the Recognition of Temporal Expressions 225
Oleksandr Kolomiyets and Marie-Francine Moens

Meta-level Information Extraction 233
Peter Kluegl, Martin Atzmueller, and Frank Puppe

Robust Processing of Situated Spoken Dialogue 241
Pierre Lison and Geert-Jan M. Kruijff

iDocument: Using Ontologies for Extracting and Annotating Information from Unstructured Text 249
Benjamin Adrian, Jörn Hees, Ludger van Elst, and Andreas Dengel

Behaviorally Flexible Spatial Communication: Robotic Demonstrations of a Neurodynamic Framework 257
John Lipinski, Yulia Sandamirskaya, and Gregor Schöner

SceneMaker: Automatic Visualisation of Screenplays 265
Eva Hanser, Paul Mc Kevitt, Tom Lunney, and Joan Condell

Knowledge Representation and Reasoning

A Conceptual Agent Model Based on a Uniform Approach to Various Belief Operations 273
Christoph Beierle and Gabriele Kern-Isberner

External Sources of Axioms in Automated Theorem Proving 281
Martin Suda, Geoff Sutcliffe, Patrick Wischnewski, Manuel Lamotte-Schubert, and Gerard de Melo

Presenting Proofs with Adapted Granularity 289
Marvin Schiller and Christoph Benzmüller

On Defaults in Action Theories 298
Hannes Strass and Michael Thielscher

Analogy, Paralogy and Reverse Analogy: Postulates and Inferences 306
Henri Prade and Gilles Richard

Cognition

Early Clustering Approach towards Modeling of Bottom-Up Visual Attention 315
Muhammad Zaheer Aziz and Bärbel Mertsching

A Formal Cognitive Model of Mathematical Metaphors 323
Markus Guhe, Alan Smaill, and Alison Pease

Hierarchical Clustering of Sensorimotor Features 331
Konrad Gadzicki

P300 Detection Based on Feature Extraction in On-line Brain-Computer Interface 339
Nikolay Chumerin, Nikolay V. Manyakov, Adrien Combaz, Johan A.K. Suykens, Refet Firat Yazicioglu, Tom Torfs, Patrick Merken, Herc P. Neves, Chris Van Hoof, and Marc M. Van Hulle

Human Perception Based Counterfeit Detection for Automated Teller Machines 347
Taswar Iqbal, Volker Lohweg, Dinh Khoi Le, and Michael Nolte

History and Philosophical Foundations

Variations of the Turing Test in the Age of Internet and Virtual Reality (Extended Abstract) 355
Florentin Neumann, Andrea Reichenberger, and Martin Ziegler

A Structuralistic Approach to Ontologies 363
Christian Schäufler, Stefan Artmann, and Clemens Beckstein

AI Viewed as a "Science of the Culture" 371
Jean-Gabriel Ganascia

Beyond Public Announcement Logic: An Alternative Approach to Some AI Puzzles 379
Paweł Garbacz, Piotr Kulicki, Marek Lechniak, and Robert Trypuz

Behavioural Congruence in Turing Test-Like Human-Computer Interaction 387
Stefan Artmann

AI and Engineering

Machine Learning Techniques for Selforganizing Combustion Control ... 395
Erik Schaffernicht, Volker Stephan, Klaus Debes, and Horst-Michael Gross

Constraint-Based Integration of Plan Tracking and Prognosis for Autonomous Production 403
Paul Maier, Martin Sachenbacher, Thomas Rühr, and Lukas Kuhn

Fault Detection in Discrete Event Based Distributed Systems by Forecasting Message Sequences with Neural Networks 411
Falk Langer, Dirk Eilers, and Rudi Knorr

Fuzzy Numerical Schemes for Hyperbolic Differential Equations 419
Michael Breuss and Dominik Dietrich

Model-Based Test Prioritizing – A Comparative Soft-Computing Approach and Case Studies 427
Fevzi Belli, Mubariz Eminov, and Nida Gokce

Automated Reasoning

Comparing Unification Algorithms in First-Order Theorem Proving 435
Kryštof Hoder and Andrei Voronkov

Atomic Metadeduction 444
Serge Autexier and Dominik Dietrich

Spatial and Temporal Reasoning

Toward Heterogeneous Cardinal Direction Calculus 452
Yohei Kurata and Hui Shi

The Scared Robot: Motivations in a Simulated Robot Arm 460
Martin V. Butz and Gerulf K.M. Pedersen

Right-of-Way Rules as Use Case for Integrating GOLOG and Qualitative Reasoning 468
Florian Pommerening, Stefan Wölfl, and Matthias Westphal

Assessing the Strength of Structural Changes in Cooccurrence Graphs 476
Matthias Steinbrecher and Rudolf Kruse

Maximum a Posteriori Estimation of Dynamically Changing Distributions 484
Michael Volkhardt, Sören Kalesse, Steffen Müller, and Horst-Michael Gross

Agents and Intelligent Virtual Environments

Kinesthetic Bootstrapping: Teaching Motor Skills to Humanoid Robots through Physical Interaction 492
Heni Ben Amor, Erik Berger, David Vogt, and Bernhard Jung

To See and to be Seen in the Virtual Beer Garden - A Gaze Behavior System for Intelligent Virtual Agents in a 3D Environment 500
Michael Wissner, Nikolaus Bee, Julian Kienberger, and Elisabeth André

Requirements and Building Blocks for Sociable Embodied Agents 508
Stefan Kopp, Kirsten Bergmann, Hendrik Buschmeier, and Amir Sadeghipour

Modeling Peripersonal Action Space for Virtual Humans by Learning a Tactile Body Schema 516
Nhung Nguyen and Ipke Wachsmuth

Hybrid Control for Embodied Agents Applications 524
Jan Miksatko and Michael Kipp

Towards System Optimum: Finding Optimal Routing Strategies in Time-Dependent Networks for Large-Scale Evacuation Problems 532
Gregor Lämmel and Gunnar Flötteröd

Formalizing Joint Attention in Cooperative Interaction with a Virtual Human 540
Nadine Pfeiffer-Leßmann and Ipke Wachsmuth

Towards Determining Cooperation Based on Multiple Criteria 548
Markus Eberling

Experience and Knowledge Management

The SEASALT Architecture and Its Realization within the docQuery Project 556
Meike Reichle, Kerstin Bach, and Klaus-Dieter Althoff

Case Retrieval in Ontology-Based CBR Systems 564
Amjad Abou Assali, Dominique Lenne, and Bruno Debray

Behaviour Monitoring and Interpretation: A Computational Approach to Ethology 572
Björn Gottfried

Automatic Recognition and Interpretation of Pen- and Paper-Based Document Annotations 581
Marcus Liwicki, Markus Weber, and Andreas Dengel

Robotics

Self-emerging Action Gestalts for Task Segmentation 589
Michael Pardowitz, Jan Steffen, and Helge Ritter

Prediction and Classification of Motion Trajectories Using Spatio-Temporal NMF 597
Julian P. Eggert, Sven Hellbach, Alexander Kolarow, Edgar Körner, and Horst-Michael Gross

A Manifold Representation as Common Basis for Action Production and Recognition 607
Jan Steffen, Michael Pardowitz, and Helge Ritter

Posters

An Argumentation-Based Approach to Handling Inconsistencies in DL-Lite 615
Xiaowang Zhang and Zuoquan Lin

Thresholding for Segmentation and Extraction of Extensive Objects on Digital Images 623
Vladimir Volkov

Agent-Based Pedestrian Simulation of Train Evacuation Integrating Environmental Data 631
Franziska Klügl, Georg Klubertanz, and Guido Rindsfüser

An Intelligent Fuzzy Agent for Spatial Reasoning in GIS 639
Rouzbeh Shad, Mohammad Saadi Mesgari, Hamid Ebadi, Abbas Alimohammadi, Aliakbar Abkar, and Alireza Vafaeenezhad

Learning Parametrised RoboCup Rescue Agent Behaviour Using an Evolutionary Algorithm 648
Michael Kruse, Michael Baumann, Tobias Knieper, Christoph Seipel, Lial Khaluf, Nico Lehmann, Alex Lermontow, Christian Messinger, Simon Richter, Thomas Schmidt, and Daniel Swars

Heuristics for Resolution in Propositional Logic 656
Manfred Kerber

Context-Aware Service Discovery Using Case-Based Reasoning Methods 664
Markus Weber, Thomas Roth-Berghofer, Volker Hudlet, Heiko Maus, and Andreas Dengel

Forward Chaining Algorithm for Solving the Shortest Path Problem in Arbitrary Deterministic Environment in Linear Time - Applied for the Tower of Hanoi Problem 672
Silvana Petruseva

Early Top-Down Influences in Control of Attention: Evidence from the Attentional Blink 680
Frederic Hilkenmeier, Jan Tünnermann, and Ingrid Scharlau

Probabilistic Models for the Verification of Human-Computer Interaction 687
Bernhard Beckert and Markus Wagner

HieroMate: A Graphical Tool for Specification and Verification of Hierarchical Hybrid Automata 695
Ammar Mohammed and Christian Schwarz

Building Geospatial Data Collections with Location-Based Games 703
Sebastian Matyas, Peter Wullinger, and Christian Matyas

Design Principles for Embodied Interaction: The Case of Ubiquitous Computing 711
Rainer Malaka and Robert Porzel

Multidisciplinary Design of Air-Launched Space Launch Vehicle Using Simulated Annealing 719
Amer Farhan Rafique, He LinShu, Qasim Zeeshan, and Ali Kamran

Stochastic Feature Selection in Support Vector Machine Based Instrument Recognition 727
Oliver Kramer and Tobias Hein

Author Index 735

Solving Fully-Observable Non-deterministic Planning Problems via Translation into a General Game

Peter Kissmann[1] and Stefan Edelkamp[2]

[1] Fakultät für Informatik, TU Dortmund, Germany
[2] Technologie-Zentrum Informatik, Universität Bremen, Germany

Abstract. In this paper, we propose a symbolic planner based on BDDs, which calculates strong and strong cyclic plans for a given non-deterministic input. The efficiency of the planning approach is based on a translation of the non-deterministic planning problems into a two-player turn-taking game, with a set of actions selected by the solver and a set of actions taken by the environment.

The formalism we use is a PDDL-like planning domain definition language that has been derived to parse and instantiate general games. This conversion allows to derive a concise description of planning domains with a minimized state vector, thereby exploiting existing static analysis tools for deterministic planning.

1 Introduction

Non-deterministic planning has been characterized as planning in an *oily world* [5], where the outcome of actions is uncertain. Uncertainty can be inherent to the unpredictable existence of nature, due to abstractions in modeling the real-world, or caused by other agents sharing the same environment and acting adversarial to the solver's choice to reach the goals. There is a tight connection between fully-observable non-determistic and adversarial planning [13]. A plan in both cases is a mapping from physical states to applicable actions. Uncertainty may be induced by the underlying dynamics.

Moreover, non-deterministic planning and two-player game playing have much in common as in the former the environment plays the role of the opponent. As a consequence, our planner's specification format refers to general game playing [16] inputs.

In this paper, we bridge the gap between general game playing, non-deterministic and adversarial planning by applying a PDDL-like input, which enables us to use existing static analysis tools for planning. The distinctive advantage we aim at is, therefore, the efficient state encoding by using multi-variate variable encodings [12], as automatically inferred by many current deterministic planners. As a consequence, we compile each non-deterministic action into two, one representing the actor's desired move, and one for the environment's response.

As non-deterministic and adversarial plans are expected to be rather large, a compact description with binary decision diagrams (BDDs) is promising [7]. BDDs [4] encode state sets space efficiently, exploiting the sharing of state vectors in a decision diagram with respect to a fixed ordering of the state variables. In the following, we thus discuss the design of the BDD-based planner GAMER, while solving non-deterministic and adversarial planning problems optimally, i. e. while computing strong and strong-cyclic plans that reflect optimal policies. Its output are optimal playing strategies in an accepted format.

B. Mertsching, M. Hund, and Z. Aziz (Eds.): KI 2009, LNAI 5803, pp. 1–8, 2009.

2 Non-deterministic Planning

Non-deterministic planning with full observability computes a conditional plan, which achieves the goal for a non-deterministic planning problem $\mathcal{P} = (\mathcal{S}, \mathcal{I}, \mathcal{A}, \mathcal{T}, \mathcal{G})$ with a finite state space (set of states) $\mathcal{S}$, an initial state $\mathcal{I} \in \mathcal{S}$, a set of goal states $\mathcal{G} \subseteq \mathcal{S}$, sets of applicable actions $\mathcal{A}(s)$ for each $s \in \mathcal{S} \setminus \mathcal{G}$, and a non-deterministic transition relation $\mathcal{T}(s, a, s) \in \mathcal{S} \times \mathcal{A} \times \mathcal{S}$.

Solutions are policies (partial functions) mapping states into actions. Let $\pi : \mathcal{S} \rightarrow \bigcup_{s \in \mathcal{S} \setminus \mathcal{G}} \mathcal{A}(s)$ be a policy, $\mathcal{S}_\pi$ the domain of definition of π, and $\mathcal{S}_\pi(s)$ the set of states reachable from s using π, then we say that π is closed with respect to s iff $\mathcal{S}_\pi(s) \subseteq \mathcal{S}_\pi$; π is proper with respect to s iff a goal state can be reached using π from all $s \in \mathcal{S}_\pi(\mathrm{s})$; π is acyclic with respect to s iff there is no trajectory $s = (s_0, \ldots, s_n)$ with i and j such that $0 \leq i < j \leq n$ and $s_i = s_j$. We also say that π is closed (resp. proper or acyclic) with respect to $\mathcal{S}' \subseteq \mathcal{S}$ if it is closed (resp. proper or acyclic) with respect to all $s \in \mathcal{S}'$.

A policy π is a *valid solution* for the non-deterministic model iff π is closed and proper with respect to the initial state $\mathcal{I}$. A valid policy π is assigned a (worst-case scenario) cost V_π equal to the longest trajectory starting at $\mathcal{I}$ and ending at a goal state. For acyclic policies with respect to $\mathcal{I}$ we have $V_\pi < +\infty$.

A policy π is *optimal* if it is a valid solution of minimum V_π value. Plans in non-deterministic domains that admit acyclic solutions have finite optimal cost, while in domains with cyclic solutions, they are often judged by the time taken to generate it.

Valid policies can be computed using the algorithms of Cimatti *et al.* [7]. These are defined in terms of state-action tables that map states to their according actions. The ultimate output is a state-action table in form of a BDD representing a plan π.

For a planning problem $\mathcal{P}$, a plan π is called *weak*, if for $\mathcal{I}$ a terminal state in $\mathcal{G}$ is reachable; *strong*, if the induced execution structure is acyclic and all its terminal states are contained in $\mathcal{G}$; and *strong cyclic*, if from every state in the plan a terminal state is reachable and every terminal state in the induced execution structure is contained in $\mathcal{G}$.

The intuition for weak plans is that the goal can be reached, but not necessarily for all possible paths, due to non-determinism. For strong plans, the goal has to be satisfied despite all non-determinism, and for strong cyclic plans all execution paths at least have the chance to reach the goal. We have

$$\mathit{WeakPreImage}(S') = \{s \in \mathcal{S} \mid \exists s' \in S', a \in \mathcal{A} : \mathcal{T}(s, a, s')\}$$

as the set of all states s that can reach a state in s' by performing action a; and

$$\mathit{StrongPreImage}(S') = \{s \in \mathcal{S} \mid \forall a \in \mathcal{A} : \emptyset \neq \mathit{Result}(s, a) \subseteq S'\}$$

as the set of all states s from which a is applicable and application of that reaches s', where $\mathit{Result}(s, a) = \{s' \in \mathcal{S} \mid \mathcal{T}(s, a, s')\}$.

While strong planning grows an initially empty plan, strong cyclic planning truncates an initially universal plan. For the strong planning algorithm, the state-action table is extended by the state-action pairs calculated by the strong pre-image of all states in π in each step. The algorithm terminates if no further change in the state-action table can be observed. An alternative stopping condition is that the initial state has been reached.

Procedure *Adversarial-Plan*(P)
  reach $\leftarrow$ *reachable*(); *front* $\leftarrow$ *winPlanner* $\leftarrow$ *loseEnv* $\leftarrow$ *reach* $\wedge \mathcal{G}$
  repeat
    to $\leftarrow$ *movePlanner* $\wedge$ *WeakPreImage*(*front*)
    winPlanner $\leftarrow$ *winPlanner* $\vee$ (*to* $\wedge$ *reach*)
    front $\leftarrow$ $\neg$*movePlanner* $\wedge$ *reach* $\wedge$ $\neg$*loseEnv* $\wedge$ *StrongPreImage*(*winPlanner*)
    loseEnv $\leftarrow$ *loseEnv* $\vee$ *front*
  until (*front* $= \bot$)

Fig. 1. Adapted Two-Player Zero-Sum Algorithm for Adversarial Planning

The strong cyclic planning algorithm of [7] starts with the universal state-action table. Iteratively, state-action pairs whose actions lead to states outside the table are pruned, followed by those that are not reachable anymore. Once a fixpoint is reached, pairs that do not provide any progress toward the goal are removed from the table.

3 Adversarial Planning

As indicated above, adversarial planning is closely related to two-player games, but is also similar to non-deterministic planning. We have two players: one (also called protagonist) is the planner and thus plays an active role; the other, its opponent (also called antagonist), is the environment and we cannot influence its moves. The planner's aim is to reach a goal for certain, i. e. it has to find a plan which will ensure it reaching the goal, no matter what the environment does. Thus, it has to find a strong plan. Extensions to strong cyclic adversarial planning are discussed by Jensen *et al.* [13].

Winning sets, i. e., sets containing all the states from which a specific player surely reaches its goal, are of fundamental importance in model checking. It has been shown that model checking with property specification in the μ-calculus is equivalent to solving a parity game, with the maximal priority roughly corresponding to the alternation depth of the μ-formula ϕ [8]. For each play there is a unique partitioning of the parity game in two winning sets, as shown for example by Emerson and Jutla [11]. Another example is an on-the-fly algorithm for solving games based on timed game automata with respect to reachability and safety properties. The algorithm of Cassez *et al.* [6] is a symbolic extension of the on-the-fly algorithm [15] for linear-time model-checking of finite-state systems. There is also a link to CTL model checking [17]. It has been observed that searching for a strong plan for goal ϕ can be casted as satisfying the CTL goal $\mathbf{AF}\,\neg\phi$. A strong cyclic plan corresponds to satisfying the CTL formula $\mathbf{AGEF}\,\neg\phi$.

To solve the adversarial planning problem, we thus start by calculating the planner's winning set (cf. Fig. 1). The algorithm we use resembles the symbolic two-player zero-sum algorithm proposed by Edelkamp [9]. Here, we calculate the winning set for the planner only, whereas the original version could calculate both players' winning sets. It is sufficient to omit the environment's winning set, as no draw states are defined and we are only interested in the states where the planner surely wins.

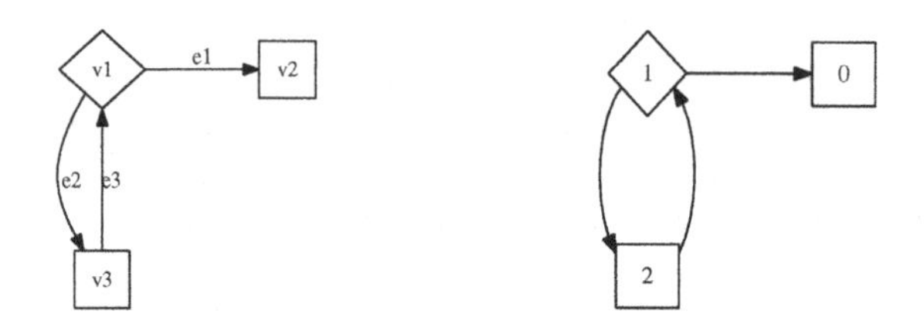

Fig. 2. Example Game Graph (left) and Backward BFS layers(right)

After calculating the set of reachable states *reach*, the algorithm starts at the goal states $\mathcal{G}$ and repeats performing a double step, until it cannot find any new clearly won states. Each double step works as follows. In the first step, the predecessor states, where the planner could choose an action (denoted by *movePlanner*), are calculated by applying the weak pre-image. In the second step, all the states won by the planner found so far (*winPlanner*) are taken and the set of predecessor states, where the environment had to choose an action, is calculated using the strong pre-image and added to the set of the environment's lost states *loseEnv*. Thus, this double step returns all the states where the environment might choose any action, but the planner can react by taking an action that will lead to the states won by it.

Note, that the calculation of the reachable states is optional. If it is omitted, in which case *reach* is set to $\top$, additionally several unreachable states might be created in the backward search. In our experiments, this enlargement of the winning sets had no effect on the runtime – this is not surprising, as we handle state-sets and the number of pre-images should remain the same. Thus, the amount of work needed for computing the reachable states might be saved, as it provides no advantage in the backward search.

A shortcoming of this algorithm is that it calculates only the winning sets, not the plans (i. e. the policies). The problem is that given only a winning set, one cannot necessarily calculate an optimal policy. Consider the graph given in Fig. 2. Let the initial state be denoted $v1$ and the goal state $v2$ and let the planner have control in $v1$ and the environment in $v2$ and $v3$. Then all states would be calculated as being won by the planner – thus, the winning set is $\{v1, v2, v3\}$. But, if the policy would assign the taking of action $e2$ when in $v1$, the planner cannot ever reach the goal and thus would not win.

To prevent this, it is sufficient to store the calculated winning sets in layers, according to the backward BFS layer they were found in (as shown in Fig. 2). Bakera *et al.* [1] have shown that always taking an action leading to a smaller backward BFS layer (i. e. to a state closer to a goal) will result in an optimal policy.

4 Transformation

In previous work [10], we have proposed an algorithm to classify general two-player turn-taking games, i. e. an algorithm to find the reward the players will get for each state, in case of optimal play. For this, we started with uninstantiated input from the game description language [16] and came up with an intermediate format quite close to instantiated PDDL, which we call GDDL [14], but additionally to normal actions also incorporating what we called multi-actions. These multi-actions consist of a global precondition *glob* and a number of precondition/effect pairs $p_1, e_1, \ldots, p_n, e_n$. Such a multi-action is interpreted as *multi-action* $:=$ *glob* $\wedge (p_1 \Leftrightarrow e_1) \wedge \ldots \wedge (p_n \Leftrightarrow e_n)$.

Due to the closeness of specifications for two-player games, non-deterministic and adversarial planning problems, we decided to use the same intermediate format. Thus, we can use the same BDD algorithms for all sources.

Recall that the formal definition of the non-deterministic version of PDDL has been enriched with an additional statement of the form (*oneof* e_1 e_2 $\ldots$ e_n), where each e_i is a PDDL effect. The semantics is that, when executing such effect, exactly one of the e_i, $i \in \{1, \ldots, n\}$, is chosen and applied to the current state.

Usually, PDDL domain input is schematic (parameterized), i. e. the finite number of domain objects for substituting parameters in predicates and actions are specified in the problem specific file. A PDDL domain in which all predicates are atoms and all actions have zero parameters is called grounded.There are many existing tools that infer a grounded PDDL representation given a schematic one. Some of them additionally provide a partitioning of atoms into mutually exclusive fact groups, which enable a concise state encoding to multi-variate variables. The inference of a minimal state encoding [12] is essential for the effectiveness of BDD-based search methods. Moreover, such state encoding leads to improved heuristics and is also referred to as SAS^+ planning.

We wrote a compiler that parses a non-deterministic domain and returns the PDDL description of a two-player turn-taking game, where the active role is played by the planner, while the non-deterministic effects in the environment are determined by the opponent. We added an additional variable to determine the player's turn and further variables to determine the action that was chosen. These variables are all mutually exclusive and thus in one fact group. Whenever the planner chooses an action, the corresponding action-variable is set. This gives control to the environment, which decides the effect to be taken. Afterwards, control is returned to the planner.

After the translation of the non-deterministic domain into PDDL representing the two-player game of planner and environment, any PDDL static analyzer can be used to instantiate the domain.

Once the algorithm terminates, extracting the conditional plan is rather obvious. If a plan is found (otherwise there is no solution), we just iterate over all elements of π. Given such an element s, we existentially quantify over the action variables $a_1, \ldots, a_m$ to get the state's description and over the state variables $v_1, \ldots, v_l$ to get the corresponding action to take. As some state variables are compiled away by our static analyzer, we need to insert these again (as being $\top$ in all states) in the policy .

When operating on the grounded input, we observed that running Cimatti *et al.*'s strong planning algorithm still yields correct classification results for adversarial planning problems. This was unexpected as the two algorithms for computing strong plans and for winning sets in adversarial planning are rather different. The one insists on strong pre-images, while the other one alternates between strong and weak ones. The reason for the equivalence is as follows. In an instantiated adversarial planning problem, only the action variable and the effect are modified in the actions of the environment.

Let us consider the strong planning algorithm in more detail. We first compute all reachable states and remove all states, in which the player moves. This is possible,

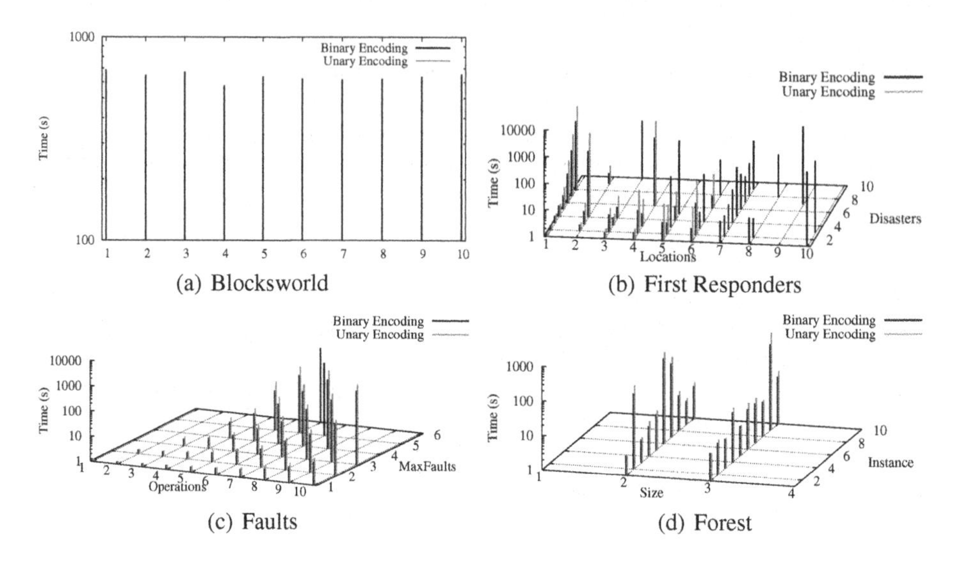

(a) Blocksworld (b) First Responders

(c) Faults (d) Forest

Fig. 3. Performance in Competition Domains

since the states of the planner and the environment are almost identical, only the action encoded in the action-variable has changed. This implies that we construct a state-action table. While chaining backwards we compute first a strong pre-image, intersect it with the subset of reachable states, and insert it into the state-action table. Then we take all states of the state-action table and remove the action variables via existential quantification, calculate the conjunction of the result with the goals and start again. This means that we always compute the strong pre-image from states without executing the according action. Hence, we have an implicit form of a weak pre-image, as used in computing winning sets for adversarial plans or for solving two-player games.

5 Results

Non-Deterministic Planning We tested the planner on some of the problems from the uncertainty part of the 2008 international planning competition[1]. The results are shown in Fig. 3. All experiments were perfomed on a machine with two Opteron processors with 2.4 GHz each. A timeout of 30 minutes and a memory limit of 2 GB were chosen.

As we could hardly find a competitor, we tested the influence of minimizing state encoding and found some differences throughout the domains.

In the Blocksworld (Fig. 3(a)) domain, the usage of the encoding found by our static analysis tool enables the planner to solve 10 out of a total of 30 instances, whereas the unary encoding of all variables prevents any solution. This might be due to the fact that in the Blocksworld domain there are lots of possible actions to take (more than 300 in the simplest instance) and thus in case of the unary encoding the BDDs quickly become too large for the given memory limit.

[1] All problems can be found on the official web-page http://ippc-2008.loria.fr

Table 1. Runtimes in seconds for the Airplane domain

number of cities	number of packages	BFS	AO* w/ FF	AO* w/Adv. FF	MBP	GAMER
3	4	0.870	0.463	0.232	1.780	0.300
3	5	5.556	1.437	0.321	9.041	1.142
3	6	87.691	16.323	1.157	44.287	4.705
4	6	–	76.718	82.701	130.064	21.072
4	7	–	373.553	99.639	–	114.164
4	8	–	–	–	–	563.381

A similar situation arises in the First Responders domain (Fig. 3(b)). The efficient binary encoding enables the planner to find 47 plans. In the unary enoding, only 26 plans can be found. For the more complex instances, the runtime using the unary encoding is a lot higher than with the binary one: in some cases, the latter is about six times as fast.

In the Faults domain (Fig. 3(c)), the usage of the binary encoding enables the planner to solve 36 instances, whereas the unary encoding yields only 34 plans. For the more complex problems it can be observed that the usage of the unary encoding takes about twice the time of the more efficient binary encoding.

Finally, in the Forest domain (Fig. 3(d)), the planner finds the same number of plans using any of the two encodings. Only, the unary encoding is somewhat slower: In most cases its runtime is about 60% higher than with the binary encoding, for the most complex instance, the runtime even doubles.

Adversarial Planning. For adversarial planning, we performed some experiments in the airplane domain presented in [2]. The possible actions for the protagonist (pilot) can load packages, unload packages, fly between cities, or smoke, while the antagonist (co-pilot) can unload packages, refuel, or make a coffee-break. The pilot's goal is to bring a number of packages to their corresponding destination.

Bercher and Mattmüller [2] solved this by using AO* with a special Adversarial FF heuristic and compared their results to simple BFS, AO* with the FF heuristic and MBP [3]. In Table 1, we additionally provide the results for the solution using the adapted two-player algorithm[2]. Given a timeout of 10 minutes, BFS is clearly inferior. Also, MBP does not perform too well, compared to the other algorithms. Using AO* with the Adversarial FF heuristic typically is a lot faster than with the FF heuristic, but both algorithms can calculate the same number of solutions. With our approach, we find the solution for another more complex instance.

6 Conclusion

We showed how to transform a non-deterministic planning problem into a two-player turn-taking game in PDDL notation, mainly in order to apply deterministic tools to infer a minimized state encoding fully automatically. Additionally, this allows to compute

[2] The first four results were calculated on 2 Quad Xeon with 2.66 GHz and 16 GB RAM; the two-player algorithm was performed on a Core 2 Duo with 2 GHz and 2 GB RAM.

disjunctive pre-images. Using a concise state encoding, we get smaller BDDs and using partitioned transition relations, we expect faster running times. As a feature, conditional plans can be inferred very naturally, and the extraction of the ASCII representation given the BDD was straight-forward. With this approach we won the 2008 international non-deterministic planning competition on fully observable planning.

It seems worthwile to investigate to what extent bidirectional search can be used to solve adversarial planning problems. For the progression part, we would use a variant of AO* search [2] over the AND/OR graph induced by the planning problem, whereas for the regression we would use a BDD based symbolic BFS, as elaborated above.

References

1. Bakera, M., Edelkamp, S., Kissmann, P., Renner, C.D.: Solving μ-calculus parity games by symbolic planning. In: MoChArt 2008. LNCS (LNAI), vol. 5348, pp. 15–33. Springer, Heidelberg (2009)
2. Bercher, P., Mattmüller, R.: A planning graph heuristic for forward-chaining adversarial planning. In: ECAI, pp. 921–922 (2008)
3. Bertoli, P., Cimatti, A., Pistore, M., Roveri, M., Traverso, P.: MBP: A model based planner. In: IJCAI Workshop on Planning under Uncertainty and Incomplete Information, pp. 93–97 (2001)
4. Bryant, R.E.: Graph-based algorithms for boolean function manipulation. IEEE Transactions on Computers 35(8), 677–691 (1986)
5. Bryce, D., Buffet, O.: 6th International Planning Competition: Uncertainty Part (2008)
6. Cassez, F., David, A., Fleury, E., Larsen, K.G., Lime, D.: Efficient on-the-fly algorithms for the analysis of timed games. In: Abadi, M., de Alfaro, L. (eds.) CONCUR 2005. LNCS, vol. 3653, pp. 66–80. Springer, Heidelberg (2005)
7. Cimatti, A., Pistore, M., Roveri, M., Traverso, P.: Weak, strong, and strong cyclic planning via symbolic model checking. Artificial Intelligence 147(1–2), 35–84 (2003)
8. Cleaveland, R., Klein, M., Steffen, B.: Faster model checking for the modal μ-calculus. Theoretical Computer Science 663, 410–422 (1992)
9. Edelkamp, S.: Symbolic exploration in two-player games: Preliminary results. In: AIPS 2002, Workshop on Model Checking, pp. 40–48 (2002)
10. Edelkamp, S., Kissmann, P.: Symbolic classification of general two-player games. In: Dengel, A.R., Berns, K., Breuel, T.M., Bomarius, F., Roth-Berghofer, T.R. (eds.) KI 2008. LNCS (LNAI), vol. 5243, pp. 185–192. Springer, Heidelberg (2008)
11. Emerson, E.A., Jutla, C.S.: Tree automata, μ-calculus and determinacy. In: Foundations of Computer Science, pp. 368–377 (1991)
12. Helmert, M.: Understanding Planning Tasks: Domain Complexity and Heuristic Decomposition. LNCS (LNAI), vol. 4929. Springer, Heidelberg (2008)
13. Jensen, R.M., Veloso, M.M., Bowling, M.H.: Obdd-based optimistic and strong cyclic adversarial planning. In: ECP, pp. 265–276 (2001)
14. Kissmann, P., Edelkamp, S.: Instantiating general games. In: IJCAI-Workshop on General Game Playing (2009)
15. Liu, X., Smolka, S.A.: Simple linear-time algorithms for minimal fixed points. In: Larsen, K.G., Skyum, S., Winskel, G. (eds.) ICALP 1998. LNCS, vol. 1443, pp. 53–66. Springer, Heidelberg (1998)
16. Love, N.C., Hinrichs, T.L., Genesereth, M.R.: General game playing: Game description language specification. Technical Report LG-2006-01, Stanford Logic Group (April 2006)
17. McMillan, K.L.: Temporal logic and model checking. In: Verification of Digital and Hybrid Systems, pp. 36–54 (1998)

Planning with h^+ in Theory and Practice

Christoph Betz and Malte Helmert

Albert-Ludwigs-Universität Freiburg, Institut für Informatik,
Georges-Köhler-Allee 52, 79110 Freiburg, Germany
{betzc,helmert}@informatik.uni-freiburg.de

Abstract. Many heuristic estimators for classical planning are based on the so-called *delete relaxation*, which ignores negative effects of planning operators. Ideally, such heuristics would compute the actual goal distance in the delete relaxation, i. e., the cost of an *optimal relaxed plan*, denoted by h^+. However, current delete relaxation heuristics only provide (often inadmissible) estimates to h^+ because computing the correct value is an NP-hard problem.

In this work, we consider the approach of planning with the actual h^+ heuristic from a theoretical and computational perspective. In particular, we provide *domain-dependent complexity results* that classify some standard benchmark domains into ones where h^+ can be computed efficiently and ones where computing h^+ is NP-hard. Moreover, we study *domain-dependent implementations* of h^+ which show that the h^+ heuristic provides very informative heuristic estimates compared to other state-of-the-art heuristics.

1 Introduction

Many algorithms for classical planning employ the approach of heuristic search based on *delete relaxation heuristics*. Given a state s of a planning task, such planners approximate the distance to the goal by solving a simplified planning task with initial state s where all undesirable effects of planning operators are ignored. For some planning formalisms, such as the STRIPS fragment of PDDL [1], such undesirable effects of planning operators can easily be identified syntactically. For more general planning formalisms like the ADL fragment of PDDL, linear-time compilation methods exist that can transform input tasks into the desired form [2].

Delete relaxation heuristics have been used quite successfully in the past. For example, they are a key component of the winners of the sequential satisficing tracks of the International Planning Competitions in 2000 (FF), 2004 (Fast Downward), 2006 (SGPlan) and 2008 (LAMA).

Once we commit to the idea of using heuristics based on delete relaxations, ideally we would like to use the cost of an *optimal relaxed plan* from a state s, denoted by $h^+(s)$, as the heuristic estimate for s [3]. However, computing $h^+(s)$ is an NP-equivalent problem [4], and good admissible approximations to h^+ are also provably hard to compute, even in a restricted propositional STRIPS formalism where each operator has only one precondition and only one effect (as shown later in this paper.) Therefore, common delete relaxation heuristics use approximations to h^+ which can differ from h^+ by an arbitrarily large multiplicative factor. Many such approaches are described in the planning literature:

B. Mertsching, M. Hund, and Z. Aziz (Eds.): KI 2009, LNAI 5803, pp. 9–16, 2009.

- The max heuristic $h^{\max}$ [5] computes the makespan of an optimal *parallel plan* for the relaxed task. The $h^{\max}$ value of the state is always a lower bound to the h^+ value, i. e., provides an admissible estimate to h^+.
- The additive heuristic h^{add} [5] computes the cost of a relaxed plan under the pessimistic assumption that there are no positive interactions between goal conditions and between operator preconditions, i. e., all conditions have to be achieved completely independently. The h^{add} value of a state is always an upper bound to the h^+ value and is in general not admissible.
- The FF heuristic h^{FF} [3] computes an actual relaxed plan for the delete relaxation, using a greedy algorithm based on backchaining in so-called relaxed planning graphs. The heuristic value is then the cost of that plan. The FF heuristic is defined procedurally, and $h^{\text{FF}}(s)$ is generally ambiguous because the precise heuristic values depend on tie-breaking behaviour in the backchaining step of the algorithm. Similar to h^{add}, the FF heuristic is generally not admissible and provides an upper bound to h^+.
- The *cost-sharing heuristic* h^{cs} and *pairwise max heuristic* h^{pmax} are other delete relaxation heuristics based on relaxed planning graphs, but using different propagation rules from h^{FF} [6]. The pairwise max heuristic is inadmissible; the cost-sharing heuristic is admissible but typically less informed than $h^{\max}$.
- The *FF/additive heuristic* $h^{\text{FF/a}}$ and *set-additive heuristic* h^{sa} [7] are variants of the FF heuristic which use different methods for computing the relaxed plans that define the heuristic value. The set-additive heuristic in particular can be considered more accurate than h^{FF} because it keeps track of positive interactions between operator preconditions in a more precise way than h^{FF}. However, theoretically, the heuristics are incomparable (that is, either can be larger than the other). Neither $h^{\text{FF/a}}$ nor h^{sa} is admissible.
- The recently introduced *local Steiner tree* heuristic h^{lst} [8] is another method for computing more accurate relaxed plans than h^{FF} in order to get closer approximations to h^+. The local Steiner tree heuristic first computes a relaxed plan using the $h^{\text{FF/a}}$ method, then reduces the size of this plan by exploiting local improvement properties of Steiner trees. Like the heuristics it builds on, it is inadmissible.
- The *LAMA heuristic* h^{LAMA} [9] counts the number of *landmarks* (from a set of possible landmarks that is precomputed prior to search) for which it can prove that they need to be achieved on the way to the goal. While it is not introduced as a kind of delete relaxation heuristic in the original paper, it can be considered such because the set of landmarks it considers are guaranteed to be landmarks of the delete relaxation, so that in particular the h^{LAMA} value for the initial state does not change when the task is replaced by its delete relaxation. The LAMA heuristic is not admissible, although it is admissible in the special case where each operator achieves at most one landmark. A family of admissible landmark heuristics built on top of LAMA has recently been introduced by Karpas and Domshlak [10].
- Finally, *additive $h^{\max}$ heuristics* are a family of admissible approximations to h^+ based on the *action cost partitioning* paradigm introduced by Haslum et al. [11] and later generalized by Katz and Domshlak [12]. Examples of additive $h^{\max}$ heuristics include the original algorithm of Haslum et al. [11] and the additive-disjunctive heuristic graphs of Coles et al. [13].

This large number of relaxation-based planning heuristics is clear evidence that delete relaxations are a very important approach to heuristic planning. Still, quite little is known about their theoretical properties, and in particular about their *limitations*. The motivation of most of the research efforts mentioned above is to find more and more precise estimates to the h^+ heuristic. However, it is not clear how good the estimates provided by h^+ *itself* actually are. The "holy grail" of delete relaxation would be an efficient heuristic estimator which provides perfect h^+ values. But would this actually be a good heuristic, compared to approaches not based on delete relaxation, such as the context-enhanced additive heuristic [14] or abstraction heuristics [15]?

Hoffmann [16] provides a partial answer to this question by showing that certain *satisficing* (suboptimal) h^+-based planners have guaranteed polynomial runtime on many classical planning benchmarks. Additionally, Helmert and Mattmüller [17] provide a theoretical analysis that shows that h^+ generally outperforms pattern database heuristics *in the limit of large problems* on typical planning domains. In this paper, we complement these results by evaluating the quality of h^+ as an admissible heuristic for *optimal planning* on *practical benchmarks*, i. e., tasks of a size for which we can actually hope to compute a solution (unlike the in-the-limit results of Helmert and Mattmüller).

Since computing h^+ is generally NP-hard and no empirically fast algorithms are known, we perform our study by designing and evaluating *domain-dependent* algorithms for h^+ in a number of classical benchmark domains. One obvious question when designing such domain-dependent h^+ implementations is whether we can come up with sub-exponential algorithms by exploiting that we only have to deal with, e. g., LOGISTICS or BLOCKSWORLD tasks. For the domains we study in this paper, we answer this question by either describing a polynomial-time algorithm or proving that computing h^+ value remains NP-hard even when restricted to tasks of the given domain. These theoretical results, discussed in Sect. 3, form the first contribution of this paper.

In addition to this theoretical study, we provide empirical results obtained by using our domain-specific h^+ implementations as a heuristic in an A*-based planner. These results, discussed in Sect. 4, form the second contribution of this paper. Of course, runtime results obtained through domain-dependent implementations cannot be directly compared to domain-independent planners (e. g., using abstraction heuristics) in order to judge which of the approaches is generally more useful. However, they can tell us what the *theoretical limits* of relaxation-based approaches to optimal planning are, so that we can give an answer whether it is actually worth working on increasingly more sophisticated methods to compute more and more accurate approximations to h^+. To anticipate our experimental results, it appears that the answer to this question is affirmative: delete relaxations compare very favourably with the state of the art, and it definitely appears to be worth looking at their application to optimal planning more closely.

2 Background

For the theoretical results of this paper, we use the propositional STRIPS formalism [4]. (Some of the planning tasks we consider go slightly beyond STRIPS by requiring *conditional effects*, but we omit the formal details for these because they are not relevant to the abbreviated proofs we can present within the limited space of this paper.)

Definition 1. *A* planning task *is a 4-tuple* $\Pi = \langle V, O, I, G \rangle$, *where*

- V *is a finite set of* propositional state variables *(also called* propositions *or* facts*),*
- O *is a finite set of* operators, *each with associated* preconditions $pre(o) \subseteq V$, add effects $add(o) \subseteq V$ *and* delete effects $del(o) \subseteq V$,
- $I \subseteq V$ *is the* initial state, *and*
- $G \subseteq V$ *is the set of* goals.

A *state* is a subset of facts, $s \subseteq V$, representing the propositions which are currently true. *Applying* an operator o in s results in state $(s \setminus del(o)) \cup add(o)$, which we denote as $s[o]$. The notation is only defined if o is *applicable* in s, i. e., if $pre(o) \subseteq s$. Applying a sequence $o_1, \ldots, o_n$ of operators to a state is defined inductively as $s[\epsilon] := s$ and $s[o_1, \ldots, o_{n+1}] := (s[o_1, \ldots, o_n])[o_{n+1}]$. A plan for a state s (s-*plan*, or *plan* when s is clear from context) is an operator sequence π such that $s[\pi]$ is defined and satisfies all goals (i. e., $G \subseteq s[\pi]$). The objective of *optimal planning* is to find an I-plan of minimal length (called an *optimal* I*-plan*) or prove that no plan exists.

Heuristic functions or *heuristics* are a key ingredient of heuristic search planners. A heuristic is a function $h : 2^V \to \mathbb{N}_0 \cup \{\infty\}$ with the intuition that $h(s)$ estimates the length of an s-plan. The *perfect heuristic* h^* maps each state to the length of an optimal s-plan (infinite if no s-plan exists). A heuristic h is *admissible* if $h(s) \leq h^*(s)$ for all states s. All common heuristic search algorithms for optimal planning require admissible heuristics. If $h(s) \geq h'(s)$ for all states s, we say that h *dominates* h'.

Relaxation heuristics estimate the distance to the goal by considering a *relaxed task* Π^+ derived from the actual planning task Π by ignoring all delete effects of operators, i. e., replacing each operator o by a new operator o^+ with the same preconditions and add effects as o and $del(o^+) = \emptyset$. The h^+ *heuristic* [3] uses the length of an optimal s-plan in Π^+ as the heuristic estimate $h^+(s)$ for a state s of the original task Π.

3 Theory: Complexity of Computing h^+

Computing h^+ estimates for states of a planning task is an NP-equivalent problem [4]. It is due to this computational complexity that h^+ has not previously been used in an actual planning system designed to solve planning tasks of interesting size. However, far from being optimal, all approximations to h^+ discussed in the introduction can actually be arbitrarily far off from the correct h^+ values, i. e., $h(s)/h^+(s)$ can be arbitrarily large for the inadmissible heuristics h discussed there, and $h^+(s)/h(s)$ can be arbitrarily large for the admissible ones. We now prove that there is a theoretical reason for this.

Theorem 2. *If* P $\neq$ NP, *then there exists no constant* $c > 0$ *and no polynomial-time algorithm for computing an admissible heuristic function* h *such that for all states* s, $h(s) \geq c \cdot h^+(s)$. *This is true even when only allowing planning tasks where each operator has only a single precondition and only a single add effect.*

Proof sketch: We present an approximation-preserving reduction (see the textbook by Ausiello et al. [18]) from MINIMUM SET COVER to planning for delete relaxations. Since MINIMUM SET COVER has no constant-factor approximations unless P = NP [18, problem SP4], the claim follows. Given a MINIMUM SET COVER instance with set

S and subsets $C_1, \dots, C_m \subseteq S$, the reduction generates operators $o_i^1, o_i^2, \dots, o_i^N$ for each subset C_i such that all these operators need to be applied (in sequence) in order to achieve a fact a_i that marks that C_i has been selected. From a_i, facts corresponding to the elements of C_i can then be directly achieved using operators which have precondition a_i and add one element of C_i at a time. To satisfy the goal, which consists of all facts corresponding to elements of S, we must select enough subsets C_i to cover S completely. By choosing N appropriately (e. g., $N = |S|$), we can ensure that the overall plan length is dominated by the number of subsets chosen, and hence short relaxed plans correspond to small set covers. ■

Theorem 2 shows that we cannot hope to find a polynomial algorithm that is guaranteed to find good approximations to h^+. However, since theoretical results of this kind tend to rely on somewhat pathological problem instances, this does not mean that computing or approximating h^+ is necessarily difficult for *practically interesting planning tasks*. Hence, to get a handle on the complexity of computing h^+ in more typical cases, we now investigate the behaviour of h^+ in *specific* planning domains used as benchmarks in the planning community, specifically the domains considered by Helmert and Mattmüller in their theoretical study of admissible planning heuristics [17].

It turns out that, at least for these domains, the situation is not quite as bleak. In all cases, we can compute constant-factor approximations to h^+ in polynomial time, and in some cases we even have polynomial algorithms for the perfect h^+ value, despite the fact that in most of these domains (all except GRIPPER and SCHEDULE [19]), computing the actual goal distance h^* is NP-hard.

For space reasons, we refer to the literature for formal definitions of these common planning benchmarks [19] and only provide very brief proof sketches. An extensive discussion of these results, including full proofs, can be found in Betz's thesis [20].

Theorem 3. *There exists a polynomial-time algorithm for computing $h^+(s)$ for arbitrary reachable states s of* BLOCKSWORLD *tasks.*
Proof sketch: The subgoal ordering issues that make optimal BLOCKSWORLD planning hard in general [21] do not exist in the delete relaxation where simple greedy criteria are sufficient to decide which blocks to pick up and, after all pick-ups have been performed, where to drop them. See Betz's thesis for details [20, Corollary 6.1]. ■

Theorem 4. *There exists a polynomial-time algorithm for computing $h^+(s)$ for arbitrary reachable states s of* GRIPPER *tasks.*
Proof sketch: Due to symmetries in GRIPPER tasks, a closed formula for h^+ can be given. This formula can be evaluated in linear time [20, Theorem 5.1]. ■

Theorem 5. *Computing $h^+(s)$ for arbitrary reachable states s of* LOGISTICS *tasks is NP-hard, but polynomial-time constant-factor approximations exist.*
Proof sketch: Hardness is proved by a reduction from SET COVER. There is one truck corresponding to each candidate subset, which is loaded with one package for each element of that subset. The instance is then constructed in such a way that a subset of trucks need to visit a special location, called the *Ω-location*, and the overall quality of a relaxed plan is determined by how many trucks visit the Ω-location. In optimal relaxed plans this subset corresponds to an optimal set cover [20, Theorem 8.3]. For the constant-factor approximation result, we refer to Betz's thesis [20, Theorem 8.5]. ■

We remark that polynomial h^+-algorithms for LOGISTICS exist if we only consider valid *initial states*, where vehicles are required to be empty [20, Theorem 8.2], and also when there is only one truck per city and only one airplane [20, Theorem 8.1].

Theorem 6. *There exists a polynomial-time algorithm for computing $h^+(s)$ for arbitrary reachable states s of* MICONIC-STRIPS *tasks.*
Proof sketch: This follows directly from the previous remark due to the similarity of MICONIC-STRIPS to LOGISTICS with only one truck [20, Theorem 3.1]. ■

Theorem 7. *Computing $h^+(s)$ for arbitrary reachable states s of* MICONIC-SIMPLE-ADL *tasks is NP-hard, but polynomial-time constant-factor approximations exist.*
Proof sketch: In MICONIC-SIMPLEADL, computing h^+ is closely related to computing h^*, and the known results for h^* [22] carry over [20, Theorem 3.2]. ■

Theorem 8. *Computing $h^+(s)$ for arbitrary reachable states s of* SATELLITE *tasks is NP-hard, but polynomial-time constant-factor approximations exist.*
Proof sketch: The proof [20, Theorem 7.1] is again based on a reduction from SET COVER and uses similar ideas to the proof that establishes NP-hardness for h^* [22]. ■

Theorem 9. *There exists a polynomial-time algorithm for computing $h^+(s)$ for arbitrary reachable states s of* SCHEDULE *tasks.*
Proof sketch: A simple algorithm that achieves the goals one object ("part") at a time is sufficient [20, Theorem 4.1]. ■

4 Practice: Using h^+ Inside an Optimal Planner

As noted in the introduction, delete relaxation heuristics are state of the art for satisficing planning. For optimal planning, however, the literature suggests that the admissible representatives of the family, $h^{\max}$ and h^{cs}, lag behind other approaches such as abstraction. For example, merge-and-shrink abstractions ($h^{\text{m\&s}}$ in the following) clearly outperform $h^{\max}$ [15], and h^{cs} is empirically even worse than $h^{\max}$ [6]. Does this indicate that delete relaxation heuristics are generally not useful for optimal planning, or is this a specific weakness of $h^{\max}$ and h^{cs}? To answer that question, we have added domain-specific implementations of the h^+ heuristic to a state-of-the-art A*-based optimal planner [15] and empirically compared it to $h^{\max}$, to see how far current admissible relaxation heuristics are from what is possible, and to $h^{\text{m\&s}}$, to see if relaxation heuristics may be competitive with the state of the art in optimal planning.

Experiments were conducted under the usual planning competition settings. Table 1 shows the results. Note that while our h^+ *implementations* are domain-dependent, the *estimates* themselves are fully domain-independent, and hence comparisons of heuristic quality (e. g., number of A* state expansions) are meaningful. We compare on all domains considered in the previous section except for those not supported by the underlying planner, MICONIC-SIMPLEADL and SCHEDULE. Note that this includes the LOGISTICS and SATELLITE domains where computing h^+ is NP-hard; in these cases, each state evaluation in our implementation can require exponential time. Table 1 indicates that the time per state expansion is indeed very high for SATELLITE, but h^+

Table 1. Experimental comparison of h^+, $h^{\mathrm{m\&s}}$ and $h^{\max}$. Parameters for $h^{\mathrm{m\&s}}$ are hand-tuned per domain. We report heuristic values for the initial state (h), number of expanded states (Exp.), and runtime in seconds (Time) for the largest tasks solved in each domain. Dashes indicate running out of time (30 minutes) or memory (2 GB). Best results for each task are highlighted in bold.

		h^+			$h^{\mathrm{m\&s}}$			$h^{\max}$		
Inst.	h^*	h	Exp.	Time	h	Exp.	Time	h	Exp.	Time
BLOCKSWORLD ($h^{\mathrm{m\&s}}$: one abstraction of size 50000)										
#9-0	30	**16**	**13215**	**0.65**	**16**	971774	191.12	9	3840589	85.00
#9-1	28	**16**	**360**	**0.02**	**16**	60311	69.25	10	1200345	32.06
#9-2	26	**17**	**594**	**0.04**	16	54583	90.12	9	1211463	32.15
#10-0	34	**18**	**241489**	**15.42**	**18**	19143580	367.82	9	—	—
#10-1	32	**19**	**29144**	**1.82**	16	12886413	316.28	8	—	—
#10-2	34	**19**	**83007**	**5.68**	18	—	—	10	—	—
#11-0	32	19	**63891**	**4.35**	**21**	7291064	199.01	8	—	—
#11-1	30	**21**	**59340**	**4.64**	17	—	—	4	—	—
#11-2	34	**19**	**53642**	**3.39**	**19**	—	—	9	—	—
#12-0	34	**22**	**58124**	**4.54**	21	—	—	10	—	—
#12-1	34	**22**	**6284**	**0.48**	21	—	—	11	—	—
#13-1	44	**25**	**9990123**	**1078.59**	23	—	—	12	—	—
#14-0	38	**25**	**100499**	**10.64**	19	—	—	10	—	—
#14-1	36	**27**	**160352**	**19.99**	20	—	—	6	—	—
#15-0	40	**28**	**3540691**	**420.91**	18	—	—	7	—	—
GRIPPER ($h^{\mathrm{m\&s}}$: one abstraction of size 5000)										
#1	11	9	82	**0.00**	**11**	**12**	**0.00**	2	208	**0.00**
#2	17	13	1249	**0.00**	**15**	**975**	0.10	2	1760	0.01
#3	23	**17**	**10304**	**0.06**	11	11506	0.34	2	11616	0.08
#4	29	**21**	**65687**	**0.44**	13	68380	1.04	2	68468	0.56
#5	35	**25**	**371726**	**2.86**	14	376510	3.59	2	376496	3.51
#6	41	**29**	**1974285**	17.79	16	1982018	**16.19**	2	1982016	21.57
#7	47	**33**	**10080252**	97.60	18	10091970	**79.83**	2	10091968	119.64
LOGISTICS-1998 ($h^{\mathrm{m\&s}}$: one abstraction of size 200000)										
#1	26	24	**8623**	**2.91**	**25**	375885	67.64	6	—	—
#5	22	**22**	**30**	**0.03**	20	527498	99.94	4	—	—
#17	42	**42**	**67**	**0.62**	—	—	—	6	—	—
#31	13	12	68	**0.02**	**13**	**14**	7.22	4	32282	0.57
#32	20	18	116	**0.02**	**20**	**21**	1.97	6	81156	1.00
#33	27	25	88629	**21.80**	**27**	**28992**	81.01	4	—	—
#35	30	**29**	**1682**	**2.65**	22	—	—	5	—	—

		h^+			$h^{\mathrm{m\&s}}$			$h^{\max}$		
Inst.	h^*	h	Exp.	Time	h	Exp.	Time	h	Exp.	Time
LOGISTICS-2000 ($h^{\mathrm{m\&s}}$: one abstraction of size 200000)										
#8-0	31	29	3269	**0.32**	**31**	**32**	26.75	6	—	—
#8-1	44	41	43665	**3.87**	**44**	**45**	28.17	6	—	—
#9-0	36	33	14090	**1.62**	**36**	**37**	37.58	6	—	—
#9-1	30	29	707	**0.10**	**30**	**31**	37.49	6	—	—
#10-0	45	41	**193846**	**26.39**	**45**	196342	79.12	6	—	—
#10-1	42	39	**165006**	**24.88**	**42**	518215	86.22	6	—	—
#11-0	48	45	156585	**28.81**	**48**	**12822**	87.99	6	—	—
#11-1	60	55	5649882	775.53	**59**	**2608870**	**187.85**	6	—	—
#12-0	42	39	**116555**	**22.25**	**42**	272878	117.98	6	—	—
#12-1	68	63	—	—	**68**	**828540**	**137.67**	6	—	—
SATELLITE ($h^{\mathrm{m\&s}}$: three abstractions of size 10000)										
#1	9	8	**10**	**0.00**	**9**	**10**	**0.00**	3	59	**0.00**
#2	13	12	**14**	0.02	**13**	**14**	0.09	3	940	**0.00**
#3	11	10	21	**0.07**	**11**	**12**	4.01	3	6822	0.11
#4	17	**17**	**26**	**0.15**	**17**	237	7.57	3	180815	3.37
#5	15	**14**	**34**	**5.02**	12	38598	44.66	3	—	—
#6	20	**18**	**526**	**16.23**	16	375938	48.73	3	10751017	371.43
#7	21	**20**	**812**	**250.37**	15	—	—	3	—	—
#9	27	**26**	**264**	**1350.72**	17	—	—	3	—	—
#10	29	**29**	**40**	**401.41**	16	—	—	3	—	—
MICONIC-STRIPS ($h^{\mathrm{m\&s}}$: one abstraction of size 200000)										
#28-4	92	**92**	**123**	**0.05**	54	—	—	3	—	—
#29-0	94	**94**	**126**	**0.06**	54	—	—	3	—	—
#29-1	91	**91**	**184**	**0.08**	53	—	—	3	—	—
#29-2	95	**95**	**155**	**0.06**	55	—	—	3	—	—
#29-3	97	**96**	**178**	**0.07**	56	—	—	3	—	—
#29-4	99	**99**	**141**	**0.05**	55	—	—	3	—	—
#30-0	95	**95**	**138**	**0.06**	55	—	—	3	—	—
#30-1	98	**98**	**150**	**0.06**	55	—	—	3	—	—
#30-2	97	**96**	**130**	**0.04**	55	—	—	3	—	—
#30-4	99	**99**	**124**	**0.05**	53	—	—	3	—	—

still scales much further than the other approaches due to the accuracy of the heuristic. Aggregating results over all domains, h^+ convincingly outperforms the other heuristics considered, including the state-of-the-art $h^{\mathrm{m\&s}}$. This suggests that the comparatively bad results obtained with earlier delete relaxation heuristics are mostly due to their inability to accurately approximate h^+ rather than a general weakness of delete relaxations.

5 Conclusion

Starting from the observation that many current planning heuristics are based on delete relaxations, we have taken a deeper look at the optimal delete relaxation heuristic, h^+, which all these heuristics strive to approximate. Theoretically, we have seen that h^+ is in general not just hard to compute (as proved already by Bylander), but also hard to approximate. However, these worst-case results do not carry over to most planning domains, for which we could show much better theoretical results – including polynomial-time algorithms for h^+ in four of the seven benchmark results considered.

Experimentally, we have shown that h^+ is very informative across a range of planning domains, improving on the state of the art in domain-independent optimal planning. Hence, it appears worth investigating practically efficient general implementations of h^+, or alternatively better admissible approximations, more closely. In our opinion, despite the multitude of existing approaches, there is still considerable scope for research on delete relaxation heuristics, in particular admissible ones. Our results presented here can serve as a useful methodological basis for such future work by allowing, for the first time, direct comparisons of practical relaxation heuristics to h^+.

Acknowledgments

This work was partly supported by the German Research Foundation (DFG) as part of the Transregional Collaborative Research Center "Automatic Verification and Analysis of Complex Systems". See `http://www.avacs.org/` for more information.

References

1. Fox, M., Long, D.: PDDL2.1: An extension to PDDL for expressing temporal planning domains. JAIR 20, 61–124 (2003)
2. Gazen, B.C., Knoblock, C.A.: Combining the expressivity of UCPOP with the efficiency of Graphplan. In: Steel, S. (ed.) ECP 1997. LNCS, vol. 1348, pp. 221–233. Springer, Heidelberg (1997)
3. Hoffmann, J., Nebel, B.: The FF planning system: Fast plan generation through heuristic search. JAIR 14, 253–302 (2001)
4. Bylander, T.: The computational complexity of propositional STRIPS planning. AIJ 69(1–2), 165–204 (1994)
5. Bonet, B., Geffner, H.: Planning as heuristic search. AIJ 129(1), 5–33 (2001)
6. Mirkis, V., Domshlak, C.: Cost-sharing approximations for h^+. In: Proc. ICAPS 2007, pp. 240–247 (2007)
7. Keyder, E., Geffner, H.: Heuristics for planning with action costs revisited. In: Proc. ECAI 2008, pp. 588–592 (2008)
8. Keyder, E., Geffner, H.: Trees of shortest paths vs. Steiner trees: Understanding and improving delete relaxation heuristics. In: Proc. IJCAI 2009 (2009)
9. Richter, S., Helmert, M., Westphal, M.: Landmarks revisited. In: Proc. AAAI 2008, pp. 975–982 (2008)
10. Karpas, E., Domshlak, C.: Cost-optimal planning with landmarks. In: Proc. IJCAI 2009 (2009)
11. Haslum, P., Bonet, B., Geffner, H.: New admissible heuristics for domain-independent planning. In: Proc. AAAI 2005, pp. 1163–1168 (2005)
12. Katz, M., Domshlak, C.: Optimal additive composition of abstraction-based admissible heuristics. In: Proc. ICAPS 2008, pp. 174–181 (2008)
13. Coles, A., Fox, M., Long, D., Smith, A.: Additive-disjunctive heuristics for optimal planning. In: Proc. ICAPS 2008, pp. 44–51 (2008)
14. Helmert, M., Geffner, H.: Unifying the causal graph and additive heuristics. In: Proc. ICAPS 2008, pp. 140–147 (2008)
15. Helmert, M., Haslum, P., Hoffmann, J.: Flexible abstraction heuristics for optimal sequential planning. In: Proc. ICAPS 2007, pp. 176–183 (2007)
16. Hoffmann, J.: Where 'ignoring delete lists' works: Local search topology in planning benchmarks. JAIR 24, 685–758 (2005)
17. Helmert, M., Mattmüller, R.: Accuracy of admissible heuristic functions in selected planning domains. In: Proc. AAAI 2008, pp. 938–943 (2008)
18. Ausiello, G., Crescenzi, P., Gambosi, G., Kann, V., Marchetti-Spaccamela, A., Protasi, M.: Complexity and Approximation. Springer, Heidelberg (1999)
19. Helmert, M.: Understanding Planning Tasks – Domain Complexity and Heuristic Decomposition. LNCS (LNAI), vol. 4929. Springer, Heidelberg (2008)
20. Betz, C.: Komplexität und Berechnung der h^+-Heuristik. Diplomarbeit, Albert-Ludwigs-Universität Freiburg (2009)
21. Gupta, N., Nau, D.S.: On the complexity of blocks-world planning. AIJ 56(2–3), 223–254 (1992)
22. Helmert, M., Mattmüller, R., Röger, G.: Approximation properties of planning benchmarks. In: Proc. ECAI 2006, pp. 585–589 (2006)

A Framework for Interactive Hybrid Planning

Bernd Schattenberg, Julien Bidot, Sascha Geßler, and Susanne Biundo

Institute for Artificial Intelligence
Ulm University, Germany
firstname.lastname@uni-ulm.de

Abstract. Hybrid planning provides a powerful mechanism to solve real-world planning problems. We present a domain-independent, mixed-initiative approach to plan generation that is based on a formal concept of hybrid planning. It allows for any interaction modalities and models of initiative while preserving the soundness of the planning process. Adequately involving the decision competences of end-users in this way will improve the application potential as well as the acceptance of the technology.

1 Introduction

As efficient automated AI planning systems may be, for many real-world domains they are neither applicable nor accepted [1]. On the one hand, human users want to be in authority for key decisions made by the systems. On the other hand, we may assume that hardly any automated system is able to take into account all relevant aspects within a particular application domain. Given these considerations, the role of users has changed from external observers to active partners involved in the planning process. Not surprisingly, the *mixed-initiative* paradigm has attracted more and more attention [2,3]. According to thisview, the best way to realize collaboration is to establish a *dialog* between the user and the planning system, which is particularly fruitful for solving a planning problem because the partners contribute in their area of expertise. Humans bring their experience and their intuition for making decisions to the table; they have a global, high-level view on the application domain. Planning systems are able to explore systematically the underlying search space with all its details. The key to success would ideally be to pass control opportunistically to that partner with the best knowledge about which decision to make next.

There exist some interactive approaches for specific planning paradigms and applications. Ai-Chang et al. have developed MAPGEN, a mixed-initiative planning system, that generates daily plans off-line for exploration rovers on Mars [4]. Tate et al. have designed and implemented the I-X planner, which is based on a standard Hierarchical Task Network approach with a constraint model [5]. Ambite et al. have proposed Heracles, a constraint-based framework that supports the interaction with a human user [6]. Ferguson and Allen have developed the planner TRIPS, which follows a dialog-driven collaborative problem-solving approach [7].

B. Mertsching, M. Hund, and Z. Aziz (Eds.): KI 2009, LNAI 5803, pp. 17–24, 2009.

In this paper, we propose a *domain-independent*, *mixed-initiative* approach to hybrid planning that is based on a formal framework for refinement-planning. Contrary to existing work, it can realize any model of initiative or interaction modality without jeopardizing the soundness of the plan generation procedure or the consistency of the results. We will conclude with an example application.

2 A Formal Framework for Refinement-Based Planning

The mixed-initiative approach is based on a hybrid planning framework that integrates partial-order causal-link planning and hierarchical planning [8]. This framework uses an ADL-like representation of states and basic actions (*primitive tasks*). States, preconditions, and effects of tasks are specified through formulae of a fragment of first-order logic. *Abstract tasks* can be refined by so-called *decomposition methods*, which provide *task networks* (*partial plans*) that describe how the corresponding task can be solved. Partial plans may contain abstract and primitive tasks. With that, hierarchies of tasks and associated methods can be used to encode the various ways to accomplish an abstract task.

A *domain model* $D = \langle \mathtt{T}, \mathtt{M} \rangle$ consists of a set of task schemata $\mathtt{T}$ and a set $\mathtt{M}$ of decomposition methods. A partial plan is a tuple $P = \langle TE, \prec, VC, CL \rangle$, which consists of a set of plan steps, a set of ordering constraints that impose a partial order on plan steps, variable constraints, and causal links. The latter are the usual means to establish and maintain causal relationships among the tasks in a partial plan. A *planning problem* $\pi = \langle D, P_{\text{init}} \rangle$ consists of a domain model D and an initial task network P_{init}. The *solution* of a planning problem is obtained by transforming P_{init} stepwise into a partial plan P that meets the following *solution criteria*: (1) all preconditions of the tasks in P are safely supported by causal links; (2) the ordering and variable constraints of P are consistent; (3) all steps in P are primitive tasks.

Transforming partial plans into their *refinements* is done by using so-called *plan modifications*. Given a partial plan $P = \langle TE, \prec, VC, CL \rangle$ and domain model D, a plan modification is defined as $\mathtt{m} = \langle E^{\oplus}, E^{\ominus} \rangle$, where $E^{\oplus}$ and $E^{\ominus}$ are disjoint sets of elementary additions and deletions of *plan elements* over P and D. Consequently, all elements in $E^{\ominus}$ are from TE, $\prec$, VC or CL, respectively, while $E^{\oplus}$ consists of new plan elements. This generic definition makes the changes explicit that a modification imposes on a plan. With that, a planning strategy is able to compare the available refinement options qualitatively and quantitatively and can choose opportunistically among them [9]. Applying a modification $\mathtt{m} = \langle E^{\oplus}, E^{\ominus} \rangle$ to a plan P returns the plan P' that is obtained from P by adding all elements in $E^{\oplus}$ and removing those of $E^{\ominus}$. Hybrid planning distinguishes various classes of plan modifications such as task expansion and task insertion.

For a partial plan P that is a refinement of the initial task network of a given problem, but is not yet a solution, so-called *flaws* are used to make the violations of the above solution criteria explicit. Flaws list those plan elements that constitute deficiencies of the partial plan. We distinguish various flaw classes including the ones for unsupported preconditions of tasks and inconsistencies of variable and ordering constraints. A large number of search strategies can be realized in this planning framework by using a *modification trigger* function that relates flaws to modifications that are suitable to eliminate them.

The planning procedure works as follows: (1) the flaws of the current plan are collected; if no flaw is detected, the plan is a solution. (2) suitable modifications are generated using the modification trigger; if for some flaws no modification can be found, the plan is discarded (dead-end). (3) selected modifications are applied and generate further refinements of the plan. (4) the next candidate plan is selected and we proceed with (1).

3 Mixed-Initiative Hybrid Planning

In the context of mixed-initiative planning, the presented *formal framework* has three essential properties that make it particularly suitable for interactive plan generation: (1) Refinement-based planning operates on the plan space, i.e., all intermediate results are partial plans and thus accessible at any time during plan generation. (2) Explicit flaws and plan modifications can explain plan development in terms of deficiencies and refinement options at any stage of the plan generation process. (3) Flexible strategies allow us to change opportunistically the focus of the plan generation process to the appropriate areas of the refinement space.

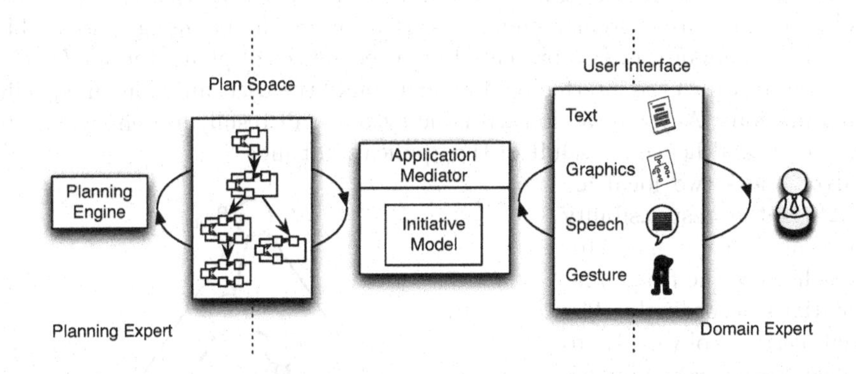

Fig. 1. The structure of the interactive framework for hybrid planning

In this setting, mixed-initiative planning is a *mediated process* between a domain-independent semi-automated plan generation and an interactive domain-specific application. The basic idea is to deploy the planning engine concurrently to the application in order to incrementally construct solutions to the given problem. At the same time, the user interactively focuses on those parts of the refinement space that are of interest to him or her, using application-specific modalities, that means, representations of (classes of) plans, their development alternatives, and interaction mechanisms.

Fig. 1 outlines the architecture: The "planning engine" is a hybrid planning system that acts as the planning expert and provides an autonomous, incremental construction of the refinement-space in which we search for solutions [10]. The perception of the plan generation process of the human user is defined by the domain-specific interface that supports multiple modalities for interacting with the application. The application mediator finally synchronizes the focus areas of the planning and the domain expert according to an application-specific model of initiative. It also provides primitives for defining user interface representations of plan components and interaction commands that imply focus changes.

Encapsulating Plan Generation. It is a consequent requirement to decouple the plan generation process, which deals with the tedious details in search of a solution, from the more long-term strategic engagements of a human user. Thus, synchronizing the construction of the refinement space with the applications' focus in an adequate manner becomes the most challenging problem in mixed-initiative planning, given that the respective "experts" differ significantly in their modus operandi. During operation, the planning system constructs the refinement space that is induced by a problem specification π up to the *planning horizon*, i.e., those plans, for which no refinement has yet been generated. The horizon has a static part of solutions and discarded inconsistent plans, about which the application mediator has perfect information by definition and which it can directly communicate on request via the interface representations. The dynamic part of the planning horizon is called the *system fringe*, which consists of all plans that are currently under investigation by the planning system. The focus of the human user is represented by a second set of plans, the *application fringe*, out of which the interface components build the user- and domain-specific representations. Any interaction with the system will result in a change of the user's focus and hence in a shift of the application fringe.

Given these two focal areas, it is the responsibility of the application mediator to synchronize the fringe manipulations accordingly. The trivial case is obviously to manage focus access on solutions and discarded plans, which does not require further actions beyond notifying the user interface. The application mediator also has to ensure that the user is always able to change focus onto any plan that is a refinement of an application fringe element.

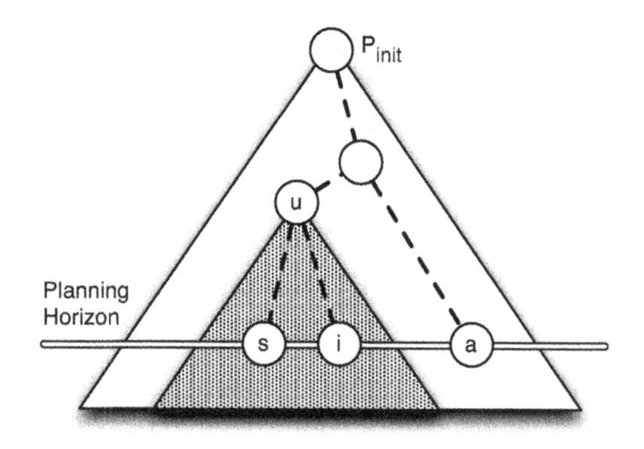

Fig. 2. Plans that are focused during mixed-initiative planning

I.e., that the system fringe has to advance such that it is a projection of the application fringe development. The mediator accomplishes this task by adapting the planning system's strategy, if the system fringe is approached too closely according to the application's representation needs (see section on representation below). Furthermore, the more the mediator knows about the refinement space ahead of the application fringe, the better informed the interactions can be processed. In particular, this means propagating information about solutions and necessary failures up to the refinement paths into the application fringe.

The planning system's focus adaptation is realized in our refinement-planning framework by means of the flexible strategy concept [9]. We deploy in fact a strategy triple that is continuously parameterized and balanced by the application mediator; Fig. 2 illustrates their co-operation in a prototypical situation during interactive planning: The figure shows the refinement-space that is induced by π. The human user's focus is represented by P_u as one plan of the application fringe and the grey area depicts the refinement subspace into which the user may focus next (prior focus areas may be revisited but are omitted here). The planning system is guided in this situation by three strategies that concurrently propose the next plan in the system fringe that is to be processed. During normal operation, an explorative general-purpose strategy is choosing a candidate P_s amongst the refinements of P_u in order to find a solution within the subspace or to reveal inconsistent refinements. This strategy may be any kind of search control that seems appropriate for the given domain and problem, it is only restricted in its input to the fringe fragment that lies inside the subspace. If the application fringe approaches the system fringe, the user-imitating strategy is preferred, which chooses a plan P_i that appears to be the most probable focus area of the human user at this point. Techniques for learning user preferences from refinement traces are out of the scope of this paper; it is, however, worth pointing out that a simple estimation like the ratio of flaw and modification classes in the refinement history provides a suitable set of candidates for stretching the planning horizon. Since the user usually needs some time to explore the new options, the system has enough time to gather alternative refinements by the general-purpose strategy. If the user's focus area is sufficiently covered, the application mediator activates an alternatives providing strategy, which selects plans P_a from system fringe candidates outside the focus subspace. Usually, this strategy is the inverse function of the user-imitating one.

Modeling Changing Initiative. The mixed-initiative paradigm emphasizes a *dynamic* notion of initiative, i.e., the roles of the experts is subject to change during the plan generation episode. In particular, mixed initiative is not necessarily reduced to a dialog scenario in a question and answer or command and response pattern. Our framework uses petri nets for modeling events that induce a change of initiative and for specifying the control flow of the application.

An example for such a trigger event is the discovery of a solution that is a refinement of a plan in the current user focus. At this occasion, the system takes the initiative in order to notify the user and passes the initiative to him/her afterwards. The user is then asked in a direct dialog to focus on the solution or to defer this for later exploration. If the system finds a solution in the alternative refinement-space, an accordingly specified rule just triggers a more unobtrusive notification, which is processed concurrently without changing the interaction mode. Our prototypical framework instances implement also a similar rule for discarded inconsistent plans: if all refinements of a plan are marked as discarded, the plan becomes marked itself. In this way, information about necessarily futile development options is propagated into the application fringe.

In addition to such generic information processing rules, mixed-initiative applications typically require more domain-specific triggers, because a typical domain expert would not be presented every refinement option but only an adequate subset. The system may retain initiative until a pre-specified set of alternatives is available. E.g., when the application fringe includes two plans with different implementations of a certain abstract task: The system then switches into a dialog mode in which the user is asked for the preferred way of implementing the task. In this application, all further details are worked out by the system autonomously.

Representing Plans and Interaction. As shown in Fig. 1, the architecture supports multiple interaction modalities, which naturally requires a broad repertoire of representation capabilities as well as mediation facilities between different representations; this includes in particular mediating between different levels of information abstraction. To this end, we have developed user-interface primitives and component structures that define visualizations (or verbalizations) for both information display and interaction metaphors. With the application mediator providing the appropriate mappings, e.g., a plan can be displayed as a Gantt-chart in which all parameter bindings are formatted in a human-readable fashion but which hides causal relationships and detailed variable constraints; the interface representation does not change unless the user does focus a plan that differs in terms of this view. The interactions however may use another modality and level of detail. In the example: The application fringe is developed automatically as long as it can be synchronized with the user focus, i.e., as long as the Gantt-chart representation is the same, until the initiative changes due to the encounter of a specific constellation of detected flaws and offered plan modification options. Depending on the trigger, the interaction may manipulate the Gantt-chart with mouse gestures, because a causal conflict makes it necessary to insert an ordering constraint. It may also use the natural language modality in a dialog metaphor, asking the user explicitly to choose between predefined modification discriminators. In this way, the user is, e.g., not presented "task expansions" that implement attending a project meeting but is instead asked, whether to "delegate the appointment" or to "attend personally". The paraphrases can be easily defined for any application domain.

It is important to point out, that the framework's representation mechanism realizes a complete encapsulation of the interface's display and interaction metaphor, including their level of detail with respect to the information conveyed and granularity in relation to the plan generation process. In particular, the user's focus and commands do not necessarily have to correspond to a single plan and plan refinements. The interface may display several plans in the user focus by one representative element from the application fringe, e.g., two plans that only differ in a hidden detail. An interface command like "delegate this task to someone else" may be translated into a sequence of several refinement steps in which the appropriate tasks are inserted, expansions are performed, and finally parameters are set. In this context of more complex focus and interaction representations, the above mentioned distance between system and application fringes, and consequently the configuration of the planning strategy components becomes an essential feature of the mixed-initiative planning framework.

4 An Interactive Planning Assistent

We instantiated the mixed-initiative framework in a domain-independent interactive planning application for explaining hybrid planning to students and for debugging domain models and problem definitions. Fig. 3 shows a screenshot from a graphical planning *editor*, i.e., it realizes a visual modality that provides a detailed representation of single plans and it interacts via the tool metaphor through mouse gestures. The application realizes an interaction model that returns the initiative to an expert user after each plan selection decision as soon as the application fringe is completely processed.

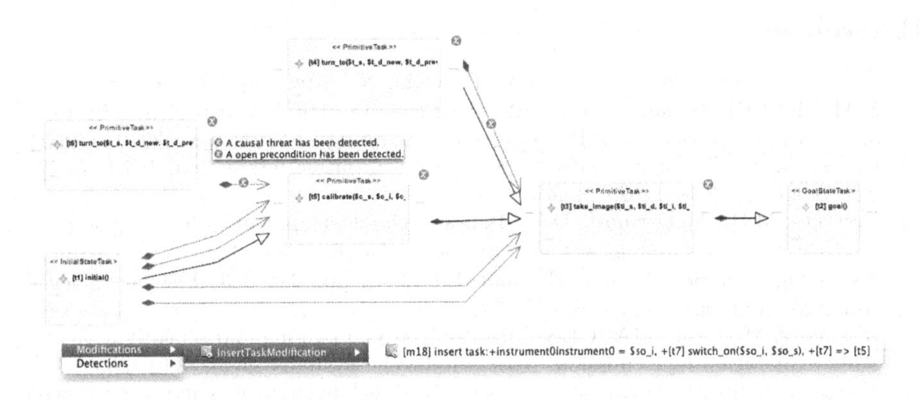

Fig. 3. An interactive planning application

The screenshot shows the graphical representation of a partial plan after applying some refinement steps for solving a planning problem in an observation-satellite domain. Each rectangle represents a primitive task, together with its name and parameter values. Black arrows symbolize ordering constraints, grey open arrows illustrate causal links (the annotated protected condition can be inspected by moving the mouse over the link). The round markings exhibit plan flaws and are placed next to the associated flawed plan elements; detailed descriptions are shown as tool-tips by clicking on the icons. E.g., a causal threat and an open precondition are associated with the task that calibrates a satellite's instrument. The open precondition exists because the calibration requires the instrument to be powered on, which is not the case in the initial state. Once the user selects a flaw to be addressed next, a set of available refinement options are shown; a menu appears that gives the options the user can select by clicking on them. E.g., once the user clicks on the open precondition flaw, a menu at the bottom of the screen appears that gives the user the possibility to insert a "switch on" task that powers on the underlying instrument.

The triggering mechanism of this application fills a list of solutions that have been found (autonomously by the planning system or guided by the user) and notifies the user, if the currently focused plan necessarily fails. Apart from the plan modification "tools", the user can undo operations by backtracking over the focus plan as well as bookmark plans for direct access.

5 Conclusion

We have presented a generic framework for mixed-initiative planning, which enables broad acceptance of the powerful and flexible hybrid-planning technology. Moreover, implementations of this approach serve as efficient debugging tools and development support for complex application domains. Future work includes the reasearch on suitable interaction modalities and on plan and plan modification representations with respect to their usability aspects.

References

1. Cesta, A., Cortellessa, G., Oddi, A., Policella, N.: Studying decision support for MARS EXPRESS planning tasks: A report from the MEXAR experience. In: Proc. of the 4th Int. Workshop for Planning and Scheduling for Space, pp. 232–240 (2004)
2. Hearst, M.A.: Trends & controversies: Mixed-initiative interaction. IEEE Intelligent Systems 14(5), 14–23 (1999)
3. Burstein, M.H., McDermott, D.V.: Issues in the development of human-computer mixed-initiative planning systems. In: Gorayska, B., Mey, J.L. (eds.) Cognitive Technology: In Search of a Humane Interface, pp. 285–303. Elsevier Science, Amsterdam (1996)
4. Ai-Chang, M., et al.: MAPGEN: Mixed-initiative planning and scheduling for the Mars exploration rover mission. IEEE Intelligent Systems 19(1), 8–12 (2004)
5. Tate, A., Dalton, J., Levine, J., Nixon, A.: Task achieving agents on the world wide web. In: Spinning the Semantic Web, pp. 431–458. MIT Press, Cambridge (2003)
6. Ambite, J.L., Knoblock, C.A., Muslea, M., Minton, S.: Conditional constraint networks for interleaved planning and information gathering. IEEE Intelligent Systems 20(2), 25–33 (2005)
7. Ferguson, G., Allen, J.F.: TRIPS: An integrated intelligent problem-solving assistant. In: AAAI/IAAI, pp. 567–572 (1998)
8. Biundo, S., Schattenberg, B.: From abstract crisis to concrete relief – A preliminary report on combining state abstraction and HTN planning. In: Cesta, A., Borrajo, D. (eds.) Proc. of the 6th European Conf. on Planning, pp. 157–168 (2001)
9. Schattenberg, B., Bidot, J., Biundo, S.: On the construction and evaluation of flexible plan-refinement strategies. In: Hertzberg, J., Beetz, M., Englert, R. (eds.) KI 2007. LNCS (LNAI), vol. 4667, pp. 367–381. Springer, Heidelberg (2007)
10. Schattenberg, B., Biundo, S.: A unifying framework for hybrid planning and scheduling. In: Freksa, C., Kohlhase, M., Schill, K. (eds.) KI 2006. LNCS (LNAI), vol. 4314, pp. 361–373. Springer, Heidelberg (2007)

A Memory-Efficient Search Strategy for Multiobjective Shortest Path Problems*

L. Mandow and J.L. Pérez de la Cruz

Dpto. Lenguajes y Ciencias de la Computación
Universidad de Málaga 29071 - Málaga, Spain
{lawrence,perez}@lcc.uma.es

Abstract. The paper develops vector frontier search, a new multiobjective search strategy that achieves an important reduction in space requirements over previous proposals. The complexity of a resulting multiobjective frontier search algorithm is analyzed and its performance is evaluated over a set of random problems.

1 Introduction

The Multiobjective Search Problem is an extension of the Shortest Path Problem that allows the evaluation of a number of distinct attributes for each path under consideration [1][2]. In multiobjective graphs arcs are labeled with vector costs. Each component in a vector cost stands for a different relevant attribute, e.g. distance, time, or monetary cost.

This paper deals with best-first algorithms like Dijkstra's and A* [3], and their multiobjective counterparts [4] [5] [6]. Frontier search [7] is a new best-first technique that exploits the monotonicity of evaluation functions to discard already explored nodes. This reduces the space complexity over standard search for certain classes of problems.

There are two previous approaches to multiobjective frontier search. *Node frontier search* [8] attempted also the elimination of explored *nodes*. However, the reduction in space requirements was limited, and incurred in an important time overhead. *Bicriterion frontier search* [9] eliminates explored *cost vectors*. The space reduction is large, but limited to uninformed bicriterion problems.

This paper extends *bicriterion frontier search* to the general uninformed multiobjective case. Section 2 highlights key issues in scalar frontier search. Sections 3, 4 discuss fundamentals and complexity results of multiobjective best-first search. Sections 5, 6 describe a new efficient pruning strategy and the scheme for vector frontier seach. Experimental results are presented in section 7. Finally, conclusions and future work are outlined.

2 Scalar Frontier Search

The shortest path problem can be stated as follows: Let G be a locally finite labeled directed graph $G = (N, A, c)$, of $|N|$ nodes, and $|A|$ arcs (n, n') labeled with positive

* This work is partially funded by/Este trabajo está parcialmente financiado por: Consejería de Innovación, Ciencia y Empresa. Junta de Andalucía (España), P07-TIC-03018.

B. Mertsching, M. Hund, and Z. Aziz (Eds.): KI 2009, LNAI 5803, pp. 25–32, 2009.

costs $c(n, n') \in \mathbb{R}$. Given a start node $s \in N$, and a set of goal nodes $\Gamma \subseteq N$, find the minimum cost path in G from s to a node in Γ.

Scalar best-first algorithms, like Dijkstra's, build a search tree rooted at the start node with the best paths found to each generated node. Nodes already expanded are called $Closed$ nodes, while those awaiting for expansion are called $Open$ nodes. Function $g(n)$ denotes the cost of the best known path reaching node n. By virtue of the optimality principle, new paths to node n are pruned whenever their cost exceeds $g(n)$.

The success of scalar frontier search is based on the combination of three different elements. First of all, frontier search assumes that evaluation functions are monotonic. Under this assumption, once a node n is selected for expansion, we can guarantee that the cost of the path reaching n is optimal. Therefore, any new paths reaching n can be discarded straightaway without further consideration. We do not even need to check the values of $g(n)$ to be sure that any new path will be suboptimal.

Secondly, frontier search keeps in memory only $Open$ nodes, eliminating $Closed$ ones. We know that new paths to closed nodes will be suboptimal, therefore all that is needed is a guarantee that once a node is closed and eliminated from memory, it will never be regenerated again. This is achieved using a special technique that keeps *used operator* vectors within each node. This idea can be applied to algorithms like Dijkstra's and A*. With monotonic evaluation functions, and the same tie-breaking rule, frontier search algorithms mimic the sequence of node expansions performed by their standard best-first counterparts. Therefore, they are admissible under the same assumptions, and terminate returning the goal node and its optimal cost in the same number of steps.

In exponential search spaces, like trees with constant branching factor b, the $Open$ list is generally *larger* than the $Closed$ list. Therefore, frontier search can reduce memory requirements by a constant factor $1/b$. However, in polynomial search spaces, like m-dimensional square grids, the frontier is typically an order of magnitude *smaller* than the set of expanded nodes. In this case, worst-case memory complexity is reduced from $O(d^m)$ to $O(d^{m-1})$ for square grids with depth d and m dimensions.

In standard search, the solution path is recovered tracing back pointers kept in closed nodes. However, these are missing in frontier search. The third element of frontier search is a recursive formulation that reconstructs the solution path after the goal node is reached. The idea is to identify some *midline node* n' traversed by the solution, and then call frontier search recursively from the start node to n', and from n' to the goal.

3 Fundamentals of Multiobjective Search

The multiobjective search problem can be stated as follows: Let G be a locally finite labeled directed graph $G = (N, A, \mathbf{c})$, of $|N|$ nodes, and $|A|$ arcs (n, n') labeled with positive vectors $\mathbf{c}(n, n') \in \mathbb{R}^q$. Given a start node $s \in N$, and a set of goal nodes $\Gamma \subseteq N$, find the set of all *non-dominated* cost paths in G from s to nodes in Γ.

The main distinguishing feature of multiobjective problems is the fact that cost vectors $\mathbf{c}(n, n')$ induce only a partial order preference relation $\prec$ called *dominance*,

$$\forall \boldsymbol{g}, \boldsymbol{g}' \in \mathbb{R}^q \quad \boldsymbol{g} \prec \boldsymbol{g}' \quad \Leftrightarrow \quad \forall i \quad g_i \leq g'_i \wedge \boldsymbol{g} \neq \boldsymbol{g}' \tag{1}$$

where g_i denotes the i-th element of vector $\boldsymbol{g}$.

Given two vectors, it is not always possible to rank one as better than the other, e.g. none of $(2,3)$, $(3,2)$ dominates each other, but both are dominated by vector $(1,1)$.

Definition 1. *Given a set of vectors X, we shall define $nd(X)$ the set of non-dominated vectors in set X in the following way, $nd(X) = \{\boldsymbol{x} \in X \mid \nexists \boldsymbol{y} \in X \quad \boldsymbol{y} \prec \boldsymbol{x}\}$.*

Most multiobjective best-first algorithms follow a similar scheme [4] [6]. These build an acyclic search graph SG rooted at s to store all nondominated paths found to each node. Whenever a new path is found to a known node n, its cost is compared to those already reaching n. Nondominated paths are kept, and dominated ones are pruned. Two different sets are associated to each node: $G_{op}(n)$ denotes the set of cost vectors of paths reaching n that can be further explored, while $G_{cl}(n)$ denotes the set of those that have already been expanded. The set of $OPEN$ *paths* in SG that can be further explored is made up of all labels $(n, \boldsymbol{g})$ where $\boldsymbol{g} \in G_{op}(n)$.

In this paper we shall consider only uninformed search. Particularly, we shall look for alternatives to the same standard multiobjective (MO) algorithm used by [9] (see table 1). In order to guarantee admissibility, MO search needs to select for expansion an open path P (i.e. a label $(n, \boldsymbol{g})$) with a cost vector g nondominated in $OPEN$. Two sets, $GOALN$ and $COSTS$, keep track of goal nodes and costs of non-dominated solutions found so far. Each time a new solution is found, dominated alternatives are filtered from $OPEN$ and search proceeds until $OPEN$ is empty.

In multiobjective search many different non-dominated paths may reach a given node. Therefore, it is the number of cost vectors stored in $G_{op}(n)$ and $G_{cl}(n)$ that dominates space requirements, while the number of nodes plays only a minor role. Regrettably, the expansion of a non-dominated path to node n does not prevent other non-dominated paths to enter $G_{op}(n)$ at later stages of the search, even with monotonic cost functions [5](lemma 20). Standard MO search algorithms keep vector costs of expanded paths in the $G_{cl}(n)$ sets in order to prune new dominated paths reaching already

Table 1. A standard multiobjective search algorithm (blind NAMOA*)

1. Initialize an acyclic search graph SG with root in s, a set of labels $OPEN = \{(s, \boldsymbol{g}(s))\}$, and two empty sets, $GOALN$, $COSTS$.
2. If $OPEN$ is empty, return paths in SG with costs in $COSTS$ reaching nodes in $GOALN$.
3. Remove from $OPEN$ a non-dominated label $(n, \boldsymbol{g}^n)$.
4. If n is a goal node, then: (a) Put n in $GOALN$ and $\boldsymbol{g}^n$ in $COSTS$; (b) Eliminate from $OPEN$ all labels $(x, \boldsymbol{g}^x)$ such that $\boldsymbol{g}^n \prec \boldsymbol{g}^x$; (c) Go back to step 2
5. If n is not a goal node, then
   (a) Move $\boldsymbol{g}^n$ from $G_{op}(n)$ to $G_{cl}(n)$.
   (b) For all successors nodes m of n, such that $\boldsymbol{g}^m = \boldsymbol{g}^n + \boldsymbol{c}(n,m)$ is not dominated by $COSTS$, and that do not produce cycles in SG do:
      - If m is new, then put $(m, \boldsymbol{g}^m)$ in $OPEN$, and g^m in $G_{op}(m)$ pointing to n. else,
         i. Remove from $G_{op}(m)$ and $G_{cl}(m)$ all vectors dominated by $\boldsymbol{g}^m$.
         ii. If $\boldsymbol{g}^m$ is not dominated by $G_{op}(m)$ nor by $G_{cl}(m)$, put it in $G_{op}(m)$ with a pointer to n, and put label $(m, \boldsymbol{g}^m)$ in $OPEN$.
      - Go back to step 2.

explored nodes. In order to reduce the space requirements of multiobjective search the size of the $G_{cl}(n)$ sets should be reduced as much as possible.

4 Complexity Issues in Multiobjective Search

Memory is an important practical limiting factor of multiobjective best-first search. The algorithms need to store every known node n, as well as the set of non-dominated cost vectors of paths reaching n, i.e. the $G_{op}(n)$ and $G_{cl}(n)$ sets. The number of distinct non-dominated costs is known to grow exponentially with node depth in the worst case for the general case [4]. However, the number of costs to any node grows only polynomially with depth in the worst case assuming arc costs are integer, positive, and bounded.

Let us assume that all components of each arc cost are integers in the interval $[c_i, c_a]$, where $c_i, c_a > 0$, and denote the range of distinct integer values for each arc cost component by $r = c_a - c_i + 1$. What is then the maximum number of distinct non-dominated cost vectors that can reach any node n at depth d? Let us consider the cost of paths to n depending on their length. Let us also consider for simplicity the two-dimensional case. The shallowest paths reaching n have depth (length) d. Each component of the cost vectors will have a value in the range $[d \times c_i, d \times c_a]$. By definition of node depth, there will always be at least one path with length d. Therefore, no path with cost dominated by $\boldsymbol{u} = (d \times c_a, d \times c_a)$ will ever take part in the Pareto front of n (i.e. the set of nondominated costs to n), regardless of the path's length. This is illustrated graphically in figure 1(left).

There are also possibly paths of lengths greater than d reaching n. For paths of length $d+1$ each component of the cost vectors will have a value in the range $[(d+1) \times c_i, (d+1) \times c_a]$. Figure 1(center) shows the feasible cost space for paths of length $d+1$ reaching a node n at depth d. Obviously, for some length $d + k$ $(k > 0)$ all path costs will be dominated by $\boldsymbol{u}$. This is the case where $(d+k) \times c_i \geq d \times c_a$, i.e. $k \geq d \times (c_a - c_i)/c_i$. Therefore, the feasible cost space for any node is bounded and finite. Figure 1(right) shows such space for a particular case ($d = 3$, costs in $[1, 2]$), and the largest possible Pareto front that fits into that space.

Since the feasible cost space for any given node is bounded and finite, it is also possible to bound the size of the largest Pareto front that fits into that space. For a node at depth d and q objectives, the maximum size of the Pareto front is $O((dr)^{q-1})$. This is also the worst-case number of cost vectors stored in any node at depth d.

Now, what is the worst-case number of iterations (or cost-vectors) in a polynomial state space? When monotonic evaluation functions are used, each new label $(n, \boldsymbol{g})$ selected for expansion must stand for a new non-dominated path to n. Therefore, the worst-case number of iterations amounts to the addition of the number of all non-dominated cost vectors reaching every node in the graph.

Let us consider the case of an m-dimensional square grid that extends uniformly from a central node s outwards up to some depth D. The number of nodes in the grid is $O(D^m)$, i.e., the size of the state space grows polynomially with depth. As a particular case, consider a bidimensional grid. There are exactly $4d$ nodes at depth d, and in a complete grid with depth D there are exactly $1 + \sum_{d=1}^{D} 4d$ nodes, i.e. $O(D^2)$.

For example, in a complete bidimensional grid with two objectives there are exactly $4d$ nodes at depth d and, a Pareto front of size $O(dr)$ at each node. Therefore, in the worst case search will explore $O(d^2r)$ cost vectors at depth d. The number of such vectors in the complete grid grows with depth D as $O(D^3r)$. With three objectives, the largest Pareto front at depth d has $O((dr)^2)$ optimal vectors, and the number in a complete bidimensional grid becomes $O(D^4r^2)$. For a general m-dimensional grid with q objectives, the number of iterations (or cost vectors) in the worst case is $O(D^{m+q-1}r^{q-1})$.

5 Efficient Path Pruning in Multiobjective Search

In this section, we shall develop a new way to test the optimality principle that does *not* require all non-dominated cost vectors typically stored in the $G_{cl}(n)$ sets.

Standard MO search algorithms apply the optimality principle to prune non-promising paths (see step 5(b)ii, table 1). When monotonic evaluation functions are used, and a label $(n, \boldsymbol{g})$ is selected for expansion, we can be sure $\boldsymbol{g}$ is a non-dominated cost to n. Accordingly, $\boldsymbol{g}$ is placed in $G_{cl}(n)$. Whenever a new path P reaches some node n with cost $\boldsymbol{g}''$, and $\boldsymbol{g}''$ is dominated by some vector in $G_{cl}(n)$, then P can be safely pruned and its cost vector discarded straightaway. Therefore, all standard MO search algorithms keep large G_{cl} sets to guarantee efficient pruning.

We shall make the following assumption regarding MO search algorithms,

Assumption 1. *The path selection procedure is monotonic nondecreasing in the first component of cost vectors, i.e. if label $(n, \boldsymbol{g})$ is selected for expansion before label $(n', \boldsymbol{g}')$, then $g_1 \leq g'_1$, where g_1 is the first component of vector $\boldsymbol{g}$.*

Definition 2. *Given a vector $\boldsymbol{v} = (v_1, v_2, \ldots v_n)$, we shall define its truncated vector $\boldsymbol{t}(\boldsymbol{v})$ as vector $\boldsymbol{v}$ without its first component, i.e. $\boldsymbol{t}(\boldsymbol{v}) = (v_2, \ldots v_n)$.*

In order to apply the optimality principle under assumption 1, we do *not* need to check the cost of every new path reaching some node n against the whole $G_{cl}(n)$ set. The reason is illustrated graphically in figure 2 for the three objective case. Let us assume three different optimal paths have been found to some node n with non-dominated cost vectors $\boldsymbol{x} = (3, 3, 2)$, $\boldsymbol{y} = (4, 2, 3)$, and $\boldsymbol{z} = (6, 1, 1)$. Therefore, $G_{cl}(n) = \{\boldsymbol{x}, \boldsymbol{y}, \boldsymbol{z}\}$.

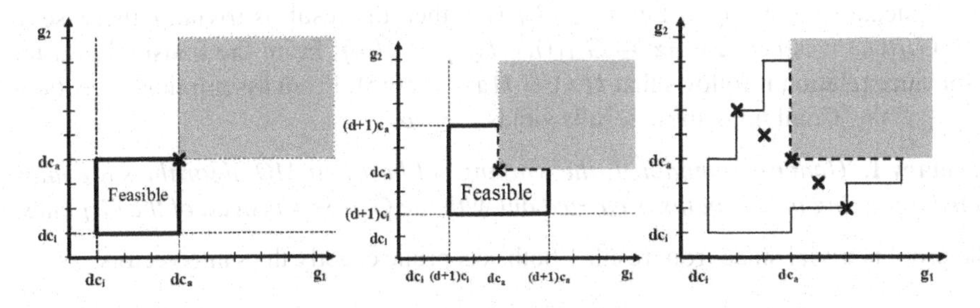

Fig. 1. Left: feasible cost space for paths of length d, vector $\boldsymbol{u} = (d \times c_a, d \times c_a)$ is marked with a cross. Center: feasible cost space for paths of length $d + 1$. Right: complete feasible cost space for depth 3, costs in $[1, 2]$, and largest Pareto front (crosses).

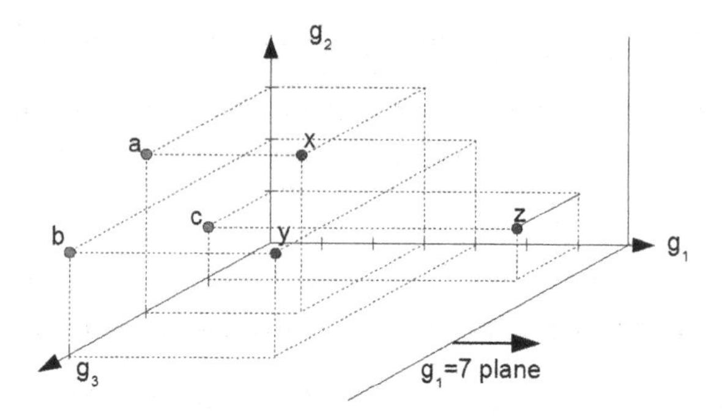

Fig. 2. Sample G_{cl} set for some node n

According to assumption 1, these costs will have been selected exactly in this order, and all paths with cost $\boldsymbol{v}$ such that $v_1 < 6$ have already been explored. Let us further assume some new path with cost $\boldsymbol{w} = (w_1, w_2, w_3)$ reaches node n. Since w_1 is greater or equal than the first component of all previously selected cost vectors, in order to dermine if $\boldsymbol{w}$ should be pruned we need to check only if $\boldsymbol{t}(\boldsymbol{w})$ is dominated by some of $\boldsymbol{t}(\boldsymbol{x})$, $\boldsymbol{t}(\boldsymbol{y})$, or $\boldsymbol{t}(\boldsymbol{z})$. These two-dimensional cost vectors are shown in figure 2 as $\boldsymbol{a}$, $\boldsymbol{b}$, and $\boldsymbol{c}$. Notice that in two-dimensional cost space, $\boldsymbol{a}$ and $\boldsymbol{b}$ are dominated by $\boldsymbol{c}$. Therefore, in order to determine if $\boldsymbol{w}$ is dominated by G_{cl} we *only* need to check if $\boldsymbol{t}(w)$ is dominated by $\boldsymbol{t}(\boldsymbol{z})$.

Definition 3. *Let* $X(n) = nd\{\boldsymbol{t}(\boldsymbol{x}) \,|\, \boldsymbol{x} \in G_{cl}(n)\}$ *be the set of non-dominated truncated closed vectors to node* n*. Let* $G_{cl1}(n)$ *and* $G_{cl2}(n)$ *be defined such that* $G_{cl1}(n) \cup G_{cl2}(n) = G_{cl}(n)$*,* $G_{cl1}(n) \cap G_{cl2}(n) = \emptyset$*, and* $G_{cl1}(n) = \{\boldsymbol{y} \,|\, \boldsymbol{y} \in G_{cl}(n) \wedge \boldsymbol{t}(\boldsymbol{y}) \in X(n)\}$.

In figure 2 we have $G_{cl1}(n) = \{\boldsymbol{z}\}$ and $G_{cl2}(n) = \{\boldsymbol{x}, \boldsymbol{y}\}$.

Result 1. *Let* $(n, \boldsymbol{w})$ *be a new label selected for expansion. Under assumption 1,* $\exists \boldsymbol{x} \in G_{cl}(n) \,|\, \boldsymbol{x} \preceq \boldsymbol{w}$ *if and only if* $\exists \boldsymbol{y} \in G_{cl1}(n) \,|\, \boldsymbol{y} \preceq \boldsymbol{w}$.

Proof. The implication to the left is trivial, since $G_{cl1}(n) \subseteq G_{cl}(n)$. Let us consider the implication to the right. Let $\boldsymbol{x} \in G_{cl1}(n)$, then the result is trivial. Otherwise, if $\boldsymbol{x} \notin G_{cl1}(n)$ it is because $\exists \boldsymbol{z} \in G_{cl1}(n) \,|\, \boldsymbol{t}(\boldsymbol{z}) \preceq \boldsymbol{t}(\boldsymbol{x})$. From the transitivity of the dominance relation it follows that $\boldsymbol{t}(\boldsymbol{z}) \preceq \boldsymbol{t}(\boldsymbol{x}) \preceq \boldsymbol{t}(\boldsymbol{w})$. From assumption 1 we know that $z_1 \leq w_1$. Combining these results yields $\boldsymbol{z} \preceq \boldsymbol{w}$.

Corolary 1. *Under assumption 1, the workings of best-first MO algorithms are unaffected if pruning of new paths is carried out with the* G_{cl1} *sets instead of the* G_{cl} *ones.*

The proof is trivial since from result 1 both sets prune exactly the same vectors.

6 Vector Frontier Search

The development of a vector frontier search (VFS) algorithm is simple from the previous section under assumption 1. Each time a new label $(n, \boldsymbol{g})$ is selected for expansion, we update the $G_{cl1}(n)$ set. Vectors that would fall into $G_{cl2}(n)$ can be simply removed from memory since they are not necessary for pruning and, in practice, are only useful in the algorithms for path recovery purposes.

Notice that, as described in section 4, the size of $G_{cl}(n)$ for a node of depth d with q objectives is $O((dr)^{q-1})$. However, truncating cost vectors reduces the worst-case size of the Pareto front to that of a cost space with $q-1$ objectives, and therefore, the size of $G_{cl1}(n)$ is just $O((dr)^{q-2})$. This effectively reduces the space requirements in polynomial state spaces to $O(D^{m+q-2}r^{q-2})$. The approach is valid for any $q \geq 2$.

A bicriterion frontier search algorithm (BCFS) was presented in [9]. BCFS uses a lexicographic order to select $Open$ paths, which is monotonic in the first component of cost vectors, as required by assumption 1. For each node n BCFS replaces closed cost vectors by an upper bound b on the second component. All new cost vectors reaching n whose second component are greater than the bound can be shown to be dominated. Truncating cost vectors $(g_1^i, g_2^i), i = 1..k$ results in a set of scalar values g_2^i. The optimal of these values is precisely the single scalar bound $b = min_{i=1}^{k} g_2^i$ used in [9]. Therefore, the BCFS algorithm is exactly the particular case of VFS for two objectives.

Finally, since VFS simply replaces the standard multiobjective pruning procedure with an equivalent memory-efficient one, it follows that with the same tie-breaking rule, VFS will expand the same labels in the same order as standard MO search. Therefore, it will reach the same goal nodes finding the same sets of non-dominated costs. A second recursive search phase is needed for path recovery. The same recovery procedures developed for *node frontier search* [10] can be applied to VFS as well.

7 Experimental Tests

Experimental results were presented in [9] for BCFS applied to two-objective problems. This section uses a similar experimental setup to confirm reductions in space requirements for VFS in problems with three objectives. A set of two-dimensional square grids of increasing size was generated with uniformly distributed random integer costs in the range $[1, 10]$. For each grid, search was conducted from one corner to the opposite.

Figure 3(left) reveals similar time requirements for standard MO search and VFS (excluding path recovery) as a function of grid size. Figure 3(right) shows the maximum

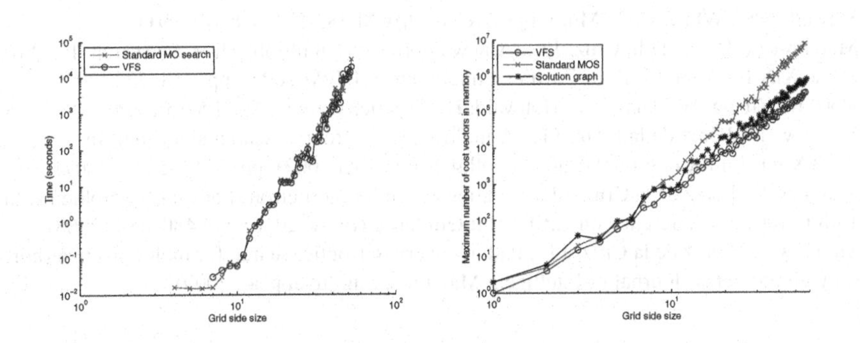

Fig. 3. Results of experimental tests

number of cost vectors simultaneously stored in memory by standard MO search, VFS, and the size of the completely labeled solution subgraph. For the analyzed state space the problem of space requirements seems technically solved, since the memory needed to run VFS is actually *smaller* than the size of the solution subgraph. The experiments show a dramatic reduction in space requirements for vector frontier search. In a grid of size 58 (solution depth 116), standard search stored 8912157 cost vectors, while VFS stored only 368178 (4'13 %). Standard MO search exhausted memory for the first time while searching a 59 x 59 grid. However, using vector frontier search harder problems can be solved, for example, VFS could explore a 70×70 grid (solution depth 140) in 14'5 hours, storing only 539504 cost vectors. Although standard search could not solve this problem we can estimate its space requirements from the number of iterations of VFS which was 14877038, i.e. VFS stored only 3'62 % of the cost vectors that standard search would have required.

8 Conclusions and Future Work

The paper describes vector frontier search, a new general framework for multiobjective search that achieves a reduction in worst-case memory complexity for polynomial state spaces with bounded positive integer costs. This result is validated experimentally over a set of random grid search problems with three objectives. Bicriterion frontier search is shown to be a particular case of vector frontier search. For the class of problems analyzed time is now the major practical limiting factor. Future research involves improving time requirements and analyzing the time needed for solution recovery.

References

1. Ehrgott, M.: Multicriteria Optimization. Springer-Verlag New York, Inc., Heidelberg (2005)
2. Raith, A., Ehrgott, M.: A comparison of solution strategies for biobjective shortest path problems. Comput. Oper. Res. 36(4), 1299–1331 (2009)
3. Hart, P., Nilsson, N., Raphael, B.: A formal basis for the heuristic determination of minimum cost paths. IEEE Trans. Systems Science and Cybernetics SSC-4 2, 100–107 (1968)
4. Hansen, P.: Bicriterion path problems. In: LNEMS 177, pp. 109–127. Springer, Heidelberg (1979)
5. Stewart, B.S., White, C.C.: Multiobjective A*. JACM 38(4), 775–814 (1991)
6. Mandow, L., Pérez de la Cruz, J.L.: A new approach to multiobjective A* search. In: Proc. of the XIX Int. Joint Conf. on Artificial Intelligence (IJCAI 2005), pp. 218–223 (2005)
7. Korf, R., Zhang, W., Thayer, I., Hohwald, H.: Frontier search. JACM 52(5), 715–748 (2005)
8. Mandow, L., Pérez de la Cruz, J.L.: A multiobjective frontier search algorithm. In: Proc. of the XX Int. Joint Conf. on Artificial Intelligence (IJCAI 2007), pp. 2340–2345 (2007)
9. Mandow, L., Pérez de la Cruz, J.L.: Frontier search for bicriterion shortest path problems. In: 18th European Conference on Artificial Intelligence (ECAI 2008), pp. 480–484 (2008)
10. Mandow, L., Pérez de la Cruz, J.L.: Path recovery in frontier search for multiobjective shortest path problems. Journal of Intelligent Manufacturing (to appear, 2009)

Perfect Hashing for State Spaces in BDD Representation

Martin Dietzfelbinger[1,*] and Stefan Edelkamp[2,**]

[1] Faculty of Computer Science and Automation, Technische Universität Ilmenau
P.O. Box 100565, 98684 Ilmenau, Germany
martin.dietzfelbinger@tu-ilmenau.de
[2] TZI, Universität Bremen, Am Fallturm 1, 28357 Bremen, Germany
edelkamp@tzi.de

Abstract. In this paper we design minimum perfect hash functions on the basis of BDDs that represent all reachable states $S \subseteq \{0,1\}^n$. These functions are one-to-one on S and can be evaluated quite efficiently. Such hash functions are useful to perform search in a bitvector representation of the state space. The time to compute the hash value with standard operations on the BDD G is $(n|G|)$, the time to compute the inverse is $O(n^2|G|)$. When investing $O(n)$ bits per node, we arrive at $O(|G|)$ preprocessing time and optimal time $O(n)$ for ranking and unranking.

1 Introduction

In many cases, BDDs [5] provide a compact representation of state spaces [2]. For some challenges, even an exponential gap between the number of represented states and the number of BDD nodes to represent them has been formally shown [8]. In general game playing [16], BDDs have been used to count all reachable states. In smaller games like *Peg-Solitaire*, a full classification is possible using BDDs [9]. For larger games like Connect-Four, however, a bottom-up procedure posterior to the reachability analysis to determine the game theoretical value of each state fails. Multi-player games are more difficult to analyze, first approaches with BDDs are described in [9].

A minimal perfect hash function is a mapping from the state space $S \subseteq \{0,1\}^n$ to the set of indices $\{0, \ldots, |S|-1\}$ that is one-to-one on S. Recent results [3,4] show that given the explicit state space on disk, minimum perfect hash functions with a few bits per state can be constructed I/O efficiently. If the state space is provided by a BDD, to the authors' knowledge so far no algorithm has been proposed to compute a perfect hash value and its inverse efficiently. We describe several constructions. We assume that the state space S is given by some BDD with some ordering of the variables. We will never change this ordering, and for simplicity will assume it is $(s_1, \ldots, s_n)$, the natural ordering. The construction works with an arbitrary fixed ordering, and the hash function depends on the

* Research partially supported by DFG grant DI 412/10-1.
** Research partially supported by DFG grant ED 74/8-1.

B. Mertsching, M. Hund, and Z. Aziz (Eds.): KI 2009, LNAI 5803, pp. 33–40, 2009.

ordering, which means that it does not necessarily represent a natural ordering of S.

The motivation is the search in state spaces that have been compressed to a bitvector. For example, in *Peg-Solitaire* 187,636,298 states are reachable [10]. A hash function, mapping each state with $0 \leq p \leq 33$ pegs to $\{0, \ldots, \binom{33}{p} - 1\}$, requires 8,589,860,936 bits in total. For the Connect-Four problem the precise number of states reachable from the initially empty board is 4,531,985,219,092 [9] (which compares to an estimate of 70,728,639,995,483 reachable states [1]). When using BDDs, 84,088,763 nodes suffice to represent all states.

As described in [14], external two-bit breadth-first search integrates a tight compression method into an I/O-efficient algorithm. This approach for solving large-scale problems relies on applying an efficient minimum perfect hash function (ranking) and its inverse (unranking). It applies a space-efficient representation of the state space in BFS with two bits per state [15], an idea that goes back to [6]. With 2 bits per state Connect-Four requires a size of about 1 TB, which, nowadays, is a standard equipment for a PC. In case of explicit representation, for a state encoding with two bits for each cell plus one for the current player nearly 43.8 terabytes would be necessary.

In this paper we discuss an appropriate design of a rank and unrank function for satisfying assignments in a BDDs. The perfect hash value will compute a *rank* in $\{0, \ldots, \textit{sat-count}(G)-1\}$, where *sat-count*($G$) denotes the number of satisfying inputs in the BDD G. Its inverse is *unrank* with respect to lexicographic ordering of the bit strings in S and both are based on the given variable ordering. Given a BDD G with n variables and an input $s = (s_1, \ldots, s_n)$, the running time for computing the rank from scratch takes time $O(n|G|)$. Unranking, namely to compute the binary state vector given the hash value costs $O(n^2|G|)$ steps. An alternative approach that precomputes and stores values at each node takes $O(|G|)$ preprocessing time, and $O(n)$ time for each call to rank and unrank.

2 Ordinary BDD without Preprocessing

Reduced and ordered BDDs can be constructed by repeatedly applying two reduction rules, namely R1 and R2 to the (ordered) decision tree that represents the truth table of a given function: R1 omits nodes whose successors are identical and R2 merges two nodes with identical successor pairs.

Given a variable ordering that all paths must obey, the BDD representation is unique. There are many libraries (like CUDD and Buddy) for BDD manipulation that provide basic functionality including the evaluation of assignments, counting the number of satisfying assignments, applying Boolean connectors and quantifiers, computing an image wrt. a set of states and a set of operators, and routines for finite domain arithmetics. By applying iterated images BDD can be used to perform a reachability analysis of a state space [17].

Computing the Hash Function: Ranking. Let BDD G with n variables represents a set of reachable states and let a satisfying assignment $s = (s_1, \ldots, s_n)$ represent the query state. The task is to find a unique value in $\{0, \ldots, \textit{sat-count}(G) -1\}$.

Algorithm 1. Ranking a state

```
rank(s,G)
  if (eval(s,G) = 0) return Error("state not reachable")
  G' := lexicographic-split(G,s)
  return sat-count(G')-1
```

Algorithm 2. Splitting a BDD G at s (wrong version with counter-example)

```
lexicographic-split(G,s)
  start at root of G; i := 1
  while (i <= n)
    if there is a node for x_i
      if (s_i = 1)
        follow 1-edge
      if (s_i = 0)
        link 1-edge to 0-sink
        follow 0-edge
    i := i+1
```

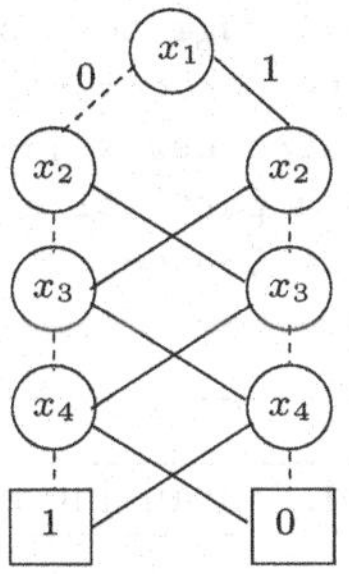

In the core routine of Alg. 1 operations *eval* and *sat-count* are available and take time $O(n)$ and $O(|G'|)$, respectively. The lexicographic split problem is defined as follows. Let $s = (s_1, \ldots, s_n)$ be the state vector and G be the input BDD. We are looking for a BDD H which represents all states s' with a lexicographic (binary) value smaller than or equal to s. The first solution shown in Alg. 2 has a subtle problem. If we take a standard BDD for the even parity function wrt. variable order $(x_1, \ldots, x_4)$ (see figure) for input state vector $s = (1001)$, the above algorithm eliminates the 1-edge emanating from node x_3 that is reached by 10. Now the satisfying input $s' = (0110)$ has been eliminated although that it is lexicographic smaller than the input $s = (1001)$.

A possible fix is Alg. 3. The BDD replicas remain untouched, BDD reduction is postponed. The blow-up in the BDD-size is at most a factor $O(n)$, as up to $O(n)$ BDDs of worst-case size $O(|G|)$ are created.

An alternative approach, also yielding $O(n|G|)$ complexity requires only operations that are provided in common libraries. Suppose that we have a BDD for *LessThanEqual*(x, y) which accepts all pairs $(x, y) \in (\{0,1\}^n)^2$ with x being lexicographically smaller than y and a BDD representing $y = s$. Then we can compute a BDD for the predicate $H(x) = (\exists y.\mathit{LessThanEqual}(y, x) \wedge (y = s)) \wedge G(x)$. Fortunately $\mathit{LessThanEqual}_s(x) = (\exists x.\mathit{LessThanEqual}(x, y) \wedge (y = s))$ has a linear structure that can be constructed directly. This BDD has one node for each x_i with $s_i = 0$. The 1-edge out of this node leads to the 0-sink, the 0-edge to the next (existing) node in the variable ordering. The conjunction of $\mathit{LessThanEqual}_s(x)$ with $G(x)$ yields the same BDD $H(x)$ as the explicit ranking algorithm above.

Unranking One solution, shown in Alg. 4, calls the rank function in a binary search manner. We first observe that ranking can be extended to non-satisfying

Algorithm 3. Splitting a BDD G along s (corrected version)

```
lexicographic-split(G,s)
  start at root of G; i := 1
  while (i <= n)
    if there is a node for x_i
      if (s_i = 1)
        link 0-edge to a COPY of the BDD of the 0-successor
        follow 1-edge
      if (s_i = 0)
        link 1-edge to 0-sink
        if the 0-successor is an x_k node
          insert new x_j-nodes on the path, i<j<k, for s_j=0,
          with parallel edges
        follow 0-edge
    i := i+1
```

Algorithm 4. Unranking based on Ranking

```
unrank(G,r)
 s := (s_1,...,s_n) with s_1 := ... := s_n := 1
 for i := 1 to n do
  s_i := 0;
  if (rank(G,s) < r) then s_i := 1
 return s
```

assignments. The invariant is that the target vector t with $\mathit{rank}(G,t) = r$ satisfies $s_j = t_j$ for $1 \leq j < i$. This is true initially since the condition is void. Now consider round i. From the invariant it follows that $\mathit{rank}(G,(s_1,\ldots,s_{i-1},0,\ldots,0)) \leq r \leq \mathit{rank}(G,(s_1,\ldots,s_{i-1},1,\ldots,1))$. If $\mathit{rank}(G,(s_1,\ldots,s_{i-1},0,1,\ldots,1) < r)$ we can conclude that $t_i = 1$, otherwise $t_i = 0$, hence the conditioned assignment makes sure that the invariant is satisfied when the next round begins. At the end the invariant implies that $s = t$.

As rank is called n times, the algorithm runs in time $O(n^2|G|)$. Not only is this quite a large value, but the frequent changes of the BDD structure are a big problem for practical applications, as they dynamically allocate substantial space, which is assumed to be limited in bitvector state space search. We first give a solution for a simpler structure, before we get back to ordinary BDDs.

3 Quasi-Reduced BDDs with Preprocessing

In a BDD with all variables appearing on every path, sat-counting is rather trivial. Note that a BDD variable is eliminated only in reduction rule R1. If we do not apply R1, or reinsert eliminated nodes pointing to the same successor (bottom-up) along the edges and prior to the root, then we have a structure that

Algorithm 5. Ranking based on Preprocessed Quasi-Reduced BDD

```
lexicographic-count(G*,s,v)
  if v is 0-sink return 0
  if v is 1-sink return 1
  if v is node labeled x_i with 0-succ. u and 1-succ. w
    if (s_i = 1) return sat-count(u) + lexicographic-count(G*,s,w)
    if (s_i = 0) return lexicographic-count(G*,s,u)
rank(G*,s)
  return lexicographic-count(G*,s,root of G*)-1
```

Algorithm 6. Unranking based on Preprocessed Quasi-Reduced BDD

```
unrank(G*,r)
 i := 1
 start at root of G
 while (i <= n)
  at node v for x_i with 0-succ. u and 1-succ. w
    if (r >= sat-count(u))
      r := r - sat-count(u)
      follow 1-edge to w, record s_i := 1
    else follow 0-edge to u, record s_i := 0
  i := i+1
```

Holzman and Puri [13] called *Minimized Automaton.* We call it a quasi-reduced BDD [18].

The blow-up between an ordinary BDD G and a quasi-reduced BDD G^* is at most a factor of $n+1$, i.e., $|G^*| \in O(n|G|)$. The worst-case example is the trivial 0 or 1 function with one sink, which is enlarged to $n+1$ nodes.

Sat-count works as follows. The 1-sink gets a *sat-count*-value of 1 and the 0-sink gets a *sat-count*-value of 0. We propagate the following rule bottom-up: if a x_i labeled node v has an x_{i+1} labeled node u as 0-successor and another x_{i+1} labeled node w as 1-successor we compute *sat-count*(v) = *sat-count*(u) + *sat-count*(w) until we reach the root, which yields the return value. The algorithm requires time and space of at most $O(|G^*|)$.

Ranking. As the BDD for the perfect hash function will not be changed we conserve all possible *sat-count* values. Alg. 5 computes the lexicographic count, given the sat-count values stored in the extended BDD G^* and the query state $s = (s_1, ..., s_n)$. The algorithm runs in time linear to the number of variables n.

Unranking If the rank r is given and the satisfying assignment s is sought, provided that the sat-count values are stored and made accessible, we can select the edges. In difference to the above unranking procedure, Alg. 6 works directly on the annotated BDD. Again, the sat-count annotation of G^* runs in $O(|G^*|)$, while unrank itself runs in $O(n)$ for each call. Let $t = (t_1, \ldots, t_n)$ be the correct

Algorithm 7. Ranking based on Preprocessed Ordinary BDD

```
lexicographic-count(G,s,v)
  if v is 0-sink return 0
  if v is 1-sink return 1
  if v is an x_i node with 1-succ. w for x_k and 0-succ u for x_j
    if (s_i = 1)
       r_1 := lexicographic-count(G,s,w)
       d_1 := binary value of (s_{i+1},...,s_{k-1})
       return 2^{j-i-1} * sat-count(u) + d_1 * sat-count(w) + r_1
    if (s_i = 0)
       r_0 := lexicographic-count(G,s,u)
       d_0 := binary value of (s_{i+1},...,s_{j-1})
       return d_0 * sat-count(u) + r_0
rank(G,s)
  if root is labeled with x_i
  d := binary value of (s_1,...,s_{i-1})
  return (d+1) * lexicographic-count(G,s,root) - 1
```

input with rank r. The invariant in the algorithm is that r is the rank of the input $(t_i, \ldots, t_n)$ among all *sat-count*(v) inputs $(s_i, \ldots, s_n)$ that are accepted when starting at node v (which is an x_i-node). This is true at the beginning since r is given. When we consider node v with label x_i, *sat-count*(u) many of the inputs $(0, s_{i+1}, ..., s_n)$ will be accepted, and *sat-count*(w) many of the inputs $(1, s_{i+1}, ..., s_n)$ will be accepted. The test $r \geq$ *sat-count*(u) finds out whether the input with rank $r \in \{0, \ldots,$ *sat-count*$(u) +$ *sat-count*$(w)\}$ starts with a 1.

4 Ordinary BDDs with Preprocessing

The above algorithm can be adapted to reduced BDDs. The core change is to take care of nodes that have disappeared and thus represent don't cares covering both possible values assignments $s_i = 0$ and $s_i = 1$. This induces some book-keeping and multiplications with factor 2 in the sat-count procedure.

Recall the sat-count algorithm for BDDs, as e.g. already provided in [5]. This is achieved by a DFS in G that associates a value *sat-count*(v) with every node v. If we reach a 1-sink v we set *sat-count*$(v) = 1$, and if we reach a 0-sink v we set *sat-count*$(v) = 0$ For an internal node v labeled with x_i and 0-successor u (labeled with x_j) and 1-successor w (labeled with x_k). we compute *sat-count*$(v) = 2^{j-i-1} \cdot$ *sat-count*$(u) + 2^{k-i-1} \cdot$ *sat-count*(w). If we arrive at the root labeled x_i, we multiply its sat-count value with 2^{i-1} and return the result. *Ranking* Alg. 7 computes the lexicographic count of some BDD G wrt. state $s = (s_1, \ldots, s_n)$. The time and space complexity for pre-processing the graph is $O(|G|)$, while each rank computation then requires $O(n)$ time. Each node is annotated with an n-bit number, which requires $O(n|G|)$ extra bits of space.

Algorithm 8. Unranking based on Preprocessed Ordinary BDD

```
unrank(G,r)
  start at root of G
  if root of G is labeled x_i
     (s_1...s_{i-1}) := binary representation of r div sat-count(root)
     r := r mod sat-count(root)
  i := 1
  while (i <= n)
    if v is an x_i node with 1-succ. w for x_k and 0-succ u for x_j
      if (r >= 2^{j-i-1} * sat-count(u))
        record s_i := 1; r := r - 2^{j-i-1} * sat-count(u)
        (s_{i+1}...s_{k-1}) := binary representation of r div sat-count(w)
        r := r mod sat-count(w)
        follow 1-edge to w
     if (r < 2^{j-i-1} * sat-count(u))
        record s_i := 0
        (s_{i+1}...s_{j-1}) := binary value of r div sat-count(u)
        r := r mod sat-count(u)
        follow 0-edge to u
     i := i + 1
```

Unranking. The inverse is analogous, starting with value r at the root. Again, one can perform the correct decision and determine the value to be subtracted from r to continue. This leads to Alg. 8. The time and space complexies are the same as for ranking. Pre-processing the graph amounts to $O(|G|)$ for performing sat-count and storing values at each node. Each unrank then requires $O(n)$ time.

5 Conclusion

We have shown how to exploit BDDs to provide a minimum perfect hash function for satisfying assignments and their inverse. Thereby we provide an important link from a symbolic to a bitvector representation of the state space. Given the BDD of all reachable states, two-bit BFS can compute the BFS-level of each state. As BDD are constructed via BFS, this is of limited interest on the first glance. However, for the construction of pattern databases [7] backward BFS levels are needed such that a forward BFS traversal with BDDs can be exploited to define the bit-vector for the backward BFS.

Besides two-player game playing, there are many different applications for such perfect hash functions. First there are semi-external and flash-memory based algorithms that have, e.g., been applied in model checking [11,12]. As the state vector has to be processed anyway, the time $O(n)$ for ranking and unranking is fact optimal. In terms of flash media like solid state disks, we highlight that only n random read accesses are needed to rank and unrank.

It is an open question, if the additional storage of $O(n|G|)$ bits can be avoided.

Acknowledgment. The authors thank the referees for their helpful hints.

References

1. Allis, L.V.: A knowledge-based approach to Connect-Four. The game is solved: White wins. Master's thesis, Vrije Universiteit, The Netherlands (1998)
2. Ball, M., Holte, R.C.: The compression power of BDDs. In: ICAPS, pp. 2–11 (2008)
3. Botelho, F.C., Pagh, R., Ziviani, N.: Simple and space-efficient minimal perfect hash functions. In: Dehne, F., Sack, J.-R., Zeh, N. (eds.) WADS 2007. LNCS, vol. 4619, pp. 139–150. Springer, Heidelberg (2007)
4. Botelho, F.C., Ziviani, N.: External perfect hashing for very large key sets. In: Conference on Information and Knowledge Management, pp. 653–662 (2007)
5. Bryant, R.E.: Symbolic manipulation of boolean functions using a graphical representation. In: Design Automation Conference (DAC), pp. 688–694 (1985)
6. Cooperman, G., Finkelstein, L.: New methods for using Cayley graphs in interconnection networks. Discrete Applied Mathematics 37/38, 95–118 (1992)
7. Culberson, J.C., Schaeffer, J.: Pattern databases. Computational Intelligence 14(4), 318–334 (1998)
8. Edelkamp, S., Kissmann, P.: Limits and possibilities of BDDs in state space search. In: AAAI, pp. 46–53 (2008)
9. Edelkamp, S., Kissmann, P.: Symbolic classification of general two-player games. In: Dengel, A.R., Berns, K., Breuel, T.M., Bomarius, F., Roth-Berghofer, T.R. (eds.) KI 2008. LNCS (LNAI), vol. 5243, pp. 185–192. Springer, Heidelberg (2008)
10. Edelkamp, S., Messerschmidt, H., Sulewski, D., Yücel, C.: Solving games in parallel with linear-time perfect hash functions. Technical report, TZI, University of Bremen (2009)
11. Edelkamp, S., Sanders, P., Simecek, P.: Semi-external LTL model checking. In: Gupta, A., Malik, S. (eds.) CAV 2008. LNCS, vol. 5123, pp. 530–542. Springer, Heidelberg (2008)
12. Edelkamp, S., Sulewski, D.: Flash-efficient LTL model checking with minimal counterexamples. In: Conference on Software Engineering and Formal Methods (SEFM), pp. 73–82 (2008)
13. Holzmann, G.J., Puri, A.: A minimized automaton representation of reachable states. International Journal on Software Tools for Technology Transfer 2(3), 270–278 (1999)
14. Korf, R.E.: Minimizing disk I/O in two-bit-breath-first search. In: AAAI, pp. 317–324 (2008)
15. Kunkle, D., Cooperman, G.: Twenty-six moves suffice for Rubik's cube. In: International Symposium on Symbolic and Algebraic Computation (ISSAC), pp. 235–242 (2007)
16. Love, N.C., Hinrichs, T.L., Genesereth, M.R.: General game playing: Game description language specification. Technical Report LG-2006-01, Stanford Logic Group (2006)
17. McMillan, K.L.: Symbolic Model Checking. Kluwer Academic Press, Dordrecht (1993)
18. Wegener, I.: Branching Programs and Binary Decision Diagrams — Theory and Applications. Society of Industrial and Applied Mathematics (2000)

On the Benefit of Fusing DL-Reasoning with HTN-Planning*

Ronny Hartanto[1,2] and Joachim Hertzberg[2]

[1] Bonn-Rhein-Sieg Univ. of Applied Sciences,
53757 Sankt Augustin, Germany
[2] University of Osnabrück, 49069 Osnabrück, Germany

Abstract. Keeping planning problems as small as possible is a must in order to cope with complex tasks and environments. Earlier, we have described a method for cascading Description Logic (DL) representation and reasoning on the one hand, and Hierarchical Task Network (HTN) action planning on the other. The planning domain description as well as the fundamental HTN planning concepts are represented in DL and can therefore be subject to DL reasoning. From these representations, concise planning problems are generated for HTN planners. We show by way of case study that this method yields significantly smaller planning problem descriptions than regular representations do in HTN planning. The method is presented through a case study of a robot navigation domain and the blocks world domain. We present the benefits of using this approach in comparison with a pure HTN planning approach.

1 Introduction

AI action planning algorithms have gained much in efficiency in recent years, as can be seen, for example, in planning system contests like International Planning Competition (IPC). Still, planning remains computationally intractable in theory, depending on the problem description size. Now consider a planning application in robot control, where a planner should generate a high-level plan for execution by a robot working in, say, a six-storey office building, performing delivery and manipulation tasks. No single plan in this scenario would typically be of impressive length, involving in fact only a tiny fraction of all the rooms, locations, or objects that the robot has to know about in its environment. Only a small fraction of individuals are normally *relevant* for the given planning problem at hand. Yet, the planner has no chance of knowing or estimating what is relevant until it has generated a plan, as it is unable to reason about the domain in any other way, bogged down by the full domain representation.

Gil [1] stated that a challenging area of future research is to integrate existing planning algorithms with rich representations of domain-specific knowledge about planning tasks and objectives, actions and events. In her survey [1], she presented previous work on combining planning techniques with Description Logics (DL) to reason about tasks, plans, and goals. Although the surveyed

* Thanks to Iman Awaad for her support on this work.

B. Mertsching, M. Hund, and Z. Aziz (Eds.): KI 2009, LNAI 5803, pp. 41–48, 2009.

approaches and systems have used DL, it has not been incorporated in state-of-the-art planning algorithms or systems.

In previous work [2], we have described a method for cascading DL representation and reasoning on the one hand, and HTN action planning on the other. The idea is to deliver to the planner a substantially filtered version of the complete domain representation to work on for any individual planning problem. The planner type used is an HTN planner. The environment is primarily represented in terms of a DL ontology, as are basic HTN planning notions. The problem description for a concrete planning problem is generated algorithmically from the DL representation by expanding in terms of the DL ontology the concepts (methods, operators, state facts) that are part of the start and goal formulations in DL. In most cases, this process extracts only a small fraction of the full set of domain individuals to send to the planner, and it does so efficiently.

In the current paper, we compare the pure HTN planning approach and our approach in two examples: a robot navigation domain and then the blocks world. We analyse how irrelevant states affect the overall planning performance. The rest of the paper is organised as follows: First the DL and HTN planning background is sketched. Then we recapitulate our approach from [2]. After that, an implementation and some experiments are described. Finally, results and conclusions are presented.

2 Description Logics and HTN Planning

Description Logics (DL) [3] is a family of state-of-the-art formalisms in knowledge representation. It is inspired by the representation and reasoning capabilities of classical semantic networks and frame systems. Yet, it is a well-defined, decidable subset of first order logic, where some exemplars are even tractable. A DL knowledge base or *ontology* consists of two components, namely, TBox and ABox: The TBox (*terminology*) defines objects, their properties and relations; the ABox (*assertions*) introduces individuals into the knowledge base. Typical reasoning tasks in DL are consistency checks, entailment, satisfiability checks, and subsumption.

Planning systems [4] also define a family of knowledge representation formalisms, yet one of a different ontological slicing than DLs. They model state facts, actions, events and state transition functions as a planning domain. The reasoning performed by a planning system is mainly to search the domain for a set and ordering of actions whose execution would lead to a goal state. Apart from algorithmic differences, planners differ in the assumptions that they make about planning domains and their representations. HTN planning [4, Ch. 11] is able to exploit procedural knowledge coded into the action representation. Standard procedures of elementary actions in the domain are modeled in a hierarchical way into compound tasks, which have to be disassembled during planning. The objective then is to perform some set of tasks, rather than achieving a set of goals. HTN planning has been used widely in applications, owing to this option of pre-coding procedural domain knowledge into the task hierarchy [4].

However, note that there is no way to do DL-type reasoning in an HTN formalism, as these deal with incommensurable aspects of representing a domain. Yet, it turns out that HTN planning can profit from DL reasoning in the domain in the sense that DL reasoning is able to determine that some subset of objects is irrelevant for a concrete planning problem. Therefore, we will recapitulate in the next section our approach to cascading HTN planning and DL reasoning [2].

3 General Approach and Implementation

3.1 Integration of DL and HTN

The basic idea is to split tasks between DL reasoning and HTN planning as shown in Fig. 1. The planning domain, as well as basic HTN planning concepts, are modeled in DL. DL reasoning is then used to automatically extract a tailored representation for each and every concrete planning problem for the planning system. Note that the DL model can be much larger than the proper planning domain. It may include any information that we want our system to have; in particular, information that we would otherwise model in separate planning domain descriptions for planning efficiency reasons. HTN planning is dedicated to finding a solution within the representation extracted by DL, which is reduced by facts, operators, and events found irrelevant by the previous DL reasoning.

[2] details the modelling of HTN Planning in DL. In addition to the DL reasoner, our approach uses two algorithms for extracting the filtered planning problem from the inferred DL model. Their complexity for extracting the initial states (s_0) is linear in the size of the sets of methods/operators and of the ontology [2]: Let $\mathbb{M}$ be set of method instances and $\mathbb{O}$ the set of operator instances in DL; let $m = (|\mathbb{M}|+|\mathbb{O}|)$; let n be the cardinality of concepts in the ontology except those that represent the planning concepts (e.g., operators, planning problem). Then the run time complexity is $O(mn)$. The complexity for generating the planning domain ($\mathcal{D}$) is linear in m. For details see [2]. The current paper examines the performance gain of our approach compared to standard HTN planning.

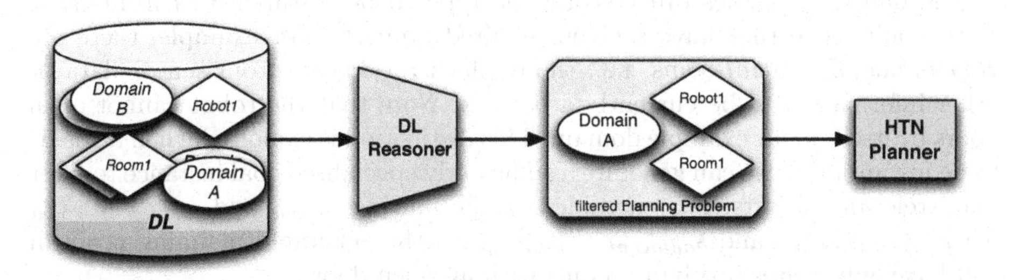

Fig. 1. The schema of how a full domain representation in DL is filtered into a planning problem description, and then passed to the HTN planner

3.2 Implementation

We have used the Web Ontology Language (OWL) to implement the DL formalism; in particular, we have used the OWL-DL variant [5,6] for representing knowledge, to achieve the compromise between expressiveness and reasoning complexity. In our implementation, we use Pellet (an open source OWL-DL reasoner in Java) [7] as the inference engine with OWL-DL. Other inference engines may also be used such as FaCT++ [8] or RacerPro [9].

For the HTN component of our system, we use JSHOP2, the open source Java variant of SHOP2. The Java implementation is a major reason for choosing JSHOP2, because we need to integrate planning and reasoning into a coherent system. In addition, JSHOP2 compiles each domain description into separate domain-specific planners, thus providing a significant performance increase in the speed of planning [10]. Other HTN planners might be used alternatively.

4 Experiments

For the case studies in this paper, we use a mobile robotic navigation domain and the blocks world domain.

Navigation Domain. The goal here is to find a sequence of actions to bring the robot to some target destination. Domain concepts need to be defined accordingly in the DL system. There are two basic concepts, namely, *Actor* and *Fixed-Object*. In our case, *Robot* is an actor; *Building*, *Room*, *Corridor* are fixed objects.

We use the following concepts for modeling the robot's opportunities for action: *Door*, *OpenDoor*, *DriveableRoom*, and *DriveableRoomInBuilding*. Once these concepts are defined, we can assert data for the system. As usual, DL reasoning can deduce inferred instances, in this case, *DriveableRoom*, *Driveable RoomInBuilding* and *OpenDoor* instances are automatically inferred.

The navigation domain has two methods, namely, m_{navi1} and m_{navi2}. They serve the same purpose, but differ in parameters. m_{navi1} expects three parameters: actor, current-location, and destination. m_{navi2} is an abstraction of m_{navi1}, requiring only two parameters: actor and destination.

The ontology enables objects of some type to be distinguished from those of the same type that have a given required property. For example, there are *Rooms* and *DrivableRooms*, i.e., those which are known from sensor data or other information to be currently accessible. Note that the robot cannot open doors by itself in our example domain. This distinction is exploited when a goal is to be instantiated; we can generate a sufficiently specialised goal instance to filter out irrelevant objects. For example, by specializing s_{room} and $s_{adjacent-room}$ with $s_{driveableroom}$ and $s_{adjacent-driveableroom}$, the generated planning problem will have only states involving rooms with an open door.

In our DL navigation domain we had 12 methods, 8 operators and 3 planning problems for the test. A user can choose which domain or goal to use for planning, as described above. Choosing $d_{navi-domain}$ or m_{navi2} will always produce

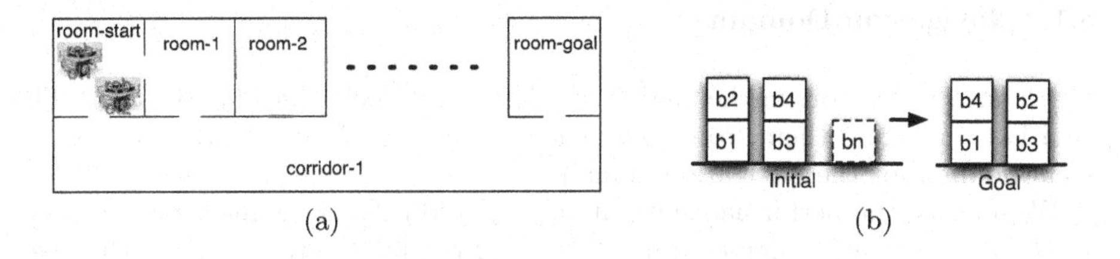

Fig. 2. Test instances for (a) the navigation domain and (b) the blocks world domain

a correct planning problem description, which has only 2 methods and 3 operators. However, m_{navi1} produces a smaller planning problem description with 1 method and 3 operators, yet it allows valid plans to be generated.

Blocks World Domain. We are using blocks world as the well-known benchmark for investigating the complexity of domains [11]. The goal here is to arrange blocks in a configuration as defined in the initial task network w. We adopt the "small problem" for blocks world from an example in the regular JSHOP2 distribution package. The domain has to be modelled in the DL representation in order to ensure that the DL reasoner can deduce a valid planning domain and its problem description.

The *BlocksWorld* concept represents all individuals that are involved in the domain. Blocks are defined either as members of *Block* or *UsedBlock*. The *UsedBlock* concept contains blocks that are explicitly defined in the initial task network w. Any other blocks should be defined under the *Block* concept.

The individual concepts *InvolvedBlock*, *UsedBlockOn*, *UsedBlockOnTable*, and *UsedBlockClear* are automatically inferred by applying the DL reasoner to the domain representation. *InvolvedBlock* is a concept for inferring blocks that are not explicitly defined in initial task network w, but are deduced to be necessary for producing a valid problem description. If, for example, the goal state is changed to block $b1$ on block $b3$, we need only to define state s_{b1} and state s_{b3} in the initial task network w since the DL reasoner will automatically deduce state s_{b2} and state s_{b4} (see Fig. 2 (b)) as members of the *InvolvedBlock* concept. These blocks are required to represent a valid model such that the generated plan is itself valid.

The *UsedBlockOn*, *UsedBlockOnTable* and *UsedBlockClear* concepts are helper concepts in the blocks world domain. They are needed for generating certain blocks' states such as s_{on}, $s_{on-table}$ and s_{clear} in the initial state s_o.

5 Results

In this section, we present empirical comparisons between our approach and a pure HTN representation for the navigation and the blocks world domains. They demonstrate the system's performance when the domain is enlarged to include additional individuals.

5.1 Navigation Domain

We defined two actors in the navigation domain, namely *robot1* and *robot2*. The goal is to navigate them from *room-start* to *room-goal.* Fig. 2 (a) shows a test scenario map for the navigation domain.

We analyse the performance iteratively by adding new rooms between *room-start* and *room-goal* as depicted in Fig. 2 (a). Initially, the test starts with *room-start*, *corridor-1* and *room-goal.* At each step, a room is added to the model. It has two closed doors, one is connected to the previous room and the other is connected to *corridor-1.*

To guarantee that the DL-filtered and the unfiltered HTN planning problem descriptions differ only in the domain filtering, we generate them both automatically from the DL domain representation: one time transforming strictly all domain individuals into the planning problem description, the other time applying the domain filtering, as described.

The number of different plans that exist for the problem schema shown in Fig. 2(a) is $(n + 2)^2$, where n is the number of added rooms, i.e., $n = 0$ for no added room. The HTN planner working on the filtered planning problem descriptions produces constantly four plans, independent of rooms added to the DL domain representation; in contrast, the pure HTN planning approach generates the full set of $(n + 2)^2$ plans.

Fig. 3 shows the required time for generating the plans in relation to the number of rooms. In the pure HTN approach, the planning time increases commensurately with the number of rooms in the problem description. The planning time is less than 2 *s* for less than 25 rooms. It is 193 *s* for 83 rooms and for more than this, the planner has failed to generate a plan.

In our approach, the overall planning time consists of DL-reasoning time plus HTN-planning time. Therefore, for $n < 25$, our approach requires more time than pure HTN planning. However, the average overall time for up to 200 rooms is 3.5 *s*.

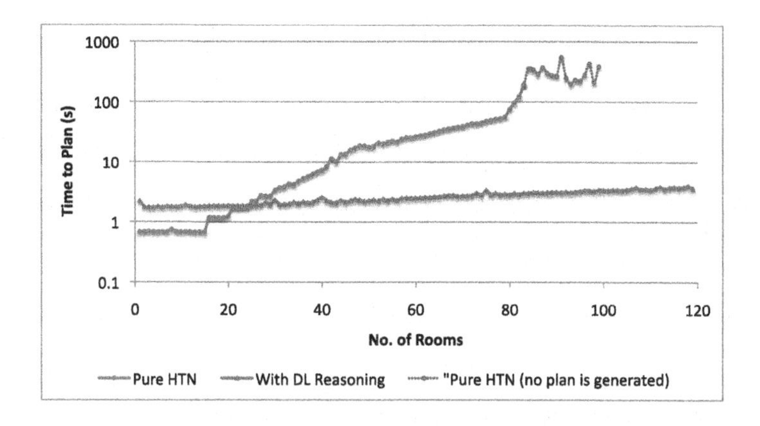

Fig. 3. Computation time (logarithmic scale) in the navigation domain

The overall time increases in average around 17 ms for each added room. This is less than 0.5% of the overall time and shows the scalability of the approach.

5.2 Blocks World Domain

In the blocks world domain, we use four blocks in the initial task network w, namely s_{b1}, s_{b2}, s_{b3} and s_{b4}. These are modelled as instances of $UsedBlock$. Fig. 2(b) depicts the blocks world domain test scenario.

Initially, only these blocks are modeled in the representation. Additional blocks are added at each step. In our approach, the added blocks are modelled as instances of the *Block* concept. In the pure HTN approach, these blocks are added directly to the basic planning problem description. Each added $block_i$ is modelled as $s_{on-table\ blocki}$, s_{blocki} and $s_{clear\ blocki}$ in the HTN problem description.

The pure HTN planning approach produces $2 * (n + 2)!$ plans, where n is the number of added blocks. The basic JSHOP2 implementation has successfully planned for up to 5 additional blocks. As in the navigation example, the filtering method makes the HTN planning independent of the added blocks. Our approach produces a constant number (four) of plans regardless of the number of blocks in the DL model.

The computation time for the blocks world problem is shown in Fig. 4. As previously mentioned in the navigation domain results, our approach needs extra computation time for the DL reasoner. This can be clearly seen in this figure. For $n < 4$, computation time in our approach takes longer than for the pure approach. Additional blocks increase the overall computation time slightly. However, this increase is relatively small: for up to 50 blocks the overall time is 2 s. The average overall time is 2.1 s. The average extra time for each additional block is 1 ms. This is, on average less than 0.2% of the overall computation time.

The pure HTN approach needs 19 s to extract the plan when 5 blocks are added into the problem description. It needs 104 s to plan with 6 additional

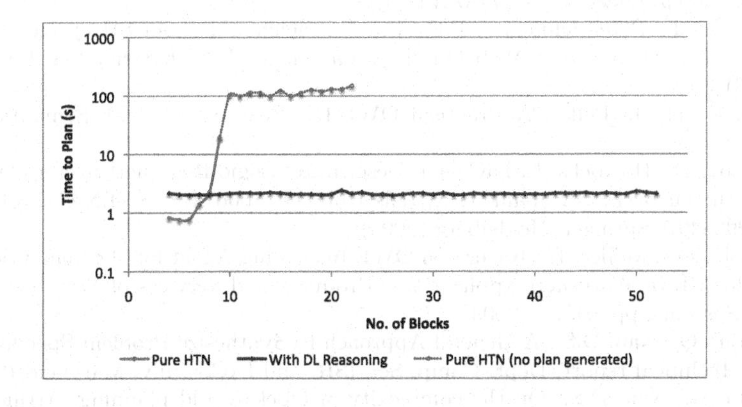

Fig. 4. Computation time (logarithmic scale) in the blocks world domain

blocks, but returns no plan due to high memory consumption. This can be seen by the dashed line in Fig. 4 (b).

6 Conclusions

We have presented benchmark results for our previously [2] described approach for amalgamating DL reasoning and HTN planning to enhance the planning process by reducing problem size. Domain concepts are represented in DL rather than in some special-purpose planning domain representation format like PDDL. Using DL inference, the information provided for the HTN planner is filtered, allowing more compact planning problems to be generated, and thereby reducing the planner's search space.

This has been exemplified by two case studies where adding irrelevant objects to the domains increases the overall planning time. Our approach can filter the domain representations and generate smaller planning problem descriptions. Moreover, additional information can be kept in the DL domain representation without affecting the overall planner performance. We also use this approach in the Johnny Jackanapes robot to plan some tasks in the RoboCup@Home competitions scenarios (e.g., fetch & carry, demo challenge); the robot has scored first in the 2009 RoboCup in Graz.

References

1. Gil, Y.: Description Logics and Planning. AI Magazine 26(2), 73–84 (2005)
2. Hartanto, R., Hertzberg, J.: Fusing DL Reasoning with HTN Planning. In: Dengel, A.R., Berns, K., Breuel, T.M., Bomarius, F., Roth-Berghofer, T.R. (eds.) KI 2008. LNCS (LNAI), vol. 5243, pp. 62–69. Springer, Heidelberg (2008)
3. Baader, F., et al.: The Description Logic Handbook: Theory, Implementation, and Applications. Cambridge University Press, Cambridge (2003)
4. Ghallab, M., Nau, D., Traverso, P.: Automated Planning: Theory and Practice. Morgan Kaufmann, San Francisco (2004)
5. Dean, M., et al.: OWL web ontology language reference. W3C Recommendation (2004), http://www.w3.org/TR/owl-ref/
6. Horrocks, I., Patel-Schneider, P.F., van Harmelen, F.: From SHIQ and RDF to OWL: The Making of a Web Ontology Language. J. Web Semantics 1(1), 7–26 (2003)
7. Sirin, E., et al.: Pellet: A Practical OWL-DL Reasoner. J. Web Semantics 5(2) (2007)
8. Tsarkov, D., Horrocks, I.: FaCT++ Description Logic Reasoner: System Description. In: Furbach, U., Shankar, N. (eds.) IJCAR 2006. LNCS (LNAI), vol. 4130, pp. 292–297. Springer, Heidelberg (2006)
9. Haarslev, V., Möller, R.: Racer: An OWL Reasoning Agent for the Semantic Web. In: Proc. Intl. Workshop Applications, Products and Services of Web-based Support Systems, pp. 91–95 (2003)
10. Ilghami, O., Nau, D.S.: A General Approach to Synthesize Problem-Specific Planners. Technical report, Dept. Comp. Sc., ISR, and IACS Univ. Maryland (2003)
11. Gupta, N., Nau, D.S.: On the complexity of blocks-world planning. Artificial Intelligence 56(2-3), 223–254 (1992)

Flexible Timeline-Based Plan Verification

A. Cesta[1], A. Finzi[2], S. Fratini[1], A. Orlandini[3], and E. Tronci[4]

[1] ISTC-CNR, Via S.Martino della Battaglia 44, I-00185 Rome, Italy
[2] DSF "Federico II" University, Via Cinthia, I-80126 Naples, Italy
[3] DIA "Roma TRE" University, Via della Vasca Navale 79, I-00146 Rome, Italy
[4] DI "La Sapienza" University, Via Salaria 198, I-00198 Rome, Italy

Abstract. Flexible temporal planning is a general technique that has demonstrated wide applications possibilities in heterogeneous domains. A key problem for widening applicability of these techniques is the robust connection between plan generation and execution. This paper describes how a model-checking verification tool, based on UPPAAL-TIGA, is suitable for verifying flexible temporal plans. Moreover, we further investigate a particular perspective, i.e., the one of verifying dynamic controllability before actual plan execution.

1 Introduction

Timeline-based planning has been shown well suited for applications, e.g., [9]. A problem for a wider diffusion of such technology stems in the limited community that has been studying formal properties of this approach to planning. In a preliminary work [1], we have listed several directions for contamination between timeline-based planning and standard techniques for formal validation and verification, then we have started addressing properties to develop a robust connection between plan generation and execution. In this context, plan verification is a crucial verification task most of the V&V tasks considered in [1] rely on. In particular, our approach allows to deploy general purpose techniques, such as model checking, to this aim. Moreover, in real domains, a generated temporally flexible plan is to be executed by an *executive* system that manages controllable processes in the presence of exogenous events. In this scenario, the duration of the execution process is not completely under the control of the executive. Thus, verifying a temporal flexible plan before its actual execution requires particular attention and major efforts. In this paper, we present a formalization used for addressing the flexible plan verification problem that makes use of Timed Game Automata [6] and UPPAAL-TIGA [4], a well known model-checking tool for verification. Other works addressed similar problems [10,2] but, up to our knowledge, the present paper is the first dealing with flexible plan verification. Moreover, we show how plan verification can also be exploited to address the *controllability problem* [11,7]. Finally, we discuss some experimental results collected using the verification tool.

2 Timeline-Based Planning and Execution

Timeline-based planning is an approach to temporal planning [9] where the generated plans are represented by sets of timelines. Each timeline denotes the evolution of a

B. Mertsching, M. Hund, and Z. Aziz (Eds.): KI 2009, LNAI 5803, pp. 49–56, 2009.

particular feature in a dynamic system. A planning domain encodes the possible evolutions of the timelines whose time points have to satisfy temporal constraints, usually represented as Simple Temporal Problem (STP) restrictions.

Here, we assume that the timelines in a planning domain are incarnations of multi-valued *state variables* as in [9]. A state variable is characterized by a finite set of values describing its temporal evolutions, and by minimal and maximal duration for each value. More formally, a state variable is defined by a tuple $\langle \mathcal{V}, \mathcal{T}, \mathcal{D} \rangle$ where: (a) $\mathcal{V} = \{v_1, \ldots, v_n\}$ is a finite set of *values*; (b) $\mathcal{T} : \mathcal{V} \rightarrow 2^{\mathcal{V}}$ is the *value transition* function; (c) $\mathcal{D} : \mathcal{V} \rightarrow \mathbb{N} \times \mathbb{N}$ is the *value duration* function, i.e. a function that specifies the allowed duration of values in $\mathcal{V}$ (as an interval $[lb, ub]$). Given a state variable, its associated *timeline* is represented as a sequence of values in the temporal interval $\mathcal{H} = [0, H)$. Each value satisfies previous (a-b-c) specifications and is defined on a set of not overlapping time intervals contained in $\mathcal{H}$. We suppose that adjacent intervals present different values. A timeline is said *completely specified* over the temporal horizon $\mathcal{H}$ when a sequence of non-overlapping valued intervals exists and its union is equal to $\mathcal{H}$. A timeline is said *time-flexible* when is completely specified and transition events are associated to temporal intervals (lower and upper bounds are given for them), instead of exact temporal occurrences. In other words, a time-flexible timeline represents a set of timelines, all sharing the same sequence of values. It is worth noting that not all the timelines in this set are valid (satisfies a-b-c). The process of *timeline extraction* from a time-flexible timeline is the process of computing (if exists) a valid and completely specified timeline from a given time-flexible timeline. In timeline-based planning, a *planning domain* is defined as a set of state variables $\{\mathcal{SV}_1, \ldots, \mathcal{SV}_n\}$ that cannot be considered as reciprocally decoupled. Then, a *domain theory* is defined as a set of additional relations, called *synchronizations*, that model the existing temporal constraints among state variables. A synchronization has the form $\langle \mathcal{TL}, v \rangle \longrightarrow \langle \{\mathcal{TL}'_1, \ldots, \mathcal{TL}'_n\}, \{v'_1, \ldots, v'_{|\mathcal{TL}'|}\}, \mathcal{R} \rangle$ where: $\mathcal{TL}$ is the reference timeline; v is a value on $\mathcal{TL}$ which makes the synchronization applicable; $\{\mathcal{TL}'_1, \ldots, \mathcal{TL}'_n\}$ is a set of target timelines on which some values v'_j must hold; and $\mathcal{R}$ is a set of *relations* which bind temporal occurrence of the *reference* value v with temporal occurrences of the *target* values $v'_1, \ldots, v'_{|\mathcal{TL}'|}$. A *plan* is defined as a set of timelines $\{\mathcal{TL}_1, \ldots, \mathcal{TL}_n\}$ over the same interval for each state variable. A plan is *valid* with respect to a domain theory if every temporal occurrence of a reference value implies that the related target values hold on target timelines presenting temporal intervals that satisfy the expected relations. A plan is *time flexible* if $\exists \mathcal{TL}_i \in \{\mathcal{TL}_1, \ldots, \mathcal{TL}_n\}$ such that $\mathcal{TL}_i$ is time flexible.

At execution time, an executive cannot completely predict the behavior of the controlled physical system because the duration of certain processes or the timing of exogenous events is outside of its control. In these cases, the values for the state variables that are under the executive scope should be chosen so that they do not constrain uncontrollable events. This *controllability problem* is defined, e.g. in [11] where *contingent* and *executable* processes are distinguished. The contingent processes are not controllable, hence with uncertain durations, instead the executable processes are started and ended by the executive system. Controllability issues have been formalized and investigated for the Simple Temporal Problems with Uncertainty (STPU) in [11] where basic formal notions are given for *dynamic* controllability (see also [8]). In the timeline-based framework, we introduce the same controllability concept defined on STNU as follows. Given a plan as a set of flexible timelines $\mathcal{PL} = \{\mathcal{TL}_1, \ldots, \mathcal{TL}_n\}$, we call *projection* the set of flexible timelines $\mathcal{PL}' = \{\mathcal{TL}'_1, \ldots, \mathcal{TL}'_n\}$ derived from $\mathcal{PL}$ setting to a fixed value the temporal occurrence of each uncontrollable timepoint. Considering N as the set of controllable flexible timepoints in $\mathcal{PL}$, a *schedule* T is a mapping $T : N \rightarrow \mathbb{N}$ where $T(x)$ is called *time* of timepoint x. A *schedule* is *consistent* if all value durations and synchronizations are satisfied in $\mathcal{PL}$. The *history* of a timepoint x w.r.t. a schedule T, denoted by $T\{\prec x\}$, specifies the time of all uncontrollable timepoints that occur prior to x. An *execution strategy* S is a mapping $S : \mathcal{P} \rightarrow \mathcal{T}$ where $\mathcal{P}$ is the set of projections and $\mathcal{T}$ is the set of schedules. An execution strategy S is viable if $S(p)$ (denoted also S_p) is consistent for each projection p. Thus, a flexible plan $\mathcal{PL}$ is *dynamically controllable* if there exists a viable execution strategy S such that $S_{p1}\{\prec x\} = S_{p2}\{\prec x\} \Rightarrow S_{p1}(x) = S_{p2}(x)$ for each controllable timepoint x and projections $p1$ and $p2$.

3 Timed Game Automata

A fundamental concept in Timed Automata is time. First, we give the formal definition of clocks and relations that can be defined over them. We call *clock* a nonnegative, real-valued variable. Let X be a finite set of clocks. We denote with $C(X)$ the set of constraints Φ generated by the grammar: $\Phi ::= x \sim c \mid x - y \sim c \mid \Phi \wedge \Phi$, where $c \in Z, x, y \in X$, and $\sim \in \{<, \leq, \geq, >\}$. We denote by $B(X)$ the subset of $C(X)$ that uses only constraints of the form $x \sim c$.

Definition 1. *A* **Timed Automaton (TA)** *[3] is a tuple* $\mathcal{A} = (Q, q_0, \text{Act}, X, \text{Inv}, E)$, *where:* Q *is a finite set of* locations, $q_0 \in Q$ *is the* initial location, Act *is a finite set of* actions, X *is a finite set of clocks,* $\text{Inv} : Q \rightarrow B(X)$ *is a function associating to each location* $q \in Q$ *a constraint* $\text{Inv}(q)$ *(the* invariant *of* q*),* $E \subseteq Q \times B(X) \times \text{Act} \times 2^X \times Q$ *is a finite set of* transitions *and each transition* (q, g, a, Y, q') *is noted* $q \overset{g,a,Y}{\rightarrow} q'$.

A *valuation* of the variables in X is a mapping v from X to the set $R_{\geq 0}$ of nonnegative reals. We denote with $R^X_{\geq 0}$ the set of valuations on X and with $\mathbf{0}$ the valuation that assigns the value 0 to each clock. If $Y \subseteq X$ we denote with $v[Y]$ the valuation (on X) assigning the value 0 ($v(z)$) to any $z \in Y$ ($z \in (X - Y)$). For any $\delta \in R^{\geq 0}$ we denote with $(v + \delta)$ the valuation such that, for each $x \in X$, $(v + \delta)(x) = v(x) + \delta$. Let $g \in C(X)$ and v be a valuation. We say that g satisfies v, notation $v \models g$ if constraint g evaluated on v returns true. A *state* of TA $\mathcal{A}$ is a pair (q, v) that $q \in Q$ and v is a valuation (on X). We denote with S the set of states of $\mathcal{A}$. An *admissible* state for $\mathcal{A}$ is a state (q, v) that $v \models \text{Inv}(q)$. A *discrete transition* for $\mathcal{A}$ is 5-tuple $(q, v) \overset{a}{\rightarrow} (q', v')$ where $(q, v), (q', v') \in S$, $a \in \text{Act}$ and there exists a transition $q \overset{g,a,Y}{\rightarrow} q' \in E$ that $v \models g$, $v' = v[Y]$ and $v' \models \text{Inv}(q')$. A *time transition* for $\mathcal{A}$ is 4-tuple $(q, v) \overset{\delta}{\rightarrow} (q, v')$ where $(q, v) \in S$, $(q, v') \in S$, $\delta \in R_{\geq 0}$, $v' = v + \delta$, $v \models \text{Inv}(q)$ and $v' \models \text{Inv}(q)$. A *run* of a TA $\mathcal{A}$ is a finite or infinite sequence of alternating time and discrete transitions of $\mathcal{A}$. We denote with $\text{Runs}(\mathcal{A}, (q, v))$ the set of runs of $\mathcal{A}$ starting from state (q, v) and write

Runs($\mathcal{A}$) for Runs($\mathcal{A}$, $(q, \mathbf{0})$). If ρ is a finite run we denote with last(ρ) the last state of run ρ. A ***network* of TA (nTA)** is a finite set of TA evolving in parallel with a CSS style semantics for parallelism. Formally, let $\mathcal{F} = \{\mathcal{A}_i \mid i = 1, \ldots n\}$ be a finite set of automata with $\mathcal{A}_i = (Q_i, q_i^0, \text{Act}, X, \text{Inv}_i, E_i)$ for $i = 1, \ldots n$. Note that the automata in $\mathcal{F}$ have all the same set of actions and clocks and disjoint sets of locations. The *network* of $\mathcal{F}$ (notation $||\mathcal{F}$) is the TA $\mathcal{P} = (Q, q^0, \text{Act}, X, \text{Inv}, E)$ defined as follows. The set of locations Q of $\mathcal{P}$ is the Cartesian product of the locations of the automata in $\mathcal{F}$, that is $Q = Q_1 \times \ldots Q_n$. The *initial state* q^0 of $\mathcal{P}$ is $q^0 = (q_1^0, \ldots q_n^0)$. The *invariant* Inv for $\mathcal{P}$ is $\text{Inv}(q_1, \ldots q_n) = \text{Inv}_1(q_1) \wedge \ldots \text{Inv}_n(q_n)$. The *transition relation* E for $\mathcal{P}$ is the synchronous parallel of those of the automata in $\mathcal{F}$. That is, E consists of the set of 5-tuples (q, g, a, Y, q') satisfying the following conditions: 1. $q = (q_1, \ldots q_n)$, $q' = (q_1', \ldots q_n')$; 2. There are $i \leq j \in \{1, \ldots n\}$ such that for all $h \in \{1, \ldots n\}$, if $h \neq i, j$ then $q_h = q_h'$. Furthermore, if $i = j$ then action a occurs only in automaton $\mathcal{A}_i$ of $\mathcal{F}$. 3. Both automata $\mathcal{A}_i$ and $\mathcal{A}_j$ can make a transition with action a. That is, $q_i \overset{g_i, a, Y_i}{\longrightarrow} q_i' \in E_i$, $q_j \overset{g_j, a, Y_j}{\longrightarrow} q_j' \in E_j$, $g = g_i \wedge g_j$, $Y = Y_i \cup Y_j$.

Definition 2. *A* **Timed Game Automaton (TGA)** *is a TA where the set of actions* Act *is split in two disjoint sets:* Act_c *the set of controllable actions and* Act_u *the set of* uncontrollable *actions.*

The notions of network of TA, run, configuration are defined in a similar way for TGA.

Given a TGA $\mathcal{A}$ and three symbolic configurations *Init*, *Safe*, and *Goal*, the *reachability control problem* or reachability game *RG*($\mathcal{A}$, *Init*, *Safe*, *Goal*) consists in finding a *strategy* f such that starting from *Init* and executing f, $\mathcal{A}$ stays in *Safe* and reaches *Goal*. More precisely, a strategy is a partial mapping f from the set of runs of $\mathcal{A}$ starting from *Init* to the set $\text{Act}_c \cup \{\lambda\}$ (λ is a special symbol that denotes "do nothing and just wait"). For a finite run ρ, the strategy $f(\rho)$ may say (1) no way to win if $f(\rho)$ is undefined, (2) do nothing, just wait in the last configuration ρ if $f(\rho) = \lambda$, or (3) execute the discrete, controllable transition labeled by l in the last configuration of ρ if $f(\rho) = l$. A strategy f is *state-based* or *memory-less* whenever its result depends only on the last configuration of the run.

4 Building TGA from Timeline-Based Planning Specifications

The main contribution of this work is the description of how flexible timeline-based plan verification can be performed solving a Reachability Game using UPPAAL-TIGA. To this end, this section describes how a flexible timeline-based plan, state variables and domain theory can be formalized as an adequate nTGA. Timelines and state variables are mapped into TGA. In addition, a *Observer* TGA checks for both illegal values occurrences and synchronizations violations. Here, we distinguish between controllable and uncontrollable state variables/timelines to simplifying the formalization.

Given a flexible plan $\mathcal{P} = \{\mathcal{TL}_1, \ldots, \mathcal{TL}_n\}$, we define a TGA for each $\mathcal{TL}_i$. For each valued interval in the timeline (also called plan step), we consider a location in the automaton. An additional final location, labeled *goal*, is considered. We consider a unique *plan clock* c_p over all the timelines automata. Then, for each planned flexible timeline $\mathcal{TL}$, we define a Timed Game Automaton $\mathcal{A}_{\mathcal{TL}} = (Q_{\mathcal{TL}}, q_0, \mathrm{Act}_{\mathcal{TL}}, X_{\mathcal{TL}}, \mathrm{Inv}_{\mathcal{TL}}, E_{\mathcal{TL}})$ as follows. For each i-th valued interval in $\mathcal{TL}$, we consider l_i in $Q_{\mathcal{TL}}$, plus the final location l_{goal} (q_0 is l_0). For each allowed value $v \in \mathcal{SV}_i$, we consider an action a_v. If the related state variable is controllable (uncontrollable) we add a_v in $\mathrm{Act}_{c\mathcal{TL}}$ ($\mathrm{Act}_{u\mathcal{TL}}$). The overall plan clock c_p is considered in $X_{\mathcal{TL}}$. For each i-th valued interval in $\mathcal{TL}$ and the related value v_p associated with the flexible interval timepoint $[lb, ub]$, we define $\mathrm{Inv}_{\mathcal{TL}}(l_i) := c_p \leq ub$ and we define a transition $e = q \overset{g,a,Y}{\rightarrow} q'$ in $E_{\mathcal{TL}}$, where $q = l_i$, $q' = l_{i+1}$, $g = c_p \geq lb$, $a = v_p!$, $Y = \emptyset$. Finally, we define a final transition $e = q \overset{g,a,Y}{\rightarrow} q'$ in $E_{\mathcal{TL}}$, where $q = l_{pl}$ (where pl is the plan length), $q' = l_{goal}$, $g = \emptyset$, $a = \emptyset$, $Y = \emptyset$. The set $Plan = \{\mathcal{A}_{\mathcal{TL}_1}, ..., \mathcal{A}_{\mathcal{TL}_n}\}$ represents the planned flexible timelines description as a nTGA.

For each state variable $\mathcal{SV}$, we have a one-to-one mapping into a Timed Game Automaton $\mathcal{A}_{SV} = (Q_{SV}, q_0, \mathrm{Act}_{SV}, X_{SV}, \mathrm{Inv}_{SV}, E_{SV})$. In fact, for each allowed value v in $\mathcal{V}$, we consider a location l_v in Q_{SV} (q_0 is set according to the initial value of the related planned timeline). Then, for each allowed value $v \in \mathcal{V}$, we consider an action a_v. If the state variable is controllable (uncontrollable), we consider a_v in Act_{cSV} (Act_{uSV}). A local clock c_{sv} is considered in X_{SV}. Finally, for each allowed value $v \in \mathcal{V}$ and both the associated $\mathcal{T}(v) = \{vs_1, ..., vs_n\}$ and $\mathcal{D}(v) = [l_b, u_b]$, we define $\mathrm{Inv}_{SV}(v) := c_{sv} \leq u_b$ and we define a transition $e = q \overset{g,a,Y}{\rightarrow} q'$, where $q = l_v$, $q' = l_{vs_i}$ in E_{SV}, $g = c_{sv} \geq l_b$, $a = a_{vs_i}?$, $Y = \{c_{sv}\}$. The set $SV = \{\mathcal{A}_{SV_1}, ..., \mathcal{A}_{SV_n}\}$ represents the State Variables description. Note that the use of actions as transitions label implements the synchronization between state variables and planned timelines.

A last TGA is the *Observer* that monitors synchronizations and values over SV and $Plan$. Basically, two locations are considered to represent *correct* and *error* status. For each possible cause of error, an appropriate transition is defined, forcing the Observer to hold the error location. In this sense, we define a TGA $\mathcal{A}_{Obs} = (Q_{Obs}, q_0, \mathrm{Act}_{Obs}, X_{Obs}, \mathrm{Inv}_{Obs}, E_{Obs})$ as follows. We consider $Q_{Obs} = \{l_{ok}, l_{err}\}$ (q_0 is l_{ok}), $\mathrm{Act}_{uObs} = \{a_{fail}\}$, $X_{Obs} = \{c_p\}$. Inv_{Obs} is undefined. For each pair plan step and associated planned value (s_p, v_p) for each timeline $\mathcal{TL}$ and the related variable $\mathcal{SV}$, we define an uncontrollable transition $e = q \overset{g,l,r}{\rightarrow} q'$ in E_{Obs}, where $q = l_{ok}$, $q' = l_{err}$, $g = \mathcal{TL}_{s_p} \wedge \neg \mathcal{SV}_{v_p}$, $l = a_{fail}$, $r = \emptyset$. Moreover, for each synchronization $\langle \mathcal{TL}, v \rangle \longrightarrow \langle \{\mathcal{TL}'_1, \ldots, \mathcal{TL}'_n\}, \{v'_1, \ldots, v'_n\}, \mathcal{R} \rangle$, we define an uncontrollable transition $e = q \overset{g,a,Y}{\rightarrow} q'$ in E_{Obs} where $q = l_{ok}$, $q' = l_{err}$, $g = \neg \mathcal{R}(\mathcal{TL}_v, \mathcal{TL}'_{1v'_1}, \ldots, \mathcal{TL}'_{nv'_n})$, $a = a_{fail}$, $Y = \emptyset$.

The nTGA $\mathcal{PL}$ composed by the set of automata $PL = SV \cup Plan \cup \{\mathcal{A}_{Obs}\}$ encapsulates Flexible plan, State Variables and Domain Theory descriptions.

5 Verifying Time Flexible Plans

Given the nTGA $\mathcal{PL}$ defined above, we can define a Reachability Game that ensures, if successfully solved, plan validity. In particular, we define *RG*($\mathcal{PL}$, *Init*, *Safe*, *Goal*) by considering *Init* as the set of initial locations of each automaton in $\mathcal{PL}$, *Safe* = $\{l_{ok}\}$ and *Goal* as the set of goal locations of each $\mathcal{TL}_i$ in $\mathcal{PL}$. By construction, it is possible to show that we use a one-to-one mapping between flexible behaviors defined by $\mathcal{P}$ and automata behaviors defined by *Plan* ∪ *SV*. While, the Observer holds the error location iff either an illegal value occurs or a synchronization is violated. Then, $\mathcal{PL}$ adequately represents all and only the behaviors defined by the flexible plan $\mathcal{P}$.

In order to solve such a reachability game, we use UPPAAL-TIGA [4]. This tool extends UPPAAL [5] providing a toolbox for the specification, simulation, and verification of real-time games. If there is no winning strategy, UPPAAL-TIGA gives a counter strategy for the opponent (environment) to make the controller lose. Given a nTGA, a set of goal states (*win*) and/or a set of bad states (*lose*), four types of winning conditions can be issued [4]. We ask UPPAAL-TIGA to solve the $RG(\mathcal{PL}, Init, Safe, Goal)$ checking the formula $\Phi = A\ [\ Safe\ U\ Goal]$ in $\mathcal{PL}$. In fact, this formula means that along all the possible paths, $\mathcal{PL}$ stays in *Safe* states until *Goal* states are reached. In other words, winning the game corresponds to ask UPPAAL-TIGA to find a strategy that, for each possible evolution of uncontrollable state variables, ensures goals to be reached and errors to be avoided. Thus, verifying with UPPAAL-TIGA the above property implies validating the flexible temporal plan.

Moreover, we show the feasibility and effectiveness of our verification method by addressing the relevant issue of plan controllability. In fact, we may notice that each possible evolution of uncontrollable automata corresponds to a timeline projection p. Each strategy/solution for the RG corresponds to a schedule T. And a set of strategy represents an execution strategy S. Thus, the winning strategies produced by UPPAAL-TIGA constitute a viable execution strategy S for the flexible timelines. The use of forward algorithms [4] guarantees that S is such that $S_{p1}\{\prec x\} = S_{p2}\{\prec x\} \Rightarrow S_{p1}(x) = S_{p2}(x)$ for each controllable timepoint x and projections $p1$ and $p2$. That is, the flexible plan is dynamically controllable.

6 Case Study and Preliminary Experiments

Figure 1 sketches a generic space domain in which the main timeline for a remote space agent should be synthesized. The planning goal is to allocate the temporal occurrences of science and maintenance operations as well as the agent's ability to communicate. There are two uncontrollable state variables pos and ava both representing orbit events. Variable pos models the position of the spacecraft with respect to a (given) deep space planet and takes values: "PERI" (pericentre, i.e. the orbital position closest to the planet) and "APO" (apocentre, i.e. the orbital position more far away from the planet). Variable ava (taking values *Available* and *Unavailable*) models the availability of the ground station opportunity windows, i.e., the visibility of the spacecraft with respect to Earth. The controllable state variable modeling the timeline should be synchronized with the above described uncontrollable variables.

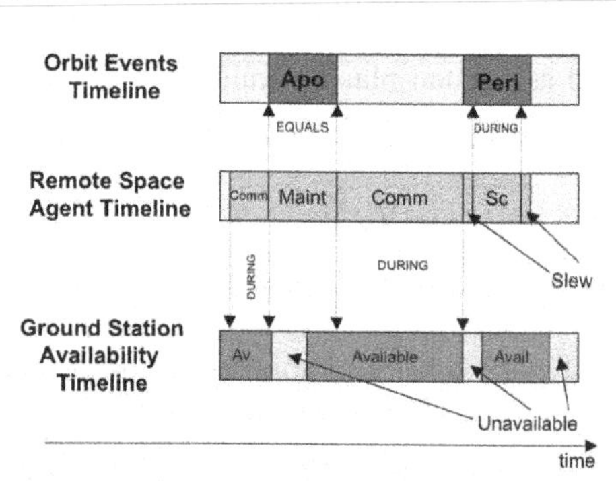

Fig. 1. Timeline synchronizations in a plan

This domain is a nondeterministic version of a real domain authors are working with [1], where the uncertainty is associated with the ground station availability. Any valid plan needs synchronizations among the agent controllable timeline (Figure 1, middle) and the uncontrollable timelines (see dotted arrows in Figure 1): science operations must occur during Pericentres, maintenance operations must occur in the same time interval as Apocentres and communications must occur during ground station visibility windows. In addition to those synchronization constraints, the operative mode timeline must respect transition constraints among values and durations for each value specified by the domain for the agent.

Here, we present some experimental results on preliminary tests focusing on the analysis of the dependency of plan verification performance from the degree of *flexibility*. We generate a flexible plan by introducing flexibility into a completely instantiated plan. This is done by replacing a time point $t = \tau$ in the instantiated plan with a time interval $t \in [\tau - \Delta, \tau + \Delta]$ in the flexible plan.

Φ \ plan size	10	20	35
3	35.6 ±0.8	36.6±1.7	37.4 ±0.5
6	35.2 ±0.4	36 ±0	37.4 ±0.5
9	36 ±1.8	36.2 ±0.4	39.2 ±1.9
12	34.8 ±0.4	36.4 ±0.5	37.8 ±0.4
15	35 ±0	36.2 ±0.4	43.6 ±10.2
18	35 ±0	40 ±8	39 ±0

Fig. 2. Experimental results collected varying plan length and the number of flexible time points (Timings in msecs)

The main parameters we consider are: the number Φ of time points that are replaced with time intervals and the width (*duration*) Δ of such intervals.

Φ \ Δ	1	5	10	15	20
3	40±6	37.4±0.5	37.8±0.4	51±7.8	37.8±1
6	38.4±0.5	38.6±1.2	38±0	44.4±8.5	38.2±0.4
9	38.4±0.5	38±0	39.2±1.9	39±0	38.8±0.4
12	52.4±10.3	38.8±0.4	38.4±0.5	39±0	39.4±0.5
15	39.2±0.4	52±13	39.2±0.4	39.2±0.4	39.8±0.4
18	39.6±0.5	39.6±0.8	40.4±1.5	48.8±9.1	40±0.6

Fig. 3. Experimental results collected with a fixed plan length (Timing in msecs.)

We perform two kind of experiments. First, keeping Δ constant ($\Delta = 10$), we study how plan verification time depends on the plan size (i.e., the number of plan time points) and on the number of flexible time points Φ. Second, keeping constant the plan size (to 35 time points), we study how plan verification time depends on the number of flexible time points Φ and on the duration Δ. Given Φ and Δ, an experiment consists in choosing at random Φ plan time points, replacing such chosen time points with time intervals of duration Δ, running the UPPAAL-TIGA verifier and, finally, measuring the verification time. For each configuration we repeat our experiment 5 times and compute the mean value (in msecs.) and variance ($\pm var$) for the verification time[1]. We note that not all experiments relative to given values for Φ and Δ yield a satisfiable flexible temporal plan. In fact, since the plan is only flexible at certain time points, the degrees of freedom may not suffice to recover from previously delayed (or anticipated) actions. Of course this is particularly the case when Φ is small with respect to the plan size. Accordingly, our verification times refer to passing (i.e., the given flexible temporal plan is dynamically controllable) as well as failing (i.e., the given flexible temporal plan is not dynamically controllable) experiments. Figure 2 shows our results for the first kind of experiments depicting the homogeneous performances of the verification tool over

[1] We run experiments on a linux workstation with 64-bit AMD Athlon 3.5GHz and 2GB RAM.

all the configurations. In Figure 3, the second kind of experiments shows that the verification tool handles well flexible plan with higher and higher degrees of flexibility both in terms of Φ and Δ.

7 Conclusion

In this paper, we presented a method to represent and verify dynamic controllability of flexible plans using TGA and UPPAAL-TIGA. In particular, the paper describes the verification method, detailing the formal representation and the modeling methodology. Preliminary tests shows the feasibility of the approach. Few related approaches have been proposed in literature. For instance, [10] proposes a mapping from *Contingent Temporal Constraint Networks* (a generalization of STPUs) to TGA which is analogous to the one exploited here. In this work, the use of a model checker is only suggested to obtain a more compact representation and not to verify plan properties. Closely related is the work [2] which proposes a mapping from temporal constraint-based planning problems into UPPAAL-TIGA game-reachability problems and presents a comparison of the two planning approaches. Here, the approach to problem modeling is similar, however, in that work the flexibility issue remains open.

Acknowledgments. Cesta, Fratini, Orlandini and Tronci are partially supported by the EU project ULISSE (Call "SPA.2007.2.1.01 Space Science". Contract FP7.218815).

References

1. Cesta, A., Finzi, A., Fratini, S., Orlandini, A., Tronci, E.: Validation and Verification Issues in a Timeline-based Planning System. In: Knowledge Engineering Review (accepted, 2009)
2. Abdedaim, Y., Asarin, E., Gallien, M., Ingrand, F., Lesire, C., Sighireanu, M.: Planning Robust Temporal Plans: A Comparison Between CBTP and TGA Approaches. In: Proceedings of ICAPS, pp. 2–10 (2007)
3. Alur, R., Dill, D.L.: A Theory of Timed Automata. Theoretical Computer Science 126, 183–235 (1994)
4. Behrmann, G., Cougnard, A., David, A., Fleury, E., Larsen, K.G., Lime, D.: UPPAAL-TIGA: Time for Playing Games! In: Damm, W., Hermanns, H. (eds.) CAV 2007. LNCS, vol. 4590, pp. 121–125. Springer, Heidelberg (2007)
5. Larsen, K.G., Pettersson, P., Yi, W.: UPPAAL in a Nutshell. International Journal on Software Tools for Technology Transfer 1(1-2), 134–152 (1997)
6. Maler, O., Pnueli, A., Sifakis, J.: On the Synthesis of Discrete Controllers for Timed Systems. In: Mayr, E.W., Puech, C. (eds.) STACS 1995. LNCS, vol. 900, pp. 229–242. Springer, Heidelberg (1995)
7. Morris, P.H., Muscettola, N., Vidal, T.: Dynamic Control of Plans with Temporal Uncertainty. In: Proc. of IJCAI 2001, pp. 494–502 (2001)
8. Morris, P.H., Muscettola, N.: Temporal Dynamic Controllability Revisited. In: Proc. of AAAI 2005, pp. 1193–1198 (2005)
9. Muscettola, N.: HSTS: Integrating Planning and Scheduling. In: Zweben, M., Fox, M.S. (eds.) Intelligent Scheduling. Morgan Kaufmann, San Francisco (1994)
10. Vidal, T.: Controllability Characterization and Checking in Contingent Temporal Constraint Networks. In: Proceedings of KR 2000 (2000)
11. Vidal, T., Fargier, H.: Handling Contingency in Temporal Constraint Networks: from Consistency to Controllabilities. JETAI 11(1), 23–45 (1999)

Solving Non-deterministic Planning Problems with Pattern Database Heuristics

Pascal Bercher[1,*] and Robert Mattmüller[2,**]

[1] Institut für Künstliche Intelligenz, Universität Ulm, Germany
pascal.bercher@uni-ulm.de
[2] Institut für Informatik, Albert-Ludwigs-Universität Freiburg, Germany
mattmuel@informatik.uni-freiburg.de

Abstract. Non-determinism arises naturally in many real-world applications of action planning. Strong plans for this type of problems can be found using AO* search guided by an appropriate heuristic function. Most domain-independent heuristics considered in this context so far are based on the idea of ignoring delete lists and do not properly take the non-determinism into account. Therefore, we investigate the applicability of pattern database (PDB) heuristics to non-deterministic planning. PDB heuristics have emerged as rather informative in a deterministic context. Our empirical results suggest that PDB heuristics can also perform reasonably well in non-deterministic planning. Additionally, we present a generalization of the pattern additivity criterion known from classical planning to the non-deterministic setting.

Keywords: Heuristic search, non-deterministic planning, PDB heuristics.

1 Introduction

Non-deterministic planning problems arise naturally as soon as the agent seeking a goal is confronted with an environment that may have an unpredictable influence on action outcomes. Specifically, in this work, we are concerned with finding strong plans [1] for non-deterministic planning tasks in fully observable and static environments. In general, approaches to tackle non-determinism include planning as model checking [1,2], QBF-based approaches, and heuristically guided explicit state techniques [3,4,5,6]. Here, we follow the latter approach and compute strong plans by explicitly constructing the relevant portion of the AND/OR graph encoding the dynamics of the world, and returning the plan corresponding to a solution subgraph. The construction of the graph follows the

* This work was partly supported by the German Research Council (DFG) as part of the Transregional Collaborative Research Center "Eine Companion-Technologie für kognitive technische Systeme" (SFB/TRR 62).

** This work was partly supported by the German Research Council (DFG) as part of the Transregional Collaborative Research Center "Automatic Verification and Analysis of Complex Systems" (SFB/TR 14 AVACS). See www.avacs.org for more information.

B. Mertsching, M. Hund, and Z. Aziz (Eds.): KI 2009, LNAI 5803, pp. 57–64, 2009.

AO* algorithm [7] and is guided by a pattern database (PDB) heuristic that estimates the cost of the solution subgraph rooted at a given node [8]. Given an informative node evaluation function, more promising parts of the graph are likely to be expanded before the less promising ones, resulting in a relatively low number of node expansions before a solution has been found.

As has been shown in earlier work [3,4,5], using the AO* algorithm in conjunction with an informative heuristic can be an efficient way to find plans of high quality. Since the evaluation functions employed so far rarely take non-determinism into account properly, in this work we investigate the use of PDB heuristics, since the abstractions underlying the PDBs can be built in a way that preserves the non-determinism of the original problem.

PDB heuristics have been studied before extensively, both for deterministic and non-deterministic problems, and both with problem-specific patterns and with problem-independent pattern selection techniques. In this work, we describe how to use given PDBs in a generic non-deterministic planner. Problem-independent ways to come up with patterns are left for future work. Our main practical contribution is the development of a non-deterministic planner for static and fully-observable problems based on AO* search guided by PDB heuristics that is domain-independent except for the lack of an automated pattern selection. On the theoretical side, we present a generalization of the additivity criterion for sets of patterns known from classical planning [9] to non-deterministic planning.

2 Non-deterministic Planning

We consider non-deterministic planning problems under full observability. In contrast to classical planning, the actions can have several outcomes, only one of which takes effect non-deterministically.

Formally, a *non-deterministic planning problem* $\mathcal{P}$ consists of a finite set *Var* of *state variables*, a finite set A of *actions*, an *initial state* s_0 and a *goal description* $G \subseteq \mathit{Var}$. The set of states $S = 2^{\mathit{Var}}$ is the set of all valuations of the state variables, and a state s is a *goal state* iff $G \subseteq s$. Each action $a \in A$ is a pair consisting of a set of *preconditions* $\mathit{pre}(a) \subseteq \mathit{Var}$ and a set $\mathit{eff}(a)$ of non-deterministic *effects* $\langle \mathit{add}_i, \mathit{del}_i \rangle$, $i = 1, \dots, n$, each consisting of *add and delete lists* $\mathit{add}_i, \mathit{del}_i \subseteq \mathit{Var}$. We call the set of all variables mentioned in the add and delete lists of an action a its *effect variables* and denote them as $\mathit{effvar}(a) := \bigcup_{i=1}^{n} (\mathit{add}_i \cup \mathit{del}_i)$. An action a is *applicable* in a state s if $\mathit{pre}(a) \subseteq s$ and its application leads to the *successor states* $\mathit{app}(s, a) := \{ \ (s \setminus \mathit{del}) \cup \mathit{add} \mid \langle \mathit{add}, \mathit{del} \rangle \in \mathit{eff}(a) \ \}$.

We want to find a strategy that is guaranteed to transform the initial state into an arbitrary goal state within a finite number of steps no matter what the outcomes of the non-deterministic actions are. Such a strategy is a partial function mapping states to applicable actions. Cimatti et al. [1] call such a strategy a *strong plan*. We formalize strong plans by means of solution graphs as follows. A planning problem $\mathcal{P}$ induces an *AND/OR graph* $\mathcal{G} = \langle V, C \rangle$, where $V = S$ is the set of *nodes* and C is the set of *connectors*. For each non-goal node $v \in V$ and $a \in A$ with $\mathit{pre}(a) \subseteq v$, there is a connector $c = \langle v, \mathit{app}(v, a) \rangle \in C$. We call *pred*($c$) $:= v$ the *predecessor* of c and *succ*(c) $:= app(v, a)$ the *successors* of c. The AND/OR graph of $\mathcal{P}$ is defined as the connected component of $\mathcal{G}$ which contains s_0. A *solution graph* $\mathcal{G} = \langle V, C\rangle$ is a connected acyclic subgraph of the AND/OR graph of $\mathcal{P}$ which contains s_0, where for all $v \in V$, either v is a goal state or there is exactly one $c \in C$ such that *pred*(c) $= v$, and where for all $c \in C$, *pred*(c) $\in V$ and *succ*(c) $\subseteq V$.

We use solution graphs to define the cost value of states. Since we prefer strong plans with a low worst-case number of action applications along the way to a goal state, we define the cost of a state $s \in S$ as $cost^*(s) := \min_{\mathcal{G}} depth(\mathcal{G})$, where $\mathcal{G}$ ranges over all solution graphs rooted at s.

Example 1. As an example, consider the following problem with variables a, b, c, d, and e, actions $a_1, \ldots, a_9$, $s_0 = \{a\}$, and $G = \{b, c, d, e\}$, where $a_1 = \langle a, b \wedge \neg a \,|\, c \wedge \neg a\rangle$, $a_2 = \langle b, e \,|\, d\rangle$, $a_3 = \langle c, e \,|\, d\rangle$, $a_4 = \langle b \wedge d, c\rangle$, $a_5 = \langle c \wedge d, b\rangle$, $a_6 = \langle b \wedge e, c\rangle$, $a_7 = \langle c \wedge e, b\rangle$, $a_8 = \langle b \wedge c \wedge d, e\rangle$, and $a_9 = \langle b \wedge c \wedge e, d\rangle$. Preconditions and effects are written in logical notation and different non-deterministic effects are separated by vertical bars. Fig. 1a shows the AND/OR graph of $\mathcal{P}$. There exists only one subgraph encoding a strong plan, shown in Fig. 1b.

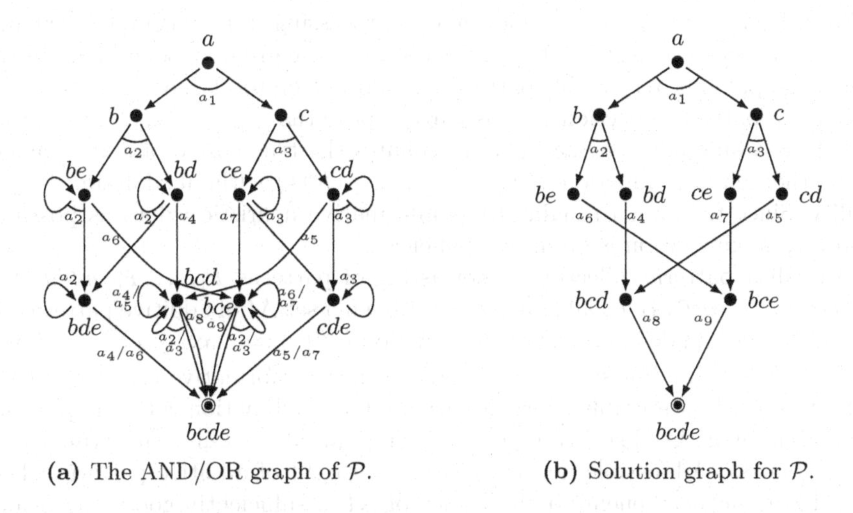

(a) The AND/OR graph of $\mathcal{P}$. **(b)** Solution graph for $\mathcal{P}$.

Fig. 1. The AND/OR graph and a solution graph for $\mathcal{P}$

3 Search Algorithm

We use AO* [7] graph search to traverse the AND/OR graph induced by a planning problem. AO* starts with the empty graph and successively expands it until a solution graph has been found or the AND/OR graph of $\mathcal{P}$ has been completely generated. The performance of AO* heavily relies on the quality of the heuristic function applied to the fringe nodes.

4 Pattern Database Heuristics

Pattern database heuristics are a special case of abstraction heuristics. The basic idea is to obtain heuristic values by optimally solving abstractions of the planning problem and using abstract costs as heuristic values. The abstractions are precomputed before the actual search is performed. During the search, no costly calculations are necessary. The heuristic values are merely retrieved from the pattern database, in which the cost values of the abstract states have been stored during the preprocessing stage. Each abstraction can be regarded as a simplification of the planning problem obtained by restraining it to a given pattern, i.e., a subset of the state variables.

Formally, an abstraction $\mathcal{P}^i$ of $\mathcal{P}$ with respect to a pattern $P_i \subseteq \mathit{Var}$ is the planning problem with variables $\mathit{Var}^i = P_i$ and states $S^i = 2^{\mathit{Var}^i}$, where all conditions and effects are restricted to P_i. More precisely, for $\mathit{var} \subseteq \mathit{Var}$, let $\mathit{var}^i := \mathit{var} \cap P_i$. Then $\mathcal{P}^i$ contains an action a^i for each $a \in A$, where $\mathit{pre}(a^i) = \mathit{pre}(a)^i$, and $\mathit{eff}(a^i)$ contains pairs $\langle \mathit{add}^i, \mathit{del}^i \rangle$ of add and delete lists for all pairs $\langle \mathit{add}, \mathit{del} \rangle$ in $\mathit{eff}(a)$. Finally, $s_0^i = s_0 \cap P_i$ and $G^i = G \cap P_i$. Given a pattern P_i, we define the heuristic function h^i by $h^i(s) := \mathit{cost}_i^*(s) := \mathit{cost}_i^*(s^i)$, where the abstract costs are defined analogously to the concrete costs of a state, i.e., as the depth of a depth-minimizing (abstract) solution graph rooted at s^i.

We calculate the heuristic values in a preprocessing step by complete exhaustive search. Since the size of the abstract state space grows exponentially in the size of the pattern, reasonable patterns should not be too large.

Given a pattern collection P consisting of patterns P_i, $i = 1, \ldots, k$, such that each h^i is admissible, i.e., never overestimates the true cost of a state, we can define the heuristic function $h^P(s) := \max_{P_i \in P} h^i(s)$ without violating admissibility. Since we want to maintain as informative heuristic values as possible, however, maximization is often not sufficient.

We call a pattern collection P consisting of patterns $P_1, \ldots, P_k$ *additive* if $\sum_{i=1}^k h^i(s) \leq \mathit{cost}^*(s)$ for all states $s \in S$. Given a set $\mathcal{M}$ of additive pattern collections P, we can define the heuristic function $h^{\mathcal{M}}(s) := \max_{P \in \mathcal{M}} \sum_{P_i \in P} h^i(s)$. While $h^{\mathcal{M}}$ is still admissible, it is in general more informative than any of the heuristics h^P. If admissible heuristics are used in combination with an appropriate search algorithm like LAO* [6], one can guarantee to find an optimal plan (if one exists). Admissible heuristics using a set $\mathcal{M}$ of additive pattern collections have another benefit: If the choice of $\mathcal{M}$ is sufficiently good, the resulting heuristic values can be even more appropriate than those of non-admissible heuristics. In classical planning, Edelkamp [9] provides a general criterion for testing whether a pattern collection is additive. It is easy to see that the analogous criterion also holds for the non-deterministic setting, and in particular, that every single h^i is admissible.

Theorem 1. *A pattern collection P is additive if for all actions $a \in A$ and for all patterns $P_i \in P$, if $P_i \cap \mathit{effvar}(a) \neq \emptyset$, then $P_j \cap \mathit{effvar}(a) = \emptyset$ for all $j \neq i$.*

Proof. The proof is by induction on the true cost $\mathit{cost}^*(s)$ of s. The base case for $\mathit{cost}^*(s) = 0$ is trivial (since in this case, all abstractions of s are abstract goal states and we only sum up costs of zero for all abstractions). For the inductive case, consider a concrete solution graph $\mathcal{G}$ minimizing $cost^*(s)$. Let c be the root connector of $\mathcal{G}$, a an action inducing c, and s' a successor of s in $\mathcal{G}$, $s' \in succ(c)$, along a cost-maximizing path. In the following, for any concrete state $\hat{s}$ and pattern P_i, $cost_i^*(\hat{s})$ denotes the abstract cost of $\hat{s}^i$, i.e., the depth of a depth-minimizing abstract solution graph rooted at $\hat{s}^i$. Without loss of generality, assume that there is a pattern, say P_1, in P such that $P_1 \cap effvar(a) \neq \emptyset$. Then, by assumption, $P_j \cap effvar(a) = \emptyset$ for $j = 2, \ldots, k$. Now let s'' be a successor of s, $s'' \in succ(c)$, such that $(s'')^1$ maximizes the $cost_1^*$ among the abstract successors of s^1. For all patterns P_i other than P_1, we have $(s'')^i = s^i$. Applying the induction hypothesis to s'', we obtain

$$\sum_{i=1}^{k} cost_i^*(s'') \leq cost^*(s'') \ , \tag{1}$$

whereas by assumption and by the definition of $cost^*$, we get

$$cost^*(s'') + 1 \leq cost^*(s') + 1 = cost^*(s) \ . \tag{2}$$

On the other hand,

$$\sum_{i=1}^{k} cost_i^*(s) \leq cost_1^*(s'') + 1 + \sum_{i=2}^{k} cost_i^*(s'') = \sum_{i=1}^{k} cost_i^*(s'') + 1 \ . \tag{3}$$

Taking all this together, by (3), (1), and (2) we obtain that

$$\sum_{i=1}^{k} cost_i^*(s) \leq \sum_{i=1}^{k} cost_i^*(s'') + 1 \leq cost^*(s'') + 1 \leq cost^*(s') + 1 = cost^*(s) \ .$$

Since the heuristic values $h^i(s)$ are defined as $cost_i^*(s)$, this is the same as $\sum_{i=1}^{k} h^i(s) \leq cost^*(s)$. □

Corollary 1. *The heuristic h^i is admissible for any pattern P_i.* □

Finding an appropriate pattern collection $P_1, \ldots, P_k$ is in itself a challenging problem. Different approaches include clustering variables with strong interaction into one pattern, or to perform local search in the space of pattern collections [10,11], starting from singleton patterns for all goal variables and extending the collection until an evaluation function estimating the number of node expansions in a search using the current pattern collection reaches a local optimum. In the following, however, we will focus on how to use a given pattern collection during search, not on how to obtain it in the first place.

Example 2. Consider the problem from Example 1 and its two abstractions $\mathcal{P}^1$ and $\mathcal{P}^2$ with respect to the patterns $P_1 = \{a, b, c\}$ and $P_2 = \{d, e\}$. Then $s_0^1 = \{a\}$, $s_0^2 = \emptyset$, $G^1 = \{b, c\}$, $G^2 = \{d, e\}$, and the abstract actions are

$a^1_1 = \langle a, b \wedge \neg a \,|\, c \wedge \neg a\rangle$, $a^1_2 = \langle b, \top\rangle$, $a^1_3 = \langle c, \top\rangle$, $a^1_4 = \langle b, c\rangle$, $a^1_5 = \langle c, b\rangle$, $a^1_6 = \langle b, c\rangle$, $a^1_7 = \langle c, b\rangle$, $a^1_8 = \langle b \wedge c, \top\rangle$, and $a^1_9 = \langle b \wedge c, \top\rangle$ for pattern P_1, and $a^2_1 = \langle \top, \top\rangle$, $a^2_2 = \langle \top, e \,|\, d\rangle$, $a^2_3 = \langle \top, e \,|\, d\rangle$, $a^2_4 = \langle d, \top\rangle$, $a^2_5 = \langle d, \top\rangle$, $a^2_6 = \langle e, \top\rangle$, $a^2_7 = \langle e, \top\rangle$, $a^2_8 = \langle d, e\rangle$, and $a^2_9 = \langle e, d\rangle$ for pattern P_2, respectively.

The two graphs to the right show the optimal solution graphs for both abstract problems (left: $\mathcal{P}^1$, right: $\mathcal{P}^2$). We get a more accurate heuristic value by summing over all patterns, since the pattern collection $\{P_1, P_2\}$ is additive. E.g.,

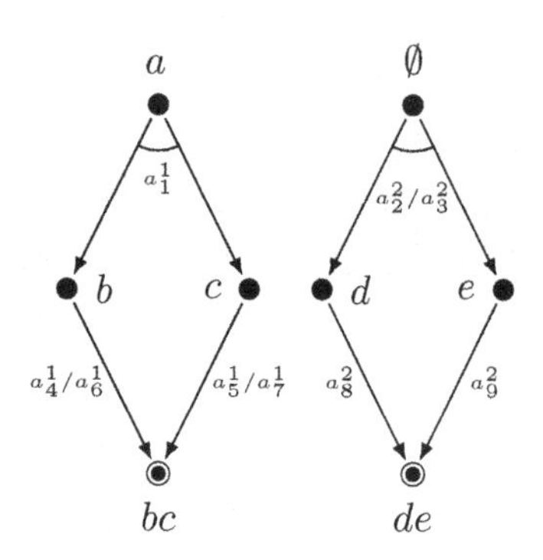

$$\begin{aligned} h(\{a\}) &= cost^*_1(\{a\}^1) + cost^*_2(\{a\}^2) \\ &= cost^*_1(\{a\}) + cost^*_2(\emptyset) \\ &= 2 + 2 = 4 = cost^*(\{a\}). \end{aligned}$$

Hence, in this case, the heuristic even computes the true cost value.

5 Experimental Results

We have encoded several instances of non-deterministic planning problems and solved them with our AO*-based planner and two different heuristics (FF heuristic [12,5] and PDB heuristics), as well as with Gamer [2], the winner of the fully observable non-deterministic (FOND) track of the uncertainty part of the International Planning Competition 2008 (IPPC'08). For the comparison, we did not use the domains from the IPPC'08, since those problems only allow for strong cyclic plans which our planner cannot find. The results, which were obtained on a machine with an AMD Turion 64 X2 processor (1600 MHz per core), and 1500 MB memory, are summarized in Tables 1 and 2.

In the first domain, Chain of Rooms [13], a number of rooms are sequentially connected by doors. A robot starting in the leftmost room has to visit each room at least once. It can move between rooms if the connecting door is open. Before a door can be passed or opened, the robot has to observe whether the door is open or closed. This observation action is modeled as turning on a light which changes the state of the door from undefined to either open or closed non-deterministically.

In the second domain, Coin Flip, there are n coins, which are initially contained in a bag. The coins have to be tossed exactly once each, in an arbitrary order. Tossing results in the coin showing heads or tails non-deterministically. After a coin has been tossed, it can be turned from heads to tails or vice versa, depending on which side is currently up. The goal is to have all coins on the table showing heads.

For the PDB heuristics and the Chain of Rooms domain, we used one pattern collection for each instance, where each pattern holds all state variables that belong to four neighboring rooms (24 Boolean variables). We omitted one room

Table 1. Experimental results from the Chain of Rooms domain. The numbers given in the columns *pre*, *search*, and *sum* are preprocessing, search, and overall times in seconds, *mem* denotes RAM in MB (for PDB plus concrete search space), |PDB| the overall number of PDB entries, *nodes* the number of generated nodes in the AND/OR graph, and |BDD| is the number of BDD nodes. Dashes indicate that the time-out of 30 minutes or the memory bound of 1500 MB was exceeded.

#rooms	AO* Planner, FF			AO* Planner, PDB						Gamer	
	search	mem	nodes	pre	search	sum	mem	|PDB|	nodes	search	|BDD|
20	2	3	236	4	1	5	3	315	274	6	16699
40	15	13	873	7	2	9	17	679	1138	23	130657
60	69	33	1909	16	6	22	46	1043	2602	—	—
80	250	69	3346	26	15	41	100	1407	4666	—	—
100	655	133	5183	41	33	74	190	1771	7330	—	—
120	1497	214	7419	61	73	134	319	2135	10594	—	—
140	—	—	—	91	117	208	495	2499	14458	—	—
160	—	—	—	127	194	321	736	2863	18922	—	—
180	—	—	—	177	314	491	1050	3227	23986	—	—

Table 2. Experimental results from the Coin Flip domain

#coins	AO* Planner, FF			AO* Planner, PDB						Gamer	
	search	mem	nodes	pre	search	sum	mem	|PDB|	nodes	search	|BDD|
20	—	—	—	2	1	3	4	60	1125	—	—
40	—	—	—	3	4	7	27	120	4645	—	—
60	—	—	—	4	19	23	85	180	10565	—	—
80	—	—	—	7	60	67	180	240	18885	—	—
100	—	—	—	11	217	328	366	300	29605	—	—
120	—	—	—	18	306	324	597	360	42725	—	—
140	—	—	—	23	533	556	905	420	58245	—	—
160	—	—	—	34	908	942	1307	480	76165	—	—

between each group of four rooms to meet the condition of Theorem 1, thus dividing a problem with n rooms into $\lceil n/5 \rceil$ subproblems. In the Coin Flip domain, we represented each of n coins by a corresponding pattern (giving us n patterns containing 3 Boolean variables each). Since those subproblems are completely independent, we achieve perfect heuristic values and no unnecessary nodes are expanded.

It is worth mentioning that Gamer will always find optimal solutions, whereas our AO*-based planner in general only finds suboptimal solutions.

6 Conclusion and Future Work

We have presented and evaluated a planner for fully-observable non-deterministic planning problems based on AO* search guided by PDB heuristics. Additionally, we have shown a generalization of the pattern additivity criterion known from classical planning, which allows for more informative heuristics while maintaining admissibility. The experimental results show that PDB heuristics are a promising tool to guide heuristic search algorithms for non-deterministic planning problems.

So far, the patterns are still selected manually. Obviously, a reasonable automated pattern selection technique is necessary to obtain a truly domain-independent planner. In classical planning, local search in the space of pattern collections [10,11] is often used to automatically select patterns. We believe that this approach will result in good patterns for the non-deterministic case, too. Besides pattern selection, future work includes the adaptation of our implementation to multi-valued state variables and the generalization of the algorithm from strong planning to strong cyclic planning [1].

Acknowledgments. We want to thank Peter Kissmann for providing us with the Gamer planning system and his help in using it.

References

1. Cimatti, A., Pistore, M., Roveri, M., Traverso, P.: Weak, strong, and strong cyclic planning via symbolic model checking. Artificial Intelligence 147(1–2), 35–84 (2003)
2. Edelkamp, S., Kissmann, P.: Solving fully-observable non-deterministic planning problems via translation into a general game. In: Proc. 32nd German Annual Conference on Artificial Intelligence, KI 2009 (2009)
3. Bryce, D., Kambhampati, S., Smith, D.E.: Planning graph heuristics for belief space search. Journal of Artificial Intelligence Research 26, 35–99 (2006)
4. Hoffmann, J., Brafman, R.I.: Contingent planning via heuristic forward search with implicit belief states. In: Proc. 15th International Conference on Automated Planning and Scheduling (ICAPS 2005), pp. 71–80 (2005)
5. Bercher, P., Mattmüller, R.: A planning graph heuristic for forward-chaining adversarial planning. In: Proc. 18th European Conference on Artificial Intelligence (ECAI 2008), pp. 921–922 (2008)
6. Hansen, E.A., Zilberstein, S.: LAO*: A heuristic search algorithm that finds solutions with loops. Artificial Intelligence 129(1–2), 35–62 (2001)
7. Nilsson, N.J.: Principles of Artificial Intelligence. Springer, Heidelberg (1980)
8. Bercher, P.: Anwendung von Pattern-Database-Heuristiken zum Lösen nicht-deterministischer Planungsprobleme. Diplomarbeit, Albert-Ludwigs-Universität Freiburg im Breisgau (2009)
9. Edelkamp, S.: Planning with pattern databases. In: Proc. 6th European Conference on Planning (ECP 2001), pp. 13–24 (2001)
10. Edelkamp, S.: Automated creation of pattern database search heuristics. In: Edelkamp, S., Lomuscio, A. (eds.) MoChArt IV. LNCS (LNAI), vol. 4428, pp. 35–50. Springer, Heidelberg (2007)
11. Haslum, P., Botea, A., Bonet, B., Helmert, M., Koenig, S.: Domain-independent construction of pattern database heuristics for cost-optimal planning. In: Proc. 22nd AAAI Conference on Artificial Intelligence (AAAI 2007), pp. 1007–1012 (2007)
12. Hoffmann, J., Nebel, B.: The FF planning system: Fast plan generation through heuristic search. Journal of Artificial Intelligence Research 14, 253–302 (2001)
13. Rintanen, J.: Constructing conditional plans by a theorem-prover. Journal of Artificial Intelligence Research 10, 323–352 (1999)

An Exploitative Monte-Carlo Poker Agent

Immanuel Schweizer, Kamill Panitzek, Sang-Hyeun Park, and Johannes Fürnkranz

TU Darmstadt, Knowledge Engineering Group,
D-64289 Darmstadt, Germany

Abstract. We describe the poker agent AKI-REALBOT which participated in the 6-player Limit Competition of the third Annual AAAI Computer Poker Challenge in 2008. It finished in second place, its performance being mostly due to its superior ability to exploit weaker bots. This paper describes the architecture of the program and the Monte-Carlo decision tree-based decision engine that was used to make the bot's decision. It will focus the attention on the modifications which made the bot successful in exploiting weaker bots.

1 Introduction

Poker is a challenging game for AI research because of a variety of reasons [3]. A poker agent has to be able to deal with *imperfect* (it does not see all cards) and *uncertain* information (the immediate success of its decisions depends on random card deals), and has to operate in a *multi-agent* environment (the number of players may vary). Moreover, it is not sufficient to be able to play an optimal strategy (in the game-theoretic sense), but a successful poker agent has to be able to exploit the weaknesses of the opponents. Even if a game-theoretical optimal solution to a game is known, a system that has the capability to model its opponent's behavior may obtain a higher reward. Consider, for example, the simple game of *rock-paper-scissors* aka *RoShamBo* [1], where the optimal strategy is to randomly select one of the three possible moves. If both players follow this strategy, neither player can gain by unilaterally deviating from it (i.e., the strategy is a *Nash equilibrium*). However, against a player that always plays *rock*, a player that is able to adapt its strategy to always playing *paper* can maximize his reward, while a player that sticks with the "optimal" random strategy will still only win one third of the games. Similarly, a good poker player has to be able to recognize weaknesses of the opponents and be able to exploit them by adapting its own play. This is also known as *opponent modeling*.

In every game, also called a *hand*, of fixed limit Texas Hold'em Poker, there exist four game states. At the *pre-flop* state, every player receives two *hole cards*, which are hidden to the other players. At the *flop*, *turn* and *river* states, three, one and one *community cards* are dealt face up respectively, which are shared by all players. Each state ends with a *betting* round. At the end of a hand (the *showdown*) the winner is determined by forming the strongest possible five-card poker hand from the players's hole cards and the community cards. Each game begins by putting two forced bets (*small blind* and *big blind*) into the pot, where the big blind is the minimal betting amount, in this context also called *small bet* (SB). In pre-flop and flop the betting amount is restricted to SBs, whereas on turn and river one has to place a *big bet* (2× SB). At every turn, a player can either *fold*, *check/call* or *bet/raise*.

B. Mertsching, M. Hund, and Z. Aziz (Eds.): KI 2009, LNAI 5803, pp. 65–72, 2009.

In this paper, we will succinctly describe the architecture of the AKI-REALBOT poker playing engine (for more details, cf. [7]), which finished second in the AAAI-08 Computer Poker Challenge in the 6-player limit variant. Even though it lost against the third and fourth-ranked player, it made this up by winning more from the fifth and sixth ranked player than any other player in the competition.

2 Decision Engine

2.1 Monte-Carlo Search

The Monte Carlo method [6] is a commonly used approach in different scientific fields. It was successfully used to build AI agents for the games of bridge [5], backgammon [8] and Go [4]. In the context of game playing, its key idea is that instead of trying to completely search a given game tree, which is typically infeasible, one draws random samples at all possible choice nodes. This is fast and can be repeated sufficiently frequently so that the average over these random samples converges to a good evaluation of the starting game state.

Monte-Carlo search may be viewed as an orthogonal approach to the use of evaluation functions. In the latter case, the intractability of exhaustive search is dealt with by limiting the search depth and the use of an evaluation function at the leaf nodes, whereas Monte Carlo search deals with this problem by limiting the search breadth at each node and the use of random choice functions at the decision nodes. A key advantage of Monte Carlo search is that it can deal with many aspects of the game without the need for explicitly representing the knowledge. Especially for poker, these include *hand strength, hand potential, betting strategy, bluffing, unpredictability* and *opponent modeling* [2]. These concepts, for which most are hard to model explicitly, are considered implicitly by the outcome of the simulation process.

In each game state, there are typically three possible actions, *fold, call* and *raise*.[1] AKI-REALBOT uses the simulated expected values (EV) for them to evaluate a decision. These EVs are estimated by applying two independent Monte-Carlo searches, one for the call action and the other one for the raise action (cf. Figure 1). Folding does not have to be simulated, since the outcome can be calculated immediately. Then at some point, these search processes are stopped by the Time Management component [7], which tries to utilize the available time as effective as possible. Since an increase in the number of simulated games also increases the quality of the EVs and therefore improves the quality of the decision, a multi-threading approach was implemented.

Our Monte-Carlo search is not based on a uniformly distributed random space but the probability distribution is biased by the previous actions of a player. For this purpose, AKI-REALBOT collects statistics about each opponent's probabilities for folding (f), calling (c), and raising (r), thus building up a crude opponent model. This approach was first described in [2] as selective sampling. For each played hand, every active opponent player is assigned hole cards. The selection of the hole cards is influenced by the opponent model because the actions a player takes reveal information about the strength of his cards, and should influence the sample of his hole cards. This selection

[1] The AAAI rules restrict the number of bets to four, so that a *raise* is not always possible.

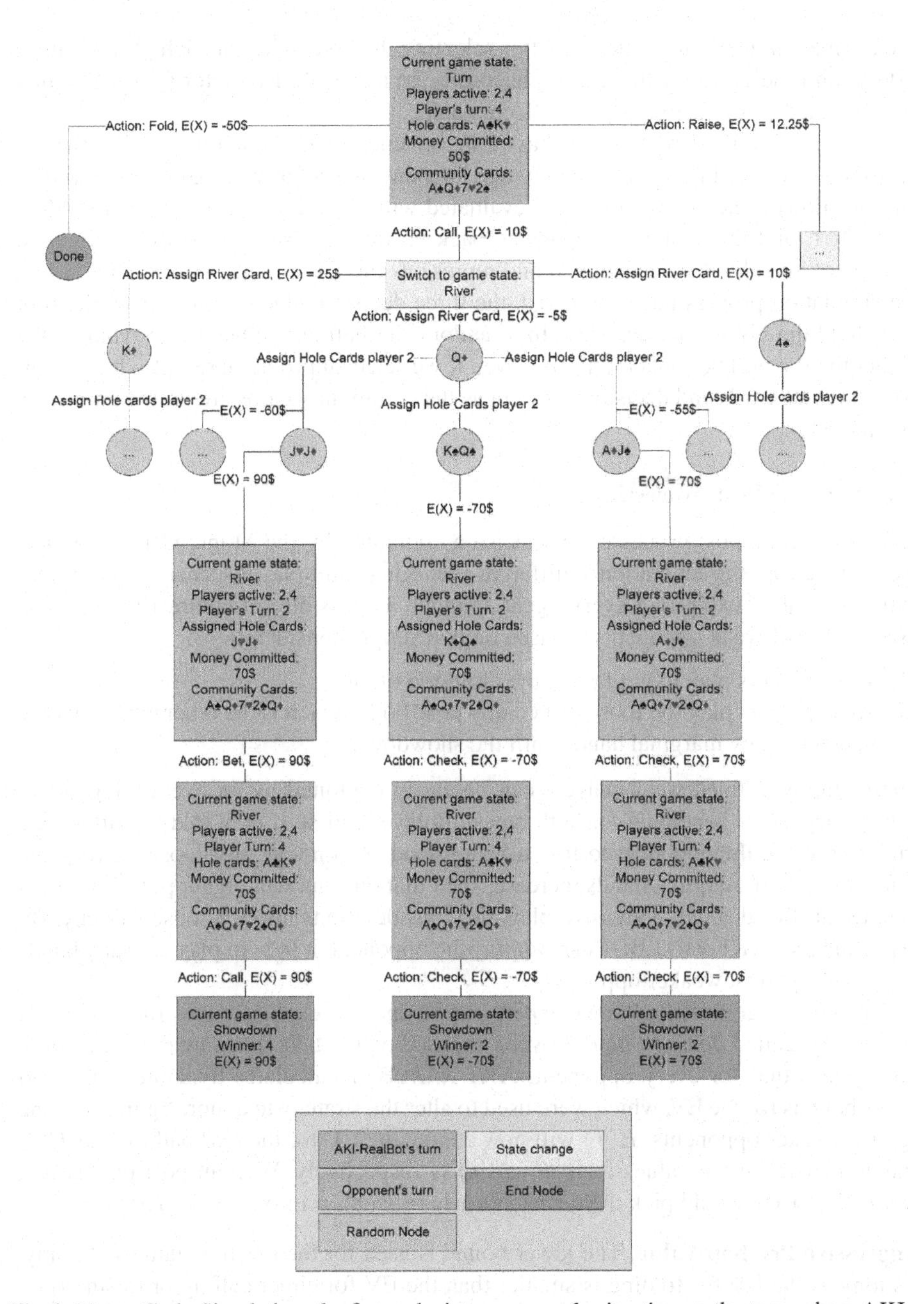

Fig. 1. Monte Carlo Simulation: the figure depicts an example situation on the turn, where AKI-REALBOT is next to act (top). The edges represent in general the actions of players or that of the chance player. For the decisions *call* or *raise* (middle and right path), two parallel simulations are initiated. The path for the *call* decision for example (in the middle), simulates random games until the showdown (the river card is Qd, the opponent cards are estimated as KsQs, and both players check on the river.) and the estimated loss of 70$ is backpropagated along the path.

is described in detail in Section 3. After selecting the hole cards, at each player's turn, a decision is selected for this player, according to a probability vector (f, c, r), which are estimated from the previously collected data.

Each community card that still has to be unveiled is also randomly picked whenever the corresponding game state change happens. Essentially the game is played to the showdown. The end node is then evaluated with the amount won or lost by AKI-REALBOT, and this value is propagated back up through the tree. At every edge the average of all subtrees is calculated and represents the EV of that subtree. Thus, when the simulation process has terminated, the three decision edges coming from the root node hold the EV of that decision. In a random simulation, the better our hand is, the higher the EV will be. This is still true even if we select appropriate samples for the opponents' hole cards and decisions as long as the community cards are drawn uniformly distributed.

2.2 Decision Post-processing

AKI-REALBOT post-processes the decision computed by the Monte-Carlo search in order to increase the adaptation to different agents in a multiplayer scenario even further with the goal of exploiting every agent as much as possible (in contrast to [2]). The exploitation of weak opponents is based on two simple considerations:

1. Weak players play too tight, i.e. they fold too often
2. Weak players play too loose (especially post-flop), which is the other extreme: they play too many marginal hands until the showdown

These simply defined weak players can be easily exploited by an overall aggressive play strategy. It is beneficial for both types of players. First, if they fold too often, one can often bring the opponent to fold a better hand. Second, against loose players, the hand strength of marginal hands increase, such that one can win bigger pots with them than usual. Besides the aggressive play, the considerations imply a loose strategy. By expecting that AKI-REALBOT can *outplay* the opponent, it tries to play as many hands as possible against weaker opponents.

This kind of commonly known expert-knowledge was explicitly integrated. For this purpose, so-called *decision bounds* were imposed on the EVs given by the simulation. This means that for every opponent, AKI-REALBOT calculates dynamic upper and lower bounds for the EV, which were used to alter the strategy to a more aggressive one against weaker opponents. $E(f)$ will now denote the EV for the *fold* path, while $E(c)$ and $E(r)$ will be the values for *call* and *raise* respectively. Without post-processing, AKI-REALBOT would pick the decision x where $E(x) = \max_{i=\{f,c,r\}} E(i)$.

Aggressive Pre-flop Value. The lower bound is used for the pre-flop game state only. As long as the EV for folding is smaller than the EV for either calling or raising (i.e., $E(f) < \max(E(c), E(r))$), it makes sense to stay in the game. More aggressive players may even stay in the game if $E(f) - \delta < \max(E(c), E(r))$ for some value $\delta > 0$. If AKI-REALBOT is facing a weak agent W it wants to exploit its weakness. This means that AKI-REALBOT wants to play more hands against W. This can be achieved by setting $\delta > 0$. We assume that an agent W is weak if he has lost money against AKI-REALBOT over a fixed period of rounds. For this purpose, AKI-REALBOT maintains

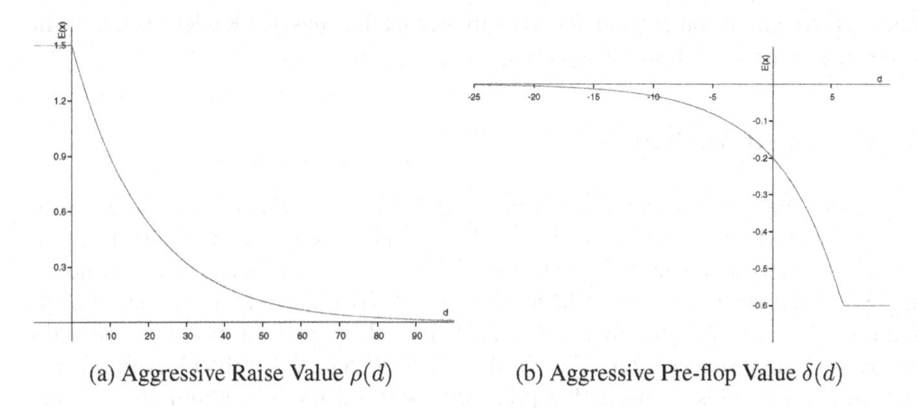

(a) Aggressive Raise Value $\rho(d)$

(b) Aggressive Pre-flop Value $\delta(d)$

Fig. 2. Dynamic Decision Bounds

a statistic over the number of small bets (SB) d, that has been lost or won against W in the last $N = 500$ rounds. For example, if W on average loses 0.5 SB/hand to AKI-REALBOT then $d = 0.5 \times 500 = 250$ SB. Typically, d is in the range of $[-100, 100]$. Then, the *aggressive pre-flop value* δ for every opponent is calculated as

$$\delta(d) = \max(-0.6, -0.2 \times (1.2)^d)$$

Note that $\delta(0) = -0.2$ (SB), and that the value of maximal aggressiveness is already reached with $d \approx 6$ (SB). That means, that AKI-REALBOT already sacrifices in the initial status $d = 0$ some EV (maximal -0.2 SB) in the pre-flop state, in the hope to outweight this drawback by *outplaying* the opponent post-flop. Furthermore, if AKI-REALBOT has won in the last 500 hands only more than 6 SB against the faced opponent, it reaches its maximal *optimism* by playing also hands which EVs were simulated as low as ≈ -0.6 SB. This makes AKI-REALBOT a very aggressive player pre-flop, especially if we consider that δ for more than one active opponent is calculated as the average of their respective δ values.

Aggressive Raise Value. The upper bound is used in all game states and makes AKI-REALBOT aggressive on the other end of the scale. As soon as this upper bound is reached, it will force AKI-REALBOT to raise even if $E(c) > E(r)$. This will increase the amount of money that can be won if AKI-REALBOT is very confident about his hand strength. This upper bound is called the *aggressive raise value* ρ.

$$\rho(d) = \min(1.5, 1.5 \times (0.95)^d)$$

Here, the upper bound returns $\rho(0) = 1.5$ for the initial status $d = 0$, which is 1.5 times the SB and therefore a very confident EV. In fact, it is so confident that this is also the maximum value for ρ. The aggressive raise value is not influenced if we lose money against a player. If, on the other hand, AKI-REALBOT wins money against an agent W, it will slowly converge against zero, resulting in a more and more aggressive play.

As said before, the value of d is calculated based on a fixed amount of past rounds. It is therefore continuously changing with AKI-REALBOT's performance over the past rounds. The idea is to adapt dynamically to find an optimal strategy against any single player. On the other hand, it is easy to see that this makes AKI-REALBOT highly vulnerable against solid, strong agents.

3 Opponent Modeling

In general, the opponent modeling of the AKI-REALBOT, which is used to adjust the implemented Monte-Carlo simulation to the individual behavior of the other players, considers every opponent as a straight-forward player. That means, we assume that aggressive actions indicate a high hand strength and passive actions a low hand strength. Within the simulation, the opponent's hand strength is guessed based on the action he takes. So if a player often folds in the pre-flop phase but calls or even raises in one special game this means he has probably a strong hand. In addition the opponent modeling tries to map cards to the actions every player takes.

AKI-REALBOT has two different routines that enables it to guess hole cards according to the opponent model. Which routine is used depends on the game state.

Pre-Flop State: In pre-flop, we assume that the actions of a player are only based on his hole cards. He is either confident enough to raise or to make a high call, whereas making a small call may indicate a lower confidence in his hand. A high call is indicated by committing more than a big bet. In either case his observed fold ratio f and call ratio c are used to calculate an upper and lower bound for the set of hole cards. It is common to divide the set of possible hole cards into *buckets*, where each bucket consists of hole cards of similar hand strength, to reduce the space of hole card combinations. We used five buckets, U_0 being the weakest bucket and U_4 (e.g. containing the cards AA) the strongest. The buckets have the following probability distribution: $p(U_0) = 0.65$, $p(U_1) = 0.14$, $p(U_2) = 0.11$, $p(U_3) = 0.07$, $p(U_4) = 0.03$.

In the first case of the above example, the upper bound U is set to the maximum possible bucket value ($U_h = 4$) because high confidence was shown. The lower bound is calculated by taking $l = c + f$ and relating this to the bucket. That means, the lower bound is set to exclude the hole cards, for which the player would only call or fold. If for example $f = 0.71$ and $c = 0.2$, the player raises only in 9% of cases. Since we assumed a straight-forward or honest player, we imply that he only does this with the top 9% of hole cards. So, the lower bound is set to U_3. Then, the hole cards for that player are selected randomly from the set of hole cards which lie between the bounds.

Post-Flop State. The second routine for guessing the opponents' hole cards is used when the game has already entered a post-flop state. The main difference is that the actions a player takes are now based on both hidden (his hole cards) and visible information (the board cards). Therefore, AKI-REALBOT has to estimate the opponent's strength also by taking the board cards into account. It estimates how much the opponent is influenced by the board cards. This is done by considering the number of folds for the game state flop. If a player is highly influenced by the board he will fold often on the flop and only play if his hand strength has increased with the board cards or if his starting hand was irrespectively very strong.

This information is used by AKI-REALBOT to assign hole cards in the post-flop game state. Two different methods are used here:

- *assignTopPair*: increases the strength of the hole cards by assigning the highest rank possible, i.e., if there is an ace on the board the method will assign an ace and a random second card to the opponent.
- *assignNutCard*: increases the strength of the hole cards even more by assigning the card that gives the highest possible poker hand using all community cards i.e. if there is again an ace on the board but also two tens the method will assign a ten and a random second card.

These methods are used for altering one of the player's hole card on the basis of his fold ratio f on the flop. We distinguish among three cases based on f, where probability values p_{Top} and p_{Nut} are computed.

(1) $f < \frac{1}{3} \Rightarrow p_{Top} = 3(f)^2 \in [0, \frac{1}{3}[\quad p_{Nut} = 0$

(2) $\frac{1}{3} \leq f < \frac{2}{3} \Rightarrow p_{Top} = \frac{1}{3} \quad p_{Nut} = \frac{1}{3}(3f - 1)^2 \in [0, \frac{1}{3}[$

(3) $f \geq \frac{2}{3} \Rightarrow p_{Top} = \frac{1}{3} \quad p_{Nut} = f - \frac{1}{3} \in [\frac{1}{3}, \frac{2}{3}]$

To be clear, *assignTopPair* is applied with a probability of p_{Top}, *assignNutCard* is applied with a probability of p_{Nut} and with a probability of $1 - (p_{Top} + p_{Nut})$ the hole cards are not altered. As one can see in the formulas, the higher f is, the more likely it is that the opponent will be assigned a strong hand in relation to the board cards. Note that for both methods the second card is always assigned randomly. This will sometimes strongly underestimate the cards e.g. when there are three spade cards on the board *assignNutCard* will not assign two spade cards.

4 AAAI-08 Computer Poker Competition Results

AKI-REALBOT participated in the 6-player Limit competition part of the Computer Poker Challenge at the AAAI-08 conference in Chicago. There were six entries: HYPERBOREAN08_RING aka POKI0 (University of Alberta), DCU (Dublin City University), CMURING (Carnegie Mellon University), GUS6 (Georgia State University), MCBOTULTRA and AKI-REALBOT, two independent entries from TU Darmstadt.

Among these players, 84 matches were played with different seating permutations so that every bot could play in different positions. Since the number of participants were exactly 6, every bot was involved in all 84 matches. In turn, this yielded 504000 hands for every bot. In that way, a significant result set was created, where the final ranking was determined by the accumulated win/loss of each bot over all matches.[2]

Table 1 shows the results over all 84 matches. All bots are compared with each other and the win/loss statistics are shown in SBs. Here it becomes clear that AKI-REALBOT exploits weaker bots because the weakest bot, GUS6, loses most of it's money to AKI-REALBOT. Note, that GUS6 lost in average more than 1.5 SBs per hand, which is a worse outcome than by folding every hand, which results in an avg. loss of 0.25 SB/Hand. Although AKI-REALBOT loses money to DCU and CMURING it manages to rank second, closely behind POKI0, because it is able to gain much higher winnings against the weaker players than any other player in this field. Thus, if GUS6 had not participated in this competition, AKI-REALBOT's result would have been much worse.

[2] The official results can be found at
http://www.cs.ualberta.ca/~pokert/2008/results/

Table 1. AAAI-08 Poker Competition Results: pairwise and overall performance of each entry

	POKI0	AKI-REAL	DCU	CMURING	MCBOT	GUS6
POKI0		65,176	2,655	18,687	29,267	214,840
AKI-REALBOT	-65,176		-15,068	-2,769	30,243	348,925
DCU	-2,665	15,068		7,250	16,465	90,485
CMURING	-18,687	2,769	-7,250		7,549	92,453
MCBOTULTRA	-29,267	-30,243	-16,465	-7,549		16,067
GUS6	-214,840	-348,925	-90,485	-92,453	-16,067	
Total	330,822	296,293	126,657	76,848	-67,529	-763,091
avg. winnings/game	3934	3579	1512	939	-800	-9042
SB/Hand	0.656	0.588	0.251	0.152	-0.134	-1.514
Place	1.	2.	3.	4.	5.	6.

5 Conclusion

We have described the poker agent AKI-REALBOT that finished second in the AAAI-08 Poker Competition. Its overall performance was very close to the winning entry, even though it has lost against three of its opponents in a direct comparison. The reason for its strong performance was its ability to exploit weaker opponents. In particular against the weakest entry, it won a much higher amount than any other player participating in the tournament. The key factor for this success was its very aggressive opponent modeling approach, due to the novel adaptive post-processing step, which allowed it to stay longer in the game against weaker opponents as recommended by the simulation.

Based on these results, one of the main further steps is to improve the performance of AKI-REALBOT against stronger bots. An easy way would be to adopt the approaches of the strongest competitors of the competition, for which there exists a multitude of publications. But, we see also yet many possible improvements for our exploitative approach, which we elaborate in [7] and are currently working on.

References

1. Billings, D.: Thoughts on RoShamBo. ICGA Journal 23, 3–8 (2000)
2. Billings, D., Castillo, L.P., Schaeffer, J., Szafron, D.: Using probabilistic knowledge and simulation to play poker. In: AAAI/IAAI, pp. 697–703 (1999)
3. Billings, D., Davidson, A., Schaeffer, J., Szafron, D.: The challenge of poker. Artif. Intell. 134(1-2), 201–240 (2002)
4. Bouzy, B.: Associating domain-dependent knowledge and Monte Carlo approaches within a Go program. In: Joint Conference on Information Sciences, pp. 505–508 (2003)
5. Ginsberg, M.L.: Gib: Steps toward an expert-level Bridge-playing program. In: Proc. of the 16th International Joint Conference on Artificial Intelligence (IJCAI 1999), pp. 584–589 (1999)
6. Metropolis, N., Ulam, S.: The Monte Carlo method. J. Amer. Stat. Assoc. 44, 335–341 (1949)
7. Schweizer, I., Panitzek, K., Park, S.-H., Fürnkranz, J.: An exploitative Monte-Carlo poker agent. Technical Report TUD-KE-2009-02, TU Darmstadt, Knowledge Engineering Group (2009), `http://www.ke.informatik.tu-darmstadt.de/publications/reports/tud-ke-2009-02.pdf`
8. Tesauro, G.: Temporal difference learning and TD-Gammon. Commun. ACM 38(3), 58–68 (1995)

Interaction of Control and Knowledge in a Structural Recognition System

Eckart Michaelsen, Michael Arens, and Leo Doktorski

Research Institute for Optronics and Pattern Recognition FGAN-FOM
Gutleuthausstr. 1, 76275 Ettlingen, Germany
{michaelsen,arens,doktorski}@fom.fgan.de

Abstract. In this contribution knowledge-based image understanding is treated. The knowledge is coded declaratively in a production system. Applying this knowledge to a large set of primitives may lead to high computational efforts. A particular accumulating parsing scheme trades soundness for feasibility. Per default this utilizes a bottom-up control based on the quality assessment of the object instances. The point of this work is in the description of top-down control rationales to accelerate the search dramatically. Top-down strategies are distinguished in two types: (i) Global control and (ii) localized focus of attention and inhibition methods. These are discussed and empirically compared using a particular landmark recognition system and representative aerial image data from GOOGLE-earth.

1 Introduction

Structural recognition of patterns or objects is an option in cases where the structure of the target patterns is the property which distinguishes them from clutter or background best, i. e., for patterns and objects where no obvious or simple numerical features are at hand which promise satisfying recognition performance of machine learning or statistical methods. Structural recognition of patterns or objects should also be considered if the structure of the targets is the desired output, i. e., if recognition is meant as automatic pattern understanding. Last but not least structural recognition of patterns or objects may be beneficial if the number of training examples is low – or none are given at all – while there is knowledge accessible on the constructive elements and rules that define the targets, i. e., if the learning data consist of handbooks, CAD-models, thesauri, or ontologies respectively. Production systems give an approved formalism for structural recognition of patterns or objects.

The idea to employ high-level knowledge for the guidance or control of low-level image analysis processes has ever been conceived since the days of early work of Kanade [5] and, e. g., Tenenbaum and Barrow [13]. Performed on single images – often termed image understanding, compare [12] – the work of Neumann [4] and colleagues can be seen as a prominent and current work in the field of top-down control of low-level image analysis. Work on exploitation of high-level knowledge in form of expectations or anticipations on time-varying imagery – i. e.

B. Mertsching, M. Hund, and Z. Aziz (Eds.): KI 2009, LNAI 5803, pp. 73–80, 2009.

videos – have been reported by the groups of Dickmanns [3] and Nagel [1], to name only two. Declarative production rules as means to code knowledge for aerial image understanding has been introduced, e. g., in the SIGMA system [8]. This includes a discussion on intelligent control of the search – being aware of the dangerous combinatorics inherent in such formulations.

2 Knowledge-Based Recognition by Production Systems

Context-free constrained multi-set grammars are discussed in [7] particularly with regard to graphical languages and computer interfaces. The basic idea is generalizing the generative string grammars by replacing the concatenation constraint by a more general constraint to be tested in higher dimensional space – such as a picture. This generalized grammar can then be employed to parse an image – taking basic image features such as lines as input and trying to derive complex structures: elements of language defined by the grammar. Next to their symbolic name the instances then have attributes – such as locations, orientations, etc. Basically, we have a finite set of non-terminals N and terminals T and a finite set of production rules $p : \mathtt{A} \rightarrow \Sigma$ where only context free forms are allowed, i. e., $\mathtt{A} \in N$ and $\Sigma = \mathtt{BC}$, $\mathtt{Bc}$, or $\mathtt{bc}$ with $\mathtt{B}, \mathtt{C} \in N$ and $\mathtt{b}, \mathtt{c} \in T$. In [10] recurrent subsystems of the form $\{p : \mathtt{A} \rightarrow \mathtt{bb}, q : \mathtt{A} \rightarrow \mathtt{Ab}\}$ are approximated by short-cut productions $s : \mathtt{A} \rightarrow \mathtt{b} \ldots \mathtt{b}$ for particular classes of constraints – such as adjacency or collinearity. Here clustering techniques or accumulator methods such as the Hough transform are included in a declarative way into the knowledge representation. Actually, this can not only be done on the terminal level but also with non-terminals yielding the form $s : \mathtt{A} \rightarrow \mathtt{B} \ldots \mathtt{B}$. The constraint is tested on a set of instances of the object class in Σ where the size of this set is not fixed. In order to distinguish this form we call it *cluster_form*, while the classical production are said to have *normal_form*. An example production system is given in Table 1. For the classical language of such systems P one root object $\mathtt{R} \in N$ must be reduced from a set of primitive instances S:

$$L_{reduce} = \{S : \{R\} \xrightarrow{*}_P S\} \tag{1}$$

where the asterisk has the usual meaning of successive right-to-left reduction using productions from P. In recognition tasks from out-door scenery usually

Table 1. Example production system for recognizing bridges over highways

	left-hand	constraint	right-hand
$p_{prolong}$	longline	collinear and overlapping	line ... line
$p_{stripe1}$	road	parallel and $6.75m < d < 11.25m$	longline longline
$p_{stripe2}$	halfhighway	parallel and $16m < d < 19m$	longline longline
$p_{dstripe}$	highway	parallel and $6.0m < d < 8.4m$	halfhighway halfhighway
p_{cross}	bridge	crossing the T	road highway

clutter objects have to be tolerated. Thus the language is defined as the set of all primitive object sets S where a set X is reducible that contains a root object:

$$L_{cluttered} = \{S : R \in X \xrightarrow{*}_P S\}. \tag{2}$$

2.1 Recognition Using the Approximate Any-Time Interpreter

The language definitions given in Section 2 are of combinatorial nature. To the best of our knowledge there exists no interpretation algorithm of polynomial computational complexity for such systems. In order to assure feasibility in the presence of input data containing several thousands of primitives – as they are commonly segmented from input images or videos – approximate interpretation is crucial. Algorithm 1 adapted from [6] gives such method in particular for $L_{cluttered}$. It works on a constantly growing database (DB) of object instances which is initialized by the set of primitive terminal instances S. It associates working hypotheses (o, p, h, α) with each instance o in the DB where p is the partner class (i.e. $\Sigma = op$ or po), h is the hypothesis type (left-hand side) and α is the priority. All newly inserted instances cause a working hypothesis initially with $p, h = nil$ and α set from the quality of the object instance o. When such a hypothesis is encountered appropriate clones are constructed with p and h according to the productions in the system. At this point the priority can be changed.

As can be seen from Algorithm 1 production of *normal_form* are treated in the usual combinatorial way leading to a branching search tree. Productions of *cluster_form* are treated differently: Only the maximal set is reduced. All sub-sets which also may fulfill the constraint are not reduced. This violates the soundness. Here the algorithm only gives an approximate solution. For the justification of this see [10] and the definition of the maximal meaningful gestalt in [2]. Cluster analysis is often iterative. Here this means that for a *cluster_form* production $s : \mathtt{A} \rightarrow \mathtt{b} \ldots \mathtt{b}$ hypotheses $h_1 = (\mathtt{b}, \mathtt{b}, \mathtt{A}, \alpha)$ and $h_2 = (\mathtt{A}, \mathtt{b}, \mathtt{A}, \alpha)$ are permitted – the first for triggering objects b and the latter for triggering objects A. For justification see again [10]. An example is given with $p_{prolong}$ in Table 1. For each object *longline* a hypothesis is formed for further prolonging it. When such hypothesis triggers, a more accurate query can be posed giving a larger set of line objects to feed the production.

2.2 Re-evaluation Strategies Control the Search

For all kinds of search control it is demanded that every object concept should have a quality attribute with it assessing the saliency, relevance, or importance respectively. This allows comparing instances of the same object concept. In order to control the search using Algorithm 1 the importance of hypotheses interrelated with instances of different object concepts have to be compared, e. g., by weighting factors. So much for the data-driven bottom-up control. But more can be done. The importance of the hypotheses can be re-assessed with respect to the state of the DB reached. We distinguish two different classes of such importance calculation.

```
repeat
    sort(queue);
    set_of_hypothesis = choose_best_n(queue);
    foreach trigger_hypo ∈ set_of_hypothesis do
        if p=nil then
            foreach q where trigger_obj ∈ right-hand side do
                adjust_priority(q);
                append_queue(trigger_elem, q, new_priority);
            end
        else
            actual_query = construct_query(trigger_hypo);
            candidate_set = select_DB(actual_query);
            switch p of type do
                case normal_form
                    foreach partner ∈ candidate_set do
                        p:new_elem ← (trigger_elem, partner);
                        insert_DB(new_elem);
                        construct_null_hypo(new_elem);
                    end
                end
                case cluster_form
                    p:new_elem ← candidate_set;
                    insert_DB(new_elem);
                    construct_null_hypo(new_elem);
                end
            end
        end
        remove_queue(trigger_hypo);
    end
    foreach newly inserted element do
        re_evaluate all hypotheses;
    end
until root R found OR timeout OR queue=∅ ;
```

Algorithm 1. Approximate Any-time Interpreter for Production Systems

A very simple example for **global priority control** is delaying hypotheses (o, p, h, α) for triggering objects o until the set of partner objects p in the DB is not empty anymore. In the example given in Section 3 this is included in all variants. When there are multiple hypotheses (o, p, h_1, α), ..., (o, p, h_k, α) with the same triggering instance o there often will be a preference for one of these inhibiting the others. For the system given in Table 1 actually three hypotheses are formed for objects *longline*, namely for $p_{stripe1}$, $p_{stripe2}$ and also for $p_{prolong}$ (see Section 2.1). According to the principle of maximal meaningful gestalts [2] the hypotheses for further prolonging has high preference over the others. Only after the prolongation production failed to produce any new instance the other hypotheses regain their original priority. This is included in all our systems.

Furthermore, in production systems with a hierarchy on the non-terminals higher priorities α can be chosen with rising hierarchy. In the example reported in Section 3 the priorities were chosen as linear functions of the quality α_{qual} in overlapping intervals with ascending hierarchy *hie* using appropriate offset v and factor w:

$$\alpha_{min}(hie) \leq \alpha = v + (w\alpha_{qual}) \leq \alpha_{max}(hie) \tag{3}$$

This leads to a depth first search characteristic. More dynamically, a histogram of instance numbers for each symbol of the DB can be acquired with low computational effort. Hypotheses with h being a frequent object type may be punished and rare ones rewarded. In the beginning of an interpretation run this also leads to a depth-first search characteristic. Later with rising computation time the search will get broader. For the sake of simplicity such control has not been used in the present system.

Local priority control: With every hypothesis tested there is a triggering object instance, and this instance is located somewhere in the corresponding attribute space. Also during the last execution of the *foreach* trigger_hypo block new object instances have been constructed. Of course while inserting them to the DB it was tested whether they are not already present there. The re-evaluation is based on the relation between the newly inserted objects and the triggering objects of the hypotheses. Since here all pairs of new objects and hypotheses have to be considered this causes considerable computational effort. In Table 2 these extra costs are shown in the column titled *top down*. The first possibility for such control is *general local inhibition*: other hypotheses of the same production with their corresponding triggering object located close to the instance at hand will lead to similar queries. Such control has been preliminarily investigated with a very simple toy production system in [11] using a smooth re-evaluation function. A similar strategy is appositely termed as *anticipation* or focus of attention: It uses declarative knowledge from productions that contain both, the newly built object and the object resulting from the triggered hypothesis. Figure 1 gives an example: On the abscissa here we have distance of a location from a newly built object *long-line*. All hypotheses building long-line objects are re-evaluated. If the corresponding objects are very close their priority will be multiplied by 0.1. The focus of attention is at a particular distance interval. Here the weighting factor will be 2. Otherwise, their priority will not be changed. Qualitatively, this is similar to the smooth re-evaluation functions used in [11], but has more parameters and is more focused – recall that this has to be defined for all combinations. Interesting here is that the particular distance and the width must be taken from "higher" productions. In the example the priority of hypotheses with triggering objects *line* is doubled because an object *longline* has been constructed in the vicinity and another such object is needed here in order to form an object *road*. The construction of the focus of attention uses the constraint predicate of the production forming road-stripe objects. This means that declarative knowledge and control knowledge interact – a dangerous but successful thing.

Table 2. Number of fully/partial successful cases **with** and **without** anticipation control strategy and average runtimes in seconds

	success	partial	time out	failure	production	sorting	top down	complete
with	15	13	6	7	8.22	8.18	0	17.25
without	22	11	2	6	2.03	2.80	3.76	8.89

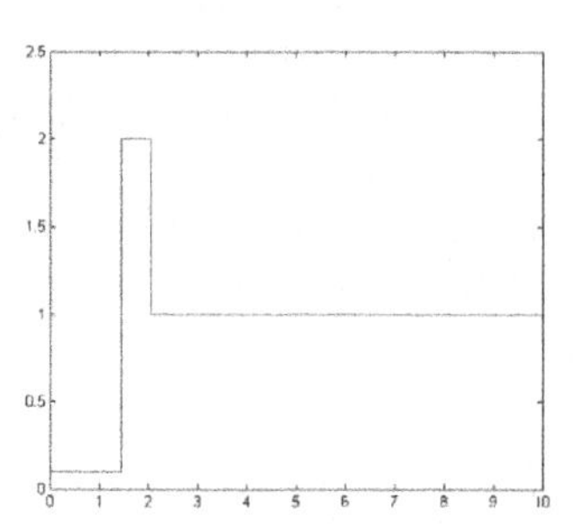

Fig. 1. Local re-evaluation function automatically derived from 'higher' productions

3 Experiments

The simple production system used here (compare Table 1) is designed to recognize bridges over highways. Knowledge – such as the expected width of single carriage ways used in production $p_{stripe2}$ (and in the localized top-down control) – can be obtained from Wikipedia. Experiments where made with a set of images taken during an evaluation run with a test-bed based on Google earth [9]. These images where picked blindly, i. e., the operator sees on the map layer of Google maps that there should be a landmark consistent with the model at that location. One of these 41 images actually doesn't contain a bridge – it is present in the map, but in the corresponding image there is a construction under way with the bridge removed. Some other pictures turn out to be difficult: Constructions under way, low contrasts, shadows, similar structures running parallel to the autobahn, etc. Not every failure of the system may be crucial depending on the task at hand. We distinguish for classes of recognition behaviour: (1) full success – the model has been instantiated properly with corresponding contours in the image, the location is precise; (2) semi success – the model has been instantiated partially correct, the localization is good enough for the task at hand; (3) no instantiation of the model could be achieved in the time limit – set in this experiment to 60 seconds; (4) false instantiation of the model to contours that are really caused by something else. Table 2 gives the success rates with and without the localized anticipation top-down control function automatically derived from higher productions and indicated in Figure 1 (right). In particular in time critical applications we have fewer problems with recognition failure due to time out constraints. The anticipation control produces only one such error – the other time-out is correct because that is the image which contains no bridge. Whereas, without this control five time-out failures are produced – and also one correct time-out. All times have been obtained on the same server machine running eight processors at 3GHz. The instantiation of the productions is run in parallel threads – thus the time spent on them is not the major part. The administration of the process queue – in particular sorting and the control handling – is done serially. We observed one more failure of type (4) with the default control than with the anticipation control. This cannot be regarded as significant given the small test sample set. Because this is the most important type of failure for the assessment of the robustness of the system for the task at hand we would still like to learn something on this topic from this experiment. We therefore give two examples, where one control works and the other one fails. Example results can be seen in Figure 2. The sets of primitive objects are displayed in the left column (usually between 10.000 and 20.000 instances per picture). The upper example is from a forested area, the lower from a urban environment. In the forest example the south eastern contour of the bridge is very weak. Accordingly, the default control concentrates on the strong autobahn contours – until the time is over. This example counts as a failure of type (3) in table 1. On the same data the anticipation first instantiates the major contours of the autobahn. Then it spreads its focus of attention more into the forest. It also instantiates the critical weak contour of the bridge. However, before this can trigger any hypotheses other parallel structures in the forest are found to be consistent with the model. We counted this as a type (4) failure. Thus we see that the anticipation automatic is not always necessarily better in every single

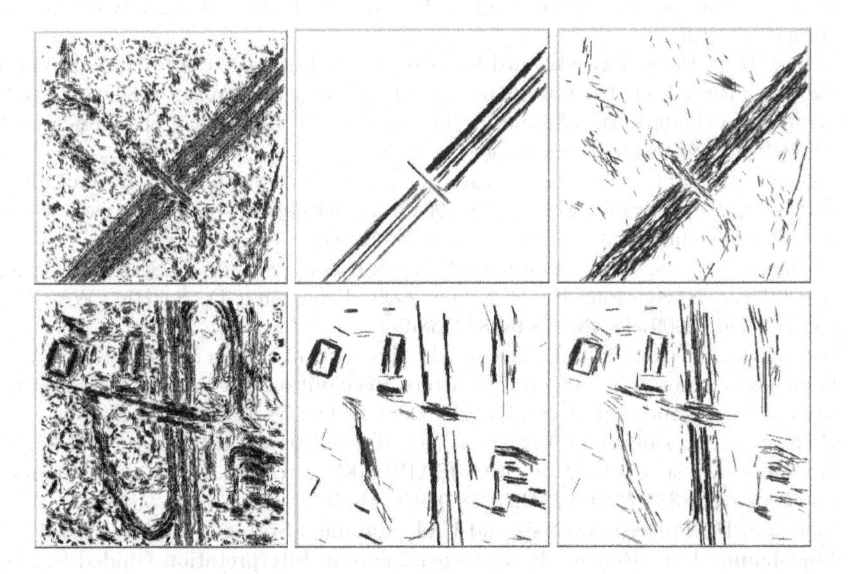

Fig. 2. Left terminal primitives for this image, center result with bottom up control, right result with localized top down control, upper row image 17 lower row image 4

example. The lower example contains many adjacent salient structures that may well be confused with the landmark. Important contours of the autobahn are much weaker than some other contours caused by buildings. Such structures at the south-western part of the image happen to fit the autobahn model. Some other road structure then forms the desired "crossing the T". This is a clear type (4) failure. Nothing here corresponds correctly and the landmark position is localized completely wrong. On the very same data the anticipation control gives a full type (1) success. A fairly tight cluster of landmark instances is found in the correct position (in fact some of them also refer to the shadow of the bridge – which we tolerate as still correct).

References

1. Arens, M., Nagel, H.-H.: Quantitative Movement Prediction based on Qualitative Knowledge about Behavior. In: KI – Künstliche Intelligenz 2/2005, pp. 5–11 (2005)
2. Desolneux, A., Moisan, L., Morel, J.-M.: From Gestalt Theory to Image Analysis. Springer, Berlin (2008)
3. Dickmanns, E.: Expectation-based Dynamic Scene Understanding. In: Blake, A., Yuille, A. (eds.) Active Vision, pp. 303–335. MIT Press, MA (1993)
4. Hotz, L., Neumann, B., Terzic, K.: High-Level Expectations for Low-Level Image Processing. In: Dengel, A.R., Berns, K., Breuel, T.M., Bomarius, F., Roth-Berghofer, T.R. (eds.) KI 2008. LNCS (LNAI), vol. 5243, pp. 87–94. Springer, Heidelberg (2008)
5. Kanade, T.: Model Representations and Control Structures in Image Understanding. In: Reddy, R. (ed.) Proc. 5th Int. Joint Conf. on Artificial Intelligence (IJCAI 1977), Cambridge, MA, USA, August 1977, pp. 1074–1082. William Kaufman, San Francisco (1977)
6. Lütjen, K.: BPI: Ein Blackboard-basiertes Produktionssystem für die automatische Bildauswertung. In: Hartmann, G. (ed.) Mustererkennung 1986, 8. DAGM-Symposium, Paderborn, September 30 – October 2. Informatik Fachberichte 125, pp. 164–168. Springer, Heidelberg (1986)
7. Marroitt, K., Meyer, B. (eds.): Visual Language Theory. Springer, Berlin (1998)
8. Matsuyama, T., Hwang, V.S.-S.: SIGMA a Knowledge-Based Aerial Image Understanding System. Plenum Press, New York (1990)
9. Michaelsen, E., Jäger, K.: A GOOGLE-Earth Based Test Bed for Structural Image-based UAV Navigation. In: FUSION 2009, Proc. on CD, Seattle, WA, USA, pp. 340–346 (2009) ISBN 978-0-9824438-0-4
10. Michaelsen, E., Doktorski, L., Arens, M.: Shortcuts in Production Systems – A way to include clustering in structural Pattern Recognition. In: Proc. of PRIA-9-2008, Nischnij Nowgorod, vol. 2, pp. 30–38 (2008) ISBN 978-5-902390-14-5
11. Michaelsen, E., Doktorski, L., Arens, M.: Making Structural Pattern Recognition Tractable by Local Inhibition. In: VISAPP 2009, Proc. on CD, Lisboa, Portugal, vol. 1, pp. 381–384 (2009) ISBN 978-989-8111-69-2
12. Niemann, H.: Pattern Analysis and Understanding. Springer, Berlin (1989)
13. Tenenbaum, J.M., Barrow, H.G.: Experiments in Interpretation Guided Segmentation. Artificial Intelligence Journal 8(3), 241–274 (1977)

Attention Speeds Up Visual Information Processing: Selection for Perception or Selection for Action?

Katharina Weiß and Ingrid Scharlau

University of Paderborn, Department of Psychology, Warburgerstr.100
33098 Paderborn
{katharina.weiss,ingrid.scharlau}@upb.de

Abstract. Attention speeds up information processing. Although this finding has a long history in experimental psychology, it has found less regard in computational models of visual attention. In psychological research, two frameworks explain the function of attention. *Selection for perception* emphasizes that perception- or consciousness-related processing presupposes selection of relevant information, whereas *selection for action* emphasizes that action constraints make selection necessary. In the present study, we ask whether or how far attention, as measured by the speed-up of information processing, is based on selection for perception or selection for action. The accelerating effect was primarily based on selection for perception, but there was also a substantial effect of selection for action.

Keywords: visuo-spatial attention, prior entry, selection for action, selection for perception.

1 Introduction

Most theories of attention assume that attention is beneficial for information processing by yielding more accurate and detailed processing of information, see e.g. [4, 6, 9, 13]. A wide body of empirical research demonstrated that attention is a pre-condition for complex object representations and awareness of these objects [16, 26, 30, 31].

Two interesting phenomena, which support the assumption that attention is a pre-condition for awareness, are *inattentional blindness* [11, 27] and *change blindness* [16]. Simons and Chabris (1999) had observers watch a film of two teams playing basketball. They directed observers' attention by the experimental task to one of the teams: counting the ball passes of this team. During the basketball play, a person in a gorilla costume respectively a woman with an umbrella walked through the scene. Almost half of the observers were "inattentional" and overlooked these unexpected, but very prominent events. Studies on *change blindness* [16, 17] demonstrated that observers have difficulties to detect salient changes in static scenes if these changes co-occur with dynamic events, such as eye blinks, saccades, or flicker on the screen. Change blindness can, however, be attenuated by attention: If a cue directs observer's attention to the part of scene that will change, observers can detect the changes more quickly.

B. Mertsching, M. Hund, and Z. Aziz (Eds.): KI 2009, LNAI 5803, pp. 81–88, 2009.

In this paper we focus on a less familiar, though not less important, consequence of attention on information processing: Attention accelerates information processing, e.g., it speeds up detection [15]. This acceleration means, for example, that in comparison to an unattended stimulus, an attended stimulus is perceived earlier or as appearing earlier. Although this notion of "prior entry" of attended stimuli has been studied for more than 150 years in experimental psychology [26, 28, 29] it was little noticed by models and theories of attention.

Prior entry is especially interesting and valuable because it allows a direct and easily understandable quantification of the effect of attention on information processing. The speed-up is the temporal interval by which an attended stimulus can trail an unattended stimulus and still be perceived as simultaneous. Psychophysics allows to measure it with temporal order judgments.

In the present paper, we use prior entry to distinguish between attention as selection for action and attention as selection for perception. Selection for perception is the more established framework, a sort of common sense, which has prevailed in psychology and modeling of attention. But the less known selection for action framework has some very interesting aspects, especially, if attention is used in controlling autonomous agents.

The underlying idea of *selection for perception* is that attention is needed for coping with the information overload of the sensory system by *selecting* relevant information and rejecting irrelevant information for further processing, e.g .,[4, 10], for reviews see [14]. Selection for perception focuses on the input-level of information processing and assumes attention-mediated selection as precondition for high-level-processing. This perspective is supported by a large body of research: Empirical findings, for example, indicate that visual attention speeds up detection [15] and finding of targets in an area of distractors, for an overview see [33]. Furthermore, attention plays an important role for the integration of features, represented in different visual modules, into object files [30]. Ultimately, attention may lead to conscious perception. For example, the *asynchronous-updating model* [13] assumes that attention is a necessary, but not a sufficient condition for conscious perception.

By contrast, *selection for action* [2] assumes attention as selection mechanism not on the input, but on the output level of information processing: Due to the constraints of effector systems - humans, for example, have only two arms and hands -, humans can direct an action only at one, at most at a few, objects. Although humans can, for example, see many apples on the tree, at the same moment, they can pick up at most two apples at a given time. Therefore, relevant spatial parameters of the action target must be provided to the motor system by excluding effects of action-irrelevant distractors. This is selection for action, and attention is necessary to execute this selection.

Selection for perception and selection for action were seen as alternative frameworks: A dissociation between perception and action is, for example, demonstrated by brain damage participants, which have selectively disrupted perception to visual stimuli, but not disrupted action (or vice versa). Additionally some visual illusions are larger if measured by perceptual judgments than by actions, e.g., [1].

But recently evidence for an interaction of selection for perception and for action arose, demonstrating that actions have immense influence on information processing, often on early processing stages. On the one hand, action-relations between objects (e.g., a bottle and a glass correctly positioned to action or not) affect selection for perception: Brain damage patients with *visual extinction* (difficulty of identifying objects on contralesional side, if they appear simultaneously with another object on the ipsilesional side) showed reduced extinction, if two objects were correctly positioned for action [18]. On the other hand, perception is affected by actions: Deubel, Schneider and colleagues showed, e.g., [25] that identification of a stimulus is improved, if observers point at the location where the stimulus appears in comparison to pointing at another location nearby. Due to the described interaction between selection for perception and selection for action, it seems interesting to assess both mechanisms in the same experimental paradigm.

In this study, we assess the contribution of both mechanisms to *prior entry*. We controlled visuo-spatial attention by peripheral *cues*. Peripheral cues are visual abrupt-onset stimuli appearing at a non-foveated location. Several studies showed that cues orient attention towards a specific location [15, 35]. According to current accounts, this orientation is not "automatic", "bottom-up", but rather contingent on intentions of the observer [3, 7].

As in earlier studies, e.g., [19, 20, 21], we used a cue which was not consciously perceivable because it was masked by the target trailing at its location. Under certain spatial and temporal conditions, a later visual stimulus can render an earlier stimulus completely invisible. For details concerning this so called backward masking see [24]. In the present context, directing attention by non-conscious cues has a practical, rather than theoretical, reasons. Mainly, it is more cautious: If they do not perceive the cue, observers cannot confuse cue and target stimuli or mistake the cue for a target. Still, the control of attention by information which is itself not consciously available is an interesting topic which is currently much debated in psychology. To state the most interesting finding, this control of attention is not automatic (like parallel information processing on saliency maps). Several studies show that it depends on the current intentions of the observer [3, 7, 22].

2 Experiment

The present experiment explores the accelerating effect of attention by means of visual prior entry. We attempt to separate this acceleration into a part that is due to selection for perception and a part that is due to selection for action. Observers judged the temporal order of two visual stimuli: an attended target, preceded by a cue, and an unattended target. The two targets were presented with variable temporal intervals. The cue had either the shape of the attended target (*shape-congruent cueing*) or the shape of the unattended target, which appeared at a different location (*shape-incongruent cueing*). This manipulation allows separating the accelerating effect on information processing of selection for action from the effect which is due to selection for perception: Observers indicate which of the two targets appeared first (square and diamond) by pressing different buttons. Although the cue cannot be consciously perceived, it may specify a corresponding motor response, e. g. pressing the button for "square first", if it was a square. This direct specification of response parameters reflects selection for action.

With a shape-congruent cue, prior entry can either be caused by speeded perception or by the motor response specified by the cue (selection for action). With a shape-incongruent cue, by contrast, the motor response (selection for action) would indicate that the uncued (unattended) target, which has the same shape as the cue in the incongruent case, was perceived first. Any prior entry found in this condition must thus be a true effect of selection for perception. Attention as selection for action can be estimated as the difference between prior entry by shape-congruent cueing and prior entry by shape-incongruent cueing.

2.1 Participants, Apparatus, Stimuli, Procedure

Sixteen voluntary naïve participants, with normal or corrected-to-normal visual acuity (10 female, mean age = 22.38) gave their informed consent and received € 6 or course credits.

Participants sat in a dimly lit room. Viewing distance was fixed at 57 cm by a chin rest. The centre of the monitor was at eye level. Stimuli were presented on a 17 in. colour monitor with a refresh rate of 60 Hz.

Targets were a square and diamond (edge length 2.1° of visual angle). Cues were smaller replicas of the targets which fitted into their inner contours (edge length 1.6°). In each trial, two targets, appeared at two of four possible locations, either below or above a fixation cross, which was in the middle of the screen. The horizontal distance was 6.3° of visual angle from fixation. In half of the trials, a cue preceded one of the targets (attended or cued target). Target intervals (onset asynchronies between the two targets) were 68, 51, 34, 17, and 0 ms. The cue appeared, if presented, 68 ms before onset of the cued target. There were two experimental blocks. In one block, cueing

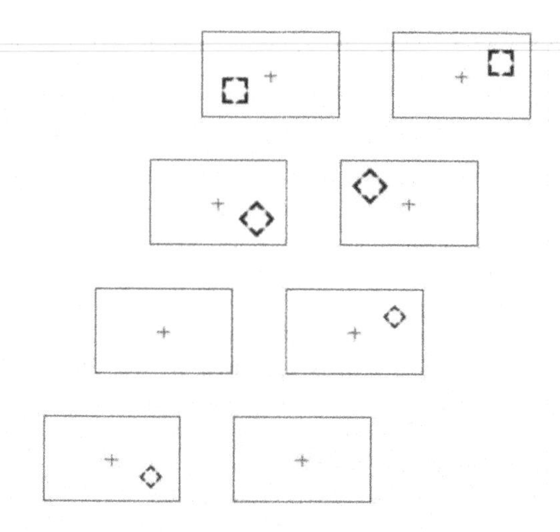

Fig. 1. Shows an example of a congruent (left) and an incongruent cued trial (right). The trials started with the frames on the bottom of the figure.

was shape-congruent (cue and cued target had the same shape), in the other block, cueing was shape-incongruent (cue and uncued target had the same shape). Order of the blocks was balanced over participants. See Figure 1 for an example of a congruent and an incongruent cued trial. Targets and cues were deleted after 34 ms.

The participants fixated a central cross throughout each trial. They judged the temporal order of the targets by indicating which appeared first[1]. The instruction emphasized accuracy; there was no time pressure. Every 40 trials, a break was made automatically.

2.2 Results and Discussion

Acceleration values (prior entry values) were calculated separately for shape-congruent and shape-incongruent cues (for details see [20, 21, 32]). As expected, both cueing conditions revealed a speed-up by attention. This acceleration is, as to be expected from earlier studies, primarily due to selection for perception, but in a smaller amount due to selection for action.

Acceleration was 45 ms in the shape-congruent condition and 38 ms in the shape-incongruent condition. Note that these values are relative to an upper value: Since the cue preceded the target by only 68 ms, the maximum speed-up by attention is this same value of 68 ms. Of course, with larger intervals between cue and cued target, attention has more time to operate and the respective gains might be much larger [23].

The difference of 7 ms between congruent and incongruent cueing can be seen as quantitative estimate of selection for action. Prior entry for incongruent cueing can be seen as a rather pure (but also conservative) estimate of the acceleration of perception by the cue (selection for perception). See Figure 2 for an illustration of prior entry effects. Thus, we find a very strong effect of attention, which is most easily interpreted as selection for perception, but a much smaller effect of selection for action.

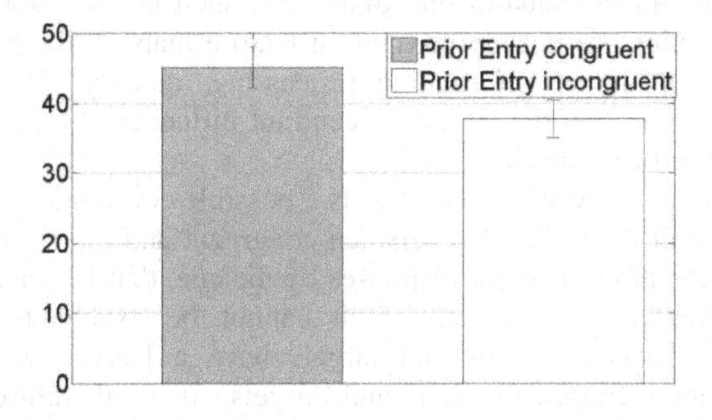

Fig. 2. Shows Prior Entry effects for both cueing conditions (congruent vs. incongruent) in ms. Errorbars indicate the standard error of the mean.

---

[1] Participants could also judge that both stimuli appeared simultaneously or that they were uncertain about temporal order, but these categories are not of interest for the present paper and therefore not analyzed here.

In statistical terms, there was a temporal advantage for cued stimuli (shape-congruent-cueing: $t(15) = 15.5$, $p < .001$, $d = 4.14$; shape-incongruent cueing = $t(15) = 13.64$, $p < .001$, $d = 3.65$). Furthermore, as assumed prior entry was larger for *shape-congruent cueing* (45 ms) than for *shape-incongruent-cueing* (38ms; $t(15) = 2.76$, $p < .01$, $d = 0.73$, one-tailed).

3 General Discussion

As demonstrated in the large body of literature on prior entry [26, 28, 29, 34], we found that perception of an attended (visual) stimulus is accelerated in comparison to an unattended stimulus. The main purpose of the present paper was to provide a quantification of selection for action and selection for perception in prior entry: To this aim, we tested whether the cue draws attention to its location in a perceptual manner or whether this effect is related to the motor relevance of the cue. We found a substantial accelerating of perception by 38 ms (selection for perception), and a smaller effect for specification of motor responses by the cue of 7 ms (selection for action).

Prior entry, the accelerating effect of attention on perception, is well documented in experimental psychology. Such a speed-up possibly is an important part of prioritized processing of attended information: Information is transferred faster to higher levels, such as, internal models or consciousness [12]. To our knowledge, it is only rarely entailed in computational models of visual attention. One computational model, which incorporates prior entry, is the *systems-model of visual attention* by Fred Hamker, which can explain prior entry by masked peripheral cues with the help of early top-down influences [8]. Bröckelmann et al. showed, for example, that target-like cues attracted attention more effectively than distractor-like, and thus irrelevant, cues.

It is worth noticing that we found that attention was differentially controlled by stimuli of different shape. Squares and diamonds, such as we used them here, are *complex* features which are not represented on feature maps. Still, they were able to influence attention-mediated information processing, quickly. This finding is in accordance with current theories on early reentrant influences by target templates or intentions in visual processing [8].

Finally, we want to draw attention towards a possible confound in our experiment. Although we interpret the difference between congruent and incongruent cueing as a difference in specification of motor responses by the cue, that is, selection for action, there is an alternative explanation, which cannot be excluded by the present experiment. Shape-congruent cues and targets have a larger overlap of sensory features than shape-incongruent cues and targets. It is therefore possible that difference between prior entry resulting from congruent and incongruent cueing, represents, at least partially, *sensory priming* (sensory detectors which are relevant for detecting the cued target were pre-activated by the cue). This question should be investigated in further experiments by varying feature overlap between cues and targets and task implications of the cues more independently.

To summarize, attention, which is controlled by non-conscious information accelerates information processing. This acceleration is due to selection for perception *and* selection for action. This finding emphasizes the relevance of action for information processing: to modulate attention-mediated specification of motor-responses (selection for action) seems, especially, interesting for research on the control of autonomous agents.

Acknowledgments

Supported by DFG grant Ne 366/ 7-2.

References

1. Aglioti, S., DeSouza, J.F.X., Goodale, M.A.: Size contrast illusions deceive the eye but not the hand. Curr. Biol. 5, 679–685 (1995)
2. Allport, A.: Attention and Selection-For-Action. In: Perspectives on Perception and Action, pp. 395–419. Lawrence Erlbaum, Hillsdale (1987)
3. Ansorge, U., Neumann, O.: Intentions Determine the Effect of Invisible Metacontrast-Masked Primes: Evidence for Top-Down Contingencies in a Peripheral Cuing Task. J. Exp. Psychol. Human. 31, 762–777 (2005)
4. Broadbent, D.E.: Perception and Communication. Pergamon Press, London (1958)
5. Bröckelmann, A.K., Junghöfer, M., Scharlau, I., Hamker, F.H.: Reentrant processing from attentional task sets: Converging support from magnetoencephalography and computational modeling (in preparation)
6. Deutsch, J.A., Deutsch, D.: Attention: Some theoretical considerations. Psychol. Rev. 70, 80–90 (1963)
7. Folk, C.L., Remington, R.W., Johnston, J.C.: Involuntary Covert Orienting is Contingent on Attentional Control Settings. J. Exp. Psychol. Human. 18, 1030–1044 (1992)
8. Hamker, F.H.: A Dynamic Model of How Feature Cues Guide Spatial Attention. Vision Res. 44, 501–521 (2004)
9. Itti, L., Koch, C.: Computational Modeling of Visual Attention. Nature Reviews Neuroscience 2, 194–203 (2001)
10. Kahneman, D.: Treisman: The Cost of Visual Filtering. J. Exp. Psychol. Human. 9, 510–522 (1983)
11. Mack, A., Rock, I.: Inattentional Blindness. Oxford University Press, Oxford (1998)
12. Neumann, O., Scharlau, I.: Visual Attention and the Mechanism of Metacontrast. Psychol. Res. – Psych. Fo. 71, 626–633 (2007)
13. Neumann, O., Scharlau, I.: Experiments on the Fehrer-Raab-Effect and on the "Weather-Station-Model" of visual backward masking. Psych. Fo. 71, 667–677 (2007)
14. Pashler, H.E.: The Psychology of Attention. MIT Press, Cambridge (1998)
15. Posner, M.I.: Orienting of Attention. Q. J. Exp. Psychol. 32, 3–25 (1980)
16. Rensink, R.A.: Change Detection. Annu. Rev. Psychol. 53, 245–277 (2002)
17. Rensink, R.A., O'Regan, J.K., Clark, J.J.: To See or not to See: The Need for Attention to Perceive Changes in Scenes. Psychol. Sci. 8, 368–373 (1997)
18. Riddoch, M.J., Humphreys, G.W., Edwards, S., Baker, T., Willson, K.: Seeing the Action: Neuropsychological Evidence for action-based effects on object selection. Nat. Neurosci. 6, 82–89 (2003)

19. Scharlau, I.: Leading, but not Trailing, Primes Influence Temporal Order Perception: Further Evidence for an Attentional Account of Perceptual Latency Priming. Percept. Psychophys. 64, 1346–1360 (2002)
20. Scharlau, I.: Evidence Against Response Bias in Temporal Order Tasks with Attention Manipulation by Masked Primes. Psychol. Res. – Psych. Fo. 68, 224–236 (2004)
21. Scharlau, I.: Temporal Processes in Prime-Mask Interaction: Assessing Perceptual Consequences of Masked Information. Adv. Cognitive Psychol. 3, 241–255 (2007)
22. Scharlau, I., Ansorge, U.: Direct Parameter Specification of an Attention Shift: Evidence from Perceptual Latency Priming. Vision Res. 43, 1351–1363 (2003)
23. Scharlau, I., Ansorge, U., Horstmann, G.: Latency Facilitation in Temporal Order Judgments: Time Course of Facilitation as Function of Judgment Type. Acta Psychol. 122, 129–159 (2006)
24. Scharlau, I., Neumann, O.: Perceptual Latency Priming by Masked and Unmasked Stimuli: Evidence for an Attentional Explanation. Psychol. Res. – Psych. Fo. 67, 184–197 (2003)
25. Schneider, W.X., Deubel, H.: Selection-for-Perception and Selection-for-spatial-motor-action are coupled by visual attention: a review of recent findings and new evidence from stimulus-driven-saccade control. In: Prinz, W., Hommel, B. (eds.) Attentenion and Performance XIX. Common mechanisms in perception and action, pp. 609–627. Oxford University Press, Oxford (2002)
26. Shore, D.I., Spence, C., Klein, R.M.: Visual Prior Entry. Psychol. Sci. 12, 205–212 (2001)
27. Simons, D.J., Chabris, C.F.: Gorillas in our Midst: Sustained Inattentional Blindness for Dynamic Events. Perception 28, 1059–1074 (1999)
28. Stelmach, L.B., Herdman, C.M.: Directed Attention and Perception of Temporal Order. J. Exp. Psychol. Human. 17, 539–550 (1991)
29. Titchener, E.M.: Lectures on the Elementary Psychology of Feeling and Attention. MacMillan, New York (1908)
30. Treisman, A., Gelade, G.: A Feature Integration Theory of Attention. Cognitive Psychol. 12, 97–136 (1980)
31. Treisman, A., Schmidt, H.: Illusory Conjunctions in the Perception of Objects. Cognitive Psychol. 14, 107–141 (1982)
32. Weiß, K., Scharlau, I.: Simultaneity and Temporal Order Perception: Different Sides of the Same Coin? Evidence from a Visual Prior Entry Study (submitted)
33. Wolfe, J.M.: Guided Search 2.0. A Revised Model of Visual Search. Psychon. B. Rev. 1, 202–238 (1994)
34. Wundt, W.: Grundzüge der physiologischen Psychologie [Main Features of Physiological Psychology]. Engelmann, Leipzig (1887)
35. Yantis, S., Jonides, J.: Abrupt Visual Onsets and Selective Attention: Voluntary versus Automatic Allocation. J. Exp. Psychol. Human. 16, 121–134 (1990)

Real-Time Scan-Line Segment Based Stereo Vision for the Estimation of Biologically Motivated Classifier Cells

M. Salah E.-N. Shafik and Bärbel Mertsching

GET Lab, University of Paderborn
Pohlweg 47-49, 33098 Paderborn, Germany
{shafik,mertsching}@get.uni-paderborn.de
http://getwww.uni-paderborn.de

Abstract. In this paper we present a real-time scan-line segment based stereo vision for the estimation of biologically motivated classifier cells in an active vision system. The system is challenged to overcome several problems in autonomous mobile robotic vision such as the detection of incoming moving objects and estimating its 3D motion parameters in a dynamic environment. The proposed algorithm employs a modified optimization module within the scan-line framework to achieve valuable reduction in computation time needed for generating real-time depth map. Moreover, the experimental results showed high robustness against noises and unbalanced light condition in input data.

1 Introduction

During the last decade, the importance of depth estimation from stereo images has been extensively increased active vision applications e.g. autonomous mobile robots. The research presented in this paper is a building block of a comprehensive system of multi-object motion analysis in robotic vision which is intended to be included in active vision applications [1,2]. In such systems the challenge of avoiding multiple moving obstacles overcoming the ego-motion problem of the mobile robot is raised. In this context, priority is given to the fast detection and 3D motion estimation of incoming and outgoing objects towards the mobile robot. Hence, a biologically motivated classifier approach presented in [3] has been improved to integrate depth information generated from a developed stereo matching algorithm. The need for developing a real-time stereo approach is not only constrained by the available computation time in the system, but also by the available hardware and its power consumption, especially if there is no specialized hardware or graphical processing unit (GPU) at hand. Hence, every millisecond counts.

Stereo algorithms can be classified based on the taxonomy proposed by Scharstein and Szelinski [4] into two major categories: local methods and global methods. While most of the global methods attempt to minimize an energy function computed for the entire image area, local methods determine the stereo correspondence between two frames by minimizing a matching cost using an aggregation window. Local algorithms in general are suitable for real-time applications. The main problem is to determine the suitable size and shape of the aggregation window in order to avoid crossing depth discontinuities while covering sufficient intensity variation. Such problems results in noisy

B. Mertsching, M. Hund, and Z. Aziz (Eds.): KI 2009, LNAI 5803, pp. 89–96, 2009.

disparities in texture-less regions and blurred object boundaries. On the other hand, global methods such as Dynamic Programming (DP) [5], Belief Propagation (BP) [6] and Graph Cut (GC) [7] make explicit smoothness assumptions on the disparity map. A major problem in DP approaches is the requirement of enforcing the monotonicity, i. e., the relative ordering of pixels on a scan-line remains the same between two frames, which cause errors in case of narrow foreground objects. Although BP and GC algorithms gives impressive results by enforcing the optimization in two-dimensions, they are computationally expensive and are not suitable for real-time applications as most of the global algorithms. Between local and global approaches, there are some algorithms which are based on the minimization of an energy function computed over a subset of the whole image area or along the epipolar lines. The implemented minimization strategy is usually based on Dynamic Programming or Scan-line Optimization (SO) [8] techniques and recently on line segmentation [9].

In this paper, we developed fast SO-based optimization technique for a line segment stereo approach. The generated depth map is used to update and optimize an active weight function $\omega(p)$ in order to estimate biologically motivated classifier cells $(c - cells)$.

The remainder of the paper is organized as follows: section 2 gives an account of the related work, while section 3 describes the proposed algorithm and the structure of the $c - cells$. Section 4 discusses the results of experiments and evaluates the proposed method, while finally, section 5 concludes the paper.

2 Related Work

Recent stereo algorithms have significantly advanced the state-of-the-art in terms of quality. However, in terms of speed, the best stereo algorithms typically take from several seconds to several minutes to generate a single disparity map ([10] gives a performance evaluation of cost aggregation strategies proposed for stereo matching). There are many interesting applications in which depth maps at video rate are needed such as (autonomous mobile robots, augmented-reality and automatic vehicle guidance). Hence, the importance of real-time stereo algorithms increases as in [11] where an adapted recursive formulation is proposed to reduce the computing cost of SAD cost function of a local approach which in turn inherits the ambiguity problem from the local algorithms. As a solution to overcome this problem, they implement a post processing filter application at the last phase of the algorithm which is considered as an overhead to the computation time. On the other hand, [12] overcomes the ambiguity problem in low textured areas by replacing estimates in texture-less regions with fitting planes. The algorithm starts with window-based multi-view stereo matching followed by the application of consistency fusion module. Afterword, a plane-fitting phase is applied by using color segmentation, where a plane is adjusted for each segment. In order to correct the potential errors and enhance the overall computation time, a GPU hierarchical recursive belief propagation approach is implemented. In spite of using a GPU, the BP-based plane-fitting stereo pipeline runs at about 1 frame per second. The use of a GPU has been introduced before in global approaches with hierarchical BP [13] and DP based on adaptive cost aggregation [14]. As the use of a GPU due to hardware constraints is not applicable on some platforms, solving the low texture problem using an effective variable support based on image segmentation within the SO framework has been addressed in [15]. While the result is promising, the performance is far from being real-time (i. e. some minutes). The computational time has been improved by using line segment techniques and tree dynamic programming as in [9]. The segmentation module there contains three steps: computing the initialization marks, repositioning marks, and removing isolated marks. In order to extract linear planes, a parameter estimation approach is used for fitting planes on sparse correspondence. Afterward, dynamic programming is used on the constructed tree to minimize the energy function. The algorithm has performed well with an Intel Pentium IV 2.4 GHz processor (processing time for "tsukkuba" [4] is about 160ms). However, the requirement of enforcing the monotonicity inherited from the DP techniques still cause the thin foreground objects problems. On the other hand, a biologically motivated classifier can be derived from neurophysiological studies [16], which found that there are two classes of neurons in the middle temporal area (MT) of the mammalian visual cortex which are considered to encode motion information from the retinal signal: direction selective D-cells and velocity selective S1-cells. Both cells covering small receptive fields of three types of neurons which have been identified in the medial superior temporal area (MTS): R-cells detecting rotation, S-cells for expansion, and D-cells for translation. Furthermore, cells which combine all three motion types give evidence for the composition of elementary motions into more complex motion patterns similar to motion templates [17]. In this module, motion-sensitive cells for the preferential direction will be constructed in order to measure the response of the sensitive cells to a corresponding motion which helps predicting hazardous scenarios such as collisions. The generation of the corresponding map requires calculating a radial symmetric weight function for each cell which is in term of computation time is very expensive. Developing a fast connection weight function based on the depth information reduce the time consumption dramatically without the use of a GPU power as in [18] where a biologically motivated classifier and feature descriptors are designed for execution on single instruction multi data hardware using the programmable GPU.

3 Proposed Algorithm

The goal of stereo algorithms is to establish pixel correspondences between the left image I_l and the right image I_r. In order to achieve reasonable results, two geometric constraints are used: first on the imaging systems, i. e. the input stereo images are *rectified* where the epipolar lines are aligned with corresponding scan-lines. And second on the scene, i. e. the *smoothness assumption* where the disparity map is smooth almost everywhere except at the border of objects assuming that scene is composed of smooth structures which in the case of autonomous mobile robots applications is not granted.

The first step of the proposed technique is the line segmentation of the reference image I_r, in which the epipolar line epl_y is segmented into different labels $l_i(epl_y) \in \Gamma$ in the label space Γ based on the color differences between a seed pixel $P_s^{l_i}$ and the neighborhood pixels of the same epipolar line:

$$P_r(x+k, y) \in l_i(epl_y) \qquad \forall \, |P_s^{l_i} - P_r(x+k, y)| < \tau_c \tag{1}$$

where τ_c is the Euclidean color distance threshold. The correspondence problem is formulated as an energy minimization function between segments of the input images.

$$E(d_\Gamma) = argmin(E^{l_i}_{data}(d_\Gamma) + E^{l_{(i,d)\in K}}_{smooth}(d_\Gamma)) \tag{2}$$

where $E(d_\Gamma)$ is the disparity map of line segments and K is an adjacent segment in a particular epipolar line epl_y. The data term $E^{l_i}_{data}(d_\Gamma)$ of the energy function is the matching cost between a segment $l_i(epl_y)$ in the reference image and the opponent segments $l_{i,d}(epl_y)$ in the target image. The smoothness term $E^{l_{(i,d)\in K}}_{smooth}(d_\Gamma)$ encodes the smoothness assumption.

3.1 Matching Cost and Optimization

In order to reduce the complexity of calculations, a matching cost C_M based on the absolute color difference between the points of the current segment in the reference image and all disparity hypotheses is used to evaluate the data term

$$C_M(P_r, P_{l,d}) = \sum_{c \in \Re} |P^c_r(x, y) - P^c_{l,d}(x + d, y)| \tag{3}$$

where $\Re$ is the RGB color space, $P_r(x, y) \in I_r$, $P_{l,d}(x, y) \in I_l$, and d is the hypothesized disparity value.

The data term is computed from the sum of the matching cost along the segment points

$$E^{l_i}_{data}(d_\Gamma) = \sum_{P=P_s}^{P^N_r} C_M(P_r, P_{l,d}) \qquad \forall\ P_r \in l_i(epl_y) \tag{4}$$

where P_s is the starting seed pixel of a line segment and P^N_r is the last point in the same segment.

In order to enhance the optimization performance, we propose an effective and simplified smoothness term within the SO-framework.

$$E^{l_{(i,d)\in K}}_{smooth}(d_\Gamma) = \lambda(\ell_{l_i}) \;.\; |d_{l_i} - d_{l_K}| \tag{5}$$

where $\lambda(\ell_{l_i})$ is an ascending function to the length of the current segment ℓ_{l_i} used to penalize depth discontinuities. The concept behind the function is to balance the relation between the disparity of a segment and the sum of the matching cost of the segment points. While the matching cost is affected by the length of the segment, only one disparity value is assigned to all the segment points and the best value is chosen within a winner take all (WTA) scheme. Considering the inter-scan-line smoothness resulting from line segmentation leads to overcome the ambiguity problem without the use of a recursive smoothing function as in BP approaches or facing narrow front objects problem as in DP algorithms.

3.2 Biologically Motivated Classifier

In [3], a model neuron for the biologically motivated classifier cell $c_i(p_0)$ represents the instantaneous velocity in the preferential direction of the cell

$$c_i(p_0) = \frac{1}{|V|} \sum_{p \in V} \frac{\omega(p_0 - p)\psi_i^L(p)}{|\psi_i^L|^2} v(p) \quad (6)$$

where $v(x, y)$ is a motion vector (MV) at an image point (x, y) generated by a motion template ψ_i^L which represents the six infinitesimal generators of a 2D vector field (translation in x,y,z and rotation around x,y,z axis). V denotes the set of MVs detected in the image. Instead of using a radial symmetric weight function $\omega(p) = \frac{1}{2\pi\sigma^2} exp(-\frac{1}{2}\frac{|p|^2}{\sigma^2})$ to create a local neighborhood around p_0, we use a connection weight function $\varpi(p)$ based on the depth information generated from the previous stereo module

$$\varpi(p_0 - p) = \begin{cases} 1 & \forall\ p \in \Im(p_0) \\ 0 & otherwise \end{cases} \quad (7)$$

where $\Im(p_0)$ is the segment label off point p_0. The new connection weight function enhances the overall computation time as well as it overcomes the blurring effect of the $\omega(p)$ function especially at the edges of an object. Moreover, considering only the points that belongs to the same depth level improves the estimation process overcoming the ambiguous interpretation problem.

Generating a response map requires a validation function to measure how well is the correspondence of a *c-cell* to a preferential motion.

$$\xi_i(p_0) = \frac{1}{|V|} \sum_{p \in V} |\varpi(p_0 - p)(v(p) - c_i(p_0)\psi_i^L(p))| \quad (8)$$

where $c_i(p_0)$ is the estimated *c-cell*, and $\varpi(p_0 - p)$ the updated connection weight function between a point p_0 and its neighbor point p.

4 Results and Discussion

In this section, the results of applying the proposed algorithm on different images will be presented and discussed. In order to analyze the output of our approach in a scalable complex scene, a virtual environment for simulating a mobile robot platform (SIMORE) is used [19,20]. The simulator provides a possibility to test the outcome of the proposed algorithm in different ideal scenarios, where complex 3D motions of multiple objects are not easily available in a real-time. In this environment, the simulated robot is moving forward and steering towards the left in front of a stable cube, a moving cone, and a size-changing ball. The experiments have been performed on a P4 personal computer (PC) with 3.00 GHz speed and 2 GB RAM.

In an effort to reduce the overall computation time, the depth from the stereo approach is applied without a refinement step depending on the enhancement done by the modified smoothing function. Fig. 1 represents a qualitative comparison of the proposed algorithm to the ground truth of the Middlebury data-set [4,21]. The second row depicts

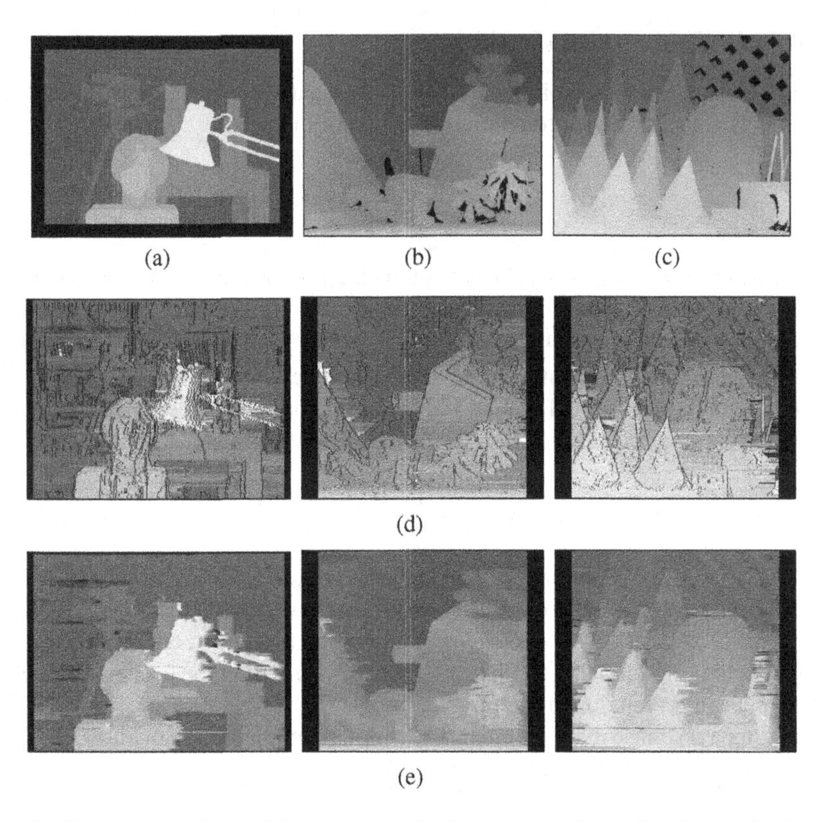

Fig. 1. Qualitative comparison of the generated depth map: (a-c) Ground truth data for the three images from the Middle-bury data-set (Tsukkuba, Teddy, and Cones). Result of the proposed line segment based stereo algorithm (d) without and (e) with the use of the modified smoothing function.

the generated depth map without the use of the smoothing function, while the third row represents the result of the stereo approach using the smoothing function. The result shows that when the smoothing function is applied, it provides a better quality. However, the use of the smoothing function increases the processing time by about 5∼9 ms which is not very critical for real-time application. On the other hand, the result without the smoothing function is affected by noise but this is not very critical to the depth perception.

The second result as shown in fig. 2 represents the response of the motion sensitive cells using image sequence from SIMORE. The generated depth maps are implemented in the estimation of the *e cells* process, where the first and the second row represent the response of the motion sensitive cells in the preferential direction of the translation in x, y, and z and the rotation around x, y, and z axis, respectively. The response maps are valuable in tackling typical problems in autonomous mobile robotic vision such as the detection and estimation of a specific motion pattern of multiple moving objects in a dynamic environment, and the ego-motion problem resulted from the motion of the mobile robot.

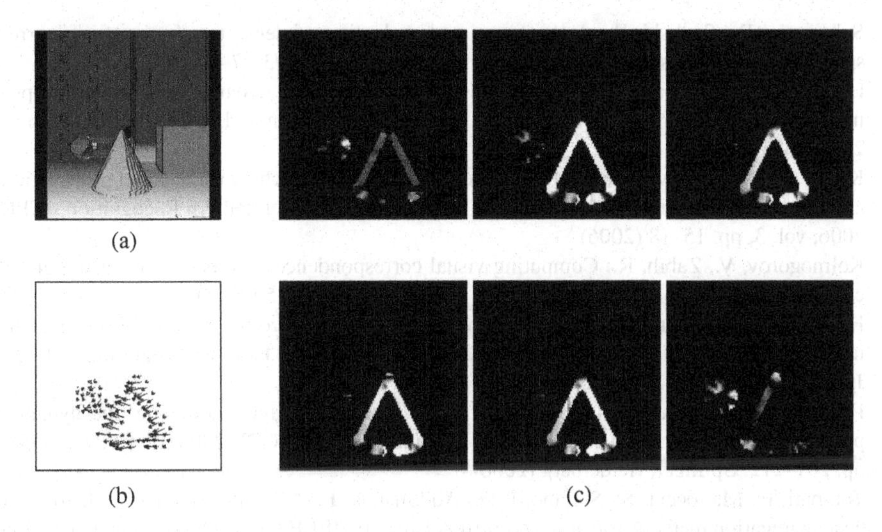

Fig. 2. Response maps of motion sensitive neuron using image sequence from SIMORE. (a) Generated depth map. (b) Generated MVF. (c) Response map of (*e cell*)in the preferential directions, up: in the translation in the x, y, and z axis, down: around the x, y, and z axis.

The overall computation time used for generating the proposed line segment based depth approach from the Tsukkuba image is about 60∼70 ms without any use of the GPU due to the energy constraint of the real platform (tele-operated mobile robot). The computation time enhancement provided by the proposed algorithm is vital to speed-up the overall estimation process of the biologically motivated classifier cells, which is intended to be implemented in the tele-sensory mobile robot.

5 Conclusion

In this paper, we have presented a real-time scan-line segment based stereo vision employing a modified optimization module within the scan-line framework. The achievement of reducing the computation time provides the ability to implement the generated depth map into the estimation process of biologically motivated classifier. Moreover, updating the connection weight function of the motion sensitive cells using the depth information helps in the fast detection of incoming objects in a dynamic environment.

References

1. Shafik, M., Mertsching, B.: Enhanced motion parameters estimation for an active vision system. Pattern Recognition and Image Analysis 18(3), 370–375 (2008)
2. Aziz, Z., Mertsching, B.: Fast and robust generation of feature maps for region-based visual attention. IEEE Trans. on Image Processing (5), 633–644 (2008)
3. Massad, A., Jesikiewicz, M., Mertsching, B.: Space-variant motion analysis for an active-vision system. In: Advanced Concepts for Intelligent Vision Systems, Ghent, Belgium (2002)

4. Scharstein, D., Szeliski, R.: A taxonomy and evaluation of dense two-frame stereo correspondence algorithms. Int. Journal of Computer Vision 47(1/2/3), 7–42 (2002)
5. Lei, C., Selzer, J., Yang, Y.: Region-tree based stereo using dynamic programming optimization. IEEE Computer Society Conf. on Computer Vision and Pattern Recognition 2, 2378–2385 (2006)
6. Klaus, A., Sormann, M., Karner, K.: Segment-based stereo matching using belief propagation and a self-adapting dissimilarity measure. In: 18th Int. Conf. on Pattern Recognition, ICPR 2006, vol. 3, pp. 15–18 (2006)
7. Kolmogorov, V., Zabih, R.: Computing visual correspondence with occlusions using graph cuts. In: Proc. Int. Conf. Computer Vision (ICCV), pp. 508–515 (2001)
8. Hirschmuller, H.: Stereo vision in structured environments by consistent semi-global matching. In: IEEE Computer Society Conf. on Computer Vision and Pattern Recognition, vol. (2), June 2006, pp. 2386–2393 (2006)
9. Deng, Y., Lin, X.: A fast line segment based dense stereo algorithm using tree dynamic programming. In: Leonardis, A., Bischof, H., Pinz, A. (eds.) ECCV 2006. LNCS, vol. 3953, pp. 201–212. Springer, Heidelberg (2006)
10. Tombari, F., Mattoccia, S., Stefano, L.D., Addimanda, E.: Classification and evaluation of cost aggregation methods for stereo correspondence. In: IEEE Int. Conf. on Computer Vision and Pattern Recognition (CVPR), Florida, USA (December 2008)
11. Foggia, P., Limongiello, A., Vento, M.: A real-time stereo-vision system for moving object and obstacle detection in avg and amr applications. In: Proc. of the Seventh Int. Workshop on Computer Architecture for Machine Perception (CAMP), Washington, DC, USA, pp. 58–63 (2005)
12. Yang, Q., Engels, C., Akbarzadeh, A.: Near real-time stereo for weakly-textured scenes. In: British Machine Vision Conference (BMVC), Leeds, UK (2008)
13. Yang, Q., Wang, L., Yang, R., Wang, S., Liao, M., Nistér, D.: Real-time global stereo matching using hierarchical belief propagation. In: BMVC, pp. 989–998 (2006)
14. Wang, L., Liao, M., Gong, M., Yang, R., Nister, D.: High-quality real-time stereo using adaptive cost aggregation and dynamic programming. In: Int. Symposium on 3D Data Processing Visualization and Transmission, pp. 798–805 (2006)
15. Mattoccia, S., Tombari, F., Stefano, L.D.: Stereo vision enabling precise border localization within a scanline optimization framework. In: Yagi, Y., Kang, S.B., Kweon, I.S., Zha, H. (eds.) ACCV 2007, Part II. LNCS, vol. 4844, pp. 517–527. Springer, Heidelberg (2007)
16. Albright, T.D., Desimone, R., Gross, C.G.: Columnar organisation of directionally selective cells in visual area mt of the macaque. J. Neurophysiol 51, 16–31 (1984)
17. Duffy, C.J., Wurtz, R.H.: Sensitivity of mst neurons to optic flow stimuli. i. a continuum of response selektivity to large-field stimuli. J. Neurophysiol. 65, 1329–1345 (1991)
18. Woodbeck, K., Roth, G., Chen, H.: Visual cortex on the GPU: Biologically inspired classifier and feature descriptor for rapid recognition. In: CVPRW, June 2008, pp. 1–8 (2008)
19. Mertsching, B., Aziz, Z., Stemmer, R.: Design of a simulation framework for evaluation of robot vision and manipulation algorithms. In: Proceedings of Asia Simulation Conference (ICSC), pp. 494–498 (2005)
20. Kutter, O., Hilker, C., Simon, A., Mertsching, B.: Modeling and simulating mobile robots environments. In: 3rd Int. Conf. on Computer Graphics Theory and Applications, Funchal, Madeira, Portugal (January 2008)
21. Scharstein, D., Szeliski, R.: High-accuracy stereo depth maps using structured light. In: IEEE Computer Society Conf. on Computer Vision and Pattern Recognition, June 2003, vol. 1, pp. 195–202 (2003)

Occlusion as a Monocular Depth Cue Derived from Illusory Contour Perception

Marcus Hund and Bärbel Mertsching*

University of Paderborn, Dept. of Electrical Engineering, GET-Lab
Pohlweg 47-49, D-33098 Paderborn, Germany
hund@get.upb.de, mertsching@get.upb.de

Abstract. When a three dimensional scene is projected to the two dimensional receptive field of a camera or a biological vision system, all depth information is lost. Even without a knowledgebase, i. e. without knowing what object can be seen, it is possible to reconstruct the depth information. Beside stereoscopic depth cues, also a number of moncular depth cues can be used. One of the most important monocular depth cues ist the occlusion of object boundaries. Therefore one of the elaborated tasks for the low level image processing stage of a vision system is the completion of cluttered or occluded object boundaries and the depth assignment of overlapped boundaries. We describe a method for depth ordering and figure-ground segregation from monocular depth cues, namely the arrangement of so-called illusory contours at junctions in the edge map of an image. Therefore, a computational approach to the perception of illusory contours, based on the tensor voting technique, is introduced and compared with an alternative contour completion realized by spline interpolation. While most approaches assume, that the position of junctions and the orientations of associated contours are already known, we also consider the preprocessing steps that are necessary for a robust perception task. This implies the anisotropic diffusion of the input image in order to simplify the image contents while preserving the edge information.

1 Introduction

For technical as well as for biological vision systems, the completion of object boundaries that are interrupted due to occlusions or low luminance contrast is essential. This implies the distinction of completions of foreground object boundaries, called modal completions, and those of occluded background object boundaries, called amodal completions. The entirety of these completions are referred to as virtual contours. Neurophysical studies have shown that the perception of modal completions, called illusory contours, can be found in mammals, birds and insects [10]. The importance of virtual contour perception for object boundary completion becomes obvious regarding the fact that the brains of these animals

* We gratefully acknowledge partial funding of this work by the Deutsche Forschungsgemeinschaft under grant Me1289/7-1.

B. Mertsching, M. Hund, and Z. Aziz (Eds.): KI 2009, LNAI 5803, pp. 97–105, 2009.

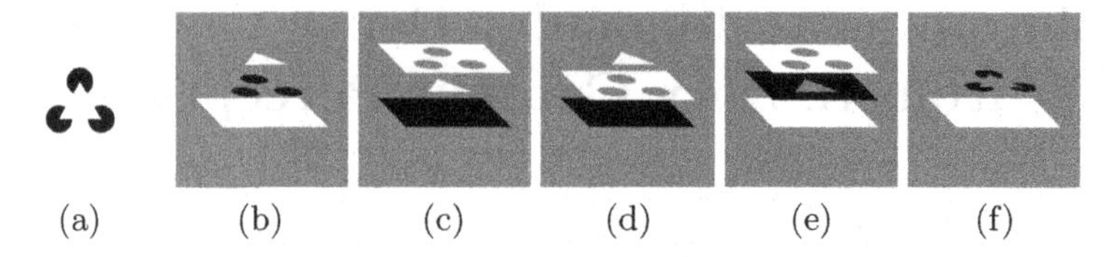

Fig. 1. Some possible depth interpretations for the Kanizsa triangle shown in (a)

have developed independently throughout evolution. Also for technical vision systems the perception of complete object boundaries is an important issue, as well for depth perception as for object recognition tasks.

Fig. 1(a) shows the Kanizsa triangle, which is the standard example for an optical illusion caused by illusory contour perception and for depth ordering derived from monocular depth cues. Even though there exists a multitude of correct scene interpretations leading to the same retinal image (Fig. 1(b)-(f)), the most often perceived depth ordering is the one shown in Fig. 1(b). Consequently, the aim of a computational approach to scene interpretation cannot be to find the correct depth ordering but to find the most probable scene interpretation. One possibility for figure-ground segregation is to consider contours that are interrupted by other contours. Such T-junctions can be interpreted as contours that are interrupted by a foreground object boundary and therefore belong to the background. But, as is stated in [2], where the figure-ground segregation was based on this principle, "such a simple mechanism can be fooled". For example, the stripes of a zebra produce T-junctions at the "object boundary", which would lead the mechanism to regard the zebra as the background. In such a case it is impossible to distinguish a hole from a textured object. In the following we will show that the ambiguities can be reduced if the possible completions of the interrupted contour are taken into account.

2 Related Work

The majority of approaches or models dealing with the problem of spatial contour integration use some kind of bipole connection scheme, as introduced by Grossberg and Mingolla [1]. This perceptual grouping kernel usually consists of two symmetric lobes encoding the connection strength and orientation. For a detailed overview of contour integration approaches, see [1] or [9]. In [9], emphasis is placed on models including illusory contour perception, namely the model of Heitger et al. [2], and Neumann and coworkers [1] as a neurophysical inspired computational model while the approach of Zweck and Williams [13] models the Brownian motion of a particle from source to sink.

For the task of determining the connection path for illusory contours we present and compare two alternative methods: spline interpolation and the tensor voting technique introduced by Medioni et al. [8]. Tensor voting was applied successfully to contour inference problems on synthetic and binary input images in [8]. In [6], this approach was extended to grayscale images as input, using gabor filtering as a preceding image processing step. In the tensor voting framework the grouping kernel, called stick voting field, is orientational, i.e. with angles from 0° to 180° and designed for contour inference. Considering illusory contour perception, the use of this stick voting field could make sense in the context of a unified treatment of all contour elements, but would lead to interference of contour elements, especially in the case of amodal completions behind a textured foreground. What is needed for illusory contours including amodal completions is a directional communication pattern (with angles from 0° to 360°), e.g. one lobe, which was already used in [7], but addressed to spontaneously splitting figures and binary input images. Relatively little computational work has been done so far on figure-ground segregation derived from illusory contour cues. As mentioned above, Heitger et al. have implemented a merely rudimentary mechanism for figure-ground segregation [2]. In [5], the phenomena of unit formation and depth ordering are treated on a theoretical basis with focus on the neurophysical view. Furthermore the "identity hypothesis" is proposed, stating that the mechanisms for modal and amodal completions are identical and that the depth ordering can be separated from the contour completion process. Regarding the monocular case in a low level processing step, we will give hints that, for this early stage, the depth ordering indeed can be realized with the contour completion being identical for modal and amodal completions.

3 Preprocessing

For the determination of the gradient magnitude and orientation we use a Gaussian based filterbank, which is also used for subsequent preprocessing steps. As a further preprocessing step we use a threshold-free edge detection and a subsequent corner detection [3]. This is achieved by applying an edge tracking over edge candidates. The edge candidates are not only local extrema of the used filterbank, but also local extrema of a corner shaped orientation selective filtermask and zero crossings of a center-surround filter with a LoG-like behaviour.

Regarding the aim of a robust illusory contour perception, also noisy input images have to be taken into account. While smoothing of the input image interferes with the edge information, anisotropic diffusion methods are capable of an edge preserving smoothing. Furthermore, anisotropic diffusion provides the possibility to scale the level of detail of the input image.

In our approach the discrete image is regarded as a vector $\xi_{\mathbf{0}} \in \mathbb{R}^{(\mathbf{m} \cdot \mathbf{d})}$ with m being the total number of pixels in the image and d the number of colors used. The components $\xi_{(x,k)_0}$ of the vector represent the grayscale values for $d = 1$ or RGB channel values for $d = 3$. Here, x is the image coordinate and k is the color plane. Two assumptions are made about the cost function [4]. First, it should cause costs if the image elements $\xi_{(x,k)}$ differ from the initial pixel values $\xi_{(x,k)_0}$. Second, neighboring elements must satisfy a continuity constraint. This leads to a cost function of the form

$$P(\xi) = \frac{1}{2} \cdot \sum_{k=1}^{d} \sum_{i=1}^{m} (\xi_{(i,k)} - \xi_{(i,k)_0})^2 + \frac{c}{4} \cdot \sum_{k=1}^{d} \sum_{i=1}^{m} \sum_{j \in U_i} w_{ij} (\xi_{(i,k)} - \xi_{(j,k)})^2 \quad (1)$$

with $\xi = (\xi_{(1,1)}, \ldots, \xi_{(m,1)}, \xi_{(1,2)}, \ldots, \xi_{(m,d)})^T$ with a smoothness or scaling factor c. Here, U_x is the neighborhood of a given pixel x with a weighting factor $w_{xi} > 0$ with $\sum_{i \in U_x} w_{xi} \leq Q := \sum_{i \in U_x} 1$. Note that $w_{xi} \neq w_{ix}$ is allowed. Due to the cost function there must exist a minimum. The minimum ξ must satisfy $\nabla P(\xi) = 0$, which leads to a system of linear equations $\nabla P(\xi) = \mathbf{A} \cdot \xi - \xi_0 = 0$ with the elements of $\nabla P(\xi)$ given by $\frac{\partial}{\partial \xi_{(x,k)}} P(\xi) = \xi_{(x,k)} - \xi_{(x,k)_0} + \frac{c}{2} \big(\sum_{j \in U_x} w_{xj} (\xi_{(x,k)} - \xi_{(j,k)}) + \sum_{\{i | x \in U_i\}} w_{ix} \left(\xi_{(x,k)} - \xi_{(i,k)} \right) \big)$. Consequently, the matrix $\mathbf{A} = (a_{i,j})_{i,j} \in \mathbb{R}^{(m \cdot d) \times (m \cdot d)}$ is given by

$a_{i,j} = \begin{cases} 1 + \frac{c}{2} (\sum_k w_{ik} + \sum_k w_{ki}) & \text{for} \quad i = j \\ \frac{c}{2} (w_{ji} + w_{ij}) & \text{else} \end{cases}$ with $w_{kl} = 0$ for $l \notin U_k$. Obviously, $\mathbf{A}$ is symmetric and positive definite. Therefore, an inverse $\mathbf{A}^{-1}$ must exist and we have a unique solution of the equation system. For the numeric solution, we use the gradient descent method. Starting with $\xi^0 \in \mathbb{R}^{(m \cdot d)}$ the vector ξ^k is iteratively updated via $\xi^{k+1} = \xi^k - \lambda_k \nabla P(\xi^k)$, $\lambda_k > 0$. Instead of using the local optimal step size, which has to be computed for each iteration and therefore causes high computational costs, we use a constant step size λ. It can be shown that $\lambda := \frac{1}{1 + c \cdot Q}$ ensures the convergence [4]. We receive a step size that is dependent on the scale, i. e. the smoothness factor c and the maximum number of pixels Q in the neighborhood U. An important advantage of our problem formulation in (1) is the fact that we are free to choose the weighting factors w_{ij} without loosing existence or uniqueness of the solution. To achieve an anisotropic diffusion behaviour, we use the following definition for the local support area. The orientation angle associated with a pixel is derived by applying the orientation selective Gaussian based filterbank to the input image. Due to the given orientation angle of a pixel the coordinates of its eight-point neighborhood are rotated into the coordinates (x, y) and the corresponding weighting factors w_{ij} are determined from the function $f(x, y) = \begin{cases} 0 & \text{for } y > 0.7 \\ 1.0 + \cos\left(\frac{\pi \cdot y}{0.7}\right) & \text{else} \end{cases}$. This ensures a strict diffusion along one direction. For a more generous behaviour, the constants 0 and 0.7 have to be changed. It is clear that $Q_i = \frac{1}{2} \sum_k (w_{ik} + w_{ki}) < Q$ and hence $f(x, y)$ is well posed. As it is mentioned above, we use a Gaussian based filterbank to derive the orientation angle we need to determine the weighting factors w_{ij}. Unfortunately gradient based filters are very sensitive to noise. In order to overcome this problem, we apply the anisotropic regularization to the edge map that is generated by the Gaussian based filterbank, i. e. we apply eq. (1) for $d = 2$ to the edge map that is defined by $(\xi_{(x,1)}, \xi_{(x,2)}) = |\nabla g_x| (cos(\alpha_x), sin(\alpha_x))$ with $|\nabla g_x|$ being the gradient magnitude and α_x the associated orientation at image position x. During the iterative process, eq. (1) is reformulated for each iteration step, since the weighting factors depend on the edge values that are modified. That is, the orientaion of an edgel determines the weighting factors, that are used for one iteration step. This step changes the edgel and therefore its orientation. This leads to a self-organization of the edge map, enhancing salient contours. Note that this proceeding no longer guarantees the uniqueness, the iteration converges to a solution that depends on the starting vector.

4 Illusory Contour Perception

In [8], Medioni, Lee and Tang describe a framework for feature inference from sparse and noisy data called tensor voting. The most important issue is the representation of edge elements as tensors. In the 2D-case, a tensor over $\mathbb{R}^2$ can be denoted by a symmetric 2×2 matrix T with two perpendicular eigenvectors $\boldsymbol{e}_1, \boldsymbol{e}_2$ and two corresponding real eigenvalues $\lambda_1 > \lambda_2$. A tensor can be visualized as an ellipse in 2-D with the major axis representing the estimated tangent direction $\boldsymbol{e}_1$ and its length λ_1 reflecting the saliency of this estimation. The length λ_2 assigned to the perpendicular eigenvector $\boldsymbol{e}_2$ encodes the orientation uncertainty. Note that for junctions or corners neither the tensor representation suffices to encode the at least two different orientations nor is the ball saliency a trustable measure for junctions, since it is highly dependent on the orientations of incoming edges. Grouping can be formulated as the combination of elements according to their so-called stick-saliency $(\lambda_1 - \lambda_2)$ or to their ball-saliency λ_2. In stick-voting, for each oriented input token the grouping kernel called stick-voting-field is aligned to the eigenvector $\boldsymbol{e}_1$. In the following the input tokens consist of detected junctions and their associated directions.

The crossing of two possible completions can lead to an undesired interference of voting fields. This is shown in Fig. 2(c). The directions of completion, that are generated by the completion of the black bars in Fig. 2(a) lead to Voting fields that are perpendicular to each other. Consequently, they produce a high orientation uncertainty. This leads to a corrupted contour completion, as shown in Fig. 2(d).

To overcome this problem, we also implemented an alternative contour completion in form of a spline interpolation. Here, the crossing of completed contours leads to no problems, as can be seen in Fig. 2(e).

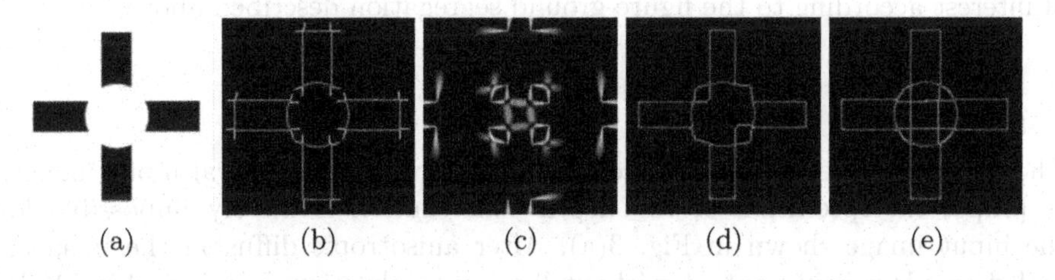

Fig. 2. Superposition of two bars and a circle: (a) input image, (b) detected edges with diirection of possible completions, (c) stick salliency after voting process, (d) contour completions after tensor voting and (e) after spline interpolation

5 Figure-Ground Segregation and Depth Ordering

The figure-ground segregation for the results shown in the next section is based on two simple rules for the crossing of two contours:

If a contour changes its state from real to illusory at a point of interest and if a crossing contour does not, then the first contour belongs to the background and the second contour belongs to an occluding foreground surface.

If a contour changes its state from real to illusory and if the illusory part lies within the bounded region of a closed crossing contour, then the first contour belongs to the background and the second contour is the closed boundary of an occluding object.

If both contours are closed and both contours have illusory parts that lie within the bounded region of the other contour, no reasonable decision can be made. This would be the case of spontaneously splitting figures.

More complex graph based labeling schemes can be found in [12] and [11], but these two rules already suffice to interpret a variety of situations in an accurate way, e.g. the second rule is sufficient to describe the complex depth ordering in the Kanizsa triangle. While the second rule describes a more complex scene understanding mechanism, the first rule is even simpler and just requires the treatment of local image structure. To ensure that the depth assignment is continuous even for contours that bound surfaces that change from back- to foreground and vice versa, we realize the depth ordering with a diffusion processthat is similar to the one presented above: Let x_n be the relative depth of a contour element with neighbors x_{n+1}, x_{n+2}, x_{n-1} and x_{n-2}. Depth of a contour means the depth of the occluding surface if the contour marks the boundary of a surface. Given a smoothing factor a, the cost term $C = \frac{1}{2}\sum_n a \cdot (((x_{n+1} - x_n) - (x_n - x_{n-1}))^2 + (x_n - x_{n+1})^2 + (x_n - x_{n-1})^2) + \begin{cases} (x_n - x_0)^2 & \text{if } x_n \text{ belongs to a junction} \\ 0 & \text{otherwise} \end{cases}$ enforces the smoothness of neighboring values as well as of the first derivative. Using the gradient descent method to minimize the cost function leads to $-\frac{\partial}{\partial x_n} C = -a \cdot (6x_n - 4x_{n+1} - 4x_{n-1} + x_{n+2} + x_{n-2} + 4x_n - 2x_{n+1} - 2x_{n-1}) - \begin{cases} 2(x_n - x_0) & \text{if } x_n \text{ belongs to a junction} \\ 0 & \text{otherwise} \end{cases}$.

Then, with λ being the step width, the update rule for x_n is given by $x_n^{(k+1)} = x_n^{(k)} - \lambda \cdot \frac{\partial}{\partial x_n} C$. During the dynamic process we update the x_0 values at points of interest according to the figure-ground segregation described above.

6 Results

The impact of preprocessing becomes obvious regarding natural input images. A proper inetrpretation of the depth constellation is merely impossible for the input image shown in Fig. 3(a). After anisotropic diffusion the image is scaled to it's salient content without loosing localization accuracy (Fig. 3(c)). In Fig. 3(d)the results for the figure-ground and depth ordering mechanisms are shown. The contour elements bounding a foreground surface are displayed lighter while edges being assigned a higher distance to the viewer are displayed

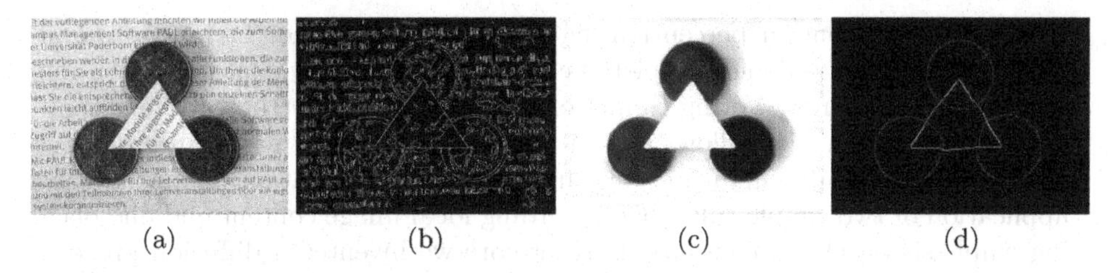

(a) (b) (c) (d)

Fig. 3. Three coins on a sheet of paper: (a) original image, (b) result of edge detection applied to the original image, (c) input image after anisotropic diffusion mit with thesholded weighting factors und (d) result of perceptual organization with completed contours and depth ordering

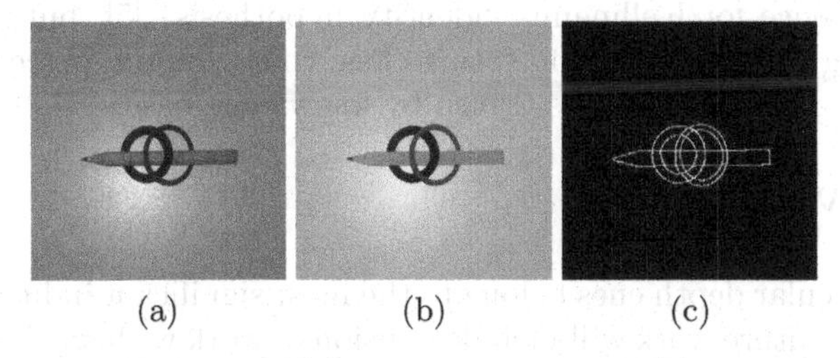

(a) (b) (c)

Fig. 4. Pencil with gasket rings: (a) original image and (b) reasult after anisotropic diffusion with thesholded weighting factors and (c) results of contour completion

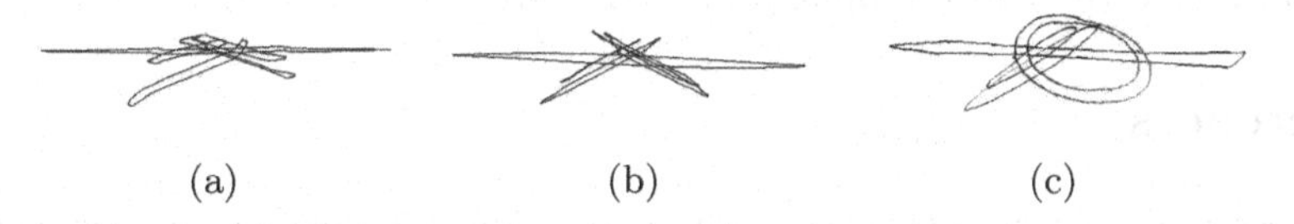

(a) (b) (c)

Fig. 5. Pencil image:(a) lateral view of depth ordering results without post processing, (b) lateral view with post processing assuming that depth values of one contour lie on an a plane and (c) isometric view of depth values with post processing

darker.This illustrates that the second rule presented in the previous section lets the mechanism interpret the depth ordering that is identical to human perception. Also for the pencil image (Fig. 4) the edge and corner detection on the original image would create problems that can be overcome by applying the anisotropic diffusion as a preprocessing step. In 4(d) the figure-ground segregation results from the first rule. As can be seen, the dynamic depth ordering ensures that even with key points changing from figure to ground a continuous depth signal can be produced.

7 Conclusion

A computational approach was presented for the depth ordering at surface boundaries in monocular images. The approach includes corner- and edge detection, illusory contour perception, figure-ground segregation at junctions and depth ordering for real and completed contours. The illusory contour perception is based on the Tensor Voting framework, for what we constructed a new communication pattern that allows a clear distinction beween the distance along a circular path and its curvature. The figure-ground segregation is done by the application of two simple rules, one regarding local image content and one placing emphasis on closed contours. Furthermore we invented a diffusion process, in which the figure-ground segregation at key points is carried to the rest of the contour. It was shown that this proceeding even works with two surfaces occluding one another. Consequently, this approach allows an assumption about an object's relative position in space, not just a distinctuion between figure and ground. The promising results presented for these mechanisms may be considered as an evidence for Kellmanns "identity hypothesis" [5], but on the other hand, regarding a larger scale, it seems clear that contour perception is also influenced by top down processes driven by knowledge base.

8 Future Work

Obviously, binocular depth cues belong to the most significant inducers for depth perception. Our future work will include a fusion of work we have done on stereoscopic algorithms with the mechanisms presented here. On the one hand, the depth ordering presented here can be used as an input to stereoscopic disparity estimation, on the other hand the results of a stereoscopic algorithm can be feeded back to the figure-ground segregation step.

References

1. Hansen, T.: A neural model of early vision: Contrast, contours, corners and surfaces. PhD thesis, Universität Ulm (2003)
2. Heitger, F., von der Heydt, R., Peterhans, E., Rosenthaler, L., Kübler, O.: Simulation of neural contour mechanisms: representing anomalous contours. Image Vision Comput. 16(6-7), 407–421 (1998)
3. Hund, M., Mertsching, B.: A Computational Approach to Illusory Contour Perception Based on the Tensor Voting Technique. In: Sanfeliu, A., Cortés, M.L. (eds.) CIARP 2005. LNCS, vol. 3773, pp. 71–80. Springer, Heidelberg (2005)
4. Hund, M., Mertsching, B.: Anisotropic Diffusion by Quadratic Regularization. In: 3rd International Conference on Computer Vision Theory and Applications (VISAPP 2008) (January 2008)
5. Kellman, P.J., Guttman, S.E., Wickens, T.D.: Geometric and neural models of object perception. In: Shipley, T.F., Kellman, P.J. (eds.) From fragments to objects: Segmentation and grouping in vision. Elsevier Science, Oxford (2001)
6. Massad, A., Babos, M., Mertsching, B.: Perceptual grouping in grey level images by combination of gabor filtering and tensor voting. In: Kasturi, R., Laurendeau, D., Suen, C. (eds.) ICPR, vol. 2, pp. 677–680 (2002)
7. Massad, A., Medioni, G.: 2-D Shape Decomposition into Overlapping Parts. In: Arcelli, C., Cordella, L.P., Sanniti di Baja, G. (eds.) IWVF 2001. LNCS, vol. 2059, pp. 398–409. Springer, Heidelberg (2001)

8. Medioni, G., Lee, M.-S., Tang, C.-K.: A Computational Framework for Segmentation and Grouping. Elsevier Science, Amsterdam (2000)
9. Neumann, H., Mingolla, E.: Computational neural models of spatial integration in perceptual grouping. In: Shipley, T., Kellman, P. (eds.) From fragments to units: Segmentation and grouping in vision, pp. 353–400. Elsevier Science, Oxford (2001)
10. Nieder, A.: Seeing more than meets the eye: processing of illusory contours in animals. Journal of Comparative Physiology A: Sensory, Neural, and Behavioral Physiology 188(4), 249–260 (2002)
11. Saund, E.: Perceptual organization of occluding contours of opaque surfaces. Comput. Vis. Image Underst. 76(1), 70–82 (1999)
12. Williams, L.R., Hanson, A.R.: Perceptual completion of occluded surfaces. Computer Vision and Image Understanding: CVIU 64(1), 1–20 (1996)
13. Zweck, J.W., Williams, L.R.: Euclidean group invariant computation of stochastic completion fields using shiftable-twistable functions. In: Vernon, D. (ed.) ECCV 2000. LNCS, vol. 1843, pp. 100–116. Springer, Heidelberg (2000)

Fast Hand Detection Using Posture Invariant Constraints

Nils Petersen and Didier Stricker*

DFKI GmbH
Kaiserslautern, Germany
{nils.petersen,didier.stricker}@dfki.de

Abstract. The biggest challenge in hand detection and tracking is the high dimensionality of the hand's kinematic configuration space of about 30 degrees of freedom, which leads to a huge variance in its projections. This makes it difficult to come to a tractable model of the hand as a whole. To overcome this problem, we suggest to concentrate on posture invariant local constraints, that exist on finger appearances. We show that, besides skin color, there is a number of additional geometric and photometric invariants. This paper presents a novel approach to real-time hand detection and tracking by selecting local regions that comply with these posture invariants. While most existing methods for hand tracking rely on a color based segmentation as a first preprocessing step, we integrate color cues at the end of our processing chain in a robust manner. We show experimentally that our approach still performs robustly above cluttered background, when using extremely low quality skin color information. With this we can avoid a user- and lighting-specific calibration of skin color before tracking.

1 Introduction

Hand gestures are an intuitive modality of communication. While already used to control Virtual Reality environments, the interest in harnessing them as a mean of human-machine-communication has been recently increasing, with consumer electronics getting more and more aware of gestures. This ranges from finger gestures to control touch tables and smart phones up to hand/arm gestures used by gaming consoles.

All the devices mentioned above use specialized hardware connected with the user to recognize hand and finger gestures, like data gloves, touchscreens, or hand held inertial-visual 'remote controls'. However, detecting hands and hand posture on camera images only is appealing, since the user does not need to wear or hold additional hardware. Additionally, cameras are a cheap mass market product, often already pre-installed in computers and phones and deliver more context information that can be incorporated into the interaction model.

The biggest challenge in visual hand detection is the high dimensionality of the hand's kinematic configuration space of about 30 degrees of freedom (DOFs). This makes it difficult to find hands in images, due to the huge variance in its projections.

* This work was supported in part by a grant from the Bundesministerium fuer Bildung und Forschung, BMBF (project AVILUSplus, grant number 01IM08002).

B. Mertsching, M. Hund, and Z. Aziz (Eds.): KI 2009, LNAI 5803, pp. 106–113, 2009.

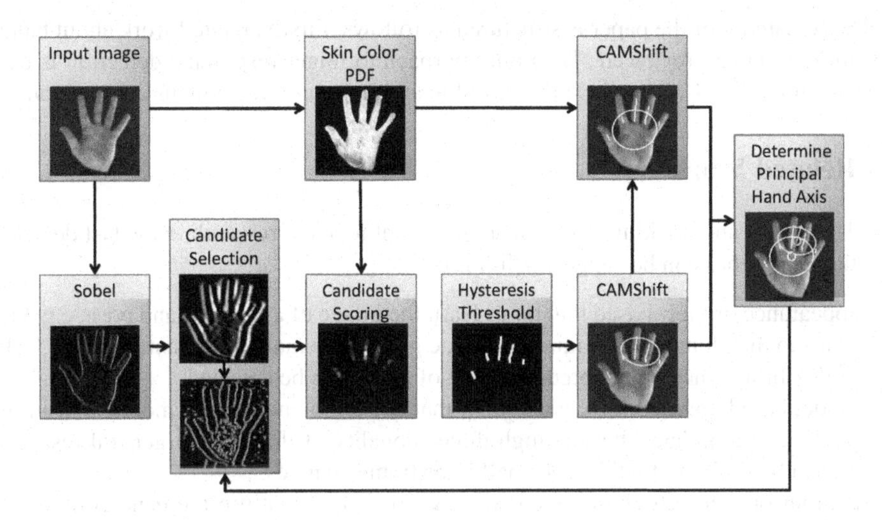

Fig. 1. Overview of our approach: First candidates are selected on geometric features only. In the succeding scoring step the local image content around each candidate is evaluated, integrating color cues. The candidates are being combined to lines using hysteresis thresholding and taken as input to a first CAMShift tracker. If the tracker converges, a second CAMShift starts from this point on the skin color probability image. The direction between the two means is taken as the hand's principal axis.

When it comes to tracking, a further challenge is, that the hand and fingers can be moved at high velocity. This leads to highly dissimilar consecutive images at standard camera frame rates around 30 fps. The use of a motion model of the hand is only of modest utility, since accelerations are likely to vary too quickly as well. This will eventually result in frequent tracking losses and makes (re-)detection an important prerequisite for practical hand tracking systems.

Constraints on finger apperance

On the projection of a finger, there exist some constraints that are invariant to the hand pose and posture. Firstly, its bounding edges are always approximately parallel with no aligned edges in between. Additionally, since the anatomical variance is limited, the distance of the bounding edges can be approximately estimated by only taking the hand's distance from the image sensor into account. Finally, unless the lighting is extremely ambient, each finger exhibits smooth shading along its profile.

If there is no occlusion by the palm itself or foreground objects, there must always be at least one finger fully visible in the image with the exception of the finger's flexion axis being aligned with the image plane. In this case the finger can occlude itself if flexed to a certain extent but would still be partly visible. So, both in 2D and 3D, the approach to detect hands by detecting fingers is feasible.

The remainder of the paper is structured as follows. Firstly, related work about hand detection is shortly reviewed. Then our approach to finger resp. hand detection is described. In section 4 we outline our algorithm and its results and conclude in section 5.

2 Related Work

Hand detection and tracking has been an important focus of research in the last decade. Existing approches can be roughly split into:

- appearance or view based that try to learn the image of a certain hand pose, e.g. [1] or try to match model and image edges, e.g. [2]. The main disadvantage is the lack of flexibility since only a predefined set of poses can be detected.
- model based approaches, e.g. [3,4,5], that use a kinematic hand model instead of predefined templates, but the high dimensionality of the hand in general restricts the applicability to tracking of small inter-frame changes.
- contour or silhouette based methods, e.g. [6,7,8,9] that use the hand contour as feature for tracking.

The method proposed in this paper is in principle contour based, but in contrast to the other methods, the contour is not calculated using a color based segmentation. Actually, in our approach, the hand contour is never computed directly and is only implicitly reconstructable from the outcome of the algorithm. Also we do not make the assumption, that our contour must also be the boundary between skin-color- and non-skin-color-pixels. The first advantage of our method is, that we do not need to rely on thresholding to detect the candidates. Moreover, since we integrate color information in a robust manner, our approach can deal with extremely low quality skin color probability images, as well as above skin colored background. A field that is quite close to our first candidate selection step is curvilinear structure extraction [10,11]. The main difference is, that we already integrate assumptions about the finger appearance in the extraction process and use a second scoring step to evaluate the local image content around a candidate.

3 Finger Detection

Besides the constraints given in section 1, we integrate another constraint on hand postures: For the outstretched hand, the pairwise maximum angle between fingers has an upper bound. In other words, the maximum angle between an outstretched finger and the principal hand axis is limited. This not only simplifies computation, but also helps to increase robustness during tracking as we will show later.

Color segmentation is typically one of the first steps in many hand tracking algorithms. Though skin color is a very decisive feature, the problem is, that a threshold parameter has to be selected to determine, if a pixel is treated as skin color or not. The success of the whole method then depends on the selection of this parameter. The same thing applies to edge extraction, which also relies on a threshold. In our approach we do not rely on any parametric preprocessing at all and only threshold the processing results. Especially, if two fingers are touching silhouette based methods fail to distinguish the fingers. Since the shared contour line typically exhibits a weak magnitude, it is also often ignored by edge extraction. We overcome this problem by looking on pairs of parallel edges in a distinct distance. If only one edge is sufficiently strong, the pairing one can be emphasized due to the distance constraint. Thus it is not necessary, that both edges exhibit a strong magnitude.

The detection works in two stages: First a number of pixel-candidates are being selected by a fast convolution-based method. These candidates are pixels that are most likely to be located on finger centers as defined through the criteria given in section 1. The selected candidates still contain a lot of false positives, since the criteria are not decisive enough. The candidates are then being scored by evaluating the image content on a scan line through the candidate. In the following subsections, we will describe this in detail, as well as a fast and robust method to determine the hand's principal axis.

3.1 Candidate Selection

The leadoff set of candidates is selected by a convolution of the edge image with a specialized filter kernel. This kernel is a function of the average finger radius expected in the image $\bar{r}$, its variance $\bar{\sigma}$, and the maximum deviation $\Delta\alpha$ from the finger orientation from the principal hand axis α. To shorten notation, we use the helper variables $r_{\Delta\alpha} = \frac{1}{2}(\bar{r} + \bar{r}/cos(\Delta\alpha))$, which denotes the width respectively $\sigma_{\Delta\alpha} = \bar{\sigma}/\sqrt{cos(\Delta\alpha)}$ which denotes the variance, both adapted to accomodate for an angular deviation $\Delta\alpha$ from the principal hand axis. The kernel function then becomes

$$f_{r_{\Delta\alpha},\sigma_{\Delta\alpha}}(x) = (\delta(|x - w_{\Delta\alpha}|) * 2\mathcal{N}_{\sigma_{\Delta\alpha}}(x)) - \frac{1}{\sigma_{\Delta\alpha}\sqrt{2\pi}}, x \in [-r_{\Delta\alpha}, r_{\Delta\alpha}], |\Delta\alpha| < \pi, \tag{1}$$

where δ denotes the Dirac-function, $\mathcal{N}_{\sigma_{\Delta\alpha}}$ the zero-mean normal distribution with variance $\sigma_{\Delta\alpha}$, and $*$ convolution. An example of this function is shown in fig. 2 We apply this kernel to the component-wise squared directional image derivative perpendicular to the principal hand axis. The filter emphasizes edges in the distance of the assumed finger width and penalizes edges in between and is similar to the sign-free first derivative of a blurred bar profile. Differences are firstly, that we crop the kernel to diminish influence from the surrounding to be able to cope with cluttered background. Secondly, we substract a discount from the kernel, to actually penalize image edges in between. The local maxima of the filter response are chosen as candidates.

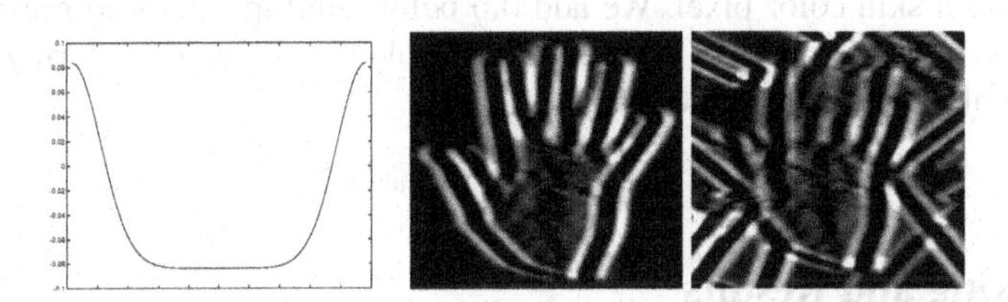

Fig. 2. Example of the kernel function for $\bar{r} = 10$, $\bar{\sigma} = 4$, and $\Delta\alpha = 45$ degrees. The graph on the left shows the corresponding kernel function. The two images show the filter response for a hand above a uniform background (middle) and cluttered background (right).

3.2 Scoring Candidates

The scoring step is used in two ways: Firstly, to find the exact center and width of the assumptive finger. Secondly, to retrieve a score for the candidate. This is done

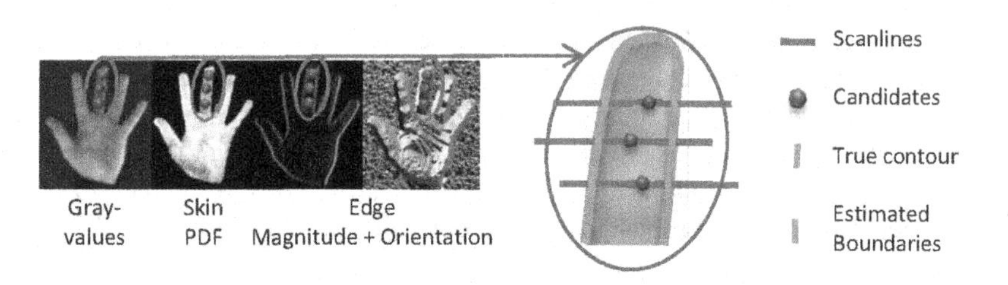

Fig. 3. Illustration of the scoring step. The image is evaluated on a scanline through the candidate point, integrating color cues. The boundaries are estimated by maximizing eq. 2.

by maximizing a scoring function of the center c and the finger radius r, evaluating the image along a scan-line through the candidate, perpendicular to the principal axis. Thus, if the candidate actually lies near a finger center, the scan-line contains the finger profile as illustrated in fig. 3. Following the intuition that the bounding edges should be the most salient ones, we chose the difference between the weaker bounding edge and the maximum in-between contrast as one quality measure: $\phi_e(c,r) = \min(\mathbf{M}(c-r), \mathbf{M}(c+r)) - \max_{x=1-c-r}^{c+r-1}(\mathbf{M}(x))$, where $\mathbf{M}(x)$ is the gradient magnitude along the scan-line. Furthermore parallelism of the edges is enforced by $\phi_o(c,r) = \mathcal{N}_{\sigma_p}(\mathbf{O}(c-r) - \mathbf{O}(c+r))$, using the local gradient orientation $\mathbf{O}(x)$. To account for the error we make, since the scan-line is generally not perpendicular to the finger axis, we formulate $\phi_w(c,r) = \mathcal{N}_{\sigma_p}(\bar{r} - r\cos(\mathbf{O}(c-r) - \alpha))$ as the distance score, with a constant variance parameter σ_p. To describe shading on a finger, we use the range of brightness values between the boundaries. Since we also penalize hard edges in between, this is an easy way to enforce smooth shading along the scan-line: $\phi_s(c,r) = \max_{x=1-c-r}^{c+r-1}(\mathbf{I}_{\text{lum}}(x)) - \min_{x=1-c-r}^{c+r-1}(\mathbf{I}_{\text{lum}}(x))$. Finally we integrate skin color information through $\phi_c(c,r) = \prod_{x=1-c-r}^{c+r-1}(p(skin|\mathbf{I}(x)) + 0.5)$, where $p(skin|\mathbf{I}(x))$ is the probability of $\mathbf{I}(x)$ being a skin color pixel. We add 0.5 before multiplication to prevent a single probability being close to zero rejecting the whole scan-line. The scoring function is the product of the parts just described:

$$(\hat{c}, \hat{r}) = \arg\max_{c,r} \phi_e \phi_w \phi_o \phi_c \phi_s. \tag{2}$$

4 Tracking and Results

4.1 Tracking Algorithm

Until now, only local information was incorporated into the generation of the candidates and scores. Since the candidates are assumed to be located at finger centers, they should

Fig. 4. Comparison between high (first row) and low (second row) quality of skin color distribution. The low quality distribution image was generated from virtually gray values only. Results are very robust even with poor skin color distribution images.

be combineable to a line along the finger. This is done using hysteresis thresholding, similar to a Canny edge extractor: Starting from candidates with scores above a certain threshold we combine neighboring candidates in direction of the candidate's orientation into lines. This is continued until the score of the next candidate is falling below a second, lower threshold. Through this step weak isolated candidates or even clusters of cluttered weak candidates can get rejected while retaining all candidates that are connected curvilinearly to strongly scored candidates. The white lines visible on the right half of fig. 4 show typical results of the hysteresis thresholding step. The region of the image with the candidates scored highest is then chosen as the hand region.

This is done using a continuously adaptive mean shift approach (CAMShift) [12] on the thresholded candidates. CAMShift is an extension of the mean shift algorithm [13], that performs hill climbing on discrete distributions, therefore appropriate for use with images. The mean shift approach iteratively centers a search window according to the first moment of the distribution calculated over this window. CAMShift additionally adapts the size of the search window according to the second moment of the distribution.

If the mean shifting does not converge or the sum of candidate scores within the converged search window lies below a certain threshold, we reset the search window to the whole image, thus trying to detect the hand in the next frame.

Tracking and principal hand axis determination. If the CAMShift on the candidates converges we use another CAMShift, this time on the skin color probability distribution image, using the outcome of the first CAMShift as initial search window. If the skin color distribution is not entirely degenerated, a mode of the distribution is located at the hand's palm, thus attracting the mean during CAMShift iterations. The direction from the local skin color mean to the local candidate mean is taken as the hand's principal axis. If the two means are diverging, i.e. the candidate pixels are completely disjunct from the color pixels set, this is taken as another criterion for tracker loss. The whole algorithm is illustrated in fig. 1.

The skin color probability for each pixel is calculated by using the euclidean distance between the pxiel's color value $\mathbf{I}(x)_{\mathrm{hsv}}$ in a HSV and a constant reference color value $\mathbf{I}_{\mathrm{ref}}$ weighted with the column vector w: $p(\mathrm{skin}|\mathbf{I}(x)) = 1.0 - ||w^T(\mathbf{I}(x)_{\mathrm{hsv}} - \mathbf{I}_{\mathrm{ref}}))||_2$, which is then simply clamped to the range $[0, 1]$.

4.2 Results

To demonstrate that the method does not substantially rely on color cues we show the outcome of the algorithm fig. 4 using different qualities of skin color distribution images. The low quality skin color distribution was generated by decreasing the weights for hue and saturation by a factor of 10, thus almost only relying on gray-values. The algorithm was applied to a sequence showing a moving hand above cluttered background and using the low quality mask. Fig. 5 shows example pictures of the sequence. The sequence consists of 140 VGA color images. For detection we assume a vertical hand axis and only slightly adapt it to the recognized one, due to the noisy outcome with this quality of skin distribution. Still, the skin color distribution CAMShift step converges approximately in the right direction, depsite the extremely imprecise distribution. This is due to the fact, that we initialize its search window on a part of the hand in an appropriate size. Therefore, the CAMShift iterations take off, focussing the hand area only and is attracted to the local mode on the hand's palm without being distracted from surrounding clutter. In the last row of fig. 5, one can see that the hand is still being tracked, though above the other hand. This is due to the principal axes of the hands are almost perpendicular, thus ignoring the finger responses of the untracked hand.

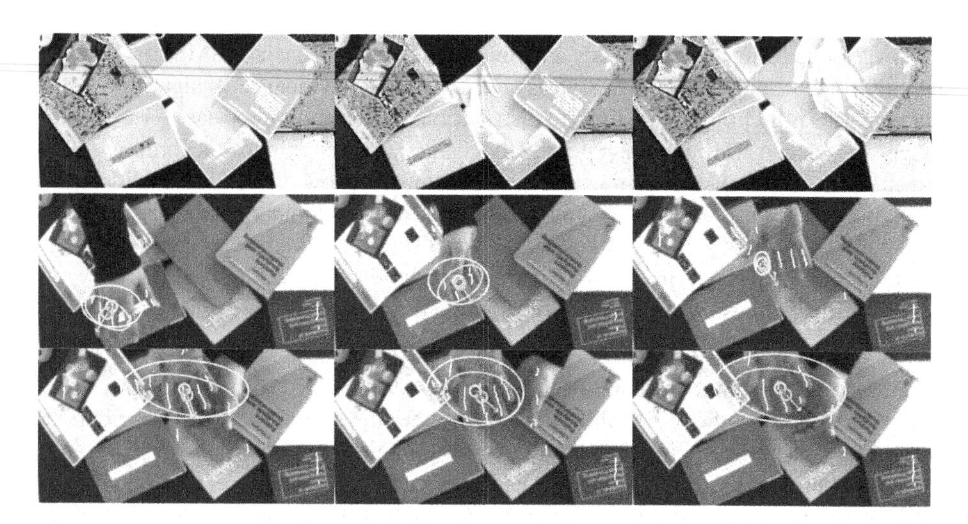

Fig. 5. The first row exemplarily shows three low quality skin color distribution images used for tracking. The six images below show tracking results above cluttered background using this imprecise distribution.

5 Conclusion and Future Work

A framework for robust hand detection tracking has been presented, that focusses on geometric hand posture invariant constraints to find hands in images. The method does not require any color or foreground based segmentation of the hand but robustly integrates color cues in a scoring step together with other geometric and photometric features. Therefore our method can deal with extremely degenerate skin color distributions and thus does not need to be calibrated to a users skin color before use. The confirmed candidates also give appoximate finger positions. A succeding posture analysis could readily incorporate this information.

Next steps include the expansion of this approach to posture classication and fitting of a hand model in the candidate set. Another promising approach would be to employ the appearance vector along the candidate's scan-line as descriptor for use in a tracking by matching approach.

References

1. Kolsch, M., Turk, M.: Robust hand detection. In: Int. Conf. on Automatic Face and Gesture Recognition, May 17–19, pp. 614–619 (2004)
2. Athitsos, V., Sclaroff, S.: Estimating 3d hand pose from a cluttered image. In: CVPR, June 2003, vol. 2, pp. II–432–II–439 (2003)
3. de La Gorce, M., Paragios, N., Fleet, D.: Model-based hand tracking with texture, shading and self-occlusions. In: CVPR (2008)
4. Rehg, J., Kanade, T.: Digiteyes: vision-based hand tracking for human-computer interaction. In: Workshop on Motion of Non-Rigid and Articulated Objects, November 1994, pp. 16–22 (1994)
5. Stenger, B., Thayananthan, A., Torr, P., Cipolla, R.: Model-based hand tracking using a hierarchical bayesian filter. PAMI 28(9), 1372 (2006)
6. von Hardenberg, C., Bérard, F.: Bare-hand human-computer interaction. In: Workshop on Perceptive User Interfaces, pp. 1–8. ACM, New York (2001)
7. Lee, T., Hollerer, T.: Handy ar: Markerless inspection of augmented reality objects using fingertip tracking. In: Int. Symp. on Wearable Computers, pp. 83–90 (2007)
8. Oka, K., Sato, Y., Koike, H.: Real-time fingertip tracking and gesture recognition. IEEE Computer Graphics and Applications 22(6), 64–71 (2002)
9. Maccormick, J., Isard, M.: Partitioned sampling, articulated objects, and interface-quality hand tracking. In: Vernon, D. (ed.) ECCV 2000. LNCS, vol. 1843, pp. 3–19. Springer, Heidelberg (2000)
10. Koller, T., Gerig, G., Szekely, G., Dettwiler, D.: Multiscale detection of curvilinear structures in 2-d and 3-d imagedata. In: ICCV, pp. 864–869 (1995)
11. Steger, C.: Extracting curvilinear structures: A differential geometric approach. In: Buxton, B.F., Cipolla, R. (eds.) ECCV 1996. LNCS, vol. 1065, pp. 630–641. Springer, Heidelberg (1996)
12. Bradski, G.R., et al.: Computer vision face tracking for use in a perceptual user interface. Intel Technology Journal 2(2), 12–21 (1998)
13. Comaniciu, D., Meer, P.: Mean shift: A robust approach toward feature space analysis. PAMI 24(5), 603–619 (2002)

A Novel and Efficient Method to Extract Features and Vector Creation in Iris Recognition System

Amir Azizi[1] and Hamid Reza Pourreza[2]

[1] Islamic Azad University Qazvin Branch
[2] Ferdowsi University of Mashhad
Amirazizi_b@Yahoo.com, hpourreza@um.ac.ir

Abstract. The selection of the optimal feature subset and the classification has become an important issue in the field of iris recognition. In this paper we propose several methods for iris feature subset selection and vector creation. In this paper we propose a new feature extraction method for iris recognition based on contourlet transform. Contourlet transform captures the intrinsic geometrical structures of iris image. For reducing the feature vector dimensions we use the method for extract only significant bit and information from normalized iris images. In this method we ignore fragile bits. At last, the feature vector is created by two methods: Co-occurrence matrix properties and contourlet coefficients. For analyzing the desired performance of our proposed method, we use the CASIA dataset, which is comprised of 108 classes with 7 images in each class and each class represented a person. Experimental results show that the proposed increase the classification accuracy and also the iris feature vector length is much smaller versus the other methods.

Keywords: Biometric, Iris Recognition, Contourlet Transform, Co-occurrence Matrix, Contourlet coefficients.

1 Introduction

There has been a rapid increase in the need of accurate and reliable personal identification infrastructure in recent years, and biometrics has become an important technology for the security. Iris recognition has been considered as one of the most reliable biometrics technologies in recent years [1, 2]. The human iris is the most important biometric feature candidate, which can be used for differentiating the individuals. For systems based on high quality imaging, a human iris has an extraordinary amount of unique details. Features extracted from the human iris can be used to identify individuals, even among genetically identical twins [3]. Iris-based recognition system can be noninvasive to the users since the iris is an internal organ as well as externally visible, which is of great importance for the real-time applications [4]. Based on the technology developed by Daugman [3, 5, 6], iris scans have been used in several international airports for the rapid processing of passengers through the immigration which have pre registered their iris images.

B. Mertsching, M. Hund, and Z. Aziz (Eds.): KI 2009, LNAI 5803, pp. 114–122, 2009.

1.1 Related Works

The usage of iris patterns for the personal identification began in the late 19th century; however, the major investigations on iris recognition were started in the last decade. In [9], the iris signals were projected into a bank of basis vectors derived by the independent component analysis, and the resulting projection coefficients were quantized as Features. A prototype was proposed in [10] to develop a 1D representation of the gray-level profiles of the iris. In [11], biometrics based on the concealment of the random kernels and the iris images to synthesize a minimum average correlation energy filter for iris authentication were formulated. In [5, 6, 12], the Multiscale Gabor filters were used to demodulate the texture phase structure information of the iris. In [13], an iris segmentation method was proposed based on the crossed chord theorem and the collarette area. An interesting solution to defeat the fake iris attack based on the Purkinje image was depicted in [16]. An iris image was decomposed in [17] into four levels by using the 2D Haar wavelet transform, the fourth-level high-frequency information was quantized to form an 87-bit code, and a modified competitive learning neural network (LVQ) was adopted for classification.

Fourth-level high-frequency information was quantized to form an 87-bit code, and a modified competitive learning neural network (LVQ) was adopted for classification. In [18], a modification to the Hough transform was made to improve the iris segmentation, and an eyelid detection technique was used, where each eyelid was modeled as two straight lines. A matching method was implemented in [19], and its performance was evaluated on a large dataset. In [20], a personal identification method based on the iris texture analysis was described. The remainder of this paper is organized as follows: Section 2 deals with proposed method. Section 3 deals with Feature Extraction method discussion. Section 4 deals with feature subset selection and vector creation techniques, Section 5 shows our experimental results and finally Section 6 concludes this paper.

2 Proposed Method: The Main Steps

First the image preprocessing step performs the localization of the pupil, detects the iris boundary, and isolates the collarette region, which is regarded as one of the most important areas of the iris complex pattern. The collarette region is less sensitive to the pupil dilation and usually unaffected by the eyelids and the eyelashes [8]. We also detect the eyelids and the eyelashes, which are the main sources of the possible occlusion. In order to achieve the invariance to the translation and the scale, the isolated annular collarette area is transformed to a rectangular block of fixed dimension. The discriminating features are extracted from the transformed image and the extracted features are used to train the classifiers. The optimal features subset is selected using several methods to increase the matching accuracy based on the recognition performance of the classifiers.

2.1 Iris Normalization

We use the rubber sheet model [12] for the normalization of the isolated collarette area. The center value of the pupil is considered as the reference point, and the radial vectors are passed through the collarette region. We select a number of data points along each radial line that is defined as the radial resolution, and the number of radial lines going around the collarette region is considered as the angular resolution. A constant number of points are chosen along each radial line in order to take a constant number of radial data points, irrespective of how narrow or wide the radius is at a particular angle. We build the normalized pattern by backtracking to find the Cartesian coordinates of data points from the radial and angular positions in the normalized pattern [3, 5, and 6]. The normalization approach produces a 2D array with horizontal dimensions of angular resolution, and vertical dimensions of radial resolution form the circular-shaped collarette area (See Fig.1I). In order to prevent non-iris region data from corrupting the normalized representation, the data points, which occur along the pupil border or the iris border, are discarded. Fig.1II (a) (b) shows the normalized images after the isolation of the collarette area.

3 Feature Extraction and Encoding

Only the significant features of the iris must be encoded so that comparisons between templates can be made. Gabor filter and wavelet are the well-known techniques in texture analysis [5, 19, 20, 21]. In wavelet family, Haar wavelet [22] was applied by Jafer Ali to iris image and they extracted an 87-length binary feature vector. The major drawback of wavelets in two-dimensions is their limited ability in capturing Directional information. The contourlet transform is a new extension of the wavelet transform in two dimensions using Multi scale and directional filter banks. The feature representation should have information enough to classify various irises and be less sensitive to noises. Also in the most appropriate feature extraction we attempt to extract only significant information, more over reducing feature vector dimensions, the processing lessened and enough information is supplied to introduce iris feature vectors classification.

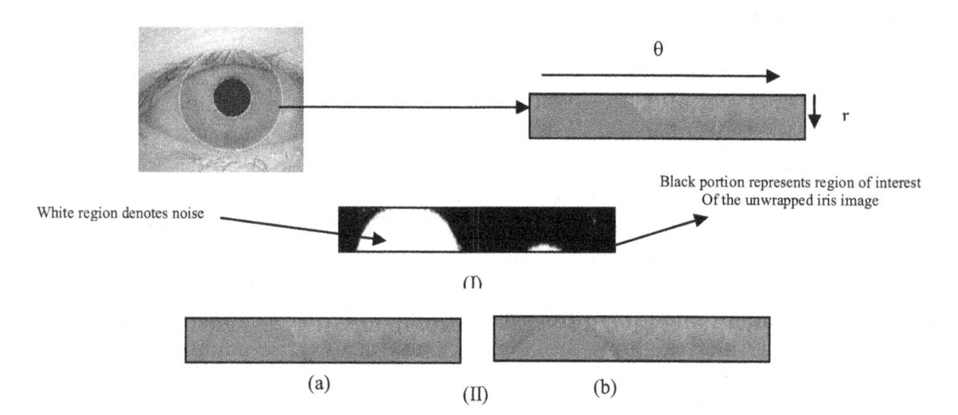

Fig. 1. (I) shows the normalization procedure on CASIA dataset; (II) (a), (b) Show the normalized images of the isolated collarette regions

3.1 Contourlet Transform

Contourlet transform (CT) allows for different and flexible number of directions at each scale. CT is constructed by combining two distinct decomposition stages [33], a multistage decomposition followed by directional decomposition. The grouping of wavelet coefficients suggests that one can obtain a sparse image expansion by applying a multi-scale transform followed by a local directional transform. It gathers the nearby basis functions at the same scale into linear structures. In essence, a wavelet-like transform is used for edge (points) detection, and then a local directional transform for contour segments detection. A double filter bank structure is used in CT in which the Laplacian pyramid (LP) [23] is used to capture the point discontinuities, and a directional filter bank (DFB) [24] to link point discontinuities into linear structures. The combination of this double filter bank is named pyramidal directional filter bank (PDFB) as shown in Fig.4.Benefits of Contourlet Transform in the Iris Feature Extraction to capture smooth contours in images, the representation should contain basis functions with variety of shapes, in particular with different aspect ratios. A major challenge in capturing geometry and directionality in images comes from the discrete nature of the data; the input is typically sampled images defined on rectangular grids.

3.2 The Best Bit in an Iris Code

Biometric systems apply filters to iris images to extract information about iris texture. Daugman's approach maps the filter output to a binary iris code. The fractional Hamming distance between two iris codes is computed and decisions about the identity of a person are based on the computed distance. The fractional Hamming distance weights all bits in an iris code equally. However, not all the bits in an iris code are equally useful. For a given iris image, a bit in its corresponding iris code is defined as "fragile" if there is any substantial probability of it ending up a 0 for some images of the iris and a 1 for other images of the same iris. According to [25] the percentages of fragile bits in each row of the iris code, Rows in the middle of the iris code (rows 5 through 12) are the most consistent.

4 Feature Vector Creation in Proposed Method

According to the method mentioned in section 3.2, we concluded the middle band of iris normalized images have more important information and less affected by fragile bits, so for introducing iris feature vector based on contourlet transform the rows between 5 and 12 in iris normalize image are decomposed into eight directional sub-band outputs using the DFB at three different scales and extract their coefficients. In our method we use using the Grey Level Co-occurrence Matrix (GLCM) and Contourlet Coefficient. Various textural features have been defined based on the work done by Haralick [26]. These features are derived by weighting each of the co-occurrence matrix values and then summing these weighted values to form the feature value. The specific features considered in this research are defined as follows:

1) Energy $= \sum_i \sum_j p(i,j)^2$ 2) Contrast $= \sum_{n=0}^{N_g-1} n^2 \left[\sum_{i=1}^{N_g} \sum_{j=1}^{N_g} P(i,j) \| i-j \| = n \right]$

3) Correlation $= \dfrac{\sum_i \sum_j (ij)P(i,j) - \mu_x \mu_y}{\sigma_x \sigma_y}$

4) Homogeneity $= \sum_i \sum_j \dfrac{1}{1+(i-j)^2} P(i,j)$

5) Autocorrelation $= \sum_i \sum_j (ij)P(i,j)$ 6) Dissimilarity $= \sum_i \sum_j |i-j| . P(i,j)$

7) Inertia $= \sum_i \sum_j (i-j)^2 P(i,j)$

Here $\mu_x, \mu_y, \sigma_x, \sigma_y$ are mean and standard deviation along x and y axis .as stated in the previous section level 2 sub bands are extracted and according to the Following Rule are modified into binary mode:

If Coeff (i)>=0 then NewCoeff (i) =1
Else NewCoeff (i) =0

And with hamming distance between the vectors of the generated coefficients is calculated. Numbers ranging from 0 to 0.5 for inter-class distribution and 0.45 and 0.6 for intra-class distribution are included. In total 192699 comparisons inter-class and 1679 comparisons intra-class are carried out. In Fig.2 you can see inter-class and intra-class distribution. In implementing this method, we have used point 0.42 the inter-class and intra-class separation point.

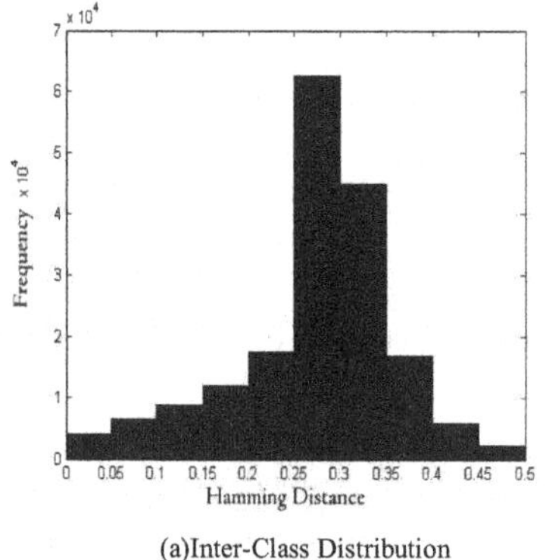

(a)Inter-Class Distribution

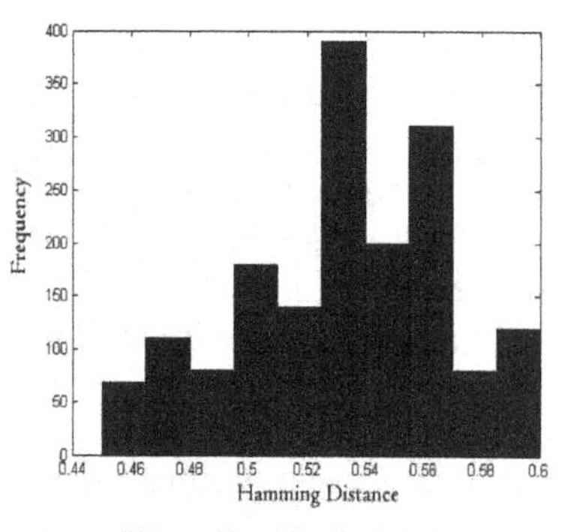

(b)Inter-Class Distribution

Fig. 2. Inter and Intra Class Distribution

5 Experimental Results

For creating iris feature vector we by GLCM carried out the following steps:

1) Iris normalized image (Rows in the middle of the iris code (rows 5 through 12)) is decomposed up to level two.(for each image ,at level one, 2 and at level two, 4 sub band are created) .
2) The sub bands of each level are put together, therefore at level one a matrix with 4*120 elements, and at level two a matrix with 16*120 elements is created. We named these matrixes: Matrix1 and Matrix 2.
3) By putting together Matrix1 and Matrix 2, a new matrix named Matrix3 with 20*120 elements is created. The co-occurrence of these three matrixes with offset one pixel and angles 0, 45, 90 degree is created and name this matrix: CO1, CO2 and CO3.in this case for each image 3 co-occurrence matrixes with 8*8 dimensions are created.
4) According to the Haralick's [26] theory the co-occurrence matrix has 14 properties, of which in iris biometric system we used 7 properties which are used for 3 matrixes , so the feature vector is as follow:

F=[En1,Cont1,cor1,hom1,Acor1,dis1,ine1, En2,Cont2,cor2, hom2,Acor2,dis2,ine2 En3,Cont3,cor3,hom3,Acor3,dis3,ine3] In other word the feature vector in our method has only 21 elements. Also for improving results, for each sub bands and scale we create a feature vector by using GLCM.in other words for each eight sub bands in level 3 of Contourlet transform we computed GLCM properties, separately and then by combining these properties the feature vector is created. In this case the feature vector has 56 elements. In Table 1 and Table 2 you can see the result of implementing our proposed methods:

Table 1. Result of Implementing Proposed Method(GLCM)

The Number Of Classes	The Correct of Percentage Classification (%)		
	KNN Classifier	SVM Classifier(Kernel 1)	SVM Classifier(Kernel 2)
20	96.6	100	100
40	88.3	94.3	96.3
60	90.8	91.6	95.6
80	89.3	90.1	95.8
100(GLCM)	88.5	90.07	94.2
100(GLCM (Combining Sub bands)	87.5	91.3	96.3

Table 2. Result of Implementing Proposed Method (Binary Vector)

separation point	FAR%	FRR%
0.2	99.047	0
0.25	82.787	0
0.30	37.880	0
0.35	5.181	0
0.4	2.492	0
0.42	1.2	2.3
0.45	0.002	7.599
0.5	0	99.499

In Table 3 we compared our proposed method with some other well known methods from 2 view points: feature vector length and the correct of percentage classification:

Table 3. Comparison Between Our Proposed Method and Some well- known Method

Method	The Correct Of Percentage Classification (%)	The Feature Vector Length(Bit)
Dugan[3]	100	2048
Lim[17]	90.4	87
Ma[20]	959	1600
Jafar Ali[22]	92.8	87
	Our Methods	
(GLCM)	94.2	21
GLCM (Combining Sub bands)	96.3	56
Binary Vector	96.5	2520

6 Conclusions

In this paper we proposed an effective algorithm for iris feature extraction using contourlet transform Co-occurrence Matrix have been presented. The GLCM and binary Coefficients proved to be a good technique as it provides reasonable accuracy and is invariant to iris rotation. For Segmentation and normalization we use Daugman methods. Our proposed method can classify iris feature vector properly. The rate of expected classification for the fairly large number of experimental date in this paper verifies this claim. In the other words our method provides a less feature vector length with an insignificant reduction of the percentage of correct classification.

References

1. Wildes, R.P.: Iris recognition: an emerging biometric technology. Proceedings of the IEEE 85(9), 1348–1363 (1997)
2. Jain, A., Bolle, R., Pankanti, S.: Biometrics: Personal Identification in a Networked Society. Kluwer Academic Publishers, Norwell (1999)

3. Daugman, J.: Biometric personal identification system based on iris analysis, US patent no. 5291560 (1994)
4. Mansfield, T., Kelly, G., Chandler, D., Kane, J.: Biometric product testing. Final Report, National Physical Laboratory, Middlesex, U.K (2001)
5. Daugman, J.G.: High confidence visual recognition of persons by a test of statistical independence. IEEE Transactions on Pattern Analysis and Machine Intelligence 15(11), 1148–1161 (1993)
6. Daugman, J.: Demodulation by complex-valued wavelets for stochastic pattern recognition. International Journal of Wavelets, Multiresolution and Information Processing 1(1), 1–17 (2003)
7. CASIA, Chinese Academy of Sciences – Institute of Automation. Database of 756 Grayscale Eye Images. Versions 1.0 (2003), `http://www.sinobiometrics.com`
8. He, X., Shi, P.: An efficient iris segmentation method for recognition. In: Singh, S., Singh, M., Apte, C., Perner, P. (eds.) ICAPR 2005. LNCS, vol. 3687, pp. 120–126. Springer, Heidelberg (2005)
9. Bae, K., Noh, S., Kim, J.: Iris feature extraction using independent component analysis. In: Kittler, J., Nixon, M.S. (eds.) AVBPA 2003. LNCS, vol. 2688, pp. 1059–1060. Springer, Heidelberg (2003)
10. Boles, W.W., Boashash, B.: A human identification technique using images of the iris and wavelet transform. IEEE Transactions on Signal Processing 46(4), 1185–1188 (1998)
11. Chong, S.C., Teoh, A.B.J., Ngo, D.C.L.: Iris authentication using privatized advanced correlation filter. In: Zhang, D., Jain, A.K. (eds.) ICB 2005. LNCS, vol. 3832, pp. 382–388. Springer, Heidelberg (2005)
12. Daugman, J.: Statistical richness of visual phase information: update on recognizing persons by iris patterns. International Journal of Computer Vision 45(1), 25–38 (2001)
13. He, X., Shi, P.: An efficient iris segmentation method for recognition. In: Singh, S., Singh, M., Apte, C., Perner, P. (eds.) ICAPR 2005. LNCS, vol. 3687, pp. 120–126. Springer, Heidelberg (2005)
14. Jeong, D.S., Park, H.-A., Park, K.R., Kim, J.: Iris recognition in mobile phone based on adaptive Gabor filter. In: Zhang, D., Jain, A.K. (eds.) ICB 2005. LNCS, vol. 3832, pp. 457–463. Springer, Heidelberg (2005)
15. Vijaya Kumar, B.V.K., Xie, C., Thornton, J.: Iris verification using correlation filters. In: Kittler, J., Nixon, M.S. (eds.) AVBPA 2003. LNCS, vol. 2688, pp. 697–705. Springer, Heidelberg (2003)
16. Lee, E.C., Park, K.R., Kim, J.: Fake iris detection by using purkinje image. In: Zhang, D., Jain, A.K. (eds.) ICB 2005. LNCS, vol. 3832, pp. 397–403. Springer, Heidelberg (2005)
17. Lim, S., Lee, K., Byeon, O., Kim, T.: Efficient iris recognition through improvement of feature vector and classifier. Electronics and Telecommunications Research Institute Journal 23(2), 61–70 (2001)
18. Liu, X., Bowyer, K.W., Flynn, P.J.: Experiments with an improved iris segmentation algorithm. In: Proceedings of the 4th IEEE Workshop on Automatic Identification Advanced Technologies (AUTO ID 2005), Buffalo, NY, USA, October 2005, pp. 118–123 (2005)
19. Liu, X., Bowyer, K.W., Flynn, P.J.: Experimental evaluation of iris recognition. In: Proceedings of the IEEE Computer Society Conference on Computer Vision and Pattern Recognition (CVPR 2005), San Diego, Calif, USA, June 2005, vol. 3, pp. 158–165 (2005)
20. Ma, L., Tan, T., Wang, Y., Zhang, D.: Personal identification based on iris texture analysis. IEEE Transactions on Pattern Analysis and Machine Intelligence 25(12), 1519–1533 (2003)

21. Daugman, J.: How Iris Recognition works. IEEE Transactions on Circuits and systems for video Technology 14(1), 21–30 (2004)
22. Ali, J.M.H., Hussanien, A.E.: An Iris Recognition System to Enhance E-security Environment Based on Wavelet Theory. AMO - Advanced Modeling and Optimization 5(2) (2003)
23. Do, M.N., Vetterli, M.: The contourlet transform: An Efficient directional multiresolution image representation. IEEE Transactions on Image Processing 14(12), 2091–2106
24. Burt, P.J., Adelson, E.H.: The Laplacian pyramid as a compact image code. IEEE Trans. Commun. 31(4), 532–540 (1983); Bamberger, R.H., Smith, M.J.T.: A filter bank for the directional decomposition of images: Theory and design. IEEE Trans. Signal Proc. 40(4), 882–893 (1992)
25. Hollingsworth, K.P., Bowyer, K.W.: The Best Bits in an Iris Code. IEEE Transactions on Pattern Analysis and Machine Intelligence (PAMI) (April 2008)
26. Haralick, R.M., Shanmugam, K., Din, L.: Textural features for image classification. IEEE Transactions on Systems, Man, and Cybernetics 3(6), 610–621 (1973)

What You See Is What You Set – The Position of Moving Objects

Heinz-Werner Priess and Ingrid Scharlau

Department of Cultural Sciences,
University of Paderborn,
Warburger Straße 100
33098 Paderborn, Germany

Abstract. Human observers consequently misjudge the position of moving objects towards the direction of motion. This so called flash-lag effect is supposed to be related to very basic processes such as processing latencies in the human brain. In our study we show that this effect can be inversed by changing the task-set of the observer. A top-down change of the observers attentional set leads to a different perception of otherwise identical scenes. Cognitive theories regard the misperception of the moving object as an important feature of attention-mediated processing, because it reflects the prioritized processing of important objects.

1 Where Do We See Moving Objects?

If human observers look at moving objects and judge their position at a specified point in time, they misjudge their positions towards the directory of motion. This effect is called flash-lag and known since 1931 [1]. Most theories agree that this effect is related to processing latency. The human visual system has a processing latency of about 80 ms [2]. We do not see the world as it is, but as it was about 80 ms ago. In a static world this would not be much of a problem, but in a dynamic world, objects can change positions during this time. If, for example, the tennis player Andy Roddick is to serve, he accelerates the ball up to 250 km/h [3]. In 80 ms, this ball travels 5.5 meters. Seeing the ball 5.5 meters displaced is no beginning for a successful return. A more up-to-date representation of the environment would enhance the chance to return the ball.

The same is true for technical systems. A football robot consequently misjudging the position of the ball with about 80 ms could only interact with very slowly moving footballs. Real-time representations of the outer world would enhance attempts to interact with the world. On the other hand, real-time processing is not possible, neither for robots nor for humans. How does the human visual system cope with a considerable slower processing speed? Attentional accounts answer: by having a clue what is important. If attention is deployed to a specified area in the visual field, objects inside this particular area are processed faster than objects outside this area [4]. According to the attentional view, the misperception of the moving object is not a bug but a feature.

B. Mertsching, M. Hund, and Z. Aziz (Eds.): KI 2009, LNAI 5803, pp. 123–127, 2009.

But how does the visual system know what is important? One object could be important in one situation but totally useless in another situation. If observers have the difficult task to count passes between random moving basketball players, the ball and the players are very important. A moonwalking bear, crossing the scene would be very unimportant although moonwalking bears could be assumed to be important. If observers watch the same scene without a particular task, the player would be less important and the bear comparatively more important. Studies show that human observers consequently miss salient objects when busy with another task, but have no problem in detecting them without a demanding task [5]. In the following study, we will show that the relative importance of objects sets the perceived position of moving objects at a specified point in time.

2 Experimental Setup

We use a standard flash-lag paradigm with a moving stimulus and a static stimulus. The static stimulus is used as a time marker. Our setup looks like a clock face with a seconds hand traveling on the outer rim of the clock (see figure 1). The time marker is an inner clock hand that can be seen on four different positions, 3 o'clock, 6 o'clock, 9 o'clock and 12 o'clock. We realized three conditions in which the position of the moving object was to judge. The seconds hand was always visible and moving smoothly with 25 rpm around the clock. The inner clock hand was also moving with 25 rpm but only visible on four positions. It started at the 12 o'clock position, jumped to the 3 o'clock, the 6 o'clock, the 9 o'clock and again to the starting position.

In condition 1, the inner clock hand was only visible for 13.3 ms. The observer's task was to adjust the position of the seconds hand in a manner that it was aligned to the 3 o'clock position, the moment the inner clock hand was visible. The onset of the inner hand (the time marker) triggers the onset of the comparison task. One cannot judge the relative position of two objects when only one is visible. So the position of the moving object is relatively unimportant while it cannot be compared to the time marker. The time marker has to be attended-to first and afterwards the moving object.

In condition 2two, the task stays the same but the setup changes a bit. The inner hand is always visible and the seconds hand gets a head start of ¼ revolution. The task is to adjust the position of the moving stimulus that both objects are aligned when the inner hand catches up with the outer hand. To solve the task, one has to wait until the inner hand jumps and compare the new position of the inner hand to the actual position of the outer hand. Again, the time marker has to be attended-to first and afterwards the moving object.

In condition 3, the inner hand gets a head start of ¼ revolution. Now the seconds hand has to catch up with the inner hand. The task is to adjust the position of the seconds hand so that the inner hand jumps to the next position the moment both objects are aligned. Priority changes in this condition. The inner clock hand is unimportant until the seconds hand reaches its position. In this condition the moving object has to be attended-to first and afterwards the time marker.

In all three conditions the task remains the same. Two objects have to be aligned in a single, specified moment. But human observers use different strategies to detect this moment. In the first two conditions the task requires first to attend to the inner hand

and later to the seconds hand. In the 3rd condition attention is first deployed to the seconds hand and later to the inner hand.

3 Results

Figure 1 shows the average results of five observers (one author and four naïve observers) at Paderborn University. The seconds hand was seen as displaced in direction of motion when the inner clock hand triggered the onset of the task. The seconds hand was seen as displaced contrary to the direction of motion when the onset of the task was triggered by itself. The setup and the average results are depicted in Figure 1.

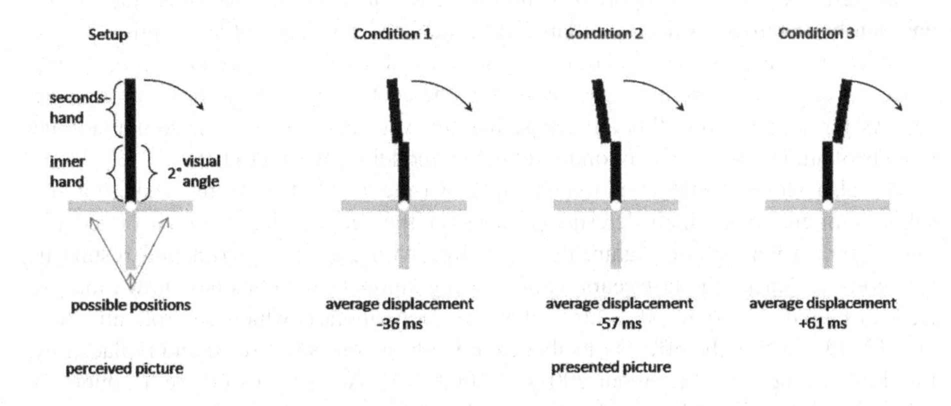

Fig. 1. In each condition the perception of the moving bars should look like the reference picture depicted under "setup". The actually presented picture did not match the subjective perceived picture. In conditions 1 and 2, the seconds hand was lagging behind the inner hand while perceived as aligned. In condition 3, the seconds hand was leading the inner hand while perceived as aligned.

These findings indicate sequential encoding of first the object that heralds the beginning of the task and second the reference object. Such a task-dependent, top-down sequence was predicted by the attentional explanation of the flash-lag effect. The point is: It is not the bottom-up properties of the presented objects which modulate the percept, but the task the observer performs.

The moving object could be seen either as leading or as lagging. Transferred to the tennis player, this would mean a more up-to-date representation of the ball if the task is returning the ball. The exact position of the opponent would be less important and could be processed later. The perception of the exact position of the tennis player has not been investigated yet. However, there are some studies on football players which are perfectly in line with our attentional explanation. If the linesman is to judge offside he has a bias towards judging offside even if there was no offside [6]. And indeed the lineman's task reminds of a flash-lag task. One has to judge the position of a moving object (the player) at a specified time (when the pass was shot). The onset of the lineman's task is triggered by the pass, so attention is first deployed to the ball and after the pass was recognized deployed to the player. Unfortunately (for the attacking team) the player has moved during this time and is being seen in an offside position after the ball was passed. Deploying attention first to the player (is he in an offside position?) and then to the ball carrier (is he passing?) should inverse the bias into less offside judgments even when the attacking player is in an offside position. Unfortunately (for the linesman) players are often in passive offside positions. In this case the linesman has to do nothing. So the strategy of first attending the player and then the ball carrier would result in much useless cognitive work.

To sum up: Our experimental results with human observers demonstrated that attention speeds up the perception of an attended object quite a bit (36 to 61 ms). Such latency effects have been reported with other tasks such as temporal-order judgments and can be regarded as a very reliable consequence of visuo-spatial attention [7, 8]. We thus can conclude that attention does not only allow for more detailed processing, object-level representations or even conscious perception, but also to faster processing [9]. As far as we know, this advantage has not yet been included in computational models of attention. An exception might be the model by Hamker [10].

We also showed that attention is top-down-mediated. It was the task, not the saliency of features which determined whether the seconds hand trailed or led the inner hand in perception. Again, this is in line with current experimental results in psychology. During the last years, an increasing amount of studies has shown that, at least in human observers, it is task-relevance, not salience, which controls attention [11, 12, 13], or that the effects of salience are at best very short-lived and replaced by top-down influences after about 200 to 300 ms [14]. According to these accounts, if salience controls attention, it does so not by default, but because salience is task-relevant in the present context. Again, this important finding has not been incorporated into computational models of attention. Although many of the current models include some top-down information, this influence is not as weighty and basic as in experimental psychology.

At first sight, attending to task-relevant objects seems to be a reasonable strategy for human observers. But what are the advantages in more detail? If processing capacity is limited, such a strategy could ensure the processing of relevant features, that is, features which are important for the current actions of tasks at hand. Another side effect of prioritization is shielding against interfering information. If for example a football robot is tracking an orange football in order to score a goal, this football has to be processed with priority. The orange t-shirt of an audience member would get less attention because it is not related to the task and although it might be equally salient) and would get less of a chance to interfere with the tracking of the football. We might also speculate that task-relevant objects very often are the objects to be acted upon – for example the object were are fixating, manipulating, tracking, grasping etc. Attention would thus serve a very important function in action control. This idea is most directly included in the premotor theory of attention [15]. In this theory covered attention shifts involve the mechanisms for saccade programming. Both, attention shifts and saccades utilize motor control circuits. All these findings an notions – attention is top-down controlled, attention is tightly coupled to the control of eye movements, attention is vital for current actions and tasks at hand – corroborate the belief that computational modeling of attention will take a major step if attention is implemented in autonomous systems.

Acknowledgments. Supported by DFG grant Scha 1515/1-1 to Ingrid Scharlau and Ulrich Ansorge.

References

1. Metzger, W.: Versuch einer gemeinsamen Theorie der Phänomene Fröhlichs und Hazelhoffs und Kritik ihrer Verfahren zur Messung der Empfindngszeit. Psychologische Forschung 16, 176–200 (1931)
2. De Valois, R.L., De Valois, K.K.: Vernier acuity with stationary moving gabors. Vision Res. 31, 1619–1626
3. Tennis Records and Statistics, `http://www.tennis-x.com/stats/tennisrecords.php`
4. Brown, J.M., Breitmeyer, B.G., Leighty, K., Denney, H.I.: The path of visual attention. Acta Psychologica 121, 199–209 (2006)
5. Most, S.B., Simons, D.J., Scholl, B.J., Jimenez, R., Clifford, E., Chabris, C.F.: How not to be seen: The contribution of similarity and selective ignoring to sustained inattentional blindness. Psych. Sci. 12(1), 9–17 (2001)
6. Gilis, B., Helsen, W., Weston, M.: Errors in judging "offside" in association football: test of the optical error versus the perceptual flash-lag hypothesis. Journal of Sport Sci. 24(5), 521–528 (2006)
7. Stelmach, L.B., Herdman, C.M.: Directed attention and perception of temporal order. Journal of Experimental Psychology: Human Perception and Performance 17, 539–550 (1991)
8. Scharlau, I.: Perceptual latency priming: A measure of attentional facilitation. Psychological Research 71, 678–686 (2007)
9. Kahneman, D., Treisman, A., Gibbs, B.J.: The Reviewing of Object Files: Object-Specific Integration of Information. Cognitive Psychology 24, 175–219 (1992)
10. Hamker, F.: The emergence of attention by population-based inference and its role in disturbed processing and cognitive control of vision. Computer Vision and Image Understanding 100, 64–106 (2005)
11. Ansorge, U., Kiss, M., Eimer, M.: Goal-driven attentional capture by invisible colours. Psychonomic Bulletin & Review (in press)
12. Ansorge, U., Neumann, O.: Intentions determine the effects of invisible metacontrast-masked primes: Evidence for top-down contingencies in a peripheral cueing task. Journal of Experimental Psychology: Human Perceptuin and Performance 31, 762–777 (2005)
13. Folk, C.L., Remington, R.W., Johnston, J.C.: Involuntary covert orienting is contingent on attentional control settings. Journal of Experimental Psychology: Human Perception and Performance 18, 1030–1044 (1992)
14. Donk, M., van Zoest, W.: Effects of Salience Are Short-Lived. Psych. Sci. 19(7), 733–739 (2008)
15. Rizzolatti, G., Riggio, L., Dascola, I., Umiltá, C.: Reorienting attention across the horicontal and vertical meridians: Evidence in favor of a premotor theory of attention. Neuropsychologica 25(1A), 31–40 (1987)

Parameter Evolution: A Design Pattern for Active Vision

Michael Müller, Dennis Senftleben, and Josef Pauli

Lehrstuhl Intelligente Systeme, Fakultät für Ingenieurwissenschaften,
Universität Duisburg-Essen, Duisburg, Germany
michael_mueller@uni-due.de

Abstract. In current robot applications the developer has to deal with changing environments, making a one-time calibration of algorithm parameters for the vision system impossible. A design pattern dealing with this problem thereby incorporating evolutionary strategies is presented and demonstrated on an example. The example shows that it is possible for the vision system to adjust its parameters automatically and to achieve results near to the optimum.

1 Introduction

Nowadays robots are utilized successfully in many areas of manufacturing and assembly. They relieve humans from monotonous or laborious tasks. At present new areas are subject to research, where robots are confronted with varying requirements and changing environmental conditions [1]. One example is the usage of a robot outdoor in contrast to an indoor-usage with safety zones and invariant light conditions. In this situation the vision system of the robot is confronted with changing conditions and it is no longer possible to adjust all parameters in advance. To overcome this problem, a design pattern is presented that summarizes the steps to build algorithms with automatically adapting parameters.

A common description of a pattern consists of a structured text. In the following the structure used by Buschmann et al. [2] is presented. First of all a pattern needs a *name* to refer to it. Then the problem to solve is explained with a small *example*. The *context* sums up the situation to which the pattern applies. Now the *problem* is explained in more detail. For this purpose the idea of conflicting forces that has to be balanced is used. Then the *solution* is stated, containing structural and dynamical aspects. An *implementation* is given as a step-by-step guide. The example from the beginning is implemented, forming an *exemplary solution*. Projects that uses the described pattern are presented as *known implementations*. Finally, advantages and disadvantages resulting from the pattern are discussed as *effects*.

2 Pattern Description: Parameter Evolution

Example: The following example is a small subtask of a robot vision system. Imagine a scenario, where a robot should search for an object to use it in another

B. Mertsching, M. Hund, and Z. Aziz (Eds.): KI 2009, LNAI 5803, pp. 128–135, 2009.

subtask. Here the robot's camera captures images showing the object, e.g. a key. A simple algorithm should separate the object in the foreground from the brighter background in two steps. First, noise is eliminated or weakened using a Gaussian filter with a parameter σ to describe the intensity of blurring. Second, the foreground/background-separation is done via a threshold operator with a parameter θ. So two parameters have to be adjusted to achieve good separation results. This is complicated by different light conditions and bright reflections on the dark object.

Context: An algorithm should process data that stems from a changing environment. The parameters of this algorithm cannot be calibrated beforehand.

Problem Description: A robot uses its sensors to gather information about its environment. Missing information is complemented with a set of assumptions. Assembling these assumptions and adjusting the robot's parameters accordingly is often called *calibration*. Hereby the robot is customized to one fixed environment. If the environment changes, the gathered information gets dissonant. A robot using a dissonant model of its surrounding world will show erratic behavior. Showing well defined behavior in a changing environment means to recalibrate the robot on-line, thereby keeping its set of assumptions up-to-date. This continuous recalibration has to be accomplished automatically to make the robot autonomous, i.e. independent from sustained human support.

The assumptions about the environment are encoded as algorithm parameters. Often the relationship between the assumptions and the algorithm parameters is not evident and has to be determined by laborious experimentation.

If the algorithm parameters are not perfectly calibrated, the quality of the results achieved will change. A quality measure would be advantageous to check the reliability of a result.

Solution: In Fig. 1(a) a control cycle performing a continuous recalibration is shown that enables a robot to keep its set of assumptions up-to-date. The *sensor* supplies the system with *data* (e.g. an image acquired by a camera). An *algorithm* analyzes this data in conjunction with some *algorithm parameters* (a-parameters) and produces a *result* (e.g. a separation between a foreground object and the background). The result can be *valuated* against some constraints given by high-level knowledge (e.g. abstract properties of the expected object), encoded in *valuation parameters* (v-parameters). The valuation parameters encode assumptions about the result of the algorithm and therefore do not depend on the current situation in the environment. A *quality* is assigned to the result by the valuation.

Some symbols in Fig. 1(a) are shown stacked to represent multiple or parallel execution. So on one set of data the algorithm can be used with different sets of algorithm parameters (e.g. one for sunlight and one for artificial light). This will yield different results, which in turn can be valuated each. The set of results and qualities are then *fused* into one final result with attached quality. For data fusion there are several possibilities. The simplest is to choose the result with the

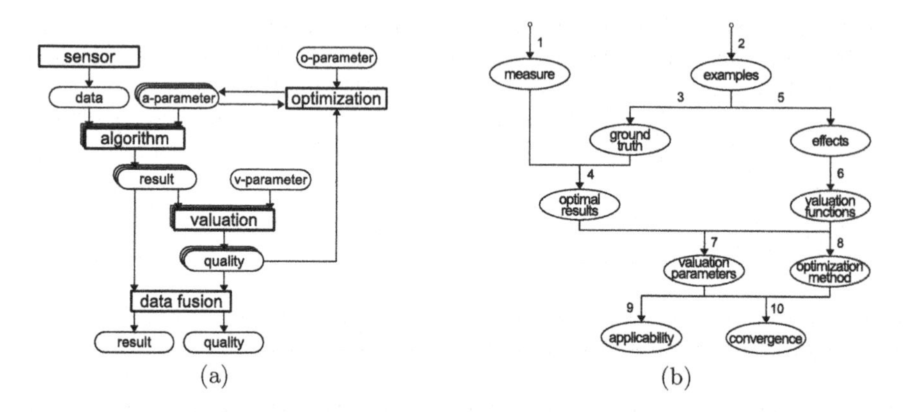

Fig. 1. (a) Data flow diagram of a control cycle for continuous recalibration of algorithm parameters. (b) Pattern implementation in ten steps.

highest quality. More complex solutions calculate the average of nearby results with high quality.

The control cycle is closed by an *optimization* that uses the quality gained by the valuation to optimize the current set of algorithm parameters. Also for the optimization several possibilities can be chosen, what is mainly affected by the shape of the quality function. Because of their formation, quality functions tend to have a quite complex shape that can be handled by *evolutionary strategies* (ES) described by Rechenberg [3] and Schwefel [4]. The *optimization parameters* (o-parameter) are farther away from the changing environment than the algorithm parameters and therefore are more likely to be situation independent. Moreover, some optimization methods, like ES, are able to optimize their own parameters.

Implementation: In Fig. 1(b) an overview of the implementation of this pattern in ten steps is given. The order of these steps is not fixed, but have to follow the depicted flow.

1. Define a measure to compare the result of the algorithm to a given ground truth.
2. Select a set of exemplary data.
3. Manually create the ground truth for the set of exemplary data. If the ground truth cannot be determined directly, in some cases it is possible to find a good estimation (see e.g. [5]).
4. For each element of the exemplary data, do an exhaustive search (with a given discretization) for the combination of algorithm parameters producing a result that fits the ground truth best. For comparison use the measure defined in Step 1. Store all results in an intermediary data structure.
5. Find out, how changing the algorithm parameters qualitatively influences the result. Distinguish between desirable and non-desirable effects.

6. Create valuation functions that take into account the found effects from Step 5. Select a method to combine the results of several valuation functions.
7. If some valuation functions requires parametrization, use the results from Step 4 as a basis for the valuation parameters.
8. For each element of the exemplary data, go through all combinations of parameters and store the combined value of all valuation functions in an adequate data structure. Combining the results of several simple valuation functions will yield a combined valuation function with a shape that is likely to be complex. Use the combined valuation function to decide on an optimization method (e.g. see Bäck [6] for a small overview). In the majority of cases an ES can be found that is able to cope with the complex shape of the found function.
9. Use the results from Step 4 and Step 7 to decide, if the method is applicable and if results based on the combined valuation function are comparable to the ground truth.
10. Investigate the convergence of the chosen optimization method (Step 7) based on the set of exemplary data.

Known Implementations: Liedtke et al. [7] introduce a system that optimizes algorithm parameters for a vision system with a set of fuzzy rules instead of valuation functions and ES. A full description is given by Pauli et al. in [8,9]. Exemplary three different tasks are solved: recognizing position and orientation of industrial objects, segmentation of cell images, and inspection of chip bonding.

For this system an expert is needed to create the necessary set of fuzzy rules. However, for many algorithms the relationship between changes of parameters and effects in the result is not obvious and therefore adequate rules cannot easily be determined. Instead, describing attributes of the desired result by means of valuation functions frees the user from figuring out these relationships. Then each individual or sub-population in a given ES will try to discover relationships on a case-by-case basis, and also those relationships can be used that are valid infrequently or are unknown to the user.

Effect: An advantage of this pattern is that it allows a vision system to work in different known environments without manual recalibration of its algorithm parameters. In moderation the system will also be able to cope with unknown environments. Another advantage is that the assumptions about the changing environment no longer have to be translated in algorithm parameters manually. Instead non-changing high-level knowledge can be encoded in valuation parameters as assumptions about the result, automatically translating these requirements into algorithm parameters by the optimization process. Finally, a quality measure is created, allowing to check the reliability of each produced result.

But there are also some disadvantages. First, optimizing the parameters needs processing time, dependent on the chosen optimization method. Second, in a fixed and known environment, the optimization process, guided by several valuation functions, will not give perfect results inevitably, but instead only good ones. Finally, the pattern can only be used, if it is possible to find adequate valuation functions.

3 Exemplary Implementation

Now the proposed pattern is implemented for the example given in the beginning of the description.

Step 1 – Defining a Measure: To assess the quality of the produced results, a comparison to a manually generated foreground/background-separation is performed. Each pixel in the result is considered to be correctly detected foreground (*true positive*, TP), wrongly detected foreground (*false positive*, FP), wrongly detected background (*false negative*, FN) or correctly detected background (*true negative*, TN). From three of these four values the recall ratio ($rec = \frac{TP}{TP+FN}$) and the precision ratio ($pre = \frac{TP}{TP+FP}$) can be calculated. Rijsbergen describes, how both measures can be combined into a F-measure [10]: $F_2 = \frac{2 \cdot pre \cdot rec}{pre+rec}$.

Step 2 – Selecting Exemplary Data: The exemplary data consists of 16 images with three varying aspects. Four keys are used with different sizes and shapes. Then two different light conditions and two different backgrounds (a uniform and a textured one) are used.

Step 3 – Manual Creation of Ground Truth: The ground truth was determined manually and stored as black/white-mask images.

Step 4 – Determining Best Combination of Parameters: The algorithm uses two parameters: σ describes the intensity of Gaussian blur to remove or weaken noise. θ sets the threshold for foreground/background-separation. The parameter σ was changed in an interval of $[0.8, 5.39]$ (corresponding to a Gaussian operator mask of 3×3 to 32×32 pixels size) The parameter θ was changed in an interval of $[0, 255]$. For both parameters a discretization of 256 steps was used. Each achieved result was compared to the ground truth from Step 3 with F_2 defined in Step 1. In Fig. 2(a), first row, the smooth quality function is shown with marked optimum area.

Step 5 – Observing Effects: In Fig. 2(b) the effects of changes to the parameters σ and θ are visualized:

1. If σ is too low, the key segment will be jagged.
2. If σ is too high, the key segment will be abraded.
3. If θ is too low, the key segment will be scattered.
4. If θ is too high, the key segment will be too large and merge with the background.

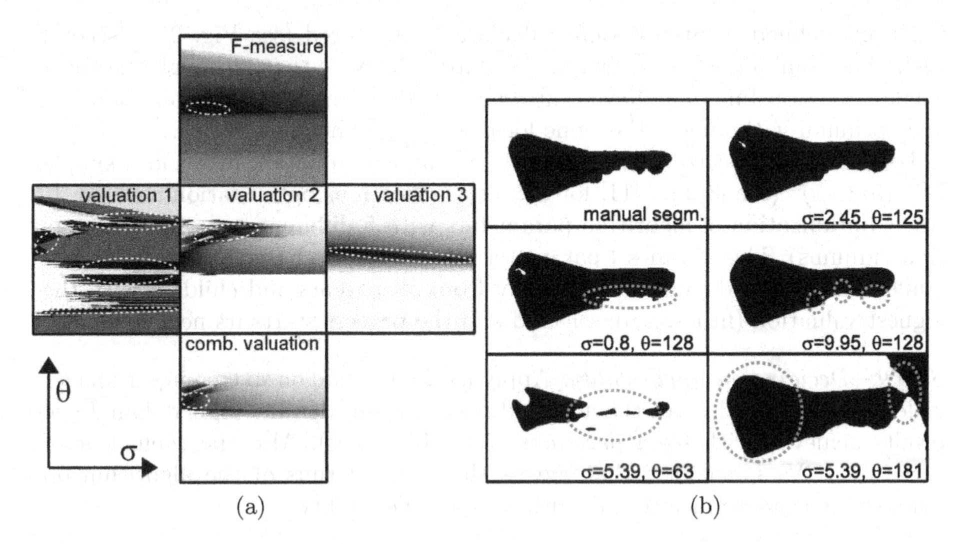

Fig. 2. (a) Quality functions. First row: F_2 for all combinations of parameters; second row: valuation functions 1-3 (no. of segments, length of boundary, area of segment); third row: combined valuation function. (b) Effects of changes to σ and θ parameter. First row: ground truth, best segmentation producible with algorithm ($F_2 = 86\%$); second row: σ is too low, too high; third row: θ is too low, too high.

Step 6 – Creating Valuation Functions: Based on Step 5 three simple valuation functions can be defined: First, the number of found segments should be small. Second, the boundary length of the largest segment should be in a range to be defined. Finally, the area of the largest segment should be in a range to be defined. This set of valuation functions is quite simple, but can be calculated easily and fast. As a model a Gaussian distribution for each valuation can be used with its mean and standard deviation as parameters (see Step 7).

To calculate the combination of several valuation functions it was chosen to use the product-t-norm. It helps to avoid passing of discontinuities from valuation functions, so that optimization can make use of a continuous combined valuation function.

Step 7 – Setting Up Valuation Parameters: Based on Step 4 the following three pairs of mean value and standard deviation can be extracted that are needed as parameters for the valuation functions. The average number of found segments is 1.55 with a standard deviation of 0.79. The average length of the key segment is 871.72 with a standard deviation of 129.35. The average area of the key segment is 23353.78 with a standard deviation of 4588.47. Note that these numbers form a model of the desired result of the algorithm, independent from the given input.

Step 8 – Choosing Optimization Method: Again, for σ values in the interval $[0.8, 5.39]$ and for θ values in the interval of $[0, 255]$ were chosen, both with a discretization of 256 steps. For each achieved result the values of the valuation functions defined in Step 6 were calculated and stored (see Fig. 2(a), second row). The combination is shown in the third row. Note that although the three single valuation functions have scattered optimum areas, the combination has one optimum with almost the same location as the one given by F_2.

Because of the scattered shape of the combined valuation function a simple ES "(5/1+5)" (see [3] and [11] for details) is chosen as optimization method. It starts optimization of algorithm parameters with 5 different sets of parameters (individuums). Then 5 times 1 parameter set (parent) is chosen randomly, copied (child) and changed (mutated). Finally from all parents and childs 5 with the highest valuation (fitness) are selected and the process starts its next iteration.

Step 9 – Deciding on Applicability: Applying the method on sixteen input images yields that averaged F_2 of the achieved result is considerably higher than F_2 of results achieved with fixed parameter sets (Fig. 3(a)). Also the coefficient of variation (CV) is smaller; that means that several runs of the algorithm on different images produce results with comparable quality.

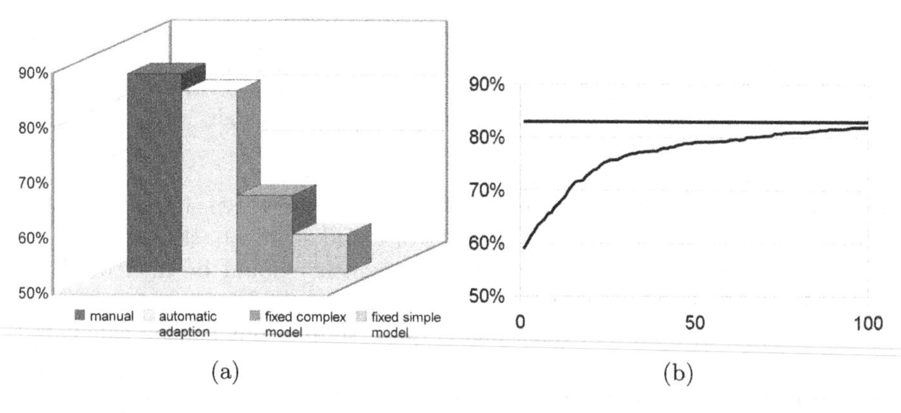

Fig. 3. (a) Comparison of F_2 for manual parameter adaption, automatic parameter adaption, fixed parameters based on several sample images, and fixed parameters based on one sample image. (b) Mean F_2 averaged over 16 images and 20 repetitions, dependent on the number of generations. The optimum of 83% is also shown for comparison.

The highest achievable F_2 for the given algorithm is 86% with a CV of 11%, incorporating an exhaustive search and a known ground truth. An automatic adaption based on valuation functions almost reaches comparable values: F_2 is 83% with a CV of 14%. Using a fixed set of parameters averaged from sixteen sample images yields a much lower F_2 of only 64% with a higher CV of 39%. Using a fixed set of parameters based on only one sample image (averaged over sixteen different sample images) yields a still lower F_2 of 57% with a high CV of 49%.

Step 10 – Investigation of Convergence: In Fig. 3(b) an overview of the convergence of the chosen optimization method is given, averaged over 16 images and

20 repetitions. After about 50 generations, this means 255 applications of the image processing algorithm for ES "(5/1+5)", the optimization method reaches a value near to the optimum. Whereas the CV has an averaged high value of 58% in the first generation, it is getting lower fast and finally reaches a value of 14%.

Note that the initial low F_2 of 59% is at the same level as F_2 of a fixed set of parameters based on one sample image (57%, see above). The automatic adaption lifts this value after some iterations to a level near to the one achieved by an exhaustive search and a known ground truth (79% compared to 86%).

50 generations or 255 applications of the image processing algorithm are acceptable numbers. Assuming that the camera takes 25 images per second and the processing system is capable to work on four parameter sets in parallel, then after only 2.55 seconds F_2 reaches a value near to the optimum.

4 Conclusion

A pattern using a control cycle to continuously recalibrate the parameters of an algorithm has been presented. The recalibration was directed by a set of simple valuation functions that represent situation independent high-level knowledge. The proposed pattern is intended to contribute to a pattern language for cognitive robotic systems.

References

1. Brugali, D.: Stable analysis patterns for robot mobility. In: Brugali, D. (ed.) Software Engineering for Experimental Robotics, vol. 30, pp. 9–30. Springer, Berlin (2007)
2. Buschmann, F., Meunier, R., Rohnert, H., Sommerlad, P.: Pattern-Oriented Software Architecture: A System of Patterns, vol. 1. Wiley&Sons, Hoboken (1996)
3. Rechenberg, I.: Evolutionsstrategie 1994. Frommann-Holzboog, Stuttgart (1994)
4. Schwefel, H.P.: Evolution and Optimum Seeking. Wiley, New York (1995)
5. Müller, M., Keimling, R., Lang, S., Pauli, J., Dahmen, U., Dirsch, O.: Estimating Blood Flow Velocity in Liver Vessels. In: Meinzer, H.P., Deserno, T., Handels, H., Tolxdorff, T. (eds.) Bildverarbeitung für die Medizin. Informatik aktuell, pp. 36–40. Springer, Heidelberg (2009)
6. Bäck, T.: Evolutionary Algorithms in Theory and Practise. Oxford University Press, New York (1996)
7. Liedtke, C.E., Blömer, A., Gahm, T.: Knowledge based configuration of image segmentation processes. IJIST 2, 285–295 (1990)
8. Pauli, J., Blömer, A., Liedtke, C.E., Radig, B.: Zielorientierte Integration und Adaption von Bildanalyseprozessen. In: KI, vol. 3, pp. 30–34 (1995)
9. Pauli, J., Radig, B., Blömer, A., Liedtke, C.E.: Integrierte, adaptive Bildanalyse. Technical Report I9204, TU München, Institut für Informatik (1992)
10. van Rijsbergen, C.J.: Information Retrieval, 2nd edn. Butterworths, London (1979)
11. Arnold, D.V.: Noisy Optimization with Evolution Strategies. Kluwer Academic Publishers, Norwell (2002)

Clustering Objects from Multiple Collections*

Vera Hollink**, Maarten van Someren, and Viktor de Boer

University of Amsterdam

Abstract. Clustering methods cluster objects on the basis of a similarity measure between the objects. In clustering tasks where the objects come from more than one collection often part of the similarity results from features that are related to the collections rather than features that are relevant for the clustering task. For example, when clustering pages from various web sites by topic, pages from the same web site often contain similar terms. The collection-related part of the similarity hinders clustering as it causes the creation of clusters that correspond to collections instead of topics. In this paper we present two methods to restrict clustering to the part of the similarity that is not associated with membership of a collection. Both methods can be used on top of standard clustering methods. Experiments on data sets with objects from multiple collections show that our methods result in better clusters than methods that do not take collection information into account.

1 Introduction

In many clustering applications we have an a priori idea about the types of clusters we are looking for. For example, in document clustering tasks we often want clusters that correspond to the documents' topics. Even though the cluster type is known, clustering is still unsupervised, as we do not know in advance *which* topics are present in the data. Clustering algorithms form clusters of objects that are similar to each other in terms of some measure of similarity. For homogeneous sets of objects standard similarity or distance measures usually lead to satisfying results. For instance, cosine distance applied to word vectors is suitable for finding topic clusters in a news archive or web site (e.g. [1,2]).

When the data comes from multiple collections, often the collections do not coincide with the type of clusters that we want to find. In this situation we know in advance that features that are related to the collections are not relevant for our clustering task. The part of the similarity which is associated with these features can hinder clustering as it causes clustering algorithms to group objects primarily by collection. For example, this problem occurs when we want to cluster pages from a number of web sites by topic using a word-based similarity measure. In terms of such a measure pages from the same web sites are usually more similar to each other than to pages from other web sites, because pages from one web

* This research is part of the project 'Adaptive generation of workflow models for human-computer interaction' (project MMI06101) funded by SenterNovem.

** Currently at Centre for Mathematics and Computer Science (CWI).

B. Mertsching, M. Hund, and Z. Aziz (Eds.): KI 2009, LNAI 5803, pp. 136–143, 2009.

site share a common terminology and often contain the same names. As a result, we get clusters of pages from the same site instead of clusters of pages on the same topic. Similarity, with image clustering, images can become clustered by illumination or background instead by the things that they depict.

The task of clustering objects from multiple collections can be formalized as follows. There is a set of objects $\mathcal{I}$ and a set of collections $\mathcal{R}$. The collection of each object is defined by a special feature $col : \mathcal{I} \rightarrow \mathcal{R}$. The value of *col* influences the value of an unknown part of the other features. There is a similarity function $sim : \mathcal{I} \times \mathcal{I} \rightarrow \mathbb{R}$, which comprises both relevant similarity and similarity associated with the value of *col*. The task is to divide $\mathcal{I}$ in a set of clusters that represent only relevant similarity. In this paper we do not consider overlapping or hierarchical clusters, but the proposed methods can be applied without modification to algorithms that create these types of clusters.

An intuitive solution for the problems caused by collections is to use only features that are relevant for the clustering task. Selecting the right features directly requires much effort and a deep understanding of the domain. E.g., specialized image analysis techniques are needed to compensate for illumination, background, etc. In this paper we provide two simple and generally applicable methods to deal with collection-related similarity. The methods do not require any domain knowledge, but only need to know which objects are from which collections. For a user, this is easy to determine, as the collections represent some obviously irrelevant feature such as where or by who the objects were created. The proposed methods can be used as an addition to standard clustering algorithms.

2 Related Work

Li and Liu [3] present a method for binary text classification on data sets where the training and the test set come from different web sites and thus have different term distributions. To compensate for this they add irrelevant documents as extra negative examples. This task is related to ours, but it is fully supervised.

Huang and Mitchell [4] address document clustering. They use an extension of EM to determine which terms are cluster-specific and which terms are shared by all clusters. This task differs from the one addressed in this work, in that they reinforce features of clusters that are found automatically instead of features of user-defined collections. If their method initially groups objects by collection, in later iterations it will assign even higher weights to collection-specific features.

Bronstein et al. [5] defined the concept of 'partial similarity'. Two images are partially similar when parts of the objects are similar while other parts differ, e.g. a picture of a horse and a picture of a centaur. The authors developed a method to identify a subset of the object features that correspond to the similar parts of the images. Unlike partial similarity the methods that we present are also applicable in situations in which there is no clear distinction between irrelevant and relevant features. This happens, for example, if we have a set of pictures taken with different lightings. Illumination influences the whole pictures and thus can not be compensated for by focussing on a part of the pixels.

3 Two Methods to Deal with Multiple Collections

3.1 The Omission Method

Similarity-based clustering algorithms compute the similarities between objects from the object features. During clustering these algorithms use only the similarities and not the raw features. The quality of a potential clustering $\mathcal{C}$ is computed as a function f of the similarities between objects (the objective function):

$$quality(\mathcal{C}) = f(\mathcal{C}, \{sim(i,j) | i,j \in \mathcal{I}\}) \tag{1}$$

For example, the average linkage algorithm [6] expresses the quality of a cluster as the average similarity between all pairs of objects in the cluster.

The omission method computes the same function, but omits the similarity of pairs of objects from the same collection, because these similarities often consist for a large part of collection-related similarity:

$$quality(\mathcal{C}) = f(\mathcal{C}, \{sim(i,j) | i,j \in \mathcal{I} \wedge col(i) \neq col(j)\}) \tag{2}$$

3.2 The Estimation Method

The estimation method estimates which part of the similarity between objects is relevant similarity and which part is caused by collections. The relevant part of the similarities is given to the clustering algorithm to cluster the objects.

We decompose the similarity between any two objects a and b into relevant similarity, $sim_{rel}(a,b)$, and collection-related similarity, $sim_{col}(a,b)$:

$$sim(a,b) = sim_{rel}(a,b) + sim_{col}(a,b) \tag{3}$$

We estimate $sim_{col}(a,b)$ as the average collection-related similarity between all pairs of objects in the collections of a and b. Therefore, the collection-related similarity between objects a and b from collections A and B can be expressed as the similarity between their collections, $sim_{col}(A,B)$:

$$\forall a \in A, b \in B : sim_{col}(a,b) = Avg_{\{a' \in A, b' \in B\}} sim_{col}(a',b') = sim_{col}(A,B) \tag{4}$$

When the target clusters (the clusters that we are trying to find) are independent of the collections, the target clusters will be more or less evenly spread over the collections. As a result, the average relevant similarity between objects from a pair of collections will be roughly the same for all pairs of collections. Consequently, we can estimate the average relevant similarity between objects from all pairs of collections as a constant φ:

$$\forall A, B \in \mathcal{R} : Avg_{\{a \in A, b \in B\}} sim_{rel}(a,b) = \varphi \tag{5}$$

The average similarity between the objects in a pair of collections A and B can now be expressed as:

$$\begin{aligned} Avg_{\{a \in A, b \in B\}} sim(a,b) &= Avg_{\{a \in A, b \in B\}} (sim_{rel}(a,b) + sim_{col}(a,b)) \\ &= Avg_{\{a \in A, b \in B\}} (sim_{rel}(a,b) + sim_{col}(A,B)) \\ &= \varphi + sim_{col}(A,B) \end{aligned} \tag{6}$$

The value of $Avg_{\{a \in A, b \in B\}} sim(a, b)$ can be computed directly. The correct value of φ is unknown. We define the collection-related similarity between the two most dissimilar collections to be 0 (other values lead to different absolute values for the relevant similarities, but do not change the differences between the relevant similarities). Now we can compute φ as:

$$\varphi = minimum_{\{A,B \in \mathcal{R}\}} Avg_{\{a \in A, b \in B\}} sim(a, b) \quad (7)$$

We use φ in Equation 6 to compute for all pairs of collections A and B the collection-related similarity, $sim_{col}(A, B)$. From this we can estimate the relevant similarity between the objects using Equation 3.

3.3 Applicability

Both the omission and the estimation method can be applied to all similarity-based clustering algorithms. Examples of such algorithms are given in the Sect. 4.

The more objects belong to the same collection, the more similarities are omitted by the omission method. This reduces the amount of data, which may lead to less accurate clusters. The estimation method bases its estimation of the collection-related similarity of a pair of collections on the average similarity between the objects in the collections. If the collections contain very few objects, the estimations become uncertain. Therefore, we expect that omission is the best choice for data sets with a large number of small collections and estimation is best for data sets with a small number of large collections. In Sect. 4 we test the effects of the number of collections in experiments.

The computational complexity of both methods is small, so that they scale very well to large clustering tasks. The extra complexity that the omission method adds to a clustering method is negligible. The only action it introduces is checking whether two objects are from the same collection. The estimation method does not change a clustering algorithm, but requires computing the relevant similarities. This can be done in two passes through the similarity matrix, so that the time complexity is $O((n-1)^2)$, where n is the number of objects.

4 Evaluation

We applied the omission and the estimation method to three commonly used clustering algorithms. The first algorithm is K-means [7], a partitional algorithm. We used a similarity-based version of K-means, which uses the average similarity between all objects in a cluster instead of the distance between objects and the cluster centre. K-means was run with 100 random initializations. The second algorithm was an agglomerative clustering algorithm: average linkage clustering [6]. This algorithm starts with each object in its own cluster and merges the two clusters with the smallest average similarity until the desired number of clusters is reached. The third algorithm was bisecting K-means [1], a divisive algorithm that starts with all objects in one cluster. In each step the cluster

Table 1. Properties of the data sets: the type of objects, the type of sets that form the collections, the type of clusters we want to find, the number of objects, the number of clusters in the gold standard and the number of collections

Data set	Object type	Col. type	Target cluster type	#Objects	#Clus	#Cols
hotel	web page	web site	topic	52	7	5
conference	web page	web site	topic	56	13	5
surf	web page	web site	topic	84	12	5
school	web page	web site	topic	105	17	5
no-flash	picture	location	item in picture	108	18	3
flash	picture	location	item in picture	108	18	6
artificial	-	-	-	400	40	1 to 10

with the largest number of objects is split into two. To split the clusters we used the similarity-based version of K-means with 20 random initializations.

The algorithms were evaluated on two types of real world data: web pages and pictures. The data sets can be downloaded from http://homepages.cwi.nl/~vera/-clustering_data. In addition, we tested the influence of various characteristics of the data using artificial data. Table 1 shows the main properties of the data sets.

For the web site data sets the task of the clustering algorithms was to find clusters of web pages about the same topic in a number of comparable web sites. We used 4 data sets with sites from different domains: windsurf clubs, schools, small hotels and computer science conferences. Each data set consisted of the pages from 5 sites (collections). We manually constructed for each domain a gold standard: a set of page clusters that corresponded to topics (e.g. *pages listing important dates*). The gold standard clusters were not evenly spread over the sites: for many topics some of the sites contained multiple pages or zero pages. We used a standard similarity measure: the cosine of word frequency vectors [8].

The image data sets consisted of pictures of 18 small items, such as toys and fruit, taken at three different backgrounds (locations). The task was to find clusters of pictures of the same item. The 'no-flash' data set contained for each item two pictures per location taken without flash. All pictures taken at one location formed one collection. The 'flash' data set contained for each item and location one picture that was taken with flash and one that was taken without flash. Each combination of lighting and background formed a collection. We used a simple pixel-based similarity function that compared the RGB-values of the pixels that were at the same position in the pictures. The similarity between two pictures was computed as 1/total_RGB_difference. Of course, more advanced image-comparison techniques exist, but the goal of this paper is not to provide an optimal solution for clustering images.

For the artificial data sets we created 400 objects that were divided evenly over a number of collections (1 to 10). We created 40 gold standard clusters, in such a way that each cluster contained objects from all collections. The similarities between the objects were drawn from a normal distribution with a mean of 0.05 and a standard deviation of 0.075. When two objects were from the same cluster, we added 0.15 to the similarity, representing relevant similarity. For objects from the same collection we added a certain amount of collection-related similarity. We called this amount ρ and experimented with various values for ρ. The mentioned values were chosen in such a way that they resembled the values that were found in the school data set. For each parameter setting 10 data sets were created. All reported numbers are averages over the 10 data sets.

4.1 Results

The standard and the enhanced clustering algorithms were applied to the data sets. The number of clusters was set equal to the actual number of clusters in the gold standards. For each cluster we counted the number of different collections from which the objects in the cluster originated. The standard clustering algorithms created clusters with very small numbers of collections: on average 1.0 to 2.3, where the gold standards had 4.1 to 6.0 collections per cluster. This confirms our hypothesis that standard algorithms tend to form clusters of objects from one collection. The enhanced methods frequently created clusters that spanned multiple collections. On average, the omission and the estimation method increased the number of collections per cluster by respectively 54% and 58%.

We evaluated the clusters through comparison with the gold standards. We counted how many of the created clusters also occurred in the gold standards and vice versa. A created cluster was considered to be the same as a cluster from the gold standard if more than 50% of the objects in the clusters were the same. The overlap of a set of clusters with a gold standard is expressed by F-measure [8].

Table 2 shows the F-measure of the standard and the enhanced clustering algorithms. On average, the omission method increased the F-measure by 93%. The estimation improved the average with 87%. The omission method gave excellent results with K-means, but performed less well with average linkage and bisecting K-means. A possible explanation for this is that divisive and agglomerative clustering algorithms are sensitive to incorrect choices that are made in the early steps of the clustering process. Omitting similarities may remove information that is essential at this stage. The estimation method led to fairly large improvements with all three clustering methods.

To test the effects of the number of collections, we generated data sets with 1 to 10 collections. ρ was 0.1. Fig. 1a shows the F-measure of the standard

Table 2. F-measure of the clusters created by the standard algorithms (std.), with the omission method (omi.) and with the estimation method (est.). Best scores are bold.

Data set	K-means			Average linkage			Bisecting K-means		
	std.	omi.	est.	std.	omi.	est.	std.	omi.	est.
hotel	0.14	**0.43**	0.29	0.00	**0.29**	0.14	0.14	**0.18**	0.14
conference	0.15	**0.61**	0.31	0.00	0.08	**0.15**	0.00	**0.15**	**0.15**
surf	0.00	**0.42**	0.00	0.00	0.00	0.00	**0.25**	0.12	**0.25**
school	0.20	**0.41**	0.38	0.12	0.08	**0.18**	0.06	**0.15**	0.12
no-flash	0.25	**0.47**	0.33	0.11	0.00	**0.13**	0.22	0.00	**0.50**
flash	0.11	0.11	**0.13**	0.06	0.06	**0.11**	0.06	0.07	**0.17**

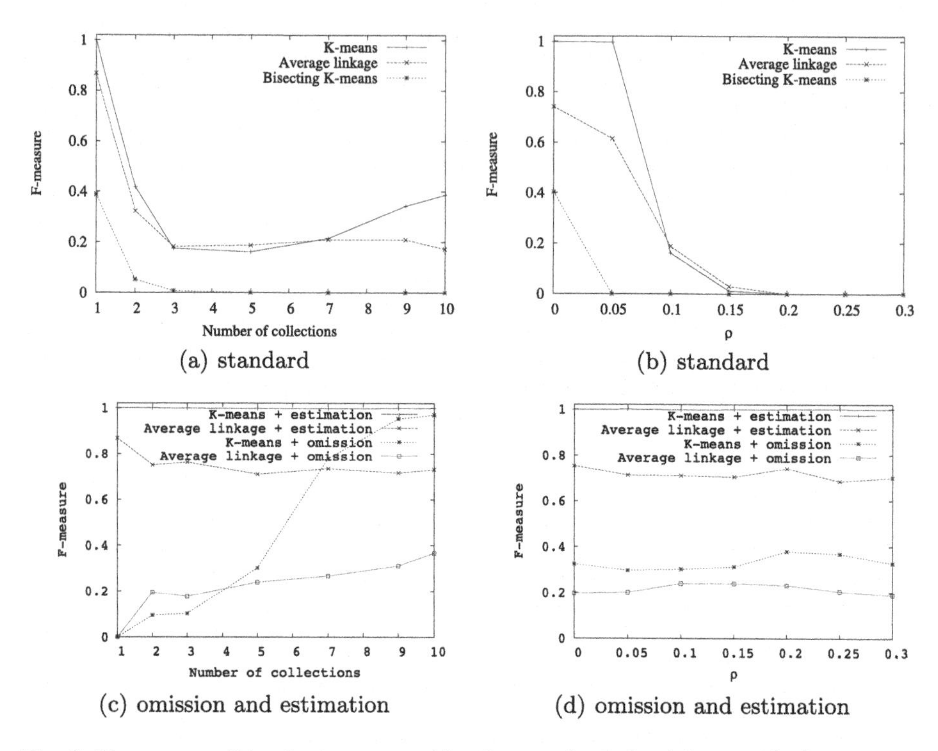

Fig. 1. F-measure of the clusters created by the standard algorithms, with the omission method and with the estimation method on artificial data sets with various numbers of collections and various amounts of collection-related similarity (ρ).

algorithms. When the objects came from several collections, the algorithms could not distinguish between collection-related similarity and relevant similarity and often produced clusters that coincided with collections. The results of the enhanced methods are shown in Fig. 1c (results of bisecting K-means are similar). In Sect. 3 we postulated the hypothesis that omission works best if we have a large number of collections. Fig. 1c shows that this is indeed the case. As expected, the performance of the estimation method deteriorated with increasing numbers of collections. However, the effect is very small. Apparently, with 10 collections, there was still enough data to accurately estimate collection-related similarity.

Next, we tested the influence of the strength of the collection-related similarity by varying the value of parameter ρ. The data sets in these experiments had 5 collections. From Fig. 1b we can see that for the standard algorithms more collection-related similarity led to lower F-measures. Fig. 1d shows that with the omission and the estimation method the performance of the algorithms was stable even when there was a large amount of collection-related similarity.

In sum, these experiments show that commonly used similarity measures do not lead to satisfying results when the data comes from multiple collections. In this situation using collection information improves clustering.

5 Conclusions

In many clustering tasks the data come from more than one collection. In this paper we showed that if we want to find clusters that span multiple collections, standard similarity measures and clustering methods are not adequate.

We provided two methods to suppress similarity associated with collections: omission, a modification to clustering algorithms and estimation, a modification to similarity measures. The methods do not require any domain-specific knowledge and can be applied on top of all clustering algorithms that use similarities between objects. Our experiments show that clustering methods enhanced with one of these methods can effectively find clusters in data from multiple collections. Compared to the standard clustering algorithms, the enhanced methods created clusters that spanned more collections and were more similar to gold standard clusters created by humans. On average, the omission and the estimation method increased the number of collections per cluster by respectively 54% and 58%. The average overlap with the gold standards was increased by 93% and 87%. Both the omission and the estimation method proved effective, but on artificial data and most of the real world data sets estimation outperformed omission. This shows that the more fined-grained analysis of the estimation method is effective despite the additional assumptions that underlie this analysis.

Future work includes generalizing our approach to situations where there is more than one type of collection. In these cases, we have multiple collection labels that all correspond to collection-related similarity. Most likely, improvements can be made by compensating the collection-related similarity of each collection type separately, instead of treating each combination of collections as one collection.

References

1. Steinbach, M., Karypis, G., Kumar, V.: A comparison of document clustering techniques. In: KDD Workshop on Text Mining, Boston, MA (2000)
2. Sahoo, N., Callan, J., Krishnan, R., Duncan, G., Padman, R.: Incremental hierarchical clustering of text documents. In: CIKM 2006, Arlington, VA, pp. 357–366 (2006)
3. Li, X., Liu, B.: Learning from positive and unlabeled examples with different data distributions. In: Gama, J., Camacho, R., Brazdil, P.B., Jorge, A.M., Torgo, L. (eds.) ECML 2005. LNCS (LNAI), vol. 3720, pp. 218–229. Springer, Heidelberg (2005)
4. Huang, Y., Mitchell, T.M.: Text clustering with extended user feedback. In: SIGIR 2006, Seattle, WA, USA, pp. 413–420 (2006)
5. Bronstein, A.M., Bronstein, M.M., Bruckstein, A.M., Kimmel, R.: Partial similarity of objects, or how to compare a centaur to a horse. International Journal of Computer Vision (in press)
6. Voorhees, E.M.: Implementing agglomerative hierarchical clustering algorithms for use in document retrieval. Information Processing and Management 22(6), 265–276 (1986)
7. MacQueen, J.B.: Some methods for classification and analysis of multivariate observations. In: 5th Berkeley Symposium on Mathematical Statistics and Probability, vol. I, pp. 281–297 (1967)
8. Salton, G., McGill, M.J.: Introduction to modern information retrieval. McGraw-Hill, New York (1983)

Generalized Clustering via Kernel Embeddings

Stefanie Jegelka[1], Arthur Gretton[1,2], Bernhard Schölkopf[1],
Bharath K. Sriperumbudur[3], and Ulrike von Luxburg[1]

[1] Max Planck Institute for Biological Cybernetics, Tübingen, Germany
[2] Carnegie Mellon University, Pittsburgh, PA 15213, USA
[3] Dept. of ECE, UC San Diego, La Jolla, CA 92093, USA

Abstract. We generalize traditional goals of clustering towards distinguishing components in a non-parametric mixture model. The clusters are not necessarily based on point locations, but on higher order criteria. This framework can be implemented by embedding probability distributions in a Hilbert space. The corresponding clustering objective is very general and relates to a range of common clustering concepts.

1 Introduction

In this paper we consider a statistical, non-parametric framework for clustering. Assuming the data points to be drawn from some underlying distribution P, we treat a cluster as a sample from a component distribution P_k. A clustering can then be described as a decomposition of the underlying distribution of the form $P = \sum_{k=1}^{K} \pi_k P_k$ with mixture weights π_k.

A common statistic to quantify the separation between clusters is the distance between the cluster means (e.g. Figure 1.a). However, separation based on "location" is not always what we want to achieve. The example in Figure 1.b is a mixture of two Gaussians with identical means, but with different variances. In this situation a decomposition is desirable where the difference between the variances of P_1 and P_2 is large. The difference between cluster means or between cluster variances are just two examples of distance functions between distributions. A straightforward generalization of the traditional clustering problem is to replace the distance between the means by a more general distance function. To avoid unnecessarily complicated solutions, we additionally require that the components P_k be "simple". This leads to the following generalized clustering problem:

> *Generalized clustering:* Decompose the density into *"simple"* components P_i, while *maximizing a given distance* function between the P_i.

A particularly suitable distribution representation and associated distance measure is given by Gretton et al. (2006). In this framework, a probability distribution P is embedded as $\mu[P]$ into a reproducing kernel Hilbert space (RKHS) $\mathcal{H}$ corresponding to some kernel function k. The Hilbert space norm $\|\mu[P]\|_{\mathcal{H}}$ can be interpreted as a "simplicity score": the smaller the norm, the "simpler"

B. Mertsching, M. Hund, and Z. Aziz (Eds.): KI 2009, LNAI 5803, pp. 144–152, 2009.

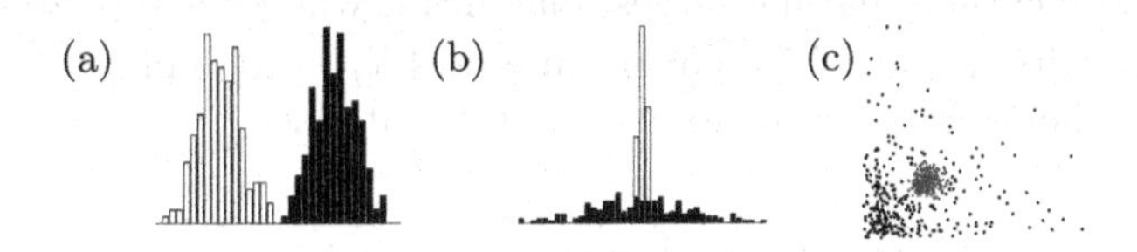

Fig. 1. (a) Traditional clustering, (b) generalized clustering; (c) example 2D result labeled by the MMD approach

the corresponding distribution (e.g., the smoother the density). The maximum mean discrepancy $\text{MMD}(P, Q)$ between two distributions P and Q is defined as the Hilbert space distance $\|\mu[P] - \mu[Q]\|_{\mathcal{H}}$ between the two embedded distributions. MMD with a dth-order polynomial kernel only discriminates between distributions based on the first d moments; with a linear kernel, it is simply the distance of means. At the most general level, i.e., with a characteristic kernel (e.g., a Gaussian kernel), *all* moments are accounted for (Sriperumbudur et al., 2008). Thus, the combination of simplicity score and distance between distributions afforded by RKHS embeddings yields a straightforward expression for the two objectives of generalized clustering. Our formulation will turn out to be very generic and to relate to many well-known clustering criteria. In this sense, the main contribution of this work is to reveal and understand the properties of the MMD approach and its relations to existing clustering algorithms. We will discuss them in Section 4 and simplicity in Section 5, after formally introducing MMD in Section 2 and the optimization problem in Section 3.

Alternative approaches to generalized clustering exist in the literature, but they are less general than the MMD approach. We summarize them in Section 4.

2 Maximum Mean Discrepancy (MMD)

We begin with a concise presentation of kernel distribution embeddings and the MMD, following Gretton et al. (2006). Given a kernel function $k : \mathcal{X} \times \mathcal{X} \to \mathbb{R}$ on some space $\mathcal{X}$, it is well known that points $x \in \mathcal{X}$ can be embedded into the corresponding reproducing kernel Hilbert space $\mathcal{H}$ via the embedding $\varphi : \mathcal{X} \to \mathcal{H}$, $x \mapsto k(x, \cdot)$. If P denotes a probability distribution on $\mathcal{X}$, one can show that the expectation $\mu[P] := \mathbb{E}_{x \sim P}[\varphi(x)]$ realizes an embedding[1] $\mu : \mathcal{P} \to \mathcal{H}$ of the space of all probability distributions $\mathcal{P}$ in $\mathcal{H}$. The Maximum Mean Discrepancy between two distributions P_1 and P_2 can be defined in two equivalent ways:

$$\text{MMD}(P_1, P_2) = \|\mu[P_1] - \mu[P_2]\|_{\mathcal{H}} \tag{1}$$

$$= \sup_{g \in \mathcal{H}, \|g\|_{\mathcal{H}} \leq 1} \left(\mathbb{E}_{x \sim P_1} g(x) - \mathbb{E}_{y \sim P_2} g(y)\right). \tag{2}$$

The first form shows that MMD is a metric. The second form shows that two probability distributions P_1 and P_2 are particularly "far" from each other if there

[1] Assume $\mathbb{E}_P[k(x,x)] < \infty$ and k is *characteristic*, then the embedding is injective (see Sriperumbudur et al. (2008) for the proof and the definition of 'characteristic').

exists a smooth function g that has largest magnitude where the probability mass of P_1 differs most from that of P_2. Given samples $\{X_i^{(1)}\}_{i=1}^n$ and $\{X_i^{(2)}\}_{i=1}^n$, the embedding and the MMD can be empirically estimated as

$$\mu[\hat{P}_1] = \frac{1}{n}\sum_{i=1}^{n}\varphi(X_i) = \frac{1}{n}\sum_{i=1}^{n}k(X_i,\cdot) \tag{3}$$
$$\widehat{\mathrm{MMD}}(P_1,P_2) = \mathrm{MMD}(\hat{P}_1,\hat{P}_2) := \|\mu[\hat{P}_1] - \mu[\hat{P}_2]\|_{\mathcal{H}}.$$

3 The Generalized Clustering Optimization Problem

We now describe how kernel distribution embeddings can be used to implement a generalized clustering algorithm. For simplicity, we focus on the case of two clusters only. Our goal is to decompose the underlying distribution P such that $\mathrm{MMD}(P_1, P_2)$ is large and $\|\mu[P_1]\|_{\mathcal{H}}$ and $\|\mu[P_2]\|_{\mathcal{H}}$ are small. With only a finite sample $\{X_i\}_{i=1}^n$, we must estimate these quantities empirically. To this end, we parameterize the empirical distributions $\hat{P}_k$ via assignments $\alpha_i^{(k)}$ of Dirac measures δ_{X_i} on the sample points X_i:

$$\hat{\pi}_k\hat{P}_k = \frac{1}{n}\sum_{i=1}^{n}\alpha_i^{(k)}\delta_{X_i} \quad \text{with} \quad \hat{\pi}_k = \frac{1}{n}\sum_{i=1}^{n}\alpha_i^{(k)}$$

for $\alpha_i^{(1)} + \alpha_i^{(2)} = 1$. For a soft clustering we allow $\alpha^{(1)}, \alpha^{(2)} \in [0,1]^n$; for a hard clustering we constrain $\alpha^{(1)}, \alpha^{(2)} \in \{0,1\}^n$. The resulting decomposition takes the form $\hat{P} = \hat{\pi}_1\hat{P}_1 + \hat{\pi}_2\hat{P}_2$. These estimates lead to the following optimization problem (note that $\alpha^{(2)} = \mathbf{1} - \alpha^{(1)}$ is determined by $\alpha^{(1)}$):

$$\max_{\alpha^{(1)}} \Psi(\alpha^{(1)}) := \max_{\alpha^{(1)}} \quad \text{MMD-Term}(\hat{P}_1,\hat{P}_2) + \lambda \cdot \text{Regularization-Term}(\hat{P}_1,\hat{P}_2).$$

Let $\mathbf{K} = (k(X_i, X_j))_{i,j=1,\ldots,n}$ denote the kernel matrix of the sample. The MMD-Term and its parametric form are then

$$\hat{\pi}_1\hat{\pi}_2\mathrm{MMD}(\hat{P}_1,\hat{P}_2) = \hat{\pi}_1\hat{\pi}_2\|\mu[\hat{P}_1] - \mu[\hat{P}_2]\|_{\mathcal{H}}^2 \tag{4}$$
$$= \frac{\hat{\pi}_2}{n^2\hat{\pi}_1}(\alpha^{(1)})^\top\mathbf{K}\alpha^{(1)} + \frac{\hat{\pi}_1}{n^2\hat{\pi}_2}(\alpha^{(2)})^\top\mathbf{K}\alpha^{(2)} - \frac{2}{n^2}(\alpha^{(1)})^\top\mathbf{K}\alpha^{(2)}.$$

The product of the cluster sizes $\hat{\pi}_1, \hat{\pi}_2$ acts as a balancing term to avoid particularly small clusters. We will call the maximization of (4) *maxMMD* and give various interpretations in Section 4. As a regularization term we use

$$\lambda_1\|\mu[\hat{P}_1]\|_{\mathcal{H}}^2 + \lambda_2\|\mu[\hat{P}_2]\|_{\mathcal{H}}^2 = \frac{\lambda_1}{n^2\hat{\pi}_1^2}(\alpha^{(1)})^\top\mathbf{K}\alpha^{(1)} + \frac{\lambda_2}{n^2\hat{\pi}_2^2}(\alpha^{(2)})^\top\mathbf{K}\alpha^{(2)}. \tag{5}$$

To avoid empty clusters, we introduce a constraint for the minimum size $\varepsilon > 0$ of a cluster. This leads to the final optimization problem

$$\max_{\alpha^{(1)} \in [0,1]^n} \Psi(\alpha^{(1)}) \qquad \text{s. t. } \varepsilon \leq \sum_i \alpha_i^{(1)} \leq (1-\varepsilon).$$

As we shall see in Section 4, maxMMD alone can be optimized efficiently via a variant of the kernel k-means algorithm ensuring the minimum size constraint. For the full objective, we used the Ipopt solver (Wächter and Biegler, 2006). Even though we evaluated our criterion and variants in many experiments, we will exclude them to save space and concentrate on the theory, our main contribution.

4 MaxMMD, Discriminability, and Related Approaches

We will now describe how the *discriminability* criterion maxMMD (Eqn. (4)) encompasses the concepts behind a number of classical clustering objectives. Figure 2 gives an overview.

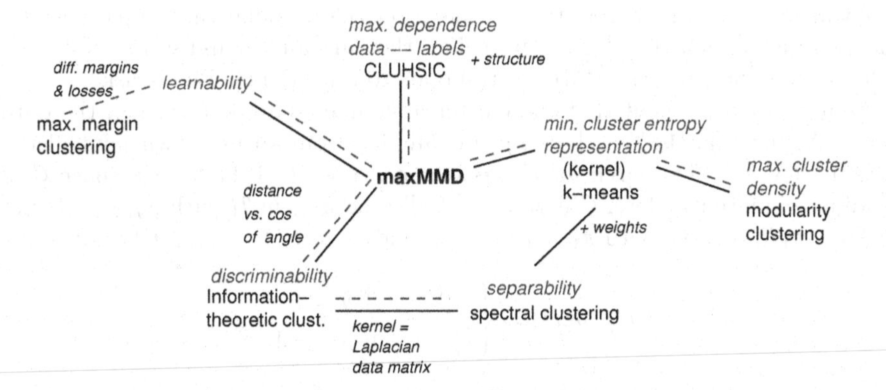

Fig. 2. Overview of connections. Dashed links are conceptual, solid links mathematical. First, *discriminability* is conceptually close to *learnability*. Learning seeks to detect patterns in the data, i.e., dependencies between data and labels (cluster assignments). Only if we capture those dependencies can we reliably predict labels. In other words, we want clusters to maximize the *dependence between data and labels*. If the label distribution is closely linked to the data, it *represents* the data well. Representation conceptually connects to compression and coding. Clusters that are less complex to describe have *lower entropy*. Small entropy means *dense clusters*, which leads back to a generalization of the k-means criterion. Extended by weights, this criterion encompasses spectral clustering, e.g., graph cut criteria. Those cuts favor sparse connections between clusters, simply another measure of discriminability. Spectral clustering also relates to discriminability via the angle of embeddings $\mu[P_1], \mu[P_2]$.

Concept 1 (Discriminability). *MaxMMD seeks dissimilar components* P_k.

Since MMD is a metric for distributions, maxMMD strives for distant component distributions by construction. Hence, it indirectly *promotes discriminability* of the clusters by their statistics.

Moreover, MaxMMD behaves similarly to the Jensen-Shannon (JS) divergence in the clustering context. For a mixture $P = \pi_1 P_1 + \pi_2 P_2$, the latter is $D_{\text{JS}}(P_1, P_2) = \pi_1 D_{\text{KL}}(P_1, P) + \pi_2 D_{\text{KL}}(P_2, P)$ (Fuglede and Topsøe, 2004). If we replace the KL divergence D_{KL} by the squared MMD, we arrive at the parallel form

$$\pi_1 \text{MMD}(P_1, P)^2 + \pi_2 \text{MMD}(P_2, P)^2 = \pi_1 \pi_2 \text{MMD}(P_1, P_2)^2. \tag{6}$$

Discriminability via projections or some moments is used by, e.g., Chaudhuri and Rao (2008), Arora and Kannan (2001), but for specific distributions. Information-theoretic clustering (Gokcay and Principe, 2002, Jenssen et al., 2004) measures discriminability by the cosine of the angle between the $\mu[P_k]$. Its motivation, however, restricts k to have a specific form, whereas MMD is more general.

Concept 2 (Learnability). *MaxMMD finds clusters that are well learnable.*

It turns out that our objective also connects unsupervised and supervised learning. In supervised learning, the Bayes risk R^* measures the difficulty of a learning problem. If R^* is large, then there is no good rule (among the simple ones we choose from) to tell the classes apart, i.e., they are almost indiscriminable. We will see that the negative MMD corresponds to a particular Bayes risk.

Assume for a moment that the cluster assignments are known, i.e., the data points X_i have labels $Y_i \in \{-1, 1\}$, all sampled from an unknown joint distribution $P(X, Y)$. We search for a classifier $g : \mathcal{X} \to \{-1, 1\}$ from a space $\mathcal{G}$ of candidate functions. Here, $\mathcal{G}$ is the set of all functions in $\mathcal{H}$ with $\|g\|_{\mathcal{H}} \leq 1$. Let $P(X) = \pi_1 P_1 + \pi_2 P_2$ with $\pi_1 = P(Y = 1)$ and $\pi_2 = P(Y = -1)$. Choosing loss

$$\ell(y, g(x)) = \begin{cases} \frac{-g(x)}{\pi_1} & \text{if } y = 1 \\ \frac{g(x)}{\pi_2} & \text{if } y = -1 \end{cases}$$

and using the definition (1) of MMD, the Bayes risk becomes

$$\begin{aligned} R^* &= \inf_{g \in \mathcal{G}} -\left(\int g dP_1 - \int g dP_2\right) = -\sup_{g \in \mathcal{G}} \left(\int g dP_1 - \int g dP_2\right) \\ &= -\text{MMD}(P_1, P_2). \end{aligned} \tag{7}$$

A large MMD hence corresponds to a low Bayes risk. The classifier g^* that minimizes the risk is (Gretton et al., 2008)

$$g^* = \arg \inf_{g \in \mathcal{H}, \|g\|_{\mathcal{H}} \leq 1} -\left(\int g dP_1 - \int g dP_2\right) = \frac{\mu[P_1] - \mu[P_2]}{\|\mu[P_1] - \mu[P_2]\|_{\mathcal{H}}}, \quad \text{i.e.,}$$

$$g^*(x) = \langle g^*, \varphi(x) \rangle_{\mathcal{H}} \propto \int k(x, x') dP_1(x') - \int k(x, x'') dP_2(x'').$$

Estimating $\mu[P_k]$ as in (3) with the assignments $\alpha_i^{(k)}$ yields a Parzen window classifier with the window function k, assuming k is chosen appropriately.

In clustering, we choose labels Y_i and hence implicitly P_1 and P_2. Maximizing their MMD defines classes that are well learnable with $\mathcal{G}$. This concept is reminiscent of Maximum Margin Clustering (MMC) that seeks to minimize a 0-1-loss (Xu et al., 2005). As opposed to MMC, however, we do not maximize a minimum separation (margin) between the cluster points in the RKHS, but strive for discrepancy of the means in $\mathcal{H}$.

Concept 3 (Dependence maximization). *MaxMMD seeks the dependence between data and cluster assignments.*

What is the aim of learning? We assume an association between the data and the labels that can be expressed by a pattern or learnable rule. Well-learnable cluster assignments are in conformity with such a dependence.

In matrix notation and with binary assignments, Criterion (4) becomes

$$\hat{\pi}_1\hat{\pi}_2\|\mu[\hat{P}_1] - \mu[\hat{P}_2]\|^2_{\mathcal{H}} = \text{tr}(\mathbf{KL}) - const = -\text{HSIC}(P, \alpha^{(1)}) - const \quad (8)$$

where $\mathbf{K}$ is the kernel matrix of the data and $\mathbf{L}$ that of the labels with entries $l_{ij} = n^{-1}\hat{\pi}_1^{-1/2}$ if $\alpha_i^{(1)} = \alpha_j^{(1)} = 1$, $l_{ij} = n^{-1}\hat{\pi}_2^{-1/2}$ if $\alpha_i^{(1)} = \alpha_j^{(1)} = 0$ and $l_{ij} = 0$ otherwise. HSIC (Gretton et al., 2005) is a measure of statistical dependence between random variables. Hence, *maxMMD is equivalent to maximizing the dependence between cluster assignments and the data distribution.*

This criterion has been used in existing clustering approaches. Song et al. (2007) exploit the HSIC formulation to introduce structural constraints. The criterion by Aghagolzadeh et al. (2007) is similar to (8) but derived as an estimate of the Mutual Information between labels and data.

Concept 4 (Representation). *MaxMMD aims to find functions $\mu[\hat{P}_k]$ that represent the data well.*

We can rewrite the maxMMD criterion as

$$2\hat{\pi}_1\hat{\pi}_2\|\mu[\hat{P}_1] - \mu[\hat{P}_2]\|^2_{\mathcal{H}} = const - \sum_{k=1}^{2}\sum_{i=1}^{n} \alpha_i^{(k)} \|\varphi_i - \mu[\hat{P}_k]\|^2_{\mathcal{H}}. \quad (9)$$

Consider a probabilistic encoding with a limited number of deterministic codebook vectors $y(l) = \mu[\hat{P}_l]$. Choose a label $l(X_i)$ for a point X_i with probability $\hat{P}(l|X_i) = \alpha_i^{(l)}$ and encode it as $y(X_i) = y(l(X_i))$ (Rose, 1994). Then Criterion (9) is an estimate of the average distortion $\mathbb{E}[D] = \int d(X, y(X))dP(X, k)$ with divergence $d(X_i, y(X_i)) = \|\varphi_i - y(X_i)\|^2_{\mathcal{H}}$. An assignment with minimal distortion, as favored by (9), is a good encoding, i.e., *it represents the data well.* Following further algebra, the minimization of the average distortion corresponds to the maximization of the sum of dot products

$$\sum_{k=1}^{2}\sum_{i=1}^{n} \alpha_i^{(k)} \langle\varphi(X_i), \mu[\hat{P}_k]\rangle_{\mathcal{H}} = \sum_{k=1}^{2}\sum_{i=1}^{n} \alpha_i^{(k)} f_{\mu[\hat{P}_k]}(X_i). \quad (10)$$

The functions $f_{\mu[\hat{P}_k]}$ have the form of a Parzen window estimator: $f_{\mu[\hat{P}_k]}(X) = \sum_{j=1}^{n} \alpha_j^{(k)} k(X, X_j)$. Criterion (10) is large if the $f_{\mu[\hat{P}_k]}$ represent the density structure of the data, i.e., if each X_i is likely under the estimator it belongs to.

Concept 5 (Entropy). *MaxMMD finds clusters with low generalized entropy.* With the definition $n\mu[\hat{P}_k] = \hat{\pi}_k^{-1} \sum_i \alpha_i^{(k)} \varphi(X_i)$, Criterion (10) becomes

$$\max \sum_{k=1}^{2} \hat{\pi}_k \|\mu[\hat{P}_k]\|_{\mathcal{H}}^2. \tag{11}$$

The related term $H_2(P_k) = -\log \|\mu[P_k]\|_{\mathcal{H}}^2$, as shown by Erdogmus and Principe (2006), is the Parzen window estimate of a generalized entropy, the Renyi entropy (Renyi, 1960). Consequently, large norms $\|\mu[\hat{P}_k]\|_{\mathcal{H}}$ result in small Renyi entropies of the $\hat{P}_k$. Thus, similar to the analogy in Equation (6), Criterion (11) parallels the JS divergence: maximizing D_{JS} minimizes the weighted sum of Shannon entropies, $D_{\mathrm{JS}}(P_1, P_2) = H_S(P) - \pi_1 H_S(P_1) - \pi_2 H_S(P_2)$ (Lin, 1991).

Criterion (9) is in fact the kernel k-means objective. While linear k-means seeks clusters with minimal variance, its kernel version generalizes to clusters with minimal entropy, as shown by Concept 5, and in line with the viewpoint of Erdogmus and Principe (2006) that entropy is a generalization of variance to non-Gaussian settings.

Entropy can be viewed as an information theoretic complexity measure; recall the association between coding length and Shannon entropy. This means that maxMMD favors less complex clusters in an information theoretic sense.

The function $\mu[P_k]$ has a larger norm if (the image of) its support is narrower. In light of this view, small entropies relate to criteria that favor small, dense clusters. For graphs, those are densely connected subgraphs. Construct a graph with vertices X_i. Let the kernel matrix $\mathbf{K}$ be its adjacency matrix (with $k(X, X) = c$ for all X) and cluster C_k have n_k nodes. Let $m(C_k)$ denote the sum of the weights of the within-cluster edges. Then maxMMD promotes the average connectedness of a vertex within its cluster:

$$\hat{\pi}_k \|\mu[\hat{P}_k]\|_{\mathcal{H}}^2 = \hat{\pi}_k^{-1} (\alpha^{(k)})^\top K \alpha^{(k)} = nc + n\frac{m(C_k)}{n_k}.$$

Edge density is vital in the modularity criterion (Newman and Girvan, 2004) that compares the achieved to the expected within-cluster connectivity.

5 Regularization

As we have seen, the maxMMD criterion is well supported by classical clustering approaches. That said, clustering by maxMMD alone can result in $\hat{P}_k$ with non-smooth, steep boundaries, as in Figure 3. Smoothness of a function f in form of "simplicity" is commonly measured by the norm $\|f\|_{\mathcal{H}}$, for example in Support Vector Machines. To avoid steep or narrow $\hat{P}_k$ — unless the data allows for that — we add a smoothness term for the estimates $f_{\mu[\hat{P}_k]}$,

$$\rho(\alpha^{(1)}, \alpha^{(2)}) := \lambda_1 \|\mu[\hat{P}_1]\|_{\mathcal{H}}^2 + \lambda_2 \|\mu[\hat{P}_2]\|_{\mathcal{H}}^2.$$

The weights $\hat{\pi}_k$ do not come into play here. If $\mu[P_k]$ is small, then its support in $\mathcal{H}$ is broader, and P_k has higher entropy (uncertainty). Thus, constraining the

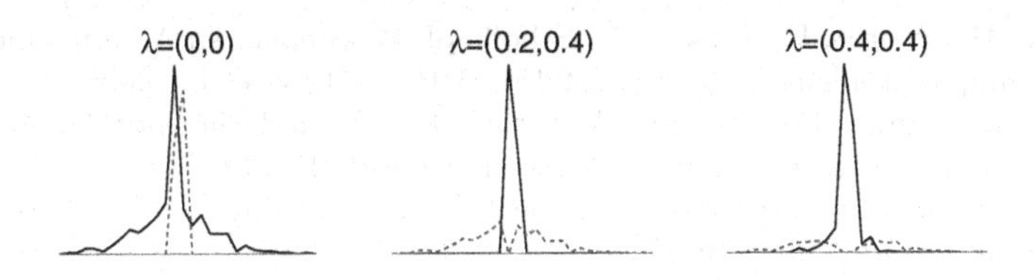

Fig. 3. Effect of increasing regularization with Gaussian kernel (estimates by binning)

norms can avoid "overfitting". Furthermore, we can introduce prior knowledge about different entropies of the clusters by choosing $\lambda_1 \neq \lambda_2$.

To enforce overlap and smoothness, more direct restriction of the assignments $\alpha^{(k)}$ is also conceivable, similar to the updates in soft k-means. This type of regularization is motivated by the analogy (9) to kernel k-means. Both ways of regularization restrict the set of candidate distributions for P_k.

6 Conclusion

The literature on clustering is overwhelming, and it is difficult even to get an overview of what is out there. We believe that it is very important to "tidy up" and discover relationships between different clustering problems and algorithms. In this paper we study a generalized clustering problem which considers clustering from a higher level point of view, based on embeddings of distributions to high dimensional vector spaces. This approach reveals connections to the concepts behind many well-known clustering criteria.

Acknowledgments. We thank Bob Williamson, Mark Reid and Dominik Janzing for discussions.

References

Aghagolzadeh, M., Soltanian-Zadeh, H., Araabi, B., Aghagolzadeh, A.: A hierarchical clustering based on maximizing mutual information. In: ICIP (2007)

Arora, S., Kannan, R.: Learning mixtures of arbitrary Gaussians. In: STOC (2001)

Chaudhuri, K., Rao, S.: Beyond Gaussians: Spectral methods for learning mixtures of heavy-tailed distributions. In: COLT (2008)

Erdogmus, D., Principe, J.C.: From linear adaptive filtering to nonlinear information processing. IEEE Signal Processing Magazine 23(6), 14–33 (2006)

Fuglede, B., Topsøe, F.: Jensen-Shannon divergence and Hilbert space embedding. In: Proc. of the Int. Symp. on Information Theory, ISIT (2004)

Gokcay, E., Principe, J.C.: Information theoretic clustering. IEEE Transactions on Pattern Analysis and Machine Intelligence 24(2), 158–170 (2002)

Gretton, A., Bousquet, O., Smola, A., Schölkopf, B.: Measuring statistical dependence with Hilbert-Schmidt norms. In: Jain, S., Simon, H.U., Tomita, E. (eds.) ALT 2005. LNCS (LNAI), vol. 3734, pp. 63–77. Springer, Heidelberg (2005)

Gretton, A., Borgwardt, K.M., Rasch, M., Schölkopf, B., Smola, A.: A kernel method for the two-sample problem. In: NIPS (2006)

Gretton, A., Borgwardt, K., Rasch, M., Schölkopf, B., Smola, A.: A kernel method for the two-sample problem. Tech. Report 157, MPI for Biol. Cyb. (2008)
Jenssen, R., Erdogmus, D., Principe, J., Eltoft, T.: The laplacian pdf distance: a cost function for clustering in a kernel feature space. In: NIPS (2004)
Lin, J.: Divergence measures based on the Shannon entropy. IEEE Transaction on Information Theory 37(1) (1991)
Newman, M.E.J., Girvan, M.: Finding and evaluating community structure in networks. Physical review E 69 (2004)
Renyi, A.: On measures of entropy and information. In: Proc. Fourth Berkeley Symp. Math., Statistics and Probability, pp. 547–561 (1960)
Rose, K.: A mapping approach to rate-distortion computation and analysis. IEEE Transactions on Information Theory 40(6), 1939–1952 (1994)
Song, L., Smola, A., Gretton, A., Borgwardt, K.: A dependence maximization view of clustering. In: 24th International Conference on Machine Learning, ICML (2007)
Sriperumbudur, B.K., Gretton, A., Fukumizu, K., Lanckriet, G., Schölkopf, B.: Injective Hilbert space embeddings of probability measures. In: COLT (2008)
Wächter, A., Biegler, L.T.: On the implementation of a primal-dual interior point filter line search algorithm for large-scale nonlinear programming. Mathematical Programming 106(1), 25–57 (2006)
Xu, L., Neufeld, J., Larson, B., Schuurmans, D.: Maximum margin clustering. In: NIPS (2005)

Context-Based Clustering of Image Search Results

Hongqi Wang[1,2], Olana Missura[1], Thomas Gärtner[1], and Stefan Wrobel[1,2]

[1] Fraunhofer Institute Intelligent Analysis and Information Systems IAIS,
Schloss Birlinghoven, D-53754 Sankt Augustin, Germany
firstname.lastname@iais.fraunhofer.de
[2] Department of Computer Science III, Rheinische Friedrich-Wilhelms-Universitt Bonn,
Römerstrasse 164, D-53117 Bonn, Germany
firstname.lastname@iai.uni-bonn.de

Abstract. In this work we propose to cluster image search results based on the textual contents of the referring webpages. The natural ambiguity and context-dependence of human languages lead to problems that plague modern image search engines: A user formulating a query usually has in mind just one topic, while the results produced to satisfy this query may (and usually do) belong to the different topics. Therefore, only part of the search results are relevant for a user. One of the possible ways to improve the user's experience is to cluster the results according to the topics they belong to and present the clustered results to the user. As opposed to the clustering based on visual features, an approach utilising the text information in the webpages containing the image is less computationally intensive and provides the resulting clusters with semantically meaningful names.

Keywords: image clustering, machine learning, image search.

1 Introduction

The information explosion brings a rapidly increasing amount of published information and makes a huge number of images available on the internet. Although search engines have made retrieval and managing of large amount of information from the internet much easier, the results from image search engines are not always satisfactory. Due to the ambiguity and context-dependency of human languages the same word can relate to wildly different things (consider, for example, that "jaguar" could refer to an animal, as well as to a car). As a result, for the same query an image search engine can return images from several different categories. In general users are interested in one particular category. Search engines such as Google [5], Yahoo! [17], and Picsearch [12] return a long list of image results, which users have to navigate by themselves by examining titles and thumbnails of these images to find the relevant results. It is a time-consuming and frustration-inducing task, especially when a large number of different topics is presented. A natural idea is to cluster the image search results returned by a search engine according to the topics they belong to.

In recent years several algorithms were developed based on Content-Based Image Retrieval (CBIR) [10,9,16,15,4,2]. Unfortunately, all of them suffer from the following problems: First, they use high dimensional visual features that are too computationally intensive to be practical for web applications. Second, the generated clusters do not have

B. Mertsching, M. Hund, and Z. Aziz (Eds.): KI 2009, LNAI 5803, pp. 153–160, 2009.

semantic names. [6] proposed IGroup as an efficient and effective algorithm. IGroup firstly builds different clusters with unique names by clustering text search results from Google Search and snippets from Picsearch using Q-grams algorithm [3]. The cluster names are then used to perform the image search on Picsearch. IGroup claims to have three unique features. First, the most representative image groups can be found with meaningful names. Second, all resulting images are taken into account in the clustering process instead of the small part. And third, this algorithm is efficient enough to be practical. In spite of these features, this algorithm, however, has some problems which have not been stated by the authors. In order to generate the cluster names, the system has to be well trained beforehand with some training data, which is improper for a real ISRC system, because for real systems query text is usually unknown and random.

In contrast to the approaches described above we propose a way to cluster the images based on the textual information contained in their referring webpages. The assumption is that given a webpage, its textual and visual content are related to each other. Furthermore, we assume that the distance within the webpage indicates the degree of relevance between the image and the text. There are two significant advantages of using clustering based on text as opposed to visual features. First, it is less computationally intensive, which is important in the context of web search, where the reaction times are on average less than a second. Second, using text features gives us an ability to construct semantically meaningful names for the clusters, which simplifies the navigation for the users.

In this paper we introduce TeBIC (Text Based Image Clustering) which clusters image search results into different category groups. We proceed as follows: In Section 2 the architecture of the TeBIC's components is described. After that the experimental results are presented in Section 3. The work is summarised and possible directions for the future work are outlined in Section 4.

2 Component Description

The images to cluster together with the information about their referring webpages are fetched from the Yahoo Image Search Engine [17]. Prior to clustering, TeBIC utilises a language filter to discard all the websites that are not written in a specified language. In our work English was used as a primary language, but any other language can be chosen due to the language independence property of the Q-gram algorithm. Then the data is extracted from the webpages using a content extractor and resulting feature vectors are clustered. After the clusters are generated, the cluster labeler assigns semantic names to the clusters according to the common topic of the images in the cluster. For each component we investigated several options, which are described below.

2.1 The Language Filter

Images of similar topics might be contained in the webpages written in different languages. A clustering algorithm using textual features is likely to cluster different languages into different clusters. To exclude this possibility TeBIC uses a language filter. The Q-gram-based text categorisation algorithm [3] is used to filter out all webpages that are not written in English. All the words of a webpage are scanned to produce a list of Q-grams. The list is sorted in a descending order of their count (Q-gram that has the largest count comes first) and stored as the document profile of the webpage. The webpage is assigned to a language such that the distance between their profiles is minimal. (For a detailed description of the algorithm and the metrics used see [3].)

2.2 The Content Extractor

Text-based. The text-based extractor takes into account only pure text information of a webpage, excluding scripts, comments and tag attributes. The resulting text document is represented by its the bag of words. The stop words are removed and the remaining words are stemmed into terms by Porter Stemmer [13]. After that the terms are weighted using the tf-idf approach with a sublinear tf scaling:

$$weight = \begin{cases} (1 + \log tf) \times \log \frac{N}{df_t} & \text{if } tf > 0; \\ 0 & \text{if } tf = 0; \end{cases} \tag{1}$$

where N is the total number of the webpages, tf is the term frequency and df_t is the inverse document frequency of each term.

DOM-based. As opposed to plain text documents, webpages have a structural organisation, which is embodied in a Document Object Model (DOM) tree. Recall that we assumed that the text located closer to the image is more relevant to the topic of the image than the text located further away. The DOM tree allows us to utilise this information by calculating the distance between the image nodes and the text nodes in the following way:

- The distance of the target image node to itself is *zero.*
- In a subtree which contains the target image node, the distance of any child node, except the target image node itself and its children, is the difference of depth between the target image node and the least common ancestor of the child node and the target image node.

The weight w_i of a text node n_i is calculated according to the idea that the closest nodes to the image are the most important ones and with the increasing distance their importance is rapidly falling:

$$w_i = \frac{1}{\sum_{i=1}^{N} e^{-\frac{d_i^2}{3}}} e^{-\frac{d_i^2}{3}}, \tag{2}$$

where d_i is the distance between the ith text node and the target image node and the constant 3 was determined empirically. The frequency of the term t is then scaled with the weight of its text node to calculate the final term weight:

$$wf_t = \sum_{i=1}^{N} ft|_{n_i} \times w_i, \tag{3}$$

where w_i is the weight of the ith text node, $ft|_{n_i}$ is the term frequency in the text node n_i.

Link-based. Another difference between a webpage and a text document is that a webpage contains hyperlinks to other webpages, which may provide additional useful information. The link-based extractor searches for all the non-nepotistic (i.e. leading to other domains) hyperlinks in all webpages returned by the search engine. Each webpage is then represented by a vector V of links according to the occurrence of the hyperlinks.

External-Page and DOM-based. Not all images have referring webpages with enough textual information (e.g. photo galleries). To overcome this drawback we designed an external-page and DOM-based extractor. It augments the word vector constructed by the DOM-based extractor with the terms mined from the webpages that the non-nepotistic hyperlinks lead to. The new terms are weighted with the the minimal weight value presented in the original word vector.

2.3 Cluster Analyser

K-means. The K-means algorithm [14] clusters n objects into k partitions, $k < n$, by minimising intra-cluster distances.

Yet another K-means. The Yet another K-means (YAK) algorithm was designed by the authors to overcome a drawback [1] of the K-means algorithm, namely the need for the parameter specifying the number of clusters upfront. Instead of a pre-defined number of clusters, YAK requires a maximum possible number of clusters k. YAK is a soft-clustering algorithm and its resulting clusters may overlap. It proceeds as follows:

1. One of the data points is selected randomly and assigned to the initial singleton cluster.
2. A data point x is assigned to each cluster c_i such that the similarity between c_i and x is above the similarity threshold s. Note that in this step the point x can be assigned to more than one cluster. If no such cluster exists, x forms a new singleton cluster. In case that the maximum cluster number k is reached, x is assigned to c_i with the maximum similarity value. The centroids of clusters are recalculated and the process is repeated until clusters no longer change.
3. Clusters which share most of their elements are merged according to a merge threshold m. After the merging process the singleton clusters are discarded.

One may say that we replaced one parameter (number of clusters) with two new ones (similarity and merge thresholds). For the number of clusters there is no way to know in advance how many topics the resulting images belong to. The thresholds, on the other hand, have universal values that can be established empirically.

Non-negative matrix factorisation (NMF). The goal of the NMF [7] is to factorise the feature space (represented by a set of feature vectors) into a k-dimensional semantic space with each dimension corresponding to a particular topic and n-dimensional weight space. Each topic represents a cluster. In the semantic space each item, in our case each referring webpage, is represented by a linear combination of the k topics. For the details about the NMF algorithm see [7].

2.4 Cluster Labeler

Information Gain (IG). The IG [18] approach calculates the importance of each term to each cluster as follows:

$$IG(T|C) = \sum_{e_c \in \{1,0\}} \sum_{e_t \in \{1,0\}} P(T = e_t, C = e_c) \log \frac{P(T = e_t, C = e_c)}{P(T = e_t)P(C = e_c)} \tag{4}$$

where T has values $e_t = 1$ when term t_i is in a webpage in the webpage pool and $e_t = 0$ when term t_i is not. Similarly C has values $e_c = 1$ when the webpage is in the cluster $_kc$ and $e_c = 0$ when the webpage is not. The terms are sorted in a descending order according to their IG values for each cluster and the top ten are used as the cluster's names.

χ^2 Test. The χ^2-test [18] approach tests the independence between each cluster and each term:

$$\chi^2(T, C) = \sum_{e_t \in 1,0} \sum_{e_c \in 1,0} \frac{(O_{e_t e_c} - E_{e_t e_c})^2}{E_{e_t e_c}} \tag{5}$$

where e_t and e_c have the same definition as in the IG approach. $O_{e_t e_c}$ and $E_{e_t e_c}$ are the *observed frequency* and the *expected frequency* that take the values of e_t and e_c.

Word Frequency (WF). The WF approach sorts the terms in a descending order according to their weight in a cluster and uses the top ten as the cluster's names.

3 Experimental Results

The goal of the evaluation was to determine the best choice of the TeBIC's components. To this purpose we tested its performance with the language filter and without, when utilising each of the described content extractors and each of the clustering algorithms. Furthermore, it was interesting to determine which of the proposed labeling methods provides semantically better cluster name. The quality of the labels was evaluated using human judgement. Two metrics were used in the evaluation, purity and Rand Index [8]:

$$purity(K, C) = \frac{1}{N} \sum_{c_i \in C} \max_j |k_j \cap c_i| \tag{6}$$

$$RI = \frac{tp + tn}{\binom{N}{2}}, \tag{7}$$

where $K = \{k_1, \cdots, k_m\}$ is the set of topics, $C = \{c_1, \cdots, c_n\}$ is the set of clusters, $|k_j \cap c_i|$ is the number of images from topic k_j that are clustered in cluster c_i, *tp* is the number of true positives, *tn* is the number of true negatives, and N is the total number of images.

We evaluated the effect that different components have on the performance of TeBIC on two query terms, "jaguar" and "apple", chosen for their ambiguity. Figure 2(a) shows

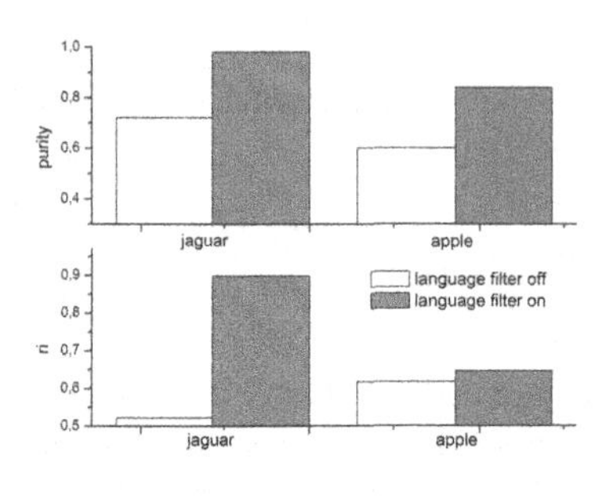

(a) With and without language filter

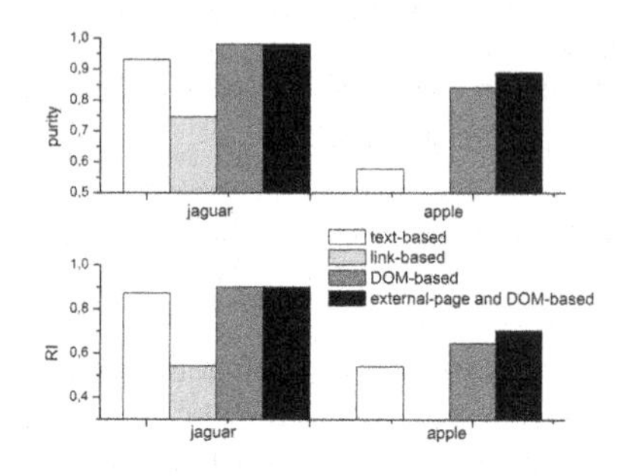

(b) With different content extractors

Fig. 1. TeBIC's performance

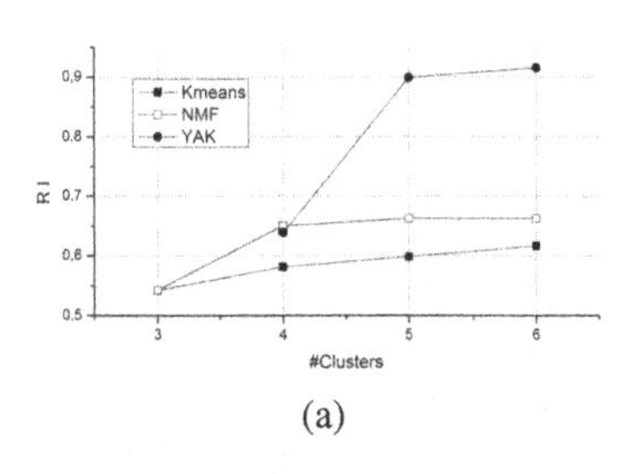

(a)

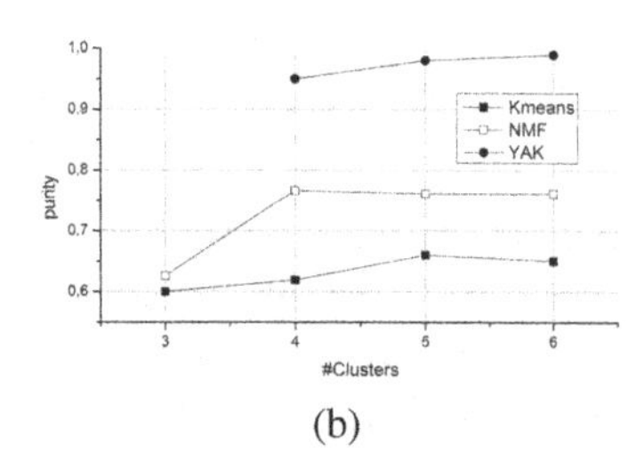

(b)

Fig. 2. Purity and RI over different number of clusters generated by different clustering algorithms. #Clusters is the actual cluster number used by the corresponding algorithm. The upper bound for YAK was set to 10.

that both purity and RI increase when the language filter is on. Figure 2(b) shows the purity and RI values for each of the proposed content extractors. The two DOM-based methods outperform the text-based extractor according to both metrics. From Figure 2 we can see that the YAK cluster analyser outperforms the other two approaches probably due to it being a soft-clustering method and removing the singletons or small clusters in the end. The comparison of the labelers is presented in Table 1. The German terms used in the labels occur in the <META> tags of some of the referring webpages even when the main text is written in English. Note that according to [18] IG and CHI scores of a term are strongly correlated. The examination of the suggested cluster names reveals that a simpler method of counting the word frequencies generates better names than the other two. It is possible that this is due to a small number of clusters produced in our experiments. [8] showed that the performance of the frequency-based selection method falls quickly when the number of clusters is larger than 10.

Based on the collected data we conclude that the best configuration from the suggested options is TeBIC consisting of the language filter, DOM-based content extractor, soft clustering algorithm YAK, and word frequency-based labeler.

Table 1. Automatically computed cluster labels for the query "jaguar"

	IG	CHI test	Word Frequency	Real Topic
1	**mieten** **leihen** hochzeitswagen ver-leih **werkstatt** film-fahrzeuge reparaturen **werkst** jaguarverkauf klassische	**mieten** **leihen** hochzeitswagen ver-leih **werkstatt** film-fahrzeuge reparaturen **werkst** jaguarverkauf klassische	type oldtimer **ausfahrten** classical **club** **veranstal-tungen** ersatzteile selber klassik fahren	jaguar club
2	**south** **animal** **mother** **america** **live** information space **rainforest** weighting prey	**south** **animal** **mother** **america** **live** information space **rainforest** weighting prey	**panthers** frazetta **animal** mayan **cats** size black walks **leopard** fear	jaguar cat
3	lancia facel tomaso **all-sportauto** lada alphine marcos **oldsmobile** tatra caterham	lancia facel tomaso **all-sportauto** lada alphine marcos **oldsmobile** tatra caterham	andros mondial louis pho-tos **allsportauto** **bagatelle** **retromobile** **sportive** masini trophe	jaguar car
4	nutz silly eclectech flappy bytemark cash tube shirts hello	nutz silly eclectech flappy bytemark cash tube shirts hello	**specs** technically **specific** **xkr** parks image arden pic-ture gray exotic	mainly cars, mixed with cats

4 Conclusion and Future Work

In this paper we proposed an approach to cluster image search results based on the textual information as a way to overcome the problems of visual features based algorithms, namely their high computational costs and lack of semantic names for the generated clusters. The preliminary results demonstrate the soundness of the idea that the text in the referring webpages provides enough information for the clustering of the images. However, further experiments are required to compare the performance of TeBIC with other approaches (e.g. IGroup [6]). In future work we also intend to conduct user studies to answer the question, whether clustering of image search results indeed improves the user experience.

Acknowledgments

We would like to thank the anonymous reviewers for their helpful and insightful comments.

References

1. Berkhim, P.: Survey of clustering data mining techniques. Tech. rep., Accrue Software, San Jose, CA (2002)
2. Cai, D., He, X., Li, Z., Ma, W.-Y., Wen, J.-R.: Hierarchical clustering of WWW image search results using visual, textual and link information. In: MULTIMEDIA 2004: Proceedings of the 12th annual ACM international conference on Multimedia, pp. 952–959. ACM, New York (2004)
3. Cavnar, W.B., Trenkle, J.M.: N-gram-based text categorization. In: Proceedings of SDAIR 1994, 3rd Annual Symposium on Document Analysis and Information Retrieval, pp. 161–175 (1994)

4. Gao, B., Liu, T.-Y., Qin, T., Zheng, X., Cheng, Q.-S., Ma, W.-Y.: Web image clustering by consistent utilization of visual features and surrounding texts. In: MULTIMEDIA 2005: Proceedings of the 13th annual ACM international conference on Multimedia, pp. 112–121. ACM, New York (2005)
5. Google Image Search, http://images.google.com
6. Jing, F., Wang, C.: IGroup: web image search results clustering. In: MULTIMEDIA 2006, pp. 377–384. ACM, New York (2006)
7. Lee, D.D., Seung, H.S.: Algorithm for non-negative matrix factorization, pp. 556–562. MIT Press, Cambridge
8. Manning, C.D., Raghvavan, P., Schütze, H.: An Introduction to Information Retrieval. Cambridge University Press, Cambridge (2008)
9. Liu, X., Croft, W.B.: Cluster-based retrieval using language models. In: SIGIR 2004: Proceedings of the 27th annual international ACM SIGIR conference on Research and development in information retrieval, pp. 186–193. ACM, New York (2004)
10. Luo, B., Wang, X., Tang, X.: World Wide Web Based Image Search Engine Using Text and Image Content Features. In: Santini, S., Schettini, R. (eds.) Society of Photo-Optical Instrumentation Engineers (SPIE) Conference Series, vol. 5018, pp. 123–130 (2003)
11. Nielson Online, http://www.nielsen-online.com
12. Picsearch, http://www.picsearch.com
13. Porter, M.F.: An algorithm for suffix stripping. Program, 313–316
14. Steinhaus, H.: Sur la division des corp materiels en parties. Bull. Acad. Polon. Sci. 1, 801–804
15. Wang, Y., Kitsuregawa, M.: Use Link-Based Clustering to Improve Web Search Results. In: WISE 2001: Proceedings of the Second International Conference on Web Information Systems Engineering (WISE 2001), Washington, DC, USA, vol. 1, p. 115 (2001)
16. Wang, X.-J., Ma, W.-Y., He, Q.-C., Li, X.: Grouping web image search result. In: MULTIMEDIA 2004: Proceedings of the 12th annual ACM international conference on Multimedia, pp. 436–439. ACM, New York (2004)
17. Yahoo! Image Search, http://images.search.yahoo.com
18. Yang, Y., Pedersen, J.O.: A Comparative Study on Feature Selection in Text Categorization. In: ICML 1997: Proceedings of the Fourteenth International Conference on Machine Learning, pp. 412–420. Morgan Kaufmann Publishers Inc., San Francisco (1997)

Variational Bayes for Generic Topic Models

Gregor Heinrich[1] and Michael Goesele[2]

[1] Fraunhofer IGD and University of Leipzig
[2] TU Darmstadt

Abstract. The article contributes a derivation of variational Bayes for a large class of topic models by generalising from the well-known model of latent Dirichlet allocation. For an abstraction of these models as systems of interconnected mixtures, variational update equations are obtained, leading to inference algorithms for models that so far have used Gibbs sampling exclusively.

1 Introduction

Topic models (TMs) are a set of unsupervised learning models used in many areas of artificial intelligence: In text mining, they allow retrieval and automatic thesaurus generation; computer vision uses TMs for image classification and content based retrieval; in bioinformatics they are the basis for protein relationship models etc.

In all of these cases, TMs learn latent variables from co-occurrences of features in data. Following the seminal model of latent Dirichlet allocation (LDA [6]), this is done efficiently according to a model that exploits the conjugacy of Dirichlet and multinomial probability distributions. Although the original work by Blei et al. [6] has shown the applicability of variational Bayes (VB) for TMs with impressive results, inference especially in more complex models has not adopted this technique but remains the domain of Gibbs sampling (e.g., [12,9,8]).

In this article, we explore variational Bayes for TMs in general rather than specific for some given model. We start with an overview of TMs and specify general properties (Sec. 2). Using these properties, we develop a generic approach to VB that can be applied to a large class of models (Sec. 3). We verify the variational algorithms on real data and several models (Sec. 4). This paper is therefore the VB counterpart to [7].

2 Topic Models

We characterise topic models as a form of discrete mixture models. Mixture models approximate complex distributions by a convex sum of component distributions, $p(x) = \sum_{k=1}^{K} p(x|z{=}k)p(z{=}k)$, where $p(z{=}k)$ is the weight of a component with index k and distribution $p(x|z{=}k)$.

Latent Dirichlet allocation as the simplest TM can be considered a mixture model with two interrelated mixtures: It represents documents m as mixtures of latent variables z with components $\vec{\vartheta}_m = p(z|m)$ and latent topics z as mixtures of words w with components $\vec{\beta}_k = p(w|z{=}k)$ and component weights $\vec{\vartheta}_m$, leading to a distribution over

B. Mertsching, M. Hund, and Z. Aziz (Eds.): KI 2009, LNAI 5803, pp. 161–168, 2009.

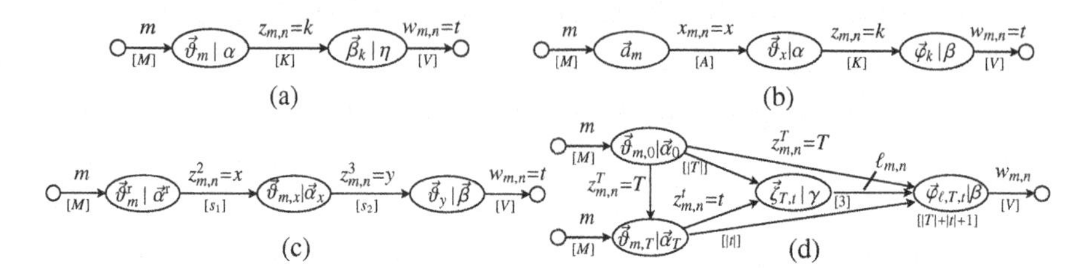

Fig. 1. Dependencies of mixture levels (ellipses) via discrete variables (arrows) in examples from literature: (a) latent Dirichlet allocation [6], (b) author–topic model (ATM [12], using observed parameters $\vec{a}_m$ to label documents, see end of Sec. 3), (c) 4-level pachinko allocation (PAM [9], models semantic structure with a hierarchy of topics $\vec{\vartheta}_m, \vec{\vartheta}_{m,x}, \vec{\vartheta}_y$), (d) hierarchical pachinko allocation (hPAM [8], topic hierarchy; complex mixture structure).

words w of $p(w|m) = \sum_{k=1}^{K} \vartheta_{m,k} \beta_{k,w}$.[1] The corresponding generative process is illustrative: For each text document m, a multinomial distribution $\vec{\vartheta}_m$ is drawn from a Dirichlet prior $\text{Dir}(\vec{\vartheta}_m|\alpha)$ with hyperparameter α. For each word token $w_{m,n}$ of that document, a topic $z_{m,n}{=}k$ is drawn from the document multinomial $\vec{\vartheta}_m$ and finally the word observation w drawn from a topic-specific multinomial over terms $\vec{\beta}_k$. Pursuing a Bayesian strategy with parameters handled as random variables, the topic-specific multinomial is itself drawn from another Dirichlet, $\text{Dir}(\vec{\beta}_k|\eta)$, similar to the document multinomial.

Generic TMs. As generalisations of LDA, topic models can be seen as a powerful yet flexible framework to model complex relationships in data that are based on only two modelling assumptions: (1) TMs are structured into Dirichlet–multinomial mixture "levels" to learn discrete latent variables (in LDA: z) and multinomial parameters (in LDA: β and ϑ). And (2) these levels are coupled via the values of discrete variables, similar to the coupling in LDA between ϑ and β via z.

More specifically, topic models form graphs of mixture levels with sets of multinomial components as nodes connected by discrete random values as directed edges. Conditioned on discrete inputs, each mixture level chooses one of its components to generate discrete output propagated to the next level(s), until one or more final levels produce observable discrete data. For some examples from literature, corresponding "mixture networks" are shown in Fig. 1, including the variant of observed multinomial parameters substituting the Dirichlet prior, which will be discussed further below.

For the following derivations, we introduce sets of discrete variables X, multinomial parameters Θ and Dirichlet hyperparameters A as model-wide quantities, and the corresponding level-specific quanitities $X^\ell, \Theta^\ell, A^\ell$ where superscript ℓ indicates the mixture level. The constraint of connecting different mixture levels (ellipses in Fig. 1) via discrete variables (arrows in Fig. 1) can be expressed by an operator $\uparrow x^\ell$ that yields all parent variables of a mixture level $\ell \in L$ generating variable x^ℓ. Here x^ℓ can refer to specific tokens $\uparrow x_i^\ell$ or configurations $\uparrow X^\ell$. Based on this and the definitions of the multinomial and Dirichlet distributions, the joint likelihood of any TM is:

[1] In example models, we use the symbols from the original literature.

$$p(X,\Theta|A)=\prod_{\ell\in L}p(X^\ell,\Theta^\ell|A^\ell,\uparrow X^\ell)=\prod_{\ell\in L}\left[\prod_i \text{Mult}(x_i|\Theta,\uparrow x_i)\prod_k \text{Dir}(\vec{\vartheta}_k|A,\uparrow X)\right]^{[\ell]} \quad (1)$$

$$=\prod_{\ell\in L}\left[\prod_i \vartheta_{k_i,x_i}\prod_k \frac{\Gamma(\sum_t \alpha_{j,t})}{\prod_t \Gamma(\alpha_{j,t})}\prod_t \vartheta_{k,t}^{\alpha_{j,t}-1}\right]^{[\ell]} ;\; k_i^\ell=g^\ell(\uparrow x_i^\ell,i),\; j^\ell=f^\ell(k^\ell)$$

$$=\prod_{\ell\in L}\left[\prod_k \frac{1}{\Delta(\vec{\alpha}_j)}\prod_t \vartheta_{k,t}^{n_{k,t}+\alpha_{j,t}-1}\right]^{[\ell]} ;\quad n_{k,t}^\ell=\left[\sum_i \delta(k_i-k)\,\delta(x_i-t)\right]^{[\ell]} . \quad (2)$$

In this equation, some further notation is introduced: We use brackets $[\cdot]^{[\ell]}$ to indicate that the contained quantities are specific to level ℓ. Moreover, the mappings from parent variables to component indices k_i are expressed by (level-specific) $k_i = g(\uparrow x_i, i)$, and $n_{k,t}^\ell$ is the number of times that a configuration $\{\uparrow x_i, i\}$ for level ℓ lead to component k^ℓ. Further, models are allowed to group components by providing group-specific hyperparameters $\vec{\alpha}_j$ with mapping $j = f(k)$. Finally, $\Delta(\vec{\alpha})$ is the normalisation function of the Dirichlet distribution, a K-dimensional beta function: $\Delta(\vec{\alpha}) \triangleq \prod_t \Gamma(\alpha_t)/\Gamma(\sum_t \alpha_t)$.

3 Variational Bayes for Topic Models

As in many latent-variable models, determining the posterior distribution $p(H,\Theta|V) = p(V,H,\Theta) \,/\, \sum_H \int p(V,H,\Theta)\,\mathrm{d}\Theta$ with hidden and visible variables $\{H,V\} = X$, is intractable in TMs because of excessive dependencies between the sets of latent variables H and parameters Θ in the marginal likelihood $p(V) = \sum_H \int p(V,H,\Theta)\,\mathrm{d}\Theta$ in the denominator. Variational Bayes [2] is an approximative inference technique that relaxes the structure of $p(H,\Theta|V)$ by a simpler variational distribution $q(H,\Theta|\Psi,\Xi)$ conditioned on sets of free variational parameters Ψ and Ξ to be estimated in lieu of H and Θ. Minimizing the Kullback-Leibler divergence of the distribution q to the true posterior can be shown to be equivalent to maximising a lower bound on the log marginal likelihood:

$$\begin{aligned}\log p(V) &\geq \log p(V) - \text{KL}\{q(H,\Theta)\,||\,p(H,\Theta|V)\}\\ &= \langle \log p(V,H,\Theta)\rangle_{q(H,\Theta)} + \text{H}\{q(H,\Theta)\} \triangleq \mathcal{F}\{q(H,\Theta)\}\end{aligned} \quad (3)$$

with entropy $\text{H}\{\cdot\}$. $\mathcal{F}\{q(H,\Theta)\}$ is the (negative) variational free energy – the quantity to be optimised using an EM-like algorithm that alternates between (E) maximising $\mathcal{F}$ w.r.t. the variational parameters to pull the lower bound towards the marginal likelihood and (M) maximising $\mathcal{F}$ w.r.t. the true parameters to raise the marginal likelihood.

Mean-field approximation. Following the variational mean field approach [2], in the LDA model the variational distribution consists of fully factorised Dirichlet and multinomial distributions [6]:[2]

$$q(\vec{z},\beta,\vartheta|\varphi,\lambda,\gamma)=\prod_{m=1}^{M}\prod_{n=1}^{N_m}\text{Mult}(z_{m,n}|\vec{\varphi}_{m,n})\cdot\prod_{k=1}^{K}\text{Dir}(\vec{\beta}_k|\vec{\lambda}_k)\prod_{m=1}^{M}\text{Dir}(\vec{\vartheta}_m|\vec{\gamma}_m) . \quad (4)$$

[2] In [6] this refers to the smoothed version; it is described in more detail in [5].

In [6], this approach proved very successful, which raises the question how it can be transferred to more generic TMs. Our approach is to view Eq. 4 as a special case of a more generic variational structure that captures dependencies $\uparrow X$ between multiple hidden mixture levels and includes LDA for the case of one hidden level ($H = \{\vec{z}\}$):

$$q(H, \Theta|\Psi, \Xi) = \prod_{\ell \in H} \Big[\prod_i \mathrm{Mult}(x_i|\vec{\psi}_i, \uparrow x_i) \Big]^{[\ell]} \prod_{\ell \in L} \Big[\prod_k \mathrm{Dir}(\vec{\vartheta}_k|\vec{\xi}_k, \uparrow X) \Big]^{[\ell]} , \quad (5)$$

where $\ell \in H$ refers to all levels that produce hidden variables. In the following, we assume that the indicator i is identical for all levels ℓ, e.g., words in documents $i^\ell = i \equiv (m, n)$. Further, tokens i in the corpus can be grouped into terms v and (observable) document-specific term frequencies $n_{m,v}$ introduced. We use shorthand $u = (m, v)$ to refer to specific unique tokens or document–term pairs.

Topic field. The dependency between mixture levels, $\uparrow x_u^\ell$, can be expressed by the likelihood of a particular configuration of hidden variables $\vec{x}_u{=}\vec{t} \triangleq \{x_u^\ell{=}t^\ell\}_{\ell \in H}$ under the variational distribution: $\psi_{u,\vec{t}} = q(\vec{x}_u{=}\vec{t}|\Psi)$. The complete structure $\underline{\psi}_u$ (the joint distribution over all $\ell \in H$ with $\Psi = \{\underline{\psi}_u\}_{\forall u}$) is a multi-way array of likelihoods for all latent configurations of token u with as many index dimensions as there are dependent variables. For instance, Fig. 1 reveals that LDA has one hidden variable with dimension K while PAM has two with dimensions $s_1 \times s_2$. Because of its interpretation as a mean field of topic states in the model, we refer to $\underline{\psi}_u$ as a "topic field" (in underline notation).

We further define $\psi^\ell_{u,k,t}$ as the likelihood of configuration (k^ℓ, t^ℓ) for document–term pair u. This "marginal" of $\underline{\psi}_u$ depends on the mappings between parent variables $\uparrow x_u$ and components k on each level. To obtain $\psi^\ell_{u,k,t}$, the topic field $\underline{\psi}_u$ is summed over all descendant paths that $x_u{=}t$ causes and the ancestor paths that can cause $k = g(\uparrow x_u, u)$ on level ℓ according to the generative process:

$$\psi^\ell_{u,k,t} = \sum_{\{\vec{t}^\ell_{\mathrm{A}}, \vec{t}^\ell_{\mathrm{D}}\}} \psi_{u;(\vec{t}^\ell_{\mathrm{A}}, k^\ell, t^\ell, \vec{t}^\ell_{\mathrm{D}})} \,; \quad \vec{t}^\ell_{\mathrm{A}} = \text{path causing } k^\ell \,, \vec{t}^\ell_{\mathrm{D}} = \text{path caused by } t^\ell \,. \quad (6)$$

Descendant paths $\vec{t}^\ell_{\mathrm{D}}$ of t^ℓ are obtained via recursion of $k = g(\uparrow x^d_u, u)$ over ℓ's descendant levels d. Assuming bijective $g(\cdot)$ as in the TMs in Fig. 1, the ancestor paths $\vec{t}^\ell_{\mathrm{A}}$ that correspond to components in parents leading to k^ℓ are obtained via $(\uparrow x^a_u, u) = g^{-1}(k)$ on ℓ's ancestor levels a recursively. Each pair $\{\vec{t}^\ell_{\mathrm{A}}, \vec{t}^\ell_{\mathrm{D}}\}$ corresponds to one element in $\underline{\psi}_u$ per $\{k^\ell, t^\ell\}$ at index vector $\vec{t} = (\vec{t}^\ell_{\mathrm{A}}, k^\ell, t^\ell, \vec{t}^\ell_{\mathrm{D}})$.

Free energy. Using Eqs. 2, 3, 5 and 6, the free energy of the generic model becomes:

$$\mathcal{F} = \sum_{\ell \in L} \Big[\sum_k \log \Delta(\vec{\xi}_k) - \log \Delta(\vec{\alpha}_j) + \sum_t \big((\textstyle\sum_u n_u \psi_{u,k,t}) + \alpha_{j,t} - \xi_{k,t} \big) \cdot \mu_t(\vec{\xi}_k) \Big]^{[\ell]}$$
$$- \sum_u n_u \sum_{\vec{t}} \psi_{u,\vec{t}} \log \psi_{u,\vec{t}} = \sum_{\ell \in L} \mathcal{F}^\ell + \mathrm{H}\{\Psi\} , \quad (7)$$

where $\mu_t(\vec{\xi}) \triangleq \Psi(\xi_t) - \Psi(\sum_t \xi_t) = \langle \log \vec{\vartheta} | \vec{\xi} \rangle_{\mathrm{Dir}(\vec{\vartheta}|\vec{\xi})} = \nabla_t \log \Delta(\vec{\xi})$, and $\Psi(\xi) \triangleq \mathrm{d}/\mathrm{d}x \log \Gamma(\xi)$ is the digamma function.[3]

[3] Note the distinction between the function $\Psi(\cdot)$ and quantity Ψ.

Variational E-steps. In the E-step of each model, the variational distributions for the joint multinomial ψ_u for each token (its topic field) and the Dirichlet parameters $\vec{\xi}_k^\ell$ on each level need to be estimated. The updates can be derived from the generic Eq. 7 by setting derivatives with respect to the variational parameters to zero, which yields:[4]

$$\psi_{u,\vec{t}} \propto \exp\left(\sum_{\ell\in L} \left[\mu_t(\vec{\xi}_k)\right]^{[\ell]}\right), \tag{8}$$

$$\xi_{k,t}^\ell = \left[\left(\sum_u n_u \psi_{u,k,t}\right) + \alpha_{j,t}\right]^{[\ell]} \tag{9}$$

where the sum $\sum_u n_u \psi_{u,k,t}^\ell$ for level ℓ can be interpreted as the expected counts $\langle n_{k,t}^\ell \rangle_q$ of co-occurrence of the value pair (k^ℓ, t^ℓ). The result in Eqs. 8 and 9 perfectly generalises that for LDA in [5].

M-steps. In the M-step of each model, the Dirichlet hyperparameters $\vec{\alpha}_j^\ell$ (or scalar α^ℓ) are calculated from the variational expectations of the log model parameters $\langle \log \vartheta_{k,t} \rangle_q = \mu_t(\vec{\xi}_k)$, which can be done at mixture level (Eq. 9 has no reference to $\vec{\alpha}_j^\ell$ across levels).

Each estimator for $\vec{\alpha}_j$ (omitting level ℓ) should "see" only the expected parameters $\mu_t(\vec{\xi}_k)$ of the K_j components associated with its group $j = f(k)$. We assume that components be associated a priori (e.g., PAM in Fig. 1c has $\vec{\vartheta}_{m,x} \sim \text{Dir}(\vec{\alpha}_x)$) and K_j is known. Then the Dirichlet ML parameter estimation procedure given in [6,10] can be used in modified form. It is based on Newton's method with the Dirichlet log likelihood function f as well as its gradient and Hessian elements g_t and h_{tu}:

$$\begin{aligned}
f(\vec{\alpha}_j) &= -K_j \log \Delta(\vec{\alpha}_j) + \textstyle\sum_t (\alpha_{j,t} - 1) \sum_{\{k:f(k)=j\}} \mu_t(\vec{\xi}_k) \\
g_t(\vec{\alpha}_j) &= -K_j \mu_t(\vec{\alpha}_j) + \textstyle\sum_{\{k:f(k)=j\}} \mu_t(\vec{\xi}_k) \\
h_{tu}(\vec{\alpha}_j) &= -K_j \Psi'(\textstyle\sum_s \alpha_{j,s}) + \delta(t-u) K_j \Psi'(\alpha_{j,t}) = z + \delta(t-u) h_{tt} \\
\alpha_{j,t} &\leftarrow \alpha_{j,t} - (\underline{H}^{-1}\vec{g})_t = \alpha_{j,t} - h_{tt}^{-1}\left(g_t - (\textstyle\sum_s g_s h_{ss}^{-1}) \big/ (z^{-1} + \sum_s h_{ss}^{-1})\right).
\end{aligned} \tag{10}$$

Scalar α (without grouping) is found accordingly via the symmetric Dirichlet:

$$\begin{aligned}
f &= -K[T \log \Gamma(\alpha) - \log \Gamma(T\alpha)] + (\alpha - 1) s_\alpha, \quad s_\alpha = \textstyle\sum_{k=1}^K \sum_{t=1}^T \mu_t(\vec{\xi}_k) \\
g &= KT[\Psi(T\alpha) - \Psi(\alpha) + s_\alpha], \quad h = KT[T\Psi'(T\alpha) - \Psi'(\alpha)] \\
\alpha &\leftarrow \alpha - g h^{-1}.
\end{aligned} \tag{11}$$

Variants. As an alternative to Bayesian estimation of all mixture level parameters, for some mixture levels ML point estimates may be used that are computationally less expensive (e.g., unsmoothed LDA [6]). By applying ML only to levels without document-specific components, the generative process for unseen documents is retained. The E-step with ML levels has a simplified Eq. 8, and ML parameters ϑ^c are estimated in the M-step (instead of hyperparameters):

$$\psi_{u,\vec{t}} \propto \exp\left(\sum_{\ell\in L\backslash c} \left[\mu_t(\vec{\xi}_k)\right]^{[\ell]}\right) \cdot \vartheta_{k,t}^c, \quad \vartheta_{k,t}^c = \langle n_{k,t}^c \rangle_q \big/ \langle n_k^c \rangle_q \propto \sum_u n_u \psi_{u,k,t}^c. \tag{12}$$

[4] In Eq. 8 we assume that $t^\ell = v$ on final mixture level(s) ("leaves"), which ties observed terms v to the latent structure. For "root" levels where component indices are observed, $\mu_t(\vec{\xi}_k)$ in Eq. 8 can be replaced by $\Psi(\xi_{k,t})$.

Moreover, as an extension to the framework specified in Sec. 2, it is straightforward to introduce observed parameters that for instance can represent labels, as in the author–topic model, cf. Fig 1. In the free energy in Eq. 7, the term with $\mu_t(\vec{\xi}_k)$ is replaced by $(\sum_u n_u \psi_{u,k,t}) \log \vartheta_{k,t}$, and consequently, Eq. 8 takes the form of Eq. 12 (left), as well.

Other variants like specific distributions for priors (e.g., logistic-normal to model topic correlation [4] and non-parametric approaches [14]) and observations (e.g., Gaussian components to model continuous data [1]), will not be covered here.

Algorithm structure. The complete variational EM algorithm alternates between the variational E-step and M-step until the variational free energy $\mathcal{F}$ converges at an optimum. At convergence, the estimated document and topic multinomials can be obtained via the variational expectation $\log \hat{\vartheta}_{k,t} = \mu_t(\vec{\xi}_k)$. Initialisation plays an important role to avoid local optima, and a common approach is to initialise topic distributions with observed data, possibly using several such initialisations concurrently. The actual variational EM loop can be outlined in its generic form as follows:

1. Repeat E-step loop until convergence w.r.t. variational parameters:
   1. For each observed unique token u:
      1. For each configuration $\vec{t}$: calculate var. multinomial $\psi_{u,\vec{t}}$ (Eq. 8 or 12 left).
      2. For each (k, t) on each level ℓ: calculate var. Dirichlet parameters $\xi^{\ell}_{k,t}$ based on topic field marginals $\psi^{\ell}_{u,k,t}$ (Eqs. 6 and 9), which can be done differentially: $\xi^{\ell}_{k,t} \leftarrow \xi^{\ell}_{k,t} + n_u \Delta\psi^{\ell}_{u,k,t}$ with $\Delta\psi^{\ell}_{u,k,t}$ the change of $\psi^{\ell}_{u,k,t}$.
   2. Finish variational E-step if free energy $\mathcal{F}$ (Eq. 7) converged.
2. Perform M-step:
   1. For each j on each level ℓ: calculate hyperparameter $\alpha^{\ell}_{j,t}$ (Eqs. 10 or 11), inner iteration loop over t.
   2. For each (k, t) in point-estimated nodes ℓ: estimate $\vartheta^{\ell}_{k,t}$ (Eq. 12 right).
3. Finish variational EM loop if free energy $\mathcal{F}$ (Eq. 7) converged.

In practice, similar to [5], this algorithm can be modified by separating levels with document-specific variational parameters $\Xi^{\ell,m}$ and such with corpus-wide parameters $\Xi^{\ell,*}$. This allows a separate E-step loop for each document m that updates ψ_u and $\Xi^{\ell,m}$ with $\Xi^{\ell,*}$ fixed. Parameters $\Xi^{\ell,*}$ are updated afterwards from changes $\Delta\psi^{\ell}_{u,k,t}$ cumulated in the document-specific loops, and their contribution added to $\mathcal{F}$.

4 Experimental Verification

In this section, we present initial validation results based on the algorithm in Sec. 3.

Setting. We chose models from Fig. 1, LDA, ATM and PAM, and investigated two versions of each: an unsmoothed version that performs ML estimation of the final mixture level (using Eq. 12) and a smoothed version that places variational distributions over all parameters (using Eq. 8). Except for the component grouping in PAM ($\vec{\vartheta}_{m,x}$ have vector hyperparameter $\vec{\alpha}_x$), we used scalar hyperparameters. As a base-line, we used Gibbs sampling implementations of the corresponding models. Two criteria are immediately useful: the ability to generalise to test data V' given the model parameters Θ, and the convergence time (assuming single-threaded operation). For the

Model:		LDA			ATM			PAM		
Dimensions {A,B}:		$K = \{25, 100\}$			$K = \{25, 100\}$			$s_{1,2} = \{(5,10),(25,25)\}$		
Method:		GS	VB_{ML}	VB	GS	VB_{ML}	VB	GS	VB_{ML}	VB
Convergence time [h]	A	0.39	0.83	0.91	0.73	1.62	1.79	0.5	1.25	1.27
	B	1.92	3.75	4.29	3.66	7.59	8.1	5.61	14.86	16.06
Iteration time [sec]	A	4.01	157.3	164.2	6.89	254.3	257.8	5.44	205.1	207.9
	B	16.11	643.3	671.0	29.95	1139.2	1166.9	53.15	2058.2	2065.1
Iterations	A	350	19	20	380	23	25	330	22	22
	B	430	21	23	440	24	25	380	26	28
Perplexity	A	1787.7	1918.5	1906.0	1860.4	1935.2	1922.8	2053.8	2103.0	2115.1
	B	1613.9	1677.6	1660.2	1630.6	1704.0	1701.9	1909.2	1980.5	1972.6

Fig. 2. Results of VB and Gibbs experiments

first criterion, because of its frequent usage with topic models we use the perplexity, the inverse geometric mean of the likelihood of test data tokens given the model: $\mathcal{P}(V') = \exp(-\sum_u n_u \log p(v'_u|\Theta')/W')$ where Θ' are the parameters fitted to the test data V' with W' tokens. The log likelihood of test tokens $\log p(v'_u|\Theta')$ is obtained by (1) running the inference algorithms on the test data, which yields Ξ' and consequently Θ', and (2) marginalising all hidden variables h'_u in the likelihood $p(v'_u|h'_u, \Theta') = \prod_{\ell \in L} [\vartheta_{k,t}]^{[\ell]}$.[5]

The experiments were performed on the NIPS corpus [11] with $M = 1740$ documents (174 held-out), $V = 13649$ terms, $W = 2301375$ tokens, and $A = 2037$ authors.

Results. The results of the experiments are shown in Fig. 2. It turns out that generally the VB algorithms were able to achieve perplexity reductions in the range of their Gibbs counterparts, which verifies the approach taken. Further, the full VB approaches tend to yield slightly improved perplexity reductions compared to the ML versions. However, these first VB results were consistently weaker compared to the baselines. This may be due to adverse initialisation of variational distributions, causing VB algorithms to become trapped at local optima. It may alternatively be a systematic issue due to the correlation between Ψ and Ξ assumed independent in Eq. 5, a fact that has motivated the collapsed variant of variational Bayes in [13]. Considering the second evaluation criterion, the results show that the current VB implementations generally converge less than half as fast as the corresponding Gibbs samplers. This is why currently work is undertaken in the direction of code optimisation, including parallelisation for multikernel CPUs, which, opposed to (collapsed) Gibbs samplers, is straightforward for VB.

5 Conclusions

We have derived variational Bayes algorithms for a large class of topic models by generalising from the well-known model of latent Dirichlet allocation. By an abstraction of these models as systems of interconnected mixture levels, we could obtain variational update equations in a generic way, which are the basis for an algorithm, that can be easily applied to specific topic models. Finally, we have applied the algorithm to a couple of example models, verifying the general applicability of the approach. So far, especially more complex topic models have predominantly used inference based on Gibbs

[5] In contrast to [12], we also used this method to determine ATM perplexity (from the $\vec{\varphi}_k$).

sampling. Therefore, this paper is a step towards exploring the possibility of variational approaches. However, what can be drawn as a conclusion from the experimental study in this paper, more work remains to be done in order to make VB algorithms as effective and efficient as their Gibbs counterparts.

Related work. Beside the relation to the original LDA model [6,5], especially the proposed representation of topic models as networks of mixture levels makes work on discrete DAG models relevant: In [3], a variational approach for structure learning in DAGs is provided with an alternative derivation based on exponential families leading to a structure similar to the topic field. They do not discuss mapping of components or hyperparameters and restrict their implementations to structure learning in graphs bipartite between hidden and observed nodes. Also, the authors of [9] present their pachinko allocation models as DAGs, but formulate inference based on Gibbs sampling. In contrast to this, the novelty of the work presented here is that it unifies the theory of topic models in general including labels, the option of point estimates and component grouping for variational Bayes, giving empirical results for real-world topic models.

Future work will optimise the current implementations with respect to efficiency in order to improve the experimental results presented here, and an important aspect is to develop parallel algorithms for the models at hand. Another research direction is the extension of the framework of generic topic models, especially taking into consideration the variants of mixture levels outlined in Sec. 3. Finally, we will investigate a generalisation of collapsed variational Bayes [13].

References

1. Barnard, K., Duygulu, P., Forsyth, D., de Freitas, N., Blei, D., Jordan, M.: Matching words and pictures. JMLR 3(6), 1107–1136 (2003)
2. Beal, M.J.: Variational Algorithms for Approximate Bayesian Inference. PhD thesis, Gatsby Computational Neuroscience Unit, University College London (2003)
3. Beal, M.J., Ghahramani, Z.: Variational bayesian learning of directed graphical models with hidden variables. Bayesian Analysis 1, 793–832 (2006)
4. Blei, D., Lafferty, J.: A correlated topic model of science. AOAS 1, 17–35 (2007)
5. Blei, D., Ng, A., Jordan, M.: Hierarchical Bayesian models for applications in information retrieval. Bayesian Statistics 7, 25–44 (2003)
6. Blei, D., Ng, A., Jordan, M.: Latent Dirichlet allocation. JMLR 3, 993–1022 (2003)
7. Heinrich, G.: A generic approach to topic models. In: ECML/PKDD (2009)
8. Li, W., Blei, D., McCallum, A.: Mixtures of hierarchical topics with pachinko allocation. In: ICML (2007)
9. Li, W., McCallum, A.: Pachinko allocation: DAG-structured mixture models of topic correlations. In: ICML (2006)
10. Minka, T.: Estimating a Dirichlet distribution. Web (2003)
11. NIPS corpus, `http://www.cs.toronto.edu/~roweis/data.html`
12. Steyvers, M., Smyth, P., Rosen-Zvi, M., Griffiths, T.: Probabilistic author-topic models for information discovery. In: ACM SIGKDD (2004)
13. Teh, Y.W., Newman, D., Welling, M.: A collapsed variational Bayesian inference algorithm for latent Dirichlet allocation. In: NIPS, vol. 19 (2007)
14. Teh, Y.W., Jordan, M.I., Beal, M.J., Blei, D.M.: Hierarchical Dirichlet processes. Technical Report 653, Department of Statistics, University of California at Berkeley (2004)

Surrogate Constraint Functions for CMA Evolution Strategies

Oliver Kramer, André Barthelmes, and Günter Rudolph

Department of Computer Science
Technische Universität Dortmund
44227 Dortmund, Germany

Abstract. Many practical optimization problems are constrained black boxes. Covariance Matrix Adaptation Evolution Strategies (CMA-ES) belong to the most successful black box optimization methods. Up to now no sophisticated constraint handling method for Covariance Matrix Adaptation optimizers has been proposed. In our novel approach we learn a meta-model of the constraint function and use this surrogate model to adapt the covariance matrix during the search at the vicinity of the constraint boundary. The meta-model can be used for various purposes, i.e. rotation of the mutation ellipsoid, checking the feasibility of candidate solutions or repairing infeasible mutations by projecting them onto the constraint surrogate function. Experimental results show the potentials of the proposed approach.

1 Introduction

Constraints typically are not considered available in their explicit formal form, but are assumed to be black boxes: a vector $\mathbf{x}$ fed to the black box just returns a numerical or boolean value. The constrained real-valued optimization problem is to find a solution $\mathbf{x} = (x_1, x_2, \ldots, x_N)^T$ in the N-dimensional solution space $\mathbb{R}^N$ that minimizes the objective function $f(\mathbf{x})$, in symbols:

$$\begin{array}{ll} f(\mathbf{x}) \rightarrow \min!, \mathbf{x} \in \mathbb{R}^N & \text{subject to} \\ \text{inequalities} \quad g_i(\mathbf{x}) \leq 0, \; i = 1, \ldots, n_1, & \text{and} \\ \text{equalities} \quad h_j(\mathbf{x}) = 0, \; j = 1, \ldots, n_2 \,. & \end{array} \tag{1}$$

The question arises how to cope with constraints that are not given explicitly. Various constraint handling methods for evolutionary computation have been proposed in the last decades. A survey is not possible due to the limited space. Coello [1] or Kramer [4] deliver a good starting point for literature research. Most methods fall into the category of penalty functions, e.g. by Kuri *et al.* [5] allowing the search process to discover the whole search space, but penalizing the infeasible part. Other approaches are based on decoders, e.g. by Michalewizc [7] or multi-objective optimization, e.g. by Jimenez [3].

Covariance Matrix Adaptation belongs to the most successful black box optimization algorithms in real-valued search spaces. The idea of covariance matrix

B. Mertsching, M. Hund, and Z. Aziz (Eds.): KI 2009, LNAI 5803, pp. 169–176, 2009.

adaptation techniques is to adapt the distribution of the mutation operator such that the probability to reproduce steps that led to the actual population increases. The CMA-ES was introduced by Ostermeier et al. [8,2]. All previous CMA-ES implementations make use of death penalty as constraint handling technique. The results of the CMA-ES on problems TR2 and 2.40[1] can be found in table 1. In comparison to a standard evolution strategy the CMA-ES is able to cope with the low success rates around the optimum of the TR-problem. The average number of infeasible solutions during the approximation is 44%. This indicates that a reasonable adaptation of the mutation ellipsoid takes place. As the TR problem exhibits a linear objective and a linear constraint function we would expect a rate of $\approx 50\%$ infeasible solutions and a decrease of step sizes for an ill-adapted mutation ellipsoid. An analysis of the angle between the main axis of the mutation ellipsoid and the constraint function converges to zero, the same do the step sizes during approximation of the optimum. On problem 2.40 about 64% of the produced solutions are infeasible.

Table 1. Experimental analysis of the CMA-ES with death penalty

CMA-ES (DP)	best	mean	worst	dev	ffe	cfe
TR2	0.0	0.0	$2.7 \cdot 10^{-15}$	$5.8 \cdot 10^{-16}$	6,754	12,019
2.40	0.0	0.0	$9.1 \cdot 10^{-13}$	$1.3 \cdot 10^{-13}$	19,019	71,241

2 A Fast Surrogate Model Estimator for Linear Constraints

To construct more powerful constraint handling methods the idea arises to build a meta-model of the constraint function. This meta-model can be used in various kinds of ways. We will use the surrogate model for the rotation of the mutation ellipsoid, checking the feasibility of mutations, and the repair of infeasible mutations of the CMA-ES. In this work we concentrate on linear constraints, but the approach can be extended to the non-linear case.

To efficiently estimate the constraint surrogate function we developed a new efficient constraint boundary estimator based on fast binary search. We assume that our constraint handling method starts with the first occurrence of an infeasible individual. The estimation algorithm for the linear constraint boundary, i.e. a linear hyper-plance h_0 works as follows.

In the first step the surrogate estimator identifies N points *on* the N-dimensional linear constraint hyper-plane h_0. For this sake the model-estimator makes use of an N-dimensional hypersphere cutting the constraint boundary. The connection between the infeasible points on the hypersphere and the center of the hypersphere will cut the constraint boundary. When the first infeasible offspring individual $(\xi)_1$ is produced, the original feasible parent $\boldsymbol{x}_f$ is the center of the

[1] Problem TR2 is the Sphere function $\boldsymbol{x}^T\boldsymbol{x}$ with the linear constraint $\sum x_i - N > 0$, problem 2.40 is a linear function with 6 linear constraints, see Schwefel [9].

corresponding meta-model estimator and the distance becomes radius r of your model-estimator. The approach uses binary search to find the cutting points and takes the following three steps:

1. Generation of random points on the surface of the hypersphere: Point $\boldsymbol{x}_f$ is the center of the hypersphere with radius r such that the constraint boundary is cut. $N-1$ additional infeasible points $\xi_i, \quad 1 \leq i \leq N-1$ have to be produced. Infeasible points are produced by sampling on its surface until a sufficient number of infeasible points are produced. The points on the surface are sampled randomly with uniform distribution using the method of Marsaglia [6]: At first, $N-1$ Gaussian distributed points are sampled $(\xi_j)_i \sim \mathcal{N}(0,1) \quad j = 1, \ldots, N$ and scaled to length 1. Further scaling and shifting yields N randomly distributed point on the hypersphere surface

$$(\xi)_i = \frac{1}{\sqrt{(\xi_1)_i^2 + \ldots + (\xi_N)_i^2}} \cdot ((\xi_1)_i, \ldots, (\xi_N)_i)^T \tag{2}$$

2. Identification of N points $s_1, \ldots, s_N$ *on* the estimated constraint boundary, see Figure 1: The line between the feasible point $\boldsymbol{x}_f$ and the i-th infeasible point $(\xi)_i$ cuts the real constraint hyper-plane h^* in point s_i^*. Binary search is a fast approach to approximate s_i^* on this segment: The center s_i of the last interval defined by the last points of the binary search becomes an estimation of point s_i^* on h^*. The real hyperplane lies between h_1^* and h_2^*. Table 2 shows the empirical number of binary search steps to achieve the angle accuracy $\phi < 1°$ with accuracy $\phi < 0.25°$ of the estimated hyperplane h_0. Each experiment was repeated 100,000 times, i.e. in each repetition a new test case with a random hyper-plane and random points $\boldsymbol{x}_f, (\xi)_1, \ldots (\xi)_N \in \mathbb{R}$ was generated. The error of the angle ϕ can be estimated by the number of binary search steps: Let ϕ_k be the average angle error after k binary steps and η be the accuracy improvement factor we are able to achieve with one binary search step. The experimental analysis yields the following linear relation between angle accuracy ϕ and binary search steps $k = j - i$:

$$\phi_i \cdot \eta^{j-i} = \phi_j \tag{3}$$

   The magnitude of the efficiency factor η, i.e. an improvement of angle accuracy, can be estimated as $0.53 \leq \eta \leq 0.57$ from our experiments, see Table 2. Note, that the number of binary search steps grows slower than the number of dimensions of the solution space.
3. Finally, we calculate the normal vector $\boldsymbol{n}_0$ of h_0 using the N points on the constraint boundary. Assuming the points $s_i, \quad 1 \leq i \leq N$, represent linearly independent vectors a successive Gram-Schmidt orthogonalization of the $(i+1)$-th vector on the i-th previously produced vectors delivers the normal vector $\boldsymbol{n}_0$ of h_0.

Note, that we estimate the normal vector $\boldsymbol{n}_0$ of the linear constraint model h_0 only one time, i.e. when the first infeasible solutions have been detected. Later,

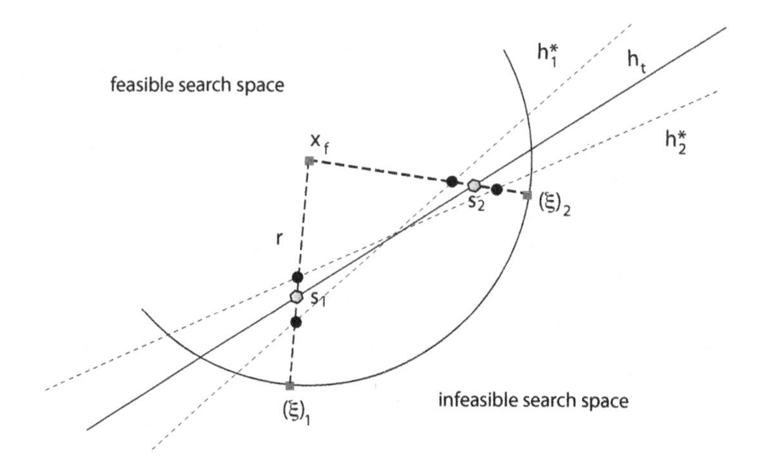

Fig. 1. Procedure to estimate the constraint boundary h_0 in two dimensions: The method performs binary search on the segments between a feasible point $\boldsymbol{x}_f$ and each two infeasible points $(\xi)_1$, $(\xi)_2$ to estimate two points s_1, s_2 on the meta-model

Table 2. Number of binary search steps and constraint function evaluations to limit the average angle accuracy in an artificial setup with one hyperplane, a feasible and an infeasible point

dimension	steps ($\phi \leq 1°$)	mean error	steps ($\phi \leq 0.25°$)	mean error	cfe
2	9	0.85	11	0.24	3.14
10	14	0.68	16	0.21	21.56
20	15	0.76	17	0.25	42.96

update steps only concern the local support point p_t of the hyper-plane[2]. At the beginning, any of the points s_i may be the support point p_0.

3 Rotation of the Covariance Matrix

Our first idea to exploit the surrogate constraint function was to adapt the covariance matrix itself. Our experiments with correlated Gaussian mutation for ES motivated that step[3]. The covariance matrix $\boldsymbol{C} \in \mathbb{R}^{N \times N}$ is a positive semi-definite matrix and thus, exhibits a decomposition into an orthonormal matrix $\boldsymbol{B}$ and a diagonal matrix $\boldsymbol{D}$, such that $\boldsymbol{C} = \boldsymbol{B}\boldsymbol{D}^2\boldsymbol{B}^T$. In case of approximating the constraint boundary $\boldsymbol{h}$ with normal vector $\boldsymbol{n}$ we adapt the covariance matrix $\boldsymbol{C}$ as follows:

[2] Hence, in iteration t the linear model h_t is specified by normal vector $\boldsymbol{n}_0$ and current support point p_t.

[3] Correlated mutation by Schwefel [9] rotates the axes of the hyper-ellipsoid to adapt to local properties of the fitness landscape by means of self-adaptation.

1. Decomposition of covariance matrix $\boldsymbol{C}$ into orthonormal matrix $\boldsymbol{B}$ and diagonal matrix $\boldsymbol{D}$.
2. Replacement of vector $\boldsymbol{c}_i$ of matrix $\boldsymbol{C}$ with the least angle d to the normal vector $\boldsymbol{n}_0$, i.e. $\boldsymbol{c}_i$ such that $\boldsymbol{c}_i\boldsymbol{n}_0$ is minimal, by normalized vector $\boldsymbol{n}_0/\|\boldsymbol{n}_0\|$.
3. Orthogonalization of $\boldsymbol{C}$ starting with $\boldsymbol{n}_0/\|\boldsymbol{n}_0\|$.
4. Scale of the $N-1$ vectors of $\boldsymbol{D}$ that are orthogonal to $\boldsymbol{n}$ with regard to a projection of $\boldsymbol{C}$ onto h, and scale of the vector parallel to $\boldsymbol{n}$ to 1/200-th of its original length.
5. Computation of $\boldsymbol{C} = \boldsymbol{B}\boldsymbol{D}^2\boldsymbol{B}^T$.

The CMA-ES runs as usual until the constraint boundary is met. The algorithm starts the constraint handling technique if 40% of the mutations are infeasible. Figure 3 shows the results of our experimental analysis on TR2 and 2.40. We can observe a slight decrease of necessary fitness function calls of about 5% in average on TR2. This slight speedup is a result of a faster adaptation of $\boldsymbol{C}$ to the constraint boundary. After the adaptation is complete, the optimization process continuous in the usual kind of way. No speedup can be observed on problem 2.40, the CMA-ES was already successful in adapting $\boldsymbol{C}$ to the constraint conditions.

Table 3. Experimental analysis of the CMA-ES using the surrogate model for adaptation of $\boldsymbol{C}$

CMA-ES (DP)	best	mean	dev	ffe	cfe
TR2	0.0	0.0	$3.1 \cdot 10^{-15}$	6,380	11,374
2.40	0.0	0.0	$1.3 \cdot 10^{-13}$	18,998	71,559

4 Checking Feasibility

In a second approach potentially feasible solutions are checked for feasibility with the surrogate model. In particular, this is recommendable if constraint function calls are expensive. For this purpose an exact estimation of the constraint boundary in necessary. Two errors for the feasibility prediction of individual $\boldsymbol{x}_t$ are possible:

1. The model predicts $\boldsymbol{x}_t$ is feasible, but it is not. Points of this category are examined for feasibility causing an additional constraint function evaluation.
2. The model predicts $\boldsymbol{x}_t$ is infeasible, but it is feasible. The individual will be discarded, but may be a very good approximation of the optimum.

We introduce a *safety margin* δ, i.e. a shift of the estimated constraint boundary into the infeasible direction. The safety margin reduces errors of type 2. We set δ to the distance d of the mutation ellipsoid center c and the estimated constraint boundary h_t. The distance between c and the shifted constraint boundary h'_t becomes $2d$. A regular update of the constraint boundary support point p_t is necessary. Again, we have to distinguish between two conditions in each iteration. Let d_{t_0} be the distance between the mutation ellipsoid center c_{t_0} and the constraint boundary h_{t_0} at time t_0 and let k be the number of binary search steps to achieve the angle accuracy of $\delta < 0.25°$.

1. The center of the mutation ellipsoid c_t approaches h_t: If distance d_t between h_t and c_{t_0} becomes smaller than $d_{t_0}/2^k$, a re-estimation of the support point p_t is reasonable.
2. The search c_t moves parallel to h_t: An exceeding of distance

$$c_{t_0} - c_t = \sqrt{\frac{1}{\tan(\phi)^2} + 4} \cdot d_{t_0} \tag{4}$$

with $\phi = 0.25 \cdot (0.57)^{3k}$ causes a re-estimation of h_t.

To find the new support point p_t we use $4k$ binary steps on the line between the current infeasible solutions and c_t. Table 4 shows significant savings of constraint evaluations with a high approximation capability using the feasibility checking approach.

Table 4. Results of the CMA-ES with meta-model feasibility checking

CMA-ES (F-Checking)	best	mean	dev	ffe	cfe
TR2	0.0	0.0	$6.9 \cdot 10^{-16}$	6,780	7,781
2.40	0.0	0.0	$1.8 \cdot 10^{-13}$	19,386	34,254

5 Repair of Infeasible Solutions

At last, we propose to repair infeasible mutations by projection onto the constraint boundary h_t. We assume the angle error ϕ that can again be estimated by the number of binary search steps k, see equation 3. In our approach we elongate the projection vector by length δ. Figure 2 illustrates the situation. Let p_t be the support point of the hyper-plane h_t at time t and let x_i be the infeasible solution. Since $a^2 + b^2 = d^2$ and $\delta/b = \tan\phi$, we obtain

$$\delta = \sqrt{a^2 - d^2} \cdot \tan\phi. \tag{5}$$

The elongation of the projection into the potentially feasible region guarantees feasibility of the repaired individuals. Nevertheless, it might prevent fast convergence, in particular in regions far away from the hyper-plane support point p_t, as δ grows with increasing length of d. In our approach we update the center of the hyper-plane for an update of accuracy every 10 generations. The results of the CMA-ES repair algorithm can be found in Table 5. We observe a significant decrease of fitness function evaluations. The search concentrates on the boundary of the infeasible search space, in particular on the feasible site. Of course, no saving of constraint function evaluations could have been expected.

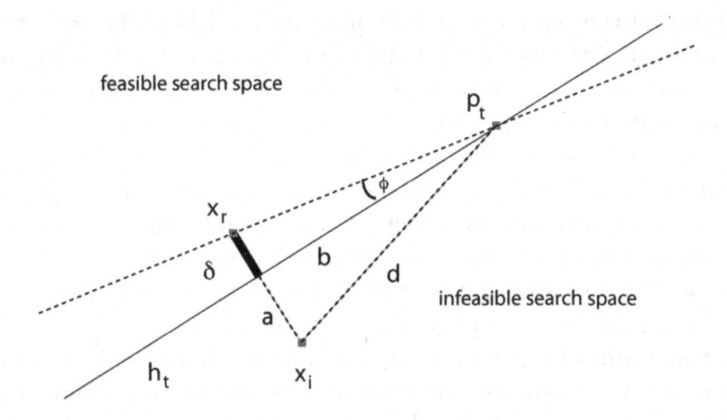

Fig. 2. Safety margin: the elongation of the projection of infeasible solution x_i onto the estimated constraint boundary h_t by length δ ensures the feasibility of the repaired point x_r

Table 5. Results of the CMA-ES with repair mechanism

CMA-ES (Repair)	best	mean	dev	ffe	cfe
TR2	0.0	0.0	$5.5 \cdot 10^{-16}$	3,432	5,326
2.40	0.0	0.0	$9.1 \cdot 10^{-14}$	16,067	75,705

6 Combination of Approaches

At last, we combine all three introduced approaches. Table 6 shows the results of the combined constraint handling techniques on problems TR2 and 2.40. In comparison to the repair approach of Section 5 the number of constraint function calls decreases significantly and the relation between fitness and constraint function evaluations reaches the level of the feasibility check technique in Section 4. In comparison to the results with death penalty, see Table 1, a significant decrease of objective and constraint function evaluations can be observed.

Table 6. Results of the CMA-ES with covariance matrix rotation, feasibility check and repair mechanism

CMA-ES (all)	best	mean	worst	dev	ffe	cfe
TR2	0.0	0.0	$8.9 \cdot 10^{-16}$	$5.1 \cdot 10^{-16}$	3,249	3,650
2.40	0.0	0.0	$9.1 \cdot 10^{-13}$	$9.1 \cdot 10^{-14}$	11,216	30,069

7 Summary and Outlook

We have proposed the first sophisticated constraint handling technique for the CMA-ES. The better the constraints are known, the more information can be investigated during the search process. Surrogate models of the constraint functions turn out to be very useful for handling of constraints. To compute a constraint surrogate function we proposed an efficient algorithm for the estimation of linear constraints based on sampling on the surface of a hypersphere and binary search on the lines into the infeasible part of the search space. Based on this model we used the surrogate constraint function for multiple purposes, i.e. rotation of the covariance matrix, checking of feasibility and repair of infeasible points. Significant savings of fitness and constraint evaluations can be observed. If the effort spent on the surrogate constraint model and the various techniques is profitable, depends on the application scenario.

As many constraints in practical applications are not linear, we plan to extend our approach to non-linear constraints. First experiments show that a linear surrogate approximation of non-linear constraints is sufficient and leads to amazingly good results. Furthermore, the approach can be extended concerning the handling of multiple constraints functions. For this purpose, it seems to be reasonable to develop a detailed case sensitive surrogate model.

References

1. Coello Coello, C.A.: Theoretical and numerical constraint handling techniques used with evolutionary algorithms: A survey of the state of the art. Computer Methods in Applied Mechanics and Engineering 191(11-12), 1245–1287 (2002)
2. Hansen, N.: The CMA evolution strategy: A tutorial. Technical report, TU Berlin, ETH Zürich (2005)
3. Jimenez, F., Verdegay, J.L.: Evolutionary techniques for constrained optimization problems. In: Zimmermann, H.-J. (ed.) 7th European Congress on Intelligent Techniques and Soft Computing (EUFIT 1999). Verlag Mainz, Aachen (1999)
4. Kramer, O.: Self-Adaptive Heuristics for Evolutionary Computation. Springer, Berlin (2008)
5. Kuri-Morales, A., Quezada, C.V.: A universal eclectic genetic algorithm for constrained optimization. In: Proceedings 6th European Congress on Intelligent Techniques and Soft Computing (EUFIT 1998), September 1998, pp. 518–522. Verlag Mainz, Aachen (1998)
6. Marsaglia, G.: Choosing a point from the surface of a sphere. The Annals of Mathematical Statistics 43, 645–646 (1972)
7. Michalewicz, Z., Fogel, D.B.: How to Solve It: Modern Heuristics. Springer, Berlin (2000)
8. Ostermeier, A., Gawelczyk, A., Hansen, N.: A derandomized approach to self adaptation of evolution strategies. Evolutionary Computation 2(4), 369–380 (1994)
9. Schwefel, H.-P.: Evolution and Optimum Seeking. Sixth-Generation Computer Technology. Wiley Interscience, New York (1995)

Rake Selection: A Novel Evolutionary Multi-Objective Optimization Algorithm

Oliver Kramer and Patrick Koch

Department of Computer Science
Technische Universität Dortmund
44227 Dortmund, Germany

Abstract. The optimization of multiple conflictive objectives at the same time is a hard problem. In most cases, a uniform distribution of solutions on the Pareto front is the main objective. We propose a novel evolutionary multi-objective algorithm that is based on the selection with regard to equidistant lines in the objective space. The so-called rakes can be computed efficiently in high dimensional objective spaces and guide the evolutionary search among the set of Pareto optimal solutions. First experimental results reveal that the new approach delivers a good approximation of the Pareto front with uniformly distributed solutions. As the algorithm is based on a $(\mu + \lambda)$-Evolution Strategy with birth surplus it can use σ-self-adaptation. Furthermore, the approach yields deeper insights into the number of solutions that are necessary for a uniform distribution of solutions in high-dimensional objective spaces.

1 Introduction

1.1 Multi-Objective Black Box Optimization

Many real-world black box optimization problems are multi-objective. Maximization of performance and minimization of costs is a typical example. Formally, the multi-objective optimization problem is defined as follows:

$$\min_{\boldsymbol{x} \in \mathbb{R}^N} \boldsymbol{f}(\boldsymbol{x}) = \min_{\boldsymbol{x} \in \mathbb{R}^N} (f_1(\boldsymbol{x}), f_2(\boldsymbol{x}), ..., f_m(\boldsymbol{x})) \tag{1}$$

with $f_i(x) : \mathbb{R}^N \rightarrow \mathbb{R}, i = 1, ..., m$. We seek for a set of Pareto optimal solutions, i.e. the Pareto set $\mathcal{P} = \{x^* \in \mathbb{R}^N | \nexists x \in \mathbb{R}^N : \forall i \in \{1, ..m\} : f_i(x) \leq f_i(x^*) \wedge \nexists j \in \{1, .., m\} : f_j(x) < f_j(x^*)\}$. Evolutionary multi-objective algorithms (EMOA) have shown great success in this field in the last decade. The SMS-EMOA [1], NSGA-II [4], and SPEA2 [7] belong to the most successful ones. In this paper we propose a new EMOA that is based on geometrical properties in the objective space. In Section 2 we propose the new multi-objective approach called Rake Selection. Afterwards, we present a first empirical study in Section 3 and discuss various aspects of the new technique like movement towards the Pareto front in Section 4.

B. Mertsching, M. Hund, and Z. Aziz (Eds.): KI 2009, LNAI 5803, pp. 177–184, 2009.

1.2 Related Work

Due to the limited space we only recapitulate the two variants that are closely related to Rake Selection[1]. The non-dominated-sorting genetic algorithm (NSGA) and its successor (NSGA-II) by Deb *et al.* [4] belong to the most frequently used algorithms in multi-objective optimization. The idea is to rank individuals of the population according to their non-domination level. All individuals that are not dominated by any others are assigned to the first rank. In a next step they are removed from the population and the non-dominance check is conducted again. These steps are repeated, until all individuals have been assigned to a rank according to their non-dominance level. Best ranked solutions with maximum crowding distance values are added to the population of the next generation.

The SMS-EMOA [1] belongs to the indicator-based EMOA using the S-metric as selection objective. The S-metric – also called hypervolume indicator – measures the size of the space dominated by at least one solution. Maximization of the S-metric among non-dominated solutions is a combinatorial problem. Hence, the SMS-EMOA applies a $(\mu+1)$ (or steady-state) selection scheme in each generation discarding the individual that contributes least to the S-metric value of the population. The invoked variation operators are not specific for the SMS-EMOA or NSGA-II, but are taken from literature [3], namely polynomial mutation and simulated binary crossover with the same parameterization as for NSGA-II.

2 Rake Selection

In this section we introduce a novel evolutionary optimization technique with a geometric based selection scheme. This scheme is designed to produce *approximately equidistant solutions on the Pareto front.* The proposed algorithm is based on evolution strategies (ES). For a comprehensive introduction to ES we refer to Beyer and Schwefel [2,6].

2.1 Idea of Rake Selection

Let m be the number of objective functions $f_1(\boldsymbol{x}), \ldots f_m(\boldsymbol{x})$ with $\boldsymbol{x} \in \mathbb{R}^N$. The intuition of rake selection is to define $k = k_0^{m-1}$ rakes – parallel straight lines in the objective space – that guide the evolutionary search and lead to a uniform, i.e. approximately equidistant, distribution on the final Pareto front $\mathcal{F}$. Parameter k_0 is the number of rakes along the connection of two extreme points on the Pareto front, see Paragraph 2.3. Based on these *edge solutions* that minimize the respective objectives, an $m-1$-dimensional hypergrid with k uniformly distributed intercept points $\boldsymbol{p}_i$, $1 \leq i \leq k_0^{m-1}$ is defined. We define *rake* lines l_j, $1 \leq j \leq k_0^{m-1}$ that cut the hypergrid in the intercept points perpendicularly. Due to the equidistant distribution of the intercept points on the hypergrid the rake lines lie equidistantly parallel to each other in the

[1] A broad list of publications in the field of evolutionary multi-objective optimization can be found at *http://delta.cs.cinvestav.mx/~ccoello/EMOO/EMOObib.html.*

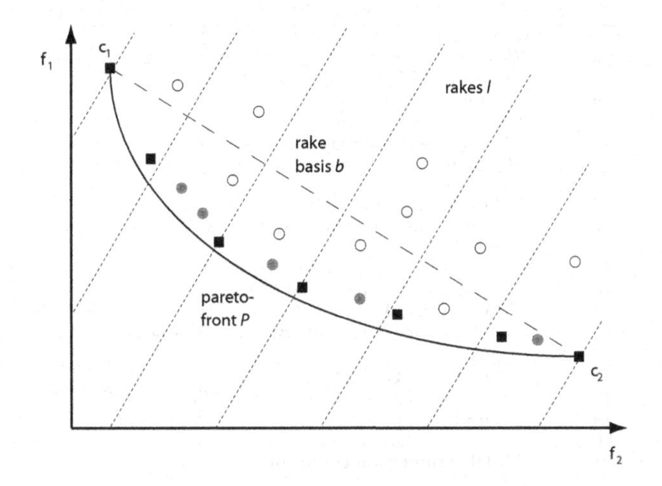

Fig. 1. Illustration of Rake Selection: The rakes lie equidistantly in the objective space and guide the evolutionary selection process. Among the set of non-dominated individuals $\mathcal{N}$ (black squares and grey circles) for each rake line l_j the closest solutions (black squares) are selected.

objective space. Consequently, they cut the Pareto front $\mathcal{F}$ equidistantly with respect to a projection of the Pareto front onto the hypergrid, and *approximately* equidistantly on the Pareto front hypersurface. Concerning a discussion about the equidistant distribution on the Pareto front surface, see Paragraph 4.1.

Figure 1 illustrates the situation for $m = 2$. Here $k = k_0$ rakes lie in the objective space perpendicularly to the rake base that connects the corner points $\boldsymbol{c}_1$ and $\boldsymbol{c}_2$. In each generation the algorithm produces λ offspring solutions. The core of Rake Selection is to select the k closest solutions (black squares) to each rake l_j among the set of non-dominated solutions (black squares and grey circles). Before each rake selects the closest solution, non-dominated sorting is applied and the set of non-dominated solutions $\boldsymbol{x}_i \in \mathcal{N}$ with $\mathcal{N} = \{\boldsymbol{x}_i | d_i = 0\}$ and dominance rank d_i is computed. Then, each rake line l_j selects the closest solution in the objective space, i.e.

$$\boldsymbol{x}_i = \arg \min_{\boldsymbol{x}_n \in \mathcal{N}} \mathrm{dist}(\boldsymbol{x}_n, l_j), \tag{2}$$

if $\mathrm{dist}(\boldsymbol{x}_n, l_j)$ measures the distance between point $\boldsymbol{x}_n$ and line l_j in $\mathbb{R}^m$. One solution may be selected multiple times by different rakes. The rakes guide the search to establish equidistant solutions in the objective space. If the number of selected solutions δ is smaller than μ, we add the $\mu - \delta$ solutions to the parental set $\mathcal{P}_{t+1}$ of solutions that are dominated least.

2.2 The Algorithm

Figure 2 shows the pseudo-code of our novel multi-objective optimization algorithm. At the beginning the single objectives have to be minimized in order to

```
1   Start
2     P_0 ← init, t ← 0;
4     minimize each f_1(x), ..., f_m(x);
5     compute corner points c_1, ..., c_m;
6     compute rake base b;
7     compute rake lines l_1, ..., l_{k-m};
8     Repeat
8       t ← t + 1;
9       For μ = 1 To λ
10        O_t ← O_t ∪ variation(P_t);
11      Next
12      N ← non-dominated_individuals (P_t ∪ O_t);
13      compute distance matrix D;
14      For rake l_j Do
15        R_t ← R_t ∪ x_i = arg min_{x_j∈N} dist(x_n, l_j);
16      Next
17      For i = 0 To |R_t| − μ
18        R_t ← R_t ∪ element(i, sort_d((P_t ∪ O_t)\R_t);
19      Next
20      P_{t+1} ← R_t
21    Until termination condition;
22  End
```

Fig. 2. Pseudocode of the novel EMOA with Rake Selection

compute the rake base $\boldsymbol{b}$ and the rake lines l_j in the objective space. In an iterative loop the individuals are produced. We use a $(\mu + \lambda)$ population scheme, i.e. in each generation the algorithm produces λ offspring solutions with intermediate recombination. For two uniformly selected parents $\boldsymbol{x}_1$ and $\boldsymbol{x}_2$, the offspring is $\boldsymbol{x}' := \frac{1}{2}\boldsymbol{x}_1 + \frac{1}{2}\boldsymbol{x}_2$. We point out that we do not use the specialized multi-objective variation operators simulated binary crossover or polynomial mutation [3] like other approaches, e.g. NSGA-II or SMS-EMOA. Instead, Rake Selection uses σ-self-adaptive Gaussian mutation, i.e. each solution $\boldsymbol{x}''$ is produced by

$$\boldsymbol{x}'' := \boldsymbol{x}' + \boldsymbol{\sigma}' \cdot \mathcal{N}(0,1). \tag{3}$$

with

$$\boldsymbol{\sigma}' := e^{(\tau_0 \mathcal{N}_0(0,1))} \cdot \left(\sigma_1 e^{(\tau_1 \mathcal{N}_1(0,1)}, \ldots, \sigma_N e^{(\tau_1 \mathcal{N}_N(0,1)}\right), \tag{4}$$

with learning rates τ_0 ad τ_1 and put into the offspring population O_t. Due to premature convergence that is discussed in Paragraph 4.3 we have to introduce a minimum step size ϵ, so that it holds $\sigma_i \geq \epsilon, i = 1 \ldots N$. Without the minimum step size the EMOA converges towards any set of non-dominated solutions with minimum distance to the rakes.

2.3 Computation of Rake-Base and Rake Lines

At the beginning the algorithm requires the calculation of a hypergrid in the objective space, i.e. the rake base and its interception points. For this sake Rake Selection has to compute the corner points c_i in the first stage by minimizing the single objectives $f_i, 2 \leq i \leq m$. If $\boldsymbol{x}_i^* \in \mathbb{R}^N$ is the optimal solution minimizing f_i, the corner points $\boldsymbol{c}_i$ are computed by insertion $\boldsymbol{c}_i = (f_1(\boldsymbol{x}_i^*), \ldots, f_i(\boldsymbol{x}_i^*), \ldots, f_m(\boldsymbol{x}_i^*))^T$. The rake base, i.e. the connection of corner points is the difference between each two corner points, e.g. for $m = 2$ we get $\boldsymbol{b} = \boldsymbol{c}_2 - \boldsymbol{c}_1$. If k is the number of rakes, we get the inner $k - 2$ rake intercept points $\boldsymbol{p}_i$ by $\boldsymbol{p}_i = \boldsymbol{c}_1 + q \cdot \frac{\boldsymbol{b}}{k-1}$ for $q = 1, \ldots, k-2$. The rakes cut the rake base perpendicularly in these intercept points. Thus, to define each rake l_j we only need to compute the vector $\boldsymbol{n}$ that is orthogonal to the rake base. For $m > 2$ the rake base is not a line, but a hyperplane with k^{m-1} intercept points, i.e. a hypergrid and the rake lines cut the hyperplane in the intercept points of the grid.

2.4 Selection with Rakes

In each generation Rake Selection performs non-dominated sorting, i.e. for each individual $\boldsymbol{x}_i$ the number d_i of solutions is computed that dominate $\boldsymbol{x}_i$. The set of non-dominated solutions $\mathcal{N} = \{\boldsymbol{x}_i \in \mathcal{P} | d_i = 0\}$ is subject to the Rake Selection procedure. For this sake the algorithm has to compute a distance matrix $\boldsymbol{D} = d_{ij}$ containing the distances d_{ij} of all non-dominated solutions $\boldsymbol{x}_i$ to the rakes l_j with $1 \leq j \leq |\mathcal{N}|$ and $1 \leq j \leq k - m$. For each rake l_j the closest solution is selected and put into the next parental population $\mathcal{P}_{t+1} = \mathcal{R}$. One solution may be selected multiple times by different rakes for $\mathcal{P}_{t+1}$. That leads to $|\mathcal{P}_{t+1}| < \mu$. In this case the population $\mathcal{P}_{t+1}$ is filled with the $\delta = |\mathcal{R}| - \mu$ solutions with minimal rank d_i.

One of the advantages of Rake Selection is its computational performance, in particular for $m > 2$. Rake base $\boldsymbol{b}$ and rake intercept points $\boldsymbol{p}_i$ are easy to compute. The normal vector $\boldsymbol{n}$ of the rake base requires the solving of m linear equations, i.e. $O(m^3)$. Each distance computation in the objective space between a solution and a rake line takes $O(m^2)$ that is based on computation of the intersection point between two lines.

3 Experimental Study

In this section we present a first experimental study of our novel approach on the multi-objective test problems ZDT1 to ZDT4, each with $m = 2$ objectives. Minimization of the single objectives yields the corner points. We use $\mu = 50$ parental solutions, $k = 50$ rakes, and the mutation settings $\tau_0 = 5.0$, $\tau_1 = 5.0$, and $\epsilon = 10^{-7}$. Figure 3 shows the experimental results[2] of typical runs of Rake Selection on the problems ZDT1 to ZDT4 with $N = 30$ dimensions ($N = 10$ for ZDT4) and random initialization after 1,000 iterations. We can observe that Rake Selection places the non-dominated solutions directly on the rakes and converges towards the Pareto front. The average distance to the rakes reaches the area of the minimum step size ϵ at the end of each run. The results are stable, i.e. the Pareto front is reached in almost every run.

[2] The figures of the experimental analysis are generated by the *matplotlib* package in *Python*.

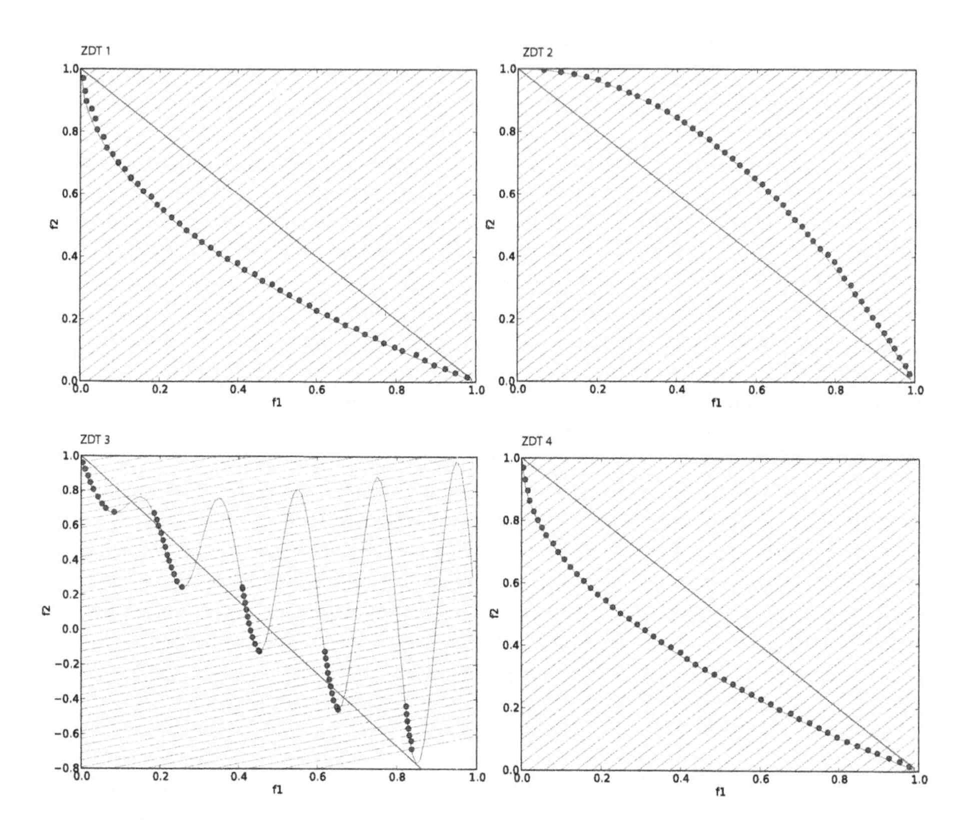

Fig. 3. Experimental results of typical runs of Rake Selection on the multi-objective problems ZDT1 to ZDT4. The rake base connects the corner points. The perpendicular lines define the rakes in the objective space. Note that the rakes don't appear to be perpendicular to the rake base in this figure due to the unequal scaling of the axes. After 1,000 iterations, i.e. 50,000 objective function evaluations the solutions lie on the rakes and on the curves of the Pareto fronts.

4 Discussion of Rake Selection

4.1 Arbitrary Placement of Rakes in the Objective Space

Does the Rake Selection approach really achieve equidistantly distributed solutions on the Pareto front? As the Pareto front is cone-convex, the condition of an approximately equidistant distribution cannot be hurt strong. E.g. the solutions are equidistantly distributed if the Pareto front is linear. In a worse case the rakes cut the Pareto front perpendicularly at one area and almost lie parallel to the Pareto front at another area. But as the search follows the rakes, the approach is quite flexible concerning the rake placement in the objective space, i.e. the intercept points $\boldsymbol{p}_i$ on the rake base $\boldsymbol{b}$ can be adapted during the run. The rake placement can be adapted automatically or by hand:

Automatic rake placement: To improve the equidistant distribution of solutions on the Pareto front, we currently experiment with an automatic adaptation of the rake-distances according to the Euclidean distances of the solutions in the objective space. If $\boldsymbol{c}_a$ and $\boldsymbol{c}_b$ are the corner points, a good linear[3] approximation could be the following: Let $\Delta = \frac{1}{\mu}\sum_{i=1}^{\mu-1} \|\boldsymbol{x}_i - \boldsymbol{x}_{i+1}\|$ be the average Euclidean distance between the solutions on the Pareto front, let b_{ij} be the distance between two solutions $\boldsymbol{x}_i$ and $\boldsymbol{x}_j$, and $r = \|\boldsymbol{b}\|/(k-1)$ be the initial distance between the rakes. Then, a good linear approximation for the rake distance between two solutions is $\frac{r}{b_{ij}} \cdot \Delta$ due to the theorem on intersecting lines. As $\sum^{k-1} b_{ij} = (k-1) \cdot \Delta$, the length of the rake base is kept constant.

Rake placement by hand: If visualized, the system could offer the practitioner the option to adapt the rakes by hand. The user can directly control the areas in the objective space where he wants to search for a Pareto optimal solution.

4.2 The Curse of Objective Space Dimensionality

For the search with two objectives, Rake Selection requires k rake lines depending on the density of solutions the practitioner desires for the Pareto set. To achieve a similar density of solutions in higher dimensional objective spaces with m objectives, a hypergrid is necessary with k^{m-1} rake lines. Note, that this exponential increase is no drawback of Rake Selection, but a – to our mind frequently neglected – consequence of the *curse of dimensionality* in multi-objective optimization. Neglect the fact that an equidistant distribution with constant distances into all directions of the objective space increases exponentially with the number of objectives, leads to a loss of density for every multi-objective approach.

4.3 Movement towards the Pareto Front

Why is Rake Selection moving the solutions to the Pareto front – and not only towards the rakes? The underlying concept of approximating the Pareto front is similar to the concept of maximizing the S-metric of the SMS-EMOA. The movement towards the Pareto front is achieved by considering only non-dominated solutions for the S-metric hypervolume. In Rake Selection the variation operators produce solutions that can dominate solutions of the previous generation and consequently, the population moves towards the Pareto front. Note, that the SMS-EMOA does not approximate the Pareto front by maximizing the absolute volume dominated in the objective space, but the relative volume with regard to the solution's neighbors – considering a reference point that is generated from the worst solution in each generation influencing the solutions at the corner of the Pareto front. In our experiments we also took into account the maximization of the distance to a dominated reference point $\boldsymbol{r}$, e.g. $\boldsymbol{r} = (1, 1)$, in the distance matrix, i.e. $d_{ij} = \text{dist}(\boldsymbol{x}_i, l_j) + \alpha\|\boldsymbol{x}_i - \boldsymbol{r}\|$, see Equation 2. We applied this update

[3] An even better equidistant distribution can be achieved with a nonlinear regression model.

rule to all rakes, and only one rake at the border of the non-dominated individuals with and without a decreasing α, and could not achieve a speedup towards the Pareto front.

5 Conclusions and Outlook

Rake Selection is a novel evolutionary multi-objective algorithm based on the selection by the distance to equidistantly distributed lines in the objective space. Birth surplus allows the application of σ-self-adaptation. Our first experimental studies have shown that the population follows the rakes and at the same times approximates the Pareto front. In the next step, a detailed experimental study and comparison to other approaches using methods from design of experiments will follow on various test problems, e.g. from the CEC 2007 [5]. Of interest are the capabilities of Rake Selection on problems with more than two objectives. Special attention will be payed on the behavior of σ-self-adaptation as not many results are available for self-adaptation in multi-objective solutions spaces. Further selection schemes will be interesting, e.g. selection of more than one closest non-dominated solutions or the consideration of the closest dominated solutions.

References

1. Beume, N., Naujoks, B., Emmerich, M.: SMS-EMOA: Multiobjective Selection based on Dominated hypervolume. European Journal of Operational Research 181(3), 1653–1669 (2007)
2. Beyer, H.-G., Schwefel, H.-P.: Evolution strategies - A comprehensive introduction. Natural Computing 1, 3–52 (2002)
3. Deb, K.: Multi-Objective Optimization using Evolutionary Algorithms. John Wiley & Sons, Chichester (2001)
4. Deb, K., Pratap, A., Agarwal, S., Meyarivan, T.: A fast and elitist multiobjective genetic algorithm: NSGA-II. IEEE Transactions on Evolutionary Computation 6(2), 182–197 (2002)
5. Huang, V.L., Qin, A.K., Deb, K., Zitzler, E., Suganthan, P.N., Liang, J.J., Preuss, M., Huband, S.: Problem definitions for performance assessment of multi-objective optimization algorithms. Technical report, Nanyang Technological University, Singapore (2007)
6. Schwefel, H.-P.: Evolution and Optimum Seeking. Sixth-Generation Computer Technology. Wiley Interscience, New York (1995)
7. Zitzler, E., Laumanns, M., Thiele, L.: SPEA2: Improving the Strength Pareto Evolutionary Algorithm. In: EUROGEN 2001. Evolutionary Methods for Design, Optimization and Control with Applications to Industrial Problems, Athens, Greece, pp. 95–100 (2002)

A Comparison of Neighbourhood Topologies for Staff Scheduling with Particle Swarm Optimisation

Maik Günther and Volker Nissen

TU Ilmenau, Information Systems in Services, D-98684 Ilmenau
volker.nissen@tu-ilmenau.de, maik.guenther@gmx.de

Abstract. The current paper uses a real-life scenario from logistics to compare various forms of neighbourhood topologies within particle swarm optimization (PSO). Overall, gbest (all particles are connected with each other and change information) outperforms other well-known topologies, which is in contrast to some other results in the literature that associate gbest with premature convergence. However, the advantage of gbest is less pronounced on simpler versions of the application. This suggests a relationship between the complexity of instances from an identical class of problems and the effectiveness of PSO neighbourhood topologies.

Keywords: staff scheduling, neighbourhood topologies, sub-daily planning, particle swarm optimization.

1 Application Problem and PSO Solution Approach

Staff scheduling involves the assignment of an appropriate employee to the appropriate workstation at the appropriate time while considering various constraints. The present staff scheduling problem originates from a German logistics service provider. This company operates in a spatially limited area 7 days a week almost 24 hours a day. The employees are quite flexible in terms of their working hours, which results in a variety of working-time models. There are strict regulations especially with regard to qualifications. The personnel demand for the workstations is subject to large variations during the day. However, today employees are manually scheduled to work at the same workstation all day, causing large phases of over- and understaffing.

The planning problem covers seven days (20 hours each), divided into 15-minute intervals. It includes 65 employees and, thus, an uncompressed total of 36,400 dimensions for the optimization problem to be solved. A staff schedule is only valid if any one employee is assigned to one workstation at a time and if absent employees are not included in the plan. These hard constraints can be contrasted with soft constraints (e.g. qualifications), which are penalised with error points that reflect that companys requirements.

B. Mertsching, M. Hund, and Z. Aziz (Eds.): KI 2009, LNAI 5803, pp. 185–192, 2009.

The problem is represented as a two-dimensional matrix of employees and time periods, where the cells are filled with workstation assignments. Each particle in the swarm has an own matrix that determines its position.

During initialisation valid assignments w.r.t. the hard constraints are created that use information from the company's current full-day staff schedule. In each iteration the new particle position is determined by traversing all dimensions and executing one of the following actions with predefined probability. The probability distribution was heuristically determined in prior tests:

- No change: The workstation already assigned remains. (prob. p1)
- Random workstation: A workstation is randomly determined and assigned. Only those assignments are made for which the employee is qualified. The probability function is uniformly distributed. (prob. p2)
- pBest workstation: The corresponding workstation is assigned to the particle dimension from pBest, the best position found so far by the particle. Through this, the individual PSO component is taken into account. (prob. p3)
- Neighbourhood workstation: The corresponding workstation is assigned to the particle dimension from the chosen neighbourhood. The social behaviour of the swarm is controlled with these types of assignments. (prob. p4)

The behaviour of the PSO-heuristic is relatively insensitive to changes of p1, p3, and p4. The optimal value for p2 depends on the problem size. Pre-tests revealed that a value of 0.3% for p2 works best for the problem investigated here. The other probabilities were set at p1=9.7%, p3=30%, and p4=60%. PSO terminates after 400,000 inspected solutions. Further details of the application and PSO-implementation can be found in [12] where PSO was compared to evolution strategies (ES).

2 Neighbourhood Topologies

The choice of an appropriate neighbourhood topology (sociometry) plays an important role in the effectiveness of PSO. A general statement on the quality of neighbourhoods cannot be made, however, because their effectiveness depends among other things on the problem to be solved. In PSO the behaviour of the particles is determined in large part by the individual influence of the particle itself as well as the social influence of "other particles" [5]. The individual influence refers to the best position found so far by that particular particle. The meaning of "other particles" depends on the neighbourhood topology of the swarm. Social behaviour here refers to the best position found by the neighbourhood. Watts showed that the information flow in social networks is influenced by various properties [15], [16]:

- Connectivity (k): the size of the particle neighbourhood,
- Clustering (C): the number of neighbours of the particle which are simultaneously neighbours of each other and
- Path length (L): the smallest average distance between two neighbours (strongly influenced by k and C).

The topologies used most often in the original form of PSO are the gBest and lBest topologies. In gBest the swarm members are connected in such a way that each particle is a neighbour of every other particle. This means that each particle immediately knows the best global value found up to that point. All particles are included in the position calculation of gBest. If the global optimum is not located near enough to the best particle, it can be difficult for the swarm to search other parts of the solution space, possibly converging instead to a local optimum [11].

Avoiding such convergence to a sub-optimum is one of the goals of the lBest topology, in which a particle is only connected to its immediate neighbours. The parameter k represents the number of neighbours of a particle. With k=2 the topology is a circle (or ring). Increasing k to particle count minus 1 yields a gBest topology. Each particle only possesses information about itself and its neighbours. The swarm converges slower than gBest but also has a higher chance of finding the global optimum [6].

Another neighbourhood form is the wheel. There exists a central particle which is connected to all other particles. These particles only have that central particle as its neighbour and are isolated from all others. This arrangement prevents a new best solution from being immediately distributed throughout the swarm.

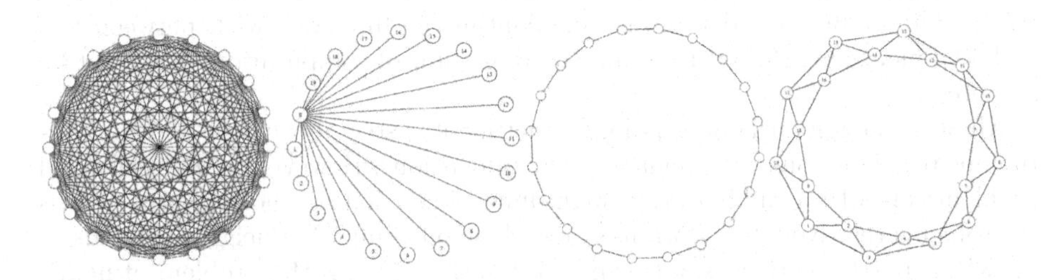

Fig. 1. Neighbourhood topologies gBest, wheel, circle (k=2) and lBest (k=4) [10]

3 Related Work

In the past, various modifications and innovations on the neighbourhood topologies have been tested to avoid premature convergence to a local optimum. The focus was especially on continuous test functions. Suganthan continuously changes the neighbourhood size [13]. He begins with an lBest topology of k = 1. Therefore, the neighbourhood initially consists of the particle itself. As time progresses, k is continuously increased until a gBest topology is attained. Using several continuous test functions with 20, 30 and 50 dimensions, generally better results were obtained compared to the standard PSO. In [7] continuous test functions (with two and 30 dimensions respectively) were thoroughly tested for various neighbourhood topologies. In addition to the topologies described above, the pyramid, the small and the von Neumann topologies were used along with randomly generated topologies. The pyramid and the von Neumann performed best, with the worst being the wheel and gBest topologies.

Blackwell and Bentley use PSO to generate music [1]. They include an additional acceleration component in the calculation of the new particle positions in order to reduce the force of attraction toward the swarm centre. The speed of convergence is slowed down by avoiding collisions. Krink, Vesterstrom and Riget follow a different approach in that they place a radius around each particle [8]. If two particles would collide, they are set apart in order to avoid the collision and thus a grouping.

In Løvbjerg and Krink's work a variable is added to each particle [9]. This variable represents a critical value which is increased when two particles come nearer to each other. If the critical value falls below a set distance, the corresponding particles are reshuffled. Better results were achieved using four continuous test functions with 30 dimensions each. Xie Zhang and Yang integrate negative entropy into the swarm by introducing chaos in the particle velocity and in the particle position [17]. Using this should help prevent convergence of the swarm to a point. This dissipative PSO is used on two continuous test functions, each with 10, 20 and 30 dimensions.

In [3] several neighbourhoods are used simultaneously. During an iteration each particle checks the fitness value of its new position for a star, circle and von Neumann topology. The particle then uses the topology yielding the best result. This hybrid fared well on nine continuous functions with between two and 30 dimensions. However, an increased amount of computation time must be accepted.

All of the neighbourhood topology experiments listed above were done on continuous test functions with relatively few dimensions. However, many real-world problems are of a combinatorial, high-dimensional nature. For this large class of problems relatively few work has been done on comparing neighbourhoods so far which partly motivates our own investigation. In [14] the problem of multiprocessor scheduling is discussed as it pertains to gBest and lBest topologies, among others. The gBest topology performed better than the lBest topology. Czogalla and Fink investigated the gBest, lBest, wheel and a hierarchical topology (H-PSO) for a combinatorial problem [2]. Experiments were done using the example of four project planning problems (RCPSP) with 30 to 120 tasks. It was shown that the lBest and H-PSO topologies yielded better results than wheel and gBest, with gBest performing the worst.

4 Experiments and Results

This section will describe experiments and results with gBest, wheel, circle (k = 2) and lBest (k = 4) topologies for the sub-daily staff scheduling problem given in [12] where a complete week is to be planned. Additionally, these neighbourhood topologies will also be used to solve simpler versions of the same problem where the week is divided into days which are then planned individually.

All test runs were conducted on a PC with an Intel 4 x 2.67 GHz processor and 4 GB of RAM. An individual run requires approx. 25 minutes with this configuration. Thirty independent runs were conducted each time for each of the experiments to allow for statistical testing. The termination criterion was 400,000 calculations of the fitness function for all tests.

Table 1 below lists the results of the week problem. The first column contains the particular PSO configuration, in which the value in brackets is the particle count in the swarm. The respective best results (mean and minimum values) are underlined for each neighbourhood topology.

Table 1. Results of various swarm sizes and neighbourhood topologies for the week problem (30 runs each)

Heuristic	Error		Number of job-changes	Wrong qualifications in minutes	Understaffing in minutes	Overstaffing in minutes	
	mean	min				demand > 0	demand = 0
PSO (20) gBest	<u>52162</u>	<u>51967</u>	1666.8	0.0	7478.5	28488.0	7265.5
PSO (40) gBest	52222	52085	1730.2	0.0	7568.6	28246.5	7339.4
PSO (100) gBest	52591	52400	1778.5	0.0	7576.3	28152.1	7542.0
PSO (200) gBest	53727	53467	2220.3	0.0	7658.5	28017.0	7916.5
PSO (20) wheel	<u>71104</u>	<u>70161</u>	2872.8	0.0	10987.5	21281.5	17981.0
PSO (40) wheel	90854	80708	2497.6	48.0	13520.0	18353.5	23441.5
PSO (100) wheel	173848	137768	1648.0	411.0	16518.5	16105.5	28688.0
PSO (200) wheel	248082	219251	1025.0	757.5	18083.5	15243.5	31115.0
PSO (20) circle	<u>52614</u>	<u>52402</u>	1925.4	0.0	7539.5	28479.5	7335.0
PSO (40) circle	52995	52719	2161.0	0.0	7566.0	28414.0	7427.0
PSO (100) circle	57841	56714	2416.8	0.0	7831.0	24618.5	11487.5
PSO (200) circle	70099	69152	1949.7	0.0	9854.0	17962.5	20166.5
PSO (20) lBest	<u>52498</u>	<u>52300</u>	1863.0	0.0	7523.5	28485.5	7313.0
PSO (40) lBest	52776	52627	2029.8	0.0	7551.5	28458.5	7368.0
PSO (100) lBest	54133	53868	2353.1	0.0	7637.5	27682.5	8230.0
PSO (200) lBest	63640	62528	2118.6	0.0	8516.5	20578.5	16213.0

All topologies demonstrate that small swarm sizes yield better results than large ones. Because of the uniform termination criterion of 400,000 fitness function calculations, more iterations can be executed with small swarms. The creation of a good schedule requires substantial changes to the initial plan, and the probability p2 for the explorative action 2 (random workstation assignment) in our PSO is rather small. Thus, it is apparently preferable to track many small changes for more iterations as compared to richer knowledge (through larger swarm size) of the solution space in each iteration.

If one compares the individual best results of the neighbourhood topologies, it can be seen that gBest performs best. This topology leads to more overstaffing in periods with demand > 0 as compared to other topologies. This is important because employees can still support each other instead of being idle when demand = 0. The lBest and circle topologies performed somewhat worse and wheel was the worst. Significantly, qualifications violations occur in the last case. The quality of the results improve as the size of the neighbourhood increases. The advantage of smaller neighbourhoods to avoid convergence to a local optimum does not carry significant weight for this application. Because of the complexity of the problem, finding "good" allocations of individual employees is difficult. If the information is not available to all particles immediately, there is a danger of its being lost. Another reason for gBest's success lies in the modified PSO procedure for this problem. The task assignment occurs randomly in one of the four actions for the new position calculation. Therefore, all particles always remain in motion and have the chance to escape from a local optimum.

PSO(20) with gbest and PSO(20) with lbest topology provided the best overall error results. With 30 independent runs for each heuristic it is possible to test the statistical significance of the performance difference between two neighbourhoods with a t-test (see table 2). A Levene-test (test level 5%) revealed the homogeniety of variances between both groups ($F = 2.995, p = 0.089$). The corresponding t-test with a 95% confidence interval confirms the better performance of gbest with a very high statistical significance ($p < 0.001$ for H_0). The results remain the same if heterogeniety of variances is assumed. Similarly, lbest outperforms circle, and circle outperforms the wheel neighbourhood on a highly significant statistical level.

Table 2. T-test results for pairwise comparison of neighbourhood topologies on the week problem for PSO(20)

H_1	T	df	significance H_0 (1-tailed)	mean difference	95% confidence intervall of differences	
					lower	upper
PSO(20)gbest < PSO(20)lbest	-12.34	58	< 0.001	-336.00	-390.50	-281.51
PSO(20)lbest < PSO(20)circle	-4.04	58	< 0.001	-116.10	-173.64	-58.56
PSO(20)circle < PSO(20)wheel	-186.09	58	< 0.001	-18489.87	-18688.80	-18291.00

Fig. 2 shows the convergence behaviour of the four neighbourhood topologies for PSO(20) in comparison. Not only does gbest generate the better final solution, but it also demonstrates a more rapid convergence towards good solutions than the other topologies an the week problem.

Next, the four neighbourhood topologies were tested on significantly smaller problem sets, that is, on the individual days of the week problem. Table 3 shows the respective mean and minimum errors for each day for the gBest, wheel, circle (k = 2) and lBest (k = 4) topologies. The results were gathered for configurations with 20 particles because small swarms also yield the best results for individual days. Comparing the mean error points of the topologies for all days in table 3 (the best values are again underlined) gBest does not always perform best. Only on five of seven days did gBest defeat the other topologies. Circle and lBest usually yielded similar values. Wheel performed quite poorly. The fact that gBest is not automatically preferred for individual days could be attributed to the less complex problem. It is much easier for the swarm to find "good" employee assignments, through which the influence of maximised information exchange throughout the swarm is reduced. The loss of "good" allocations can be compensated for in a better fashion and the advantage of the other topologies for avoiding premature convergence to a local optimum is partly visible.

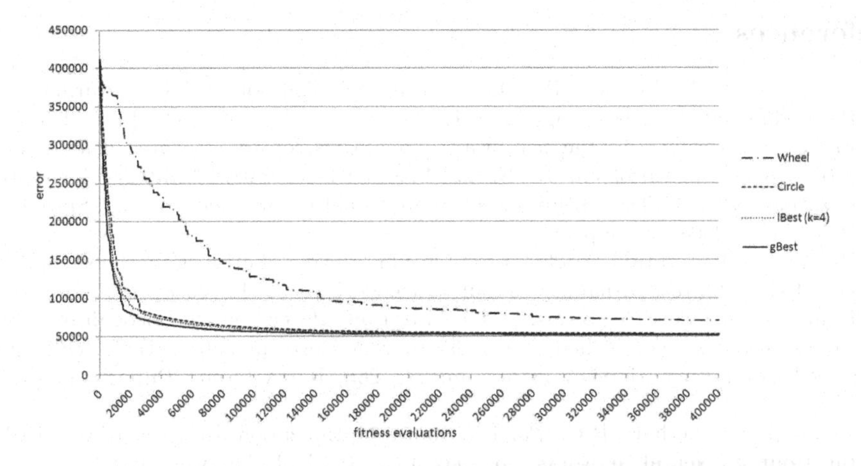

Fig. 2. Convergence chart for PSO(20) with different neighbourhoods

Table 3. Results of various neighbourhood topologies for individual days of the week problem (30 runs each)

	Monday		Tuesday		Wednesday		Thursday		Friday		Saturday		Sunday	
	mean	min.	mean	min.	mean	min.	mean	min.	mean	min.	mean	min.	mean	min.
gBest	7782	7706	5918	5904	8173	8146	8246	8230	5573	5504	8825	8817	7301	7292
Wheel	7968	7896	6205	6079	8485	8348	8423	8360	5754	5667	8958	8910	7496	7424
Circle	7746	7701	5926	5908	8194	8154	8256	8240	5528	5505	8833	8824	7308	7299
lBest	7744	7705	5928	5913	8194	8157	8254	8236	5527	5498	8834	8823	7310	7302

5 Conclusion and Future Work

For a large and highly constrained combinatorial staff scheduling problem taken from a logistics service provider it was demonstrated that the gBest topology is superior to the wheel, circle and lBest topologies and is therefore to be preferred in this context. This is in contrast to comments in the literature that gbest might lead to premature convergence on a local optimum (see section 3). After all, it also depends on other measures in the design of a PSO heuristic and the risk to loose good allocations if they are not quickly and broadly communicated in the swarm whether or not gbest actually produces premature convergence.

Investigations of seven easier versions of the same application problem were undertaken in which the advantage of gBest was only partly present. lBest performed better on some of the smaller problems, while wheel and circle were unable to perform best on any of the problems. However, the performance of circle was generally very close to lBest. This suggests a relationship between the complexity of instances from an identical class of problems and the effectiveness of PSO neighbourhood topologies.

In order to base the conclusions of this work on a wider foundation, the investigations done here are currently extended to a practical problem from the trade domain, which is even more extensive in regards to dimensions and constraints. Further neighbourhood topologies will also be assessed.

References

1. Blackwell, T.M., Bentley, P.: Don't push me! Collision-avoiding swarms. In: Proceedings of the 2002 Congress on Evol. Comp., vol. 2, pp. 1691–1696 (2002)
2. Fink, A., Czogalla, J.: Particle swarm topologies for the resource constrained project scheduling problem. In: Nature Inspired Cooperative Strategies for Optimization (NICSO 2008). Studies in Computational Intelligence, vol. 236. Springer, Heidelberg (2008) (to appear)
3. Hamdan, S.A.: Hybrid Particle Swarm Optimiser using multi-neighborhood topologies. INFOCOMP Journal of Computer Science 7(1), 36–44 (2008)
4. Kennedy, J.: Particle Swarms. Optimization Based on Sociocognition. In: De Castro, L.N., Von Zuben, F.J. (eds.) Recent Developments in Biologically Inspired Computing, Hershey et al., pp. 235–268. IDEA Group Publishing, USA (2005)
5. Kennedy, J., Eberhart, R.C.: Particle Swarm Optimization. In: Proc. of the IEEE Int. Conf. on Neural Networks, pp. 1942–1948. IEEE, Piscataway (1995)
6. Kennedy, J., Eberhart, R.C., Shi, Y.: Swarm Intelligence. Kaufmann, San Francisco (2001)
7. Kennedy, J., Mendes, R.: Population structure and particle swarm performance. In: Proceedings of the IEEE Congress on Evolutionary Computation (CEC), pp. 1671–1676 (2002)
8. Krink, T., Vesterstrom, J.S., Riget, J.: Particle swarm optimization with spatial particle extension. In: Proceedings of the 2002 Congress on Evolutionary Computation, vol. 2, pp. 1474–1479 (2002)
9. Løvbjerg, M., Krink, T.: Extending particle swarm optimisers with self-organized criticality. In: Proceedings of the 2002 Congress on Evolutionary Computation, vol. 2, pp. 1588–1593 (2002)
10. Mendes, R.: Population Topologies and Their Influence in Particle Swarm Performance, PhD Thesis, Departamento de Informática, Escola de Engenharia, Universidade do Minho (2004)
11. Mendes, R., Kennedy, J., Neves, J.: Avoiding the pitfalls of local optima: how topologies can save the day. In: Proceedings of the 12th Conference Intelligent Systems Application to Power Systems (ISAP 2003). IEEE Computer Society, Lemnos (2003)
12. Nissen, V., Günther, M.: Staff Scheduling With Particle Swarm Optimisation and Evolution Strategies. In: Cotta, C., Cowling, P. (eds.) EvoCOP, Tübingen, Germany, April 15-17. LNCS, vol. 5482, pp. 228–239. Springer, Heidelberg (2009)
13. Suganthan, P.N.: Particle Swarm Optimiser with Neighbourhood Operator. In: Proceedings of the Congress on Evolutionary Computation, Washington DC, pp. 1958–1962 (1999)
14. Sivanandam, S.N., Visalakshi, P., Bhuvaneswari, A.: Multiprocessor Scheduling Using Hybrid Particle Swarm Optimization with Dynamically Varying Inertia. International Journal of Computer Science and Applications 4(3), 95–106 (2007)
15. Watts, D.J.: Small Worlds: The Dynamics of Networks Between Order and Randomness. Princeton University Press, Princeton (1999)
16. Watts, D.J., Strogatz, S.H.: Collective dynamics of small-world networks. Nature 393, 440–442 (1998)
17. Xie, X.F., Zhang, W.J., Yang, Z.L.: Dissipative particle swarm optimization. In: Proceedings of the 2002 Congress on Evolutionary Computation, vol. 2, pp. 1456–1461 (2002)

Controlling a Four Degree of Freedom Arm in 3D Using the XCSF Learning Classifier System

Patrick O. Stalph, Martin V. Butz, and Gerulf K.M. Pedersen

University of Würzburg, Department of Psychology III
Röntgenring 11, 97070 Würzburg, Germany
{patrick.stalph,butz,gerulf}@psychologie.uni-wuerzburg.de
http://www.coboslab.psychologie.uni-wuerzburg.de/

Abstract. This paper shows for the first time that a Learning Classifier System, namely XCSF, can learn to control a realistic arm model with four degrees of freedom in a three-dimensional workspace. XCSF learns a locally linear approximation of the Jacobian of the arm kinematics, that is, it learns linear predictions of hand location changes given joint angle changes, where the predictions are conditioned on current joint angles. To control the arm, the linear mappings are inverted—deriving appropriate motor commands given desired hand movement directions. Due to the locally linear model, the inversely desired joint angle changes can be easily derived, while effectively resolving kinematic redundancies on the fly. Adaptive PD controllers are used to finally translate the desired joint angle changes into appropriate motor commands. This paper shows that XCSF scales to three dimensional workspaces. It reliably learns to control a four degree of freedom arm in a three dimensional work space accurately and effectively while flexibly incorporating additional task constraints.

Keywords: Learning Classifier Systems, XCSF, LWPR, Autonomous Robot Control, Dynamic Systems.

1 Introduction

Humans are able to solve reaching tasks in a highly flexible manner, quickly adapting to current task demands. In contrast, industrial robot applications are usually optimized to solve one particular task. Consequently, new tasks require substantial reprogramming of the system. In order to design an autonomous robotic system that can solve given tasks without predefined fixed models, we use machine learning techniques for movement control.

From a top-down view, we divide a complex reaching task into three levels: (1) high-level planning, during which a complex task is decomposed into small reaching tasks in order to avoid deadlocks and obstacles. (2) low-level control, during which high-level plan-based hand directions are translated into appropriate joint angle changes. (3) dynamic control, during which the desired joint angle changes are mapped onto corresponding forces, to, for example, activate

B. Mertsching, M. Hund, and Z. Aziz (Eds.): KI 2009, LNAI 5803, pp. 193–200, 2009.

motors or muscles. This work focuses on learning the low-level control part, that is, to derive desired *joint angle changes* given a desired *hand location.*

The challenge is to learn the forward kinematics of the arm and use the knowledge to derive the inverse kinematics for arm control. We define a forward model as a function from *posture space* Θ (or *configuration space*) to *hand space* Ξ (or *task space*):

$$f : \Theta \rightarrow \Xi, \quad f(\theta) = \xi, \tag{1}$$

where $\theta = \theta_1, \ldots, \theta_n$ are the joint angles of the n joints respectively and the Cartesian coordinates x, y, z of the robot hand are denoted as ξ. However, since the number of joints and thus the number of degrees of freedom in posture space are often greater than the three spatial dimensions, there is usually no unique solution to the inverse kinematics problem [1,2]. Thus, the inverse kinematics mapping is not a function but a one-to-many mapping, where the set of postures that corresponds to one particular hand location is often referred to as a *null-manifold* in configuration space. While this redundant mapping may be considered problematic and may be resolved by adding additional constraints during learning, our approach is to preserve the available redundancy and resolve it only during actual inverse control. In this way, the system is able to consider varying task demands online.

It was shown that a Learning Classifier System [3], namely XCSF, can be successfully applied to a kinematic arm control task [4], where a three joint planar arm reached goal locations using XCSF-based control. Most recently, this arm model was extended to a dynamic arm platform and it was shown that the learned knowledge structure can be used to flexibly resolve the arm redundancies task-dependently [5]. In this paper, we show that this approach can be extended to a dynamic 4-DOF arm in a three-dimensional workspace, while still being able to resolve varying task constraints on the fly during control.

The remainder of the paper is structured as follows. Section 2 briefly introduces the Learning Classifier System XCSF, focussing on its application to the arm control task. Section 3 describes the control approach as well as the redundancy-resolving mechanism. The XCSF-based control system is evaluated on a dynamic four degree of freedom arm in Section 4 and the paper ends with concluding remarks in Section 5.

2 Learning the Forward Kinematics

In order to control a robot arm, XCSF [6] learns to predict hand location *changes* given joint angle *changes*, as introduced in [4]. Depending on the current posture $\theta \in \Theta$, the function

$$g_\theta : \dot{\Theta} \mapsto \dot{\Xi}, \quad g_\theta(\dot{\theta}) \rightarrow \dot{\xi} \tag{2}$$

maps joint angle changes $\dot{\theta} = \dot{\theta}_1, \ldots, \dot{\theta}_n$ to hand location changes $\dot{\xi} = \dot{x}, \dot{y}, \dot{z}$, given a current posture θ. In order to approximate this function, XCSF evolves a population of rules, so called classifiers, which cover the posture space. Each classifier predicts the forward model, as defined in Equation 2 in the proximity of a posture by means of a linear approximation.

2.1 XCSF in a Nutshell

XCSF is an extension of the currently most prominent Learning Classifier System XCS [7]. XCSF learns to approximate iteratively sampled real-valued functions $f : \mathbb{R}^n \mapsto \mathbb{R}^m$ using a set of rules, so called classifiers. Each classifier consists of a condition part that specifies the active subspace and a prediction part that approximates the function in that subspace. A genetic algorithm evolves the population of classifiers with the goal to minimize the prediction error. Various condition structures are possible, including radial-basis functions acting as a kernel conditions [8]. Predictions are often linear and it was shown that recursive least squares yield very good linear approximations [9]. In our work, we use rotating hyper-ellipsoidal condition structures and recursive least squares to learn local linear approximations.

Given a function sample $x_t, f(x_t)$ at time t, the population is scanned for matching classifiers, that is, classifiers whose condition matches the input x_t. These active classifiers form the so called *match set*. Given the match set, the predicted value $\widehat{f}(x_t)$ is computed as a weighted average of the individual classifier predictions. The prediction error, that is, the difference to the actual function value $f(x_t)$ is used to refine the predictions by means of a gradient-descent technique. Furthermore, the accuracy of a classifier, which is defined as the scaled inverse of the prediction error, determines a fitness value. A steady state genetic algorithm selects two classifiers of the match set, based on their fitness, for reproduction. After mutation and crossover operators are applied, the two classifiers are inserted back into the population. If the maximal population size is exceeded, classifiers are deleted from the population, where the deletion probability is higher for overcrowded regions and lower for very accurate classifiers.

Besides the fitness pressure towards accurate predictions resulting from reproduction, a generalization pressure is added. This pressure is due to the niche-based reproduction but population-based deletion. In sum, the system evolves maximally general classifier condition structures for sufficiently accurate linear predictions. This results in an accurate, highly general piecewise overlapping solution representation. For further details the interested reader is referred to the available literature [6,8].

2.2 Predicting the Forward Model

To learn hand location changes $\dot{\xi}$ given joint angle changes $\dot{\theta}$, the function given in Equation 2 is approximated conditioned on the current posture state θ. Using pure function approximation techniques, this could, for example, be accomplished by learning the function $\Theta \rightarrow \Xi$ and computing the derivative of that function during control [10]. However, using this technique the input space would be structured based on error information with respect to the hand location given a posture and not based on the desired function, that is, the first order derivative. Another way would be to learn the augmented function $\Theta \times \dot{\Theta} \rightarrow \dot{\Xi}$, resulting in twice the dimensionality of the input space or to learn a unique inverse mapping directly with loss of redundancy. Locally Weighted Projection Regression [11] has been used to approximate the inverse mapping of a 7-degree of freedom arm directly without redundancy resolution [12]. However, as the inverse mapping is not a function, the learned mapping does not correspond to appropriate movement commands but mixes different redundant movements.

In order to avoid these problems, we (1) learn the forward model to preserve redundancy and (2) split the condition and prediction mechanisms. In particular, the condition part of a classifier specifies the context of the prediction, that is, a posture and its proximity. The linear prediction part of the classifier is fed with joint angle changes in order to predict hand location changes. Since the accuracy of the predictions is used to evolve new classifiers, the posture space is clustered based on the error information of the desired function. Thus, the resulting system learns motor-induced location changes. In the present arm control application, XCSF receives kinematic information about the current state of the arm and structures the posture space in order to optimize the linear approximation of arm location changes given joint angle changes.

3 Inverse Kinematics Control

The learned forward predictive model represented in XCSF's classifier population yields local linear predictions of hand location changes $\dot{\xi}$ given joint angle changes $\dot{\theta}$. To move the hand in a specific direction, the inverse of these predictions can be used. Given the current hand location $h \in \Xi$ and the hand goal $g \in \Xi$ the desired hand direction $\dot{\xi}$ is $g - h$. The linear prediction of the forward model is a system of three linear equations

$$w_{i1}\dot{\theta}_1 + \ldots + w_{in}\dot{\theta}_n = \dot{\xi}_i$$

where w_{ij} are the learned linear prediction weights ($i \in \{1, 2, 3\}$). Note that an offset weight is not needed, because a zero change in joint angles yields zero movement.

To derive the inverse solution, that is, $\dot{\theta}$ given desired $\dot{\xi}$, we have to solve an under-determined[1] linear system. Using Gaussian elimination with backward propagation[2] results in a solution set L, where the dimension of the solution is lower or equal $n - 3$. Any vector $\dot{\theta} \in L$ specifies a change in joint angles that corresponds to the desired change in hand location. We recall that the prediction is linear and thus the approximation is accurate for small movements but gets worse for large movements.

Resolving the Redundancy. As mentioned before, if the arm has more degrees of freedom n than the dimension of the workspace, we have to deal with redundancy. Instead of appending constraints to the linear system, such that the number of equations equals the number of joints, we derive the general solution L and select one particular solution afterwards.

Given a secondary goal s, such as a particular change in angles, we can derive the solution $\dot{\theta} \in L$ that is closest to the secondary goal. Therefore we minimize

[1] We consider arm models with redundant degrees of freedom, that is, $n > 3$.

[2] We also apply scaled partial pivoting to maintain numerical stability.

the distance to s by projecting s onto L. If the solution is one-dimensional, that is, a line, we can simply determine the point on the line with minimal distance to the secondary goal.

Adaptive PD Controllers. Since we consider a dynamic system with gravity, inertia, and friction we cannot simply translate the desired change of angles $\dot{\theta}$ into appropriate motor commands. Therefore we use classical proportional-derivative (PD) controllers independently for each joint. The controllers are extended with a gravity adaptation module as well as a velocity module in order to counteract transfer of momentum. We omit further details, since the design of PD-controllers is out of the scope of this work.

4 Experimental Evaluation

XCSF is tested on a simulated dynamic arm plant (including gravity, inertia, and friction) with four degrees of freedom in a three dimensional workspace. The idea of the robot arm model is to resemble the proportions and at least four degrees of freedom of a human arm. The model has a shoulder joint with two degrees of freedom (glenohumeral joint with flexion-extension and abduction-adduction), an elbow joint with one degree of freedom (humeroulnar joint with flexion-extension), and the wrist with one degree of freedom (distal radio-carpal joint with flexion-extension). The limbs starting from the shoulder have lengths 40, 40, and 20 cm and masses of 4, 3, and 1 kg. The range of joint rotations is restricted to $[-1.05, 2.9]$, $[-1.5, 1.5]$, $[0, 2.8]$, and $[-1.5, 1.5]$ radians and maximum torques applicable are 350, 350, 250, and 150 Nm, respectively for the four joints.

Learning. Learning was done by moving the arm to random postures and learning from observed movements. If individual movements were smaller than 0.02 radians, the learning mechanism waits for further movements until the signal is strong enough, which prevents learning from zero movements and helps on a more uniform problem space sampling. All reported results are generated from ten independent runs. The XCSF parameter settings are those used elsewhere [4].[3] The problem-dependent error threshold was set to $\varepsilon_0 = 0.001$. According to [13], we set θ_{GA} to a higher value of 200 to counteract the non-uniform sampling of the problem space.

Fig. 1(a) shows the learning performance of XCSF on the learning task: The prediction error drops from 0.006 to about 0.0015. After 180K iterations condensation with closest classifier matching [8] is applied, effectively reducing the population size by more than 80% while maintaining the target error.

XCSF-based Control. The control capabilities are tested rigorously during the learning process every 10000 iterations. We evaluate XCSF-based control throughout the complete joint angle space. Therefore we generate $4^4 = 256$ postures P by setting each joint angle to four different values from min to max

[3] These settings are $N = 6400$, $\beta = \delta = 0.1$, $\alpha = 1$, $\nu = 5$, $\chi = 1$, $\mu = 0.05$, $\theta_{\mathrm{del}} = 20$, $\theta_{\mathrm{sub}} = 20$ $r_0 = 0.5$.

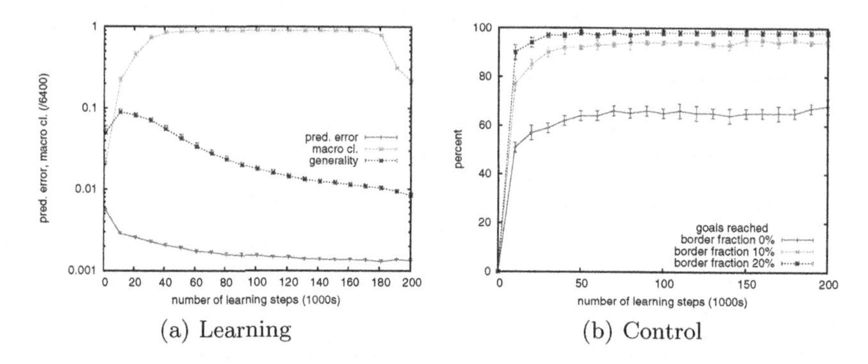

(a) Learning (b) Control

Fig. 1. (a) XCSF learning performance including average accuracy, classifier generality, and effective number of classifiers. (b) XCSF arm control performance with different border fractions—almost all goals in the inner region are reached.

value. The corresponding hand location for the i-th posture $P[i]$ is denoted as $h(P[i])$. XCSF-based control is tested by setting the initial arm posture to $P[i]$ and the hand location goal to $h(P[i+1])$, resulting in 255 start-goal combinations distributed throughout the posture space. A goal was said to be reached, if the hand was within 7.5% arm length of the goal location.

Fig. 1(b) shows the control performance of XCSF on this testbed, where available redundancy was resolved such that extreme angular values are avoided. In order to do so, we apply a minimal movement constraint when joint i is inside the 50% inner region of his movement range. If the joint moves towards the minimum or maximum we apply a quadratically increasing penalty. The border region of the joint angle space poses a problem since (1) the border region is less sampled during training, (2) fewer degrees of freedom are available at the border compared to the inner region, and (3) the effect of gravity and inertia are pronounced at the border. However, XCSF-based control is able to reach almost all goals if we omit a fraction of the border region for testing.

The interplay between XCSF's control and the PD controllers yields a somewhat bad path efficiency, that is, the distance from initial hand location to the goal divided by the traveled distance of the hand. Since the controllers of the joints are independent of each other, the system is unable to exactly reach a goal location. Additionally XCSF's approximations are linear and the resulting joint angle changes are slightly inaccurate, too. Together, the components yield an average path efficiency below 65% for all experiments. Further optimization seems necessary to yield an optimal hand to goal path.

Exploiting the Redundancy. Finally, we take a closer look at the redundancy-resolving mechanism. We compare three different constraints: a minimal movement constraint (denoted 0000), which effectively tries to reach the given goal with minimal angular changes, a natural posture constraint (denoted NNNN), which avoids extreme joint angles trying to resemble human behavior, and a favorite joint angle constraint (denoted 000J), which tries to center the wrist

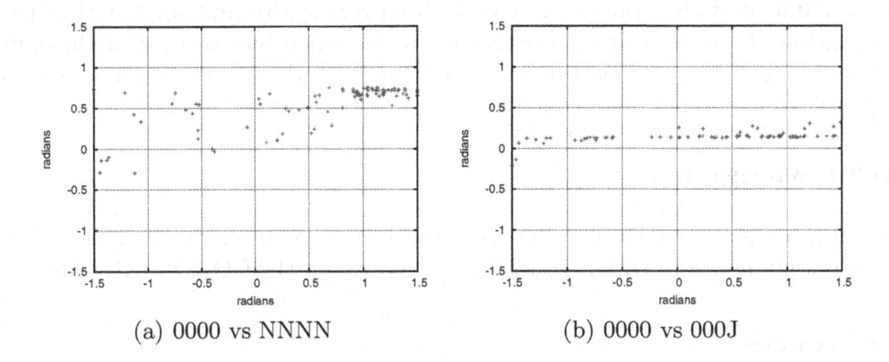

(a) 0000 vs NNNN (b) 0000 vs 000J

Fig. 2. When adding the natural posture constraint NNNN or the favorite joint angle constraint 000J, the final wrist joint angle in radians (movement range $[-1.5, 1.5]$) focuses much more on the inner joint region, compared to the minimal movement constraint 0000.

while the minimal movement constraint is applied to the other joints. In order to compare the three constraints, we generate 100 random postures and use XCSF-based control to reach a hand location at $x = 15\,\text{cm}, y = 65\,\text{cm}, z = -15\,\text{cm}$ that can be reached with a natural posture as well as with non-natural postures. After the hand location goal is reached we store the wrist joint angle.

Fig. 2(a) shows the final wrist joint angles with constraint 0000 (horizontal axis) against the final wrist joint angles with the natural posture constraint NNNN (vertical axis). Since the minimal movement constraint does not care about the absolute joint angle, the values are distributed in the full angular range of the joint, while the natural posture constraint yields values that are farther away from the borders. The effect is even stronger for constraint 000J, as shown in Fig. 2(b). The wrist is centered (angle close to 0), while other joints are not constrained. We observe that the wrist is usually slightly extended, which is due to the effect of gravity and independent PD controllers.

5 Conclusion

The evaluations of the XCSF-based control approach have shown that the system is able to control a dynamic four degree of freedom arm in a three-dimensional workspace. In an exhaustive testbed covering the complete configuration space, almost all goals are reached, while effectively and flexibly considering additional task constraints. Thus, the unique, locally-linear knowledge representation learned by XCSF can be effectively used to resolve redundancies online flexibly satisfying additional task constraints in 3D controlling four degrees of freedom. Only border regions posed a problem due to the sampling method, restricted degrees of freedom, and the pronounced effect of gravity. Future work needs to optimize the random exploration phase to improve performance also in the border regions. Further analyses with even more degrees of freedom will investigate the scalability of the approach. Overall, however, the results suggest that the application of the developed mechanisms on a real robot arm are within our grasp. The result would be a robot control system that learns from scratch a highly flexible and adaptive controller.

Acknowledgments

The authors acknowledge funding from the Emmy Noether program (German Research Foundation, DFG, BU1335/3-1) and thank the COBOSLAB team.

References

1. Craig, J.J.: Introduction to Robotics: Mechanics and Control. Addison-Wesley Longman Publishing Co., Inc., Boston (1989)
2. Baker, D.R., Wampler II., C.W.: On the inverse kinematics of redundant manipulators. The International Journal of Robotics Research 7(2), 3–21 (1988)
3. Holland, J.H., Reitman, J.S.: Cognitive systems based on adaptive algorithms. SIGART Bull. 63(63), 49 (1977)
4. Butz, M.V., Herbort, O.: Context-dependent predictions and cognitive arm control with XCSF. In: GECCO 2008: Proceedings of the 10th annual conference on Genetic and evolutionary computation, pp. 1357–1364. ACM, New York (2008)
5. Butz, M.V., Pedersen, G.K., Stalph, P.O.: Learning sensorimotor control structures with XCSF: Redundancy exploitation and dynamic control. In: GECCO 2009: Proceedings of the 11th annual conference on Genetic and evolutionary computation, pp. 1171–1178 (2009)
6. Wilson, S.W.: Classifiers that approximate functions. Natural Computing 1, 211–234 (2002)
7. Wilson, S.W.: Classifier fitness based on accuracy. Evolutionary Computation 3(2), 149–175 (1995)
8. Butz, M.V., Lanzi, P.L., Wilson, S.W.: Function approximation with XCS: Hyperellipsoidal conditions, recursive least squares, and compaction. IEEE Transactions on Evolutionary Computation 12, 355–376 (2008)
9. Lanzi, P.L., Loiacono, D., Wilson, S.W., Goldberg, D.E.: Prediction update algorithms for XCSF: RLS, kalman filter, and gain adaptation. In: GECCO 2006: Proceedings of the 8th annual conference on Genetic and evolutionary computation, pp. 1505–1512. ACM, New York (2006)
10. Salaün, C., Padois, V., Sigaud, O.: Control of redundant robots using learned models: An operational space control approach. In: IEEE/RSJ International Conference on Intelligent Robots and Systems (submitted)
11. Vijayakumar, S., D'Souza, A., Schaal, S.: Incremental online learning in high dimensions. Neural Computation 17, 2602–2634 (2005)
12. Vijayakumar, S., Schaal, S.: Locally weighted projection regression: An O(n) algorithm for incremental real time learning in high dimensional space. In: ICML 2000: Proceedings of the Seventeenth International Conference on Machine Learning, pp. 1079–1086. Morgan Kaufmann Publishers Inc., San Francisco (2000)
13. Orriols-Puig, A., Bernadó-Mansilla, E.: Bounding XCS's parameters for unbalanced datasets. In: GECCO 2006: Proceedings of the 8th annual conference on Genetic and evolutionary computation, pp. 1561–1568. ACM, New York (2006)

An Evolutionary Graph Transformation System as a Modelling Framework for Evolutionary Algorithms*

Hauke Tönnies

University of Bremen, Department of Computer Science
P.O. Box 330440, D-28334 Bremen, Germany
hatoe@informatik.uni-bremen.de

Abstract. In this paper an heuristic method for the solving of complex optimization problems is presented which is inspired equally by genetic algorithms and graph transformation. In short it can be described as a genetic algorithm where the individuals (encoding solutions of the given problem) are always graphs and the operators to create new individuals are provided by graph transformation. As a case study this method is used to solve the independent set problem.

1 Introduction

Solving combinatorial optimization problems, especially those whose decision variant belong to the complexity class of NP-complete, remains an important and highly active field of research, partly because of their importance in many industrial areas, partly because of the still unclear relationship between optimization problems that are NP-complete and optimization problems that are known to be solvable in polynomial time. A broad and diverse number of important optimization problems are NP-complete which means that in most cases realistic sizes of problem instances can only be solved by heuristic methods since an important property of NP-complete problems is that the worst case time behaviour of the currently best algorithm to solve them is exponential in the size of the problem instance.

This paper proposes to use graphs and graph transformation as a rule-based formal framework for modelling evoluionary algorithms which finally leads to a system called evolutionary graph transformation system. Graphs are quite generic data structure and therefore suitable for the modelling of a lot of interesting and complex optimization problems. The common search algorithms on graphs traverse the graphs to find either an optimal or an optimal-close solution, depending on the complexity of the problem. Graph transformation

* The author would like to acknowledge that his research is partially supported by the Collaborative Research Centre 637 (Autonomous Cooperating Logistic Processes: A Paradigm Shift and Its Limitations) funded by the German Research Foundation (DFG).

B. Mertsching, M. Hund, and Z. Aziz (Eds.): KI 2009, LNAI 5803, pp. 201–208, 2009.

works directly on the representation of the problem, i.e. the graph, and alter the graph in a way that it represents a new and maybe better solution of the problem. Since graph transformation rules perfom local changes on graphs, it is a straight-forward idea, to model a mutation operator with graph transformation. Altogether with a suitable fitness and selection function and some rules for the initialization of the graphs, a nice way of modelling evolutionary algorithms is achieved.

The paper is organized as follows. In section 2 graphs and graph transformation are formally introduced. In section 3 it is shown how an evolutionary algorithm can be modelled by an evolutionary graph transformation system to solve the maximum independet set problem. The conclusion is given in section 5.

2 Graphs and Graph Transformation for Modelling Static and Dynamic Systems

In this paper, undirected, edge-labelled graphs are assumed, although the presented method is suitable for any kind of graphs. In the following, we present some formal definitions:

Let V be some set. Then $\binom{V}{k}$ denotes the set of all subsets of V containing exactly k elements. Moreover, V^{1+2} is a notational shorthand for the union $\binom{V}{1} \cup \binom{V}{2}$.

Definition 1 (*Undirected, edge-labelled graph*)

Let Σ be an arbitrary set of labels. An undirected, edge-labelled graph is a quadruple $G = (V, E, att, l)$, where

1. *V is the set of nodes,*
2. *E is the set of edges,*
3. *$att : E \to V^{1+2}$ and*
4. *$l : E \to \Sigma$ are total functions.*

An edge e with $att(e) = \{v\}$ for one $v \in V$ is called *loop*. The sets V and E and the functions att and l of a graph G will be denoted $V(G)$, $E(G)$, att_G, l_G respectively. To cover unlabeled graphs a special element $* \in \Sigma$ is assumed, which is not drawn in the visualization of graphs. The set of all graphs over Σ is denoted by $\mathcal{G}_\Sigma$.

Definition 2 (*Subgraph*)

Let G and G' be two graphs. G is subgraph of G', if $V(G) \subseteq V(G')$ and $E(G) \subseteq E(G')$, $att(e) = att'(e)$ and $l(e) = l'(e)$ for all $e \in E(G)$.

Given a graph, a subgraph is obtained by removing some nodes and edges subject to the condition that the removal of a node is accompanied by the removal of all its incident edges. Let G be a graph and $X = (V(X), E(X)) \subseteq (V, E)$ be a pair of sets of nodes and edges. Then $G - X = (V - V(X), E - E(X), att, l)$ is only then properly defined, if the above condition is met.

Definition 3 (*Graphmorphism*)

Let G and G' be two graphs. A graph morphism $\alpha : G \to G'$ is a pair of functions $\alpha_V : V^{1+2} \to V^{1+2}$ and $\alpha_E : E(G) \to E(G')$ that are structure-preserving.

The image $g(G)$ of a graphmorphism $\alpha : G \to G'$ in G' is called *match* of G in G' and is subgraph of G'.

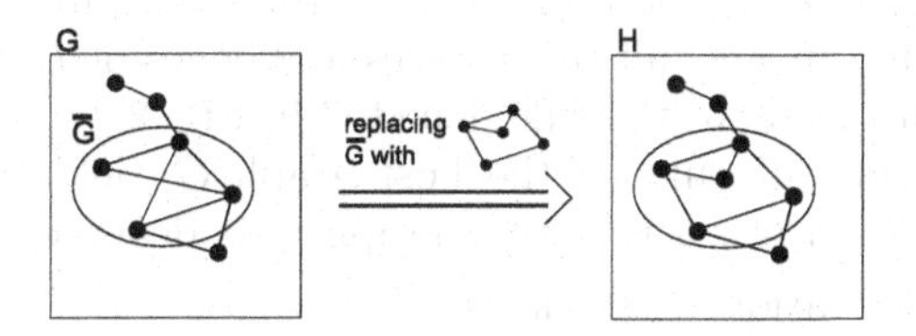

Fig. 1. Replacing subgraph $\overline{\mathrm{G}}$ by a new graph

2.1 Graph Transformation

The idea of graph transformation consists in replacing a subgraph of a given graph by another graph (see Fig. 1). Usually this replacement is based on rules, the so called *graph transformation rules.*

Definition 4 (*Graph transformation rule*)

A graph transformation rule r *consists of three graphs $L, K, R \in \mathcal{G}_\Sigma$ with $L \supseteq K \subseteq R$.*

The graph to transform is called *hostgraph.* An application of a graph transformation rule r on a hostgraph G consists of three steps:

1. A graph morphism $\alpha : L \to G$ is chosen respecting two *application conditions:*
   (a) *Contact condition:* The removal of the match $\alpha(L)$ in G must not leave edges without attachment.
   (b) *Identification condition:* If two nodes or edges of L are identified in the match $\alpha(L)$ they must be in K.
2. The match $\alpha(L)$ is removed up to $\alpha(K)$ from G yielding a new graph $Z = G - (\alpha(L) - \alpha(K))$.
3. The graph R is added to Z by gluing Z with R in $\alpha(K)$ yielding the graph $H = Z + (V(R) - V(K), E(R) - E(K), att', l')$ with $att'(e') = att_R(e')$, if $att(e') \in (V(R) - V(K))^{1+2}$ and $att'(e') = \alpha_V(att_R(e'))$ otherwise and $l'(e') = l_R(e')$ for all $e' \in E(R) - E(K)$.

Analyzing in this respect the replacement of Fig. 1, the corresponding graph transformation rule is easy to obtain. The graphs L and R are obviously the graphs H and the graph depicted above the arrow in the figure. The graph K

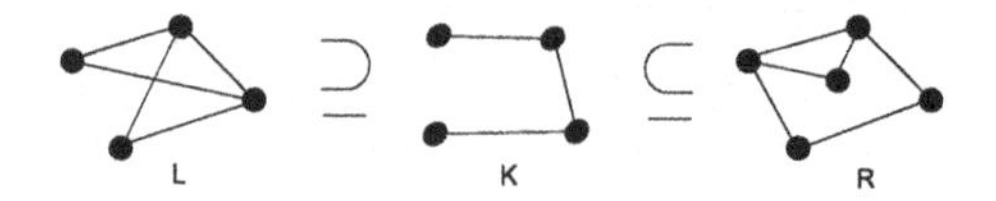

Fig. 2. Graph transformation rule

helps to unambigously embed the graph R in the hostgraph after the removal of $\alpha(L) - \alpha(K)$. That means, all the vertices and edges of L that we still need, should be in K. This leads to the rule depicted in Fig. 2.

The application of this rule on the host graph G yields the graph G'. In general, the application of a rule r to a graph G is denoted by $G \underset{r}{\Longrightarrow} G'$ and is called a direct derivation. A sequence $G_0 \underset{r_1}{\Longrightarrow} G_1 \underset{r_2}{\Longrightarrow} \ldots \underset{r_n}{\Longrightarrow} G_n$ of direct derivations can also be denoted by $G_0 \overset{n}{\underset{P}{\Longrightarrow}} G_n$, if $r_1, \ldots, r_n \in P$ or $G \overset{*}{\underset{r}{\Longrightarrow}} G'$ for $G = G_0$ and $G' = G_n$. The string $r_1, \ldots, r_n$ is called *application sequence* of the derivation $G_0 \underset{r}{\Longrightarrow} G_1 \underset{r}{\Longrightarrow} \ldots \underset{r}{\Longrightarrow} G_n$.

Graph transformation comprises a broad and wide area of application, for an overview see for example [1] . In this paper we focus on the application in the area of operations research. Before we show how graph transformation can be used to model algorithms to solve problems, we have to introduce one more important concept: the *graph transformation unit.* A graph transformation unit uses subsets of graphs specified by so called graph *class expressions* and so called *control conditions* to cut down the derivation process.

Definition 5 (*Graph class expression*)

A graph class expression is a syntactical entity X that specifies a set of graphs $SEM(X) \subseteq \mathcal{G}_\Sigma$.

A popular example of a graph class expression is a set of labels $\Delta \subseteq \Sigma$ such that

$$SEM(\Delta) = \{G \in \mathcal{G}_\Sigma \mid l(e) = \Delta \quad \forall e \in E(G)\}$$

Definition 6 (*Control condition*)

A control condition is a syntactical entity C that specifies a language $L(C)$ such that an application sequence s is permitted if and only if $s \in L(C)$.

Control conditions are useful to reduce the inherent non-determinism of applications of graph transformation rules. Without control conditions, any rule of the set P can be applied arbitrarily at any time, but sometimes it is preferable to control for example the order of the rule applications. In this case, regular expressions serve as an useful control conditions: Let $r_1 r_2^* r_3$ be a regular expression as a control condition. All application sequences must then start with applying the rule r_1, following the application of an arbitrarily number of times rule r_2 and ending with the application of rule r_3. Another useful control condition consists in applying a rule *as long as possible* before other rules can be applied. To denote this control condition, we write an exclamation mark after the rule (e.g. $r_1 r_2 ! r_3$). More about graph transformation units and further examples of graph class expressions and control conditions can be found in [2], [3], [4].

Now we have all the components to define formally a graph transformation unit:

Definition 7 (***Graph transformation unit***)

A graph transformation unit is quadruple $tu = (I, P, C, T)$ where I, T are graph class expressions, P a set of rules and C a control condition.

A graph transformation unit defines a binary relation on the set of all graphs $\mathcal{G}_\Sigma$. Intuitivly, a graph G is tu-related to G', denoted by $G \underset{tu}{\Longrightarrow} G'$, iff $G \in SEM(I), G' \in SEM(T)$ and $G \overset{*}{\underset{P}{\Longrightarrow}} G'$ respecting the control condition C. As it is shown in the following sections, graph transformation units are useful in modelling algorithms on graphs.

Sometimes it is useful not only to transform a single graph, but a set or a multiset of graphs. Let tu be a transformation unit and $M : \mathcal{G}_\Sigma \to \mathbb{N}$ a multiset of graphs. A multiset transformation $M \underset{tu}{\Longrightarrow} M'$ transform all graphs of M by applying the transformation unit tu on all graphs of M. The multiset M' consists then of all transformed graphs. More about graph multiset transformation can be found in [5].

3 Solving Combinatorial Optimization Problems with Evolutionary Graph Transformation Systems

As mentioned before, the following heuristic needs the combinatorial optimization problem to be represented as a graph. Keeping in mind that most of the interesting combinatorial problems are modelled as graphs anyways, this restriction can be seen as an advantage to avoid finding an appropriate coding. The key idea of this method consists of a so called evolutionary graph transformation system that uses graph transformation units as a mutation operator to create new graphs which encode new solutions to the given problem.

3.1 An Evolutionary Graph Transformation System for the Independent Set Problem

First of all, a formal definition of an evolutionary graph transformation system is given.

Definition 8 (***Evolutionary graph transformation system***)

Let Σ be a set of labels and $\mathcal{G}_\Sigma$ the set of all graphs over Σ according to definition 1. An evolutionary graph transformation system is a system

$$evoGTS = (init, mut, fitness, selection, T, n)$$

where

1. init *is a graph transformation unit,*
2. mut *is a graph transformation unit,*
3. $fitness : \mathcal{G}_\Sigma \to \mathbb{R}$ *is a total function,*
4. $selection : \mathcal{M}_{finite}(\mathcal{G}_\Sigma) \to \mathcal{M}_{finite}(\mathcal{G}_\Sigma)$ *is a total function, where* $\mathcal{M}_{finite}(\mathcal{G}_\Sigma)$ *denotes the set of all finite multisets of* $\mathcal{G}_\Sigma$*,*
5. $T : \mathcal{M}_{finite}(\mathcal{G}_\Sigma)^* \to BOOL$ *is a total function and*
6. $n \in \mathbb{N}$ *a natural number.*

It is now shown, how these components interplay to solve a given optimization problem. As a running example a maximum independent set in the given graph is searched.

Fig. 3. Initiation rule

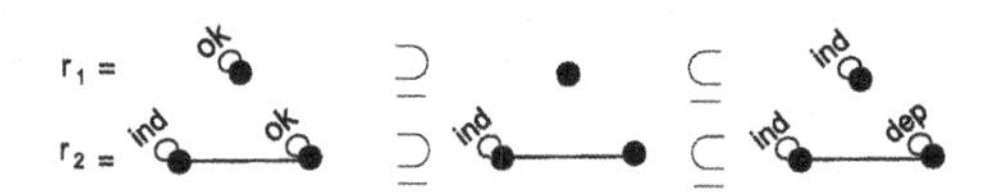

Fig. 4. Mutation rules

3.2 Solving Problems

1. The first step consists of initializing the set POP_0 through the transformation unit *init*.
$$POP_0 \underset{init}{\Longrightarrow} POP_1 \tag{1}$$
In this case, all graphs are initialized by putting an unlabeled loop to every node. The corresponding rule is depicted in Fig. 3. The control condition consists of applying this rule an arbitrary number of times and the terminal graphs, that is the succesfully initialized graphs, are exactly the graphs where every node has exactly one loop labeled with *OK*.
2. Once the multiset is initialized, the transformation unit *mut* is used to create a new multiset, which we will call, in analogy to other genetic algorithms, *children*.
$$POP_1 \underset{mut}{\Longrightarrow} children_1 \tag{2}$$
In case of the independent set problem a mutation consists of putting one random node, which has not been marked before, into the current independent set. The adjacent nodes are marked, so that the constraints of an independent set are never violated. The corresponding rule is depicted in Fig. 4 and the control condition is $(r1)(r2)!$

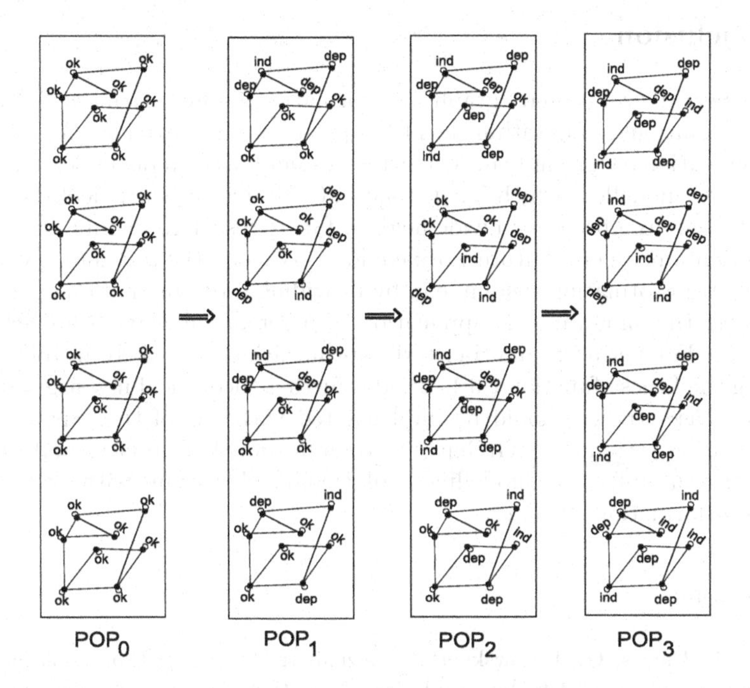

Fig. 5. Searching the maximum independent set

3. Out of the two multisets POP_1 and $children_1$ the function *selection* chooses some graphs to form the multiset POP_2(the new parent generation) usually based on the fitness values. It seems a good idea to use well-known selection functions from other evolutionary algorithm.

$$selection(POP_1 + children_1) = POP_2 \tag{3}$$

One suitable fitness function for the independent set could be the following: Let S be the current independent set of a graph G.

$$fitness(G) = \begin{cases} \infty & if\ S = \varnothing \\ \frac{\sum_{s \in S} grad(s)}{|S|} & otherwise \end{cases}$$

4. It is checked, whether the function $T(POP_1, POP_2)$ yields *true*. If it does, the best graphs (e.g. the graphs with the lowest $fitness$-value) from POP_2 are returned. If it does not, the procedure beginning from step 2 starts again, using the multiset POP_2 as the multiset to be mutated.

In Fig. 5 a possible run of the evolutionary graph transformation system is shown, where the maximum independent set can be found in only one of the graphs in the last population.

4 Conclusion

In this paper, an evolutionary graph transformation system as a modelling framework for evolutionary algorithms has been proposed. In particular, it was shown how an evolutionary graph transformation system for the independent set problem could be modelled, which has turned out suitable to heuristically solve the problem. There is a broad and wide field for future research, so only a few examples are mentioned here. Future work could also extend the presented evolutionary graph transformation system, e.g. by providing a recombination operator or by allowing the simultaneous application of a set of rules and thus achieving parallelism. Furthermore, multi-criteria optimization can easily introduced by adjusting the fitness-function ([6]). Besides further theoretical investigation, this and other case studies should be implemented with one of the existing graph transformation system, e.g. GrGen ([7]) to gain some benchmarks and to prove the practical usefulness of modelling evolutionary algorithms with evolutionary graph transformation systems.

References

1. Ehrig, H., Engels, G., Kreowski, H.J., Rozenberg, G. (eds.): Handbook of Graph Grammars and Computing by Graph Transformation. Applications, Languages and Tools, vol. 2. World Scientific, Singapore (1999)
2. Kreowski, H.J., Kuske, S., Rozenberg, G.: Graph transformation units - an overview. In: Degano, P., De Nicola, R., Meseguer, J. (eds.) Concurrency, Graphs and Models. LNCS, vol. 5065, pp. 57–75. Springer, Heidelberg (2008)
3. Kreowski, H.J., Kuske, S.: Graph transformation units with interleaving semantics. Formal Aspects of Computing 11(6), 690–723 (1999)
4. Kuske, S.: Transformation Units—A Structuring Principle for Graph Transformation Systems. PhD thesis, University of Bremen (2000)
5. Kreowski, H.J., Kuske, S.: Graph multiset transformation as a framework for massively parallel computation. In: Ehrig, H., Heckel, R., Rozenberg, G., Taentzer, G. (eds.) ICGT 2008. LNCS, vol. 5214, pp. 351–365. Springer, Heidelberg (2008)
6. Deb, K.: Multi-Objective Optimization Using Evolutionary Algorithms. John Wiley & Sons, Inc., New York (2001)
7. Geiß, R., Kroll, M.: GrGen.NET: A fast, expressive, and general purpose graph rewrite tool. In: Schürr, A., Nagl, M., Zündorf, A. (eds.) AGTIVE 2007. LNCS, vol. 5088, pp. 568–569. Springer, Heidelberg (2008)

Semi-automatic Creation of Resources for Spoken Dialog Systems

Tatjana Scheffler, Roland Roller, and Norbert Reithinger*

DFKI GmbH, Projektbüro Berlin
Alt-Moabit 91c
10559 Berlin, Germany
{firstname.lastname}@dfki.de

Abstract. The increasing number of spoken dialog systems calls for efficient approaches for their development and testing. Our goal is the minimization of hand-crafted resources to maximize the portability of this evaluation environment across spoken dialog systems and domains. In this paper we discuss the user simulation technique which allows us to learn general user strategies from a new corpus. We present this corpus, the VOICE Awards human-machine dialog corpus, and show how it is used to semi-automatically extract the resources and knowledge bases necessary in spoken dialog systems, e.g., the ASR grammar, the dialog classifier, the templates for generation, etc.

1 Introduction

The more spoken dialog systems (SDSs) are put into practice in different domains, the more efficient methods for their development and deployment are urgently needed. The project SpeechEval aims to address this need in two ways: First, by investigating the use of dialog corpora in order to automatically or semi-automatically create the resources necessary for the construction of SDSs. And second, by learning general user behavior from the same corpora, and building a flexible user simulation which can be used to test the overall usability of SDSs during development or after deployment.

Automatic testing of dialog systems is attractive because of its efficiency and cost-effectiveness. However, previous work in this area concentrated on detailed tests of individual subcomponents of the SDS (such as the ASR). In order to judge the overall usability of a system, extended testing by human callers has been necessary – a step that is usually too costly to be undertaken during the prototype stage or repeatedly after changes to the deployed system. SpeechEval intends to fill this gap. Maximum modularity of the system architecture (see [1]) as well as the (semi-)automatic techniques for the creation of the underlying resources for the user simulation (in particular, domain knowledge and user strategies) allow SpeechEval to be easily portable across different SDSs.

* This research was funded by the IBB through the ProFIT framework, grant #10140648, and sponsored by the European Regional Development Fund.

B. Mertsching, M. Hund, and Z. Aziz (Eds.): KI 2009, LNAI 5803, pp. 209–216, 2009.

In the following, we first discuss our approach to user simulation. Then we present the VOICE Awards corpus, a new dialog corpus which is the basis of our further work. The rest of the paper describes our finished and ongoing work in extracting knowledge bases for spoken dialog systems from corpora.

2 User Simulation

User simulation is used in the SDS literature for several purposes. First, for training the dialog manager of a SDS during reinforcement learning. In this case, the SDS with the learned strategy is the actual purpose of the research, whereas the user simulation is just a means to that end. Second, user simulation is used for evaluation or testing of the trained policies/dialog managers of the developed SDSs. The two types of purposes of user simulations may call for different methods. A user simulation may be used to test for general soundness of an SDS, specifically searching for errors in the design. In such a case, a random exploration may be called for [2]. A restricted random model may also perform well for learning [3].

In other cases, ideal users may be modelled so that reinforcement learning is able to learn good paths through the system's states to the goal [4]. Often (as in the previous example), a suitable user simulation is hand-crafted by the designer of the dialog system. A good overview of state-of-the-art user models for SDS training is given in [5].

Our goal is to as much as possible avoid hand-crafting the strategy (i.e., user simulation). An optimal strategy is not needed for our user simulation, neither is a random explorative strategy. Instead, the aim should be realistic user behavior. Our goal is to rapidly develop user simulations which show similar behavior (at least asymptotically) to human users in the same situations. The behavior of human callers of spoken dialog systems can be observed in our corpus, the VOICE Awards (VA) Corpus described below in section 3. We therefore define realistic user behavior in our case as user utterances that probabilistically match the ones represented in our corpus. Such probabilistic models are often used for evaluation of learned dialog managers [3].

Our current target approach is very close to the one proposed in [6] for an information state update system. At each state in the dialog, the user model choses the next action based on the transition probabilities observed in the corpus. Since some states have never or only rarely been seen in the corpus, we choose a vector of features as the representation of each dialog state. These features in our case include properties of the dialog history (such as the previous dialog act, the number of errors), the current user characteristics (expert vs. novice, for example), as well as other features such as the ASR confidence score. We estimate from the corpus the amount that each feature in the vector contributes to the choice of the next action. Thus, unseen states can be easily mapped onto the known state space as they lead to similar behavior as closely related seen states would.

The chosen next action is a dialog act type that must be enriched with content based on the goal and user characteristics. General heuristics are used to perform this operation of tying in the user simulation with the domain- and system-specific ontology.

3 A Human-Machine Dialog Corpus

For the development of a new spoken dialog system, the rules and knowledge bases must be specified by a human expert. As an alternative to hand-crafted systems, the strategies in a SDS may be learned automatically from available corpora. Much research has been done in this area recently, especially on dialog strategy optimization by reinforcement learning with (Partially Observable) Markov Decision Processes ((PO)MDPs) (see for example [7] for an overview). This approach works best for learning very specific decisions such as whether or not to ask a confirmation question or how many pieces of information to present to a user [8]. In addition, such systems must have access to large corpora of interactions with the particular system for training. Our goal, however, is to be able to interact with a new SDS in a new domain with little modification. In particular, in real applications we cannot assume the existence of a large specialized corpus of human-machine dialogs from that particular system or domain, as has been done in much of the previous literature. Therefore, we aim to learn general strategies of user behavior as well as other kinds of knowledge bases from a general dialog corpus.

Since we could not identify an appropriate human-machine dialog corpus in German, we are currently in the process of compiling and annotating the VOICE Awards (VA) corpus, which is based on the "VOICE Awards" contest. The annual competition "VOICE Awards"[1] is an evaluation of commercially deployed spoken dialog systems from the German speaking area. Since 2004, the best German SDSs are entered in this benchmarking evaluation, where they are tested by lay and expert users. We are constructing an annotated corpus of the available audio recordings from this competition, including the years 2005–2008 (recording of 2009 data is in progress).

The corpus represents a large breadth of dialog systems and constitutes a cut through the state-of-the-art in commercially deployed German SDSs. Altogether, there are more than 120 dialog systems from different domains in the corpus, with over 1500 dialogs. In each year of the competition, several lay users were asked to call the dialog systems to be tested and perform a given task in each of them. The task was pre-determined by the competition organizers according to the developers' system descriptions. After completing the task, the users filled out satisfaction surveys which comprised the bulk of the evaluation for the award. In addition, two experts interacted with each system and performed more intensive tests, specifically to judge the system's reaction to barge-ins, nonsensical input, etc. Table 1 contains a list of some of the domains represented by the dialog systems included in the VOICE Awards corpus.

[1] http://www.voiceaward.de/

Table 1. Some domains of SDSs included in the VOICE Awards corpus

public transit schedule information
banking
hotel booking
flight info confirmation
phone provider customer service
movie ticket reservation
package tracking
product purchasing

Audio data for the VA corpus is available in separate .wav files for each dialog. The transcription of the corpus, using the open source Transcriber tool, is more than 50% complete. With the transcription, a rough segmentation into turns and dialog act segments is being performed. Since more fine-grained manual timing information is very difficult and time-consuming to obtain, it is planned to retrieve word-level timing by running a speech recognizer in forced alignment mode after the transcription is completed. As a basis of our statistical analyses, the entire corpus is currently being hand-annotated with several layers of information: (1) dialog acts, (2) sources of miscommunication, (3) repetitions, and (4) task success. Since the lack of space prohibits a detailed discussion, the annotation schemes are simply listed in table 2. We are using a modified tool from the NITE XML Toolkit (NXT) that has been adapted to our needs to perform these annotations in a single step.

We have performed an evaluation of the annotation scheme on part of the available data. Two annotators independently segmented and classified the data from 4 systems (69 dialogs). This test showed very good inter-annotator agreement of Cohen's $\kappa = 0.89$, as shown in table 3. The confusion matrix between the annotators further reveals that most mismatches concern only very few dialog act types (e.g., *alternative_question* and *instruction*), suggesting that revisiting the annotation scheme for these categories could further improve the agreement.

The result will be a large corpus of human-SDS-dialogs from many different domains, covering the entire breadth of the current state-of-the-art in commercially deployed German-language SDSs. In the next section, we describe how we are using this corpus and its annotations to derive resources for the rapid development of spoken dialog systems.

4 Corpus-Assisted Creation of SDS Resources

ASR Grammar. In order to improve the coverage of an SDS's speech recognition, the recognizer's grammar must be augmented by adding both domain specific terminology as well as terms and phrases that are important in the scenario of spoken dialog systems in general. Different strategies will be used to extract both kinds of vocabulary from the VA Corpus as well as other sources.

Table 2. Hand-annotation schemes of the VOICE Awards corpus

dialog acts	errors	repetition	task_success
hello	not_understand	repeat_prompt	task_completed
bye	misunderstand	repeat_answer	subtask_completed
thank	state_error		system_abort
sorry	bad_input		user_abort
open_question	no_input		escalated
request_info	self_correct		abort_subtask
alternative_question	system_command		other_failure
yes_no_question	other_error		
explicit_confirm			
implicit_confirm			
instruction			
repeat_please			
request_instruction			
provide_info			
accept			
reject			
noise			
other_da			

Table 3. Inter-annotator agreement for the dialog act (DA) dimension

total # DAs	agree on segmentation		agree on seg & type
2375	1917		1740
	.81		.73
Chance agreement	0.16	(matching segments only)	
Cohen's kappa	0.89	(matching segments only)	

For the extraction of domain specific terminology, we have categorized the systems in the corpus along two dimensions into 24 topic domains (see table 1) and 8 interaction types (e.g., game, number entry, shopping, etc.). A simple chi-square test is used to determine whether a certain word i is significant for a domain j. Using a stop-word list of the 1000 most frequent terms in German, any word with a chi-square value greater than 3.84 is likely ($p < 0.05$) to be significant for the domain. Words which occurred less than 5 times in the corpus were discarded since the test is likely to be inaccurate. This method yielded very good results even when evaluated on a very small subcorpus. Table 4 shows the top 15 positively significant words for the banking domain, as computed on only 58 dialogs (3 systems) from the domain, and a similar amount of out-of-domain dialogs. The only false hits are words that are very suggestive of customer service SDSs in general ("möchten" / "would like"). These can be excluded by a second stop word list.

Table 4. Significant words in the banking domain

term	English	χ^2	term	English	χ^2
Kontostand	account balance	56.6	Ziffer	digit	27.6
Kontonummer	account number	54.5	Geburtsdatum	birth date	26.0
möchten	would like	44.1	Hauptmenü	main menu	23.9
Umsätze	transactions	40.7	Bankleitzahl	routing number	22.9
Konto	account	40.2	Servicewunsch	service request	21.8
Überweisung	wire transfer	32.9	beträgt	amounts to	21.3
Cent	Cent	29.1	Gutschrift	credit	20.8
minus	negative	28.1			

We are extracting SDS-specific terminology (such as "customer id", "main menu", etc.) using the same methodology. All dialogs in the VA corpus are used as the positive subcorpus. For the negative examples, we plan to use text extracted from web pages representing a similar range of topics and domains as the VA corpus. This will ensure that only terminology specific to the medium of spoken dialog systems is marked significant by the chi-square test, and not other frequent content words such as domain-specific terms.

User Characteristics. In order to perform realistic testing of dialog systems, the user simulation's behavior must be relatively varied. We aim to identify suitable user types from the VA corpus to model them in our user simulation. Broad distinctions such as expert vs. novice users are known from the literature, but aren't easily observable in the corpus, since by far most dialogs are by lay users. Thus, we instead try to distinguish objectively observable characteristics such as the user reaction time, number of barge-ins, etc. We will perform a clustering on each of these variables in order to obtain a "user properties vector" for each caller in the corpus. The obtained user characteristics then become part of the dialog state vector which determines the following user actions. This will account for the differences in behavior of different user types.

Dialog Act Segmentation and Classification. Machine learning approaches are the standard approaches to the tasks of dialog act segmentation and classification. Good results can be obtained when the number of classes is not too high, although the quality of the ASR output has a large impact on the accuracy, as well. We distinguish 18 dialog act types (see table 2). Further, the types can be grouped into a flat hierarchy of broad categories such as "request" and "answer". Thus, even in cases where an incoming dialog act has been wrongly classified, SpeechEval's reply may still be appropriate if the misclassified type is of the same super-category.

Our segmentation and classification follows closely the method developed in the AMIDA project [9]. We use the WEKA toolkit to implement separate segmentation and dialog act classification learners. As opposed to this previous work, we use the learned classification modules within an online system. This

Table 5. SMO dialog act classification results

total # DAs	**correct**	**incorrect**
1680	1396	284
	.83	.17
Cohen's kappa	0.78	

means that we cannot make use of dynamic features that require the knowledge of future assignments (as is done in the dialog act classifier). Each determined dialog act type is passed on immediately down the pipeline architecture and is acted upon in further modules.

As a first experiment we have trained a dialog act classifier using Sequential Minimal Optimization (SMO). The subcorpus for this experiment consists of 23 different spoken dialog systems with a total of 355 dialogs. For the dialog act classification, we divided the corpus into a specific training (20 systems, 298 dialogs, 9399 dialog acts) and test set (3 systems, 57 dialogs, 1680 dialog acts). The trained classifier showed very promising results, shown in table 5.

The kappa statistic of 0.78 shows relatively good agreement of the predicted dialog act with the hand-annotated one, especially considering the fact that human annotators only agree with κ=0.89 as shown above (table 3). The classification accuracy is promising since the test dialogs came from dialog systems and domains that were unseen in the training stage. This suggests that the final classifier trained on the full VA corpus will be very portable across systems and domains. In addition, we are currently using less than 25% of the available data, while the full amount of training data will increase the performance significantly. Furthermore there will be optimizations on the training data itself (for example, the treatment of overlapping segments) and the classification algorithm.

Concerning misclassifications, our evaluation so far has been very strict. The dialog act types can be grouped into a few broad super-categories (request, answer, etc.). Three super-categories are crucial for the interaction with a SDS: requests (*open_question*, *request_info*, *alternative_question*, *yes_no_question*), confirmations (*explicit_confirm*, *implicit_confirm*) and metacommunication (*instruction*, *repeat_please*). The confusion matrix shows that many misclassified instances were assigned to a dialog act class in the correct super-category. This means that the information is at least partially recoverable.

Compared to other recent work, the data reported here is very good. [10] report an accuracy of only 58.78% on dialog act classification of multi-party meeting data, even though they use a very similar feature set and a dialog act scheme of 15 types. This shows that system prompts in spoken dialog systems tend to be very schematic, and generalize well even across systems and domains. This validates our approach of extracting general system-independent knowledge bases for our user simulation.

User Utterance Templates. Our corpus shows that by far most user utterances in our corpus consist of just one word. In an initial study, only 12% of the lay user's turns contained more than one word (number sequences such as ID or telephone numbers were excluded). For genuine more-word utterances, we are exploring a grammar induction technique in order to extract possible user utterance templates from our corpus.

5 Conclusion

In this paper we presented an approach to user simulation and spoken dialog system development that allows for very rapid prototyping. We introduced our new corpus of German human-machine dialogs, and discussed the ongoing annotation effort. This corpus constitutes the basis of our statistical methods for extracting both general and domain-dependent knowledge bases for SDSs. We discuss how many resources in a user simulation for SDSs, from the ASR grammar to dialog strategy, can be derived semi-automatically from the general dialog corpus or other supplementary corpora. This ensures easy portability of the user simulation across SDSs and domains and alleviates the need for large specialized corpora or expensive human evaluators.

References

1. Scheffler, T., Roller, R., Reithinger, N.: Speecheval – evaluating spoken dialog systems by user simulation. In: Proceedings of the 6th IJCAI Workshop on Knowledge and Reasoning in Practical Dialogue Systems, Pasadena, CA, pp. 93–98 (2009)
2. Alexandersson, J., Heisterkamp, P.: Some notes on the complexity of dialogues. In: Proceedings of the 1st Sigdial Workshop on Discourse and Dialogue, Hong Kong, vol. 10, pp. 160–169 (2000)
3. Ai, H., Litman, T., Litman, D.: Comparing user simulation models for dialog strategy learning. In: Proceedings of NAACL/HLT 2007, Rochester, NY, pp. 1–4 (2007)
4. López-Cózar, R., de la Torre, A., Segura, J., Rubio, A.: Assessment of dialog systems by means of a new simulation technique. Speech Communication 40, 387–407 (2003)
5. Schatzmann, J., Weilhammer, K., Stuttle, M., Young, S.: A survey of statistical user simulation techniques for reinforcement-learning of dialogue management strategies. The Knowledge Engineering Review (2006)
6. Georgila, K., Henderson, J., Lemon, O.: Learning user simulations for information state update dialogue systems. In: Proceedings of the 9th European Conference on Speech Communication and Technology (Eurospeech), Lisbon, Portugal (2005)
7. Lemon, O., Pietquin, O.: Machine learning for spoken dialogue systems. In: Proceedings of Interspeech (2007)
8. Rieser, V., Lemon, O.: Learning dialogue strategies for interactive database search. In: Proceedings of Interspeech (2007)
9. AMIDA: Deliverable D5.2: Report on multimodal content abstraction. Technical report, DFKI GmbH, ch. 4 (2007)
10. Germesin, S., Becker, T., Poller, P.: Determining latency for on-line dialog act classification. In: MLMI 2008 (September 2008)

Correlating Natural Language Parser Performance with Statistical Measures of the Text

Yi Zhang[1] and Rui Wang[2]

[1] LT-Lab, German Research Center for Artificial Intelligence and Computational Linguistics, Saarland University
yzhang@coli.uni-sb.de
http://www.coli.uni-saarland.de/~yzhang
[2] Computational Linguistics, Saarland University
rwang@coli.uni-sb.de
http://www.coli.uni-saarland.de/~rwang

Abstract. Natural language parsing, as one of the central tasks in natural language processing, is widely used in many AI fields. In this paper, we address an issue of parser performance evaluation, particularly its variation across datasets. We propose three simple statistical measures to characterize the datasets and also evaluate their correlation to the parser performance. The results clearly show that different parsers have different performance variation and sensitivity against these measures. The method can be used to guide the choice of natural language parsers for new domain applications, as well as systematic combination for better parsing accuracy.

1 Introduction

Natural language parsing is not only one of the central tasks in the field of natural language processing, but also widely used in many other AI areas, e.g. human-computer interaction, robotics, etc. While many parsing systems achieve comparably high accuracy from application perspective [1], the robustness of parser performance remains as one of the major problems which is not only unresolved, but also less acknowledged and largely overlooked. The capability of most statistical parsing systems to produce a parse for almost any input does not entail a consistent and robust parser performance on different inputs.

For example, in robotics, the input of the parsers usually comes from an automatic speech recognition (ASR) system, which is error-prune and much worse than the human listeners [2]. More seriously, as one of the earliest components, in many applications, the unsatisfying outputs of the parsers will be propagated and the errors will be amplified through the common pipeline architecture. For example, in a popular task in Bioinformatics, protein-protein interaction extraction, [1] have shown correlation between the parse accuracy and the extraction accuracy.

B. Mertsching, M. Hund, and Z. Aziz (Eds.): KI 2009, LNAI 5803, pp. 217–224, 2009.

Through the past decade, there has been development of numerous parsing systems with different approaches using various representations, many of which are available as open source softwares and ready for use off-the-shelf. However, it is a common knowledge now that treebank-trained parsers usually perform much worse when applied onto texts of different genres from the training set. Although it is the nature of human languages to be diverse, the variation of parser performance does not always correspond to the difficulty of the text for human readers.

More recently, the problem has been studied as the task of parser domain adaptation. For instance, [3] generalized the previous approaches using a maximum a posteriori (MAP) framework and proposed both supervised and unsupervised adaptations of statistical parsers. [4] and [5] have shown that out-domain parser performance can be improved with self-training on a large amount of extra unlabeled data. The CoNLL shared task 2007 [6] has a dedicated challenge to adapt parsers trained on the newspaper texts to process chemistry and child language texts. [7] and [8] port their parsers for biomedical texts, while [9] adapts her parser for various Wikipedia biographical texts. In all, most of the studies take a liberal definition of "domain": the term is used to dub almost any dataset, either slightly or remotely different from the training set. Also, while substantial performance improvements have been reported for different parsing systems, it is not clear whether such methods are equally effective for other parsers (when intuition usually suggests the opposite).

In this paper, we present a series of experiments which correlate the performance of several state-of-the-art parsing systems (Section 3) to three very simple statistical measures of the datasets (Section 2). The result clearly shows that even for a group of datasets of similar genres, parser performance varies substantially. Furthermore, performances of different parsers are sensitive to different statistical measures (Section 4).

2 Statistical Measures for Datasets

There has been a rich literature in text classification on statistical measures that can be used to categorize documents. However, here we are not interested in differentiating the semantic contents of the texts, but in those basic measures which can be potentially correlated with the parser performance. As another related work, [10] focused on annotation differences between datasets and attributed many errors to that; while in this paper, we concern more about the basic statistical distribution of the texts itself within the datasets, without considering the syntactic annotations. When the parsers are tested on datasets with compatible and consistent annotations to the training set, the performance correlation to these measures on unannotated texts reflect the characteristics of the parsers, independent from the annotation scheme adopted.

The following three measures are used for the experiments reported in the this paper.

Average Sentence Length (ASL) is the most simple measure which can be easily calculated for any given dataset without consulting extra resources. Common intuition is that the performance of the parser is relatively worse on longer sentences than shorter ones.

Unknown Word Ratio (UWR) is calculated by counting instances of the unseen words in the training set and deviding it by the total number of word instances in the target dataset.

Unknown Part-of-Speech Trigram Ratio (UPR) is calculated by counting the instances of unseen POS trigram patterns in the training set and deviding it by the total number of trigrams in the target dataset. We also add speical sentence initial and final symbols into the POS patterns, so as to denote the rough position of the trigram in the sentence.

It should be noted that these simple measures are given here as examples to show the performance variation among different parsers. Adding further statistical measures is straightforward, and will be experiment in our future work.

3 Parser Performance Evaluation

Parser evaluation has turned out to be a difficult task on its own, especially in the case of cross-framework parser comparison. Fortunately, in this study we are not interested in the absolute scores of the parser performance, and instead, only the variation of the performance among different datasets. For this reason, we select representative evaluation metrics for each individual parsing system,

We select the following group of representative parsing systems in our experiment. All these parsers are freely available on-line. For those parsers where training is required, we use the Wall Street Journal (WSJ) section 2-21 of the Penn Treebank (PTB) as the training set. This includes both the phrase-structure trees in the original PTB annotation, and the automatically converted word-word dependency representation.

*Dan Bikel's Parser (*DBP*)*[1] [11] is an open source multilingual parsing engine. We use it to emulate Collins parsing model II [12].

*Stanford Parser (*SP*)*[2] [13] is used as an unlexicalized probabilistic Context-Free Grammar (PCFG) parser. It utilizes important features commonly expressed by closed class words, but no use is made of lexical class words, to provide either monolexical or bilexical probabilities.

*MST Parser (*MST*)*[3] [14] is a graph-based dependency parser where the best parse tree is acquired by searching for a spanning tree which maximize the score on an either partially or fully connected dependency graph.

[1] http://www.cis.upenn.edu/ dbikel/software.html
[2] http://nlp.stanford.edu/software/lex-parser.shtml
[3] http://sourceforge.net/projects/mstparser/

*Malt Parser (*MALT*)* [4] [15] follows a transition-based approach, where parsing is done through a series of actions deterministically predicted by an oracle.

*ERG+PET (*ERG*)* [5] is the combination of a large scale hand-crafted HPSG grammar for English [16], and a language independent unification-based efficient parser [17]. The statistical disambiguation model is trained with part of the WSJ data.

Although these parsers adopt different representations, making cross-framework parser evaluation difficult, here we are only interested in the relative performance variation of individual parsers. Hence, different evaluation metrics are used for different parsers. For constituent-based PCFG parsers (DBP and SP), we evaluate the labeled bracketing F-score; and for dependency parsers (MST and MALT), we evaluate the labeled attachment score. Since there is no gold HPSG treebank for our target test set, we map the HPSG parser output into a word dependency representation, and evaluate the unlabeled attachment score against our gold dependency representation.

4 Experiment Results

4.1 Datasets

As test datasets, we use the Brown Sections of the Penn Treebank. The dataset contains in total 24243 sentences with an average sentence length of 18.9 tokens. The dataset has a mixture of genres, ranging from fictions to biographies and memoires, arranged into separate sections. We further split these sections into 97 smaller datasets, and each one contains continuous texts from two adjacent files in the original corpus. The average size of 250 sentences per dataset will provide reliable parser evaluation results.

4.2 Results

All five parsers were evaluated on the 97 datasets. The performance variation is very substantial for all these parsers, although it is hard to compare on concrete numbers due to the different evaluation metrics. Figure 1 shows the correlation between parser performance and the average sentence length (ASL), unknown word ratio (UWR), and unknown POS trigram ratio (UPR)[6]. It is not surprising to observe that all the parsers' performances have negative correlations to these three measures, with some of which more significant than the others. We should note that the correlation reflects the noisiness and direction of the relation between the statistical measure and the parser performance, but not the slope of that relationship.

---

[4] http://w3.msi.vxu.se/ jha/maltparser/

[5] http://lingo.stanford.edu/erg/

[6] All these evaluation results and parser outputs will be available online.

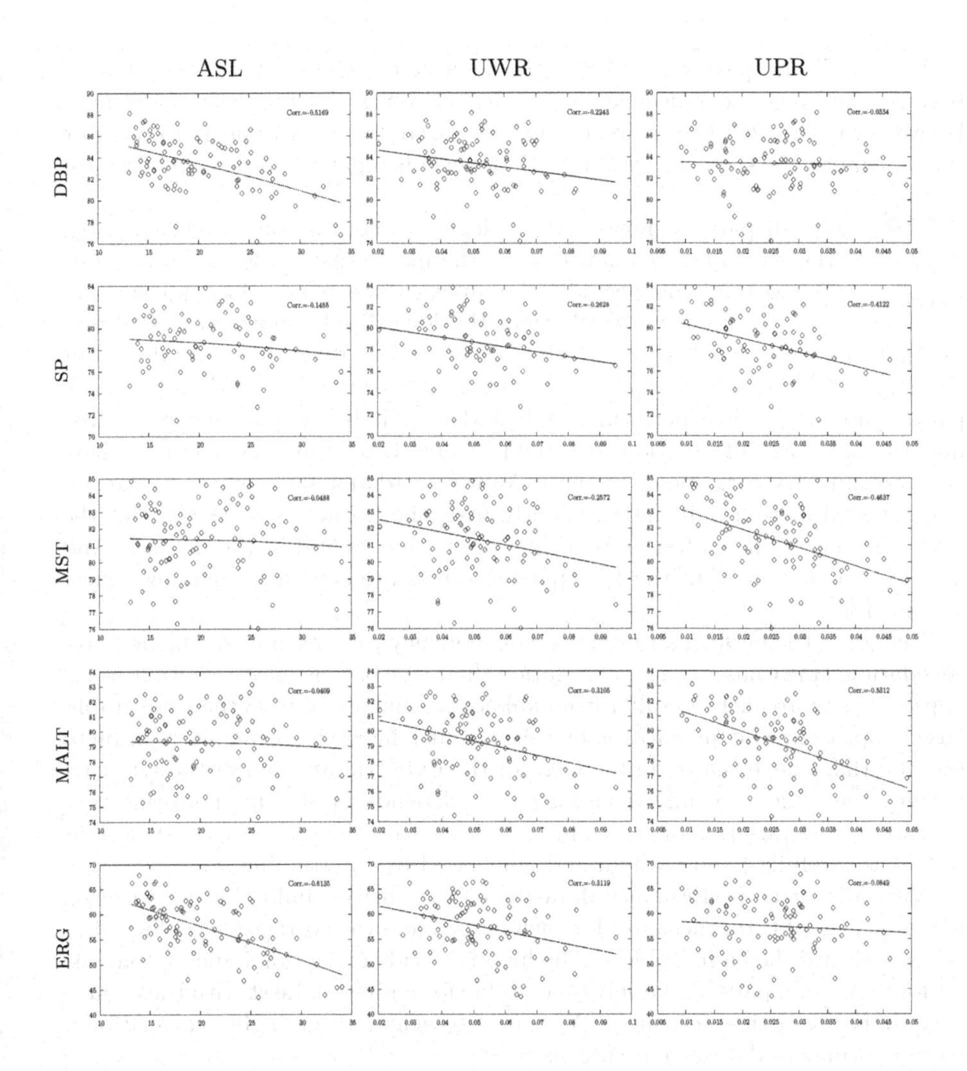

Fig. 1. Parser performance against three statistical measures, and their correlation coefficients

Table 1. Correlation coefficient of the linear regression models using all three measures

	DBP	SP	MST	MALT	ERG
Corr.	0.6509	0.5980	0.6124	0.6961	0.8102

Among all the parsers, ERG has the highest correlation with ASL. This is because the longer sentences lead to a sharp drop in parsing coverage of ERG. Between the two PCFG parsers, the unlexicalized SP parser appears to be more robust against the sentence length. Both dependency parsers appear to be robust to ASL.

With UWR, all parsers shows certain degree of correlation (coefficient from -0.22 to -0.31), with ERG and MALT being the most sensitive ones. An interesting observation is that unexpectedly the unlexicalized parser does not show to be more robust to unknown words than the lexicalized counterpart. SP's performance not only has higher negative correlation to UWR than DBP does, but also suffers a sharper performance drop with increasing UWR. Both dependency parsers also shows clear performance degradation, indicating the parser is missing critical lexical information. It should be noted that UWR as calculated here does not directly correspond to the unknown words for ERG, which contains a hand-crafted lexicon built independently from the training set (WSJ). But the UWR still reflects how often infrequent words are observed in the dataset. And it is known that most of the ERG parsing errors are caused by missing lexical entries [18].

With UPR, the performances of both dependency parsers show strong negative correlation. MALT has stronger correlation than MST because the transition-based approach is more likely to suffer from unknown sequence of POS than the graph-based approach. The unlexicalized SP shows much more significant correlation to the UPR than the lexicalized DBP does, for the POS trigrams reflect the syntactic patterns on which the unlexicalized parser depends most. ERG performs very robustly to the variation of UPR. This is because that the syntactic constructions in ERG is carefully hand-crafted, and not biased by the training set.

With all parser performance data points, we further built linear regression models using all three measures for each parser, and the correlation coefficient of the models are shown in Table 1. The high levels of correlation indicate that the performance of a parser is largely predictable for a given dataset with these three very simple statistical measures. We expect to achieve even higher correlation if more informative dataset measures is used.

5 Conclusion and Future Work

The method we proposed in this paper can be adapted to various other parsers and datasets, and potentially for different languages. The varying correlation of the proposed statistical measures with parsers' performance suggests that different parsing models are sensitive to different characteristics of the datasets.

Since all the measures are obtained from the unannotated texts, the method is not committed to specific linguistic framework or parsing algorithm. In the future we plan to experiment with more statistical measures and their combinations.

The linear regression model we built suggests one way of predicting the parser performance on unseen datasets, so that an estimation of the parser performance can be achieved without any gold-standard annotations on the target datasets. The result analysis also shows the possibility of parser combination (by either parse reranking or feature stacking) to achieve more robust performances. Furthermore, the variance of the performance and its correlation to the statistical measures can be viewed as an alternative parser evaluation metrics revealing the robustness of the parser performance, in addition to the standard accuracy measures.

Acknowledgments

The first author is grateful to DFKI and the German Excellence Cluster of Multimodal Computing and Interaction for their support of the work. The second author is funded by the PIRE PhD scholarship program.

References

1. Miyao, Y., Sagae, K., Sætre, R., Matsuzaki, T., Tsujii, J.: Evaluating Contributions of Natural Language Parsers to Protein-Protein Interaction Extraction. Journal of Bioinformatics 25(3), 394–400 (2009)
2. Moore, R.K.: Spoken language processing: piecing together the puzzle. Speech Communication: Special Issue on Bridging the Gap Between Human and Automatic Speech Processing 49, 418–435 (2007)
3. Bacchiani, M., Riley, M., Roark, B., Sproat, R.: Map adaptation of stochastic grammars. Computer speech and language 20(1), 41–68 (2006)
4. McClosky, D., Charniak, E., Johnson, M.: Reranking and self-training for parser adaptation. In: Proceedings of the 21st International Conference on Computational Linguistics and the 44th Annual Meeting of the Association for Computational Linguistics, Sydney, Australia, pp. 337–344 (2006)
5. McClosky, D., Charniak, E., Johnson, M.: When is self-training effective for parsing? In: Proceedings of the 22nd International Conference on Computational Linguistics (Coling 2008), Manchester, UK, pp. 561–568 (2008)
6. Nivre, J., Hall, J., Kübler, S., McDonald, R., Nilsson, J., Riedel, S., Yuret, D.: The CoNLL 2007 shared task on dependency parsing. In: Proceedings of EMNLP-CoNLL 2007, Prague, Czech Republic, pp. 915–932 (2007)
7. Hara, T., Miyao, Y., Tsujii, J.: Adapting a probabilistic disambiguation model of an HPSG parser to a new domain. In: Dale, R., Wong, K.-F., Su, J., Kwong, O.Y. (eds.) IJCNLP 2005. LNCS (LNAI), vol. 3651, pp. 199–210. Springer, Heidelberg (2005)
8. Rimell, L., Clark, S.: Porting a Lexicalized-Grammar Parser to the Biomedical Domain. Journal of Biomedical Informatics (in press, 2009)
9. Plank, B.: Structural Correspondence Learning for Parse Disambiguation. In: Proceedings of the Student Research Workshop at EACL 2009, Athens, Greece, pp. 37–45 (2009)

10. Dredze, M., Blitzer, J., Pratim Talukdar, P., Ganchev, K., Graca, J.a., Pereira, F.: Frustratingly hard domain adaptation for dependency parsing. In: Proceedings of the CoNLL Shared Task Session of EMNLP-CoNLL 2007. Association for Computational Linguistics, Prague, June 2007, pp. 1051–1055 (2007)
11. Bikel, D.M.: Intricacies of Collins' parsing model. Computational Linguistics 30, 479–511 (2004)
12. Collins, M.: Three Generative, Lexicalised Models for Statistical Parsing. In: Proceedings of the 35th annual meeting of the association for computational linguistics, Madrid, Spain, pp. 16–23 (1997)
13. Klein, D., Manning, C.D.: Accurate Unlexicalized Parsing. In: Proceedings of the 41st Meeting of the Association for Computational Linguistics, Sapporo, Japan, pp. 423–430 (2003)
14. McDonald, R., Pereira, F., Ribarov, K., Hajic, J.: Non-Projective Dependency Parsing using Spanning Tree Algorithms. In: Proceedings of HLT-EMNLP 2005, Vancouver, Canada, pp. 523–530 (2005)
15. Nivre, J., Nilsson, J., Hall, J., Chanev, A., Eryigit, G., Kübler, S., Marinov, S., Marsi, E.: Maltparser: A language-independent system for data-driven dependency parsing. Natural Language Engineering 13(1), 1–41 (2007)
16. Flickinger, D.: On building a more efficient grammar by exploiting types. In: Oepen, S., Flickinger, D., Tsujii, J., Uszkoreit, H. (eds.) Collaborative Language Engineering, pp. 1–17. CSLI Publications, Stanford (2002)
17. Callmeier, U.: Efficient parsing with large-scale unification grammars. Master's thesis, Universität des Saarlandes, Saarbrücken, Germany (2001)
18. Baldwin, T., Bender, E.M., Flickinger, D., Kim, A., Oepen, S.: Road-testing the English Resource Grammar over the British National Corpus. In: Proceedings of the 4th International Conference on Language Resources and Evaluation (LREC 2004), Lisbon, Portugal (2004)

Comparing Two Approaches for the Recognition of Temporal Expressions*

Oleksandr Kolomiyets and Marie-Francine Moens

Department of Computer Science, K.U.Leuven, Celestijnenlaan 200A, Heverlee, B-3001, Belgium
{Oleksandr.Kolomiyets,Sien.Moens}@cs.kuleuven.be

Abstract. Temporal expressions are important structures in natural language. In order to understand text, temporal expressions have to be extracted and normalized. In this paper we present and compare two approaches for the automatic recognition of temporal expressions, based on a supervised machine learning approach and trained on TimeBank. The first approach performs a token-by-token classification and the second one does a binary constituent-based classification of chunk phrases. Our experiments demonstrate that on the TimeBank corpus constituent-based classification performs better than the token-based one. It achieves F1-measure values of 0.852 for the detection task and 0.828 when an exact match is required, which is better than the state-of-the-art results for temporal expression recognition on TimeBank.

1 Introduction

Temporal information extraction in free text has been a research focus since 1995, when temporal expressions, sometimes also referred to as time expressions or TIMEXes, were processed as single capitalized tokens within the scope of the Message Understanding Conference (MUC) and the Named Entity Recognition task. As the demand for deeper semantic analysis tools increased, rule-based systems were proposed to solve this problem. The rule-based approach is characterized by providing decent results with a high precision level, and yields rules, which can be easily interpreted by humans. With the advent of new, annotated linguistic corpora, supervised machine-learning approaches have become the enabling technique for many problems in natural language processing. In 2004 the Automated Content Extraction (ACE) launched a competition campaign for Temporal Expression Recognition Normalization, TERN. The tasks were set to identify temporal expressions in free text and normalize them providing ISO-based date-time values.

While the ACE TERN initiative along with the provided corpus was aimed at the recognition and normalization problems, more advanced temporal processing on the same dataset was not possible. The most recent annotation language for temporal expressions, TimeML [1], and the underlying annotated corpus TimeBank [2], opens up new horizons for automated temporal information extraction and reasoning.

* This work has been partly funded by the Flemish government (through IWT) and by Space Applications Services NV as part of the ITEA2 project LINDO (ITEA2-06011).

B. Mertsching, M. Hund, and Z. Aziz (Eds.): KI 2009, LNAI 5803, pp. 225–232, 2009.

A large number of rule-based and machine learning approaches were proposed for identification of temporal expressions. Comparative studies became possible with standardized annotated corpora, such as the ACE TERN and TimeBank. While the ACE TERN corpus is very often used for performance reporting, it restricts the temporal analysis to identification and normalization. By contrast, TimeBank provides a basis for all-around temporal processing, but lacks experimental results. In this paper we describe and compare two supervised machine learning approaches for identifying temporal information in free text. Both are trained on TimeBank, but follow two different classification techniques: token-by-token following B-I-O encoding and constituent-based classifications.

The remainder of the paper is organized as follows. In Section 2 we provide the details of relevant work done in this field along with corpora and annotations schemes used. Section 3 describes the approaches. Experimental setup, results and error analysis are provided in Section 4. Finally, Section 5 gives an outlook for further improvements and research.

2 Background and Related Work

2.1 Evaluation Metrics

With the start of the ACE TERN competition in 2004, two major evaluation conditions were proposed: Recognition+Normalization (full task) and Recognition only [3]. For the recognition task, temporal expressions have to be found. The system performance is scored with respect to two major metrics: detection and exact match. For detection the scoring is very generous and implies a minimal overlap of one character between the reference extent and the system output. For an exact match, in turn, the correct extent boundaries are sought.

As the exact match evaluation appeared to be too strict for TimeBank, an alternative evaluation metric was used in [4] and called "sloppy span". In this case the system scores as long as the detected right-side boundary is the same as in the corresponding TimeBank's extent.

2.2 Datasets

To date, there are two annotated corpora used for performance evaluations of temporal taggers, the ACE TERN corpus and TimeBank [2]. Most of the implementations referred to as the state-of-the-art were developed in the scope of the ACE TERN 2004. For evaluations, a training corpus of 862 documents with about 306 thousand words was provided. Each document represents a news article formatted in XML, in which TIMEX2 XML tags denote temporal expressions. The total number of temporal expressions for training is 8047 TIMEX2 tags with an average of 10.5 per document. The test set comprises 192 documents with 1828 TIMEX2 tags [5].

The annotation of temporal expressions in the ACE corpus was done with respect to the TIDES annotation guidelines [6]. The TIDES standard specifies so-called markable expressions, whose syntactic head must be an appropriate lexical trigger, e.g. "*minute*", "*afternoon*", "*Monday*", "*8:00*", "*future*" etc. When tagged, the full extent of the tag must correspond to one of the grammatical categories: nouns, noun phrases, adjectives, adjective phrases, adverbs and adverb phrases. According to this, all pre- and postmodifiers as well as dependent clauses are also included to the TIMEX2 extent, e.g. "*five days after he came back*", "*nearly four decades of experience*".

The most recent annotation language for temporal expressions, TimeML [1], with the underlying TimeBank corpus [2], opens up new avenues for temporal information extraction. Besides a new specification for temporally relevant information, such as TIMEX3, EVENT, SIGNAL, TLINK etc, TimeML provides a means to capture temporal semantics by annotations with suitably defined attributes for fine-grained specification of analytical detail [7]. The annotation schema establishes new entity and relation marking tags along with numerous attributes for them. The TimeBank corpus includes 186 documents with 68.5 thousand words and 1423 TIMEX3 tags.

2.3 Approaches for Temporal Tagging

As for any recognition problem, there are two major ways to solve it – *rule-based* and *machine-learning methods*. As the temporal expression recognition is not only about to detect them but also to provide an exact match, machine learning approaches can be divided into *token-by-token classification* following B(egin)-I(nside)-O(utside) encoding and *binary constituent-based classification*, in which an entire chunk-phrase is considered for classification as a temporal expression.

2.3.1 Rule-Based Systems

One of the first well-known implementations of temporal taggers was presented in [8]. The approach relies on a set of hand-crafted and machine-discovered rules, based upon shallow lexical features. On average the system achieved a value of 0.832 for F1-measure against hand-annotated data for the exact match evaluation. The dataset used comprised a set of 22 New York Times articles and 199 transcripts of Voice of America. Another example of rule-based temporal taggers is Chronos, described in [9], which achieved the highest F1-scores in the ACE TERN 2004 of 0.926 and 0.878 for detection and exact match respectively.

Recognition of temporal expressions using TimeBank as an annotated corpus, is reported in [4] and based on a cascaded finite-state grammar (500 stages and 16000 transitions). A complex approach achieved an F1-measure value of 0.817 for exact match and 0.896 for detecting "sloppy" spans. Another known implementation for TimeBank is GUTime[1] – an adaptation of [8] from TIMEX2 to TIMEX3 with no reported performance level.

2.3.2 Machine Learning Recognition Systems

Successful machine learning TIMEX2 recognition systems are described in [10; 11; 12]. Proposed approaches made use of a token-by-token classification for temporal expressions represented by B-I-O encoding with a set of lexical and syntactic features, e.g., the token itself, part-of-speech tag, weekday names, numeric year etc. The performance levels are presented in Table 1. All the results were obtained on the ACE TERN dataset.

[1] http://www.timeml.org/site/tarsqi/modules/gutime/index.html

Table 1. Performance of machine learning approaches with B-I-O encoding

Approach	*F1* (detection)	*F1* (exact match)
Ahn et al. [10]	0.914	0.798
Hacioglu et al. [11]	0.935	0.878
Poveda et al. [12]	0.986	0.757

A constituent-based, also known as chunk-based, classification approach for temporal expression recognition was presented in [13]. By comparing to the previous work of the same authors [10] and on the same ACE TERN dataset, the method demonstrates a slight decrease in detection with F1-measure of 0.844 and a nearly equivalent F1-measure value for exact match of 0.787.

3 Our Approaches

The approaches presented in this section employ a supervised machine learning algorithm following a similar feature design but different classification strategies. Both classifiers implement a Maximum Entropy Model[2].

3.1 Token-Based Classification Approach

Multi-class classifications, such as the one with B-I-O encoding, are a traditional way for recognition tasks in natural language processing, for example in Named Entity Recognition and chunking. For this approach we employ the OpenNLP toolkit[3] when pre-processing the data. The toolkit makes use of the same Maximum Entropy model for detecting sentence boundaries, part-of-speech (POS) tagging and parsing tasks [14; 15]. The tokenized output along with detected POS tags is used for generating feature vectors with one of the labels from the B-I-O encoding. The feature-vector design comprises the initial token in lowercase, POS tag, character type and character type pattern. Character type and character type pattern features are implemented following Ahn et al. [10]. The patterns are defined by using the symbols X, x and 9. X and x are used to represent capital and lower-case letters respectively, 9 is used for numeric tokens. Once the character types are computed, the corresponding character patterns are produced by removing redundant occurrences of the same symbol. For example, the token "*January*" has character type "Xxxxxxx" and pattern "X(x)". The same feature design is applied to each token in the context window of three tokens to the left and to the right in the sequence limited by sentence boundaries.

3.2 Constituent-Based Classification Approach

For constituent-based classification the entire phrase is under consideration for labeling as a TIMEX or not. We restrict the classification for the following phrase types and grammatical categories: nouns, proper nouns, cardinals, noun phrases,

---

[2] http://maxent.sourceforge.net/

[3] http://opennlp.sourceforge.net/

adjectives, adjective phrases, adverbs, adverbial phrases and prepositional phrases. For each sentence we parse the initial input line with a Maximum Entropy parser [15] and extract all phrase candidates with respect the types defined above. Each phrase candidate is examined against the manual annotations for temporal expressions found in the sentence. Those phrases, which correspond to the temporal expressions in the sentence are taken as positive examples, while the rest are considered as a negative set. After that, for each candidate we produce a feature vector, which includes the following features: head phrase, head word, part-of-speech for head word, character type and character type pattern for head word as well as for the entire phrase.

4 Experiments, Results and Error Analysis

All experiments were conducted following 10-fold cross validation and evaluated with respect to the TERN 2004 evaluation plan [3].

4.1 Token-Based Classification Experiments

After pre-processing the textual part of TimeBank, we had a set of 26 509 tokens with 1222 correctly aligned TIMEX3 tags. The experimental results demonstrated the performance in detection of temporal expressions with precision, recall and F1-measure of 0.928, 0.628 and 0.747 respectively. When an exact match is required, the classifier performs at the level of 0.888, 0.382 and 0.532 for precision, recall and F1-measure respectively.

4.2 Constituent-Based Classification Experiments

After pre-processing the TimeBank corpus of 182 documents we had 2612 parsed sentences with 1224 temporal expressions in them. 2612 sentences resulted in 49 656 phrase candidates. After running experiments the classifier demonstrated the performance in detection of TIMEX3 tags with precision, recall and F1-measure of 0.872, 0.836 and 0.852 respectively. The experiments on exact match demonstrated a small decline and received scores of 0.866, 0.796 and 0.828 for precision, recall and F1-measure respectively.

4.3 Comparison and Improvements

Comparing the performance levels of the tested temporal taggers, we discovered the differences in classification results of chunk-based and token-based approaches with corresponding F1-measure values of 0.852 vs. 0.747 for detection, and 0.828 vs. 0.532 for exact match. Previous experiments on the ACE TERN corpus, especially those in [10; 13], reported the same phenomenon and a drop in F1-measure between detection and exact match, but the token-based approach delivers generally better results. For our experimental results we assume that the problem lies in the local token classification approach based on pure lexico-syntactic features alone. To prove this hypothesis, the next series of experiments is performed with an additional feature set, which contains the classification results obtained for preceding tokens, so called Maximum Entropy Markov Model, MEMM. The experimental setup varies the

Table 2. Performance levels for token-by-token classifications with MEMM

N	Detection			Exact match		
	P	*R*	*F1*	*P*	*R*	*F1*
0	0.928	0.628	0.747	0.888	0.382	0.532
1	0.946	0.686	0.793	0.921	0.446	0.599
2	0.94	0.652	0.768	0.911	0.426	0.578
3	0.936	0.645	0.762	0.905	0.414	0.566

number of previously consecutive obtained labels between 1 and 3 with the same context window size. The context is considered within the sentence only. The results of these experiments are presented in Table 2. The number of the previously obtained labels used as features is denoted by N – the order of the Markov Model, with N=0 as a baseline (see Section 3.1).

It is worth to mention that by taking into account labels obtained for preceding tokens the performance level rises and reaches the maximum at N=1 for both, the detection and exact match tasks, and decreases from N=2 onwards.

Putting these results into context, we can conclude that the chunk-based machine learning approach for temporal expression recognition performed at a comparable operational level to the state-of-the-art rule-based approach of Boguraev and Ando [4] and outperformed it in exact match. A comparative performance summary is presented in Table 3.

4.4 Error Analysis

Analyzing the classification errors, we see several causes for them. We realized that the current version of TimeBank is still noisy with respect to annotated data. An ambiguous use of temporal triggers in different context, like “*today*”, “*now*”, “*future*”, makes correct identification of relatively simple temporal expressions difficult. Apart from obviously incorrect parses, the inexact alignment between temporal expressions and candidate phrases was caused by annotations that occurred at the middle of a phrase, for example “*eight-years-long*”, “*overnight*”, “*yesterday’s*”. In total there are 99 TIMEX3 tags (or 8.1%) misaligned with the parser output, which resulted in 53

Table 3. Performance summary for the constituent-based classification (CBC) approach

	P	*R*	*F1*
	Detection		
CBC approach	0.872	0.836	0.852
	Sloppy Span		
Boguraev and Ando [4]	0.852	0.952	0.896
	Exact Match		
CBC approach	0.866	0.796	0.828
Boguraev and Ando [4]	0.776	0.861	0.817

(or 4.3%) undetected TIMEX3s. Definite and indefinite articles are unsystematically left out or included into TIMEX3 extent, which introduces an additional bias in classification.

5 Conclusion and Future Work

In this paper we presented two machine learning approaches for the recognition of temporal expressions using a recent annotated corpus for temporal information, TimeBank. Two approaches were implemented: a token-by-token and a binary constituent-based classifiers. The feature design for both methods is very similar and takes into account contentual and contextual features. As the evaluation showed, both approaches provide a good performance level for the detection of temporal expressions. The constituent-based classification outperforms the token-based one with F1-measure values of 0.852 vs. 0.747. If an exact match is required, only the constituent-based classification can provide a reliable recognition with an F1-measure value of 0.828. For the same task token-based classification reaches only 0.532 in terms of F1-measure. By employing a Maximum Entropy Markov Model, the method increases in performance and reaches its maximum, when only the classification result for the previous token is used (with F1-measures of 0.793 and 0.599 for detection and exact match respectively).

Our best results were obtained by the binary constituent-based classification approach with shallow syntactic and lexical features. The method achieved a performance level of a rule-based approach presented in [4] and for the exact match task our approach even outperforms the latter. Although a direct comparison with other state-of-the-art systems is not possible, our experiments disclose a very important characteristic. While the recognition systems in the ACE TERN 2004 reported a substantial drop of F1-measure between detection and exact match results (6.5 - 11.6%), our phrase-based detector demonstrates a light decrease in F1-measure (2.4%), whereas the precision declines only by 0.6%. This important finding leads us to the conclusion that most of TIMEX3s in TimeBank can be detected at a phrase-based level with a reasonably high performance.

References

1. Pustejovsky, J., Castaño, J., Ingria, R., Saurí, R., Gaizauskas, R., Setzer, A., Katz, G.: TimeML: Robust Specification of Event and Temporal Expressions in Text. In: Fifth International Workshop on Computational Semantics (2003)
2. Pustejovsky, J., Hanks, P., Saurí, R., See, A., Day, D., Ferro, L., Gaizauskas, R., Lazo, M., Setzer, A., Sundheim, B.: The TimeBank Corpus. In: Corpus Linguistics 2003, pp. 647–656 (2003)
3. TERN, Evaluation Plan (2004), http://fofoca.mitre.org/tern_2004/tern_evalplan-2004.29apr04.pdf
4. Boguraev, B., Ando, R.K.: TimeBank-Driven TimeML Analysis. In: Annotating, Extracting and Reasoning about Time and Events. Dagstuhl Seminar Proceedings. Dagstuhl, Germany (2005)

5. Ferro, L.: TERN Evaluation Task Overview and Corpus, http://fofoca.mitre.org/tern_2004/ferro1_TERN2004_task_corpus.pdf
6. Ferro, L., Gerber, L., Mani, I., Sundheim, B., Wilson, G.: TIDES 2003, Standard for the Annotation of Temporal Expressions (2003), http://timex2.mitre.org
7. Boguraev, B., Pustejovsky, J., Ando, R., Verhagen, M.: TimeBank Evolution as a Community Resource for TimeML Parsing. Language Resource and Evaluation 41(1), 91–115 (2007)
8. Mani, I., Wilson, G.: Robust Temporal Processing of News. In: 38th Annual Meeting on Association for Computational Linguistics, pp. 69–76 (2000)
9. Negri, M., Marseglia, L.: Recognition and Normalization of Time Expressions: ITC-irst at TERN 2004. Technical Report, ITC-irst, Trento (2004)
10. Ahn, D., Adafre, S.F., de Rijke, M.: Extracting Temporal Information from Open Domain Text: A Comparative Exploration. Digital Information Management 3(1), 14–20 (2005)
11. Hacioglu, K., Chen, Y., Douglas, B.: Automatic Time Expression Labeling for English and Chinese Text. In: Gelbukh, A. (ed.) CICLing 2005. LNCS, vol. 3406, pp. 548–559. Springer, Heidelberg (2005)
12. Poveda, J., Surdeanu, M., Turmo, J.: A Comparison of Statistical and Rule-Induction Learners for Automatic Tagging of Time Expressions in English. In: International Symposium on Temporal Representation and Reasoning, pp. 141–149 (2007)
13. Ahn, D., van Rantwijk, J., de Rijke, M.: A Cascaded Machine Learning Approach to Interpreting Temporal Expressions. In: NAACL-HLT 2007 (2007)
14. Ratnaparkhi, A.: A Maximum Entropy Model for Part-of-Speech Tagging. In: Conference on Empirical Methods in Natural Language Processing, pp. 133–142 (1996)
15. Ratnaparkhi, A.: Learning to Parse Natural Language with Maximum Entropy Models. Machine Learning 34(1), 151–175 (1999)

Meta-level Information Extraction

Peter Kluegl, Martin Atzmueller, and Frank Puppe

University of Würzburg,
Department of Computer Science VI
Am Hubland, 97074 Würzburg, Germany
{pkluegl,atzmueller,puppe}@informatik.uni-wuerzburg.de

Abstract. This paper presents a novel approach for meta-level information extraction (IE). The common IE process model is extended by utilizing transfer knowledge and meta-features that are created according to already extracted information. We present two real-world case studies demonstrating the applicability and benefit of the approach and directly show how the proposed method improves the accuracy of the applied information extraction technique.

1 Introduction

While structured data is readily available for information and knowledge extraction, unstructured information, for example, obtained from a collection of text documents cannot be directly utilized for such purposes. Since there is significantly more unstructured (textual) information than structured information e.g., obtained by structured data acquisition information extraction (IE) methods are rather important. This can also be observed by monitoring the latest developments concerning IE architectures, for example *UIMA* [1] and the respective methods, e.g., conditional random fields (CRF), support vector machines (SVM), and other (adaptive) IE methods, cf., [2,3,4].

Before IE is applied, first an IE model is learned and generated in the learning phase. Then, the IE process model follows a standard approach depicted in Figure 1: As the first step of the process itself, the applicable features are extracted from the document. In some cases, a (limited form of) knowledge engineering is used for tuning the relevant features of the domain. Finally, the generated model is applied on the data such that the respective IE method selects or classifies the relevant text fragments and extracts the necessary information.

With respect to the learning phase, usually machine-learning related approaches like candidate classification, windowing, and markov models are used. Often SVMs or CRFs are applied for obtaining the models. The advantages of CRFs are their relation to the sequence labeling problem, while they do not suffer from the dependencies between the features. However, a good feature selection step is still rather important. The advantages of SVMs are given by their automatic ranking of the input features and their ability to handle a large number of features. Therefore, less knowledge engineering is necessary. However, the IE task can also be implemented by knowledge engineering approaches applying rules or lambda expressions. In the case studies we will present a rule-based approach that is quite effective compared to the standard approaches.

B. Mertsching, M. Hund, and Z. Aziz (Eds.): KI 2009, LNAI 5803, pp. 233–240, 2009.

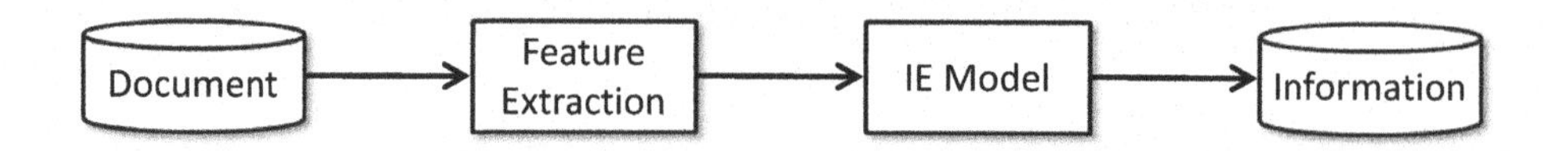

Fig. 1. Common Process Model for Information Extraction

Specifically, this paper proposes extensions to the common IE approach, such that meta-level features that are generated during the process can be utilized: The creation of the meta-level features is based on the availability of already extracted information that is applied in a feedback loop. Then repetitive information like structural repetitions can be processed further by utilizing transfer knowledge. Assuming that a document, for example, is written by a single author, then it is probably the case that the same writing and layout style is used for all equivalent structures. We present two case studies demonstrating the applicability and benefit of the approach and show how the proposed method improves the accuracy of the applied information extraction technique.

The rest of the paper is structured as follows: Section 2 presents the proposed novel process model for information extraction extending the standard process. We first motivate the concrete problem setting before we discuss the extensions in detail and specifically the techniques for meta-level information extraction. After that, Section 3 presents two real-world case studies: We demonstrate the applicability of the presented approach, for which the results directly indicate its benefit. Next, we discuss related work. Finally, Section 4 concludes with a summary of the presented approach and points at interesting directions for future research.

2 Meta-level Information Extraction

In the following, we first motivate the proposed approach by presenting two examples for which the commonly applied standard process model is rather problematic. Next, we present the process model for meta-level information extraction and discuss its elements and extensions in detail.

2.1 Problem Statement

To point out certain flaws of the standard process, we discuss examples concerning information extraction from curricula vitae (CV) and from medical discharge letters. Both examples indicate certain problems of the common process model for information extraction and lead to following claim:

> Using already extracted information for further information extraction can often account for missing or ambiguous features and increase the accuracy in domains with repetitive structure(s).

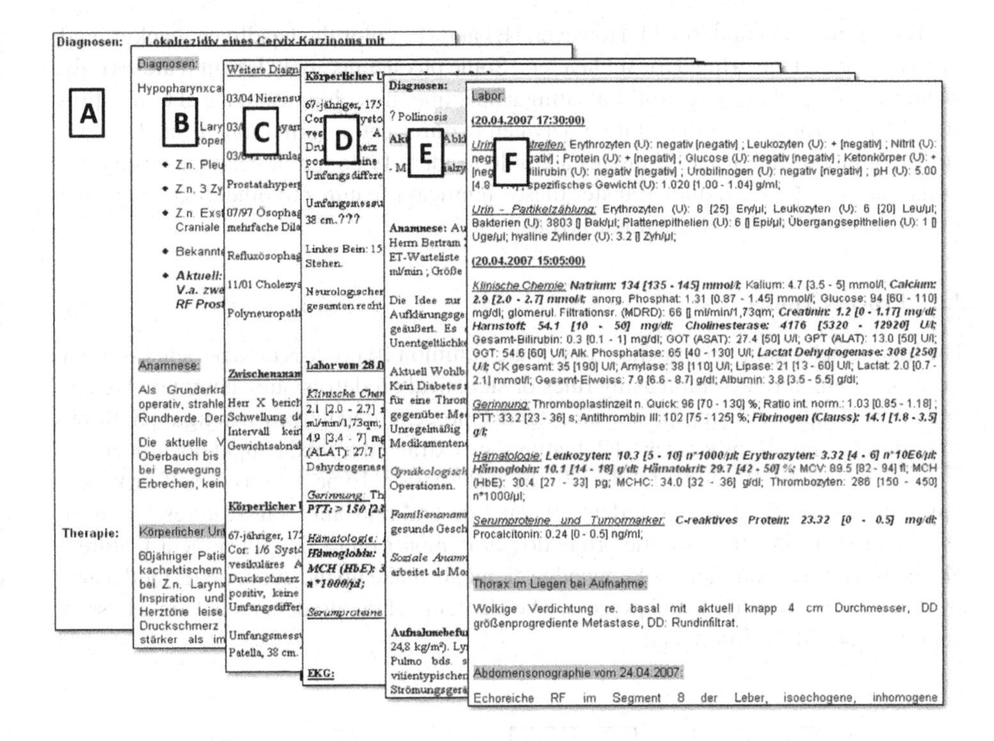

Fig. 2. Examples of different headlines in medical discharge letters

CVs. For the extraction from CV documents, a predefined template with slots for start time, end time, title, company and description is filled with the corresponding text fragments. The text segments describing experiences or projects are used to identify a template. Then, the slots of the templates are extracted. Often the company can be identified using simple features, e.g., common suffixes, lists of known organizations or locations. Yet, these word lists cannot be exhaustive, and are often limited for efficency reasons, e.g., for different countries. This can reduce the accuracy of the IE model, e.g., if the employee had been working in another country for some time. Humans solve these problems of missing features, respectively of unknown company names, by transfering already 'extracted' patterns and relations. If the company, for example, was found in the third line of ninety percent of all project sections, then it is highly probable that an 'unclear' section contains a company name in the same position.

Medical discharge letters. Medical discharge letters contain, for example, the observations, the history, and the diagnoses of the patient. For IE, different sections of the letter need to be identified: The headlines of a section cannot only help to create a segmentation of the letter, but also provide hints what kind of sections and observations are present. Since there are no restrictions, there is a variety of layout structure; Figure 2 shows some examples: Whereas the headlines in (A) are represented using a table,

(C+D) use bold and underlined. However, (B) and (F) color the headlines' background and use bold and underlined for subheaders. Some physicians apply layout features only to emphasize results and not for indicating a headline. It is obvious, that a classification model faces problems, if the relation between features and information differs for each input document. In contrast, humans are able to identify common headlines using the containing words. Then, they transfer these significant features to other text fragments and extract headlines with a similar layout.

2.2 Process Model

The human behaviour solving the flaws of the common IE process model seems straight forward, yet its formalization using rules or statistical models is quite complex. We approach this challenge by proposing an extended process model, shown in figure 3: Similar to the common IE process model, features are extracted from the input document and are used by a static IE model to identify the information. Expectations or self-criticism can help to identify highly confident information and relevant meta-features. Transfer knowledge is responsible for the projection or comparison of the given meta-features. The meta-features and transfer knowledge elements make up the dynamic IE model and are extended in an incremental process. The elements of the process model are adressed in more detail in the following:

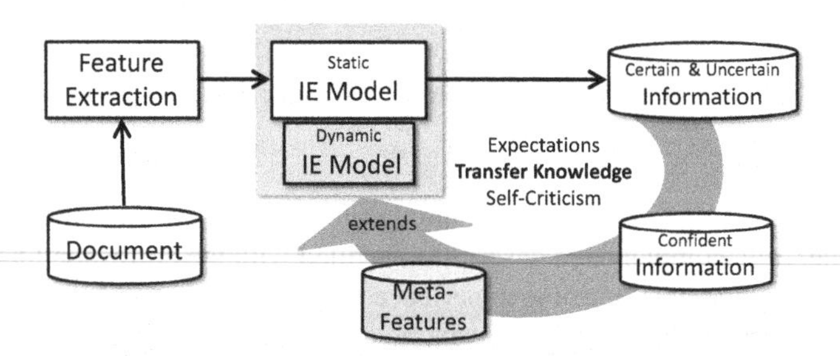

Fig. 3. Extended process model with meta-features and transfer knowledge

Meta-Features. Relations between features and information, respectively patterns, are explicitly implemented by meta-features. These are not only created for the extracted information, but also for possible candidates. A simple meta-feature, for example for the extraction of headlines, states that the bold feature indicates a headline in this document.

Expectations and Self-Criticism. Since even only a single incorrect information can lead to a potentially high number of incorrect information, the correctness and confidence of an information is essential for the meta-level information extraction. There are two ways to identify an information suitable for the extraction of meta-features. If the knowledge engineer already has some assumptions about the content of the input documents, especially on the occurence of certain information, then these expectations can be formalized in order to increase the confidence of the information. In the absence of expecations, self-criticism of the IE model using features or a confidence value can highlight a suitable information. Furthermore, self-critism can be used to reduce the incorrect transfer of meta-features by rating newly identified information.

Transfer Knowledge. The transfer knowledge models the human behaviour in practice and can be classified in three categories: *Agglomeration* knowledge processes multiple meta-features and creates new composed meta-features. Then, *projection* knowledge defines the transfer of the meta-features to possible candidates of new information. *Comparison* knowledge finally formalizes how the similarity of the meta-features of the original information and a candidate information is calculated. The usage of these different knowledge types in an actual process depends on the kind of repetitive structures and meta-features.

In section 3, specific examples of these elements are explained in the context of their application.

3 Case Studies

For a demonstation of the applicability and benefit of the approach, the two subtasks of the IE applications introduced earlier are addressed. The meta-level approach is realized with the rule-based TextMarker system and the statistical natural language processing toolkit ClearTK [5] is used for the the supervised machine learning methods CRF[1] and SVM[2]. The three methods operate in the same architecture (UIMA) and process the identical features. The same documents are applied for the training and test phase of the machine learning approaches and intentionally no k-fold cross evaluation is used, since it is hardly applicable for the knowledge engineering approach. Yet, the selected features do not amplify overfitting, e.g., no stem information is used. The evaluation of the SVM did not return reasonable values, probably because of the limited amount of documents and features in combination with the selected kernel method. Therefore, only results of the meta-level approach and CRF are presented using the F1-measure.

3.1 The TextMarker System

The TEXTMARKER system[3] is a rule-based tool for information extraction and text processing tasks [6]. It provides a full-featured development environment based on the DLTK framework[4] and a build process for UIMA Type Systems and generic UIMA Analysis Engines [1]. Different components for rule explanation and test-driven development [7] facilitate the knowledge engineering of rule-based information extraction components. The basic idea of the TEXTMARKER language is similar to JAPE [8]: rules match on combinations of predefined types of annotations and create new annotations. Furthermore, the TEXTMARKER language provides an extension mechanism for

[1] The CRF implementation of Mallet (http://mallet.cs.umass.edu/) is used.

[2] The SVM implementation of SVMLight (http://svmlight.joachims.org/) is used.

[3] http://textmarker.sourceforge.net/

[4] http://www.eclipse.org/dltk/

domain dependent language elements, several scripting functionalities and a dynamic view on the document. Due to the limited space, we refer to [6,7] for a detailed description of the system.

3.2 CVs

In this case study, we evaluate a subtask of the extraction of CV information: Companies in a past work experience of a person, respectively the employer. The corpus contains only 15 documents with 72 companies. The selected features consists of already extracted slots, layout information, simple token classes and a list of locations of one country. The meta-features are based on the position of confident information dependent on the layout and in relation to other slots. Agglomeration knowledge uses these meta-features to formalize a pattern of the common appearence of the companies. Then, projection knowledge uses this pattern to identify new information, that is rated by rules for self-criticism. In Figure 4, the results of the evaluation are listed. The meta-level approach achieved a F1-measure of 97.87% and the CRF method reached 75.00%. The low recall value of the CRF is caused by the limited amount of available features. However, the meta-level approach was able to compensate for this loss using the meta-features.

3.3 Medical Discharge Letters

A subtask of the extraction of information from medical discharge letters is the recognition of headlines. In oder to evaluate the approaches we use a corpus with 141 documents and 1515 headlines. The extracted features consist mainly of simple token classes and layout information, e.g., bold, underlined, italic and freeline. In this case study, the expectation to find a *Diagnose* or *Anamnese* headline is used to identify a confident information. Then, meta-features describing its actual layout are created and transferred by projection knowledge. Finally, comparison knowledge is used to calculate the similarity of the layout of the confident information and a candidate for a headline. The results of the evaluation are shown in figure 4: The meta-level approach was evaluated with 97.24% and the CRF method achieved a F1-measure of 87.13%. CRF extracted the same headlines as the meta-level approach in many documents. However, the conflicting layout styles of the some authors caused, as expected, a high number of false negative errors resulting in a lower recall value.

CVs	Precision	Recall	F1
CRF	93.75%	62.50%	75.00%
META	100.00%	95.83%	97.87%

Medical	Precision	Recall	F1
CRF	97.87%	78.52%	87.13%
META	99.11%	95.44%	97.24%

Fig. 4. Results of the CVs and medical discharge letters evaluation

3.4 Related Work and Discussion

In the case studies, we have seen that the proposed approach performs very promissing and achieves considerably better accuracy measures than the standard approach using machine learning techniques, that is CRF. The machine learning methods would potentially perform better using more (and/or 'better') features, however, the same is true for the meta-level IE approach. The approach is not only very effective but also rather efficient, since the proposed approach required only about 1-2 hours for formalizing the necessary meta-features and transfer knowledge, significantly less time than the time spent for the annotation of the examples.

To the best of the authors' knowledge, the approach is novel in the IE community and application. However, similar ideas to the core idea of transfering features have been adressed in the feature construction and inductive logic programming community, e.g., [9]. However, in this context there is no direct 'feedback' according to a certain process, and also no distinction between meta-features and transfer knowledge that is provided by the presented approach. According to the analogy of human reasoning, it is often easier to formalize each knowledge element separately. Especially in information extraction, there are approaches using extracted information in a meta-learning process, e.g., [10]. However, compared to our approach no meta-features dependent on extracted information and no transfer knowledge is used. The proposed approach is able to adapt to peculiarities of certain authors of the documents, similarly to the adaptation phase of common speech processing and speech understanding systems.

4 Conclusions

In this paper, we have presented a meta-level information extraction approach that extends the common IE process model by including meta-level features. These meta-features are created using already extracted information, e.g., given by repetitive structural constructs of the present feature space. We have described a general model for the application of the presented approach, and we have demonstrated its benefit in two case studies utilizing a rule-based system for information extraction.

For future work, the exchange of transfer knowledge and meta-features between documents can further enrich the process model in specific domains. We plan to extend the approach in order to incorporate the automatic acquisition of transfer knowledge. Techniques from inductive logic programming [11] can potentially provide helpful methods and support the knowledge engineer to automatically acquire the needed transfer and meta knowledge. For the automatic acquisition, self criticism capabilities and the inclusion of the expecations of the developer are essential. Finally, we are also planning to combine the approach with machine learning methods, like SVM and CRF [5]:

Acknowledgements

This work has been partially supported by the German Research Council (DFG) under grant Pu 129/8-2.

References

1. Ferrucci, D., Lally, A.: UIMA: An Architectural Approach to Unstructured Information Processing in the Corporate Research Environment. Nat. Lang. Eng. 10(3-4), 327–348 (2004)
2. McCallum, A., Li, W.: Early Results for Named Entity Recognition with Conditional Random Fields, Feature Induction and Web-Enhanced Lexicons. In: Proc. of the seventh conference on Natural language learning at HLT-NAACL 2003, pp. 188–191. Association for Computational Linguistics, Morristown (2003)
3. Turmo, J., Ageno, A., Català, N.: Adaptive Information Extraction. ACM Comput. Surv. 38(2), 4 (2006)
4. Li, D., Savova, G., Kipper-Schuler, K.: Conditional Random Fields and Support Vector Machines for Disorder Named Entity Recognition in Clinical Texts. In: Proc. of the Workshop on Current Trends in Biomedical Natural Language Processing, pp. 94–95. Association for Computational Linguistics, Columbus (2008)
5. Ogren, P.V., Wetzler, P.G., Bethard, S.: ClearTK: A UIMA Toolkit for Statistical Natural Language Processing. In: UIMA for NLP workshop at Language Resources and Evaluation Conference, LREC (2008)
6. Atzmueller, M., Kluegl, P., Puppe, F.: Rule-Based Information Extraction for Structured Data Acquisition using TextMarker. In: Proc. of the LWA 2008 (KDML Track), pp. 1–7 (2008)
7. Kluegl, P., Atzmueller, M., Puppe, F.: Test-Driven Development of Complex Information Extraction Systems using TextMarker. In: KESE at KI 2008 (2008)
8. Cunningham, H., Maynard, D., Tablan, V.: JAPE: A Java Annotation Patterns Engine, 2 edn. Research Memorandum CS–00–10, Department of Computer Science, University of Sheffield (November 2000)
9. Flach, P.A., Lavrac, N.: The Role of Feature Construction in Inductive Rule Learning. In: Raedt, L.D., Kramer, S. (eds.) Proc. ICML 2000 Workshop on Attribute-Value and Relational Learning: crossing the boundaries, 17th International Conference on Machine Learning, July 2000, pp. 1–11 (2000)
10. Sigletos, G., Paliouras, G., Spyropoulos, C.D., Stamatopoulos, T.: Meta-Learning beyond Classification: A Framework for Information Extraction from the Web. In: Proc. of the Workshop on Adaptive Text Extraction and Mining. The 14th Euro. Conf. on Machine Learning and the 7th Euro. Conf. on Principles and Practce of knowledge Discovery in Databases (2003)
11. Thomas, B.: Machine Learning of Information Extraction Procedures - An ILP Approach. PhD thesis, Universität Koblenz-Landau (2005)

Robust Processing of Situated Spoken Dialogue

Pierre Lison and Geert-Jan M. Kruijff

Language Technology Lab
German Research Centre for Artificial Intelligence (DFKI GmbH)
Saarbrücken, Germany

Abstract. Spoken dialogue is notoriously hard to process with standard language processing technologies. Dialogue systems must indeed meet two major challenges. First, natural spoken dialogue is replete with disfluent, partial, elided or ungrammatical utterances. Second, speech recognition remains a highly error-prone task, especially for complex, open-ended domains. We present an integrated approach for addressing these two issues, based on a robust incremental parser. The parser takes word lattices as input and is able to handle ill-formed and misrecognised utterances by selectively relaxing its set of grammatical rules. The choice of the most relevant interpretation is then realised via a discriminative model augmented with contextual information. The approach is fully implemented in a dialogue system for autonomous robots. Evaluation results on a Wizard of Oz test suite demonstrate very significant improvements in accuracy and robustness compared to the baseline.

1 Introduction

Spoken dialogue is one of the most natural means of interaction between a human and a robot. It is, however, notoriously hard to process with standard language processing technologies. Dialogue utterances are often incomplete or ungrammatical, and may contain numerous *disfluencies* like fillers (err, uh, mm), repetitions, self-corrections, fragments, etc. Moreover, even if the utterance is perfectly well-formed and does not contain disfluencies, the dialogue system still needs to accommodate the various *speech recognition errors* thay may arise. This problem is particularly acute for robots operating in real-world environments and dealing with complex, open-ended domains.

Spoken dialogue systems designed for human-robot interaction must therefore be robust to both *ill-formed* and *ill-recognised* inputs. In this paper, we present a new approach to address these two issues. Our starting point is the work done by Zettlemoyer and Collins on parsing using CCG grammars [10]. To account for natural spoken language phenomena (more flexible word order, missing words, etc.), they augment their grammar framework with a small set of non-standard rules, leading to a *relaxation* of the grammatical constraints. A discriminative model over the parses is coupled to the parser, and is responsible for selecting the most likely interpretation(s).

In this paper, we extend their approach in two important ways. First, [10] focused on the treatment of ill-formed input, ignoring the speech recognition issues. Our approach, however, deals with both ill-formed and misrecognized input, in an integrated fashion. This is done by augmenting the set of non-standard rules with new ones specifically tailored to deal with speech recognition errors. Second, we significantly extend the range

B. Mertsching, M. Hund, and Z. Aziz (Eds.): KI 2009, LNAI 5803, pp. 241–248, 2009.

of features included in the discriminative model, by incorporating not only *syntactic*, but also *acoustic*, *semantic* and *contextual* information into the model.

An overview of the paper is as follows. We describe in Section 2 the general architecture of our system, and discuss the approach in Section 3. We present the evaluation results on a Wizard-of-Oz test suite in Section 4, and conclude.

2 Architecture

The approach we present in this paper is fully implemented and integrated into a cognitive architecture for autonomous robots (see [4]). It is capable of building up visuo-spatial models of a dynamic local scene, and of continuously planning and executing manipulation actions on objects within that scene. The robot can discuss objects and their material- and spatial properties for the purpose of visual learning and manipulation tasks. Figure 1 illustrates the architecture for the communication subsystem.

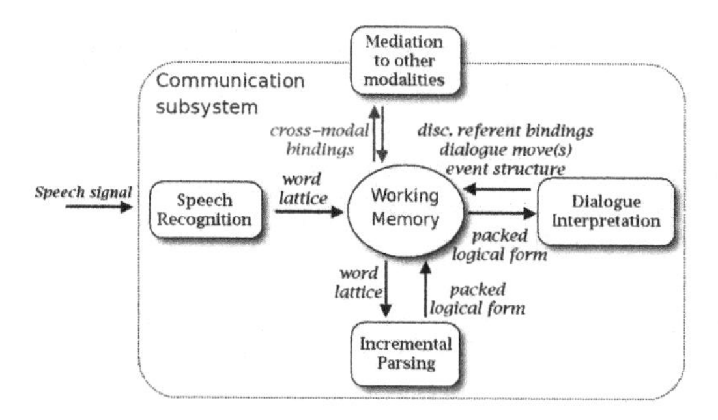

Fig. 1. Architecture schema of the communication subsystem (only for comprehension)

Starting with speech recognition, we process the audio signal to establish a *word lattice* containing statistically ranked hypotheses about word sequences. Subsequently, parsing constructs grammatical analyses for the given word lattice. A grammatical analysis constructs both a syntactic analysis of the utterance, and a representation of its meaning. The analysis is based on an incremental chart parser[1] for Combinatory Categorial Grammar [8]. These meaning representations are ontologically richly sorted, relational structures, formulated in a (propositional) description logic, more precisely in HLDS [1][2]. Finally, at the level of dialogue interpretation, the logical forms are resolved against a dialogue model to establish co-reference and dialogue moves.

3 Approach

3.1 Grammar Relaxation

Our approach to robust processing rests on the idea of **grammar relaxation**: the grammatical constraints specified in the grammar are "relaxed" to handle slightly ill-formed

[1] Built using the OpenCCG API: http://openccg.sf.net

[2] An example of such meaning representation (HLDS logical form) is given in Figure 2.

or misrecognised utterances. Practically, the grammar relaxation is done via the introduction of *non-standard CCG rules* [10]. We describe here two families of relaxation rules: the *discourse-level composition rules* and the *ASR correction rules* [5].

Discourse-level composition rules. In natural spoken dialogue, we may encounter utterances containing several independent "chunks" without any explicit separation (or only a short pause or a slight change in intonation), such as "*yes take the ball right and now put it in the box*". These chunks can be analysed as distinct "discourse units". Syntactically speaking, a discourse unit can be any type of saturated atomic categories – from a simple discourse marker to a full sentence.

The type-changing rule $\mathbf{T}_{du}$ converts atomic categories into discourse units:

$$\mathsf{A} : @_i f \Rightarrow \mathsf{du} : @_i f \qquad (\mathbf{T}_{du})$$

where A represents an arbitrary saturated atomic category (s, np, pp, etc.).

Rule $\mathbf{T}_C$ then integrates two discourse units into a single structure:

$$\mathsf{du} : @_a x \Rightarrow \mathsf{du} : @_c z \;/\; \mathsf{du} : @_b y \qquad (\mathbf{T}_C)$$

where the formula $@_c z$ is defined as:

$$@_{\{c:\text{d-units}\}}(\mathbf{list} \wedge (\langle \text{FIRST} \rangle\ a \wedge x) \wedge (\langle \text{NEXT} \rangle\ b \wedge y)) \qquad (1)$$

ASR error correction rules. Speech recognition is highly error-prone. It is however possible to partially alleviate this problem by inserting error-correction rules (more precisely, new lexical entries) for the most frequently misrecognised words. If we notice for instance that the ASR frequently substitutes the word "wrong" for "round" (because of their phonological proximity), we can introduce a new lexical entry to correct it:

$$round \vdash \mathsf{adj} : @_{attitude}(\mathbf{wrong}) \qquad (2)$$

A small set of new lexical entries of this type have been added to our lexicon to account for the most frequent recognition errors.

3.2 Parse Selection

Using more powerful rules to relax the grammatical analysis tends to increase the number of parses. We hence need a mechanism to discriminate among the possible parses. The task of selecting the most likely interpretation among a set of possible ones is called *parse selection*. Once the parses for a given utterance are computed, they are filtered or selected in order to retain only the most likely interpretation(s). This is done via a (discriminative) statistical model covering a large number of features.

Formally, the task is defined as a function $F : \mathcal{X} \rightarrow \mathcal{Y}$ where $\mathcal{X}$ is the set of possible inputs (in our case, $\mathcal{X}$ is the space of *word lattices*), and $\mathcal{Y}$ the set of parses. We assume:

1. A function $\mathbf{GEN}(x)$ which enumerates all possible parses for an input x. In our case, the function represents the admissibles parses of the CCG grammar.
2. A d-dimensional feature vector $\mathbf{f}(x,y) \in \Re^d$, representing specific features of the pair (x,y) (for instance, acoustic, syntactic, semantic or contextual features).
3. A parameter vector $\mathbf{w} \in \Re^d$.

The function F, mapping a word lattice to its most likely parse, is then defined as:

$$F(x) = \underset{y \in \mathbf{GEN}(x)}{\operatorname{argmax}} \ \mathbf{w}^T \cdot \mathbf{f}(x,y) \tag{3}$$

where $\mathbf{w}^T \cdot \mathbf{f}(x,y)$ is the inner product $\sum_{s=1}^{d} w_s \ f_s(x,y)$, and can be seen as a measure of the "quality" of the parse. Given the parameter vector $\mathbf{w}$, the optimal parse of a given word lattice x can be therefore easily determined by enumerating all the parses generated by the grammar, extracting their features, computing the inner product $\mathbf{w}^T \cdot \mathbf{f}(x,y)$, and selecting the parse with the highest score.

3.3 Learning

Training data. To estimate the parameters $\mathbf{w}$, we need a set of training examples. Since no corpus of situated dialogue adapted to our task domain is available to this day – let alone semantically annotated – we followed the approach advocated in [9] and *generated* a corpus from a hand-written task grammar. We first designed a small grammar covering our task domain, each rule being associated with a HLDS representation and a weight. Once specified, the grammar is then randomly traversed a large number of times, resulting in a large set of utterances along with their semantic representations.

Perceptron learning. The algorithm we use to estimate the parameters $\mathbf{w}$ using the training data is a **perceptron**. The algorithm is fully online - it visits each example in turn, in an incremental fashion, and updates $\mathbf{w}$ if necessary. Albeit simple, the algorithm has proven to be very efficient and accurate for the task of parse selection [3,10].

The pseudo-code for the online learning algorithm is detailed in [**Algorithm 1**].

The parameters $\mathbf{w}$ are first initialised to arbitrary values. Then, for each pair (x_i, z_i) in the training set, the algorithm computes the parse y' with the highest score according to the current model. If this parse matches the best parse associated with z_i (which we denote y^*), we move to the next example. Else, we perform a perceptron update on the parameters:

$$\mathbf{w} = \mathbf{w} + \mathbf{f}(x_i, y^*) - \mathbf{f}(x_i, y') \tag{4}$$

The iteration on the training set is repeated T times, or until convergence.

3.4 Features

As we have seen, the parse selection operates by enumerating the possible parses and selecting the one with the highest score according to the linear model parametrised by $\mathbf{w}$. The accuracy of our method crucially relies on the selection of "good" features $\mathbf{f}(x,y)$ for our model - that is, features which help *discriminating* the parses. In our model, the features are of four types: semantic features, syntactic features, contextual features, and speech recognition features.

Algorithm 1. Online perceptron learning

Require: - set of n training examples $\{(x_i, z_i) : i = 1...n\}$
- T: number of iterations over the training set
- $\texttt{GEN}(x)$: function enumerating the parses for an input x according to the grammar.
- $\texttt{GEN}(x, z)$: function enumerating the parses for an input x with semantics z.
- $L(y)$ maps a parse tree y to its logical form.
- Initial parameter vector $\mathbf{w_0}$

% Initialise
$\mathbf{w} \leftarrow \mathbf{w_0}$
% Loop T times on the training examples
for $t = 1...T$ **do**
  for $i = 1...n$ **do**
    % Compute best parse according to current model
    Let $y' = \text{argmax}_{y \in \mathbf{GEN}(x_i)} \mathbf{w}^T \cdot \mathbf{f}(x_i, y)$
    % If the decoded parse ≠ expected parse, update the parameters
    if $L(y') \neq z_i$ **then**
      % Search the best parse for utterance x_i with semantics z_i
      Let $y^* = \text{argmax}_{y \in \mathbf{GEN}(x_i, z_i)} \mathbf{w}^T \cdot \mathbf{f}(x_i, y)$
      % Update parameter vector $\mathbf{w}$
      Set $\mathbf{w} = \mathbf{w} + \mathbf{f}(x_i, y^*) - \mathbf{f}(x_i, y')$
    end if
  end for
end for
return parameter vector $\mathbf{w}$

Semantic features. Semantic features are defined on *substructures* of the logical form. We define features on the following information sources: the nominals, the ontological sorts of the nominals, and the dependency relations (following [2]). These features help us handle various forms of lexical and syntactic ambiguities.

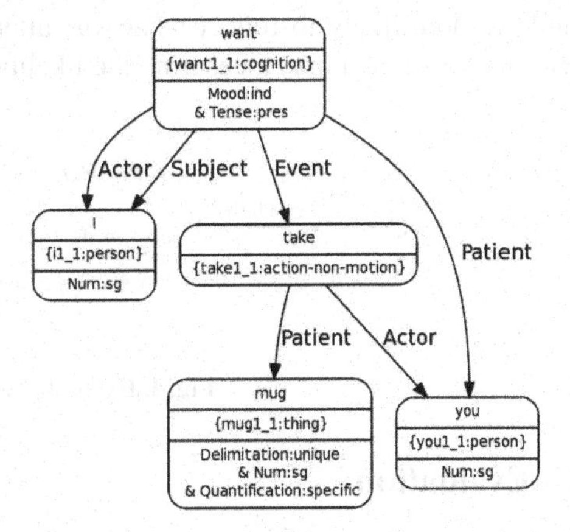

Fig. 2. Logical form for *"I want you to take the mug"*

Syntactic features. Syntactic features are features associated to the *derivational history* of a specific parse. The main use of these features is to *penalise* to a correct extent the application of the non-standard rules introduced into the grammar.

To this end, we include in the feature vector $\mathbf{f}(x, y)$ a new feature for each

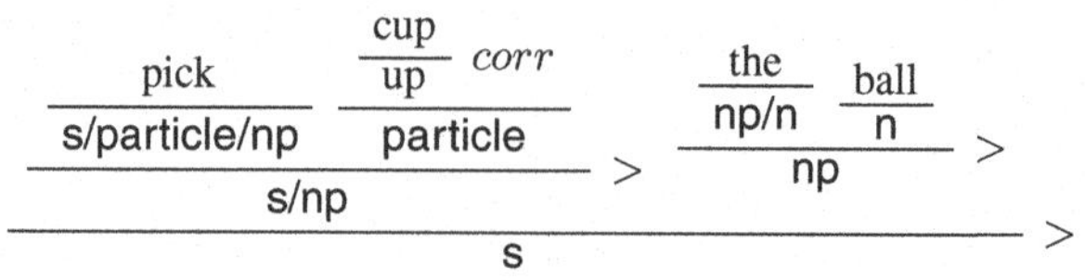

pick / s/particle/np; cup — up ($corr$) / particle; → s/np (>)
the / np/n; ball / n; → np (>)
→ s (>)

Fig. 3. CCG derivation of *"pick cup the ball"*

non-standard rule, which counts the number of times the rule was applied in the parse. In the derivation shown in Figure 3, the rule $corr$ (application of an ASR correction rule) is applied once, so the corresponding feature value is set to 1. These syntactic features can be seen as a *penalty* given to the parses using these non-standard rules, thereby giving a preference to the "normal" parses over them. This mechanism ensures that the grammar relaxation is only applied "as a last resort" when the usual grammatical analysis fails to provide a full parse.

Contextual features. One striking characteristic of spoken dialogue is the importance of context. Understanding the visual and discourse contexts is crucial to resolve potential ambiguities and compute the most likely interpretation(s) of a given utterance. The feature vector f (x, y) therefore includes various contextual features [5]. The dialogue system notably maintains in its working memory a list of contextually activated words [7]. This list is continuously updated as the dialogue and the environment evolves. For each context-dependent word, we include one feature counting its occurrence in the utterance.

Speech recognition features. Finally, the feature vector $\mathbf{f}(x, y)$ also includes features related to the *speech recognition*. The ASR module outputs a set of (partial) recognition hypotheses, packed in a word lattice. One example is given in Figure 4. To favour the hypotheses with high confidence scores (which are, according to the ASR statistical models, more likely to reflect what was uttered), we introduce in the feature vector several acoustic features measuring the likelihood of each recognition hypothesis.

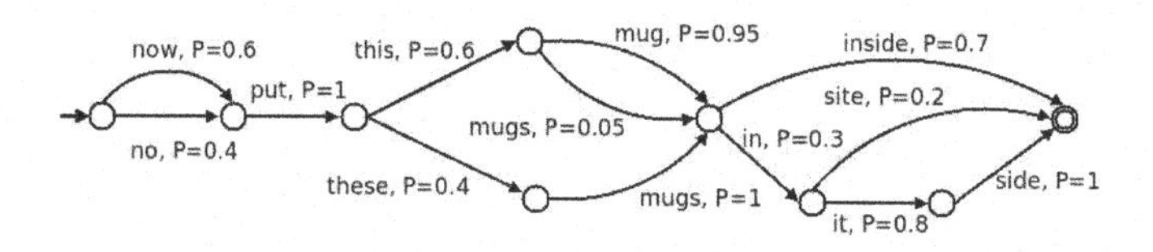

Fig. 4. Example of word lattice

4 Evaluation

We performed a quantitative evaluation of our approach, using its implementation in a fully integrated system (cf. Section 2). To set up the experiments for the evaluation, we have gathered a Wizard-of-Oz corpus of human-robot spoken dialogue for our task-domain, which we segmented and annotated manually with their expected semantic interpretation. The data set contains 195 individual utterances along with their complete logical forms.

Three types of quantitative results are extracted from the evaluation results: *exact-match, partial-match*, and *word error rate*. Tables 1, 2 and 3 illustrate the results, broken down by use of grammar relaxation, use of parse selection, and number of recognition hypotheses considered. Each line in the tables corresponds to a possible configuration. Tables 1 and 2 give the precision, recall and F_1 value for each configuration (respectively for the exact- and partial-match), and Table 3 gives the Word Error Rate [WER].

Table 1. Exact-match accuracy results (in percents)

	Size of word lattice (number of NBests)	Grammar relaxation	Parse selection	Precision	Recall	F_1-value
(Baseline)	1	No	No	40.9	45.2	**43.0**
.	1	No	Yes	59.0	54.3	56.6
.	1	Yes	Yes	52.7	70.8	60.4
.	3	Yes	Yes	55.3	82.9	66.3
.	5	Yes	Yes	55.6	84.0	66.9
(Full approach)	10	Yes	Yes	55.6	84.9	**67.2**

Table 2. Partial-match accuracy results (in percents)

	Size of word lattice (number of NBests)	Grammar relaxation	Parse selection	Precision	Recall	F_1-value
(Baseline)	1	No	No	86.2	56.2	**68.0**
.	1	No	Yes	87.4	56.6	68.7
.	1	Yes	Yes	88.1	76.2	81.7
.	3	Yes	Yes	87.6	85.2	86.4
.	5	Yes	Yes	87.6	86.0	86.8
(Full approach)	10	Yes	Yes	87.7	87.0	**87.3**

Table 3. Word error rate (in percents)

Size of word lattice (NBests)	Grammar relaxation	Parse selection	WER
1	No	No	**20.5**
1	Yes	Yes	19.4
3	Yes	Yes	16.5
5	Yes	Yes	15.7
10	Yes	Yes	**15.7**

The baseline corresponds to the dialogue system with no grammar relaxation, no parse selection, and use of the first NBest recognition hypothesis. Both the partial-, exact-match accuracy results and the WER demonstrate statistically significants improvements over the baseline. We also observe that the inclusion of more ASR recognition hypotheses has a positive impact on the accuracy results.

5 Conclusions

We presented an *integrated* approach to the processing of (situated) spoken dialogue, suited to the specific needs and challenges encountered in human-robot interaction.

In order to handle disfluent, partial, ill-formed or misrecognized utterances, the grammar used by the parser is "relaxed" via the introduction of a set of *non-standard rules* which allow for the combination of discourse fragments or the correction of speech recognition errors. The relaxed parser yields a (potentially large) set of parses, which are then retrieved by the parse selection module. The parse selection is based on a discriminative model exploring a set of relevant semantic, syntactic, contextual and acoustic features extracted for each parse.

The outlined approach is currently being extended in new directions, such as the exploitation of parse selection *during* incremental parsing to improve the parsing efficiency [6], the introduction of more refined contextual features, or the use of more sophisticated learning algorithms, such as Support Vector Machines.

References

1. Baldridge, J., Kruijff, G.-J.M.: Coupling CCG and hybrid logic dependency semantics. In: ACL 2002: Proceedings of the 40th Annual Meeting of the Association for Computational Linguistics, pp. 319–326. Association for Computational Linguistics, Philadelphia (2002)
2. Clark, S., Curran, J.R.: Log-linear models for wide-coverage ccg parsing. In: Proceedings of the 2003 conference on Empirical methods in natural language processing, pp. 97–104. Association for Computational Linguistics, Morristown (2003)
3. Collins, M., Roark, B.: Incremental parsing with the perceptron algorithm. In: ACL 2004: Proceedings of the 42nd Annual Meeting of the Association for Computational Linguistics, p. 111. Association for Computational Linguistics, Morristown (2004)
4. Hawes, N.A., Sloman, A., Wyatt, J., Zillich, M., Jacobsson, H., Kruijff, G.-J.M., Brenner, M., Berginc, G., Skocaj, D.: Towards an integrated robot with multiple cognitive functions. In: Proc. AAAI 2007, pp. 1548–1553. AAAI Press, Menlo Park (2007)
5. Lison, P.: Robust processing of situated spoken dialogue. Master's thesis, Universität des Saarlandes, Saarbrücken (2008), `http://www.dfki.de/ plison/pubs/thesis/main.thesis.plison2008.pdf`
6. Lison, P.: A method to improve the efficiency of deep parsers with incremental chart pruning. In: Proceedings of the ESSLLI Workshop on Parsing with Categorial Grammars, Bordeaux, France (in press, 2009)
7. Lison, P., Kruijff, G.-J.M.: Salience-driven contextual priming of speech recognition for human-robot interaction. In: Proceedings of the 18th European Conference on Artificial Intelligence, Patras, Greece (2008)
8. Steedman, M., Baldridge, J.: Combinatory categorial grammar. In: Borsley, R., Börjars, K. (eds.) Nontransformational Syntax: A Guide to Current Models. Blackwell, Oxford (2009)
9. Weilhammer, K., Stuttle, M.N., Young, S.: Bootstrapping language models for dialogue systems. In: Proceedings of Interspeech 2006, Pittsburgh, PA (2006)
10. Zettlemoyer, L.S., Collins, M.: Online learning of relaxed CCG grammars for parsing to logical form. In: Proceedings of the 2007 Joint Conference on Empirical Methods in Natural Language Processing and Computational Natural Language Learning (EMNLP-CoNLL), pp. 678–687 (2007)

iDocument: Using Ontologies for Extracting and Annotating Information from Unstructured Text

Benjamin Adrian[1], Jörn Hees[2], Ludger van Elst[1], and Andreas Dengel[1,2]

[1] Knowledge Management Department, DFKI
Kaiserslautern, Germany
[2] CS Department, University of Kaiserslautern
Kaiserslautern, Germany
firstname.lastname@dfki.de

Abstract. Due to the huge amount of text data in the WWW, annotating unstructured text with semantic markup is a crucial topic in Semantic Web research. This work formally analyzes the incorporation of domain ontologies into information extraction tasks in iDocument. Ontology-based information extraction exploits domain ontologies with formalized and structured domain knowledge for extracting domain-relevant information from un-annotated and unstructured text. iDocument provides a pipeline architecture, an extraction template interface and the ability of exchanging domain ontologies for performing information extraction tasks. This work outlines iDocument's ontology-based architecture, the use of SPARQL queries as extraction templates and an evaluation of iDocument in an automatic document annotation scenario.

1 Introduction

Automatically or semi-automatically extracting structured information from unstructured text is an important step towards text understanding. Existing information extraction (IE) systems are mostly specialized in limited domains. A scenario where end users may rapidly query a text base with ad hoc queries[1] is nearly impossible to implement with existing IE technologies. (e.g., *Select scientific conferences in 2009 and their deadlines for paper submissions with a focus on information extraction.*)

Knowledge engineering approaches (e.g., Textmarker [1]) suffer from the knowledge engineering bottleneck which means an expert has to provide and maintain a rule base. Existing machine learning approaches avoid these efforts. Instead they require previously annotated training corpora with expensive ground truth data [2].

Thus, common IE systems do not provide scalability, adaptability, and maintainability for being used in cost saving and generic user scenarios.

---

[1] According to Ralph Grishman (2002) about the project Proteus: *Our long-term goal is to build systems that automatically find the information you're looking for, pick out the most useful bits, and present it in your preferred language, at the right level of detail.*

B. Mertsching, M. Hund, and Z. Aziz (Eds.): KI 2009, LNAI 5803, pp. 249–256, 2009.

In order to overcome these shortcomings, we present iDocument, a flexible ontology-based information extraction (OBIE) system. It uses terminology and instance knowledge of domain ontologies that are written in RDF/S [3]. Exchanging a domain ontology customizes the system on a completely new domain of interest. Instead of using expensive IE template definition specifications, iDocument provides a generic and standardized IE template interface based on the syntax of the RDF query language SPARQL. Thus, extraction templates can be expressed by SPARQLing the ontology's RDFS scheme. As result, iDocument generates an RDF graph that contains RDF triples which were extracted from given text documents and comply with the SPARQL template.

The structure of this paper is as follows: Related work on using ontologies in information extraction is summarized in Section 2. Next, Section 3 formally describes, how RDFS ontologies may be used in IE tasks. Section 4 shows iDocument's architecture, extraction pipeline, and template population algorithm. In Section 5, an evaluation presents the quality of annotated instances and facts.

2 Related Work

iDocument extends basic IE principles by using ontologies. Comparable OBIE systems are GATE [4], SOBA [5] or SummIt-BMT [6]. In difference to these, iDocument does not use the ontology as simple input gazetteer that is a plain list of relevant labels but as model for semantic analysis such as instance disambiguation and discourse analysis. The technique of using existing domain ontologies as input for information extraction tasks and extraction results as base for ontology population was first presented by Empley in [7] by using relational database technologies. Sintek et al. [8] applied a similar scenario on Semantic Web ontologies. A bootstrapping approach was presented by Maedche, Neumann, and Staab in [9] by learning new domain ontologies semi-automatically and populating these in a final step. The previously mentioned OBIE systems do not support any template mechanisms. In the Message Understanding Conference series (MUC), extraction templates were defined by using named slots that were annotated with constraints and rules [10]. These template specifications tended to be very long and complex and were hard to create by system users. Hobbs and Israel claimed in [11] that template design is related to ontology engineering. Following this assumption, the use of extraction ontologies as sort of template for IE is presented by Labský in [12]. Here the entire ontology specifies those information elements (e.g., fields and entities) that should be extracted from text. In iDocument, SPARQL queries that are formulated by using the ontology's RDFS vocabulary [3] serve as technique for expressing extraction templates.

3 Domain Ontologies in Information Extraction Tasks

"In common sense, a domain ontology (or domain-specific ontology) models a specific domain, or part of the world. It represents the particular meanings of terms as they apply to that domain."[2] Domains may be for example: the Olympic Summer Games 2004, the Knowledge Management Department at DFKI, or a Personal Information Model [13] inside a Semantic Desktop.

By using the RDFS [3] vocabulary the main components of a domain ontology $\mathcal{O}$ can be defined as $\mathcal{O}(H_C, H_P, I, S, A)$(For the sake of simplicity, assertions about instances and ontology are not separated as done in frame logics):

Hierarchy of classes (H_C). A resource c can be defined as `rdfs:class`. A class c_1 may specialize class c_2 by expressing `rdfs:subClassOf`(c_1, c_2). The transitive closure of all `rdfs:subClassOf` expressions builds the class hierarchy H_C.

Hierarchy of properties (H_p). A property p expresses a semantic relation between two classes $p(c_1, c_2)$. A property p_1 may specialize property p_2 if `rdfs:subPropertyOf`(p_1, p_2). The transitive closure of `rdfs:subPropertyOf` expressions builds the hierarchy of properties H_P.

  Object Properties (P_O). A property $p \in P_O$ is called object property if `rdfs:range` is defined as `rdfs:range`(p, c) with $c \in H_C \backslash$`rdfs:Literal`^{T}, where `rdfs:Literal`^{T} denotes the reflexive, transitive subhierarchy of class `rdfs:Literal`.

  Datatype Properties (P_{DT}). A property $p \in P_{DT}$ is a datatype property if `rdfs:range` is defined as `rdfs:range`$(p,$ `rdfs:Literal`$)$.

Instances (I). Instances are resources i with an `rdf:type` property that is defined as `rdf:type`(i, c) with $c \in H_C \setminus \{$`rdfs:Literal`$^T\}$.

Symbols (S). Symbols are resources s with an `rdf:type` property that is defined as `rdf:type`(s, c) with `rdfs:subClassOf`$(c,$ `rdfs:Literal`$)$.

Assertions (A). Assertions are triple expressions in the form of $p(i, r)$ with $p \in H_P$ and $i, r \in H_C \cup I \cup S$.

Based on such an ontology, it is possible to define four major OBIE tasks.

Symbol Recognition. If a similarity function $sim(s, s_e)$ decides that a phrase s inside a text matches an existing symbol $s_e \in S$, s is recognized as symbol. Each datatype property $p \in P_{DT}$ inside a valid assertion $p(i, s_e) \in A$ about an instance $i \in I$ is called the symbol's type.

- If $sim(s, s_e)$ bases on content similarity, s is called *content symbol* and denoted as s_c. Each content symbol s_c has to exist in S.
  (e.g., assuming `foaf:mbox`(urn:BenjaminAdrian, "adrian@dfki.de") $\in A$, all occurrences of "adrian@dfki.de" in text will be recognized as content symbol with type `foaf:mbox`)
  In traditional IE systems this task is called *Named Entity Recognition.*
- If $sim(s, s_e)$ bases on structural similarity, s is called *structure symbol* and denoted as s_s.

[2] as described in Wikipedia, online available at `http://en.wikipedia.org/w/index.php?title=Ontology_(information_science)&oldid=284405492`, 2009-04-21

(e.g., assuming `foaf:mbox`(urn:BenjaminAdrian, "adrian@dfki.de") $\in A$, the occurrence of "dengel@dfki.de" in text will be recognized as structure symbol with type `foaf:mbox`).

In traditional IE systems this task is called *Structured Entity Recognition.*

Instance Recognition. The unification of recognized content symbols to existing instances i_{exist} is called instance recognition. For a *content symbol* s_c, i_{exist} is a recognized instance if an assertion $p(i_{exist}, s_c) \in A$ exists where at least one type of s_c matches p. Thus it is possible that for single content symbols more than one recognized instance exist (e.g., content symbols of type `foaf:firstName`).

In case of a *structure symbol* s_s, instance recognition may either be solved with unification that adds a new assertion $p(i_e, s_s) \notin A$ to an existing instance $i_e \in I$ or as instantiation of a new instance $i_{new} \notin I$ with an assertion $p(i_{new}, s_s) \notin A$, with p is a type of symbol s_s.

Traditional IE systems call this task *Template Unification* or *Template Merge.*

Fact Recognition. Assume that $P_Q \subseteq H_P$ is a set of queried properties inside the extraction template Q. Recognized facts are assertions of type $p(i_1, i_2)$ or $p(i_1, s)$ with $i_1, i_2 \in I, p \in P_Q, s \in S$. If all components i_1, i_2 or s of a recognized fact were recognized in text in the previous steps and the recognized fact is not an existing assertion $f_r \notin A$, the fact is called *extracted fact.* Otherwise if $f_r \in A$ it is called *completed fact* as the template is completed with known assertions of the domain ontology.

Traditional IE systems call this task *Fact Extraction.*

Template Population. A given extraction template Q is populated with recognized facts. It is possible that a single template is populated with multiple populations. (e.g., In case of "Select scientific conferences in 2009 and their deadlines for paper submissions with a focus on information extraction", each recognized conference with its paper due forms a single populated template.) Traditional IE systems call this task *Scenario Extraction.*

The following section describes the architecture, IE task implementations, and extraction templates of the OBIE system iDocument.

4 The OBIE System iDocument

As outlined in Figure 1, iDocument's architecture comprises the following components: a domain ontology, a text collection, SPARQL templates, an OBIE pipeline, and finally results in populated templates. A text collection contains content that is relevant for the current question and domain of interest.

4.1 Alignment Metadata

For using a domain ontology in iDocument, relevant parts for extraction purpose have to be annotated with the MOBIE vocabulary[3] (Metadata for OBIE) in a mapping file:

[3] refer to `http://ontologies.opendfki.de/repos/ontologies/obie/mobie`

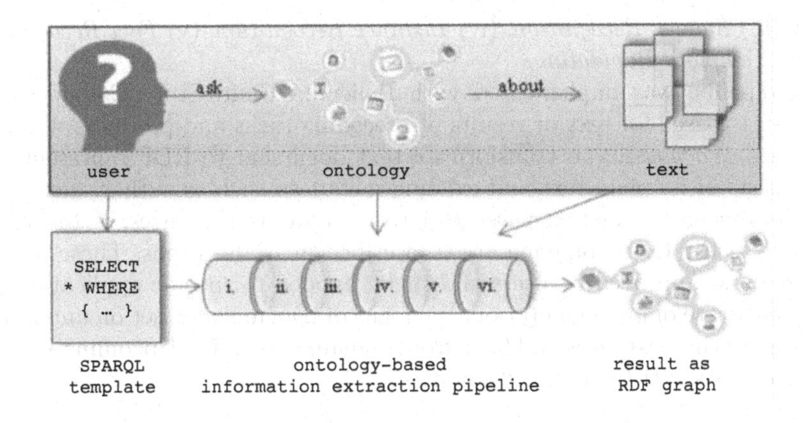

Fig. 1. OBIE scenario

`mobie:Entity` For using a class of instances for instance recognition purpose, it has to be assigned to mobie:Entity by using rdfs:subClassOf. (e.g., foaf:Person rdfs:subClassOf mobie:Entity)

`mobie:symbol` For using a datatype property for symbol recognition purpose, it has to be assigned to mobie:symbol by using rdfs:subPropertyOf. (e.g., foaf:firstName rdfs:subPropertyOf mobie:symbol)

`mobie:relates` For using an object property for fact recognition purpose, it has to be assigned to mobie:relates by using rdfs:subPropertyOf. (e.g., foaf:knows rdfs:subPropertyOf mobie:relates)

4.2 Extraction Templates

iDocument provides an extraction template interface that processes templates written in SPARQL. Users may write templates by using the ontology's RDFS vocabulary. (e.g., SELECT * WHERE {?person rdf:type foaf:Person. ?person foaf:member ?org. ?org rdf:type foaf:Organisation.} extracts persons and organisations and facts about memberships from text.)

4.3 Template Population

The algorithm for populating templates follows a most constrained variable heuristic. A SPARQL query can be represented as forest of join expressions. The algorithm transforms the forest into a list of paths from root to leaf and sorts this list in a descending order by the length of paths. Iteratively, it removes the longest path from the list and tries to populate it with recognized facts.

4.4 Extraction Pipeline

iDocument is built upon an OBIE pipeline consisting of six OBIE tasks. The first two are standard text based analyzes for instance (i) *Normalization* and (ii) *Segmentation*. Succeeding tasks are OBIE tasks as described in Section 3, namely (iii) *Symbol Recognition*, (iv) *Instance Recognition*, (v) *Fact Recognition*, and (vi) *Template Population*.

This pipeline was implemented with Believing Finite-State Cascades [14]. Each task is based on text or results of preceding tasks and produces weighted hypotheses. *Normalization* transforms a text document to RDF representation that consists of its plain text and existing metadata such as author, date, title. *Segmentation* partitions text passages into paragraphs, sentences, or tokens.

Results of the OBIE pipeline are transcribed in RDF graphs. These may be visualized and approved by users and/or be handed to specific applications. In the current state of implementation, the focus of iDocument is set on annotating text with existing instances and facts from domain ontologies. Upcoming versions will enhance its extraction functionalities.

5 Evaluation

The evaluation of iDocument was done by analyzing the quality of extracted results of instance and fact recognition tasks with precision and recall. As populating and maintaining domain ontologies is burdensome, the evaluation was designed to show how the degree of extraction quality correlates with the amount of assertions inside the domain ontology. As corpus data, the DFKI/OCAS 2008[4] corpus was used. It contains a domain ontology about the Olympic Summer games 2004 and fifty documents that were annotated with symbols, instances and facts [15]. In an adaption of leave-one-out cross validation, one document was taken as candidate, the remaining documents with their annotations built the assertions of the domain ontology. For analyzing the learning behaviour, assertions of the domain ontology were divided into two parts, i.e. training set and test set. The impact of existing knowledge about a single document's content or knowledge about other documents in the same domain was analyzed by creating four types of training sets, i.e.:

x-Doc, 0-Ont. Training set contains just a ratio of x % of the currently analyzed document's annotations.

0-Doc, x-Ont. Training set just contains a ratio of x % of the all other documents' annotations, except annotations of the currently analyzed document.

x-Doc, x-Ont. Training set contains a ratio of x % of the all existing documents' annotations.

x-Doc, 100-Onto. Training set contains a ratio of 100 % of the all other documents' annotations and a ratio of x % of the currently analyzed document's annotations.

The training sets were used as assertions inside the domain ontology. During each test run the ratio x was increased from 0 % to 100 % in steps of 10 %. The test set always contained 100 % of annotations about the currently analyzed document. The evaluation task was to check the degree of extracted instances and facts. Figure 2 shows results of the instance recognition task for instances of

[4] http://idocument.opendfki.de/wiki/Evaluation/Corpus/OlympicGames2004

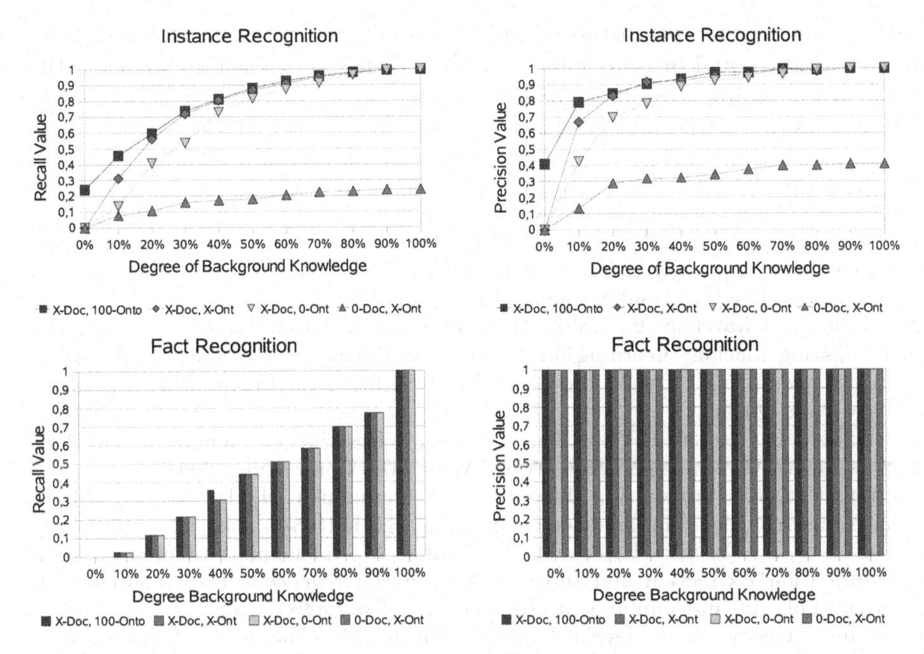

Fig. 2. Recall and precision progressions for (i) instance recognition as shown on top and (ii) fact recognition as shown on bottom

type (Person, Nation, Sport, and Discipline) in the upper row. At left it shows the progression of recall regarding an increasing amount of assertions inside the domain ontology. The logarithmic increase can be explained by the existence of multiple symbols per instance inside the ontology that were recognized during symbol recognition. If the ontology contains all relevant instances and facts that occur inside a test document, iDocument is able to recognize these with 100 % recall and precision. The recall progression of recognizing facts about a person's nationality follows a linear behavior as shown in the lower row. The step from 90 % to 100 % backround knowledge can be explained by rounding double values, as the amount of these facts per document were often below ten. The precision of recognized facts is always at 100 %, because iDocument only annotated existing facts that are asserted inside the domain ontology.

In both cases instance and fact recognition, the progression of an ontology with assertions shows, that precision does not decrease if the ontology knows more than just the document annotations.

6 Conclusion and Outlook

This work described the use of RDF/S domain ontologies for information extraction and annotation. The OBIE system iDocument was presented, including a pipeline-based architecture of OBIE tasks and a SPARQL-based template interface for defining which information to extract from text. The evaluation showed that iDocument retrieved instances and facts in text if they exist inside a domain ontology. Future and ongoing efforts in iDocument are spent on increasing the ability of extracting new information from text.

This work was financed by the BMBF project Perspecting (Grant 01IW08002).

References

1. Atzmüller, M., Klügl, P., Puppe, F.: Rule-Based Information Extraction for Structured Data Acquisition using TextMarker. In: Proc. LWA 2008 (Special Track on Knowledge Discovery and Machine Learning) (2008)
2. Ireson, N., Ciravegna, F., Califf, M.E., Freitag, D., Kushmerick, N., Lavelli, A.: Evaluating Machine Learning for Information Extraction. In: Raedt, L.D., Wrobel, S. (eds.) ICML. ACM Int. Conf. Proc. Series, vol. 119, pp. 345–352. ACM, New York (2005)
3. Brickley, D., Guha, R.V.: RDF Vocabulary Description Language 1.0: RDF Schema. W3C recommendation, World Wide Web Consortium (2004)
4. Bontcheva, K., Tablan, V., Maynard, D., Cunningham, H.: Evolving GATE to meet new challenges in language engineering. JNLE 10(3-4), 349–373 (2004)
5. Buitelaar, P., Cimiano, P., Frank, A., Hartung, M., Racioppa, S.: Ontology-based Information Extraction and Integration from Heterogeneous Data Sources. Int. Journal of Human-Computer Studies (11), 759–788 (2008)
6. Endres-Niggemeyer, B., Jauris-Heipke, S., Pinsky, M., Ulbricht, U.: Wissen gewinnen durch Wissen: Ontologiebasierte Informationsextraktion. Information - Wissenschaft & Praxis 57(1), 301–308 (2006)
7. Embley, D.W., Campbell, D.M., Smith, R.D., Liddle, S.W.: Ontology-based Extraction and Structuring of Information from Data-Rich Unstructured Documents. In: CIKM 1998: Proc. of the 7th Int. Conf. on Information and Knowledge Management, pp. 52–59. ACM, New York (1998)
8. Sintek, M., Junker, M., van Elst, L., Abecker, A.: Using Information Extraction Rules for Extending Domain Ontologies. In: Workshop on Ontology Learning. CEUR-WS.org (2001)
9. Maedche, A., Neumann, G., Staab, S.: Bootstrapping an Ontology-based Information Extraction System. In: Szczepaniak, P., Segovia, J., Kacprzyk, J., Zadeh, L.A. (eds.) Intelligent Exploration of the Web. Springer, Berlin (2002)
10. Grishman, R., Sundheim, B.: Design of the MUC-6 evaluation. In: Proc. of a workshop held at Vienna, pp. 413–422. Association for Computational Linguistics, Virginia (1996)
11. Hobbs, J., Israel, D.: Principles of Template Design. In: HLT 1994: Proc. of the workshop on HLT, pp. 177–181. ACL, Morristown (1994)
12. Labský, M., Svátek, V., Nekvasil, M., Rak, D.: The Ex Project: Web Information Extraction using Extraction Ontologies. In: Proc. Workshop on Prior Conceptual Knowledge in Machine Learning and Knowledge Discovery, PriCKL 2007 (2007)
13. Sauermann, L., van Elst, L., Dengel, A.: PIMO - a Framework for Representing Personal Information Models. In: Proc. of I-Semantics 2007, JUCS, pp. 270–277 (2007)
14. Adrian, B., Dengel, A.: Believing Finite-State cascades in Knowledge-based Information Extraction. In: Dengel, A.R., Berns, K., Breuel, T.M., Bomarius, F., Roth-Berghofer, T.R. (eds.) KI 2008. LNCS (LNAI), vol. 5243, pp. 152–159. Springer, Heidelberg (2008)
15. Grothkast, A., Adrian, B., Schumacher, K., Dengel, A.: OCAS: Ontology-Based Corpus and Annotation Scheme. In: Proc. of the HLIE Workshop 2008, ECML PKDD, pp. 25–35 (2008)

Behaviorally Flexible Spatial Communication: Robotic Demonstrations of a Neurodynamic Framework

John Lipinski, Yulia Sandamirskaya, and Gregor Schöner

Institut für Neuroinformatik
Ruhr-Universität Bochum
Bochum, Germany
Tel.: +49-234-3224201
2johnlipinski@gmail.com

Abstract. Human spatial cognitive processes provide a model for developing artificial spatial communication systems that fluidly interact with human users. To this end, we develop a neurodynamic model of human spatial language combining spatial and color terms with neurally-grounded scene representations. Tests of this model implemented on a robotic platform support its viability as a theoretical framework for flexible spatial language behaviors in artificial agents.

1 Introduction

The fluidity of human-human spatial communication suggests that human cognitive processes may provide a useful model for human-robot spatial communication systems. The complexity of human spatial cognition, however, is an obvious challenge to this approach. In developing such a human-based artificial system it is therefore useful to focus on a constrained but fundamental set of characteristics underlying human spatial communication. To this end, the current work focuses on autonomy and behavioral flexibility in spatial language processing. This limited focus provides for a manageable research agenda that still incorporates core features of human spatial cognition.

1.1 Autonomy and Flexibility

In cognition, autonomy refers to the unfolding of cognitive processes continuously in time according to past and present sensory and behavioral states [5]. In agent-based systems such as robots, autonomy further implies that behaviors are structured according to sensory information that the agent itself acquires [7]. These aspects of autonomy draw attention to two core elements of human spatial language. First, natural spatial language comprehension and production depend on the continuous integration of both visual and linguistic information over time. The inherent variability of speaker timing (e.g. slow versus rapid speech), word order, and visual context demands a system that continuously integrates this

B. Mertsching, M. Hund, and Z. Aziz (Eds.): KI 2009, LNAI 5803, pp. 257–264, 2009.

information in a manner permitting contextually adaptive behavior. Second, because spatial language often changes behavior (e.g. guiding ones attention or action to a specific location), behavioral changes needs to be fluidly coordinated with continuous linguistic input.

Flexibility, the second key characteristic, is embedded in the principle of autonomy. Flexibility is important because the same spatial language system must support many behaviors including the generation of descriptive spatial terms and the combination of the spatial and non-spatial features in processing a spatial description (i.e. "The *red* apple to the *right* of the glass"). These behaviors must also be flexibly deployed across a limitless array of visual scenes.

1.2 Dynamic Field Theory

The Dynamic Field Theory [2,6] provides an implementable theoretical framework that can capture these characteristics. Dynamical Field Theory (DFT) is a neural-dynamic approach to human cognition in which cognitive states are represented as distributions of neural activity defined over metric dimensions. These dimensions may represent perceptual features (e.g., retinal location, color, orientation), movement parameters (e.g. heading direction) or more abstract parameters (e.g. visual attributes of objects like shape or size). These continuous metric spaces represent the space of possible percepts, actions, or scenes.

Spatially continuous neural networks (neural fields) are at the heart of the DFT and were originally introduced as approximate descriptions of cortical and thalamic neuroanatomy. Neural fields are recurrent neural networks whose temporal evolution is described by iteration equations. In continuous form, these take the form of dynamical systems. The mathematics of dynamical neural fields was first analyzed by Amari [1]. Importantly, recent modeling and empirical work in spatial cognition (for review see [8]) shows how the DFT captures core characteristics of human spatial cognition. This suggests that the DFT may facilitate the development of fluid human-robot spatial communication.

To rigorously test this claim we present a new DFT-based spatial language model and implement it on a robotic platform continuously linked to real-world visual images. Our "human style"model extracts the categorical, cognitive information from the low-level sensory input through the system dynamics. Our demonstrations specifically combine visual space, spatial language, and color.

2 Modeling Neurons and Dynamical Neural Fields

2.1 Dynamical Fields

The dynamical neural fields are mathematical models first used to describe cortical and subcortical neural activation dynamics [1]. The dynamic field equation Eq. (1) is a differential equation describing the evolution of activation u defined over a neural variable(s) $\mathbf{x}$. These neural variables represent continuous perceptual (e.g. color) or behavioral (e.g. reaching amplitude) dimensions of interest that can be naturally defined along a continuous metric.

$$\tau \dot{u}(\mathbf{x}, t) = -u(\mathbf{x}, t) + h + \int f(u(\mathbf{x}', t))\omega(\Delta \mathbf{x})d\mathbf{x}' + I(\mathbf{x}, t) \quad (1)$$

Here, $h < 0$ is the resting level of the field; the sigmoid non-linearity $f(u) = 1/(1+e^{-\beta u})$ determines the field's output at suprathreshold cites with $f(u) > 0$. The field is quiescent at subthreshold cites with $f(u) < 0$. The homogeneous interaction kernel $\omega(\Delta x) = c_{exc}e^{\frac{-(\Delta x)^2}{2\sigma^2}} - c_{inh}$ depends only on the distance between the interacting cites $\Delta x = \mathbf{x} - \mathbf{x}'$. This interaction kernel is a Bell-shaped, local excitation/lateral inhibition function. The short-range excitation is of amplitude c_{exc} and spread σ. The long-range inhibition is of amplitude c_{inh}. $I(\mathbf{x}, t)$ is the summed external input to the field; τ is the time constant.

If a localized input activates the neural field at a certain location, the interaction pattern ω stabilizes a localized "peak", or "bump"solution of the field's dynamics. These activation peaks represent the particular value of the neural variable coded by the field and thus provide the representational units in the DFT. In our model, all entities having "field"in their name evolve according to Eq. (1), where $\mathbf{x}$ is a vector representing the two-dimensional visual space in Cartesian coordinates. The links between the fields are realized via the input term $I(\mathbf{x}, t)$, where only cites with $f(u) > 0$ propagate activation to other fields or neurons.

2.2 Discrete Neurons

The discrete (localist) neurons in the model representing linguistic terms can be flexibly used for either user input or response output and evolve according to the dynamic equation (2).

$$\tau_d \dot{d}(t) = -d(t) + h_d + f(d(t)) + I(t). \quad (2)$$

Here, d is the activity level of a neuron; the sigmoidal non-linearity term $f(d)$ shapes the self-excitatory connection for each discrete neuron and provides for self-stabilizing activation. The resting level is defined by h_d. The $I(t)$ term represents the sum of all external inputs into the given neuron. This summed input is determined by the input coming from the connected neural field, the user-specified language input, and the inhibitory inputs from the other discrete neurons defined for that same feature (color or space); τ is the time constant of the dynamics.

3 The Spatial Language Framework

This section outlines the overall structure (see Fig. 1) and functionality of the model. The color-space fields (Fig. 1A) are an array of several dynamical fields representing the visual scene. Each field is sensitive to a hue range which corresponds to a basic color. The color resolution was low in the presented examples because only a few colors were needed to represent the used objects. In principle, the color (hue) is a continuous variable and can be resolved more finely.

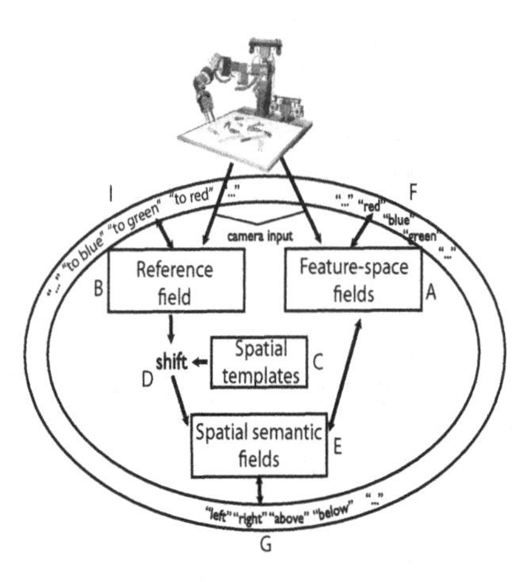

Fig. 1. Overview of the architecture

The stack of color-space fields is therefore a three-dimensional dynamic field that represents colors and locations on the sensor surface. The camera provides visual input to the color-space field which is below the activation threshold before the task is defined.

Once the language input specifies the *color* of the object, however, the resting levels of all cites of the corresponding color-space field are raised homogeneously. Because the color-space fields receive localized camera input, this uniform activation increase is summed with that input to enable the development of an instability and, ultimately, the formation of a single-peak solution. This peak is centered over the position of the object with that specified color. The *spatial* language input also influences the color-space fields' dynamics through the aligned spatial semantic fields (see below).

The reference field (Fig. 1B) is a spatially-tuned dynamic field which also receives visual input. When the user specifies the reference object color, the corresponding "reference-color" neuron becomes active and specifies the color in the camera image that provides input into the reference field. A peak of activation in the reference field specifies the location of the reference object. The reference field continuously tracks the position of the reference object. It also filters out irrelevant inputs and camera noise and thus stabilizes the reference object representation. Having a stable but updatable reference object representation allows the spatial semantics to be continuously aligned with the visual scene.

The spatial semantic templates (Fig. 1C) are represented as a set of synaptic weights that connect spatial terms to an abstract, "retinotopic" space. The particular functions defining "left", "right", "below", and "above" here were two-dimensional Gaussians in polar coordinates and are based on a neurally-inspired approach to English spatial semantic representation [3]. The shift mechanism (Fig. 1D) aligns these retinotopically defined spatial semantics with the current task space. The shift is done by convolving the "egocentric" weight matrices with the outcome of the reference field. Because the single reference object is represented as a localized activation peak in the reference field, the convolution simply centers the semantics over the reference object. The spatial terms thus become defined relative to the reference object location (for related method see [4]).

The aligned spatial semantic fields (Fig. 1E) are arrays of dynamical neurons with weak lateral interaction. They receive input from the spatial alignment or "shift" mechanism which maps the spatial semantics onto the current scene by "shifting" the semantic representation of the spatial terms to the reference object position. The aligned spatial semantic fields integrate the spatial semantic input with the summed outcome of the color-space fields and interact reciprocally with the spatial-term nodes. Thus, a positive activation in an aligned spatial semantic field increases the activation of the associated spatial term node and vice versa.

4 Demonstrations

In the presented scenarios, everyday objects (e.g. a red plastic apple, a blue tube of sunscreen) were placed in front of the robot. The visual input was formed from the camera image and sent to the reference and color-space fields. The color-space field input was formed by extracting hue value ("color") for each pixel in the image and assigning that pixel's intensity value to the corresponding location in the matching color-space field. Analogously, the reference field input was formed according to the user-specified reference object color. The reference and color-space fields receive localized inputs corresponding to the three objects in the camera image (marked with arrows, see Fig. 2(a) and Fig. 2(b)).

4.1 Demonstration 1: Describing "Where"

Demonstration 1 asks "Where is the yellow object relative to the green one?" To respond correctly, the robot must select "Right". Fig. 2(a) shows the neural field activation just before the answer is given. The task input first activates two discrete neurons, one representing "green" for the user-specified reference object color and the other "yellow" for the user-specified object color (see user inputs, top Fig. 2(a)). The reference object specification "green" leads to the propagation of the green camera input into the reference field, creating an activation bump in the reference field at the location of the green item (see Reference Field, Fig. 2(a)). The specification of the target color "yellow" increases the activation for the yellow node linked to the yellow color-space field, which raises the resting level of the associated yellow color-space field. This uniform activation boost coupled with the camera input from the yellow object induces an activation peak in the field (see "yellow" color-space field, Fig. 2(a)).

This localized target object activation is then transfered to the aligned semantic fields. The aligned semantic fields also receive input from spatial term semantic units. Critically, these semantic profiles are shifted to align with the reference object position. In the current case, the yellow target object activation therefore overlaps with the aligned "right" semantic field (see red arrow in the

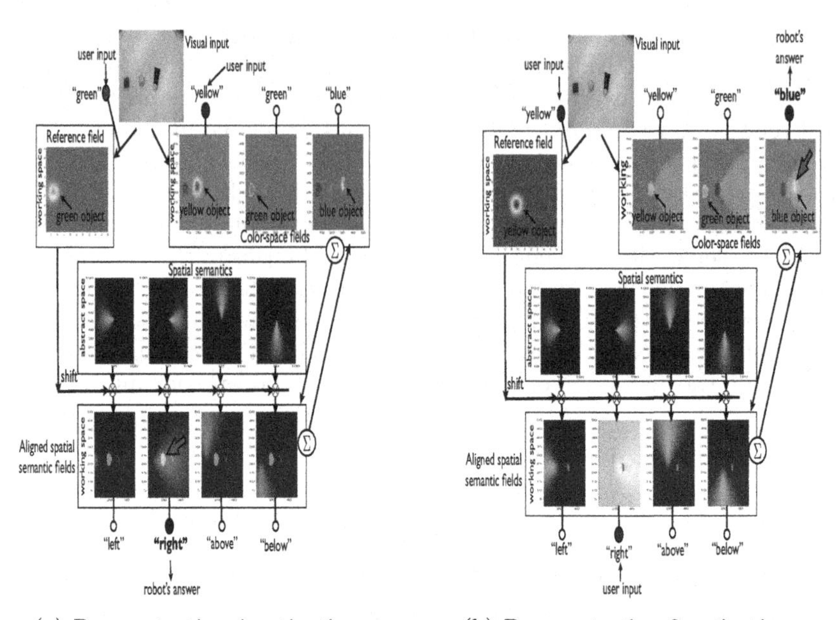

(a) Demonstration 1 activations. (b) Demonstration 2 activations.

Fig. 2. The basic behaviors of the architecture

"right" aligned spatial semantic field, Fig. 2(a)). This overlap ultimately drives the activation and selection of the "right" node.

4.2 Demonstration 2: Describing "Which"

Demonstration 2 asks "Which object is to the right of the yellow one?". To respond correctly, the robot must select "Blue". As indicated in Fig. 2(b), the task input first activates two discrete neurons, one representing the reference object color "yellow" and the other representing "right".

The reference object specification "yellow" creates an activation bump in the reference field location matching that of the yellow item (see reference field, Fig. 2(b)). The specification of "right", in its turn, increases the activation for that spatial-term node, creating a homogeneous activation boost to the "right" semantic field. This activation boost creates a positive activation in the field to the right of the yellow reference object (see "right" aligned spatial semantic field, Fig. 2(b)). This spatially-specific activation is then input into the color-space fields and subsequently raises activation at all those color-space field locations to the right of the reference object (see lighter blue color-space field regions, Fig. 2(b)). This region overlaps with the localized blue object input in the "blue" color-space field and an activation peak develops in that field (see red arrow in the "blue" object color-space field, Fig. 2(b)). This increases the activation of the "blue" color-term node, triggering selection of the answer, "blue".

4.3 Demonstration 3: Dynamically Driven Sensor Movement

Movement presents an additional set of behavioral challenges. First, movements (e.g. gaze, reaching) can be driven by cognitive states shaped by spatial language. Second, when that movement involves the sensor providing spatial information (e.g. eyes) the changing visual input can disrupt the peak dynamics driving cognitive behaviors. Robustly adaptive behavior in the context of such movement is thus a strong test of our dynamic model. Demonstration 3 addresses these challenges with the addition of a dynamic motor control module that drives sensor (camera) movement. We present the sentence "The red one to the left of the blue" in the context of two red objects. The robot's task to establish a peak at the correct object location, shifting the camera accordingly.

Fig. 3 shows the time course and summary camera movements of the task (blue reference object specified previously). The"red" color term first uniformly boosts the "red" color-space field and creates a peak for the slightly larger, but *incorrect*, red object (red apple) on the right (see Fig. 3B). The camera then shifts to the right to center that object. This in turn leads to the smearing and shift of the activation profiles across the fields (see especially Fig. 3B). Nevertheless, the peak

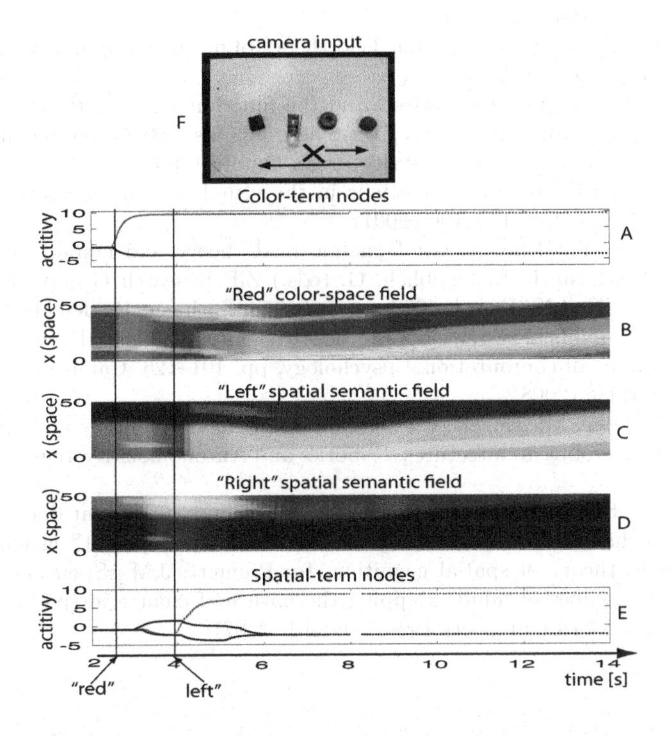

Fig. 3. Demonstration 3. Horizontal axis in all panels represents time. Vertical axis in A and E represents activation. Vertical axis in panels B-D represents the projected activation onto horizontal axis of the image. See text for additional details.

is stably maintained across the camera movement, thus tracking the red object location. When we later specify the "left" spatial relation(Fig. 3E), however, this initial peak is extinguished and a peak at the fully described location arises (see later portion of Fig. 3B). This new peak then shifts the camera dynamics and the camera centers the correct object (see shifting profiles in Fig. 3B-D). The motor behaviors were thus driven by emergent, dynamic decision processes.

5 Conclusion

The present work adopted a systems-level neural dynamic perspective to develop an implementable spatial language framework.The behavioral flexibility arising from our autonomous neurodynamic model suggests that systems-level neural dynamic theories like the DFT can aid the development of effective artificial spatial communication systems.

References

1. Amari, S.: Dynamics of pattern formation in lateral-inhibition type neural fields. Biological Cybernetics 27, 77–87 (1977)
2. Erlhagen, W., Schöner, G.: Dynamic field theory of movement preparation. Psychological Review 109, 545–572 (2002)
3. O'Keefe, J.: Vector grammar, places, and the functional role of the spatial prepositions in english. In: van der Zee, E., Slack, J. (eds.) Representing direction in language and space. Oxford University Press, Oxford (2003)
4. Salinas, E.: Coordinate transformations in the visual system. Advances in Neural Population Coding 130, 175–190 (2001)
5. Sandamirskaya, Y., Schöner, G.: Dynamic field theory and embodied communication. In: Wachsmuth, I., Knoblich, G. (eds.) ZiF Research Group International Workshop. LNCS (LNAI), vol. 4930, pp. 260–278. Springer, Heidelberg (2008)
6. Schöner, G.: Dynamical systems approaches to cognition. In: Sun, R. (ed.) The Cambridge handbook of computational psychology, pp. 101–126. Cambridge University Press, Cambridge (2008)
7. Schöner, G., Dose, M., Engels, C.: Dynamics of behavior: Theory and applications for autonomous robot architectures. Robotics and Autonomous Systems 16, 213–245 (1995)
8. Spencer, J.P., Simmering, V.S., Schutte, A.R., Schöner, G.: What does theoretical neuroscience have to offer the study of behavioral development? insights from a dynamic field theory of spatial cognition. In: Plumert, J.M., Spencer, J.P. (eds.) Emerging landscapes of mind: Mapping the nature of change in spatial cognition, pp. 320–361. Oxford University Press, Oxford (2007)

SceneMaker: Automatic Visualisation of Screenplays

Eva Hanser, Paul Mc Kevitt, Tom Lunney, and Joan Condell

School of Computing & Intelligent Systems
Faculty of Computing & Engineering
University of Ulster, Magee
Derry/Londonderry BT48 7JL
Northern Ireland
`hanser-e@email.ulster.ac.uk`,
`{p.mckevitt,tf.lunney,j.condell}@ulster.ac.uk`

Abstract. Our proposed software system, *SceneMaker*, aims to facilitate the production of plays, films or animations by automatically interpreting natural language film scripts and generating multimodal, animated scenes from them. During the generation of the story content, SceneMaker will give particular attention to emotional aspects and their reflection in fluency and manner of actions, body posture, facial expressions, speech, scene composition, timing, lighting, music and camera work. Related literature and software on Natural Language Processing, in particular textual affect sensing, affective embodied agents, visualisation of 3D scenes and digital cinematography are reviewed. In relation to other work, SceneMaker will present a genre-specific text-to-animation methodology which combines all relevant expressive modalities. In conclusion, SceneMaker will enhance the communication of creative ideas providing quick pre-visualisations of scenes.

Keywords: Natural Language Processing, Text Layout Analysis, Intelligent Multimodal Interfaces, Affective Agents, Genre Specification, Automatic 3D Visualisation, Affective Cinematography, SceneMaker.

1 Introduction

The production of movies is an expensive process involving planning, rehearsal time, actors and technical equipment for lighting, sound and special effects. It is also a creative act which requires experimentation, visualisation of ideas and their communication between everyone involved, e.g., play writers, directors, actors, camera men, orchestra and set designers. We are developing a software system, *SceneMaker*, which will provide a facility to pre-visualise scenes. Users input a natural language (NL) script text and automatically receive multimodal 3D visualisations. The objective is to give directors or animators a reasonable idea of what a scene will look like. The user can refine the automatically created output through a script and 3D editing interface, accessible over the internet and on mobile devices. Such technology could be applied in the training of those

B. Mertsching, M. Hund, and Z. Aziz (Eds.): KI 2009, LNAI 5803, pp. 265–272, 2009.

involved in scene production without having to utilise expensive actors and studios. Alternatively, it could be used for rapid visualisation of ideas and concepts in advertising agencies. SceneMaker will extend an existing software prototype, CONFUCIUS [1], which provides automated conversion of single natural language sentences to multimodal 3D animation of character actions. SceneMaker will focus on the precise representation of emotional expression in all modalities available for scene production and especially on most human-like modelling of body language and genre sensitive art direction. SceneMaker will include new tools for text layout analysis of screenplays, commonsense and affective knowledge bases for context understanding, affective reasoning and automatic genre specification. This work focuses on three research questions: How can emotional information be computationally recognised in screenplays and structured for visualisation purposes? How can emotional states be synchronised in presenting all relevant modalities? Can compelling, life-like and believable animations be achieved? Section 2 of this paper gives an overview of current research on computational, multimodal and affective scene production. In section 3, the design of SceneMaker is discussed. SceneMaker is compared to related multimodal work in section 4 and Section 5 discusses the conclusion and future work.

2 Background

Automatic and intelligent production of film/theatre scenes with characters expressing emotional states involves four development stages:

1. Detecting personality traits and emotions in the film script
2. Modelling affective 3D characters, their expressions and actions
3. Visualisation scene environments according to emotional findings
4. Intelligent storytelling interpreting the plot.

This section reviews state-of-the-art advances in these areas.

2.1 Detecting Personality and Emotions in Film Scripts

All modalities of human interaction, namely voice, word choice, gestures, body posture and facial expression, express personality and emotional states. In order to recognise emotions in text and to create life-like characters, psychological theories for emotion, mood, personality and social status are translated into computable models, e.g Ekman's 6 basic emotions [2], the Pleasure-Dominance-Arousal model (PAD) [3] with intensity values or the OCC model (Ortony-Clore-Collins) [4] with cognitive grounding and appraisal rules. Different approaches to textual affect sensing are able to recognise explicit affect phrases such as keyword spotting and lexical affinity [5], machine learning methods [6], hand-crafted rules and fuzzy logic systems [7] and statistical models [6]. Common knowledge based approaches [8,9] and a cognitive inspired model [10] include emotional context evaluation of non-affective words and concepts. The strict formatting of screenplays eases the machine parsing of scripts and facilitates the detection of semantic context information for visualisation. Through text layout analysis of capitalisation, indentation and parentheses, elements such as dialog, location, time, present actors, actions and sound cues can be recognised and directly mapped into XML-presentation [11].

2.2 Modelling Affective Embodied Agents

Research aiming to automatically animate virtual humans with natural expressions faces challenges not only in automatic 3D character transformation, synchronisation of face and body expressions with speech, path finding and collision detection, but furthermore in the refined sensitive execution of each action. The exact manner of an affective action depends on intensity, fluency, scale and timing. Various XML based scripting languages specifically cater for the modelling of affective behaviour, e.g., the Behaviour Expression Animation Toolkit (BEAT) [12], the Multimodal Presentation Mark-up Language (MPML) [13], SCREAM (Scripting Emotion-based Agent Minds) [14] and AffectML [15]. The Personality & Emotion Engine [7], a fuzzy rule-based system, combines the OCEAN personality model [16], Ekman's basic emotions [2] and story character roles to control the affective state and body language of characters mapping emotions to postural values of four main body areas. Embodied Conversational Agents (ECA) are capable of face-to-face conversations with human users or other agents, generating and understanding NL and body movement, e.g., Max [17] and Greta [18]. Multimodal annotation coding of video or motion captured data specific to emotion collects data in facial expression or body gesture databases [19]. The captured animation data can be mapped to 3D models instructing characters precisely on how to perform desired actions.

2.3 Visualisation of 3D Scenes

The composition of the 3D scene environment or set, automated cinematography, acoustic elements and the effect of genre styles are addressed in text-to-visual systems. WordsEye [20] depicts non-animated 3D scenes with characters, objects, actions and environments considering the attributes, poses, kinematics and spatial relations of 3D models. CONFUCIUS [1] produces multimodal 3D animations of single sentences. 3D models perform actions, dialogues are synthesised and basic cinematic principles determine the camera placement. ScriptViz [23] renders 3D scenes from NL screenplays, extracting verbs and adverbs to interpret events in sentences. Film techniques are automatically applied to existing animations in [21]. Reasoning about plot, theme, character actions, motivations and emotions, cinematic rules define the appropriate placement and movement of camera, lighting, colour schemes and the pacing of shots according to theme and mood. The Expression Mark-up Language (EML) [22] integrates environmental expressions like cinematography, illumination and music into the emotion synthesis of virtual humans. Films, plays or literature are classified into different genres with distinguishable presentation styles, e.g., drama or comedy. Genre is reflected in the detail of a production, exaggeration and fluency of movements, pace (shot length), lighting, colour and camerawork [24]. The automatic 3D animation production system, CAMEO [25], incorporates direction knowledge, like genre and cinematography, as computer algorithms and data. A system which automatically recommends music based on emotion [26] associates emotions and music features, chords, rhythm and tempo of songs, in movie scenes.

2.4 Intelligent Storytelling

Intelligent, virtual storytelling requires the automatic development of a coherent plot, fulfilling thematic, dramatic, consistency and presentation goals of the author, as in MINSTREL [27]. The plot development can be realised with a character-based approach, e.g., AEOPSWORLD [28], where characters are autonomous intelligent agents choosing their own actions or a script-based approach, e.g., Improv [29], where characters act out a given scene script or an intermediate variation of the two.

The wide range of approaches for intelligent storytelling and modelling emotions, moods and personality aspects in virtual humans and scene environments provide a sound basis for SceneMaker.

3 Design of SceneMaker

Going beyond the animation of explicit events, SceneMaker will use Natural Language Processing (NLP) methods for screenplays to automatically extract and visualise emotions and moods within the story or scene context. SceneMaker will augment short 3D scenes with affective influences on the body language of actors and environmental expression, like illumination, timing, camera work, music and sound automatically directed according to the genre style.

3.1 SceneMaker Architecture

SceneMakers's architecture is shown in Fig. 1. The main component is the *scene production module* including modules for understanding, reasoning and multimodal visualisation. The *understanding module* performs natural language processing and text layout analysis of the input text. The *reasoning module* interprets the context based on common, affective and cinematic knowledge bases, updates emotional states and creates plans for actions, their manners and the representation of the set environment. The *visualisation module* maps these plans to 3D animation data, selects appropriate 3D models from the graphics database, defines their body motion transitions, instructs speech synthesis, selects sound and music files from the audio database and assigns values to camera and lighting parameters. The visualisation module synchronises all modalities into an animation manuscript. The online user interface, available via computers and mobile devices, consists of the input module, assisting film script writing and editing

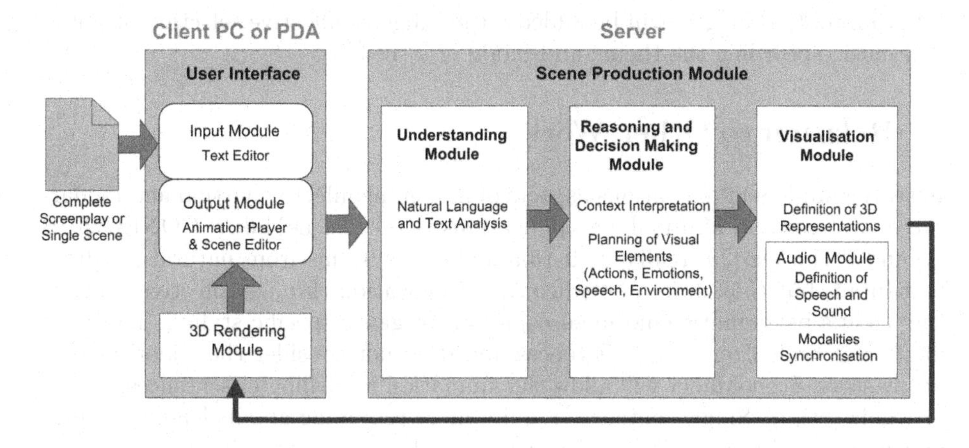

Fig. 1. SceneMaker architecture

and the output module, rendering 3D scene according to the manuscript and allowing manual scene editing to fine-tune the automatically created animations.

3.2 Implementation of SceneMaker

Multimodal systems automatically mapping text to visuals face challenges in interpreting human language which is variable, ambiguous, imprecise and relies on common knowledge between the communicators. Enabling a machine to understand a natural language text involves feeding the multimodal system with grammatical structures, semantic relations, visual descriptions and common knowledge to be able to match suitable graphics. A pre-processing tool will decompose the layout structure of the input screenplay to facilitate access to semantic information. SceneMaker's language interpretation will build upon the NLP module of CONFUCIUS [1]. The Connexor Part of Speech Tagger [30] parses the input text and identifies grammatical word types, e.g., noun, verb or adjective, and determines their relation in a sentence, e.g., subject, verb and object with Functional Dependency Grammars [31]. CONFUCIUS's semantic knowledge base (WordNet [32] and LCS database [33]) will be extended by an emotional knowledge base, e.g., WordNet-Affect [34], and context reasoning with ConceptNet [9] to enable an understanding of the deeper meaning of the context and emotions. In order to automatically recognise genre, SceneMaker will identify keyword co-occurences and term frequencies and determine the length of dialogues, sentences and scenes/shots. The visual knowledge of CONFUCIUS, such as object models and event models, will be related to emotional cues. CONFUCIUS' basic cinematic principles will be extended and classified into expressive and genre-specific categories. Resources for 3D models are H-Anim models [35] which include geometric or physical, functional and spatial properties. The speech synthesis module used in CONFUCIUS, FreeTTS [36], will be tested for its suitability with regard to effective emotional prosody. An automatic audio selection tool, as in [26], will be added for intelligent, affective selection of sound and music according the theme and mood of a scene.

4 Relation to Other Work

Research implementing various aspects of modelling affective virtual actors, narrative systems and film-making applications relates SceneMaker. CONFUCIUS [1] and ScriptViz [23] realise text-to-animation systems from natural language text input, but they do not enhance the visualisation through affective aspects, the agent's personality, emotional cognition or genre specific styling. Their animation is built from single sentences and does not consider the wider context of the story. SceneMaker will allow the animation modelling of sentences, scenes or whole scripts. Single sentences require more reasoning about default settings and more precision will be achieved from collecting context information from longer passages of text. SceneMaker will introduce text layout analysis to derive semantic content from the particular format of screenplays. Emotion cognition and display will be related to commonsense knowledge. CAMEO [25] is the only system relating specific cinematic direction for character animation, lighting and camera work to the genre or theme of a given story, but genre types are explicitly selected by the user. SceneMaker will automatically recognise genre from script text with keyword co-occurrence, term frequency and calculation of dialogue and scene length. SceneMaker will form a software system for believable affective computational animation production from NL scene scripts.

5 Conclusion and Future Work

SceneMaker contributes to believability and artistic quality of automatically produced animated, multimedia scenes. The software system, SceneMaker, will automatically visualise affective expressions of screenplays. Existing systems solve partial aspects of NLP, emotion modelling and multimodal storytelling. Thereby, this research focuses on semantic interpretation of screenplays, the computational processing of emotions, virtual agents with affective behaviour and expressive scene composition including emotion-based audio selection. In relation to other work, SceneMaker will incorporate an expressive model for multiple modalities, including prosody, body language, acoustics, illumination, staging and camera work. Emotions will be inferred from context. Genre types will be automatically derived from the scene scripts and influence the design style of the output animation. The 3D output will be editable on SceneMaker's mobile, web-based user interface and will assist directors, drama students, writers and animators in the testing of scenes. Accuracy of animation content, believability and effectiveness of expression and usability of the interface will be evaluated in empirical tests comparing manual animation, feature film scenes and real-life directing with SceneMaker. In conclusion, this research intends to automatically produce multimodal animations with heightened expressivity and visual quality from screenplay input.

References

1. Ma, M.: Automatic Conversion of Natural Language to 3D Animation. PhD Thesis, School of Computing and Intelligent Systems, University of Ulster (2006)
2. Ekman, P., Rosenberg, E.L.: What the face reveals: Basic and applied studies of spontaneous expression using the facial action coding system. Oxford University Press, Oxford (1997)
3. Mehrabian, A.: Framework for a Comprehensive Description and Measurement of Emotional States. Genetic, Social, and General Psychology Monographs. Heldref Publishing 121(3), 339–361 (1995)
4. Ortony, A., Clore, G.L., Collins, A.: The Cognitive Structure of Emotions. Cambridge University Press, Cambridge (1988)
5. Francisco, V., Hervás, R., Gervás, P.: Two Different Approaches to Automated Mark Up of Emotions in Text. In: Research and development in intelligent systems XXIII: Proceedings of AI 2006, pp. 101–114. Springer, Heidelberg (2006)
6. Strapparava, C., Mihalcea, R.: Learning to identify emotions in text. In: Proceedings of the 2008 ACM Symposium on Applied Computing. SAC 2008, pp. 1556–1560. ACM, New York (2008)
7. Su, W.-P., Pham, B., Wardhani, A.: Personality and Emotion-Based High-Level Control of Affective Story Characters. IEEE Transactions on Visualization and Computer Graphics 13(2), 281–293 (2007)
8. Liu, H., Lieberman, H., Selker, T.: A model of textual affect sensing using real-world knowledge. In: Proceedings of the 8th International Conference on Intelligent User Interfaces. IUI 2003, pp. 125–132. ACM, New York (2003)
9. Liu, H., Singh, P.: ConceptNet: A practical commonsense reasoning toolkit. BT Technology Journal. Springer Netherlands 22(4), 211–226 (2004)
10. Shaikh, M.A.M., Prendinger, H., Ishizuka, M.: A Linguistic Interpretation of the OCC Emotion Model for Affect Sensing from Text. In: Affective Information Processing, pp. 45–73. Springer, Heidelberg (2009)
11. Choujaa, D., Dulay, N.: Using screenplays as a source of context data. In: Proceeding of the 2nd ACM international Workshop on Story Representation, Mechanism and Context. SRMC 2008, pp. 13–20. ACM, New York (2008)
12. Cassell, J., Vilhjálmsson, H.H., Bickmore, T.: BEAT: the Behavior Expression Animation Toolkit. In: Proceedings of the 28th Annual Conference on Computer Graphics and interactive Techniques. SIGGRAPH 2001, pp. 477–486. ACM, New York (2001)
13. Breitfuss, W., Prendinger, H., Ishizuka, M.: Automated generation of non-verbal behavior for virtual embodied characters. In: Proceedings of the 9th International Conference on Multimodal Interfaces. ICMI 2007, pp. 319–322. ACM, New York (2007)
14. Prendinger, H., Ishizuka, M.: SCREAM: scripting emotion-based agent minds. In: Proceedings of the First International Joint Conference on Autonomous Agents and Multiagent Systems: Part 1. AAMAS 2002, pp. 350–351. ACM, New York (2002)
15. Gebhard, P.: ALMA - Layered Model of Affect. In: Proceedings of the 4th International Conference on Autonomous Agents and Multiagent Systems (AAMAS 2005), Utrecht University, Netherlands, pp. 29–36. ACM, New York (2005)
16. De Raad, B.: The Big Five Personality Factors. In: The Psycholexical Approach to Personality, Hogrefe & Huber (2000)
17. Kopp, S., Allwood, J., Grammer, K., Ahlsen, E., Stocksmeier, T.: Modeling Embodied Feedback with Virtual Humans. In: Modeling Communication with Robots and Virtual Humans, pp. 18–37. Springer, Heidelberg (2008)

18. Pelachaud, C.: Multimodal expressive embodied conversational agents. In: Proceedings of the 13th Annual ACM International Conference on Multimedia. MULTIMEDIA 2005, pp. 683–689. ACM, New York (2005)
19. Gunes, H., Piccardi, M.: A Bimodal Face and Body Gesture Database for Automatic Analysis of Human Nonverbal Affective Behavior. In: 18th International Conference on Pattern Recognition, ICPR, vol. 1, pp. 1148–1153. IEEE Computer Society, Washington (2006)
20. Coyne, B., Sproat, R.: WordsEye: an automatic text-to-scene conversion system. In: Proceedings of the 28th Annual Conference on Computer Graphics and Interactive Techniques, pp. 487–496. ACM Press, Los Angeles (2001)
21. Kennedy, K., Mercer, R.E.: Planning animation cinematography and shot structure to communicate theme and mood. In: Proceedings of the 2nd International Symposium on Smart Graphics. SMARTGRAPH 2002, vol. 24, pp. 1–8. ACM, New York (2002)
22. De Melo, C., Paiva, A.: Multimodal Expression in Virtual Humans. Computer Animation and Virtual Worlds 17(3-4), 239–348 (2006)
23. Liu, Z., Leung, K.: Script visualization (ScriptViz): a smart system that makes writing fun. Soft Computing 10(1), 34–40 (2006)
24. Rasheed, Z., Sheikh, Y., Shah, M.: On the use of computable features for film classification. IEEE Transactions on Circuits and Systems for Video Technology. IEEE Circuits and Systems Society 15(1), 52–64 (2005)
25. Shim, H., Kang, B.G.: CAMEO - camera, audio and motion with emotion orchestration for immersive cinematography. In: Proceedings of the 2008 international Conference on Advances in Computer Entertainment Technology. ACE 2008, vol. 352, pp. 115–118 (2008)
26. Kuo, F., Chiang, M., Shan, M., Lee, S.: Emotion-based music recommendation by association discovery from film music. In: Proceedings of the 13th Annual ACM international Conference on Multimedia, MULTIMEDIA 2005, pp. 507–510. ACM, New York (2005)
27. Turner, S.R.: The Creative Process: A Computer Model of Storytelling and Creativity. Lawrence Erlbaum Associates, Hillsdale (1994)
28. Okada, N., Inui, K., Tokuhisa, M.: Towards affective integration of vision, behavior, and speech processing. In: Proceedings of the Integration of Speech and Image Understanding, SPELMG, pp. 49–77. IEEE Computer Society, Washington (1999)
29. Perlin, K., Goldberg, A.: Improv: a system for scripting interactive actors in virtual worlds. In: Proceedings of the 23rd Annual Conference on Computer Graphics and interactive Techniques, SIGGRAPH 1996, pp. 205–216. ACM, New York (1996)
30. Connexor, `http://www.connexor.eu/technology/machinese`
31. Tesniere, L.: Elements de syntaxe structurale. Klincksieck, Paris (1959)
32. Fellbaum, C.: WordNet: An Electronic Lexical Database. MIT Press, Cambridge (1998)
33. Lexical Conceptual Structure Database, `http://www.umiacs.umd.edu/~bonnie/LCS_Database_Documentation.html`
34. Strapparava, C., Valitutti, A.: WordNet-Affect: an affective extension of WordNet. In: Proceedings of the 4th International Conference on Language Resources and Evaluation, LREC 2004, vol. 4, pp. 1083–1086 (2004)
35. Humanoid Animation Working Group, `http://www.h-anim.org`
36. FreeTTS 1.2 - A speech synthesizer written entirely in the JavaTM programming language: `http://freetts.sourceforge.net/docs/index.php`

A Conceptual Agent Model Based on a Uniform Approach to Various Belief Operations*

Christoph Beierle[1] and Gabriele Kern-Isberner[2]

[1] Dept. of Computer Science, FernUniversität in Hagen, 58084 Hagen, Germany
[2] Dept. of Computer Science, TU Dortmund, 44221 Dortmund, Germany

Abstract. Intelligent agents equipped with epistemic capabilities are expected to carry out quite different belief operations like answering queries and performing diagnosis, or revising and updating their own state of belief in the light of new information. In this paper, we present an approach allowing to realize such belief operations by making use of so-called c-change operations as a uniform core methodology. The key idea is to apply the binary c-change operator in various ways to create various belief operations. Ordinal conditional functions (OCF) serve as representations of epistemic states, providing qualitative semantical structures rich enough to validate conditionals in a semi-quantitative way. In particular, we show how iterated revision is possible in an OCF environment in a constructive manner, and how to distinguish clearly between revision and update.

1 Introduction

When modelling an intelligent agent, elaborate knowledge representation and reasoning facilities are required. For instance, an agent performing medical diagnosis must be able to deal with beliefs expressed by rules such as "*If symptoms A and B are present, then disease D_1 is more likely than disease D_2*" or "*If there is pain in the right lower quadrant, rebound tenderness, and leukocytosis $>$ 18.000, then it is highly plausible that the patient has appendicitis*". Furthermore, such an agent should be able to answer diagnostic questions in the presence of evidential facts like "*Given evidence for pain in right lower quadrant, rebound tenderness and low-grade fever, what is the degree of belief for inflamed and perforated appendicitis, respectively?*"

An agent capable of dealing with knowledge bases containing such default rules can be seen as an agent being able to take rules, pieces of evidence, queries, etc., from the environment and giving back sentences she believes to be true with a degree of certainty. Basically, these degrees of belief are inferred from the agent's current epistemic state which is a representation of her cognitive state at the given time. When the agent is supposed to live in a dynamic environment, she has to adapt her epistemic state constantly to changes in the surrounding world and to react adequately to new demands (cf. [1], [2]).

In order to model the semantics of qualitative default rules, we will use ordinal conditional functions (OCFs) [12], ordering possible worlds according to their implausibility. Thus, OCFs are used to represent the epistemic state of an agent carrying out knowledge

* The research reported here was supported by the Deutsche Forschungsgemeinschaft (grants BE 1700/7-1 and KE 1413/2-1).

B. Mertsching, M. Hund, and Z. Aziz (Eds.): KI 2009, LNAI 5803, pp. 273–280, 2009.

management tasks like answering queries and performing diagnosis, or revising and updating her own state of belief in the light of new information. The purpose of this paper is to study these operations involving an agent's epistemic state and to demonstrate that various belief operations can be realized in a uniform way by using the concept of c-change [9,10]. In particular, we develop a simple conceptual agent model with precisely specified belief operations such as revision, update, and diagnosis.

2 Conditionals and Epistemic States

We start with a propositional language $\mathcal{L}$, generated by a finite set Σ of atoms $a, b, c, \ldots$. The formulas of $\mathcal{L}$ will be denoted by uppercase Roman letters $A, B, C, \ldots$. For conciseness of notation, we will omit the logical *and*-connective, writing AB instead of $A \wedge B$, and overlining formulas will indicate negation, i.e. $\overline{A}$ means $\neg A$. Let Ω denote the set of possible worlds over $\mathcal{L}$; Ω will be taken here simply as the set of all propositional interpretations over $\mathcal{L}$ and can be identified with the set of all complete conjunctions (or minterms) over Σ. For $\omega \in \Omega$, $\omega \models A$ means that the propositional formula $A \in \mathcal{L}$ holds in the possible world ω. By introducing a new binary operator $|$, we obtain the set $(\mathcal{L} \mid \mathcal{L}) = \{(B|A) \mid A, B \in \mathcal{L}\}$ of conditionals over $\mathcal{L}$. $(B|A)$ formalizes "*if A then B*" and establishes a plausible connection between the *antecedent* A and the *consequent* B. Let $\mathit{Fact}_{\mathcal{U}}$ be the set of all (unquantified) propositional sentences over Σ, i.e. $\mathit{Fact}_{\mathcal{U}}$ consists of all formulas from $\mathcal{L}$. The set of all (unquantified) conditional sentences from $(\mathcal{L} \mid \mathcal{L})$ is denoted by $\mathit{Rule}_{\mathcal{U}}$, and the set of all sentences is given by $\mathit{Sen}_{\mathcal{U}} = \mathit{Fact}_{\mathcal{U}} \cup \mathit{Rule}_{\mathcal{U}}$ with elements written as A and $(B|A)$, respectively. Additionally, $\mathit{SimpleFact}_{\mathcal{U}}$ denotes the set of simple facts $\Sigma \subseteq \mathit{Fact}_{\mathcal{U}}$, i.e. $\mathit{SimpleFact}_{\mathcal{U}} = \Sigma$.

A conditional $(B|A)$ can be represented as a *generalized indicator function* [3] with $(B|A)(\omega) = 1$, if $\omega \models AB$, $(B|A)(\omega) = 0$, if $\omega \models A\overline{B}$, and $(B|A)(\omega) = u$, if $\omega \models \overline{A}$, where u stand for *unknown* or *indeterminate*. To give appropriate semantics to conditionals, they are usually considered within richer structures such as *epistemic states*. Well-known qualitative, ordinal approaches to represent epistemic states are Spohn's *ordinal conditional functions, OCFs*, also called *ranking functions*, [12], assigning degrees of plausibility to formulas and possible worlds.

An OCF is a function $\kappa : \Omega \to \mathbb{N}$ expressing degrees of implausibility of worlds: The higher $\kappa(\omega)$, the less plausible is ω. At least one world must be regarded as being completely normal; therefore, $\kappa(\omega) = 0$ for at least one $\omega \in \Omega$. Each such ranking function can be taken as the representation of a full epistemic state of an agent. Thus, the set of *ordinal epistemic states* is given by $\mathit{EpState}_{\mathcal{O}} = \{\kappa \mid \kappa : \Omega \to \mathbb{N} \text{ and } \kappa^{-1}(0) \neq \emptyset\}$ Each $\kappa \in \mathit{EpState}_{\mathcal{O}}$ uniquely determines a function (also denoted by κ) $\kappa : \mathit{Sen}_{\mathcal{U}} \to \mathbb{N} \cup \{\infty\}$ defined by $\kappa(A) = \mathit{min}\{\kappa(\omega) \mid \omega \models A\}$, if A is satisfiable, and $\kappa(A) = \infty$ otherwise, for sentences $A \in \mathit{Fact}_{\mathcal{U}}$, and by $\kappa((B|A)) = \kappa(AB) - \kappa(A)$, if $\kappa(A) \neq \infty$, and $\kappa((B|A)) = \infty$ otherwise, for conditionals $(B|A) \in \mathit{Rule}_{\mathcal{U}}$.

The beliefs of an agent being in epistemic state κ with respect to a set of default rules $\mathcal{S}$ is determined by $\mathit{belief}_{\mathcal{O}}(\kappa, \mathcal{S}) = \{S[m] \mid S \in \mathcal{S}, \kappa \models_{\mathcal{O}} S[m] \text{ and } \kappa \not\models_{\mathcal{O}} S[m+1]\}$ where the satisfaction relation $\models_{\mathcal{O}}$ for quantified sentences $(B|A)[m]$ is defined by

$$\kappa \models_{\mathcal{O}} (B|A)[m] \quad \text{iff} \quad \kappa(AB) + m < \kappa(A\overline{B}) \tag{1}$$

Thus, $(B|A)$ is believed in κ with degree of belief m if the degree of disbelief of AB (verifying the unquantified conditional) is more than m smaller than the degree of disbelief of $A\overline{B}$ (falsifying the unquantified conditional).

3 Changing Epistemic States

Belief revision, the theory of dynamics of knowledge, has been mainly concerned with propositional beliefs for a long time. The most basic approach here is the *AGM-theory* presented in the seminal paper [1] as a set of postulates outlining appropriate revision mechanisms in a propositional logical environment. Other forms of belief change operations have been discussed, too, the most prominent one being *update* [7]. It is commonly understood that revision takes place when new information about a static world arrives, whereas updating tries to incorporate new information about a (possibly) evolving, changing world. The difference between revision and update is a subtle one and will be illustrated by the following example.

Example 1. We consider the propositional atoms f - *flying*, b - *birds*, p - *penguins*, w - *winged* animals, k - *kiwis*, d - *doves*. Let the set $\mathcal{R}$ consist of the following conditionals:

r_1: $(f|b)$ *birds fly*
r_2: $(b|p)$ *penguins are birds*
r_3: $(\overline{f}|p)$ *penguins do not fly*
r_4: $(w|b)$ *birds have wings*
r_5: $(b|k)$ *kiwis are birds*

Moreover, we assume the strict knowledge *penguins, kiwis, and doves are pairwise exclusive* to hold, which amounts in considering only those worlds as possible that make at most one of $\{p, k, d\}$ true.

Suppose now that the agent finds in a book that kiwis do not have wings, a piece of information that has escaped her knowledge before. To incorporate $(\overline{w}|k)$ into her stock of beliefs, she should apply a revision operation because her beliefs about the world must change, but the world itself has not changed. On the other hand, the agent might have learned from the news, that, due to some mysterious illness that has occurred recently among doves, the wings of newborn doves are nearly completely mutilated. She wants to adopt her beliefs to the new information $(\overline{w}|d)$. Obviously, the proper change operation in this case in an update operation as the world under consideration has changed by some event (the occurrence of the mysterious illness).

The AGM framework has been widened by Darwiche and Pearl [2] for (qualitative) epistemic states and conditional beliefs. An even more general approach, unifying revision methods for quantitative and qualitative representations of epistemic states, is proposed in [8,10] and will be used in the following for a precise specification of both update and revision in the context of ordinal conditional functions. For an axiomatic approach to and in-depth discussion of revision and update, we refer to [10].

More precisely, we will understand updating as a successive process of changing the agent's epistemic state as new pieces of information referring to an evolving world arrive. Revision, on the other hand, is assumed to collect new pieces of information referring to a static context and to execute simultaneous change of the epistemic state. Formally, assume that a fully binary change operator $*$ is defined, taking epistemic states in the form of ordinal conditional functions κ and sets of conditionals $\mathcal{S}$ as information input and returning an ordinal conditional function κ^* as posterior epistemic state: $\kappa^* = \kappa * \mathcal{S}$. According to [10], revision of $\kappa * \mathcal{R}$ by $\mathcal{S}$ is then defined by $\kappa * (\mathcal{R} \cup \mathcal{S})$, while updating $\kappa * \mathcal{R}$ by $\mathcal{S}$ amounts to computing $(\kappa * \mathcal{R}) * \mathcal{S}$ as posterior epistemic state. For other recent approaches to update, see, e.g., [11,6].

Besides revision and update, more techniques for changing beliefs have been considered so far. Dubois and Prade deal with focussing as a process of applying generic knowledge to a specific situation [4]. In our framework, focussing can be realized via updating an epistemic state by evidential, i.e. factual information. Note that the nature of the basic change operator $*$ is updating, and that updating is used coherently whenever the new information refers to a shifted context, whereas revision is applied when the information about the same context is refined.

A change operator $*$ that is powerful enough to implement these ideas is the *ordinal c-change operator* [10] which transforms an ordinal conditional function κ and a set of conditional sentences $\mathcal{R}$ into a new ordinal conditional function $\kappa * \mathcal{R}$ accepting these sentences. Note that $\mathcal{R}$ may also include (plausible) facts by identifying factual information A with $(A|\mathit{true})$. C-change operations satisfy all epistemic AGM postulates, as well as the postulates for iterated revision proposed by Darwiche and Pearl (s. [9,2]). The basic idea of a c-change operation is to faithfully respect the structure of conditionals [8]. Simple c-change operations $\kappa^* = \kappa * \mathcal{R}$ of an epistemic state κ and a (finite) set of rules $\mathcal{R} = \{(B_1|A_1), \ldots, (B_n|A_n)\}$ can be obtained by adding to each $\kappa(\omega)$ a constant for each rule $r_i \in \mathcal{R}$ if ω falsifies r_i:

$$\kappa^*(\omega) = \kappa_0 + \kappa(\omega) + \sum_{1 \leqslant i \leqslant n,\ \omega \models A_i \overline{B}_i} \kappa_i^- \tag{2}$$

with a normalizing constant κ_0 and constants κ_i^- (one for each conditional) chosen appropriately to ensure that $\kappa^* \models \mathcal{R}$, for a consistent $\mathcal{R}$. The boolean function *consistencyCheck* : $\mathcal{P}(\mathit{Sen}_{\mathcal{U}}) \to \mathit{Bool}$ tests the consistency of a set of conditionals [5], where $\mathcal{P}(S)$ denotes the power set of a set S. In [10] it is shown how each κ_i^- can be chosen in a unique, computable way that ensures that κ^* from (2) accepts all conditionals in $\mathcal{R}$ while pushing up the degrees of implausibility of worlds falsifying conditionals only minimally.

Therefore, we get a function *cChange* : $\mathit{EpState}_{\mathcal{O}} \times \mathcal{P}(\mathit{Sen}_{\mathcal{U}}) \to \mathit{EpState}_{\mathcal{O}}$ yielding a c-change *cChange*$(\kappa, \mathcal{R})$ for any OCF κ and any consistent set of conditionals $\mathcal{R}$ [10].

4 A Simple Agent Model

When considering the process of iterated belief change, we may assume a discrete time scale as sketched in Figure 1. For any j, κ_j is the agent's epistemic state at time t_j, and $\mathcal{R}_j$ is the new information given to the agent at time t_j. Furthermore, for any j, the agent has to adapt her own epistemic state κ_j in the light of the new information $\mathcal{R}_j$, yielding κ_{j+1}. Now suppose that at times t_i and t_{i+k} the agent receives the information that the world has changed (i.e., a world change occurred between t_{i-1} and t_i and between $t_{i+(k-1)}$ and t_{i+k}). Thus, the changes from κ_{i-1} to κ_i and from $\kappa_{i+(k-1)}$ to κ_{i+k} are done by an update operation. Suppose further, that the world remains static in between, implying that all belief changes from κ_i up to $\kappa_{i+(k-1)}$ are revisions. Furthermore, let

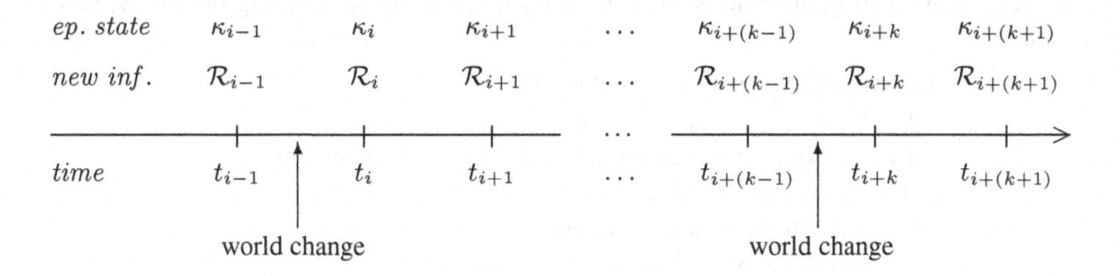

Fig. 1. Time scale for iterated belief change

the world remain static also after t_{i+k}, i.e., the belief change from κ_{i+k} to $\kappa_{i+(k+1)}$ is also achieved by a revision.

Now using a binary belief change operator $*$ as described in Sec. 3 (and that will be implemented in our conceptual agent model by a c-change), we can precisely describe how the agent's epistemic states $\kappa_i, \kappa_{i+1}, \ldots, \kappa_{i+(k-1)}, \kappa_{i+k}, \kappa_{i+(k+1)}$ are obtained:

$$\begin{array}{rlcl}
\mathit{update}: & \kappa_i & = & \kappa_{i-1} * (\mathcal{R}_{i-1}) \\
\mathit{revision}: & \kappa_{i+1} & = & \kappa_{i-1} * (\mathcal{R}_{i-1} \cup \mathcal{R}_i) \\
\mathit{revision}: & \kappa_{i+2} & = & \kappa_{i-1} * (\mathcal{R}_{i-1} \cup \mathcal{R}_i \cup \mathcal{R}_{i+1}) \\
 & \ldots & & \\
\mathit{revision}: & \kappa_{i+(k-1)} & = & \kappa_{i-1} * (\mathcal{R}_{i-1} \cup \mathcal{R}_i \cup \ldots \cup \mathcal{R}_{i+(k-2)}) \\
\mathit{update}: & \kappa_{i+k} & = & \kappa_{i+(k-1)} * (\mathcal{R}_{i+(k-1)}) \\
\mathit{revision}: & \kappa_{i+(k+1)} & = & \kappa_{i+(k-1)} * (\mathcal{R}_{i+(k-1)} \cup \mathcal{R}_{i+k})
\end{array}$$

Therefore, in our conceptual approach, it suffices to consider a simple agent model where the internal state of an agent consists of just three components:

$$\begin{array}{ll}
\mathit{currstate} & \in \mathit{EpState}_{\mathcal{O}} \\
\mathit{stateBeforeUpdate} & \in \mathit{EpState}_{\mathcal{O}} \\
\mathit{newRulesSinceUpdate} & \subseteq \mathit{Sen}_{\mathcal{U}}
\end{array}$$

At any time, *currstate* denotes the current epistemic state of the agent, and *stateBefore-Update* denotes her state before the last update took place. *newRulesSinceUpdate* is the union of all facts and rules that have been given to the agent as new information and that have been used to revise the agents epistemic state since the last update ocurred.

5 Belief Operations

When creating a new agent, at first no knowledge at all might be available. We model this situation by the ranking function *uniform* $\in \mathit{EpState}_{\mathcal{O}}$ representing complete ignorance by regarding all possible worlds as equally pausible, i.e. $\mathit{uniform}(\omega) = 0$ for all $\omega \in \Omega$. If, however, default knowledge is at hand to describe the problem area under consideration, an epistemic state has to be found to appropriately represent this background knowledge. Already here, we can use c-change since an inductive representation of $\mathcal{R}$ can be obtained by a c-change of the epistemic state *uniform* by $\mathcal{R}$. Thus, we can *initialize* the system with an epistemic state by providing a set of sentences $\mathcal{R} \subseteq \mathit{Sen}_{\mathcal{U}}$

and generating a full epistemic state from it by inductively completing the knowledge given by $\mathcal{R}$:

Initialization **Input**: rule base $R \subseteq Sen_{\mathcal{U}}$

if *consistencyCheck*(R) = *false*
then *output*("rule base for initialization is inconsistent")
else *currstate* := *cChange*(*uniform*, R)
stateBeforeUpdate := *uniform*
newRulesSinceUpdate := *rule_base*

For instance, given the rule set $\mathcal{R}$ from Ex. 1 for initialization, the epistemic state $\kappa = uniform * \mathcal{R}$ is computed and yields a model for inference and further belief operations. Note that beside the agent's *currstate*, also the book-keeping variables *stateBeforeUpdate* and *newRulesSinceUpdate* are set accordingly.

Whereas all input sentences are purely qualitative, unquantified sentences in the form of default facts and rules, the agent's answer to *queries* are believed sentences carrying a degree of belief.

Example 2. When asked the query $(f|p)$ (*"Do penguins fly?"*) in the epistemic state κ obtained in Example 1 after initialization, the agent tells us that $(f|p)$ does not belong to the set of believed sentences, i.e. $\kappa \not\models_{\mathcal{O}} (f|p)[0]$; the knowledge base used for building up κ explicitly contains the opposite rule $(\overline{f}|p)$.

On the other hand, asking $(w|k)$ (*"Do kiwis have wings?"*) and $(w|d)$ (*"Do doves have wings?"*) we get a degree of belief 0, i.e. $\kappa \models_{\mathcal{O}} (w|k)[0]$ and $\kappa \models_{\mathcal{O}} (w|d)[0]$. From their superclass *birds*, both kiwis and doves inherit the property of having wings. Note that $\kappa \models_{\mathcal{O}} (B|A)[0]$ indeed expresses the agent's basic belief in $(B|A)$ (cf. (1)).

We now want to demonstrate how our framework offers a means to provide a very clear and precise distinction between update and revision operators in a knowledge processing system. Both operators, update and revision, are implemented by c-changes of an epistemic state by a set of rules; the difference between update and revision lies in the exact specification of the parameters of the c-change operation:

Update **Input**: new rules $R \subseteq Sen_{\mathcal{U}}$

if *consistencyCheck*(R) = *false*
then *output*("rule base for update is inconsistent")
else *stateBeforeUpdate* := *currstate*
newRulesSinceUpdate := R
currstate := *cChange*(*currstate*, R)

Thus, an update operation saves the actual current state (*currstate*) to *stateBeforeUpdate*, initializes the set of new rules (*newRulesSinceUpdate*) to be taken into account for changing the epistemic state to the the new information R, and infers the new current state by a c-change of the actual current state and the new information. A consistency check ensures that the information given to the c-change is not contradictory.

Revise **Input**: new rules $R \subseteq Sen_{\mathcal{U}}$

let R_{all} = *newRulesSinceUpdate* $\cup R$ **in**
if *consistencyCheck*(R_{all}) = *false*
then *output*("rule base for revise is inconsistent")
else *newRulesSinceUpdate* := R_{all}
currstate := *cChange*(*stateBeforeUpdate*, R_{all})

A revise operation, on the other hand, computes the new current state by a c-change of the state being valid before the last update took place (*stateBeforeUpdate*) and the set of all new rules given for that last update and for all following single revision steps (*newRulesSinceUpdate*). Also here, a consistency check ensures that the information given to the c-change is not contradictory. Note also, that *stateBeforeUpdate* is not changed.

Example 3. In Example 2, we found that an agent with the given epistemic state κ believes that kiwis and doves have wings. Now we come back to the scenarios described in Example 1, to make the differences between revision and update, and their consequences, apparent. As described above, the agent first gets the information that her belief on kiwis possessing wings is wrong from a textbook describing the current world. In this case, a new inductive representation for the expanded set $\mathcal{R}' = \{(f|b), (b|p), (\overline{f}|p), (w|b), (b|k), (b|d)\} \cup \{(\overline{w}|k)\}$, i.e. a c-change of *uniform* by $\mathcal{R}'$ has to be computed: $\kappa_1^* = \mathit{uniform} * \mathcal{R}'$. Note that the new information $(\overline{w}|k)$ is not consistent with the prior epistemic state κ but with the context information $\mathcal{R}$ which is refined by $(\overline{w}|k)$.

In the other scenario, the agent wants to update κ with the new information that the wings of newborn doves are nearly completely mutilated, due to some new mysterious illness (that has occurred in the world and that therefore changed the world). The updated epistemic state $\kappa_2^* = \kappa * \{(\overline{w}|d)\}$ is a c-change of κ by $\{(\overline{w}|d)\}$. Working out the details we find that, while the revised state κ_1^*, by construction, still represents the six conditionals that have been known before (and, of course, the new conditional), the updated state κ_2^* only represents the five conditionals $(f|b)$, $(b|p)$, $(\overline{f}|p)$, and $(w|b)$, $(b|k)$, but it no longer accepts $(b|d)$ - since *birds* and *wings* have been plausibly related by the conditional $(w|b)$, the property of not having wings casts (reasonably) doubt on doves being birds. Moreover, the agent is now also uncertain about the ability of doves to fly, as $\kappa_2^*(fd) = \kappa_2^*(\overline{f}d) = 1$. This illustrates that formerly stated, explicit knowledge is kept under revision, but might be given up under update. So, revision and update can be realized within the same scenario, but yielding different results according to the different epistemic roles that have been assigned to the corresponding pieces of information in the change processes.

The process of *diagnosing* a particular case amounts to asking about the status of certain simple facts $D \subseteq \mathit{SimpleFact}_{\mathcal{U}} = \Sigma$ in an epistemic state Ψ under the condition that some particular factual knowledge E (so-called *evidential knowledge*) is given. Thus, an agent makes a diagnosis in the light of some given evidence by uttering her beliefs in the state obtained by adapting her current epistemic state by focussing on the given evidentual facts:

Diagnosis **Input**: evidence $E \subseteq \mathit{Fact}_{\mathcal{U}}$
diagnoses $D \subseteq \mathit{SimpleFact}_{\mathcal{U}}$

---

if *consistencyCheck*(E) = *false*
then *output*("evidential knowledge for diagnosis is inconsistent")
else let *focussedState* = *cChange*(*currstate*, E) **in**
output(*belief*$_{\mathcal{O}}$(*focussedState*, D))

Please note that the focussed epistemic state is only used for answering the particular diagnostic questions; specifically, the agent's current epistemic state (*currstate*) it not changed.

Example 4. Continuing Example 1, we might have the evidence for a penguin and want to ask for the diagnosis whether the penguin has wings: *evidence* $E = \{p\}$ and *diagnoses* $D = \{w\}$. Here, the agent infers $\{w[0]\}$ and therefore the degree of belief 0, i.e., with the evidence *penguin* given, she believes that *wings* also holds.

6 Conclusions and Further Work

Using the notion of the binary c-change operator transforming an epistemic state (represented by an OCF) and a set of default rules $\mathcal{R}$ as new information to an epistemic state accepting $\mathcal{R}$, we developed a model of an intelligent agent. The agent is capable to perform various belief revision operations like update, revision, and focussing (diagnosis).Our approach also contributes to the clarification of the often subtle differences between update and revision both on a conceptual and on a technical level. The constructive definitions of the agent's behaviour yield an operational model. Currently, we are extending a prototype implementation of this agent model to cover also the knowledge discovery functionality of generating rules from data, a process that is inverse to the inductive completion of knowledge given by a set of rules.

References

1. Alchourrón, C.E., Gärdenfors, P., Makinson, P.: On the logic of theory change: Partial meet contraction and revision functions. Journal of Symbolic Logic 50(2), 510–530 (1985)
2. Darwiche, A., Pearl, J.: On the logic of iterated belief revision. Artificial Intelligence 89, 1–29 (1997)
3. DeFinetti, B.: Theory of Probability, vol. 1,2. John Wiley & Sons, Chichester (1974)
4. Dubois, D., Prade, H.: Focusing vs. belief revision: A fundamental distinction when dealing with generic knowledge. In: Nonnengart, A., Kruse, R., Ohlbach, H.J., Gabbay, D.M. (eds.) FAPR 1997 and ECSQARU 1997. LNCS, vol. 1244, pp. 96–107. Springer, Heidelberg (1997)
5. Goldszmidt, M., Pearl, J.: Qualitative probabilities for default reasoning, belief revision, and causal modeling. Artificial Intelligence 84, 57–112 (1996)
6. Hunter, A., Delgrande, J.: Iterated belief change: a transition system apporach. In: Kaelbling, L.P., Saffiotti, A. (eds.) IJCAI 2005, pp. 460–465. Prof. Book Center (2005)
7. Katsuno, H., Mendelzon, A.O.: On the difference between updating a knowledge base and revising it. In: Proc. KR 1991, pp. 387–394. Morgan Kaufmann, San Mateo (1991)
8. Kern-Isberner, G.: Conditionals in nonmonotonic reasoning and belief revision. LNCS (LNAI), vol. 2087. Springer, Heidelberg (2001)
9. Kern-Isberner, G.: A thorough axiomatization of a principle of conditional preservation in belief revision. Annals of Mathematics and Artificial Intelligence 40(1-2), 127–164 (2004)
10. Kern-Isberner, G.: Linking iterated belief change operations to nonmonotonic reasoning. In: Brewka, G., Lang, J. (eds.) Proc. KR 2008, pp. 166–176. AAAI Press, Menlo Park (2008)
11. Lang, J.: Belief update revisited. In: Proceedings of the Twentieth International Joint Conference on Artificial Intelligence, IJCAI 2007, pp. 2517–2522 (2007)
12. Spohn, W.: Ordinal conditional functions: a dynamic theory of epistemic states. In: Causation in Decision, Belief Change, and Statistics, II, pp. 105–134. Kluwer Academic Publ., Dordrecht (1988)

External Sources of Axioms in Automated Theorem Proving

Martin Suda[1,⋆], Geoff Sutcliffe[2,⋆], Patrick Wischnewski[3], Manuel Lamotte-Schubert[3], and Gerard de Melo[3]

[1] Charles University in Prague, Czech Republic
[2] University of Miami, USA
[3] Max-Planck-Institut für Informatik, Germany

Abstract. In recent years there has been a growing demand for Automated Theorem Proving (ATP) in large theories, which often have more axioms than can be handled effectively as normal internal axioms. This work addresses the issues of accessing *external sources of axioms* from a first-order logic ATP system, and presents an implemented ATP system that retrieves external axioms asynchronously, on demand.

1 Introduction

In recent years there has been a growing demand for Automated Theorem Proving (ATP) in large theories. A *large theory* is one that has many functors and predicates, has many axioms of which typically only a few are required for the proof of a theorem, and many theorems to be proved using the axioms. Examples of large theories that are in (or have been translated to) a form suitable for ATP include the SUMO ontology, the Cyc knowledge base, the Mizar mathematical library, the YAGO knowledge base, WordNet, and the MeSH thesaurus. Many of these consist of lots (millions) of atomic facts, mostly positive ground facts. Large theories also arise from dynamic and computational sources of data. Such sources include online knowledge bases, computer algebra systems, lemma discovery tools, bioinformatics databases and tools, and web services. Dynamic and computational sources can provide infinitely many axioms. Large theories pose challenges for ATP, which are different from the challenges of small theory applications of ATP.

This paper addresses the issues of accessing large theories' axioms, stored as *external sources of axioms*, from a first-order ATP system. External sources of axioms provide further challenges for ATP. These include specifying what axioms are (or might be) available, building interfaces to the sources, retrieving and integrating axioms on demand during deduction, and adapting the deduction calculus to deal with axioms becoming available only after deduction has started.

⋆ Work done in the Automation of Logic group, Max-Planck-Institut für Informatik. Martin Suda supported by the Grant Agency of Charles University, grant 9828/2009. Geoff Sutcliffe supported by European Community FP7 grant PIIF-GA-2008-219982.

B. Mertsching, M. Hund, and Z. Aziz (Eds.): KI 2009, LNAI 5803, pp. 281–288, 2009.

2 Abstract System Design

This section describes design choices for a system that accesses external sources of axioms, and highlights the decisions taken for our implementation. (The design choice points are numbered, and the decisions taken are numbered correspondingly.) The general system architecture is composed of a first-order ATP system and external sources of axioms, as shown in Figure 1. The ATP system accepts a problem file containing (i) specifications for external sources of axioms, (ii) axioms to be read and stored internally, and (iii) a conjecture to be proved. It is understood that access to the external sources of axioms will be comparatively slow, and that their retrieval might be incomplete.

The Nature of External Axioms. (1) The *types of axioms* range from simple, e.g., positive ground facts, to highly expressive, e.g., full first-order formulae. The most common type is positive ground facts. (2) *Features of the axioms* can be different from those traditionally found in axiom sets. The axioms may be inconsistent, have varying epistemic and assertional status, and be temporal. (3) The *storage of axioms* can use a range of technologies, such as relational databases, semantic nets, RDF triples, executable programs, and web services.

ATP Systems' Use of External Axioms. (4) An ATP system's *interface to external axioms* has several facets. These include how their availability is specified, what roles they may have, the structure of requests for their retrieval, the format in which they are delivered, and their external manifestation. (5) The decision *when to retrieve* external axioms affects their integration into the ATP process. Advance retrieval allows the external axioms to be treated like internal axioms. Deduction-time retrieval requires that they be integrated dynamically. Deduction-time retrieval can be synchronous, or asynchronous, and can use a pull or push approach. (6) The *criteria for making a request* depend on the nature of the ATP system and its calculus. (7) There are several levels of *granularity for delivery* of external axioms. The finest grain delivery is one axiom at a time. Delivering batches of axioms is possible, but may need to be constrained to deliver less than all the axioms that match a request. Uniqueness requirements might be imposed, and the delivery order might be specified. (8) The ATP system has choices regarding the *storage and retention* of external axioms. These include where they should be stored, and whether they should be stored persistently.

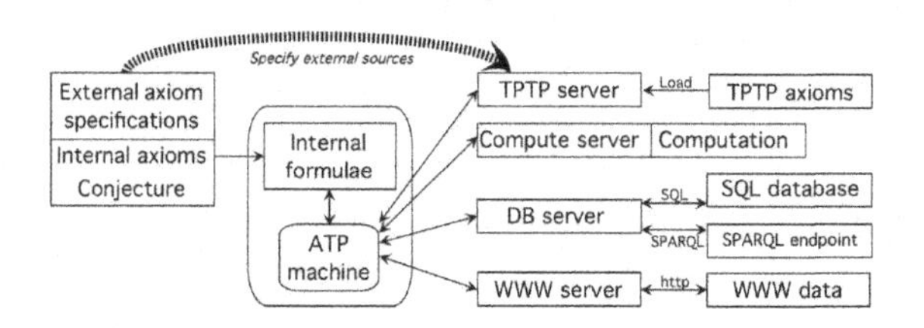

Fig. 1. System Architecture

Design Decisions. The design decisions for this work were often the simple ones, to avoid complication while developing the initial system, and also influenced by the decision to implement in the context of the SPASS ATP system [5], which is a CNF saturation-style ATP system. The choices are:

(1) The external axioms are positive ground facts (ground atoms).
(2) Positive ground facts are necessarily consistent, and it is assumed that the external axioms are consistent with the internal ones. The external axioms are also assumed to be certain, precise, and non-temporal.
(3) No constraints are imposed on the external storage technology.
(4) A TPTP-style specification of the availability of external axioms is used, and formulae with the TPTP role `axiom` are delivered. An extension of the TPTP standard for questions and answers[1] is used for requests and delivery. Initially the external manifestation will be a local process.
(5) External axioms are pulled asynchronously during deduction.
(6) External axioms are requested when a negative literal of the chosen clause (of the ATP system's saturation loop) matches an external specification.
(7) External axioms are delivered in batches, with the possibility of limiting the number of axioms that are delivered. There are no requirements for uniqueness, but no request is issued more than once.
(8) The external axioms are integrated with the internal axioms by adding them to the "Usable" list of the ATP system.

3 Concrete System Design

The availability of external axioms is specified in formulae of the form

`fof(`*external_name*`,external,`
  $\forall\exists$*template*,
  `external(`*type*`,`*access*`,`*protocol*`, [`*useful info*`]) ).`

For example, the following specifies the availability of external axioms that relate creators to their creations.

```
fof(creators,external,
    ! [Thing] : ? [Name] : s__creatorOf(Name,Thing),
    external(exec,'DBServer CreatorDB',tptp_qa,
        [xdb(limit,number(1000)),xdb(limit,cpu(4))]) ).
```

The *external_name* is an arbitrary identifier for the external specification. The `external` role separates the external specification from other types of formulae, such as `axioms` and `conjectures`. The $\forall\exists$*template* specifies the format of the external axioms. The atom is a template for the external axioms, and may contain structure including functions and constants. The universally quantified variables (`!` is the TPTP universal quantifier) must be instantiated when a request is sent, while the existentially quantified variables (`?` is the TPTP existential quantifier) can be filled in the external axioms that are delivered (see below). In the example, requests must instantiate the `Thing` for which the creators' `Names` can be

[1] `http://www.tptp.org/TPTP/Proposals/AnswerExtraction.html`

provided in the axioms delivered. The universal quantifications can be thought of as a way of providing a large (potentially infinite) number of external specifications – an external specification for each combination of values for the universally quantified variables.

The `external()` term specifies the manifestation of the external axiom source, the communication protocol, and constraints on the batch delivery. The *type* specifies the nature of the source. At this stage only locally executed processes are supported, using their `stdin` and `stdout` for requests and delivery. This is specified as `exec`, as in the example. The *access* to the source depends on the *type*, e.g., for an `exec` it provides the command that will start the process (as in the example). The *protocol* specifies the format of requests and deliveries. At this stage only the TPTP question and answer protocol is supported, as described below. This is specified by `tptp_qa`, as in the example. The `[`*useful info*`]` is an optional Prolog-like list of terms, which can be used to store arbitrary information. In particular, `xdb()` terms are used to store auxiliary information that needs to be passed to the external source when requesting axioms.

A problem file may contain multiple external specifications, specifying multiple external sources, external sources with multiple ∀∃*templates*, or the same ∀∃*template* with different `xdb()` terms.

Requests for external axioms are written as TPTP "questions", of the form

```
fof(request_name,question,
   ∃template,
   source,[useful info] ).
```

For example, the following requests axioms for the creator(s) of YAGO.

```
fof(who,question,
   ? [Name] : s__creatorOf(Name,s_YAGOOntology),
   spass,[xdb(limit,number(1000)),xdb(limit,cpu(4))] ).
```

Most of the fields of a request correspond to those of an external specification. The `question` role specifies the intended use of this formula. The *source* is used generally in the TPTP to record where a formula came from. In the example, it records that the request came from the SPASS ATP system. The `xdb()` terms in the *useful info* are copied from the external specification, and thus passed on to the external source.

External axioms are delivered as TPTP "answers", which follow the SZS standards [3]. The complete delivery package is of the form

```
% SZS status Success for source
% SZS answers start InstantiatedFormulae for source
fof(external_name,answer,
   atom,
   answer_to(request_name,[useful info]),[useful info]).
fof(external_name,answer,
   atom,
   answer_to(request_name,[useful info]),[useful info]).
  ⋮
% SZS answers end InstantiatedFormulae for source
```

For example, the following delivers axioms for the creators of YAGO.

```
% SZS status Success for spass
% SZS answers start InstantiatedFormulae for spass
fof(creators,answer,
   s__creatorOf(s_FabianMartinSuchanek,s_YAGOOntology),
   answer_to(who,[]),[]).
fof(creators,answer,
   s__creatorOf(s_GjergjiKasneci,s_YAGOOntology),
   answer_to(who,[]),[]).
% SZS answers end InstantiatedFormulae for spass
```

Most of the fields of a delivery correspond to those of an external specification. The `answer` role specifies the intended use of this formula.

The algorithm for a CNF saturation-style ATP system, integrating asynchronous requests and delivery of external axioms, is shown in Figure 2.

```
1  foreach ES ∈ ExternalSources do
2      ES.WaitingForDelivery = False;
3      ES.RequestQueue = ∅;
4  end
5  while ¬SolutionFound & (Usable | *.WaitingForDelivery | *.RequestQueue) do
6      repeat
7          foreach ES ∈ ExternalSources do
8              if ES.WaitingForDelivery & Axioms = ES.CompleteDelivery then
9                  Add Axioms to Usable;
10                 ES.WaitingForDelivery = False;
11             end
12             if ¬ES.WaitingForDelivery & ES.RequestQueue then
13                 Send(Dequeue(ES.RequestQueue));
14                 ES.WaitingForDelivery = True;
15             end
16         end
17         if ¬Usable & *.WaitingForDelivery then sleep(1);
18     until Usable | ¬*.WaitingForDelivery ;
19     if ¬Usable then break;
20     Move ChosenClause from Usable to WorkedOff;
21     foreach NL ∈ ChosenClause.NegativeLiterals do
22         if NL matches any ES ∈ ExternalSources then
23             Build Request for NL;
24             if Request not repeated then Enqueue(ES.RequestQueue,Request);
25         end
26         Do inferencing with ChosenClause and WorkedOff;
27     end
28 end
```

Fig. 2. Integrating external sources into a saturation-style ATP system

The algorithm augments a standard ATP saturation loop algorithm with steps to accept deliveries, send requests, and queue requests. The loop control is modified accordingly. Specific points to note are:

- The condition of the **while** loop is augmented to keep going if more external axioms might be delivered.
- Each time around the **while** loop, each external source is checked for axioms delivered, and for requests to send. Only one request is sent to each external source at a time, so that there is no reliance on buffering in the communication channel.
- The request queue of each external source can be a priority queue.
- All negative literals (rather than, e.g., only the selected literal in ordered resolution) are examined for a match with an external source. This causes external axioms to be requested and delivered "preemptively". It would also be acceptable to simply ensure that at least one request is enqueued.
- The "matches" condition can be as precise as desired. The intention is to identify cases when external axioms might resolve against the negative literal.
- The "not repeated" condition can be as precise as desired, ranging from syntactic equivalence through to notions of unifiability and subsumption.

The soundness of the algorithm follows from its soundness without the steps for accessing external axioms. The notion of completeness is somewhat different in this setting, but if all the external axioms used were internal axioms, and were put in the same place in the Usable list as when delievered as external axioms, the same deductions would be performed. In practice, external sources might not deliver all possible axioms, and therefore a saturation cannot be taken to mean that there is no proof.

4 Implementation and Testing

The algorithm given in Figure 2 has been implemented as SPASS-XDB, by modifying the well-known SPASS ATP system [5].[2] The implementation is based on SPASS 3.5, which can read TPTP format data. Each external source of axioms is represented by a `struct`, containing:

- The execution string and PID of the external source process, which is started using a standard `fork` and `exec` sequence.
- Unnamed pipes for the `stdin` and `stdout` of the external source process.
- A flag indicating if a request has been sent, for which there has not been a delivery.
- A queue of requests waiting to be sent.
- A list of sent requests, used to prevent sending duplicate requests.

The queue of requests is implemented as a priority queue, ordered so that lighter requests with less variables get higher priority. The effect is to give priority to

[2] An implementation based on the E prover [2] is also planned.

requests that are likely to deliver less axioms, which curbs the ATP system's search space. SPASS' constraints on the search space are relaxed in various ways, to overcome various forms of incompleteness that stem from SPASS' ignorance of the availability of external axioms, which are added during the deduction (SPASS, like most (all?) ATP systems, was designed with the assumption that all formulae are in the problem file). For the types of problems that SPASS-XDB is designed to solve, relaxing the constraints does not have a very deleterious effect, because the proofs are typically quite shallow (in contrast to the deep proofs that are common in traditional applications of ATP).

Five external sources of axioms have been implemented. The YAGO knowledge base has provided a lot (around 14.5 million) of ground facts about the world. The facts were mostly mined from Wikipedia, and were exported in TPTP format with symbols being renamed and axioms transformed to match the SUMO ontology [1]. Two implementations have been written to serve YAGO axioms. The first is a Prolog implementation that reads YAGO axioms into its database, and delivers them based on unification with a request. The second is a Java interface to a MySQL database containing YAGO axioms, with a relation for each predicate. A request is converted into a `SELECT...WHERE` statement that extracts matching tuples, which are then delivered in TPTP format.

Two computational sources of axioms have been implemented in Prolog. The first implements evaluation of ground arithmetic expressions, thus providing arithmetic capabilities to SPASS-XDB (support for arithmetic is notoriously weak in ATP systems). The second implements syntactic difference of terms, which is used to determine if two terms look different. This provides a controlled way to implement a unique names assumption.

A web service that calls the Yahoo Maps Web Service[3] has been implemented via a Prolog mediator. This provides axioms containing the latitude, longitude, official city name, and country code, for a given common city name. The mediator converts a request to an HTTP request that is posted to Yahoo, and converts Yahoo's XML result to a TPTP axiom.

4.1 Testing

SPASS-XDB has been tested on several problems. One illustrative example is given here. The problem is to name an OECD country's capital that is at the same latitude (rounded to the nearest degree) as Moscow, and is subject to flooding. The names of the 27 OECD countries are retrieved from an external source of 4.5 million YAGO entity-property axioms. The capital of each country is retrieved from an external source of YAGO axioms. The YAGO identifier for a capital is translated to its common English name using an external source of YAGO axioms for "meaning in English" - this external source has 2.7 million axioms. Given the common English name of a city, it's latitude is obtained using the Yahoo Maps source of axioms, and the real values obtained from these axioms are rounded to the nearest degree using the external source of ground

[3] `http://developer.yahoo.com/maps/`

arithmetic evaluation. Internal SUMO axioms that specify that coastal cities are near the sea, that a sea is a body of water, a body of water is a water area, things near water areas can get flooded, and class membership is inherited, are used with an external source of YAGO axioms about which cities in the world are coastal, to establish if a city is subject to flooding. This aspect requires full first-order reasoning, rather than simple data retrieval. Finally, the external source of syntactic difference axioms is consulted to prevent Moscow being reported as the OECD capital.[4] In the proof search 84 requests are queued, 52 requests were sent, and 337 external axioms are delivered. The axioms delivered break down as 27 OECD countries, 15 capital cities, 6 common English city names, 4 latitudes, 3 arithmetic evaluations, 280 coastal cities, and 2 confirmations of different city names. The proof takes 11.2 CPU seconds, 10.9s in SPASS and 0.3s in external axiom sources. 4667 clauses were derived by SPASS.

5 Conclusion

This paper has presented the design and implementation of an ATP system that retrieves axioms from external sources, asynchronously, on demand. The design decisions were taken from an analysis of many choice points. The testing shows that interesting problems can be solved, using external sources of axioms that cannot be handled effectively as normal internal axioms in an ATP system.

There is evidence that automated reasoning for large theories is growing in importance. Current classical first-order ATP systems, such as those that compete in the annual CADE ATP system competition [4], are unable to work in extremely large theories. This work provides a step forward in ATP capability – even if access to external axioms is slower than in-memory access, waiting for a proof is better than no hope for a proof at all.

References

1. de Melo, G., Suchanek, F., Pease, A.: Integrating YAGO into the Suggested Upper Merged Ontology. In: Chung, S. (ed.) Proceedings of the 20th IEEE International Conference on Tools with Artificial Intelligence, pp. 190–193 (2008)
2. Schulz, S.: E- A Brainiac Theorem Prover. AI Communications 15(2-3), 111–126 (2002)
3. Sutcliffe, G.: The SZS Ontologies for Automated Reasoning Software. In: Sutcliffe, G., Rudnicki, P. (eds.) Proceedings of the Workshop on Knowledge Exchange: Automated Provers and Proof Assistants, CEUR Workshop Proceedings, vol. 418, pp. 38–49 (2008)
4. Sutcliffe, G.: The 4th IJCAR Automated Theorem Proving System Competition - CASC. AI Communications 22(1), 59–72 (2009)
5. Weidenbach, C., Fietzke, A., Kumar, R., Suda, M., Wischnewski, P., Dimova, D.: SPASS Version 3.5. In: Schmidt, R. (ed.) Proceedings of the 22nd International Conference on Automated Deduction. LNCS (LNAI), vol. 5663, pp. 140–145. Springer, Heidelberg (2009)

[4] The answer is Copenhagen, in case you are trying to work it out.

Presenting Proofs with Adapted Granularity*

Marvin Schiller[1] and Christoph Benzmüller[2]

[1] German Research Center for Artificial Intelligence (DFKI), Bremen, Germany
[2] International University in Germany, Bruchsal, Germany
Marvin.Schiller@dfki.de, c.benzmueller@googlemail.com

Abstract. When mathematicians present proofs they usually adapt their explanations to their didactic goals and to the (assumed) knowledge of their addressees. Modern automated theorem provers, in contrast, present proofs usually at a fixed level of detail (also called granularity). Often these presentations are neither intended nor suitable for human use. A challenge therefore is to develop user- and goal-adaptive proof presentation techniques that obey common mathematical practice. We present a flexible and adaptive approach to proof presentation based on classification. Expert knowledge for the classification task can be hand-authored or extracted from annotated proof examples via machine learning techniques. The obtained models are employed for the automated generation of further proofs at an adapted level of granularity.

Keywords: Adaptive proof presentation, proof tutoring, automated reasoning, machine learning, granularity.

1 Introduction

A key capability trained by students in mathematics and the formal sciences is the ability to conduct rigorous arguments and proofs and to present them. The presentation of proofs in this context is usually highly adaptive as didactic goals and the (assumed) knowledge of the addressee are taken into consideration. Modern theorem proving systems, in contrast, do often not sufficiently address this common mathematical practice. In particular automated theorem provers typically generate and present proofs only using very fine-grained and machine-oriented calculi. Of course, some theorem proving systems exists — amongst them prominent interactive theorem provers such as Isabelle/HOL[1], HOL[2], Coq[3], and Theorema[4] — that provide means for human-oriented proof presentations. Nevertheless the challenge of supporting user- and goal-adapted proof presentations has been widely neglected in the past. This constitutes an unfortunate gap, in particular since mathematics and the formal sciences are

* This work was supported by a grant from *Studienstiftung des Deutschen Volkes e. V.*
[1] http://www.cl.cam.ac.uk/research/hvg/Isabelle/
[2] http://hol.sourceforge.net/
[3] http://coq.inria.fr/
[4] http://www.risc.uni-linz.ac.at/research/theorema/

B. Mertsching, M. Hund, and Z. Aziz (Eds.): KI 2009, LNAI 5803, pp. 289–297, 2009.

increasingly targeted as promising application areas for intelligent tutoring systems. We briefly illustrate the challenge with an example. In the elementary proof in basic set theory in Fig. 1 in Bartle & Sherbert's introductory textbook [1], intermediate proof steps are skipped when this seems appropriate: whereas most of the proof steps consist of the application of exactly one mathematical fact (e.g., a definition or a lemma, such as the distributivity of *and* over *or*), the step from assertion 9 to assertion 10 applies several inference steps at once, namely the application of the definition of $\cap$ twice, and then using the distributivity of *and* over *or*.

Similar observations were made in the empirical studies within the DIALOG project [2], where tutors (mathematicians hired to help simulate the dialog system) identified limits for how many inference steps are to be allowed at once. An example from our DIALOG corpus [3] for a correct but unacceptably large student step that was rejected by the tutor is presented to the right.

student: $(x, y) \in (R \circ S)^{-1}$
tutor: `Now try to draw inferences from that!`
student: $(x, y) \in S^{-1} \circ R^{-1}$
tutor: `One cannot directly deduce that.`

The challenge thus consists in (i) developing means to model and assess different levels of proof granularity, (ii) providing support for the interactive or even automated acquisition of such models from well chosen proof examples, and (iii) combining these aspects with natural language (NL) generation techniques to present machine generated proofs at adaptive levels of granularity to humans.

Related work has addressed this challenge only to a moderate extent. The ΩMEGA system [4], for example, provides a hierarchically organized proof data structure that allows to represent proofs at different levels of granularity which are maintained simultaneously in the system. And ΩMEGA's proof explanation system P.rex [5] was able to generate adapted proof presentations by moving up or down these layers on request. The problem remains, however, of how to identify a particular level of granularity, how to model it, and how to ensure that this level of granularity is appropriate. A similar observation applies to the Edinburgh HiProofs system [6]. One particular level of proof granularity has been proposed by Autexier and Fiedler [7], which, in brief, refers to assertion level proofs where all assertion level inference steps are spelled out explicitly and refer only to facts readily available from the assertions or the previous inference

1 Let x be an element of $A \cap (B \cup C)$, 2 then $x \in A$ and $x \in B \cup C$. 3 This means that $x \in A$, and either $x \in B$ or $x \in C$. 4 Hence we either have (i) $x \in A$ and $x \in B$, or we have (ii) $x \in A$ and $x \in C$. 5 Therefore, either $x \in A \cap B$ or $x \in A \cap C$, so 6 $x \in (A \cap B) \cup (A \cap C)$. 7 This shows that $A \cap (B \cup C)$ is a subset of $(A \cap B) \cup (A \cap C)$. 8 Conversely, let y be an element of $(A \cap B) \cup (A \cap C)$. 9 Then, either (iii) $y \in A \cap B$, or (iv) $y \in A \cap C$. 10 It follows that $y \in A$, and either $y \in B$ or $y \in C$. 11 Therefore, $y \in A$ and $y \in B \cup C$ 12 so that $y \in A \cap (B \cup C)$. 13 Hence $(A \cap B) \cup (A \cap C)$ is a subset of $A \cap (B \cup C)$. 14 In view of Definition 1.1.1, we conclude that the sets $A \cap (B \cup C)$ and $(A \cap B) \cup (A \cap C)$ are equal.

Fig. 1. Proof of the statement $A \cap (B \cup C) = (A \cap B) \cup (A \cap C)$, reproduced from [1]

steps (*what-you-need-is-what-you-stated* granularity). However, they conclude that even the simple proof in Fig. 1 cannot be fully captured by their rigid notion of proof granularity.

In this paper we present a flexible and adaptive framework for proof presentation (currently used for formal proofs, not diagrammatic proofs etc.). Our approach employs rule sets (granularity classifiers) to model different levels of proof granularity. These rule sets are employed in a straightforward proof assessment algorithm to convert machine generated proofs (in our ΩMEGA system) into proofs at specific levels of detail. Both the granularity rules and the algorithm are outlined in Sect. 2. In Sect. 3 we show that our approach can successfully model the granularity of our running example proof in Fig. 1. Different models for granularity can either be hand-coded or they may be learned from samples using machine learning techniques. Ideally, the latter approach, which is described in Sect. 4, helps reducing the effort of adapting the system to new application and user contexts, and, in particular, to train the system by domain experts who are not familiar with expert systems.

2 An Adaptive Model for Granularity

We treat the granularity problem as a classification task: given a proof step, representing one or several assertion applications[5], we judge it as either *appropriate*, *too big* or *too small*. As our feature space we employ several mathematical and logical aspects of proof steps, but also aspects of cognitive nature. For example, we keep track of the background knowledge of the user in a basic (overlay) student model. We illustrate our approach with a proof step from Fig. 1: 10 is derived from 9 by applying the definition of $\cap$ twice, and then using the distributivity of *and* over *or*. In this step (which corresponds to multiple assertion level inference steps) we make the following observations:

(i) involved are two concepts: def. of $\cap$ and distributivity of *and* over *or*,
(ii) the total number of assertion applications is three,
(iii) all involved concepts have been previously applied in the proof,
(iv) all manipulations apply to a common part in 9,
(v) the names of the applied concepts are not explicitly mentioned, and
(vi) two of the assertion applications belong to *naive set theory* (def. of $\cap$) and one of them relates to the domain of propositional logic (distributivity).

These observations are represented as a feature vector, where, in our example, the feature "distinct concepts" receives a value of "2", and so forth. Currently, our system computes the following set of features for each (single- or multi-inference) proof step:

[5] We use the notion of *assertion application* for inference steps that are justified by a mathematical fact (such as a definition, theorem or a lemma).

total: the total number of (assertion level) inference steps combined into one proof step,
conceptsunique: the number of different concepts applied within the proof step,
mastered-concepts-unique (m.c.u.): the number of different employed mathematical facts assumed to be known to the user according to the very basic user model (which is updated in the course of the proof[6]),
unmastered-concepts-unique (unm.c.u.): the number of different employed mathematical facts assumed to be unknown to the user,
verb: whether the step is accompanied by a verbal explanation,
unverbalized-unknown: the number of assertions not accompanied by a verbal description and not known to the user,
lemmas: the number of employed assertions that are lemmas (in contrast to basic definitions),
hypintro: indicates whether a (multi-inference) proof step introduces a new hypothesis,
subgoals: indicates whether (and how many) new subgoals have been introduced,
same-subformula: indicates whether all manipulations apply to a common formula part,
newinst: indicates whether a variable has been instantiated,
close: indicates whether a branch of the proof has been finished,
parallelism: indicates when it is possible to apply the same assertion several times, but it is applied only on fewer occasions than possible,
forward: indicates inference applications in forward direction,
backward: indicates inference applications in backward direction,
direction-change: indicates whether the direction of inferences has changed w.r.t. the previous step,
step-analog: indicates whether the assertions applied within the current step have been applied before in the proof, in the same order and as a single step,
multi-step-analog: indicates whether the assertions applied within the current step have been applied before in the proof, in the same order, but not necessarily within a single step,
settheory, relations, topology, geometry, etc.: the number of inference applications from each (mathematical) domain,
$\cap$-Defn, $\cup$-Defn, eq-Defn, etc.: indicator feature for each concept.

These features were motivated by the corpora obtained from the experiments in the DIALOG project (cf. [2] and [3]) and discussions with domain experts. We express our models for classifying granularity as rule sets (cf. Fig. 4), which associate specific combinations of feature values to a corresponding granularity verdict ("appropriate", "too big" or "too small"). Our straightforward algorithm

[6] All concepts that were employed in an "appropriate" or "too small" proof step obtain the status of being known in the subsequent proof steps/proofs.

$$
\begin{array}{rcccl}
 & x \in \mathbf{S} \vdash x \in \mathbf{S} & & y \in \mathbf{T} \vdash y \in \mathbf{T} & \\
\hline
\textsc{Def}\cup\ (8) & (x \in (A \cap B) \vee x \in (A \cap C)) \vdash x \in \mathbf{S} & & (y \in A \wedge y \in (B \cup C)) \vdash y \in \mathbf{T} & \textsc{Def}\cap\ (15) \\
\hline
\textsc{Def}\cap\ (7) & (x \in (A \cap B) \vee x \in A \wedge x \in C) \vdash x \in \mathbf{S} & & (y \in A \wedge (y \in B \vee y \in C)) \vdash y \in \mathbf{T} & \textsc{Def}\cup\ (14) \\
\hline
\textsc{Def}\cap\ (6) & (x \in A \wedge x \in B \vee x \in A \wedge x \in C) \vdash x \in \mathbf{S} & & (y \in A \wedge y \in B \vee y \in A \wedge y \in C) \vdash y \in \mathbf{T} & \text{DISTR}\ (13) \\
\hline
\text{DISTR}\ (5) & (x \in A \wedge (x \in B \vee x \in C)) \vdash x \in \mathbf{S} & & (y \in A \wedge y \in B \vee y \in (A \cap C)) \vdash y \in \mathbf{T} & \textsc{Def}\cap\ (12) \\
\hline
\textsc{Def}\cup\ (4) & (x \in A \wedge x \in (B \cup C)) \vdash x \in \mathbf{S} & & (y \in (A \cap B) \vee y \in (A \cap C)) \vdash y \in \mathbf{T} & \textsc{Def}\cap\ (11) \\
\hline
\textsc{Def}\cap\ (3) & (x \in (A \cap (B \cup C))) \vdash x \in \mathbf{S} & & (y \in ((A \cap B) \cup (A \cap C))) \vdash y \in \mathbf{T} & \textsc{Def}\cup\ (10) \\
\hline
\textsc{Def}\subseteq\ (2) & \vdash (A \cap (B \cup C)) \subseteq \mathbf{S} & & \vdash ((A \cap B) \cup (A \cap C)) \subseteq \mathbf{T} & \textsc{Def}\subseteq\ (9) \\
\hline
\textsc{Def Eq}\ (1) & \multicolumn{3}{c}{\vdash \underbrace{(A \cap (B \cup C))}_{\mathbf{T}} = \underbrace{((A \cap B) \cup (A \cap C))}_{\mathbf{S}}} &
\end{array}
$$

Fig. 2. Assertion level proof for the statement $A \cap (B \cup C) = (A \cap B) \cup (A \cap C)$

for granularity-adapted proof presentation takes two arguments, a granularity rule set and an assertion level proof[7] as generated by ΩMEGA.

The assertion level proof generated by ΩMEGA for our running example is given in Fig. 2; this proof is represented as a tree (or acyclic graph) in sequent-style notation and the proof steps are ordered. Currently we only consider plain assertion level proofs, and do not assume any prior hierarchical structure or choices between proof alternatives (as is possible in ΩMEGA). Our algorithm performs an incremental categorization of steps in the proof tree (where $n = 0, \ldots, k$ denotes the ordered proof steps in the tree; initially n is 1):

while there exists a proof step n **do**

evaluate the granularity of the compound proof step n (i.e., the proof step consisting of all assertion level inferences performed after the last step labeled "appropriate with explanation" or "appropriate without explanation" — or the beginning of the proof, if none exists yet) with the given rule set under the following two assumptions: (i) the involved concepts are mentioned in the presentation of the step (an *explanation*), and (ii) only the resulting formula is displayed.

1. **if** n is appropr. w. expl.
   then label n as "appropr. w. expl."; set $n := n+1$;
2. **if** n is too small w. expl., but appropr. wo. expl.
   then label n as "appropr. wo. expl."; set $n := n+1$;
3. **if** n is too small both w. and wo. explanation
   then label n as "too small"; set $n := n+1$;
4. **if** n is too big **then** label $n-1$ as "appropr. wo. expl." (i.e. consider the previous step as appropr.), unless $n-1$ is labeled "appropr. w. expl." or "appropr. wo. expl." already or n is the first step in the proof (in this special case label n as "appropr. w. expl." and set $n := n+1$).

[7] In principle, our approach is not restricted to assertion level proofs and is also applicable to other proof calculi. However, in mathematics education we consider single assertion level proof steps as the finest granularity level of interest. We gained evidence for this choice from the empirical investigations in the DIALOG project (cf. [2] and [3]).

We thereby obtain a proof tree with labeled steps (labeled nodes) which differentiates between those steps that are categorized as appropriate for presentation and those which are considered too fine-grained. Proof presentations are generated by walking through the tree,[8] skipping the steps labeled *too small*.[9]

3 Modeling the Granularity of Our Example Proof

We exemplarily model the granularity of the textbook proof in Fig. 1. Starting point is the initial assertion level proof from Fig. 2. This proof assumes the basic definitions and concepts in naive set theory (such as equality, subset, union, intersection and distributivity) and first-order logic. Notice that the Bartle & Sherbert proof in Fig. 1 starts in 1 with the assumption that an element x is in the set $A \cap (B \cup C)$. The intention is to show the subset relation $A \cap (B \cup C) \subseteq (A \cap B) \cup (A \cap C)$, which is not explicitly revealed until step 7, when this part of the proof is already finished. The same style of *delayed* justification for prior steps is employed towards the end of the proof, where statements 13 and 14 justify (or recapitulate) the preceding proof. For the comparison of proof step granularity in this paper, however, we consider a re-ordered variant of the steps in Fig. 1, which is displayed in Fig. 3 (a).[10] We now employ suitable granularity rule sets to automatically generate a proof presentation from our ΩMEGA assertion level proof which exactly matches the twelve steps of the Bartle & Sherbert proof, skipping intermediate proof steps according to our feature-based granularity model. Fig. 4 shows two sample rule sets which both lead to the automatically generated proof presentation in Fig. 3 (b). For instance, the three assertion level steps (11), (12) and (13) in the initial assertion level proof are combined into one single step from 9. to 10. in the proof presentation in Fig. 3 (b), like in the textbook proof. The rule set in Fig. 4 (a) was generated by hand, whereas the rule set in Fig. 4 (b) was automatically learned[11] (cf. Sect. 4). The rules are ordered by utility for conflict resolution. Note that rules 4–6 in Fig. 4 (a) express the relation between the appropriateness of steps and whether the employed concepts are mentioned verbally (feature *verb*), e.g. rule 6) enforces that the definition of equality is explicitly mentioned (as in step 1. in Fig 3 (b)).

[8] In case of several branches, a choice is possible which subtree to present first, a question which we do not address in this paper.

[9] Even though the intermediate steps which are *too small* are withheld, the presentation of the output step reflects the results of all intermittent assertion applications, since we include the names of all involved concepts whenever a (compound) step is appropriate with explanation.

[10] Note that step (1) in the re-ordered proof corresponds to the statements 7, 13 and 14 in the original proof which jointly apply the concept of set equality. The ordering of proof presentations can be dealt with using dialog planning techniques, as explored in [5].

[11] The sample proof was used to fit a rule set to it via C5.0 machine learning [8]. All steps in the sample proof were provided as training instances with label *appropriate*, all tacit intermediate assertion level steps were labeled as *too small*, and always the next bigger step to each step in the original proof was provided as a *too big* step.

1. In view of Definition 1.1.1, we [show] that the sets $A \cap (B \cup C)$ and $(A \cap B) \cup (A \cap C)$ are equal. 14 [First we show] that $A \cap (B \cup C)$ is a subset of $(A \cap B) \cup (A \cap C)$. 7 [Later we show] $(A \cap B) \cup (A \cap C)$ is a subset of $A \cap (B \cup C)$. 13
2. Let x be an element of $A \cap (B \cup C)$, 1
3. then $x \in A$ and $x \in B \cup C$. 2
4. This means that $x \in A$, and either $x \in B$ or $x \in C$. 3
5. Hence we either have (i) $x \in A$ and $x \in B$, or we have (ii) $x \in A$ and $x \in C$. 4
6. Therefore, either $x \in A \cap B$ or $x \in A \cap C$, 5
7. so $x \in (A \cap B) \cup (A \cap C)$. 6
8. Conversely, let y be an element of $(A \cap B) \cup (A \cap C)$. 8
9. Then, either (iii) $y \in A \cap B$, or (iv) $y \in A \cap C$. 9
10. It follows that $y \in A$, and either $y \in B$ or $y \in C$. 10
11. Therefore, $y \in A$ and $y \in B \cup C$, 11
12. so that $y \in A \cap (B \cup C)$. 12

(a)

1. We show that $((A \cap B) \cup (A \cap C) \subseteq A \cap (B \cup C))$ and $(A \cap (B \cup C) \subseteq (A \cap B) \cup (A \cap C))$ ...because of definition of equality
2. We assume $x \in A \cap (B \cup C)$ and show $x \in (A \cap B) \cup (A \cap C)$
3. Therefore, $x \in A \wedge x \in B \cup C$
4. Therefore, $x \in A \wedge (x \in B \vee x \in C)$
5. Therefore, $x \in A \wedge x \in B \vee x \in A \wedge x \in C$
6. Therefore, $x \in A \cap B \vee x \in A \cap C$
7. We are done with the current part of the proof (i.e., to show that $x \in (A \cap B) \cup (A \cap C)$). [It remains to be shown that $(A \cap B) \cup (A \cap C) \subseteq A \cap B \cup C$]
8. We assume $y \in (A \cap B) \cup (A \cap C)$ and show $y \in A \cap (B \cup C)$
9. Therefore, $y \in A \cap B \vee y \in A \cap C$
10. Therefore, $y \in A \wedge (y \in B \vee y \in C)$
11. Therefore, $y \in A \wedge y \in B \cup C$
12. This finishes the proof. Q.e.d.

(b)

Fig. 3. Comparison between (a) the (re-ordered) proof by Bartle and Sherbert [1] and (b) the proof presentation generated with our rule set from the ΩMEGA proof in Fig. 2

All other cases, which are not covered by the previous rules, are subject to a default rule. Natural language is produced here via simple patterns and more exciting natural language generation is easily possible with Fiedler's mechanisms [5]. The rule sets in Fig. 4 can be successfully reused for other examples in the domains as well (as demonstrated with a different proof exercise in [9]).

4 Learning from Empirical Data

We employ off-the-shelf machine learning tools to learn granularity rule sets (classifiers) from annotated examples (*supervised* learning), i.e. proof steps with the labels *appropriate*, *too small* or *too big*. Currently, our algorithm calls the C5.0 data mining tools [8]. To assess the performance of learning classifiers from human judgments, we have conducted a study where a mathematician (with tutoring experience) judged the granularity of 135 proof steps. These steps were presented to the mathematician via an ΩMEGA-assisted environment which computed the feature values for granularity classification in the background. The steps were (with some exceptions) generated at random step size, such that each presented step corresponded to one, two, or three assertion level inference steps (we also included a few single natural deduction (ND) steps for comparison[12]).

[12] We found that, unlike the assertion-level steps, single natural deduction steps were mostly rated as "too small" by the expert.

(a)

1) hypintro=1 $\wedge$ total> 1 $\Rightarrow$ too-big
2) $\cup$-Defn$\in\{1,2\}\wedge\cap$-Defn$\in\{1,2\}$ $\Rightarrow$ too-big
3) $\cap$-Defn< 3 $\wedge$ $\cup$-Defn=0 $\wedge$ m.c.u.=1 $\wedge$ unm.c.u.=0 $\Rightarrow$ too-small
4) total<2 $\wedge$ verb=true $\Rightarrow$ too-small
5) m.c.u.<3 $\wedge$ unm.c.u.=0 $\wedge$ verb=true $\Rightarrow$ too-small
6) eq-Defn>0 $\wedge$ verb=false $\Rightarrow$ too-big
7) _ $\Rightarrow$ app.

(b)

1) conceptsunique$\in\{0,1\}$ $\wedge$ eq-Defn=0 $\wedge$ verb=true $\Rightarrow$ too-small
2) hypintro=0 $\wedge$ eq-Defn=0 $\wedge$ $\cup$-Defn=0 $\wedge$ verb=true $\Rightarrow$ too-small
3) conceptsunique $\in\{2,3,4\}$ $\wedge$ $\cup$-Defn $\in\{1,2,3\}$ $\Rightarrow$ too-big
4) hypintro $\in\{1,2,3,4\}$ $\wedge$ conceptsunique $\in\{2,3,4\}$ $\Rightarrow$ too-big
5) unm.c.u.=0 $\wedge$ total $\in\{0,1,2\}$ $\cap$-Defn $\in\{1,2\}$ $\wedge$ close=false $\Rightarrow$ too-small
6) eq-Defn$\in\{1,2\}$ $\wedge$ verb=false $\Rightarrow$ too-big
7) eq-Defn$\in\{1,2\}$ $\wedge$ verb=true $\Rightarrow$ app.
8) eq-Defn=0 $\wedge$ verb=false $\Rightarrow$ app.
9) _ $\Rightarrow$ app.

Fig. 4. Rule sets for our running example: (a) rule set generated by hand, (b) rule set generated the using C5.0 data mining tool (ordered by the rules' confidence values)

The presented proofs belonged to one exercise in naive set theory and three different exercises about binary relations. We used the Weka suite[13] to compare the performance of the PART classifier [10] which is inspired by Quinlan's C4.5 to the support vector machines implementation SMO [11], resulting in 86.9% and 85.4% of correct classification and Cohen's (unweighted) $\kappa = 0.65$ and $\kappa = 0.61$, respectively, in 10-fold cross validation, using only the 130 steps that were generated from assertion-level inferences (excluding the single ND steps). The results were achieved after we excluded some of the attributes (in particular those that refer to the use of specific concepts, i.e., Def. of $\cap$, Def. of $\circ$, etc.), which were relevant only in some of the exercises (possibly hampering generalizability of the learned classifiers), otherwise we obtained slightly worse 85.4% of correct classification and $\kappa = 0.61$ with PART.

5 Conclusion

Granularity has been a challenge in AI for decades [12,13]. Here we have focused on adaptive proof granularity, which we treat as a classification problem. We model different levels of granularity using rule sets, which can be hand-authored or learned from sample proofs. Our granularity classifiers are applied dynamically to proof steps, taking into account changeable information such as the user's familiarity with the involved concepts. Using assertion level proofs as the basis for our approach is advantageous for the generation of natural language output, and the relevant information for the classification task (e.g., the concept names) is easily read off the proofs. Future work consists in further empirical evaluations of the learning approach — to address the questions: (i) what are the most useful features for judging granularity, and are they different among distinct experts and

[13] `http://www.cs.waikato.ac.nz/~ml/weka/`

domains, and (ii) what is the inter-rater reliability among different experts and the corresponding classifiers generated by learning in our framework? The resulting corpora of annotated proof steps and generated classifiers can then be used to evaluate the appropriateness of the proof presentations generated by our system.

Acknowledgments. We thank four anonymous reviewers for their useful comments, and Marc Wagner and Claus-Peter Wirth for internal review.

References

1. Bartle, R.G., Sherbert, D.: Introduction to Real Analysis, 2nd edn. Wiley, Chichester (1982)
2. Benzmüller, C., Horacek, H., Kruijff-Korbayová, I., Pinkal, M., Siekmann, J.H., Wolska, M.: Natural language dialog with a tutor system for mathematical proofs. In: Lu, R., Siekmann, J.H., Ullrich, C. (eds.) Cognitive Systems: Joint Chinese-German Workshop. LNCS (LNAI), vol. 4429, pp. 1–14. Springer, Heidelberg (2007)
3. Benzmüller, C., Horacek, H., Lesourd, H., Kruijff-Korbayová, I., Schiller, M., Wolska, M.: A corpus of tutorial dialogs on theorem proving; the influence of the presentation of the study-material. In: Proc. Intl. Conference on Language Resources and Evaluation (LREC 2006), Genoa, Italy, ELDA (2006)
4. Autexier, S., Benzmüller, C., Dietrich, D., Meier, A., Wirth, C.P.: A generic modular data structure for proof attempts alternating on ideas and granularity. In: [14], pp. 126–142
5. Fiedler, A.: P.rex: An interactive proof explainer. In: Goré, R.P., Leitsch, A., Nipkow, T. (eds.) IJCAR 2001. LNCS (LNAI), vol. 2083, pp. 416–420. Springer, Heidelberg (2001)
6. Denney, E., Power, J., Tourlas, K.: Hiproofs: A hierarchical notion of proof tree. In: Proc. of the 21st Annual Conf. on Mathematical Foundations of Progamming Semantics (MFPS XXI). ENTCS, vol. 155, pp. 341–359. Elsevier, Amsterdam (2006)
7. Autexier, S., Fiedler, A.: Textbook proofs meet formal logic - the problem of underspecification and granularity. In: [14], pp. 96–110
8. Quinlan, J.R.: C4.5: Programs for Machine Learning. Morgan Kaufmann, San Francisco (1993)
9. Schiller, M., Benzmüller, C.: Granularity-adaptive proof presentation. Technical report, SEKI Working-Paper (2009), `http://arxiv.org/pdf/0903.0314v4`
10. Frank, E., Witten, I.H.: Generating accurate rule sets without global optimization. In: Proc. 15th Intl. Conf. on Machine Learning, pp. 144–151. Morgan Kaufmann, San Francisco (1998)
11. Platt, J.C.: Fast training of support vector machines using sequential minimal optimization. In: Schoelkopf, B., Burges, C., Smola, A. (eds.) Advances in Kernel Methods - Support Vector Learning, pp. 185–208. MIT Press, Cambridge (1998)
12. Hobbs, J.R.: Granularity. In: Joshi, A.K. (ed.) Proc. of the 9th Int. Joint Conf. on Artificial Intelligence (IJCAI), pp. 432–435. Morgan Kaufmann, San Francisco (1985)
13. McCalla, G., Greer, J., Barrie, B., Pospisil, P.: Granularity hierarchies. In: Computers & Mathematics with Applications, vol. 23(2-5), pp. 363–375. Elsevier, Amsterdam (1992)
14. Kohlhase, M. (ed.): MKM 2005. LNCS (LNAI), vol. 3863. Springer, Heidelberg (2006)

On Defaults in Action Theories

Hannes Strass and Michael Thielscher

Department of Computer Science
Dresden University of Technology
{hannes.strass,mit}@inf.tu-dresden.de

Abstract. We study the integration of two prominent fields of logic-based AI: action formalisms and non-monotonic reasoning. The resulting framework allows an agent employing an action theory as internal world model to make useful default assumptions. We show that the mechanism behaves properly in the sense that all intuitively possible conclusions can be drawn and no implausible inferences arise. In particular, it suffices to make default assumptions only once (in the initial state) to solve projection problems.

1 Introduction

This paper combines works of two important areas of logic-based artificial intelligence: we propose to enrich formalisms for reasoning about actions and change with default logic. The present work is not the first to join the two areas—non-monotonic logics have already been used by the reasoning about actions community in the past. After McCarthy and Hayes discovered the fundamental problem of determining the non-effects of actions, the frame problem [1], it was widely believed that non-monotonic reasoning were necessary to solve it. Then Hanks and McDermott gave an example of how straightforward use of non-monotonic logics in reasoning about actions and change can lead to counter-intuitive results [2]. When monotonic solutions to the frame problem were found [3,4], non-monotonic reasoning again seemed to be obsolete.

In this paper, we argue that utilizing default logic still is of use when reasoning about actions. We will not use it to solve the frame problem—the solution to the frame problem we use here is monotonic and similar to the one of [4]—, but to make useful default assumptions about states. We consider action theories with deterministic actions where all effects are unconditional and a restricted form of default assumptions. The main reasoning task we are interested in is the projection problem, that is, given an initial state and a sequence of actions, the question whether a certain condition holds in the resulting state. As the main result of this paper, we show that restricting default application to the initial state not only guarantees a maximal set of states that are reachable, but also all possible inferences about these states.

The rest of the paper is organized as follows. The next section introduces the two areas this work is concerned with. Section 3 then combines the two fields and develops the main results. In the last section, we shortly sketch the limitations of our approach, outline directions for further work, and conclude.

B. Mertsching, M. Hund, and Z. Aziz (Eds.): KI 2009, LNAI 5803, pp. 298–305, 2009.

2 Background

This section presents the formal underpinnings of the paper. In the first subsection we acquaint the reader with a unifying action calculus that we use to logically formalize action domains, and in the second subsection we recall the notions of default logic [5].

2.1 The Unifying Action Calculus

The stated objective of introducing the unifying action calculus (UAC) in [6] was to provide a universal framework for research in reasoning about actions. Since we want to formulate our results in a most general manner, we adopt the UAC for the present work.

The most notable generalization established by the UAC is its abstraction from the underlying time structure: it can be instantiated with formalisms using the time structure of situations (as the Situation Calculus [7] or the Fluent Calculus [4]), as well as with formalisms using a linear time structure (like the Event Calculus [8]).

The UAC uses only the sorts FLUENT, ACTION, and TIME along with the predicates $<$: TIME $\times$ TIME (denoting an ordering of time points), *Holds* : FLUENT $\times$ TIME (stating whether a fluent evaluates to true at a given time point), and *Poss* : ACTION $\times$ TIME $\times$ TIME (indicating whether an action is applicable for particular starting and ending time points). Uniqueness-of-names is assumed for all (finitely many) functions into sorts FLUENT and ACTION.

The following definition introduces the most important types of formulas of the unifying action calculus: they allow to express properties of states and applicability conditions and effects of actions.

Definition 1. *Let $\boldsymbol{s}$ be a sequence of variables of sort* TIME.

- *A* state formula $\Phi[\boldsymbol{s}]$ in $\boldsymbol{s}$ *is a first-order formula with free variables $\boldsymbol{s}$ where*
  - *for each occurrence of $Holds(\varphi, s)$ in $\Phi[\boldsymbol{s}]$ we have $s \in \boldsymbol{s}$ and*
  - *predicate Poss does not occur.*

Let s, t be variables of sort TIME *and A be a function into sort* ACTION.

- *A* precondition axiom *is of the form*

$$Poss(A(\boldsymbol{x}), s, t) \equiv \pi_A[s] \tag{1}$$

  where $\pi_A[s]$ is a state formula in s with free variables among $s, t, \boldsymbol{x}$.
- *An* effect axiom *is of the form*

$$Poss(A(\boldsymbol{x}), s, t) \supset (\forall f)(Holds(f, t) \equiv (\gamma_A^+ \vee (Holds(f, s) \wedge \neg\gamma_A^-))) \tag{2}$$

  where

$$\gamma_A^+ = \bigvee_{0 \le i \le n_A^+} f = \varphi_i \text{ and } \gamma_A^- = \bigvee_{0 \le i \le n_A^-} f = \psi_i$$

  and the φ_i and ψ_i are terms of sort FLUENT *with free variables among $\boldsymbol{x}$.*

Readers may be curious as to why the predicate *Poss* carries two time arguments instead of just one: $Poss(a, s, t)$ is to be read as "action a is possible starting at time s and ending at time t." The formulas γ_A^+ and γ_A^- enumerate the positive and negative effects of the action, respectively. This definition of effect axioms is a restricted version of the original definition of [6]—it only allows for deterministic actions with unconditional effects.

Definition 2. *A* (UAC) domain axiomatization *consists of a finite set of foundational axioms Ω (that define the underlying time structure and do not contain the predicates Holds and Poss), a set Π of precondition axioms (1), and a set Υ of effect axioms (2); the latter two for all functions into sort* ACTION.

A domain axiomatization is progressing, *if*

- $\Omega \models (\exists s : \text{TIME})(\forall t : \text{TIME}) s \leq t$
- $\Omega \cup \Pi \models Poss(a, s, t) \supset s < t$

A domain axiomatization is sequential, *if it is progressing and*

$$\Omega \cup \Pi \models Poss(a, s, t) \wedge Poss(a', s', t') \supset \\ (t < t' \supset t \leq s') \wedge (t = t' \supset (a = a' \wedge s = s'))$$

That is, a domain axiomatization is progressing if there exists a least time point and time always increases when applying an action. A sequential domain axiomatization furthermore requires that no two actions overlap.

Since we are mainly interested in the projection problem, our domain axiomatizations will usually include a set Σ_0 of state formulas in the least time point that characterize the initial state.

To illustrate the intended usage of the introduced notions, we make use of a variant of the example of [2], the Yale Shooting scenario.

Example 1. Consider the domain axiomatization $\Sigma = \Omega_{sit} \cup \Pi \cup \Upsilon \cup \Sigma_0$[1]. The precondition axioms say that the action Shoot is possible if the gun is loaded and the actions Load and Wait are always possible.

$$\Pi = \{Poss(\mathsf{Shoot}, s, t) \equiv (Holds(\mathsf{Loaded}, s) \wedge t = Do(\mathsf{Shoot}, s)), \\ Poss(\mathsf{Load}, s, t) \equiv t = Do(\mathsf{Load}, s), Poss(\mathsf{Wait}, s, t) \equiv t = Do(\mathsf{Wait}, s)\}$$

With these preconditions and foundational axioms Ω_{sit}, the domain axiomatization is sequential. The effect of shooting is that the turkey ceases to be alive, loading the gun causes it to be loaded, and waiting does not have any effect. All effect axioms in Υ are of the form (2), we state only the $\gamma^\pm$ different from the empty disjunction: $\gamma^+_{\mathsf{Load}} = (f = \mathsf{Loaded})$, $\gamma^-_{\mathsf{Shoot}} = (f = \mathsf{Alive})$. Finally, we state that the turkey is alive in the initial situation, $\Sigma_0 = \{Holds(\mathsf{Alive}, S_0)\}$. We can now employ logical entailment to answer the question whether the turkey is still alive after applying the actions Load, Wait, and Shoot, respectively. With the notation $Do([a_1, \ldots, a_n], s)$ as abbreviation for $Do(a_n, Do(\ldots, Do(a_1, s)\ldots))$, it is easy to see that $\Sigma \models \neg Holds(\mathsf{Alive}, Do([\mathsf{Load}, \mathsf{Wait}, \mathsf{Shoot}], S_0)$.

[1] Ω_{sit} denotes the foundational axioms for situations. They have been omitted from the presentation due to a lack of space and can be found in [9].

2.2 Default Logic

Introduced in the seminal work by Reiter [5], default logic has become one of the most important formalisms for non-monotonic reasoning. The semantics for supernormal defaults used here is taken from [10], which is itself an enhancement of a notion developed in [11].

Definition 3. *A* supernormal default rule, *or, for short,* default, *is a closed first-order formula. Any formulas with occurrences of free variables are taken as representatives of their ground instances.*

For a set of closed formulas S, we say the default δ is active in S *if both $\delta \notin S$ and $\neg\delta \notin S$.*

A (supernormal) default theory *is a pair $(W, \mathcal{D})$, where W is a set of sentences and $\mathcal{D}$ a set of default rules.*

A default rule can thus also be seen as a hypothesis that we are willing to assume, but prepared to give up in case of contradiction. A default theory then adds a set of formulas, the *indefeasible knowledge*, that we are not willing to give up for any reason.

Definition 4. *Let $(W, \mathcal{D})$ be a default theory where all default rules are supernormal and $\prec\!\!\!\prec$ be a total order on $\mathcal{D}$. Define $E_0 := Th(W)$ and for all $i \geq 0$,*

$$E_{i+1} = \begin{cases} E_i & \text{if no default is active in } E_i \\ Th(E_i \cup \{\delta\}) & \text{otherwise, where } \delta \text{ is the } \prec\!\!\!\prec\text{-} \\ & \text{minimal default active in } E_i. \end{cases}$$

Then the set $E := \bigcup_{i \geq 0} E_i$ is called the extension generated by $\prec\!\!\!\prec$.

A set of formulas E is a preferred extension *for $(W, \mathcal{D})$ if there exists a total order $\prec\!\!\!\prec$ that generates E. The set of all preferred extensions for a default theory $(W, \mathcal{D})$ is denoted by $Ex(W, \mathcal{D})$.*

An extension for a default theory can be seen as a way of assuming as many defaults as possible without creating inconsistencies. It should be noted that, although the definition differs, our extensions are extensions in the sense of [5].

Based on extensions, one can define skeptical and credulous conclusions for default theories: skeptical conclusions are formulas that are contained in every extension, credulous conclusions are those that are contained in at least one extension.

Definition 5. *Let $(W, \mathcal{D})$ be a supernormal default theory and Ψ be a first-order formula.*

$$W \mathrel{|\!\!\approx}_{\mathcal{D}}^{skept} \Psi \stackrel{\text{def}}{\equiv} \Psi \in \bigcap_{E \in Ex(W,\mathcal{D})} E, \quad W \mathrel{|\!\!\approx}_{\mathcal{D}}^{cred} \Psi \stackrel{\text{def}}{\equiv} \Psi \in \bigcup_{E \in Ex(W,\mathcal{D})} E$$

3 Domain Axiomatizations with Defaults

The concepts established up to this point are now easily combined to the notion of a domain axiomatization with defaults, our main object of study. It is essentially a default theory having an action domain axiomatization as indefeasible knowledge.

Definition 6. *A* domain axiomatization with defaults *is a pair* $(\Sigma, \mathcal{D}[s])$*, where* Σ *is a UAC domain axiomatization and* $\mathcal{D}[s]$ *is a set of supernormal defaults of the form* $Holds(\varphi, s)$ *or* $\neg Holds(\varphi, s)$ *for a fluent* φ*.*

We next define what it means for a time point to be reachable in an action domain. Intuitively, it means that there is a sequence of actions that leads to the time point when applied in sequence starting in the initial time point.

Definition 7. *Let* Σ *be a domain axiomatization and* $\mathcal{D}[s]$ *be a set of defaults.*

$$\begin{aligned} Reach(r) &\stackrel{\text{def}}{=} (\forall R)(((\forall s)(Init(s) \supset R(s)) \\ &\qquad \wedge (\forall a, s, t)(R(s) \wedge Poss(a, s, t) \supset R(t))) \supset R(r)) \qquad (3) \\ Init(t) &\stackrel{\text{def}}{=} \neg(\exists s) s < t \qquad (4) \end{aligned}$$

A time point τ *is called*

- finitely reachable in Σ *if* $\Sigma \models Reach(\tau)$*;*
- finitely, credulously reachable in $(\Sigma, \mathcal{D}[\sigma])$[2], *if* σ *is finitely reachable in* Σ *and for some extension* E *for* $(\Sigma, \mathcal{D}[\sigma])$ *we have* $E \models Reach(\tau)$*;*
- finitely, weakly skeptically reachable in $(\Sigma, \mathcal{D}[\sigma])$*, if* σ *is finitely reachable in* Σ *and for all extensions* E *for* $(\Sigma, \mathcal{D}[\sigma])$*, we have* $E \models Reach(\tau)$*;*
- finitely, strongly skeptically reachable in $(\Sigma, \mathcal{D}[\sigma])$*, if* σ *is finitely reachable in* Σ *and there exist ground actions* $\alpha_1, \ldots, \alpha_n$ *and time points* $\tau_0, \ldots, \tau_n$ *such that* $\Sigma \approx^{skept}_{\mathcal{D}[\sigma]} Poss(\alpha_i, \tau_{i-1}, \tau_i)$ *for all* $1 \leq i \leq n$*,* $\tau_0 = \sigma$*, and* $\tau_n = \tau$*.*

With situations as underlying time structure, weak and strong skeptical reachability coincide. This is because the foundational axioms for situations [9] entail that situations have unique predecessors.

Example 1 (continued). We add a fluent Broken that indicates if the gun does not function properly. Shooting is now only possible if the gun is loaded *and* not broken:

$$Poss(\mathsf{Shoot}, s, t) \equiv (Holds(\mathsf{Loaded}, s) \wedge \neg Holds(\mathsf{Broken}, s) \wedge t = Do(\mathsf{Shoot}, s))$$

Unless there is information to the contrary, it should be assumed that the gun has no defects. This is expressed by the set of defaults $\mathcal{D}[s] = \{\neg Holds(\mathsf{Broken}, s)\}$. Without the default assumption, it cannot be concluded that the action Shoot is

[2] By $\mathcal{D}[\sigma]$ we denote the set of defaults in $\mathcal{D}[s]$ where s has been instantiated by the term σ.

possible after performing Load and Wait since it cannot be inferred that the gun is not broken. Using the abbreviations $S_1 = Do(\mathsf{Load}, S_0)$, $S_2 = Do(\mathsf{Wait}, S_1)$, and $S_3 = Do(\mathsf{Shoot}, S_2)$, we illustrate how the non-monotonic entailment relation defined earlier enables us to use the default rule to draw the desired conclusion:

$$\Sigma \approx^{skept}_{\mathcal{D}[S_0]} \neg Holds(\mathsf{Broken}, S_2),$$
$$\Sigma \approx^{skept}_{\mathcal{D}[S_0]} Poss(\mathsf{Shoot}, S_2, S_3), \text{ and}$$
$$\Sigma \approx^{skept}_{\mathcal{D}[S_0]} \neg Holds(\mathsf{Alive}, S_3).$$

The default conclusion that the gun works correctly, drawn in S_0, carries over to S_2 and allows to conclude applicability of Shoot in S_2 and its effects on S_3.

In the example just seen, default reasoning could be restricted to the initial situation. As it turns out, this is sufficient for the type of action domain considered here: effect axiom (2) never "removes" information about fluents and thus never makes more defaults active after executing an action. This observation is formalized by the following lemma. It essentially says that to reason about a time point in which an action ends, it makes no difference whether we apply the defaults to the resulting time point or to the time point when the action starts. This holds of course only due to the restricted nature of effect axiom (2).

Lemma 1. *Let $(\Sigma, \mathcal{D}[s])$ be a domain axiomatization with defaults, α be a ground action such that $\Sigma \models Poss(\alpha, \sigma, \tau)$ for some σ, τ :* TIME, *and let $\Psi[\tau]$ be a state formula in τ. Then*

$$\Sigma \approx^{skept}_{\mathcal{D}[\sigma]} \Psi[\tau] \text{ iff } \Sigma \approx^{skept}_{\mathcal{D}[\tau]} \Psi[\tau]$$

The next theorem says that all *local* conclusions about a finitely reachable time point σ (that is, all conclusions about σ using defaults from $\mathcal{D}[\sigma]$) are exactly the conclusions about σ that we can draw by instantiating the defaults only with the least time point.

Theorem 1. *Let $(\Sigma, \mathcal{D}[s])$ be a progressing domain axiomatization with defaults, λ its least time point, σ :* TIME *be finitely reachable in Σ, and $\Psi[\sigma]$ be a state formula. Then*

$$\Sigma \approx^{skept}_{\mathcal{D}[\sigma]} \Psi[\sigma] \text{ iff } \Sigma \approx^{skept}_{\mathcal{D}[\lambda]} \Psi[\sigma]$$

It thus remains to show that local defaults are indeed exhaustive with respect to local conclusions. The next lemma takes a step into this direction: it states that action application does not increase default knowledge about past time points.

Lemma 2. *Let $(\Sigma, \mathcal{D}[s])$ be a domain axiomatization with defaults, α be a ground action such that $\Sigma \models Poss(\alpha, \sigma, \tau)$ for some σ, τ :* TIME, *and let $\Psi[\rho]$ be a state formula in $\rho :$* TIME *where $\rho \leq \sigma$. Then*

$$\Sigma \approx^{skept}_{\mathcal{D}[\tau]} \Psi[\rho] \text{ implies } \Sigma \approx^{skept}_{\mathcal{D}[\sigma]} \Psi[\rho]$$

The converse of the lemma does not hold, since an action effect might preclude a default conclusion about the past. Using the above lemma and simple induction on the length of action sequences, one can establish the following.

Theorem 2. *Let* $(\Sigma, \mathcal{D}[s])$ *be a progressing domain axiomatization with defaults, let* $\Psi[s]$ *be a state formula,* $\sigma < \tau$ *be time points, and* σ *be finitely reachable in* Σ*. Then*

$$\Sigma \approx^{skept}_{\mathcal{D}[\tau]} \Psi[\sigma] \text{ implies } \Sigma \approx^{skept}_{\mathcal{D}[\sigma]} \Psi[\sigma]$$

The next theorem, our first main result, now combines Theorems 1 and 2 and tells us that default instantiation to the least time point subsumes default instantiation in any time point in the future of the time point we want to reason about.

Theorem 3. *Let* $(\Sigma, \mathcal{D}[s])$ *be a progressing domain axiomatization with defaults,* λ *be its least time point,* $\Psi[s]$ *be a state formula, and* $\sigma < \tau$ *be terms of sort* TIME *where* σ *is finitely reachable in* Σ*. Then*

$$\Sigma \approx^{skept}_{\mathcal{D}[\tau]} \Psi[\sigma] \text{ implies } \Sigma \approx^{skept}_{\mathcal{D}[\lambda]} \Psi[\sigma]$$

Proof. $\Sigma \approx^{skept}_{\mathcal{D}[\tau]} \Psi[\sigma]$ *implies* $\Sigma \approx^{skept}_{\mathcal{D}[\sigma]} \Psi[\sigma]$ *by Theorem 2. By Theorem 1, this is the case iff* $\Sigma \approx^{skept}_{\mathcal{D}[\lambda]} \Psi[\sigma]$.

What this theorem misses out, however, are time points that are not finitely reachable in Σ only, but where some action application along the way depends crucially on a default conclusion. To illustrate this, recall Example 1: the situation $Do([\mathsf{Load}, \mathsf{Wait}, \mathsf{Shoot}], S_0)$ is not reachable in Σ, because the necessary precondition that the gun is not broken cannot be inferred without the respective default.

The following theorem, our second main result, now assures sufficiency of instantiation with the least time point also for time points that are only reachable by default.

Theorem 4. *Let* $(\Sigma, \mathcal{D}[s])$ *be a progressing domain axiomatization with defaults,* λ *its least time point,* σ *be a time point that is finitely reachable in* Σ*,* $\Psi[s]$ *be a state formula, and* τ *be a time point that is finitely, strongly skeptically reachable in* $(\Sigma, \mathcal{D}[\sigma])$*. Then*

1. τ *is finitely, strongly skeptically reachable in* $(\Sigma, \mathcal{D}[\lambda])$*, and*
2. $\Sigma \approx^{skept}_{\mathcal{D}[\sigma]} \Psi[\tau]$ *iff* $\Sigma \approx^{skept}_{\mathcal{D}[\lambda]} \Psi[\tau]$.

An immediate consequence of this result is that instantiation in the least time point also provides a "maximal" number of reachable time points: default instantiation with a later time point might potentially prevent actions in the least time point, which in turn might render yet another time point unreachable.

4 Conclusions and Future Work

We have presented an enrichment of action theories with a well-known non-monotonic logic, Raymond Reiter's default logic. To the best of our knowledge, this is the first time this field is explored in the logic-based AI community. The approach has been shown to behave well (although no proofs could be included due to space limitations)—by the restrictions made in the definitions, defaults persist over time and it thus suffices to apply them only once (namely to the initial state).

With respect to further generalizations of our proposal, we remark that both allowing for disjunctive defaults and allowing for conditional effects causes un-intuitive conclusions via the employed solution to the frame problem. Future research in this topic will therefore be devoted to generalizing both the defaults (from supernormal to normal) and the considered actions (from deterministic with unconditional effects to non-deterministic with conditional effects).

References

1. McCarthy, J., Hayes, P.J.: Some Philosophical Problems from the Standpoint of Artificial Intelligence. In: Machine Intelligence, pp. 463–502. Edinburgh University Press (1969)
2. Hanks, S., McDermott, D.: Nonmonotonic Logic and Temporal Projection. Artificial Intelligence 33(3), 379–412 (1987)
3. Reiter, R.: The Frame Problem in the Situation Calculus: A Simple Solution (Sometimes) and a Completeness Result for Goal Regression. In: Artificial Intelligence and Mathematical Theory of Computation – Papers in Honor of John McCarthy, pp. 359–380. Academic Press, London (1991)
4. Thielscher, M.: From Situation Calculus to Fluent Calculus: State Update Axioms as a Solution to the Inferential Frame Problem. Artificial Intelligence 111(1–2), 277–299 (1999)
5. Reiter, R.: A Logic for Default Reasoning. Artificial Intelligence 13, 81–132 (1980)
6. Thielscher, M.: A Unifying Action Calculus. Artificial Intelligence (to appear, 2009)
7. McCarthy, J.: Situations and Actions and Causal Laws, Stanford Artificial Intelligence Project: Memo 2 (1963)
8. Kowalski, R.A., Sergot, M.J.: A Logic-based Calculus of Events. New Generation Computing 4(1), 67–95 (1986)
9. Pirri, F., Reiter, R.: Some Contributions to the Metatheory of the Situation Calculus. Journal of the ACM 46(3), 325–361 (1999)
10. Brewka, G., Eiter, T.: Prioritizing Default Logic: Abridged Report. In: Festschrift on the occasion of Prof. Dr. W. Bibel's 60th birthday. Kluwer, Dordrecht (1999)
11. Brewka, G.: Preferred Subtheories: An Extended Logical Framework for Default Reasoning. In: Proceedings of the Eleventh International Conference on Artificial Intelligence, pp. 1043–1048 (1989)

Analogy, Paralogy and Reverse Analogy: Postulates and Inferences

Henri Prade[1] and Gilles Richard[1,2]

[1] IRIT, Université Paul Sabatier, 31062 Toulouse Cedex 09, France
prade@irit.fr
[2] BITE, 258-262 Avicenna House London, E7 9HZ UK
richard@irit.fr

Abstract. Analogy plays a very important role in human reasoning. In this paper, we study a restricted form of it based on analogical proportions, i.e. statements of the form *a is to b as c is to d*. We first investigate the constitutive notions of analogy, and beside the analogical proportion highlights the existence of two noticeable companion relations: one that is just reversing the change from c to d w. r. t. the one from a to b, while the last one called *paralogical* proportion expresses that *what a and b have in common, c and d have it also.* Characteristic postulates are identified for the three types of relations allowing to provide set and Boolean logic interpretations in a natural way. Finally, the solving of proportion equations as a basis for inference is discussed, again emphasizing the differences between analogy, reverse analogy, and paralogy, in particular in a three-valued setting, which is also briefly presented.

1 Introduction

Analogical reasoning is a kind of reasoning that is commonly used in the day to day life as well as in scientific discoveries. For these reasons, analogical reasoning has raised a considerable interest among researchers in philosophy, in human sciences (cognitive psychology, linguistics, anthropology,...), in computer sciences (especially in artificial intelligence). In particular, diverse formalizations [1,2,3] and implementations of analogical reasoning [4,2,5,3] have been proposed. The underlying models usually rely on highly structured representations (graphs for instance) with substitution operations, sometimes expressed in first order logic [3], or even in second order logic [5] using unification/anti-unification.

However, a particular form of analogical reasoning is based on simple and compact statements called analogical proportions, which are statements of the form *a is to b as c is to d*, usually denoted $a : b :: c : d$. Case-based reasoning (CBR) [6], generally regarded as an elementary form of analogical reasoning, exploits the human brain ability to parallel "problems" a and b, and then "deduce" that if c is a solution to the first problem a, then some d, whose relation to b is similar to the relation between a and c, could be a solution to the second problem b. Obviously this setting might be viewed as an analogical proportion

B. Mertsching, M. Hund, and Z. Aziz (Eds.): KI 2009, LNAI 5803, pp. 306–314, 2009.

reasoning through the solving of the equation $a : b :: c : x$, or also of the equation $a : c :: b : x$, depending if we emphasize the relations problem-problem and solution-solution, or the relation problem-solution, in the comparison. However, as we shall see in section 4, analogy-based reasoning (ABR) goes beyond CBR. Strangely enough, the formal study of analogical proportions has been considered only by a few researchers [7][8] and these works have remained incomplete in several respects. That is why our paper proposes a systematic analysis at a formal level of analogical proportion. In the next section, we investigate constitutive notions of analogy and we highlight the existence of two relations beside standard analogical proportion, namely *paralogical proportion* and *reverse analogical proportion*, exhibiting other kinds of relation between entities. More precisely, starting from the idea that analogy is a matter of similarity and difference, we identify three types of proportion that can relate four entities and we state their respective postulates. In Section 3, we briefly develop the set and Boolean logic interpretations for our 3 proportions. In Section 4, we examine the problem of finding the last missing item to build up a complete proportion starting with a, b and c: different conditions of existence of solutions appear depending on the type of proportion at hand. Section 5 outlines a multiple-valued interpretation for the three types of proportion leading to different solutions. Finally we survey the related works and conclude. The main purpose of our works is to reveal that analogy has three faces, altogether providing a potentially increased inferential power which may be useful in different AI areas such as machine learning or natural language processing for instance.

2 Analogy, Reverse Analogy and Paralogy: A 3-Sided View

Analogy puts two situations in parallel (a "situation" may be, e.g. the description of a problem with its solution), and compares these situations by establishing a correspondence between them. In a simple form, each situation involves two entities, say a, c on the one hand, and b, d on the other hand. The comparison then bears on the pair (a, b), and on the pair (c, d) and lead to consider what is common (in terms of properties) to a and b (let us denote it $com(a, b)$), and what is specific to a and not shared by b (we denote it $spec(a, b)$). Clearly $com(a, b) = com(b, a)$ and generally, $spec(a, b) \neq spec(b, a)$. With this view, a is represented by the pair $(com(a, b), spec(a, b))$, b by the pair $(com(a, b), spec(b, a))$, c by the pair $(com(c, d), spec(c, d))$ and d by the pair $(com(c, d), spec(d, c))$.

Then, an analogical proportion, denoted $a : b :: c : d$, expressing that *a is to b as c is to d* (in the parallel between a, c and b, d), amounts to state that the way a and b differ is the same as the way c and d differ, namely

$$spec(a, b) = spec(c, d) \text{ and } spec(b, a) = spec(d, c) \quad (1)$$

enforcing symmetry in the way the comparison is done. In $a : b :: c : d$, when going from a to b (or from c to d), $spec(a, b)$ is changed into $spec(b, a)$ or equivalently $spec(c, d)$ is changed into $spec(d, c)$, while $com(a, b)$ and $com(c, d)$ are the respective common parts of the pairs (a, b) and (c, d).

If we now compare a represented as $(com(a,b), spec(a,b))$ to c represented as $(com(c,d), spec(c,d))$, it appears due to (1) that their common part is $spec(a,b) = spec(c,d)$, while $com(a,b)$ is changed into $com(c,d)$. In other words, $com(a,b)$ plays the role of $spec(a,c)$ and $com(c,d)$ plays the role of $spec(c,a)$. Similarly, when comparing b and d, the common part is $spec(b,a) = spec(d,c)$ due to (1) and $com(a,b)$ is again changed into $com(c,d)$ when going from b to d (i.e. $com(a,b)$ plays the role of $spec(b,d)$, and $com(c,d)$ the role of $spec(d,b)$). This amounts to write $spec(a,c) = spec(b,d)$, and $spec(c,a) = spec(d,b)$. Altogether this exactly means $a : \mathbf{c} :: \mathbf{b} : d$. Thus we have retrieved the central permutation postulate that most authors associate with analogical proportion (together with the symmetry postulate already mentioned). Namely, analogical proportion, also referred to as "analogy" in the following, satisfies the postulates:

- $a : b :: a : b$ (and $a : a :: b : b$) hold (reflexivity).
- if $a : b :: c : d$ holds then $a : c :: b : d$ should hold (central permutation)
- if $a : b :: c : d$ holds then $c : d :: a : b$ should hold (symmetry).

$a : b :: b : a$ cannot hold since $spec(a,b) \neq spec(b,a)$. Such a postulate is more in the spirit of reverse analogy that we introduce now. Still focusing on specificities, we can consider a new proportion denoted "!" where $a\ !\ b\ !!\ c\ !\ d$ holds as soon as $spec(a,b) = spec(d,c)$ and $spec(b,a) = spec(c,d)$. It expresses the *reverse* analogy *a is to b as d is to c*, and obeys the postulates:

- $a\ !\ b\ !!\ b\ !\ a$ (and $a\ !\ a\ !!\ b\ !\ b$) should hold (reverse reflexivity)
- if $a\ !\ b\ !!\ c\ !\ d$ holds then $c\ !\ b\ !!\ a\ !\ d$ should hold (odd permutation)
- if $a\ !\ b\ !!\ c\ !\ d$ holds then $c\ !\ d\ !!\ a\ !\ b$ should hold (symmetry).

Except if $a = b$, $a\ !\ b\ !!\ a\ !\ b$ will not hold in general. Having investigated all the possibilities from the viewpoint of specificities, it seems natural to focus on the shared properties, which leads to introduce a new kind of proportion denoted ";". Then, we have no other choice than defining $a\ ;\ b\ ;;\ c;\ d$ as $com(a,b) = com(c,d)$. We decide to call this new proportion *paralogical proportion*. It states that *what a and b have in common, c and d have it also.* Obviously, this proportion does not satisfy the permutation properties of its two sister relations, but rather:

- $a\ ;\ b\ ;;\ a\ ;\ b$ and $a\ ;\ b\ ;;\ b\ ;\ a$ always holds (bi-reflexivity)
- if $a\ ;\ b\ ;;\ c\ ;\ d$ holds $b\ ;\ a\ ;;\ c\ ;\ d$ should hold (even permutation)
- if $a\ ;\ b\ ;;\ c\ ;\ d$ holds then $c\ ;\ d\ ;;\ a\ ;\ b$ should hold (symmetry).

Note that symmetry applies here to the comparison between two pairs of items (a,b) and (c,d), but not internally since $a : b$ cannot be replaced with $b : a$.

3 Set and Boolean Logic Interpretations

When the entities at hand are subsets of a referential X (e.g. an entity can be represented as a subset of properties that hold in a given situation), our initial analysis can be easily translated in terms of set operations. Common properties between two sets a and b can be defined via *set intersection* $a \cap b$ and specificities via *set difference* $a \setminus b = a \cap \overline{b}$ where $\overline{b}$ denotes $X \setminus b$. In other words, we characterize a set a w.r.t. a set b via the pair $(a \cap b, a \cap \overline{b})$. Then, focusing on specificities, the analogical proportion $a : b :: c : d$, is then stated through the constraints:

$$a \cap \overline{b} = c \cap \overline{d} \text{ and } \overline{a} \cap b = \overline{c} \cap d \quad (A_{Set})$$

Similarly, reverse analogy is defined in the set framework by the constraints:

$$a \cap \overline{b} = \overline{c} \cap d \text{ and } \overline{a} \cap b = c \cap \overline{d} \quad (R_{Set})$$

Focusing on common features appreciated through properties that hold for a and b and properties that do not hold neither for a nor b, leads to define paralogy as $a \cap b = c \cap d$ and $\overline{a} \cap \overline{b} = \overline{c} \cap \overline{d}$, which is equivalent to:

$$a \cap b = c \cap d \text{ and } a \cup b = c \cup d \quad (P_{Set})$$

It is easy to switch to the Boolean lattice $\mathbb{B} = \{0, 1\}$ with the standard operators $\vee, \wedge, \neg$, and $u \to v$ defined as $\neg u \vee v$, and $u \equiv v$ is short for $u \to v \wedge v \to u$. We follow here the lines of [9]. These logical definitions come from a direct translation of (A_{Set}), (R_{Set}) and (P_{Set}), where $\cup$ is replaced by $\vee$, $\cap$ by $\wedge$, complementarity by $\neg$, and introducing the implication in the two first expressions:

- $a : b :: c : d$ iff $((a \to b \equiv c \to d) \wedge (b \to a \equiv d \to c)) \quad (A_{Bool})$
- $a \,!\, b \,!!\, c \,!\, d$ iff $((a \to b \equiv d \to c) \wedge (b \to a \equiv c \to d)) \quad (R_{Bool})$
- $a \,;\, b \,;;\, c \,;\, d$ iff $((a \wedge b \equiv c \wedge d) \wedge (a \vee b \equiv c \vee d)) \quad (P_{Bool})$

Note that $a : b :: \overline{b} : \overline{a}$ holds, which appears now reminiscent of the logical equivalences between $a \to b$ and $\overline{b} \to \overline{a}$ (and $b \to a$ and $\overline{a} \to \overline{b}$). In the context of the Boolean interpretation, the most visual way to understand the difference between analogy, reverse analogy and paralogy is to examine their truth tables when they are considered as Boolean operators. Table 1 provides the truth values making the relations to hold (among $2^4 = 16$ possibilities). Except for the first

Table 1. Analogy, Reverse analogy and Paralogy truth tables

Analogy	Reverse	Paralogy
0 0 0 0	0 0 0 0	0 0 0 0
1 1 1 1	1 1 1 1	1 1 1 1
0 0 1 1	0 0 1 1	1 0 0 1
1 1 0 0	1 1 0 0	0 1 1 0
0 1 0 1	0 1 1 0	0 1 0 1
1 0 1 0	1 0 0 1	1 0 1 0

two lines where the three proportions are identical, it appears that only two of the three relations can hold simultaneously. Indeed the patterns 1001 and 0110 are rejected by analogy since the changes from 1 to 0 and from 0 to 1 are not in the same sense. Patterns 0101 and 1010 where changes are in the same sense are rejected by reverse analogy. Patterns 0011 and 1100, expressing the absence of common features (what a and b have in common, c and d do not have it and vice versa), do not agree with paralogy. The property below holds for set and Boolean interpretations, and provides an explicit link between the three proportions:

Property 1. *$a : b :: c : d$ holds iff $a \,!\, b \,!!\, d \,!\, c$ holds and $a : b :: c : d$ holds iff $a \,;\, d \,;;\, c \,;\, b$ holds.*

4 Inferences

Let us now turn to an effective way to use these proportions. Let S be a set of vectors of Boolean features describing available cases. Then proportions between 4-tuples of vectors are defined componentwise. We are faced with a learning-like task when we want to predict for a new, only partially informed, case x (i.e. only a subvector $k(x)$ of features is known), the values of the missing features $u(x)$. The basic idea is that if the *known* part of x, $k(x)$, is *in proportion of some type denoted* $|$, with the corresponding parts of 3 cases a, b, and c belonging to S i.e. $k(a)|k(b)||k(c)|k(x)$ holds *componentwise*, then the *unknown* part of x, $u(x)$ should be in a proportion of the same type with the corresponding part of a, b and c, and thus should be solution of an equation of the form $u(a)|u(b)||u(c)|z$, where $z = u(x)$ is the unknown. This is obviously a form of reasoning that is beyond classical logic, but which may be useful for trying to guess unknown values. Given a proportion $|$, the problem of finding x such that $a|b||c|x$ holds will be referred as "equation solving". This is why this problem has to be investigated as soon as we build up a model for exploiting the different types of proportion: analogy, reverse analogy and paralogy. Despite the proportions have close relationships, the solutions they provide can be quite different as we shall see.

Let us examine a practical example of analogical proportion to support intuition. Consider a database of cars and trucks. Each data is represented as a vector of 5 binary properties (car (1) or truck (0), diesel (0 or 1), 4wd (0 or 1), "green" (small carbon footprint) (0 or 1) and expensive (0 or 1)) with obvious meaning. In our data, there are 3 items (2 cars a and b and a truck c), each represented by a feature vector: $a = (1, 1, 0, 0, 1)$, $b = (1, 0, 1, 1, 0)$, $c = (0, 1, 0, 0, 1)$. We assume a new item d with only partial information, $d = (0, 0, 1, s, t)$: using our notation, we note that $k(a) : k(b) :: k(c) : k(d)$ holds for the first three features in the set theoretical framework and we are looking for the unknown attributes $u(d) = (s, t)$ in order to guess if the new item is green and/or expensive. We start with the set interpretation (the first of the three results below is in [8]):

Property 2. *The analogical equation $a : b :: c : x$ has a solution iff $b \cap c \subseteq a \subseteq b \cup c$. In that case, the solution is unique and given by $x = ((b \cup c) \setminus a) \cup (b \cap c)$.*

The reverse analogical equation $a\ !\ b\ !!\ c\ !\ x$ is solvable iff $a \cap c \subseteq b \subseteq a \cup c$. In that case, the solution is unique and given by $x = ((a \cup c) \setminus b) \cup (a \cap c)$.

The paralogical equation $a; b;; c; x$ has a solution iff $a \cap b \subseteq c \subseteq a \cup b$. In that case, the solution is unique and given by $x = ((a \cup b) \setminus c) \cup (a \cap b)$.

The above results have a more compact formulation in terms of Boolean logic:

Property 3. *The analogical equation $a : b :: c : x$ is solvable iff $((a \equiv b) \vee (a \equiv c)) = 1$. In that case, the unique solution is $x = a \equiv (b \equiv c)$.*
The reverse analogical equation $a\ !\ b\ !!\ c\ !\ x$ is solvable iff $((b \equiv a) \vee (b \equiv c)) = 1$. In that case, the unique solution is $x = b \equiv (a \equiv c)$.
The paralogical equation $a; b;; c; x$ is solvable iff $((c \equiv b) \vee (c \equiv a)) = 1$. In that case, the unique solution is $x = c \equiv (a \equiv b)$.

Applying this result to solve our "truck" example, we get $u(d) = (s, t) = (1, 0)$ with $s = (0 \equiv (1 \equiv 0)) = 1$ and $t = (1 \equiv (0 \equiv 1)) = 0$: the new truck is "green" and cheap. Due to the associativity of the equivalence connective, the solution of the analogical, paralogical, and reverse analogical proportions have the same expression in the Boolean case, namely $x = c \equiv (a \equiv b)$. However, the conditions of existence of the solution are *not the same* for the three types of proportion. Indeed the vectors $(a, b, c, x) = (1, 1, 0, x)$ and $(a, b, c, y) = (0, 0, 1, y)$ cannot be completed in a paralogical equation (while the solutions are $x = 0$ and $y = 1$ for analogy and reverse-analogy). Similarly, $(a, b, c, x) = (1, 0, 1, x)$ and $(a, b, c, y) = (0, 1, 0, y)$ have no solution in reverse analogy, and $(a, b, c, x) = (1, 0, 0, x)$ and $(a, b, c, y) = (0, 1, 1, y)$ have no solution in analogy. As we shall see, the expression of the solutions (if any) will be no longer necessarily the same when moving to a tri-valued setting, or more generally to multiple-valued or even numerical models. As illustrated by this example, ABR simultaneously exploits the differences and similarities between 3 completely informed cases, while CBR considers the completely informed cases one by one.

5 Tri-valued Interpretation

A strict Boolean interpretation does not give space for graded properties where, for instance, we have to deal with properties whose satisfaction is described in terms of three levels, e.g. *low, medium or high,* (instead of the usual *yes* or *no*). Then we need the introduction of a 3rd value to take into account this fact. We now use the set $\mathbb{T} = \{-1, 0, 1\}$ where our items a, b, c and d can take their value, for encoding these three levels[1] (be aware that the bottom element is now -1). We have now to provide definitions in line with our analysis of sections 2 and 3. Let us start from the set interpretation. It seems natural to substitute the set difference $a \setminus b$ with $a - b$. It leads to the constraint $a - b = c - d$ for analogical proportion (the other part of A_{set} is not necessary since $a - b = c - d$ is equivalent to $b - a = d - c$). The fact that $a - b$ may no longer be in $\mathbb{T}$ does not matter here. What is important is to notice that the corresponding equation where d is unknown has a solution in $\mathbb{T}$ only in 19 cases among 27. It can be checked that only these 19 cases correspond to analogical proportion situations where both the sense and the intensity (1 or 2 steps in scale $\mathbb{T}$) changes are preserved.

Thanks to Property 1, we get the definition for reverse analogy and paralogy. These definitions (which subsume the Boolean ones) are given in Table 2. It is quite easy to prove that these definitions satisfy the required postulates for analogy, reverse analogy and paralogy over $\mathbb{T}$. As in the previous interpretations, analogical, reverse analogical and paralogical equations cannot necessarily be simultaneously solved. The Boolean interpretation is a particular case of the

[1] More generally, we can work with the unit interval scale as in fuzzy logic, and using multiple-valued logic operators, by writing counterparts of the constraints describing each type of proportion. The full study of the extension of our approach to fuzzy set connectives in the realm of appropriate algebras is left for further studies.

Table 2. 3-valued models for analogy, reverse analogy and paralogy

Analogy	**Reverse analogy**	**Paralogy**
$a - b = c - d$	$a - b = d - c$	$a + b = c + d$

tri-valued one. It appears that, when the three equations are simultaneously solvable, they may provide distinct solutions, as shown in the following example. Let us consider a simplistic dating illustrative example where individuals are identified via the 5 criteria education, sport, religion, age and income. Each criterium can be evaluated as low (-1), medium (0), high (+1) with intuitive meaning. We are looking for somebody to be associated to a new customer c profiled as $(1, 0, 0, 0, 0)$ i.e. high income, who likes sport but does not practice, who is religious without practicing, with medium age and medium education. Let us suppose now that we have in our database a couple (a, b) represented w.r.t our criteria as $(0, -1, 1, 0, 0)$ and $(0, 0, 0, -1, 1)$ of previously successfully paired profiles. In that context, two options seem attractive: either to consider paralogy in order to emphasize the common features and common rejections among couples of people, or focusing on differences as in analogical reasoning. Table 3 exhibits the solutions got by paralogy, analogy and reverse analogy.

Table 3. 3-valued model: an example

	income	*sport*	*relig*	*age*	*educ*
a	0	-1	1	0	0
b	0	0	0	-1	1
c	1	0	0	0	0
d_{para}	-1	-1	1	-1	1
d_{ana}	1	1	-1	-1	1
d_{rev}	1	-1	1	1	-1

As can be seen, we are in a particular context where it is possible to solve all the equations corresponding to the three types of proportions for all the features. The three solutions are distinct. According to intuition, paralogy or analogy may provide solutions which may appear for some features a bit risky (or innovative), and other time cautious (or conservative). It is not obvious that all the features should be handled in the same way: for instance, it appears reasonable that incomes compensate inside a couple as done by paralogy, while such a compensation seems less appropriate for education for instance (then calling for analogy for this feature). Clearly a customer c may be associated with different pairs (a, b), leading to different solutions (even if we are using only one type of reasoning). It is not coming as a surprise since it is likely that there is more than one type of individuals to be paired with another person: the repertory may contain examples of that. A proposal to gather the different obtainable solutions may be found in [10] for analogical proportions.

6 Related Works and Directions for Further Research

Analogy has been used and studied in AI for a long time [1,2], ranging from analogy-finding programs, to structure-mapping modeling, and including the heuristic use of analogical reasoning at the meta-level for improving deductive provers. In parallel, the use of analogy in cognition and especially in learning has been identified and discussed [11],[3]. Using the conceptual graphs encoding Sowa and Majumdar [4] have implemented the VivoMind engine providing an automatic analogy discovering system. Such works do not consider the analogical proportion per se. Dealing only with analogical proportions, Lepage [8] has been the first to discuss postulates, through a set interpretation analysis. However his 6 postulates are redundant in this setting, while the 3 ones we give are enough. In terms of interpretation, [12] provides a large panel of algebraic models which are likely to be subsumed in [5] by the use of second order logic substitutions.

Recently [13] has used analogical proportions for machine learning purposes, focusing on classification problems, using a k-NN-like algorithm by means of "analogical dissimilarities". Our simple examples above somehow suggest the potential of the discussed ideas for learning purposes: in particular, we may wonder if our different proportions may play specific roles in learning problems. In [14], it is shown that fuzzy CBR techniques (which parallel two cases rather than four) can be favorably compared to k-NN algorithm, thus it will be interesting to consider our more general framework in that respect.

7 Conclusion

Our analysis of core properties of analogy has introduced two new kinds of relations, called paralogy and reverse analogy, capturing intuitions other than the one underlying analogical proportion. While analogical and reverse analogical proportions focus on dissimilarities, paralogical proportions privilege a reading in terms of similarities. Then going to set and Boolean logic interpretations, we have reset definitions in a unified setting. Finally, the tri-valued interpretation introduces gradualness in the proportions and highlights the fact that they capture distinct inferential principles since they may provide distinct solutions.

References

1. Winston, P.H.: Learning and reasoning by analogy. Com. ACM 23, 689–703 (1980)
2. Gentner, D.: The Mechanisms of Analogical Learning. In: Similarity and Analogical Reasoning, pp. 197–241. Cambridge University Press, Cambridge (1989)
3. Davies, T.R., Russell, S.J.: A logical approach to reasoning by analogy. In: IJCAI 1987, pp. 264–270. Morgan Kaufmann, San Francisco (1987)
4. Sowa, J.F., Majumdar, A.K.: Analogical reasoning. In: Ganter, B., de Moor, A., Lex, W. (eds.) ICCS 2003. LNCS, vol. 2746, pp. 16–36. Springer, Heidelberg (2003)
5. Schmid, U., Gust, H., Kühnberger, K., Burghardt, J.: An algebraic framework for solving proportional and predictive analogies. In: Eur. Conf. Cogn. Sci., pp. 295–300 (2003)

6. Aamodt, A., Plaza, E.: Case-based reasoning; foundational issues, methodological variations, and system approaches. AICom 7(1), 39–59 (1994)
7. Klein, S.: Culture, mysticism & social structure and the calculation of behavior. In: Proc. Europ. Conf. in AI (ECAI), pp. 141–146 (1982)
8. Lepage, Y.: Analogy and formal languages. In: Proc. FG/MOL 2001, pp. 373–378 (2001), http://www.slt.atr.co.jp/~lepage/pdf/dhdryl.pdf.gz
9. Miclet, L., Prade, H.: Logical definition of analogical proportion and its fuzzy extensions. In: Proc. NAFIPS. IEEE, New York (2008)
10. Prade, H.: Proportion analogique et raisonnement à partir de cas. In: Rencontres Francophones sur la Logique Floue et ses Applications, Cépaduès, pp. 296–303 (2008)
11. Gentner, D., Holyoak, K.J., Kokinov, B. (eds.): The Analogical Mind: Perspectives from Cognitive Sciences. MIT Press, Cambridge (2001)
12. Stroppa, N., Yvon, F.: Analogical learning and formal proportions: Definitions and methodological issues. ENST Paris report (2005)
13. Miclet, L., Bayoudh, S., Delhay, A.: Analogical dissimilarity: definition, algorithms and two experiments in machine learning. JAIR 32, 793–824 (2008)
14. Hullermeier, E.: Case-Based Approximate Reasoning. Theory and Decision Library. Springer, New York (2007)

Early Clustering Approach towards Modeling of Bottom-Up Visual Attention

Muhammad Zaheer Aziz and Bärbel Mertsching*

GET Lab, University of Paderborn, Pohlweg 47-49,
33098 Paderborn, Germany,
{last name}@get.upb.de
http://getwww.upb.de

Abstract. A region-based approach towards modelling of bottom-up visual attention is proposed with an objective to accelerate the internal processes of attention and make its output usable by the high-level vision procedures to facilitate intelligent decision making during pattern analysis and vision-based learning. A memory-based inhibition of return is introduced in order to handle the dynamic scenarios of mobile vision systems. Performance of the proposed model is evaluated on different categories of visual input and compared with human attention response and other existing models of attention. Results show success of the proposed model and its advantages over existing techniques in certain aspects.

1 Introduction

During the recent decade emphasis has increased towards building machine vision systems according to the role model of natural vision. The intelligence, robustness, and adaptability of natural vision have foundations in visual attention, which is a phenomenon that selects significant portions of a given scene where computational resources of the brain should concentrate for detailed analysis and recognition. Many computational models have emerged for artificial visual attention in order achieve capability of autonomous selection of important portions of a given scene in machine vision. Mobile vision systems especially need such a mechanism because they have limitations on the amount of computing equipment that they can carry due to restrictions of payload and energy consumption. An additional demand of such applications is the speedup in attention processes and their compatibility with other vision modules in order to benefit from the outcome of selective attention for improving efficiency of the overall vision system.

This paper is concerned with the efforts of accelerating the attention procedures and bringing them in harmony with rest of the computer vision so that visual attention could play an effective and active role in practical vision systems. A model for bottom-up visual attention system is presented that incorporates

* We gratefully acknowledge the funding of this work by the Deutsche Forschungsgemeinschaft (German Research Foundation) under grant Me 1289 (12 - 1)(AVRAM).

B. Mertsching, M. Hund, and Z. Aziz (Eds.): KI 2009, LNAI 5803, pp. 315–322, 2009.

early pixel clustering to investigate its effects in terms of the said requirements. The basic conceptual difference of the proposed model from contemporary models is reduction in the number of items to be processed prior to attention procedures by clustering pixels into segments instead of using individual pixels during computation of attention. The proposed model is evaluated using human response of attention as benchmark and compared with other existing models to estimate the progress in the state of the art made by the proposal. Prospective applications of such systems include exploration and surveillance machines able to select relevant visual data according to human interest from a bulk of input.

2 Related Literature

In context of human vision, the features that can be detected before diverting the focus of attention are summarized in [1] where color, orientation, motion and size are reported as the confirmed channels while other fourteen probable and possible features are also listed. The mathematical models of natural attention to combine the feature channels proposed in [2] suggest a multiplication-style operation in the visual cortex while it is modeled as square of sum in [3]. In order to suppress the already attended locations to avoid continuous or frequent focusing on the same object(s) inhibition of return (IOR) takes place in terms of both location and object features [4].

In context of artificial visual attention, existing models of visual attention propose different methodologies for computing saliency with respect to a selected set of features and suggest methods to combine the feature saliency maps in order to obtain a resultant conspicuity map. Pop-out is determined using a peak selection mechanism on the final conspicuity map and each model then utilizes some mechanism for applying IOR.

The model presented in [5] and [6] first normalizes the feature maps of color contrast, intensity, and orientation and then applies a simple weighted sum to obtain the input for the saliency map which is implemented as a 2D layer of leaky integrate-and-fire neurons. A Winner Take All (WTA) neural network ensures only one occurrence of the most active location in the master saliency map. In this model inhibition of return is implemented by spatially suppressing an area in the saliency map around the current focus of attention (FOA) while feature-based inhibition is not considered. The elementary units of computation are pixels or small image neighborhoods arranged in a hierarchical structure.

The model proposed in [7] applies independent component analysis algorithm to determine relative importance of features whereas weighted sum of feature maps is used to obtain a combined saliency map. An adaptive mask suppresses the recently attended object for performing IOR. The model of [8] also computes a weighted sum of individual feature maps for obtaining an integrated attention map but introduces a manipulator map which is multiplied to the sum.

The model presented in [9] includes the aspect of tracking multiple objects while focusing attention in a dynamic scene. A separate map is used for IOR where the visited locations are marked as highly active. This activity inhibits the master map of attention to avoid immediate revisiting of the attended location. The activity of the inhibition map decays slowly in order to allow revisiting of the location after some time. Another model presented in [10] uses the direct sum of the feature channels to compute a two-dimensional saliency map but they introduce an anisotropic Gaussian as the weighting function centered at the middle of the image. They have not reported any indigenous inhibition mechanism.

All of the above mentioned techniques have used a pixel-based approach in which points, at either original scale of the input or its downscaled version, are used during computations. Working with individual pixels at input's original scale hinders consideration of global context and demands more computational resources. On the other hand, use of downscaled versions of the input cause cloudiness in the final output leading to total loss of shape information about the foci of attention and, most of the times, dislocation of the saliency clusters resulting in inaccuracy in localizing the exact position of the FOA.

The object-based method proposed in [11] implements a hierarchical selectivity process using a winner-take-all neural network. They apply a top-down influence to increase or decrease the baseline of neural activity of the most prominent feature channel. Their model deals with so called 'groupings' of pixels. This model expects the object construction module to build hierarchical structure from visual input, which is computationally very expensive with the current state of the art of computer vision.

A method of finding saliency using early segmentation can be found in [12]. They use roughly segmented regions as structural units to compute the probability of having a high saliency for each region. Although this method has similarity with the proposed solution in terms of the basic concept but they use a statistical approach mainly concentrating only on the global context.

3 Proposed Early-Clustering Approach

The proposed model groups pixels of the visual input possessing similar color attributes into clusters using a robust illumination tolerant segmentation method [13]. Architecture of the model at abstract level is sketched in figure 1. The primary feature extraction function F produces a set of regions $\Re$ consisting of n regions each represented as R_i $(1 \leq i \leq n)$ and feeds each R_i with data regarding location, bounding rectangle, and magnitudes of each feature ϕ_i^f $(f \in \Phi)$. As five channels of color, orientation, eccentricity, symmetry and size are considered in the current status of our model, we have $\Phi = \{c, o, e, s, z\}$. Each R_i also contains a list of pointers to its immediate neighbors in $\Re$ denoted by η_i. Computation of the bottom-up saliency using the rarity criteria is performed by the process S. The output of S is combined by W that applies weighted fusion of these maps to formulate a resultant bottom-up saliency map. The function P applies peak selection criteria to choose one pop-out at a time. The focus of attention obtained at a particular time t is stored in the inhibition memory using which the process of IOR suppresses the already attended location(s) at $t+1$. The memory management function M decides whether to place the recent FOA in

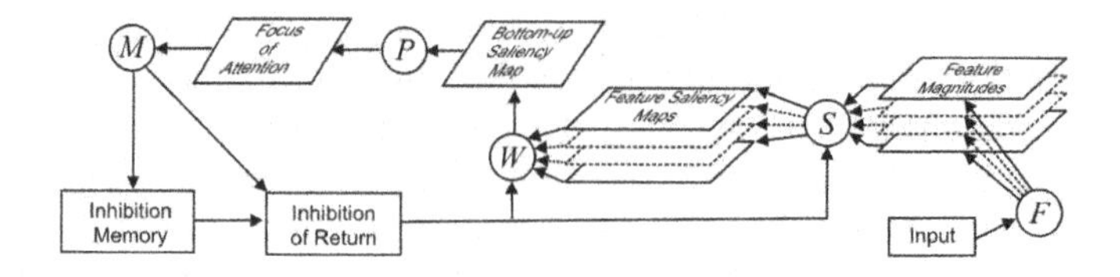

Fig. 1. Architecture of the proposed region-based bottom-up attention model

inhibition memory or excitation memory according to the active behavior and adjusts the weights of inhibition accordingly. Details of the feature extraction module F and the saliency computation step S are comprehensively described in [14] hence here the discussion will be focused on the next steps including feature map fusion W, pop-out detection P, and memory based IOR controlled by M.

The weight of each feature channel f, $W_f^{bu}(t)$, is first initialized (at $t = 0$) such that the color map gets the highest weight because it plays a major role in attention, the size map gets the lowest weight because it is effective only when other channels do not contain significant saliency, and the others a medium one. As we use a multiplicative scheme for combining the feature channels, the minimum weight that can be assigned to a channel can be a unit value, e.g. $W_z^{bu}(0) = 1$ for size channel.

Before summing up the feature saliencies, the weights are adjusted such that the feature map offering the sharpest peak of saliency contributes more in the accumulated saliency map. It is done by finding the distance, Δ_f, between the maximum and the average saliency value in each map. The feature map with the highest Δ_f is considered as most active and its weight is increased by a multiplicative factor δ (we take $\delta = 2$). Now the total bottom-up saliency $\beta_i(t)$ of a region R_i at time t is computed using all feature saliencies $S_f^i(t)$:

$$\beta_i(t) = \sum_{f \in \Phi} \left(W_f^{bu}(t) S_f^i(t) \right) / \sum_{f \in \Phi} \left(W_f^{bu}(t) \right) \tag{1}$$

The pop-out detection becomes simple after going through this procedure as the region having the highest value of $\beta_i(t)$ will be the winner to get the focus of attention. The FOA at time t is inhibited while finding pop-out at time $t+1$. Most of the existing models of attention implement either the spatial inhibition, in which a specific area around the point of attention is inhibited, or the feature-map based inhibition in which the weight of the wanted feature channel is adjusted to obtain required results. We consider three types of inhibition mechanisms namely feature-map based, spatial, and feature based.

The feature-map based inhibition is modeled for preventing a feature map from gaining extra ordinary weight so that other features do not get excluded from the competition. When a weight $W_f^{bu}(t)$ becomes equal to $max(W_f^{bu}(t) \forall f \in \Phi)$ then it is set back to its original value that was assigned to it during the initialization step. This mechanism keeps the weights of feature maps in a cycle.

In order to deal with dynamic visual input the spatial and feature-based IOR are applied using an inhibition memory, denoted by M^I, because a series of recently attended locations have to be inhibited. The inhibition memory is designed to remember the last m foci of attention:

$$M^I = \{M_k^I\},\ k \in \{1, 2, ..m\}$$

For the most recent item M_k^I stored in M^I, the value of k is set to 1. k increases with the age of M_k^I in the memory. At time $t+1$, the resultant saliency of a region $S_f^i(t+1)$ after spatial and feature-based inhibitions is computed as follows:

$$S_f^i(t+1) = S_f^i(t)\Im^s(R_i, M_k^I, k)\Im^f(R_i, M_k^I, k)\ \forall\ k \in \{1, 2, ..m\} \tag{2}$$

$\Im^s(R_i, M_k^I, k)$ finds appropriate amount of inhibition in spatial context and $\Im^f(R_i, M_k^I, k)$ inhibits already attended features after comparing R_i with all m locations stored in M^I. Decreasing suppression takes effect with the increasing value of k, i.e., less suppression will be applied when age of the memory item becomes older.

4 Results and Evaluation

The proposed attention system was tested using a variety of visual input including images and video data taken from camera of robot head, synthetic environment of the simulation framework, self created benchmark samples, and other images collected from internet image databases. In order to validate output of the proposed model it was compared with human response of attention. Data of human fixations was obtained using an eye tracking equipment. The images used as input for the model testing were shown to human subjects (13 in total). Fixation scan paths were tracked and the locations where the eyes focused longer than 200 ms were taken as the fixation points. Figure 2(a) shows a sample from the test data. Figure 2(b) demonstrates fixation spot for one of the subjects whereas figure 2(c) is combination of attended locations by all subjects. Recording of fixations was stopped after the first attention point in the experiment presented here. Salient locations in perspective of human vision in each test image were obtained using results of these psychophysical experiments.

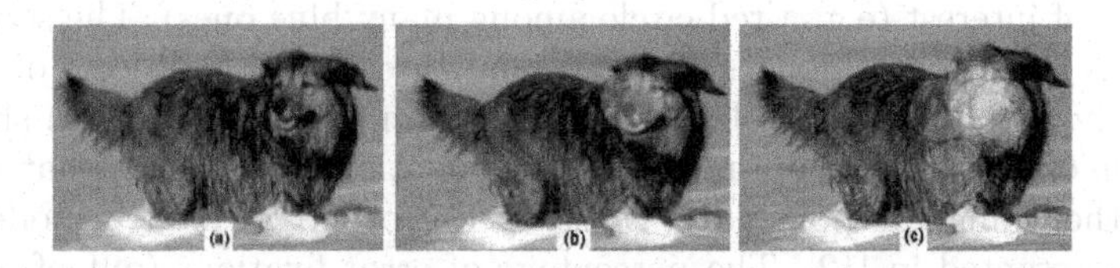

Fig. 2. Results of psychophysical experiments using eye tracking equipment. (a) A sample from input data. (b) Fixation by one subject. (c) Combined view of fixations by all subjects of the experiment.

The proposed model was executed on the same test data to obtained its focused locations. Figure 3(a) shows the saliency map produced by the proposed model at time of first focus of attention on the sample data shown in figure 2(a) whereas figure 3(b) demonstrates the first five fixations made by the model. In order to evaluate the standing of the proposed model among the contemporary models, other models were executed on the same set of input images and their fixation data was recorded. Figures 3(c) and 3(d) demonstrate first five foci of attention by the models of [5] and [12], respectively, on the same sample.

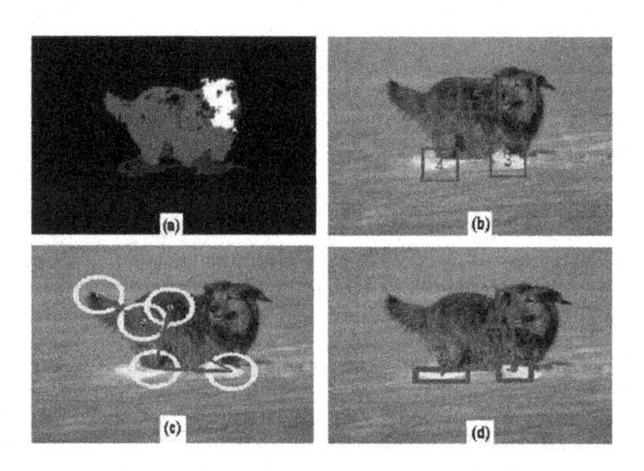

Fig. 3. Results of fixations by the computational models of attention on the input given in figure 2(a). (a) Saliency map for the first focus of attention produced by the proposed model. (b) First five foci of attention by the proposed model. (c) and (d) First five fixations by the model of [5] and [12] respectively.

Figure 4 demonstrates results of the proposed system in a dynamic scenario experimented in a virtual environment using the robot simulation framework developed in our group [15]. Figure 4 (a) shows the environment in which the simulated robot drives on the path marked by the red arrow while figure 4 (b) shows the scene viewed through the camera head of the robot. Figure 4 (c) to (g) present the output of bottom-up attention for five selected frames each picked after equal intervals of time. The current focus of attention is marked by a yellow rectangle whereas blue ones mark the inhibited locations.

For the purpose of evaluation we categorized the input into four levels of complexity. First level consisted of images on which the human vision found only one object of interest (e.g. a red circle among many blue ones). The second and third levels consisted of images with two and three possible objects of interest, respectively, while the fourth level had more than three. Figure 5(a) shows the percentage of fixations (using maximum five fixations per image) that matched between the human response and those by the proposed model, model of [5], and that presented in [12]. The percentage of error fixations (out of first five) performed by the three models that were outside the salient spots are plotted in figure 5(b). The error fixations are independent of matching fixations because a model may repeat its fixation on already attended objects.

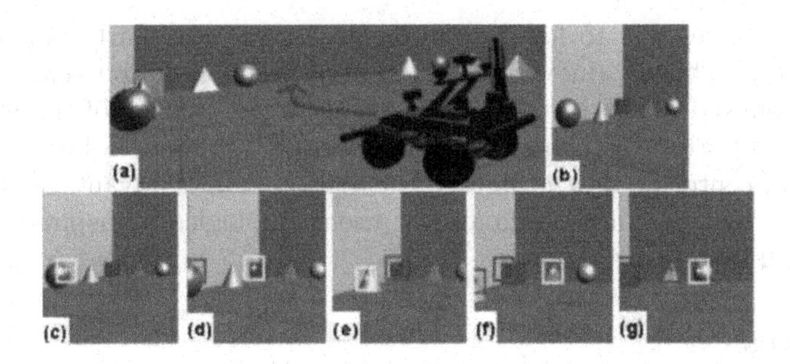

Fig. 4. Results in dynamic scenario using a simulated mobile vision system. (a) Simulated robot moving in virtual environment (along the path marked by the red arrow) (b) Scene viewed through left camera of the robot. (c) to (g) Fixated locations are indicated by yellow marks and inhibited locations by blue ones.

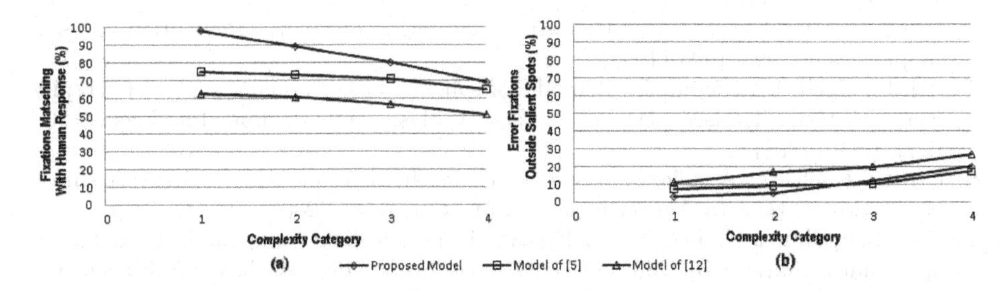

Fig. 5. (a) Percentage of fixations by the proposed model and the models by [5] and [12] matching with human response. (b) Percentage of fixation falling outside salient areas by the three models.

5 Discussion

It is noticeable in the evaluation data that the proposed model has valid results to a level comparable to the existing models. The proposed model performs much better than the other models on the simple input categories with one or two objects of interest because it works with greater number of feature channels. The drop of performance by the proposed model in case of complex scenes happens firstly because of region segmentation errors in natural scenes. Secondly, individual priorities in humans when real life scenes are shown to them make it hard to predict saliency only on basis of feature contrast. Performance of the proposed model remains equivalent to the other better performing model on the categories offering complexity. Such level of performance of the proposed system is coupled with some advantages. Firstly, the model produced output faster than other models even when it computes more feature channels to find saliency. Secondly, the foci of attention produced by the proposed model are reusable by high level machine vision algorithms as shape and already computed features of the attended region remain available. On the other hand, other models loose the shape and features due to low resolution processing and fairly blur output in the final stage. Thirdly, due to preservation of high resolution shape and position, the proposed model is able to provide accurate location of the FOA. Such accuracy is importance when the vision system is required to bring the object of interest in center of view using overt attention. These advantages makes the proposed model a leading candidate for use in biologically inspired intelligent mobile vision systems.

References

1. Wolfe, J.M., Horowitz, T.S.: What attributes guide the deployment of visual attention and how do they do it? Nature Reviews, Neuroscience 5, 1–7 (2004)
2. Neri, P.: Attentional effects on sensory tuning for single-feature detection and double-feature conjunction. In: Vision Research, pp. 3053–3064 (2004)
3. Koch, C.: Biophysics of Computation. Oxford University Press, New York (1999)
4. Weaver, B., Lupianez, J., Watson, F.L.: The effects of practice on object-based, location-based, and static-display inhibition of return. Perception & Psychophysics 60, 993–1003 (1998)
5. Itti, L., Koch, U., Niebur, E.: A model of saliency-based visual attention for rapid scene analysis. Transactions on Pattern Analysis and Machine Intelligence 20, 1254–1259 (1998)
6. Itti, L., Koch, C.: A saliency based search mechanism for overt and covert shifts of visual attention. In: Vision Research, pp. 1489–1506 (2000)
7. Park, S.J., Shin, J.K., Lee, M.: Biologically inspired saliency map model for bottom-up visual attention. In: Bülthoff, H.H., Lee, S.-W., Poggio, T.A., Wallraven, C. (eds.) BMCV 2002. LNCS, vol. 2525, pp. 418–426. Springer, Heidelberg (2002)
8. Heidemann, G., Rae, R., Bekel, H., Bax, I., Ritter, H.: Integrating context-free and context-dependant attentional mechanisms for gestural object reference. In: Crowley, J.L., Piater, J.H., Vincze, M., Paletta, L. (eds.) ICVS 2003. LNCS, vol. 2626, pp. 22–33. Springer, Heidelberg (2003)
9. Backer, G., Mertsching, B., Bollmann, M.: Data- and model-driven gaze control for an active-vision system. Transactions on Pattern Analysis and Machine Intelligence 23, 1415–1429 (2001)
10. Meur, O.L., Callet, P.L., Barba, D., Thoreau, D.: A coherent computational approach to model bottom-up visual attention. Transactions on Pattern Analysis and Machine Intelligence 28, 802–817 (2006)
11. Sun, Y., Fischer, R.: Object-based visual attention for computer vision. Artificial Intelligence 146, 77–123 (2003)
12. Avraham, T., Lindenbaum, M.: Esaliency - a stochastic attention model incorporating similarity information and knowledge-based preferences. In: WRUPKV-ECCV 2006, Graz (2006)
13. Aziz, M.Z., Mertsching, B.: Color segmentation for a region-based attention model. In: Workshop Farbbildverarbeitung (FWS 2006), Ilmenau, Germany, pp. 74–83 (2006)
14. Aziz, M.Z., Mertsching, B.: Fast and robust generation of feature maps for region-based visual attention. Transactions on Image Processing 17, 633–644 (2008)
15. Kutter, O., Hilker, C., Simon, A., Mertsching, B.: Modeling and simulating mobile robots environments. In: 3rd International Conference on Computer Graphics Theory and Applications (GRAPP 2008), Funchal, Portugal (2008)

A Formal Cognitive Model of Mathematical Metaphors

Markus Guhe, Alan Smaill, and Alison Pease*

School of Informatics, Informatics Forum, 10 Crichton Street, Edinburgh EH8 9AB
{m.guhe,a.smaill,a.pease}@ed.ac.uk

Abstract. Starting from the observation by Lakoff and Núñez (2000) that the process for mathematical discoveries is essentially one of creating metaphors, we show how Information Flow theory (Barwise & Seligman, 1997) can be used to formalise the basic metaphors for arithmetic that ground the basic concepts in the human embodied nature.

1 The Cognition of Mathematics

Mathematics is most commonly seen as the uncovering of eternal, absolute truths about the (mostly nonphysical) structure of the universe. Lakatos (1976) and Lakoff and Núñez (2000) argue strongly against the 'romantic' (Lakoff and Núñez) or 'deductivist' (Lakatos) style in which mathematics is usually presented.

Lakatos's philosophical account of the historical development of mathematical ideas demonstrates how mathematical concepts are not simply 'unveiled' but are developed in a process that involves, among other things, proposing formal definitions of mathematical concepts (his main example are polyhedra), developing proofs for properties of those concepts (e.g. Euler's conjecture that for polyhedra $V - E + F = 2$, where V, E, F are the number of vertices, edges and faces, respectively) and refining those concepts, e.g. by excluding counterexamples (what Lakatos calls *monster barring*) or widening the definition to include additional cases. Thus, the concept *polyhedron* changes in the process of defining its properties, and this casts doubt on whether ployhedra 'truly exist'.

Lakoff and Núñez (2000) describe how mathematical concepts are formed (or, depending on the epistemological view, how mathematical discoveries are made). They claim that the human ability for mathematics is brought about by two main factors: embodiment and the ability to create and use metaphors. They describe how by starting from interactions with the environment we build up (more and more abstract) mathematical concepts by processes of metaphor. They distinguish grounding from linking metaphors (p. 53). In *grounding* metaphors one domain is embodied and one abstract, e.g. the four grounding metaphors for arithmetic, which we describe below. In *linking* metaphors, both domains are abstract, which allows the creation of more abstract mathematical concepts, e.g. on the having established the basics of arithmetic with grounding metaphors this knowledge is used to create the concepts *points in space* and *functions* (p. 387).

* The research reported here was carried out in the *Wheelbarrow* project, funded by the EPSRC grant EP/F035594/1.

B. Mertsching, M. Hund, and Z. Aziz (Eds.): KI 2009, LNAI 5803, pp. 323–330, 2009.

Even if the universal truths view of mathematics were correct, the way in which people construct, evaluate and modify mathematical concepts has received little attention from cognitive science and AI. In particular, it has not yet been described by a computational cognitive model. As part of our research on understanding the cognition of creating mathematical concepts we are building such a model (Guhe, Pease, & Smaill, 2009; Guhe, Smaill, & Pease, 2009), based on the two streams of research outlined above. While there are cognitive models of learning mathematics (e.g. Lebiere 1998; Anderson 2007), there are to our knowledge no models of how humans *create* mathematics.

2 Structure Mapping, Formal Models and Local Processing

Following Gentner (1983; see also Gentner & Markman, 1997, p. 48) we assume that metaphors are similar to analogies. More specifically, there is no absolute distinction between metaphor and analogy; they rather occupy different but overlapping regions of the same space of possible mappings between concepts, with analogies comparatively mapping more abstract relations and metaphors more properties. According to Gentner's (1983, p. 156) structure mapping theory the main cognitive process of analogy formation is a mapping between the (higher-order) relations of conceptual structures.

Although we use this approach for creating computational cognitive models of mathematical discovery (see Guhe, Pease, & Smaill, 2009 for an ACT-R model using the path-mapping approach of Salvucci & Anderson, 2001), in this paper we use a formal model extending the one presented in Guhe, Smaill, and Pease (2009) that specifies the grounding metaphors for arithmetic proposed by Lakoff and Núñez (2000). Formal methods are not commonly used for cognitive modelling, but we consider it a fruitful approach to mathematical metaphors, because, firstly, mathematics is already formalised, and, secondly, if Lakoff and Núñez are correct that metaphors are a general cognitive mechanism that is applied in mathematical thinking, formal methods are an adequately high level of modelling this general purpose mechanism. Furthermore, collecting empirical data on how scientific concepts are created is difficult. This is true for case studies as well as laboratory settings. Case studies (e.g. Nersessian, 2008) are not reproducible (and therefore anecdotal), and they are usually created in retrospect, which means that they will contain many rationalisations instead of an accurate protocol of the thought processes. Laboratory settings in contrast (cf. Schunn & Anderson, 1998) require to limit the participants in their possible responses, and it is unclear whether or how this is different from the unrestricted scientific process.

Using the formal method, we take the metaphors used in the discovery process and create a formal cognitive model of the mapping of the high-level relations. The model is cognitive in that it captures the declarative knowledge used by humans, but it does not simulate the accompanying thought processes. For the formalisation we use Information Flow theory (Barwise & Seligman, 1997, see section 4), because it provides means to formalise interrelations (flows of information) between different substructures of a knowledge representation, which we called *local contexts* in Guhe (2007). A local context contains a subset of the knowledge available to the model that is relevant with respect to a particular *focussed element*. In cognitive terms, a focussed element is an item (concept, percept) that is currently in the focus of attention.

The ability to create and use suitable local contexts is what sets natural intelligent systems apart from artificial intelligent systems and what allows them to interact efficiently and reliably with their environment. The key benefit of this ability is that only processing a local context (a subset of the available knowledge) drastically reduces the computational complexity and enables the system to interact with the environment under real-time constraints.

3 Lakoff and Núñez's Four Basic Metaphors of Arithmetic

Lakoff and Núñez (2000, chapter 3) propose that humans create arithmetic with four different grounding metaphors that create an abstract conceptual space from embodied experiences, i.e. interactions with the real world. Since many details are required for describing these metaphors adequately, we can only provide the general idea here. Note that the metaphors are not interchangeable. All are used to create the basic concepts of arithmetic.

1. **Object Collection.** The first metaphor, *arithmetic is object collection*, describes how by interacting with objects we experience that objects can be grouped and that there are certain regularities when creating collections of objects, e.g. by removing objects from collections, by combining collections, etc. By creating metaphors (analogies), these regularities are mapped into the domain of arithmetic, for example, collections of the same size are mapped to the concept of number and putting two collections together is mapped to the arithmetic operation of addition.
2. **Object Construction.** Similarly, in the *arithmetic is object construction* metaphor we experience that we can combine objects to form new objects, for example by using toy building blocks to build towers. Again, the number of objects that are used for the object construction are mapped to number and constructing an object is mapped to addition.
3. **Measuring Stick.** The *measuring stick* metaphor captures the regularities of using measuring sticks for the purpose of establishing the size of physical objects, e.g. for constructing buildings. Here numbers correspond to the physical segments on the measuring stick and addition to putting together segments to form longer segments.
4. **Motion Along A Path.** The *motion along a path* metaphor, finally, adds concepts to arithmetic that we experience by moving along straight paths. For example, numbers are point locations on paths and addition is moving from point to point.

4 Information Flow

This section gives a short introduction to Information Flow theory. We focus on the aspects needed for our formalisation, see Barwise and Seligman (1997) for a detailed discussion. We need three notions for our purposes: *classification*, *infomorphism* and *channel*.

Classification. A *classification* $\boldsymbol{A}$ consists of a set of tokens $\mathrm{tok}(\boldsymbol{A})$, a set of types $\mathrm{typ}(\boldsymbol{A})$ and a binary classification relation $\models_{\boldsymbol{A}}$ between tokens and types. In this way, the classification relation classifies the tokens, for example, for a token $a \in \mathrm{tok}(\boldsymbol{A})$ and a type $\alpha \in \mathrm{typ}(\boldsymbol{A})$ the relation can establish $a \models_{\boldsymbol{A}} \alpha$. Graphically, a classification is usually depicted as in the left of figure 1, i.e. with the types on top and the tokens on the bottom.

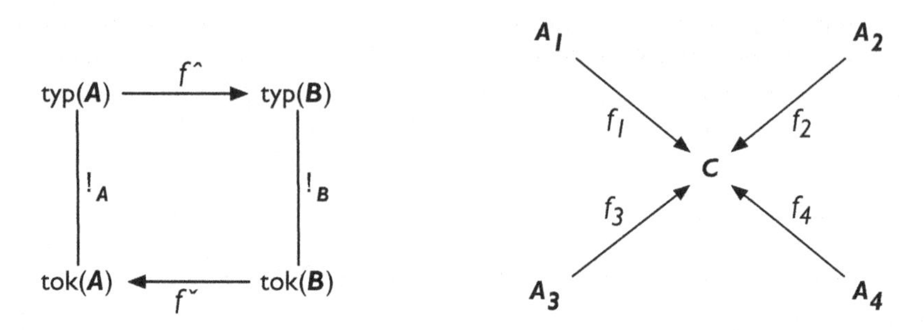

Fig. 1. Left: Two classifications (*A* and *B*) and an infomorphism (*f*) in Information Flow; Right: Channel $C = \{f_1, f_2, f_3, f_4\}$ and its core $\boldsymbol{C}$

Infomorphism. An *infomorphism* $f : \boldsymbol{A} \rightleftarrows \boldsymbol{B}$ from a classification $\boldsymbol{A}$ to a classification $\boldsymbol{B}$ is a (contravariant) pair of functions $f = \langle f^{\wedge}, f^{\vee} \rangle$ that satisfies the following condition:

$$f^{\vee}(b) \models_{\boldsymbol{A}} \alpha \quad \text{iff} \quad b \models_{\boldsymbol{B}} f^{\wedge}(\alpha)$$

for each token $b \in \text{tok}(\boldsymbol{B})$ and each type $\alpha \in \text{typ}(\boldsymbol{A})$, cf. figure 1.

Note that the type relation $f^{\wedge}$ and the token relation $f^{\vee}$ point in opposite directions. A mnemonic is that the $^{\wedge}$ of $f^{\wedge}$ points upwards, where the types are usually depicted.

Channel. A *channel* is a set of infomorphisms that have a common codomain. For example, the channel C shown in figure 1 consists of a family of four infomorphisms f_1 to f_4 that connect the four classifications $\boldsymbol{A}_1$ to $\boldsymbol{A}_4$ to the common codomain $\boldsymbol{C}$. The common codomain is the *core* of the channel. Note that the infomorphisms are all pairs of functions, i.e. $f_1 = \langle f_1^{\wedge}, f_1^{\vee} \rangle$, etc.

The core is the classification that contains the information connecting the tokens in the classifications $\boldsymbol{A}_1$ to $\boldsymbol{A}_4$. For this reason, the tokens of $\boldsymbol{C}$ are called *connections*. In our application to arithmetic the core is the arithmetic knowledge that represents what is common to the different source domains.

Channels and cores are the main way in which Information Flow achieves a distributed, localised representation of knowledge. In other words, this is the property of the Information Flow approach that fits to the localised representation and processing found in cognition. At the same time, infomorphisms provide a principled way of representing the connections between the different local contexts.

5 Formalisation of the Metaphors for Arithmetic

The basic idea of how to apply Information Flow theory to the four basic metaphors of Lakoff and Núñez (2000) is that each domain (object collection, object construction, measuring stick, motion along a path and arithmetic) is represented as a classification and the metaphors between the domains are infomorphisms.

Information Flow captures regularities in the distributed system (Barwise & Seligman, 1997, p. 8). So, the infomorphisms between the four source domains and the core (arithmetic) capture the regularities that link these domains to arithmetic and the knowledge shared by these domains. (A full arithmetic classification contains more than these commonalities – think of arithmetic concepts arising by linking metaphors like the concept of zero – but for our current purposes it suffices to think of it this way.)

In Guhe, Smaill, and Pease (2009) we present the object collection and object construction metaphors. We will, therefore, present the remaining two metaphors here. Note that we are only formalising the basics of the metaphors here. However, we chose the formalisation with respect to the required extensions, e.g. the iteration extension of the measuring stick metaphor, which should be rather straightforward.

5.1 Measuring Stick

The measuring stick metaphor, cf. table 1, is formalised with vectors, which facilitate the extensions proposed by Lakoff and Núñez (2000), e.g. for the metaphors creating the notion of square roots (p. 71). We define a classification ***MS*** where the tokens of the measuring stick domain are actual physical instances of measuring sticks encountered by the cognitive agent. We name the elements $\vec{s}_A$, $\vec{s}_B$, etc. The tokens are classified by their length, i.e. types are a set equivalent to the set of natural numbers $\mathbb{N}$. The classification relation $\models_{MS}$ classifies the measuring sticks by their length. Given this classification, we can now assign a type to each token, e.g. $\vec{s}_A \models_{MS} 3, \vec{s}_B \models_{MS} 5$.

Table 1. The measuring stick metaphor

measuring stick	arithmetic
physical segments	numbers
basic physical segment	one
length of the physical segment	size of the number
longer	greater
shorter	less
acts of physical segment placement	arithmetic operations
putting physical segments together end to end	addition
taking a smaller physical segment from a larger one	subtraction

- **Physical segments (consisting of parts of unit length):** A physical segment (of parts of unit length) is a vector (of multiples of the unit vector, i.e. the vector of length 1). All vectors in this classification are multiples of the unit vector and all vectors have the same orientation and direction.
- **Basic physical segment:** The basic physical segment is a vector $\vec{a}$ with $|\vec{a}| = 1$.
- **Length of physical segment:** The length of a physical segment is the length of the vector: $length_{MS}(\vec{s}) = |\vec{s}|$.
- **Longer/shorter:**

$$longer_{MS}(\vec{s}_A, \vec{s}_B) = \begin{cases} true, \text{ if } length_{MS}(\vec{s}_A) > length_{MS}(\vec{s}_B), \\ false, \text{ if } length_{MS}(\vec{s}_A) \leq length_{MS}(\vec{s}_B) \end{cases}$$

$shorter_{MS}$ is defined analogously.

- **Acts of physical segment placement:** Operations on vectors. Results of the segment placements (operations) are physical segments (vectors).
- **Putting physical segments together end-to-end with other physical segments:** This operation is defined as vector addition:

$$putTogether_{MS}(\vec{s}_A, \vec{s}_B) = \vec{s}_A + \vec{s}_B$$

  As all vectors have the same orientation, the length of the vector created by the vector addition is the sum of the lengths of the two original vectors.
- **Taking shorter physical segments from larger physical segments to form other physical segments:** This operation is also defined as vector addition, but this time the shorter vector ($\vec{s}_B$) is subtracted from the longer vector ($\vec{s}_A$):

$$takeSmaller_{MS}(\vec{s}_A, \vec{s}_B) = \vec{s}_A - \vec{s}_B$$

  As above, the vectors have the same orientation, so the lengths 'add up'. The corresponding type is the difference of the two lengths.

5.2 Motion along a Path

The motion along a path metaphor, cf. table 2, is formalised as sequences (lists), which we write as $\langle \ldots \rangle$. The elements of the sequences are $\bullet$ (any symbol will do here). The classification is called ***MP***. The sequences are named $path_A$, $path_B$, etc. Again the tokens are classified by their length; the type set is $\mathbb{N}$.

- **Acts of moving along the path:** Operations on sequences.
- **The origin, the beginning of the path:** The empty sequence: $\langle\rangle$. Because $\mathbb{N}$, with which model the arithmetic domain, does not contain 0, origins are not part of the model. We consider this the first (very straightforward) extension of the basic metaphors, which is reflected in the fact that historically 0 was a rather late invention.
- **Point-locations on a path:** Sequences of particular lengths, and the result of operations on sequences. The length of a sequence is given by the function $length_{MP}$, which returns the number of elements in the sequence.
- **The unit location, a point-location distinct from the origin:** The sequence with one element: $\langle\bullet\rangle$.

Table 2. The arithmetic is motion along a path metaphor

motion along a path	arithmetic
acts of moving along the path	arithmetic operations
a point location on the path	result of an operation; number
origin; beginning of the path	zero
unit location, a point location distinct from the origin	one
further from the origin than	greater
closer to the origin than	less
moving away from the origin a distance	addition
moving toward the origin a distance	subtraction

– **Further from the origin than/closer to the origin than:**

$$further_{MP}(path_A, path_B) = \begin{cases} true, \text{ if } length_{MP}(path_A) > length_{MP}(path_B), \\ false, \text{ if } length_{MP}(path_A) \leq length_{MP}(path_B) \end{cases}$$

$closer_{MP}$ is defined analogously.

– **Moving from a point-location A away from the origin, a distance that is the same as the distance from the origin to a point-location B:** Moving a distance B away from a point A that is different from the origin is the concatenation of the sequences:

$$moveAway_{MP}(path_A, path_B) = path_A \circ path_B$$

where $\circ$ concatenates two sequences in the usual fashion:

$$\langle \bullet_1 \bullet_2, \ldots \rangle \circ \langle \bullet_I, \bullet_{II}, \ldots \rangle = \langle \bullet_1 \bullet_2, \ldots, \bullet_I, \bullet_{II}, \ldots \rangle$$

The corresponding type is the addition of the two lengths.

– **Moving toward the origin from A, a distance that is the same as the distance from the origin to B:** For this operation to be possible, it must be the case that $length(path_A) > length(path_B)$.

$$towardOrigin_{MP}(path_A, path_B) = path_C, \text{where } path_A = path_B \circ path_C$$

The corresponding type is the difference of the two lengths.

5.3 Arithmetic

The classification ***AR*** representing arithmetic consists of the abstract natural numbers as tokens and concrete instances of natural numbers as types. The smallest number is 1. The arithmetic operations are defined as usual[1].

5.4 Metaphor Infomorphisms and Channel

Given these three classifications we can now define the channel C as the set of two infomorphisms ($C = \{f_{MS}, f_{MP}\}$) between the two classifications representing the source domains (***MS*** and ***MP***) and the core ***AR***.

Infomorphism from Measuring Stick to Arithmetic. The infomorphism linking the measuring stick collection domain to arithmetic is defined as $f_{MS} : \boldsymbol{MS} \rightleftarrows \boldsymbol{AR}$. The relation between types is then $f\hat{}_{MS}(n) = length_{MS}(\vec{s}_A)$, where $n \in \mathbb{N}$ and $length_{MS}(\vec{s}_A) = n$, the relation between tokens $f\check{}_{MS}(n) = \vec{s}_A$ where $n \in \mathbb{N}$ and $\vec{s}_A$ is a representation of the physical measuring stick that the number refers to.

The shortest measuring stick is $f\check{}_{MS}(\vec{s}_A) = 1$, where $length_{MS}(\vec{s}_A) = 1$ and the comparisons and operations are mapped as $f\hat{}_{MS}(longer_{MS}) = >$, $f\hat{}_{MS}(shorter_{MS}) = <$, $f\hat{}_{MS}(+) = +$, $f\hat{}_{MS}(-) = -$. Recall that the types for $putTogether_{MS}$ and $takeSmaller_{MS}$ are the sum/difference of the lengths of the vectors representing the tokens.

[1] Note that types can be structured. The natural way to look at >, for example, is to look at the sum of two classifications, each of which is some version of number: $\mathbb{N} \to \mathbb{N} + \mathbb{N} \leftarrow \mathbb{N}$. The tokens of $\mathbb{N} + \mathbb{N}$ are pairs of tokens (tk_1, tk_2) from the two copies, and the type that judges whether one token is 'greater' than the other returns a Boolean dependent on the token components tk_1, tk_2.

Infomorphism from Motion along a Path to Arithmetic. Similar to the definition above, the informorphism $f_{MP} : \boldsymbol{MP} \rightleftarrows \boldsymbol{AR}$ relates types as $f^{\wedge}{}_{MP}(n) = length_{MP}(path_A)$, where $n \in \mathbb{N}$ and $length_{MP}(path_A) = n$, and tokens as $f^{\vee}{}_{MP}(n) = path_A$, where $n \in \mathbb{N}$ and $path_A$ represents a path being referred to by the number. The other properties are defined as: $f^{\wedge}{}_{MP}(bigger_{MP}) = >$, $f^{\wedge}{}_{MP}(smaller_{MP}) = <$, $f^{\wedge}{}_{MP}(+) = +$, $f^{\wedge}{}_{MP}(-) = -$. As above, for $putTogether_{MP}$ and for $takeSmaller_{MP}$ the corresponding types are the sum/difference of the lengths of the sequences representing the tokens.

6 Conclusions and Future Work

Based on the work by Lakatos (1976) and by Lakoff and Núñez (2000) we argued that mathematics is not the uncovering of unchanging, eternal truths, but at its core it is a cognitive process. This means, that even if such truths exist, humans still discover them using cognitive processes. We demonstrated how Information Flow can be used to formalise the basic metaphors for arithmetic, i.e. to create a high-level cognitive model of these metaphors, and presented two of them (see Guhe, Smaill, & Pease, 2009 for the other two). We are currently working on an implementation of this formalisation that additionally uses quotient classifications (Barwise & Seligman, 1997, sec. 5.2).

References

Anderson, J.R.: How can the human mind occur in the physical universe? Oxford University Press, Oxford (2007)

Barwise, J., Seligman, J.: Information flow: The logic of distributed systems. Cambridge University Press, Cambridge (1997)

Gentner, D.: Structure-mapping: A theoretical framework for analogy. Cognitive Science 7(2), 155–170 (1983)

Gentner, D., Markman, A.B.: Structure mapping in analogy and similarity. American Psychologist 52(1), 45–56 (1997)

Guhe, M.: Incremental conceptualization for language production. Erlbaum, Mahwah (2007)

Guhe, M., Pease, A., Smaill, A.: A cognitive model of discovering commutativity. In: Proc. of the 31st Annual Conf. of the Cognitive Science Society (2009)

Guhe, M., Smaill, A., Pease, A.: Using information flow for modelling mathematical metaphors. In: Proc. of 9th Intern. Conf. on Cognitive Modeling (2009)

Lakatos, I.: Proofs and refutations: The logic of mathematical discovery. Cambridge University Press, Cambridge (1976)

Lakoff, G., Núñez, R.E.: Where mathematics comes from: How the embodied mind brings mathematics into being. Basic Books, New York (2000)

Lebiere, C.: The dynamics of cognition: An ACT-R model of cognitive arithmetic. Unpublished doctoral dissertation, CMU Computer Science Department (1998)

Nersessian, N.J.: Creating scientific concepts. MIT Press, Cambridge (2008)

Salvucci, D.D., Anderson, J.R.: Integrating analogical mapping and general problem solving: The path-mapping theory. Cognitive Science 25(1), 67–110 (2001)

Schunn, C.D., Anderson, J.R.: Scientific discovery. In: Anderson, J.R., Lebiere, C. (eds.) The atomic components of thought, pp. 385–427. Erlbaum, Mahwah (1998)

Hierarchical Clustering of Sensorimotor Features

Konrad Gadzicki

University of Bremen
28359 Bremen, Germany
konny@informatik.uni-bremen.de

Abstract. In this paper a method for clustering patterns represented by sets of sensorimotor features is introduced. Sensorimotor features as a biologically inspired representation have proofed to be working for the recognition task, but a method for unsupervised learning of classes from a set of patterns has been missing yet. By utilization of Self-Organizing Maps as a intermediate step, a hierarchy can be build with standard agglomerative clustering methods.

1 Motivation

The task of unsupervised discovering of meaningfull structures in data sets – formally known as clustering – is probably among the most covered in computer science. Without having any (or much) knowledge about the underlying structure of the data, one hopes that partition of class can be extracted from the similarity of patterns.

The goal of this work is investigate in how far semantic information can be found in pattern represented by sensorimotor features. Sensorimotor features are defined as two distinct points associated with some sensory data and connected by some relation. In the scope of this work, it means a saccade-like representation, but it is also possible to use it for instance in the spatial domain with two locations in space being connected by some motor actions of an agent[1,2].

The target output is a hierarchy of classes. The goal is thus not only to partition the space meaningfully but also to obtain a memory structure for further recognition tasks. The usage of a hierarchical memory is not only usefull from a computational point of view, but also agrees with cognitive memory structures. Psychological experiments give evidence for the existence of hierarchical propositional networks [3], sequence planning and execution [4], cognitive maps [5], memorizing of sequences of symbols [6] or hierarchical mental imagery[7].

In this work, the clustering of sensorimotor features will be performed on image sets.

The following section 2 will explain how images are represented by sensorimotor features and what those features actually consists of. Furthermore, Self-Organizing Maps and Agglomerative Hierarchical Clustering will be explained briefly in general since both are used in the overall clustering task. Section 3 presents the actual approach to clustering sensorimotor features and section 4 shows the results.

B. Mertsching, M. Hund, and Z. Aziz (Eds.): KI 2009, LNAI 5803, pp. 331–338, 2009.

2 Methods

2.1 Image Representation

Images are represented by a set of "eye movements" [8], mimicking the biological way of recognition of views by performing saccades. The *foveal spot* in the human eye – the only part with a high optical resolution – covers only a very narrow part of the view. Still humans are able to perceive the environment in detail by performing rapid eye movements, thus shifting the fixation location of the *fovea* around.

The data structure used to store an "eye movement" is called a sensorimotor feature. It consists of a triple $< feature, action, feature >$ (FAF) where the *features* contain some sensory data at an image location and the *action* stores the relative position change from the first to the second feature. The features are locally limited, resembling the narrow angle of the view field covered by the *foveal spot* during a fixation. The action corresponds with the relative shift of gaze.

Feature Extraction and Representation. The extraction of interesting fixation location is based on the concept of intrinsic dimensionality. This concept relates the degrees of freedom provided by a domain to the degrees of freedom actually used by a signal and states that the least redundant (the most informative) features in images are intrinsically two-dimensional signals (i2D-signals) [9,10].

For the actual extraction, nonlinear i2D-selective operators are applied to the image to find the fixation locations. Afterwards the feature vectors describing the local characteristics are generated by combining the outputs of linear orientation selective filters [8]. As a result, the foveal feature data structure stores the local opening angle, the orientation of the angle opening with respect to the image and the color.

The action structure stores the relation between two foveal features, containing the distance, the difference angle (the difference between the opening angles of those features) and the relation angle.

An image as a whole is represented as a set of such sensorimotor features. For a given number of fixation locations, each pair of *foveal features* can be connected by two *actions* since a sensorimotor feature is directional. In terms of graph theory this makes a fully connected directional graph with a cardinality of $|E| = |V| \cdot (|V| - 1)$ where an edge $|E|$ is an action and a vertex $|V|$ is a fixation location. The number of sensorimotor features in total is equivalent to the number of edges[1].

2.2 Self-Organizing Maps

A Self-Organizing Map (SOM)[11] is a competitive neural network, developed by Teuvo Kohonen in the 1980s [12].

[1] In practice, an 512x512 image has roughly 40–50 extracted foveal features.

A SOM realizes a mapping from a higher-dimensional input space to a two-dimensional grid while preserving the original topological information of the input space. The inspiration for these networks comes from topological structures in the human brain which are spatially organized according to the sensory input [13] (see [14,15] for an overview of research results).

Working Principle. The basic principle of the SOM is to adopt a neuron and its local area in order to make it fit better to a specific input, thus specializing individual nodes to particular inputs. The map as a whole converges to a state where certain areas of the map are specialized to certain parts of input space, so that the topological relations of the input space are preserved in the output space.

The training algorithm for the map is a iterative process during which the best-matching-unit (neuron with minimum distance to current input pattern) is found first. Afterwards the unit and its surrounding neurons are adapted to the current input by changing the weight vector of a neuron according to (1)

$$\mathbf{m}_n(t+1) = \mathbf{m}_n(t) + \alpha(t)\ h_{c_n}[\mathbf{x}(t) - \mathbf{m}_n(t)] \qquad (1)$$

where t denotes time. $\mathbf{x}(t)$ is an input vector at time t, the neighborhood kernel around the best-matching unit c is given as h_{c_n} and finally, the learning rate $\alpha(t)$ at a specific time.

2.3 Hierarchical Clustering

Classically clustering algorithms can be divided into hierarchical and partitioning ones. While the partitioning ones produce one, flat set of partitions, hierarchical construct nested partitions, resulting in a dendrogram. In a dendrogram each node represents a cluster; the original patterns are the leafs of the tree and thus singleton clusters.

Agglomerative hierarchical algorithms start with each pattern in a singleton cluster and merge them iteratively until only one cluster is left. The merging process is basically driven by linkage rules which define which two clusters will be merged in each step, and the similarity measure which is calculated between individual patterns that populate clusters. Both, linkage and similarity have a high impact on cluster quality.

Similarity Measures. Similarity is expressed within the range $[0\ldots1]$ where 1 states equality and 0 nothing in common at all.

Among the possible similarity measures, *Euclidean distance*-based measures seem to have the highest popularity. It can be derived from the more general L_p-norm (or *Minkowski* norm)

$$d_p(x,y) = \left(\sum_{i=1}^{n} |x_i - y_i|^p\right)^{\frac{1}{p}} . \qquad (2)$$

With $p = 2$ the *Minkowski* distance results in *Euclidean*.

The cosine measure is especially popular in document clustering. The similarity is expressed by the cosine of the angle between two vectors and measures similarity directly. It is given by

$$s_{cosine}(x_i, x_j) = \frac{x_i \cdot x_j}{\|x_i\|_2 \cdot \|x_j\|_2} \tag{3}$$

where $x_i \cdot x_j$ is the *dot*-product and $\|x\|_2$ is the Euclidean norm (p-norm with $p = 2$) of the vector.

The *Tanimoto* similarity captures the degree of overlap between two sets and is computed as the number of shared attributes.

$$s_{Tanimoto}(x_i, x_j) = \frac{x_i \cdot x_j}{\|x_i\|^2 + \|x_j\|^2 - x_i \cdot x_j} \tag{4}$$

with $x_i \cdot x_j$ being the *dot*-product and $\|x\|_2$ is the *Euclidean* norm of the vector. This measure is also referred to as *Tanimoto coefficient* (*TC*).

[16] gives a comparison about the behavior of the three above-mentioned measures.

Linkage Rules. Out of the number of available linkage rules, single and complete linkage are two extremes. Single linkage defines the distance between two clusters as the minimum distance between the patterns of these clusters. In contrast complete linkage takes the maximum distance between patterns. Both produce clusters of different shape: single linkage produces chain-like clusters which is usefull for filamentary data sets; complete linkage produces sphere-like clusters which works well with compact data.

In practice data sets often are not structured in a way suitable for the beforementioned linkage rules. Average based linkage rules incorporate all patterns from a pair of clusters in the distance calculation. There are several variants like "Average Linkage Between Groups" which takes the average distance between all pairs of patterns or "Average Linkage Within Group" which takes the average distance between all possible pairs of patterns if they formed a single cluster. "Ward's method" [17] aims at minimizing the intra-cluster variance by forming a hypothetical cluster and summing the squares of the within-cluster distances to a centroid. The average linkage rules, especially "Ward's" linkage work pretty robust on arbitrary data.

3 Hierarchical Image Clustering

3.1 Comparison of Sensorimotor Features and Images

The problem of comparing pairs of sensorimotor features brings up some problems. In [8], where the task was image recognition, those features were treated as symbols with a particular *FAF* giving evidence for a set of scenes. The initial sensory measurements – angles, distances and colors – are numerical values which can be measured rather precisely. Still 1:1 comparison of such values turned out to be not very robust to slight transformations. The mapping of the original values to intervals leads to a more robust representation with slightly different values being mapped to the same interval.

The calculation of similarity between images with features treated as symbols allows only the usage of usage of similarity measures based on set operations. Initial tests with such similarity measures produced only mediocre results and led to the idea of a numerical representation. Similar to approaches from document clustering [18] a vector representation for an image was used. Such a representation allows for the usage of numerical similarity measures like those mentioned above.

3.2 Obtaining a Numerical Image Representation

For obtaining the numerical representation, firstly the sensorimotor features were extracted from a given set of images. A Self-Organizing Map was then trained with the entire set of features. That way the system learned how to group sets of features for a given output size (the map size) which corresponds with the number of components of the vector used for image representation.

The actual representation for a particular image is obtained by presenting the sensorimotor features associated with the image to the SOM. For each feature, a particular map node will be activated, being the Best-Matching-Unit. By counting the activations of map nodes, a histogram is generated for each image which serves as the vector for similarity measurement (fig. 1).

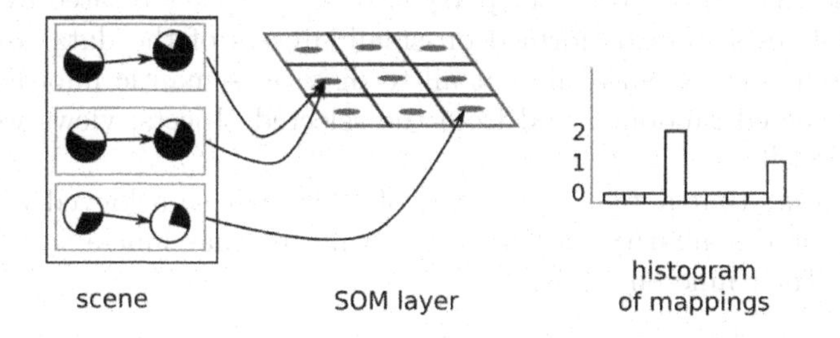

Fig. 1. From sensorimotor representation of image to histogram. The features representing an image are mapped to a previously trained SOM. A histogram representation is obtained by counting those mappings.

3.3 Hierarchical Clustering and Quality Assessment

The actual generation of the hierarchy was with agglomerative hierarchical clustering (see 2.3), with all combinations of the five linkage rules and the three similarity measures.

For the assessment of the quality of the produced hierarchy, the *f-measure* is a suitable measure[19]. It is computed from two other measures, *precision* and *recall*. *Precision* states the fraction of a cluster that belongs to a particular class and *recall* expresses the fraction of objects from a particular class found in a particular cluster out of all objects of that class.

The *f-measure* is a combination of precision and recall. It states to which extend a cluster x contains only and all objects of class c and is given by

$$F(x,c) = \frac{2 \cdot precision(x,c) \cdot recall(x,c)}{precision(x,c) + recall(x,c)} . \quad (5)$$

The values produced by the measure are in the range of $[0 \dots 1]$. A value of 1 means that a cluster is entirely populated by objects from one class and no other. If computed for a hierarchy the *f-measure* is computed for every node of the tree for a given class and the maximum is returned.

The interpretation of a value of 1 is that there is a particular subtree of the hierarchy with objects of class c only and no other. The overall *f-measure* of a hierarchy is then basically the sum of weighted *f-measures* for all classes

$$F = \sum_c \frac{n_c}{n} F(c) = \sum_c \frac{n_c}{n} \max_x(F(x,c)) \quad (6)$$

with n_c being the number of objects in cluster c and n is total number objects.

4 Results and Discussion

The performance has been tested with the "Columbia Object Image Library" (COIL-20)[20]. This image database consists of 20 objects photographed under stable conditions from 72 different perspectives, each view rotated by 5°.

The initial tests were performed on small subsets of the database in order to see whether the system is able at all to discover semantic information. Objects were selected randomly and from the selected objects, views were picked randomly as well.

Table 1 shows that a *f-measure* value of 1.000 can be achieved with certain linkage rules and similarity measures which means that image were separated according to their inherent class.

Table 1. *F-measure* for small COIL-20 subset for combinations of linkage rules and similarity measures

	Euclidean	Cosine	Tanimoto
Single	0.656	0.794	0.861
Complete	0.675	0.905	0.915
Average between groups	0.656	0.915	1.000
Average within group	0.656	0.915	0.915
Ward	0.675	1.000	1.000

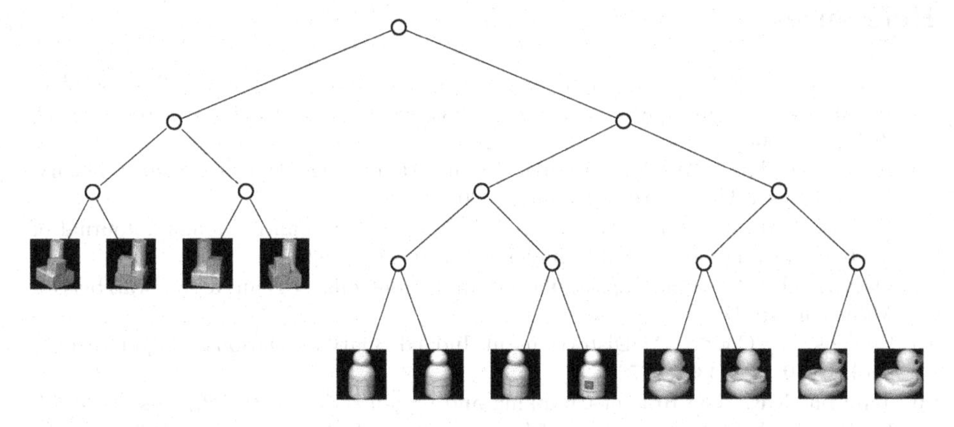

Fig. 2. Generated hierarchy with Tanimoto distance and Ward's linkage

Figure 2 shows the hierarchy generated

When tested with the full COIL-20 set, consisting of 1440 image, the figures drop significantly. With *Tanimoto* similarity measure and *Ward's* linkage rule, a *f-measure* of 0.358 can be obtained. Inspecting the generated hierarchy shows that the rough similarities have been captured. For instance, round-shaped patterns populate a particular subtree, but patterns from different classes are mixed.

Based on the results above, you can say that this method is able to capture semantic information in small scale, but on a large scale, the separation does not work well enough.

5 Summary

The hierarchical clustering approach introduced works on patterns represented by sets of sensorimotor features. By mapping the sensorimotor features to a Self-Organizing Map, a fixed-size vector representation of patterns is produced. The mapping of the set of sensorimotor features of a particular pattern to the output layer of the SOM generates a histogram of SOM activations which serves as a representation of the pattern. Being a numerical representation, it can be further processed with standard agglomerative hierarchical clustering methods in order to produced the hierarchy.

Acknowledgments

Supported by SFB/TR8.

References

1. Zetzsche, C., Wolter, J., Schill, K.: Sensorimotor representation and knowledge-based reasoning for spatial exploration and localisation. Cognitive Processing 9, 283–297 (2008)
2. Reineking, T.: Active Vision-based Localization using Dempster-Shafer Theory. Mastersthesis, University of Bremen (2008)
3. Collins, A.M., Quillian, M.R.: Retrieval Time from Semantic Memory. Journal of Verbal Learning and Verbal Behaviour 8, 240–247 (1969)
4. Oesterreich, R.: Handlungsregulation und Kontrolle. Urban & Schwarzberger, München (1981)
5. Stevens, A., Coupe, P.: Distortions in Judged Spatial Relations. Cognitive Psychology 10, 422–437 (1978)
6. Johnson, N.F.: The Role of Chunking and Organization in the Process of Recall. In: Bower, G. (ed.) Psychology of Language and Motivation, vol. 4 (1970)
7. Reed, S.K.: Structural Descriptions and the Limitations of Visual Images. Memory and Cognition 2, 329–336 (1974)
8. Schill, K., Umkehrer, E., Beinlich, S., Krieger, G., Zetzsche, C.: Scene analysis with saccadic eye movements: Top-down and bottom-up modeling. Journal of Electronic Imaging 10, 152–160 (2001)
9. Zetzsche, C., Nuding, U.: Nonlinear and higher-order approaches to the encoding of natural scenes. Network: Computation in Neural Systems 16, 191–221 (2005)
10. Zetzsche, C., Krieger, G.: Intrinsic dimensionality: nonlinear image operators and higher-order statistics. In: Nonlinear image processing, pp. 403–441. Academic Press, Orlando (2001)
11. Kohonen, T.: Self-Organizing Maps. Springer, Berlin (2001)
12. Kohonen, T.: Self Organized Formation of Topologically Correct Feature Maps. Biological Cybernetics 43, 59–69 (1982)
13. Kangas, J.: On the Analysis of Pattern Sequences by Self-Organizing Maps. PhD thesis, Helsinki University of Technology (1994), `http://nucleus.hut.fi/~jari/papers/thesis94.ps.Z`
14. Knudsen, E.I., du Lac, S., Esterly, S.D.: Computational maps in the brain. Annu. Rev. Neurosci. 10, 41–65 (1987)
15. Anderson, J.A., Pellionisz, A., Rosenfeld, E.: Neurocomputing: directions for research, vol. 2. MIT Press, Cambridge (1990)
16. Ghosh, J., Strehl, A.: Similarity-Based Text Clustering: A Comparative Study. In: Kogan, J., Nicholas, C., Teboulle, M. (eds.) Grouping Multidimensional Data. Springer, Berlin (2006)
17. Ward, J.H.: Hierarchical Grouping to Optimize an Objective Function. Journal of the American Statistical Association 58, 236–244 (1963)
18. Salton, G., Wong, A., Yang, C.S.: A Vector Space Model for Automatic Indexing. Communications of the ACM 18, 613–620 (1975)
19. Larsen, B., Aone, C.: Fast and effective text mining using linear-time document clustering. In: KDD 1999 Proceedings of the fifth ACM SIGKDD international conference on Knowledge discovery and data mining, pp. 16–22 (1999)
20. Nene, S.A., Nayar, S.K., Murase, H.: Columbia Object Image Library (COIL-20). Technical Report CUCS-005-96 (1996)

P300 Detection Based on Feature Extraction in On-line Brain-Computer Interface

Nikolay Chumerin[1,*], Nikolay V. Manyakov[1,*], Adrien Combaz[1,*], Johan A.K. Suykens[2,**], Refet Firat Yazicioglu[3], Tom Torfs[3], Patrick Merken[3], Herc P. Neves[3], Chris Van Hoof[3], and Marc M. Van Hulle[3]

[1] Laboratorium voor Neuro- en Psychofysiologie, K.U. Leuven, Herestraat 49, bus 1021, 3000 Leuven, Belgium
[2] ESAT-SCD, K.U. Leuven, Kasteelpark Arenberg 10, 3001 Heverlee, Belgium
[3] IMEC, Kapeldreef 75, 3001 Leuven, Belgium
{Nikolay.Chumerin,NikolayV.Manyakov,Adrien.Combaz, Marc.VanHulle}@med.kuleuven.be,
Johan.Suykens@esat.kuleuven.be,
{Firat,TorfsT,Patrick.Merken,Chris.VanHoof,Herc}@imec.be

Abstract. We propose a new EEG-based wireless brain computer interface (BCI) with which subjects can "mind-type" text on a computer screen. The application is based on detecting P300 event-related potentials in EEG signals recorded on the scalp of the subject. The BCI uses a simple classifier which relies on a linear feature extraction approach. The accuracy of the presented system is comparable to the state-of-the-art for on-line P300 detection, but with the additional benefit that its much simpler design supports a power-efficient on-chip implementation.

1 Introduction

Research on *brain computer interfaces* (BCIs) has witnessed a tremendous development in recent years (see, for example, the editorial in Nature [1]), and is now widely considered as one of the most successful applications of the neurosciences. BCIs can significantly improve the quality of life of neurologically impaired patients with pathologies such as: amyotrophic lateral sclerosis, brain stroke, brain/spinal cord injury, cerebral palsy, muscular dystrophy, etc.

* NC is supported by the European Commission (STREP-2002-016276), NVM and AC are supported by the European Commission (IST-2004-027017), MMVH is supported by research grants received from the Excellence Financing program (EF 2005) and the CREA Financing program (CREA/07/027) of the K.U.Leuven, the Belgian Fund for Scientific Research – Flanders (G.0234.04 and G.0588.09), the Interuniversity Attraction Poles Programme – Belgian Science Policy (IUAP P5/04), the Flemish Regional Ministry of Education (Belgium) (GOA 2000/11), and the European Commission (STREP-2002-016276, IST-2004-027017, and IST-2007-217077).

** JAKS acknowledges support of the Fund for Scientific Research Flanders (G.0588.09).

B. Mertsching, M. Hund, and Z. Aziz (Eds.): KI 2009, LNAI 5803, pp. 339–346, 2009.

Brain computer interfaces are either *invasive* (intra-cranial) or *noninvasive.* The first ones have electrodes implanted usually into the premotor- or motor frontal areas or into the parietal cortex (see review in [2]), whereas the non-invasive ones mostly employ *electroencephalograms* (EEGs) recorded from the subject's scalp. The noninvasive methods can be further subdivided into three groups. The first group explores *visually evoked potentials* (VEPs) and they can be traced back to the 70s, when Jacques Vidal constructed the first BCI [3]. The second group of noninvasive BCIs rely on the detection of imaginary movements of the right or the left hand [4]. The third noninvasive group are the BCIs that rely on the 'oddbal' evoked potential in the parietal cortex, and is the topic of this article.

An *event-related potential* (ERP) is a stereotyped electrophysiological response to an internal or external stimulus [5]. One of the most known and explored ERPs is the P300. It can be detected while the subject is classifying two types of events with one of the events occurring much less frequently than the other (rare event). The rare events elicit ERPs consisting of an enhanced positive-going signal component with a latency of about 300 ms [6].

In order to detect the ERP in the signal, one trial is usually not enough and several trials must be averaged. The averaging is necessary because the recorded signal is a superposition of *all* ongoing brain activities as well as noise. By averaging the recordings, the activites that are time-locked to a known event (*e.g.*, onset of attended stimulus) are extracted as ERPs, whereas those that are not related to the stimulus presentation are averaged out. The stronger the ERP signal, the fewer trials are needed, and *vice versa.*

A number of off-line studies have been reported that improve the classification rate of the P300 speller [7, 8, 9], but not much work has been done on on-line classification. To the best of our knowledge, the best on-line classification rate for mind-typers is reported in [10]. For a decent review of BCIs, which is out of the scope of this study, see [11].

The BCI system descibed in this article is an elaboration of the P300-based BCI but with emphasis on a simple design for a power-efficient on-chip implementation.

2 Methods

2.1 Acquisition Hardware

The EEG recordings were performed using a prototype of an ultra low-power 8-channel wireless EEG system (see Fig. 1). This system was developed by IMEC partner and is built around their ultra-low power 8-channel EEG amplifier chip [12]. The EEG signals are μV-range low-frequency signals that are correlated with a large amount of common-mode interference. This requires the use of a high performance amplifier with low-noise and high *common-mode rejection ratio* (CMRR). IMEC's proprietary 8-channel EEG ASIC consumes only 300 μW from a single 2.7 – 3.3 V supply. Each channel of the ASIC consists of an AC coupled chopped instrumentation amplifier, a chopping spike filter and

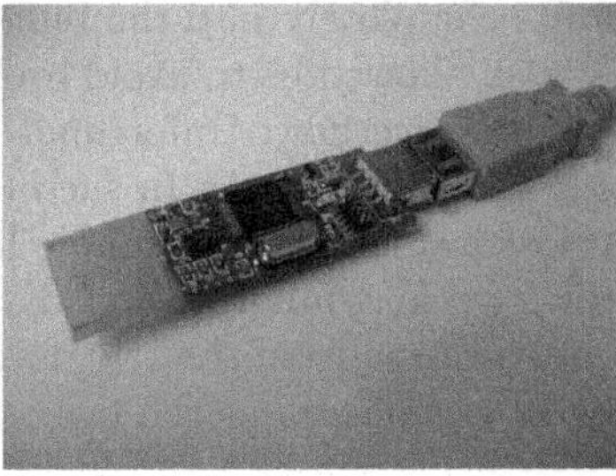

Fig. 1. Wireless 8 channel EEG system: amplifier and transmitter (left) and USB stick receiver, plugged into the extension cable (right)

a variable gain stage, and achieves 80 $nV/Hz^{\frac{1}{2}}$ input referred noise density and 130 dB CMRR at 50 Hz, while consuming 8.5 µA. The gain of the amplifier is digitally programmable between 2000 and 9000 and the higher cut-off bandwidth is digitally programmable between 52 Hz and 274 Hz. The input impedance exceeds 100 MΩ.

The wireless system uses a low-power microcontroller (Texas Instruments MSP430) combined with a low-power 2.4 GHz radio (Nordic nRF2401). In operational mode, the EEG signal is sampled at 1000 Hz with 12 bit resolution. These samples are collected in packets and transmitted in bursts at a data rate of 1 Mb/s to the receiver, which is connected through a USB interface to the PC. The average total power consumption of the system in operational mode is 18 mW (6 mA at 3 V). This implies it can be operated for more than one week continuously on two AAA batteries before the battery voltage drops below specification. At start-up, the system's parameters (such as gain and bandwidth settings) can be changed wirelessly.

The system also includes an electrode impedance measurement circuit that can measure the electrode impedances in the range of 1 kΩ to 6 MΩ. The system is designed for use with Ag/AgCl electrodes. For these experiments we have used a braincap with large filling holes and sockets for ring electrodes.

2.2 Acquisition Procedure

Recordings were collected from eight electrodes in the parietal and occipital areas, namely in positions Cz, CPz, P1, Pz, P2, PO3, POz, PO4, according to the international 10–20 system. The reference electrode and ground were linked to the left and right mastoids.

Each experiment started with a pause (approximately 90 second) needed for the stabilization of the EEG acquisition device. During this period, the EEG device transmits data but it is not recorded. The data for each symbol presentation was recorded in one session. As the duration of the session was known *a-priori*, as well as the data transfer rate, it was easy to estimate the amount of data transmitted during a session. We used this estimate, increased by a 10% margin, as the size of the serial port buffer. To make sure that the entire recording session for one symbol fits completely into the buffer, we cleared the buffer just before recording. This trick allowed us to avoid broken/lost data frames, which usually occur due to a buffer overflow. Unfortunately, sometimes data frames are still lost because of a bad radio signal. In such cases, we used the frame counter to reconstruct the lost frames, using a simple linear interpolation.

The overhead for one-symbol-session data reading from the buffer and the EEG signal reconstruction from the raw data, appeared to be negligible in Matlab, thus the latter was chosen as the main development environment.

2.3 Data-Stimuli Synchronization

Unlike a conventional EEG systems, the system we used does not have any external synchronization inputs. We tried to use one of the channels for this purpose (connecting it to a photo-sensor attached to the screen), but this scheme was not stable enough for long recording times. Finally, we came up with an "internal" synchronization scheme based on high-precision (up to hectananosecond) timing[1].

For the synchronization, we saved the exact time stamps of the start and end points of the recording session, as well as the time stamps of stimulus onsets and offsets. Due to the fact that the reconstructed EEG signal has a constant sampling rate, it is possible to find very precise correspondences between time stamps and data samples. We used this correspondence mapping for partitioning the EEG signal into signal tracks, for further processing.

2.4 Experiment Design

Four healthy male subjects (aged 23–36 with average age of 31, three righthanded and one lefthanded) participated in the experiments. Each experiment was composed of one training- and several testing stages.

We used the same visual stimulus paradigm as in the first P300-based speller, introduced by Farwell and Donchin in [13]: a matrix of 6×6 symbols. The only (minor) difference was in the type of symbols used, which in our case was a set of 26 latin characters, eight digits and two special symbols '_' (used instead of space) and '¶' (used as an *end of input* indicator).

During the training and testing stages, columns and rows of the matrix were intensified in a random manner. The intensification duration was 100 ms, followed by a 100 ms of no intensification. Each column and each row flashed only once during one trial, so each trial consisted of 12 stimulus presentations (6 rows and 6 columns).

As it was mentioned in introduction, one trial is not enough for robust ERP detection, hence, we adopted the common practise of averaging the recordings over several trials before performing the classification of the (averaged) recordings.

---

[1] TSCtime high-precision time library by Keith Wansbrough
`http://www.lochan.org/2005/keith-cl/useful/win32time.html`

During the training stage, all 36 symbols from the typing matrix were presented to the subject. Each symbol had 10 trials of intensification for each row/column (10-fold averaging). The subject was asked to count the number of intensifications of the attended symbol. The counting was used only for keeping the subject's attention to the symbol. The recorded data were filtered in the $0.5 - 15$ Hz frequency band with a fourth order zero-phase digital Butterworth filter, and cut into signal tracks. Each of these tracks consisted of 1000 ms of recording, starting from stimulus onset. Note that subsequent tracks overlap in time, since the time between two consequent stimuli onsets is 200 ms. Then, each of these tracks was downsampled to 30 tabs and assigned to one of two possible groups: *target* and *nontarget* (according to the stimuli, which they were locked to).

After training the classifier, each subject performed several test sessions and where he was asked to *mind-type* a few words (about 30–50 symbols), the performance of which was used for estimating the classification accuracy. The number of trials k that was used for averaging varied from 1 to 10. For each value of k the experiment was repeated and the classification accuracy was (re)measured. This experiment design differs from the one proposed in [10], where recordings and on-line classification were done only for $k = 10$ and the evaluation of the method for the cases $k = 1, \ldots, 9$ was done off-line using the same data. Our design is more time consuming, but it provided us with more independent data for analisys and evaluation, which is important in experiments with limited number of subjects.

2.5 Classification

In the proposed system the training stage of the classifier differs from its testing stage not only by the classification step, but also by the way of grouping the signal tracks. During the training, the system "knows" exactly which one of 36 possible symbols is attended by the subject at any moment of time. Based on this information, the collected signal tracks can be grouped into only two categories: target (attended) and non-target (not attended). However, during testing, the system does not know which symbol is attended by the subject, and the only meaningful way of grouping is by stimulus type (which in the proposed paradigm can be one of 12 types: 6 rows and 6 columns). So, during the testing stage, for each trial, we had 12 tracks (from all 12 types) of 1000 ms EEG data recorded from each electrode. The averaged (along trials) EEG response for each electrode was determined for each group. Then all 12 averaged tracks were sequentially fed to the feature extractor (see section 2.6), which extracted a scalar feature y_i for each track i. In order to decide which symbol was attended, the classifier selected the best "row candidate" and the best "column candidate" among features $(y_1, \ldots, y_{12})$ of all tracks, thus the row index i_r and the column index i_c of the classified symbol were calculated as:

$$i_r = \arg\max_{i=1,\ldots,6}\{y_i\}, \text{ and } i_c = \arg\max_{i=7,\ldots,12}\{y_i\} - 6.$$

The symbol on the intersection of the i_r-th row and i_c-th column in the matrix, was then taken as the result of the classification and presented, as a feedback, to the subject.

2.6 Feature Extraction

In order to classify averaged and subsampled EEG recordings into target and nontarget classes we used the one-dimensional version of a linear *feature extraction* (FE) approach proposed by Leiva-Murillo and Artés-Rodríguez in [14]. As all linear FE methods, this method consider as features projections of the centered input vectors $\mathcal{X} = \{\mathbf{x}_i : \mathbf{x}_i \in \mathbb{R}^D\}$ onto appropriate d-dimensional subspace ($d < D$), and the task is to find this subspace. The method searhes for the "optimal" subspace maximizing (an estimate of) mutual information between the set of projections $\mathcal{Y} = \{\mathbf{W}^T\mathbf{x}_i\}$ and the set of corresponded labels $\mathcal{C} = \{\mathsf{c}_i\}$. In the proposed system the class labels set consists of only two classes, thus we set $\mathcal{C} = \{-1, +1\}$. The dimensionality d of the desired subspace was set to 1, because it appeared to be enough to achieve a robust separation of the two classes, but compared to higher dimensionalities (we also tried $d = 2, 3$), it is computationally much cheaper. In the case of $d = 1$ the subspace is represented only by one vector $\mathbf{w} \in \mathbb{R}^D$ and projections are scalars $y_i = \mathbf{w}^T\mathbf{x}_i \in \mathbb{R}^1$. According [14], the mutual information between the set of projections $\mathcal{Y}$ and the set of corresponded labels $\mathcal{C}$ can be estimated as:

$$I(\mathcal{Y}, \mathcal{C}) = \sum_{p=1}^{N_c} p(\mathsf{c}_p) \left(J(\mathcal{Y}|\mathsf{c}_p) - \log \sigma(\mathcal{Y}|\mathsf{c}_p) \right) - J(\mathcal{Y}),$$

with $N_c = 2$ the number of classes, $\mathcal{Y}|\mathsf{c}_p$ the projection of the p-th class' data points onto the direction $\mathbf{w}$, $\sigma(\,\cdot\,)$ the standard deviation and $J(\,\cdot\,)$ the negentropy estimated using Hyvärinen's robust estimator [15].

The input vector $\mathbf{x}$ is constructed as a concatenation of the subsampled and filtered EEG data: $\mathbf{x} = (x_{11}, x_{12}, \ldots, x_{1K}, x_{21}, x_{22}, \ldots, x_{2K}, \ldots, x_{N1}, \ldots, x_{NK})^T$, where K is number of subsamples, and N is number of channels. In our experiments we used $K = 30, N = 8$ and $D = KN = 240$.

We also tried FE method proposed by Torkkola in [16], which in the one-dimensional case is almost equivalent to the considered one, but it is slightly more computationally expensive. In multi-dimensional cases, the method by Leiva-Murillo's outperforms Torkkola's method quantatively (in terms of the mutual information between classes and projections), as well as computationally (in terms of the number of floating point operations) as was shown by us in an earlier study [17].

3 Results and Discussion

The performance of each subject in mind-typing with our system is displayed in Fig. 2, where the percentage of correctly-typed symbols is plotted versus the

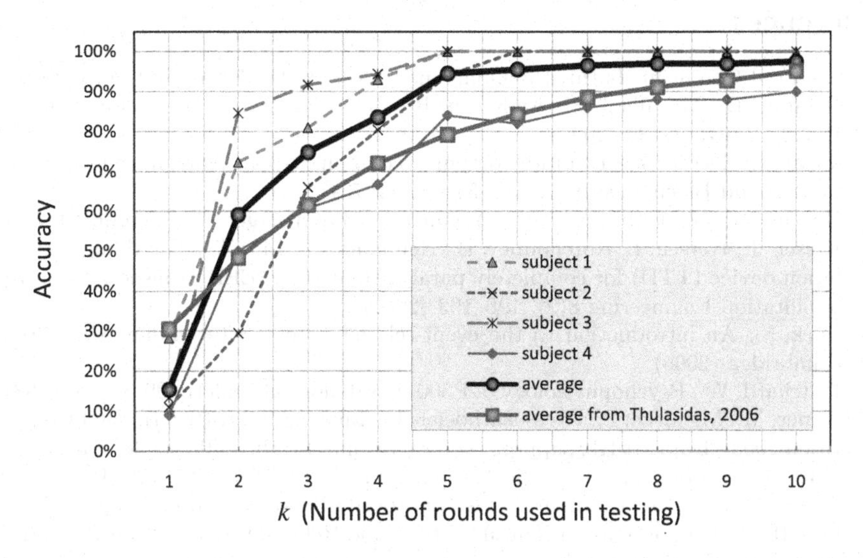

Fig. 2. Accuracy of classification for different subjects as a function of the number of trials used in testing. Averaged result and result from [10] are also plotted.

number of trials k used for averaging. The average performance of all subjects, as well as the average performance of the best mind-typing system described in the literature [10], are also plotted. It should be mentioned that the mind-typing system of Thulasidas and co-workers is based on a *support-vector machine* (SVM) classifier with a Gaussian kernel, which is usually trained using a grid-search procedure for optimal parameter selection. The training of the SVM classifier with nonlinear kernel takes substantially more time than the training of the FE-based linear classifier used in our system.

Another consideration is that the on-chip implementation of the SVM classifier is more complex than our solution, due to the presence of nonlinearity in the kernel-based function.

As it is clear from Fig. 2, the performance strongly depends on the subject. However, we hasten to add that in order to draw any statistically-grounded conclusions from only four subjects, many more experiments need to be performed.

4 Conclusions

The brain-computer interface (BCI) presented in this article allows the subject to type text by detecting P300 potentials in EEG signals. The proposed BCI consists of a EEG system and a classifier which is based on linear feature extraction. The simplicity of the proposed system supports an efficient on-chip implementation (*e.g.*, on an ASIC).

The results of this study shows that, in the field of BCIs based on event-related potentials, even simple solutions can successfully compete with the state-of-the-art systems.

References

[1] Editorial Comment: Is this the bionic man? Nature 442(7099), 109 (July 2006)
[2] Pesaran, B., Musallam, S., Andersen, R.: Cognitive neural prosthetics. Current Biology 16(3), 77–80 (2006)
[3] Vidal, J.: Toward direct brain-computer communication. Annual review of Biophysics and Bioengineering 2(1), 157–180 (1973)
[4] Birbaumer, N., Kubler, A., Ghanayim, N., Hinterberger, T., Perelmouter, J., Kaiser, J., Iversen, I., Kotchoubey, B., Neumann, N., Flor, H.: The thought translation device (TTD) for completely paralyzedpatients. IEEE Transactions on Rehabilitation Engineering 8(2), 190–193 (2000)
[5] Luck, S.: An introduction to the event-related potential technique. MIT Press, Cambridge (2005)
[6] Pritchard, W.: Psychophysiology of P300. Psychological Bulletin 89(3), 506 (1981)
[7] Kaper, M., Meinicke, P., Grossekathoefer, U., Lingner, T., Ritter, H.: BCI Competition 2003 - Data set IIb: support vector machines for the P300 speller paradigme. IEEE Transactions on Biomedical Engineering 51(6), 1073–1076 (2004)
[8] Serby, H., Yom-Tov, E., Inbar, G.: An improved P300-based brain-computer interface. IEEE Transactions on Neural Systems and Rehabilitation Engineering 13(1), 89–98 (2005)
[9] Xu, N., Gao, X., Hong, B., Miao, X., Gao, S., Yang, F.: BCI Competition 2003 - Data set IIb: enhancing P300 wave detection using ICA-based subspace projections for BCI applications. IEEE Transactions on Biomedical Engineering 51(2), 1067–1072 (2004)
[10] Thulasidas, M., Guan, C., Wu, J.: Robust classification of EEG signal for brain-computer interface. IEEE Transactions on Neural Systems and Rehabilitation Engineering 14(1), 24–29 (2006)
[11] Dornhege, G., Millán, J., Hinterberger, T., McFarland, D., Sejnowski, T., Muller, K.: Toward brain-computer interfacing. The MIT Press, Cambridge (2007)
[12] Yazicioglu, R., Merken, P., Puers, R., Van Hoof, C.: Low-power low-noise 8-channel EEG front-end ASIC for ambulatory acquisition systems. In: Proceedings of the 32nd European Solid-State Circuits Conference. ESSCIRC 2006, pp. 247–250 (2006)
[13] Farwell, L., Donchin, E.: Talking off the top of your head: toward a mental prosthesis utilizing event-related brain potentials. Electroencephalography and clinical Neurophysiology 70(6), 510–523 (1988)
[14] Leiva-Murillo, J., Artes-Rodriguez, A.: Maximization of mutual information for supervised linear feature extraction. IEEE Transactions on Neural Networks 18(5), 1433–1441 (2007)
[15] Hyvärinen, A.: New approximations of differential entropy for independent component analysis and projection pursuit. In: Proceedings of the 1997 conference on Advances in neural information processing systems 10 table of contents, pp. 273–279. MIT Press, Cambridge (1998)
[16] Torkkola, K., Campbell, W.: Mutual information in learning feature transformations. In: Proceedings of the 17th International Conference on Machine Learning (2000)
[17] Chumerin, N., Van Hulle, M.: Comparison of Two Feature Extraction Methods Based on Maximization of Mutual Information. In: Proceedings of the 2006 16th IEEE Signal Processing Society Workshop on Machine Learning for Signal Processing, pp. 343–348 (2006)

Human Perception Based Counterfeit Detection for Automated Teller Machines

Taswar Iqbal[1], Volker Lohweg[1], Dinh Khoi Le[2], and Michael Nolte[2]

[1] Institut Industrial IT, Ostwestfalen-Lippe University of Applied Science, 32657 Lemgo, Germany
[2] Wincor Nixdorf International GmbH, Heinz-Nixdorf-Ring 1, 33106 Paderborn, Germany
{taswar.iqbal,volker.lohweg}@hs-owl.de, {dinh-khoi.le,michael.nolte}@wincor-nixdorf.com

Abstract. A robust vision system for the counterfeit detection of bank ATM keyboards is presented. The approach is based on the continuous inspection of a keyboard surface by the authenticity verification of coded covert surface features. For the surface coding suitable visual patterns on the keyboard are selected while considering constraints from the visual imperceptibility, robustness and geometrical disturbances to be encountered from the aging effects. The system's robustness against varying camera-keyboard distances, lighting conditions and dirt-and-scratches effects is investigated. Finally, a demonstrator working in real-time is developed in order to publicly demonstrate the surface authentication process.

Keywords: ATMs, human perception, counterfeit resistance, digital authentication, surface coding, pattern recognition.

1 Motivation and Problem Formulation

The sensitive nature of the activities handled by the widely used ATMs and access of the adversaries to the advanced technologies have posed new challenges for the ATM industry. One such new problem is the usage of counterfeited parts by the adversaries to get the user Personal Identification Numbers (PIN) or some other important information [1], [4], [6]. Different business models (e.g. integrating more services) and varying working conditions (e.g. semi-secure service points) for ATMs have also set new requirements. In [4] foreign objects such as the spy cameras attached to the ATM or in neighborhood region to get user PIN are detected by subtracting the background image. It is mentionable that when the attached objects are not hidden these would be visible.

In this paper a new concept is proposed for an intelligent security system combining data hiding and pattern recognition techniques under the constraints of human perception is proposed. The term human perception in this context means that the coded surface is imperceptible by the unaided human visual system. A security system, which continuously authenticates the target surface by checking the patterns

B. Mertsching, M. Hund, and Z. Aziz (Eds.): KI 2009, LNAI 5803, pp. 347–354, 2009.

for the invisible coded symbols on the metallic keyboard surface and/or the neighboring regions is highlighted.

This paper is organized as follows: In Sect. 2 a surface coding technique for metal surfaces is given. Section 3 deals with surface authentication technique. Section 4 discusses the performance of the counterfeit detection system. Finally, conclusions are given in Sect. 5.

2 ATM Keyboard Surface Coding

While considering metal keyboard surface coding, the hard-stamping process and laser engraving techniques for metal surfaces result in constant-tone surface patterns, such as text on the keyboard, background patterns and logo images. Such image patterns shrink the domain for the potential feature modifications, which are necessary for surface coding.

2.1 Data Encoding Technique for Keyboard Surface

According to this technique the data is encoded in ATM keyboard surfaces by using the visual patterns (characters, numerals and symbols) on the surface by imperceptibly modifying a given feature. The data encoding features consist of two types of patterns: data encoding symbols (CS) and reference marks (RM). In the data encoding process, a CS is shifted by a small step-size (imperceptible) with respect to its nearby RM [5], which is left unmodified and is used afterwards in the data decoding process.

The data encoding process for a feature set **W**, according to the shift-based (SH) feature modification technique, can be formally described as follows:

$$\forall i,\, i=1\ldots N,\; \boldsymbol{b}=[101\ldots011],$$
$$w'(i)=\Psi\left(w(i),dt,dt_{\text{mod}}\right),\; dt=n\cdot\Delta T,\; dt_{\text{mod}}=\lambda \text{ for } b(i)=0 \text{ or } 1. \quad (1)$$

Here, $w'(i)$ is the resulting encoded feature. The variable $b(i)$ is i-th bit of the bit-stream vector $\boldsymbol{b}$. The function $\Psi(\cdot)$ defines the data encoding process. The vector **W** is the feature set to be used for data encoding with elements selected either randomly or deterministically. The single element $w(i)$ of the feature set **W** consists of two randomly selected keys on the keyboard surface. The parameter dt is the total amount of shift to be applied and it is dependent on minimum step-size ΔT and n is the number of steps to be applied. The step-size is selected in such a way that the resulting change is imperceptible by the unaided human eyes and robust against the digital-to-analog (D/A) and analog-to-digital (A/D) conversion processes, which means change ΔT can be detected reliably from the digitized surface image. The variable dt_{mod} is the modification type to encode a single bit value as 0 or 1 and its parameter λ takes the values 1 or 2. When $\lambda=1$, the first element of given feature $w(i)$ is taken for applying a modification; if $\lambda=2$, the second element of $w(i)$ is considered.

To select the data encoding feature set, the candidate key must have two types of symbols (RM and CS) to qualify as an element of feature set ***W***. The goal behind using the integrated CS and RM symbols is to tackle the potential geometrical disturbances in the keys encountered from the aging effects.

2.2 Data Encoding Mechanism vs. Surface Visual Quality

In the concept for a single data bit encoding two different keys (let us say key-1 and key-2) are used. For encoding bit "1" key-1 is moved up by a step size ΔT and key-2 is left unchanged. Similarly, for encoding bit "0" key-2 is moved up and key-1 is left unchanged. The single bit data encoding mechanism is illustrated in Fig. 1.

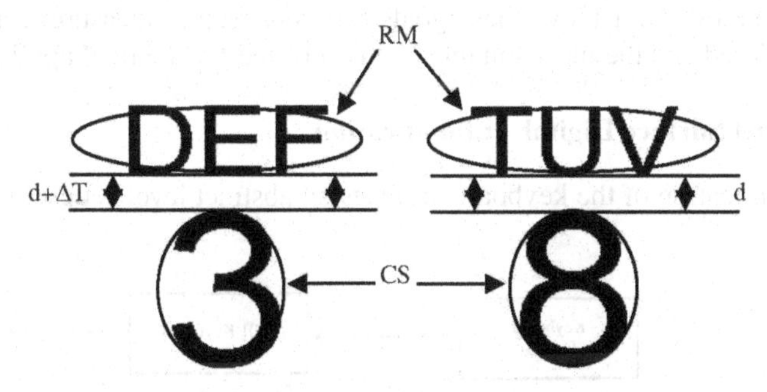

Fig. 1. Encoding single data bit "1" by using two keys

The goal behind using two keys instead of one is to minimize the visual distortion caused by the data encoding process so that coded and un-coded surfaces are visually identical. The visual distortion of the surface coding is further camouflaged by using one of the two keys that are selected pseudo-randomly (i.e. using non-neighboring keys) among all the keys. The pseudo-random key selection uniformly distributes the artifacts from the coding process over the entire surface. The value of step size ΔT is 169 micrometers.

2.3 Data Encoding Capacity for Keyboard Surface

In general the capacity of the encoded keyboard is given by:

$$C = \frac{1}{2} \cdot \sum_{i=1}^{N} \eta_i \cdot \lambda_i \cdot n_i \,. \tag{2}$$

Here, N is number of different types of features to be used for data encoding. The parameter η_i is the number of characters or symbols belonging to the same type of features that are to be used for data encoding. The parameter λ_i is the number of steps for the data encoding feature of type i and n_i is the number of modifications in the data encoding features of type i. The above mentioned Eq. (2) is formulated for a general case of the keyboard surface coding using rotation, squeezing, stretching, shifting and their combination operations for data encoding. Considering this particular scenario of a particular surface coding technique the following parameters were chosen: $N=1, n_i=2, \lambda_i=1$ and $\eta_i=6$. Therefore, the Eq. (2) results in 6 bits. Here the data encoding technique is designed for the minimum visual impact at the cost of optimal capacity and higher robustness against different types of artifacts encountered in the surface authentication. For robust surface authenticity verification even single bit information encoding would be sufficient.

3 Surface Authentication Process

In digital authentication processes, the aim is to differentiate between the coded and un-coded keys by measuring the distance between the RM and the CS position up to sub-pixel accuracy. To achieve these goals different feature measurement techniques [7] are considered and the most suitable one is selected for the final application.

3.1 Keyboard Surface Digital Authentication

Digital authentication of the keyboard surfaces on abstract level is described in Fig. 2.

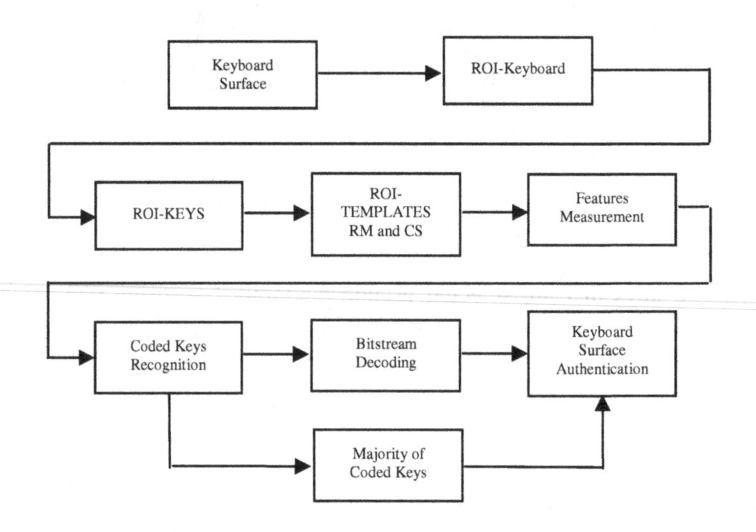

Fig. 2. Abstract view of digital authentication of keyboard surface

It begins with selecting the region of interest (ROI-Keyboard) for the keyboard surface while considering the entire digital image captured by the camera. The ROI(s) corresponding to individual keys (ROI-Keys) are selected within the ROI of the keyboard surface. The ROIs for the templates corresponding to the RM and coded CS are found for each key of interest. The distance between two templates is measured up to sub-pixel level accuracy to classify the given key as coded or un-coded. Finally, the keyboard surface is classified as coded or un-coded based on the decoded bit-stream or using the majority of the coded or un-coded keys.

3.2 Feature Measurement for Coded Key Recognition

In order to classify the coded key as coded/un-coded, a sub-pixel level feature measurement technique [2], [3] is used, which measures distance *d* up to sub-pixel level accuracy between the CS and the RM for the given key. A coded key is differentiated from an un-coded key if the distance is above a given threshold level. The selection of the feature measurement technique also plays a crucial role and impacts the computational time, distance measurement accuracy, and robustness of the measured distance. A technique selected after considering many techniques [7] for sub-pixel accurate distance measurement is described in the following.

The sub-pixel accurate distance measurement between two points begins by computing the normalized cross-correlation coefficient between the key under consideration, referred to as key template and RM template, using the following Eq. (3) [2]:

$$r(u,v)=\frac{\sum_{x,y}\left[f(x,y)-\bar{f}_{u,v}\right]\left[T(x-v,y-u)-\bar{T}\right]}{\sqrt{\sum_{x,y}\left[f(x,y)-\bar{f}_{u,v}\right]^2\cdot\sum_{x,y}\left[T(x-v,y-u)-\bar{T}\right]^2}}, \tag{3}$$

where f is the image of the Key-ROI, $\bar{T}$ is the mean of the template ROI-RM and $\bar{f}_{u,v}$ is the mean of $f(x,y)$ in the region under the template ROI-RM.

By using Eq. (3), only one point for the feature measurement is found at pixel level accuracy. The second point for the distance measurement can also be found using Eq. (3) while replacing the template ROI-RM with the CS region of interest (ROI-CS). For sub-pixel level accurate feature matching, different techniques are given. All have in common that the correlation coefficient matrix is interpolated for the selected region, corresponding to the maximum value of the correlation coefficient. This fact is used for sub-pixel accurate template matching. To measure sub-pixel level accurate distance a second order polynomial (cf. Eq. (4)) is fitted to the normalized correlation coefficient matrix using two immediate neighboring (top and bottom) points of the maximum correlation coefficient value and the point corresponding to the maximum correlation coefficient value in the interpolated matrix is found:

$$f(x)=a_0+a_1x+a_2x^2. \tag{4}$$

Finally, the sub-pixel level accurate distance *d* between two points, representing the best matching points for two templates, is computed using following relation:

$$d=\left|y_{cs}-y_{rm}\right|, \tag{5}$$

where y_{cs}, and y_{rm} are the y-coordinates of the points P_{cs} and P_{rm} correspond to the maximum values after the polynomial fitting. Figure 3 shows two keys (un-coded and coded) and the distance between these keys at sub-pixel level accuracy.

Fig. 3. Sub-pixel level distance d for the un-coded (left) and coded (right) keys (images in application resolution)

4 Performance Evaluation

The experimental setup for the surface authentication consists of a digital camera mounted on movable stand. The camera is located on top of the keyboard surface at a fixed distance L and the value of parameter L is selected by taking into account a real application scenario. Two light sources, located at a fixed distance from the keyboard surface, are used for the keyboard surface illumination. A digital USB 2.0 camera offering a resolution of 1200 by 1600 pixels, which was selected in advance by the project partner, is used for capturing digital images continuously at preselected time intervals. All parameters for the experimental setup are given in Table 1.

Table 1. Parameters for experimental setup

Distance	Image Color	Camera Model	Resolution (Pixels)	Light Sources
460 mm	greyscale	uEye USB 2.0	1200 by 1600	2 x 18 W

For a performance evaluation the real keyboard surface is simulated on a paper surface as shown in Fig. 4 and used for data encoding. Both the coded and un-coded surface images printed on white paper are found completely indistinguishable when shown to independent observers due to the imperceptible feature modification made to the coded image surface. Next, a randomly selected test surface is placed under the camera at a fixed position as a keyboard surface and classified coded/un-coded under different conditions. The performance is evaluated by running different authentication tests for long time periods (days) with randomly selected test surfaces. The surfaces have been classified correctly in all cases. With the aim to focus first on maximum robustness and real-time processing constraints, the coded/un-coded surface classification decision is made on majority basis, while considering only the coded keys. A coded bit-stream extraction would only require to detect twice the number of coded/un-coded keys. The large number of tests is made due to the constraint that the test surface (keyboard) shall be continuously authenticated against the un-coded surface detection.

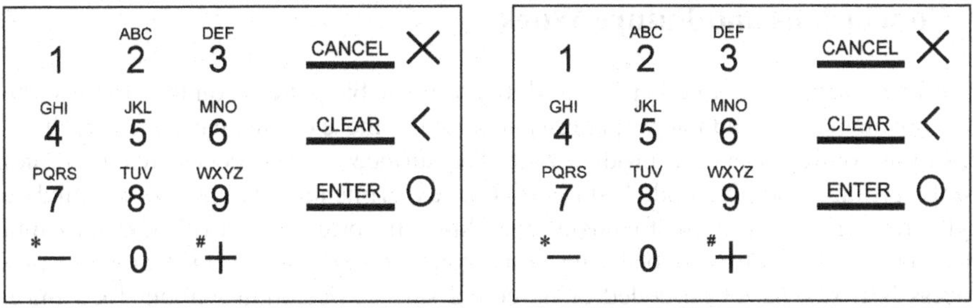

Fig. 4. Simulated ATM keyboard surfaces: left un-coded, right coded

Robustness Against Varying Lighting Conditions. As varying lighting conditions have high impact on the performance of a given surface authentication system its robustness is investigated considering different scenarios. In one scenario only the camera lights (2x18W) are considered in absence of any external light sources. The authentication is found successful. Next, the distance of the lighting sources from the surface is varied from normal (150 mm) to maximum (460 mm). The surface authentication process again has been successful. In another scenario room lights and camera lights are considered together, a scenario resulting in a relatively higher illuminated surface and the surface authentication is again successful. Furthermore, there has been no performance degradation due to the 50/100 Hz disturbances from the room lights or highly illuminated surface.

Next, the surface authentication is checked in the absence of camera illuminations, but in presence of room lights or daylight or a combination of both daylight and room lights. The tests have been successful. For daylight conditions it has been successful, however it is observed that classification results depend on the daytime and weather conditions: sunshine, cloudy, shadowing or daytime. In such cases errors have been encountered. It is mentionable that considering only room lights a small number of errors (~10 errors per 1000 analyzed images, approx 1%), due to the high frequency camera noise, were observed.

Equivalence of Metallic and Simulated Surfaces. For the metallic surface authentication, its potential impacts i) robustness of the surface coding process against laser engraving and ii) coded metallic surface authentication have been considered. The surface coding should not pose any problem as the laser engraving technology to be used is able to accomplish surface patterns modifications much more precisely (i.e. at higher resolution) as compared to the laser print process used for the simulated surface coding. The coded *metallic* surface authentication is more challenging and special care must be taken for the surface illumination. The light sources are to be positioned so that the target metallic surface is uniformly illuminated and there is a good contrast level between the visible surface patterns and the background.

Computational Time. Another important performance parameter is the computational time, which requires that the authenticity verification system works in a real-time mode and that the time must be short enough in order to detect any copying attack. The computational time is less than one second, which can be further reduced by programming code optimizations. The computational time is mainly dependent on the interpolation technique of the sub-pixel RM and the CS templates matching.

5 Conclusions and Future Work

A surface coding technique is presented that uses visible patterns on the keyboard and does not demand to add new patterns or modify the existing patterns, which is a constraint coming from end-product cost. The surface coding process results in non-visible artifacts on the coded surface. For experimental evaluation, a simulated keyboard surface is used for data encoding. In order to fulfill the maximum robustness and real-time processing constraints, a keyboard surface authentication process is firstly based on coded or un-coded surface detection without focusing on the bit-stream decoding and it is found successful. For the keyboard surface authentication, the constraints arise from the practical application scenario: non-contacting camera position and varying lighting conditions are taken into account and it is found robust against such disturbances.

In this paper the results for the first phase of the project are given. The dummy metal keyboard surface coding is in process for evaluation before it will be applied to the actual ATM keyboard surface. Experiments considering the camera at an angular position and the wear-and-tear disturbances have been conducted for the simulated surface and are not discussed here due to space limitations. They shall be disclosed in future publications along with the metallic keyboard surface authentication. Fuzzy pattern matching techniques shall be considered for performance comparison.

Acknowledgements

This work was granted by Wincor Nixdorf International GmbH, Paderborn.

References

1. Krumm, D.: Data theft at ATMs and POS terminals by internationally-organized criminal groups. Federal Office of Criminal Investigation (SO 42-3), Germany (unpublished report)
2. Lewis, J.P.: Fast Normalized Cross-Correlation. In: Proc. of Vision Interface (1995)
3. Haralick, R.M., Shapiro, L.G.: Computer and Robot Vision, vol. II, pp. 316–317. Addison-Wesley, Reading (1992)
4. Sako, H.: Self-Defence Technologies for Automated Teller Machines. In: IEEE Int. Machine Vision and Image Processing Conference (2007)
5. Iqbal, T.: High Capacity Analog Channels for Smart Documents. Ph.D. thesis, ch. 6 (pp. 125–152). University Duisburg-Essen, Fac. of Engineering, Germany (2006)
6. Enright, J.M., et al.: System and method for capturing and searching image data associated with transactions. US Patent 6, 583, 813 B1, January 24 (2003)
7. Barbara, B., Flusser, J.: Image registration methods: a survey. Elsevier Image Vision Computing (2003)

Variations of the Turing Test in the Age of Internet and Virtual Reality
(Extended Abstract*)

Florentin Neumann**, Andrea Reichenberger, and Martin Ziegler***

University of Paderborn, University of Bochum, and Vienna University of Technology

Abstract. Inspired by Hofstadter's *Coffee-House Conversation* (1982) and by the science fiction short story *SAM* by Schattschneider (1988), we propose and discuss criteria for non-mechanical intelligence:

We emphasize the practical requirements for such tests in view of massively multiuser online role-playing games (MMORPGs) and virtual reality systems like Second Life. In response to these new needs, two variations (Chomsky-Turing and Hofstadter-Turing) of the original Turing-Test are defined and reviewed. The first one is proven undecidable to a deterministic interrogator. Concerning the second variant, we demonstrate Second Life as a useful framework for implementing (some iterations of) that test.

1 The Turing Test: A Challenge

Artificial Intelligence is often conceived as aiming at simulating or mimicking human intelligence. A well-known criterion for the success of this endeavour goes back to Alan Turing: In [12] he described an *Imitation Game* in which an interrogator gets into a dialogue with a contestant solely through a teletypewriter and has to find out whether the contestant is human or not[1]. Turing forecasts:

> *I belive that in about fifty years' time it will be possible to programme computers, with a storage capacity of about* 10^9*, to make them play the imitation game so well that an average interrogator will not have more than 70 per cent chance of making the right identification after five minutes of questioning.*

It is impressive how accurate the first part of this prediction has turned out valid, whereas the Loebner Prize, announced for a computer to succeed with the second part of the prediction has not been achieved so far. The present-time

* The full version of this work is available at arxiv.org/abs/0904.3612

** Corresponding author fneumann@mail.upb.de The present work is based on the first author's bachelor thesis [3] prepared under the supervision of the other two.

*** Supported by DFG grant Zi 1009/2-1 and generously hosted by Prof. Karl Svozil.

[1] [8] points out that the original paper's intention may have been different; however, we adhere to what has become the standard interpretation of the Turing Test.

B. Mertsching, M. Hund, and Z. Aziz (Eds.): KI 2009, LNAI 5803, pp. 355–362, 2009.

world record holder `Elbot`, a popular chatterbot, having convinced 3 out of 12 interrogators of being human. Nevertheless, it is considered only a matter of years until the first system passes the test.

2 Problems with Mechanical Avatars

Science Fiction regularly adresses putative and philosophical problems that may arise with artificial agents in the future; cmp e.g. "*I, Robot*" by ISAAC ASIMOV or "*Do Androids Dream of Electric Sheep*" by PHILIP K. DICK which the movie "*Blade Runner*" is based on. In the present section we point out that the virtual reality of the internet has already turned the problem of (automatically) recognizing mechanical[2] avatars into a strongly present and practical one. We briefly mention four examples for this development.

Instant messenger protocols, services, and clients like ICQ, Jabber, or AIM nicely complement email as a means of electronic communication: They are designed for immediate, interactive, low-overhead and informal exchange of relatively short messages. Many messenger clients provide plugins for chatterbots to jump in, if the human user is unavailable. Presently, we are encountering a trend to synthesize (email-) spam and instant messenging to so-called *malicious chatterbots*: automated electronic advertisement, promotion, and luring with a new degree of interactivity that email is lacking and thus raising a very practical urge to detect and quell them in order to protect the user from such nuissance and danger. Like with email, such a detector should preferably work mechanically (i.e. no human interrogator), but unlike email, the aspect of interactivity in instant messenging prohibits any form of offline filter.

Massively Multiplayer Online *Role-Playing Games* (MMORPGs) have presently gained immense popularity. They generally provide the user with a choice of goals and challenges of various degrees of difficulty. The more advanced ones require special items (virtual tools, weapons, skills, money) that can gradually be acquired and traded in-game. Now some players lack the patience of first passing through all the initial levels—and have their client mechanically repeat a more or less simple sequence of movements, each raising a little amount of money and items. Such a client extension is called a Game Bot and well-known for many MMORPGs such as Diablo II, Ultima Online etc—and an obvious thorn in the side of the MMORPG's operator who would very much like to (automatically) identify and, say, temporarily suspend such a user's avatar.

Gold Farming is the answer to the attempt of the MMORPG developers to prevent the use of Game Bots as described above by including a source of randomness into their game play. However, their (simplified) virtual economies[3] have spurred real-life counterparts: 'Items' (must not, but) can be purchased for actual money, e.g. on `eBay`. Computer kids, mostly in China and South Korea but also elsewhere, take over the role of the (insufficiently intelligent) Game Bots and perform (still mostly mechanical and repetitive) moves with their own avatar

[2] We avoid the term '*Artificial Intelligence*' because of its philosophical ambiguities.

[3] `http://ezinearticles.com/?id=1534647`

to gain virtual wealth and then sell it for real money. MMORPGs' operators are therefore faced with the challenge of identifying and suspending those accounts. Such abuse has led regular World of Warcraft users to involve 'suspicious' avatars into an online game chat: definitely a variant of the Turing Test!

In Second Life, online virtual reality system, a distinction between non-player characters (NPCs) and player characters is largely removed. Here, NPCs are entirely missing and so is, at least initially, any form of detailed scenery. Instead users may freely construct items and buildings that can be placed and even sold on purchased virtual estate. Moreover, in striking contrast to MMORPGs, these objects can be *user*-programmed to perform actions on their own. When facing another avatar in Second Life, it is therefore not clear whether this constitutes indeed another user's virtual representation or rather a literally 'animated' object; we consider the problem of distinguishing the latter two cases as another variant of the Turing Test.

The above examples suggest to reverse the focus of the Turing Test: from devising a mechanical system to pass the test, towards devising test variants that reliably do distinguish (at least certain) mechanical systems from (other) human ones.

3 Variations of the Turing Test

In Section 2 new challenges and applications which are not covered by the original Turing Test have been discussed. We report now on some known aspects as well as on some new deficiencies of this test. Then we discuss the Hofstadter-Turing Test as a stronger variant. Finally, we prove the weaker Chomsky-Turing Test undecidable to a deterministic interrogator.

3.1 Deficiencies of the Original Test

The Turing Test can be seen as having initiated Artificial Intelligence as a science. Nevertheless it is subject to various criticism and objections, some raised by Turing himself; cf. e.g. [4,6].

Restricted Interaction and the Total Turing Test: Turing restricted communication and interaction in his test, so-to-speak *projecting* attention solely to the intellectual (if not platonic) features of the contestant. He originally ignores most human senses entirely—channels of perception which in the 1950ies were indeed unforseeable to artificial simulation. However, this situation has changed dramatically with the rise of multimedia and virtual reality systems (recall Section 2). In order to take this into account, the so-called *Total* Turing Test has been proposed as an advanced goal of Artificial Intelligence [1].

Physical Reality and Threat: Extending the above reproaches, one may argue that mere interaction with the interrogator—even if sensual—does not suffice to classify as intelligent. A considerable part of human (as opposed to animal) existence arises from, and evolves around constructing new objects, tools, and weapons within and from its natural physical environment as a means to meet with everyday challenges and threats. This is an aspect not covered even by the Total Turing Test—but well present in the Hofstadter-Turing Test described in Section 3.2 below.

Anthropocentrism is in our opinion the most serious deficiency, in fact of Artificial Intelligence as a whole: it starts with the (usually implicit) hypothesis that humans *are* intelligent, and then proceeds in *defining* criteria for intelligence based on resemblance to various aspects of humans. Anthropocentrism is known to have caused many dangerous and long-lasting errors throughout human history. Of course we have no simple solution to offer either [6, p.509], other than to constantly remain open and alert against such fallacies.

3.2 Hofstadter-Turing Test

In 1988, DR. PETER SCHATTSCHNEIDER published a science fiction short story "*SAM*" [5] in the series of the computer magazine c't. It begins from the point of view of an unspecified 'being' finding itself wandering an increasingly complex environment, later revealed to be controlled by a 'programmer', and eventually arriving at a computer terminal where it starts setting up a similar virtual environment and wanderer, thus passing what is revealed (but not specified any further) as the *Hofstadter Test.* This story may have been inspired by DOUGLAS R. HOFSTADTER's *Coffee-House Conversation* [2] of three students, Chris (physics), Pat (biology), and Sandy (philosophy) ending with the following lines:

> CHRIS: *If you could ask a computer just one question in the Turing Test, what would it be?*
> SANDY: *Uhmm...*
> PAT: *How about this:* "If you could ask a computer just one question in the Turing Test, what would it be?"

Observe the recursive self-reference underlying both, this last question and Schattschneider's story "*SAM*" as well as Turing's famous article [10] proving by diagonalization and self-reference that the question of whether a Turing machine eventually terminates (i.e. the Halting problem) is undecidable to a Turing machine. Picking up [5], we arrive at the following:

Definition 1 (Hofstadter-Turing Test). *For an entity to pass the Hofstadter-Turing Test means to devise*

i) a virtual counterpart resembling its own environment and
ii) a computer program which succeeds in recognizing itself as an entity within this virtual environment and
iii) in turn passes the Hofstadter-Turing Test.

The seemingly circular definition can be made logically sound in terms of a fixpoint sequence: 'unfolding' the condition in requiring the existence of a countably infinite hierarchy of virtual environments and entities such that the $(n+1)$-st are created by and within the n-th. This does not provide any way of operationally performing this test but at least makes the definition valid. In practice and pragmatically, once the first few n levels have succeeded in creating their successor $n+1$, one would likely be content to abort any further recursion and state with sufficient conviction that the initial entity has passed the test.

We shall return to this remark in Section 5.

We readily admit that the Hofstadter-Turing Test does not provide the ultimate solution to the problems of Artificial Intelligence and mention three reproaches.

Anthropocentrism is, again, present in its strongest form by requiring our human physical world to be the first and thus modelled by all iterated virtual counterparts according to Condition i) environments. In fact it seems that the common conception of a 'virtual environment' is highly biased and restricted. Even more, questions of intelligence and consciousness are irrelevant within the abstract 'world' of programs; they only arise through the sociocultural interface of virtual role-playing systems.

The Problem of Other Minds is a well-known philosophical issue which arises here, too: Is the contestant (sequence!) required to exhibit and visualize the virtual environment he/she has created? Can we even comprehend it, in case it is a purely digital one? How about patients with locked-in syndrome: does the single direction of their communication capabilities disqualify them from being intelligent? In final consequence, one arrives at the well-known problems of Behaviorism.

"Humans are intelligent" used to be the primary axiom of any test for true intelligence. But is actually any person able to pass (pragmatically at least some initial levels of) the Hofstadter-Turing Test? Recall he has to succeed in creating a virtual entity of true intelligence.

3.3 Chomsky-Turing Test

As pointed out, several applications prefer an automated form of Turing-like tests. The present section reveals that this is, unfortunately, infeasible in a strong sense. Referring to the Theory of Computation we formally prove that even powerful oracle Turing machines (capable of solving e.g. the Halting problem) cannot distinguish a human contestant from the simplest (abstract model of a) computing device in Chomsky's hierarchy, namely from a finite automaton.

Finite state machines and transducers (Chomsky level 3) are models of computation with immensely limited capabilities. Turing machines are located at the other end of the hierarchy (level 0). If we wish to include non-mechanical language processors, humans could be defined to reside at level -1. In which case the goal of (Artificial Intelligence and) the Turing Test amounts to (separating and) distinguishing level 0 from level -1. The present section goes for a much more modest aim:

Definition 2 (Chomsky-Turing Test). *The goal of the Chomsky-Turing Test is to distinguish Chomsky level 3 from level -1.*

We establish that such a test cannot be performed mechanically. A first result in this direction is a well-known consequence of Rice's Theorem in computability theory, namely that the language REGULARTM, defined as

$$\{\langle M\rangle : \text{the language } L(M) \subseteq \{0,1\}^* \text{ accepted by Turing machine } M \text{ is regular}\},$$

is undecidable to any Turing machine; cf. e.g. [7, THEOREM 5.3]. It is, however, decidable by an appropriate oracle machine, namely taking REGULARTM itself as the oracle. Moreover, the above mathematical claim does not quite apply to the setting we are interested in: It supposes the encoding (Gödel index) of a contestant Turing machine M to be given and to decide whether M acts as simple as (but of course not is) a finite state machine; whereas in Turing-like tests, the contestant may be human and is accessible only via dialogue. Such a dialogue amounts to a sequence $(x_1, y_1, x_2, y_2, \ldots, x_n, y_n, \ldots)$ of finite strings x_i entered by the interrogator and answered by the contestant in form of another finite string y_i upon which the interrogator adaptively enters x_{i+1}, the contestant replies y_{i+1}, and so on round by round.

Proposition 3

i) It is impossible for any deterministic interrogator to recognize with certainty and within any finite number of communication rounds $(x_1, y_1, x_2, y_2, \ldots, x_n, y_n)$ *that the answers* x_i *provided by the contestant arise from a transducer (Chomsky level* $= 3$*).*

ii) It is equally impossible in the same sense to recognize with certainty that the answers arise from a device on any Chomsky level < 3*.*

Thus, we have two separate (negative) claims, corresponding to (lack of) both recognizability and co-recognizability [7, THEOREM 4.16]. More precisely, the first part only requires the interrogator to report "`level` = 3" within a finite number of communication rounds in case that the contestant is a transducer but permits the dialogue to go on forever, in case it is another device; similarly for the second part. Also, the condition of a deterministic interrogator is satisfied even by oracle Turing machines. Hence, this can be called a strong form of undecidability result.

Proof (Proposition 3). Both claims are proven indirectly by 'tricking' a putative interrogator I. For i) we first face I with some transducer T; upon which arises by hypothesis a finite dialogue $(x_1, y_1, \ldots, x_n, y_n)$ ending in I declaring the contestant to be a transducer ("`level` = 3"). Now this transducer T can be simulated on any lower Chomsky level by an appropriate device D exhibiting an input/output behavior identical to T. Since I was supposed to behave deterministically, the very same dialogue $(x_1, y_1, \ldots, x_n, y_n)$ will arise when presenting to I the contestant D—and end in I erroneously declaring it to be a transducer.
The proof of Claim ii) proceeds similarly: first present to I some device D; which by hypothesis leads to a finite dialogue $(x_1, y_1, \ldots, x_n, y_n)$ and the report "`level` < 3". Now it is generally impossible to simulate D on a transducer since Chomsky's Hierarchy is strict [7]. However any fixed finite dialogue can be hard-coded

into some transducer T; and repeating the interrogation with this T results by determinism of I in the same, but now wrong, answer "`level` < 3". □

4 Implementing the Hofstadter Test in Second Life

Recall (Section 3.2, Definition 1) that the goal of the test is to implement a virtual reality system and an artificial entity therein which in turn passes the Hofstadter test. Fortunately, there is a variety of virtual reality systems available, so the first level of the test can be considered accomplished. However, in order to proceed to the next level, this system has to be freely programmable on the virtual level—and to the best of our knowledge, this is presently only supported by Second Life. From a technical point of view, the backbone of Second Life and the Linden Scripting Language, respectively, can actually provide the unlimited computational resources required for a truely Turing-complete environment as shown in [3].

We have succeeded in implementing within Second Life the following virtual scenario: a keyboard, a projector, and a display screen. An avatar may use the keyboard to start and play a variant of the classic arcade game Pac-Man, i.e. control its movements via arrow keys.[4] (For implementation details, please refer to [3, SECTION 4].) With some generosity, this may be considered as 2.5 levels of the Hofstadter-Turing Test:

1st: The human user installs Second Life on his computer and sets up an avatar.
2nd: The avatar implements the game of Pac-Man within Second Life.
3rd: Ghosts run through the mace on the virtual screen.

Observe that the ghosts indeed contain some (although admittedly very limited) form of intelligence represented by a simple strategy to pursue pacman.

5 Concluding Remarks

We have suggested variations of the (standard interpretation of) the Turing Test for the challanges arising from new technologies such as internet and virtual reality systems. Specifically to the operators of MMORPGs and of Second Life, the problem of distinguishing mechanical from human-controlled avatars is of strong interest in order to detect putative abuse and to answer the old philosophical question of ontology: Are the digital worlds of World of Warcraft and Second Life 'real'?

A particular feature of the Hofstadter-Turing Test is that the iterated levels of virtual reality require to be created one within another. Each one of these iterated levels can be seen as an encapsulated virtual reality, transparently contained by the one at the next higher level, like the skin layers of an onion. Similarly, behind

[4] Available within Second Life at the coordinates:
`http://slurl.com/secondlife/Leiplow/176/136/33`

the visible virtual reality of Second Life consisting of 'idealized' (i.e. platonic) geometric/architectural objects and avatars, there lies hidden the invisible and abstract virtual reality of programs and scripts that control and animate these objects.

This suggests an interesting (new?) approach to the philosophical problem of ontology: maybe reality should generally be considered composed of layers [9]. This includes the very interesting question of whether and how different such levels may interact: concerning the possibility and means to break through the confinements of one's reality. That a higher level can influence a lower one, should be pretty obvious from the above examples: the short story SAM, the Hofstadter-Turing Test, MMORPGs, and Second Life. But careful reconsideration reveals also effects in the converse direction:

- A video game addict socially isolating himself, loosing his job and/or health.
- Virtual items being sold for real money as in Gold Farming.
- Actual law-suits for breach of 'virtual' laws and unfair trade practices (see Bragg vs. Linden Lab).

References

1. Harnad, S.: Other Bodies, Other Minds: A Machine Incarnation of an Old Philosophical Problem. Minds and Machines 1, 43–54 (1991)
2. Hofstadter, D.R.: The Turing Test: A Coffee-House Conversation. In: Hofstadter, D.R., Dennett, D. (eds.) The Mind's I: Fantasies and Reflections on Self and Soul, pp. 69–95. Penguin Books, London (1982)
3. Neumann, F.: Der Hofstadter-Turing-Test als anwendungsorientiertes philosophisches Problem am Beispiel von Second Life. Bachelor thesis, University of Paderborn, http://www.upb.de/cs/ag-madh/PapersPostscript/fneumann.2009.pdf
4. Oppy, G., Dowe, D.: The Turing Test. Stanford Encyclopedia of Philosophy (May 2008), http://plato.stanford.edu/entries/turing-test/
5. Schattschneider, P.: SAM. c't Magazin 6, 214–220 (1988)
6. Saygin, A.P., Cicekli, I., Akman, V.: Turing Test: 50 Years Later. Minds and Machines 10(4), 463–518 (2000)
7. Sipser, M.: Introduction to the Theory of Computation. PWS Publishing (1997)
8. Sterrett, S.G.: Turing's Two Tests for Intelligence. Minds and Machines 10, 541–559 (2000)
9. Svozil, K.: How real are virtual realities, how virtual is reality? The constructive re-interpretation of physical undecidability. Complexity 1, 43–54 (1996)
10. Turing, A.M.: On Computable Numbers, with an Application to the Entscheidungsproblem. Proc. London Math. Soc. 42(2), 230–265 (1936)
11. Turing, A.M.: Systems of Logic Based on Ordinals. Proc. London Math. Soc. 45, 161–228 (1939)
12. Turing, A.M.: Computing Machinery and Intelligence. Mind 59, 433–460 (1950)

A Structuralistic Approach to Ontologies

Christian Schäufler[1], Stefan Artmann[2], and Clemens Beckstein[1,2]

[1] Artificial Intelligence Group, University of Jena, Germany
[2] Frege Centre for Structural Sciences, University of Jena, Germany

Abstract. It is still an open question how the relation between ontologies and their domains can be fixed. We try to give an account of semantic and pragmatic aspects of formal knowledge by describing ontologies in terms of a particular school in philosophy of science, namely *structuralism.* We reconstruct ontologies as empirical theories and interpret expressions of an ontology language by semantic structures of a theory. It turns out that there are relevant aspects of theories which cannot as yet be taken into consideration in knowledge representation. We thus provide the basis for extending the concept of ontology to a theory of the use of a language in a community.

1 Introduction

In this paper we address the representation of generalized, scientific knowledge. We focus our investigation on the terminological part of knowledge bases, namely *ontologies.* We will argue that theories and ontologies are strongly corresponding because ontologies can be understood as a special case of scientific theories. For this purpose it will be necessary to commit to a precise definition of the notions of theory and ontology. Both concepts can be given a generic and a special meaning. For ontologies we commit to ontologies specified by description logics as the general meaning, and to ontologies of sophisticated formalisms as the special one. In philosophy of science a widespread definition of the general meaning of theory is that of a set of sentences describing a certain domain. For the special meaning of theory we commit to the structuralistic theory concept [1].

We start the paper with a very short sketch of the state of the art: What are the central epistemological approaches in knowledge representation and philosophy of sciences and what attempts have been undertaken to utilize answers from philosophy of science for knowledge representation and vice versa (section 2)? As the structuralistic conception of scientific theories is not widely spread, so that it can be presupposed, we then give an introduction to its key ideas (sec. 3). This lays the ground for section 4 where we conceptually compare ontologies with structuralistic theories.

2 Related Work

It is the effort of philosophy of science to describe the *structure* and the genesis of scientific knowledge. It has been pointed out that a single homogeneous set of

B. Mertsching, M. Hund, and Z. Aziz (Eds.): KI 2009, LNAI 5803, pp. 363–370, 2009.

axioms is an insufficient representation of a substantial theory. The important aspects of scientific knowledge – like inter-theoretical links, an empirical claim, a signature – are combined in the *structuralistic* theory conceptualization. It was conceived by of J.D. Sneed [2] and advanced by W. Stegmüller [3], W. Balzer, and C.U. Moulines [1].

In knowledge representation two ways of answering what an ontology is, can be found. The philosophical one is given by stating what the purpose of an ontology is. T. Gruber, e.g., has given the most popular characterisation of this type [4]. Other attempts along this line are made by J. Sowa in [5] and B. Smith [6]. The technical answer consists of giving a formal description of the structure of an ontology. There are different conceptualisations of ontologies (for an overview of "knowledge representation ontologies" see [7]), headed by the family of description logics [8], frames formalisms [9] and semantic web formalisms.

Techniques for dealing with conceptual knowledge are highly dependent on the concrete formalism used for ontology notation. Successful methods for the most important applications like *ontology merging* and *mapping* may moreover indicate which additional aspects of conceptual evolution need to be represented by an ontology. D. Lenat experienced with EURISCO, that a frame based representation of concepts is superior to a unstructured one. In another procedure for automated discovery, Wajnberg et al. already used a structuralistic conceptualisation for the internal representation of theories [10].

3 Structuralistic Conceptualization of Scientific Theories

We will now give a short introduction to the design of structuralistic theories. This will be done by a walk through the different parts that make up a structuralistic theory description. Along with the introduction of each part we will also shed some light on the role that this part plays in the development of the overall theory. The parts themselves will be described in the style of [3,1].

Structuralism does not restrict the (object) language used to describe domain theories as long as this language has a well defined extensional semantics. The (meta) language used to express propositions about structuralistic theories is typically expressed in set theory. This is due to the structuralists' model theoretic, extensional view on axiomatization: Instead of treating an axiomatization as a set Γ of axioms (*statement view*), it is treated as the set M, $M = \{I : I \models \Gamma\}$ of models I complying to Γ (*non-statement view*). And where the non-structuralist talks about the inference of a theorem G from a set of axioms Γ, $\Gamma \models G$, the structuralist states a subset relation $M \subseteq M_G$ between the corresponding sets of models M and $M_G = \{I : I \models G\}$.

3.1 Theory Element

In order to avoid any confusion with competing understandings of theories, structuralism does not use the term *theory* at all. The central structuralistic concept, resembling most closely the traditional concept of *theory* is the *theory element.*

A theory element T is comprised of six parts:

$$T = \langle M_p(T), M(T), M_{pp}(T), C(T), L(T), I(T) \rangle.$$

Each component of this 6-tupel represents a specific facet of the theory element T. The components are sets, namely sets of models or sets of sets of models. In the following we describe the intended meaning of each component in more detail:

The set M_p of potential models. This component of a theory element represents the used terms along with their types, i.e., its *signature*. According to the non-statement view the signature is a set of *potential models* of T. Signatures are expressed using a specific set of theory-dependent symbols. These symbols can be distinguished by their logical category as designating either a set or a relation. A potential model of a theory is a structure of the form $\langle D_1, \ldots, D_m; R_1, \ldots, R_n \rangle$.

Typically the domain of a theory is represented by one or more sets D_j which are called the *universes of discourse.* Properties of domain elements are expressed by relations R_i on the domains D_j.

The set M of actual models. In statement view, the set of actual models would be described by the set of empirical axioms of T. These axioms would then model the intrinsic laws holding between the objects of the domain. In the model theoretical view, these axioms restrict the set of potential models to the models being empirically plausible: $M \subseteq M_p$.

The set M_{pp} of partial potential models. A scientist who wants to validate a theory has to decide whether the models of the theory match reality. This match requires that the observed quantities satisfy the relations belonging to these models. In order to avoid cycles, the validation of the match should not presuppose any assumptions of the theory. This is achieved via a distinction between relations that — according to the theory — are directly measurable and those that can only be measured with respect to other relations stated by the theory. The latter ones are also called *theoretical* with respect to the given theory.

Given a potential model $m_p \in M_p$, the model m_{pp} which results from m_p by removing all theoretical relations is called the *partial model* belonging to m_p. This removal procedure on models in M_p induces a *restriction function* $r : M_p \mapsto M_{pp}$, which maps each model in M_p to its corresponding partial model. The image $r(M_p)$ of M_p under r is the set M_{pp} of the partial potential models belonging to the theory.

The set C of constraints. In the model theoretical view, each application of a theory to a system is depicted by a single model. Such a model is then said to cover the system. The application of a theory to an aggregated system (a system consisting of subsystems) is described by an application of the theory to each of its subsystems. The idea of structuralism is to make explicit which of these subsystems is covered by which model of the theory. This allows for a precise description of the relation between subsystems.

Linking subsystems is not possible in terms of additional M-axioms, but by axioms relating different models of the same theory. This linking of models for an aggregated system $\mathcal{S}$ is realized by specifying a set C of potential aggregated models. C is a set of sets $C \subseteq \mathcal{P}(M_p)$, where each set $\mathcal{A} \in C$ stands for a *constraint*, i.e. a compatible collection of models covering the subsystems of $\mathcal{S}$.

The set L of links. Systems often cannot be described with a single, homogenous theory. In order to adequately describe such systems it is necessary to apply a whole set $\mathcal{T} = \{T_1, \ldots, T_n\}$ of theory elements. In structuralistic language such sets of theories are called *holons*. Each theory element in $\mathcal{T}$ describes the system from a different point of view or at a different level of generality. Because not every combination of potential models from the involved theory elements is desirable, structuralism allows us to specify compatible combinations of the respective models via the *structural setting* $\mathcal{L} \subseteq M_p(T_1) \times \ldots \times M_p(T_n)$.

For any theory element T_i of the holon $\mathcal{T}$, the projection $L(T_i)$ of $\mathcal{L}$ on $M_p(T_i)$ according to

$$m_i \in L(T_i) \quad \leftrightarrow \quad \forall j \neq i \; \exists m_j \in M_p(T_j) : \langle m_1, m_2, \ldots, m_n \rangle \in \mathcal{L}$$

defines the set of *links* of T_i with respect to $\mathcal{T}$.

The domain I of intended applications. According to structuralism a theory is not completely specified without a description of the domain I of the systems the theory is applicable to. I can be informally specified by enumerating paradigmatic examples of the intended applications of T. For a formal specification, I is given as a set of models. These models are obtained in two steps: 1. conceptualization of the application fields as a set I' of models conforming to the signature M_p, 2. empirical restriction of the models in I' to their partial models with respect to the theory T. Hence $I = r(I') \subseteq M_{pp}$. The models from I are also called *intended models*.

3.2 The Empirical Claim

The development of a theory should be driven by the applications it is intended for. These applications are the guideline that decides about the success or failure of the development process. The criteria for this decision, taken together, constitute the *empirical claim* of a theory. A theory is considered to be falsified if at least one of these criteria is violated. In structuralistic language the criteria are 1. $I \subseteq r(M)$, 2. $I \in \{r(c) \mid c \in C\}$ and 3. $I \subseteq r(L)$. Therefore (1) every model from the domain must be an empirically restricted actual model, (2) the domain must be an empirically restricted constraint of the theory, and (3) all the models from the domain must be contained in the empirical part of the theory's link. The empirical claim can be expressed equivalently with the following more compact formula:

$$I \in \mathrm{r}\big(\mathcal{P}(M) \cap C \cap \mathcal{P}(L)\big). \qquad (1)$$

For a traditional theory in statement view the empirical claim would be just $I \subseteq M$ (criterion 1 without an explicit reference to the purely empirical part of the involved models).

4 Ontologies in the Structuralistic View

In this section we describe a reconstruction and extension of the concept of ontology as a structuralist theory. By ontology we mean, in accordance with Gruber's definitions, "a specification of a conceptualisation", and by conceptualization, "an abstract, simplified view of the world." [4] From this perspective an ontology is an artifact, designed by a knowledge engineer, that provides a terminology for the description of knowledge shared by a community.

In order not to get lost in the details of different formalisms we proceed in two steps. First we define, in structuralistic terms, the conceptual core of ontologies that can be considered as the common foundation of all formalisms. Second we describe, again in structuralistic terms, extensions of the conceptual core.

4.1 Reconstructing the Conceptual Core of Ontologies

By "conceptual core" we designate the common structure on which the languages of description logic are based. Description logics have a sound semantics and a comprehensible notation, and make up the logical foundation of many ontology formalisms [8].

In description logic, an ontology is a set of axioms for the description of concepts and roles (i.e. terms). There is a strict distinction between the terminological part of knowledge bases (TBoxes) and the assertional part (ABoxes). By analogy to the semantics of predicate logic we define the semantics of ontologies as follows:

Definition 1. *(After [11]) Let $\mathcal{D}$, the domain, be an arbitrary set. Let $\mathcal{N}_C$ and $\mathcal{N}_R$ be sets of atomic concepts and atomic roles. Let $\mathcal{I}$, the interpretation function, be a function*

$$\mathcal{I} : \begin{cases} \mathcal{N}_C \rightarrow 2^{\mathcal{D}} \\ \mathcal{N}_R \rightarrow 2^{\mathcal{D} \times \mathcal{D}}. \end{cases} \tag{2}$$

A pair $\mathcal{M} = \langle \mathcal{D}, \mathcal{I} \rangle$, called interpretation, *is a model of an ontology O, written $\mathcal{M} \models O$, if and only if it satisfies all axioms of O.*

In order to define a corresponding theory element T, we supplement the definition above, which is equivalent to the definition of the set M of actual models of T, with the following components:

- The set M_p of potential models of T is given by function (2): Any structure that satisfies (2), is a potential model of T with respect to ontology O.
- We assume the set M_{pp} of partial potential models of T to be equal to the set M_p of potential models of T. This means that all terms of an ontology can be seen as non-theoretical terms. Otherwise, we would need a criterion to distinguish between theoretical and non-theoretical terms, yet such a criterion is not derivable from the conceptual core of an ontology.
- Theory elements allow for the description of an aggregate of subsystems by repeatedly applying the same theory element. The relation between the

single applications of the theory element are given by constraints. There are two possibilities to define the set C of constraints: first, $C = \mathcal{P}(M_p)$ (open world assumption); second $C = \{\{m\} \mid m \in M_p\}$ (closed world assumption).
- Def. 1 does not allow an ontology to refer to another one. The set L of links is thus equal to M_p.
- The set I of intended models is empty since def. 1 does not contain any information about applications of an ontology.

This list shows that the description-logical concept of ontology is much weaker than the structuralistic concept of theory and that there is room for extentending the classical concept of ontologies with structuralistic features.

4.2 Reconstruction of Typical Extensions of the Conceptual Core

Existing extensions of the conceptual core of ontologies that are already widely used in ontology formalisms, can be partitioned into two classes: first, extensions to the classical description logics and, second, extensions which transcend the framework of first-order predicate logic.

DL-extensions: n-ary roles, functions, concrete domains. Weak extensions of basic description logic, such as n-ary roles, functions and concrete domains, still belong to the greater family of description logics. They share a Tarski-style semantics:

- *Relations of arbitrary arity* as used in, e.g., $\mathcal{DLR}$ [8], can be understood structuralistically as relations used in the definition of the set M_p of partial models of the theory.
- *Functions* can be defined as single-valued relations.
- *Domains* in description logical languages like $\mathcal{SHIQ}(\mathcal{D}_n)^-$ are not restricted to the intended universe of discourse, but may include additional *concrete* domains besides [12]. They provide mathematico-logical domains such as numbers and strings. They can be integrated into theory elements from the outset. Structuralists call them "auxiliary sets" and explicitly enumerated in the set M_p of potential models of a theory element.

Higher-order classes and roles. In an ontology, concepts and roles are defined intensionally by properties of individuals. Some ontology formalisms have a feature called "reification" which makes it possible to consider concepts and roles themselves as individuals so that one can express propositions about these concepts and roles. In a weak form that can be used to annotate concepts, classes and ontologies. Reification is part of most formalisms that are practically applied.

Full reification transcends the expressive power of first-order predicate logic and can therefore not be expressed in todays description logics. The weak form of reification, however, can be integrated into a structuralistic theory element by including the set $\mathcal{N}$ of names of classes, of roles and of ontologies as a new domain $\mathcal{N}_C \cup N_R \cup \mathcal{T}$. Annotations are thus handled as binary relations (roles) between names and a domain of strings.

Since classes are not objects of an empirical domain, and annotations do not refer directly to an empirical domain, both can be excluded from the empirical claim of a theory element in the structuralistic framework. They are theoretical terms and do not belong to the set M_{pp} of partial potential models.

Import of ontologies. In order to standardize and modularize ontologies, one can group ontologies according to different levels of abstractions (top, upper and domain level). Also, many ontology formalisms allow the knowledge engineer to combine concept definitions from different ontologies into new concept definitions. For structuralists, these are undesirable ways of reusing ontologies: models of a theory element T_1 should not contain terms of another theory element T_2. Instead, intertheoretical laws should be expressed by the structural setting $\mathcal{L} \subseteq M_p(T_1) \times M_p(T_2)$. In particular, this can be shown as follows:

1. $M_p(T_1)$ does only contain domains and relations of T_1.
2. External references, i.e. concepts and roles of other theories used in the axioms of $M(T_1)$, are removed for the moment. Syntactically, this is done by substituting the universal concept $\top$ for concepts of $M_p(T_2)$ and the universal role $\top \times \top$ for roles of $M_p(T_2)$.
3. The original axiom that contained the external references, is then reconstructed as a cartesian product of potential models of T_1 and T_2, i.e. as a link $M_p(T_1) \times M_p(T_2)$. Of course, the concepts and roles of $M_p(T_1)$ must now be substituted for by $\top$ and $\top \times \top$.

5 Perspectives and Conclusion

This paper described ontologies in terms of the structuralistic philosophy of science. It could be shown that a standard notion of ontology is easily integrated into a theoretical framework which allows for the extension of the conceptual core of an ontology. The result is a general and precise definition of ontology that abstracts from particular ontology formalisms. We belief that this structuralistic concept of a theory element provides the basis further extensions to the conceptual core of an ontology:

- *Distributed world descriptions.* Compatible world descriptions can be grouped in subsets of the set C of constraints. Constraints open up the possibility to explicitly specify the distributed application of an ontology. That allows for, e.g., handling of inconsistencies arising between models of different subsystems.
- *Specifications of intended domains.* The determination of a set I of intended applications of a theory makes it possible to check measurements against the predictions of the theory. We propose an analogy to this idea for the application of ontologies. From our point of view, an ontology is a theory of language usage in a community, more precisely: it is a formal specification of reprensentative data of that usage.

- *Empirical claim of ontologies.* The connection between the conceptualization and the specification, which defines an ontology according to Gruber, can be expressed as the empirical claim of a theory (see equation1). By "conceptualization" we mean the transformation of the intended applications of an ontology, i.e. its domain, into the set I of intended models of a theory. By "specification" we mean the axiomatical definition of the set M of actual models of the theory.

The next steps are therefore, first, to give a formal explication of these enhanced structuralistic features and, second, to show how a real world ontology can be presented in our framework.

References

1. Balzer, W., Moulines, C.U., Sneed, J.D.: An architectonic for science - the structuralist program. Reidel, Dordrecht (1987)
2. Sneed, J.D.: The logical structure of mathematical physics, 2nd edn. Reidel, Dordrecht (1979)
3. Stegmüller, W.: Neuer intuitiver Zugang zum strukturalistischen Theorienkonzept. Theorie-Elemente. Theoriennetze. Theorienevolution, vol. 2. Springer, Heidelberg (1986)
4. Gruber, T.R.: A translation approach to portable ontology specifications. Knowledge Acquisition 5(2), 1–26 (1993)
5. Sowa, J.F.: Knowledge representation: Logical, philosophical and computational foundations. Brooks/Cole, Bangalore (2000)
6. Smith, B., Welty, C.: Ontology: Towards a new synthesis. In: Proceedings of the international conference on formal ontology in information systems (FOIS), Ogunquit, Maine, pp. 3–9. ACM Press, New York (2001)
7. Gömez-Pérez, A., Fernández-López, M., Corcho, O.: Ontological engineering: with examples from the areas of knowledge management, e-commerce and the semantic web. Springer, Berlin (2004)
8. Baader, F., Calvanese, D., McGuinness, D.L., Nardi, D., Patel-Schneider, P.F. (eds.): The description logic handbook: Theory, implementation and applications. Cambridge University Press, Cambridge (2004)
9. Chaudhri, V.K., Farquhar, A., Fikes, R., Karp, P.D., Rice, J.P.: Open knowledge base connectivity 2.0. Knowledge Systems Laboratory, Stanford (1998)
10. Wajnberg, C.D., Corruble, V., Ganascia, J.G., Moulines, C.U.: A structuralist approach towards computational scientific discovery. In: Suzuki, E., Arikawa, S. (eds.) DS 2004. LNCS (LNAI), vol. 3245, pp. 412–419. Springer, Heidelberg (2004)
11. Nebel, B.: Reasoning and Revision in Hybrid Representation Systems, 2nd edn. LNCS (LNAI), vol. 422. Springer, Heidelberg (1990)
12. Haarslev, V., Möller, R.: Description logic systems with concrete domains: Applications for the semantic web. In: Int. Workshop on KR meets Databases. CEUR-WS.org, vol. 79, pp. 1–12 (2003)

AI Viewed as a “Science of the Culture”

Jean-Gabriel Ganascia

LIP6 - University Pierre and Marie Curie, Paris, France

Abstract. Last twenty years, many people wanted to improve AI systems by making computer models more faithful to the reality. This paper shows that this tendency has no real justification, because it does not solve the observed limitations of AI. It proposes another view that is to extend the notion of “Sciences of the Artificial”, which has been introduced by Herbert Simon, into to a new “Science of the Culture”. After an introduction and a description of some of the causes of the present AI limitations, the paper recalls what the “Sciences of the Artificial” are and presents the “Sciences of the Culture”. The last part explains the possible consequences of such an extension of AI.

1 Introduction

This paper attempts to question the AI philosophical foundations and to show that there is a possible misunderstanding about its status. Our starting point concerns its epistemological status. At first sight, it appears that many contemporaneous scientists tend to build AI on a model analogous to the one which the physical or the life sciences are based on. See for instance the recent project of Marcus Hutter [1] who pretends to scientifically ground *Universal Artificial Intelligence* on the Kolmogorov information theory. In the past, AI has often been understood as a “Science of the nature”. Let us recall that the program of the very famous Dartmouth College Summer Research Project on Artificial Intelligence was based on “*the conjecture that every aspect of learning or any other feature of intelligence can in principle be so precisely described that a machine can be made to simulate it.*”. Such strong groundings would have been reassuring. Nevertheless, they are not totally satisfying and there were also some attempts to put in light other dimensions of AI than this reduction to a “Science of the nature”. For instance, the notion of “Sciences of the Artificial” introduced by Herbert Simon [2] opened many perspectives to AI. Do those “Sciences of the Artificial” differ from the traditional “Sciences of the Nature” or do they extend and renew them with new contemporaneous tools? On the one hand, most of the time, even for Simon, the aim assigned to AI is to naturalize – or to computerize, but this is more or less equivalent from a philosophical point of view – social and psychological phenomenon, i.e. to reduce social and psychological phenomenon to mechanical processes that can be simulated on digital computers. On the other hand, the notion of knowledge that was introduced in AI seems not to be reducible to mechanical processes on physical symbols. The so-called *activity* of

B. Mertsching, M. Hund, and Z. Aziz (Eds.): KI 2009, LNAI 5803, pp. 371–378, 2009.

the "Knowledge Level" that was emphasized by Alan Newell in his famous paper [3] seems to express this irreducibility.

To enlighten the epistemological nature of AI we refer to the opposition between the "Sciences of the Nature" and the "Sciences of the Culture", i.e. the humanities, that was introduced in the first half of the 20th century by some German Neo-Kantian philosophers among which the most famous were Heinrich Rickert (1863–1936) and Ernst Cassirer (1874–1945). We claim that the conceptual apparatus they have developed may be successfully applied to AI. Our working hypothesis is that AI can neither be fully reduced to a "Science of the Nature" nor to a "Science of the culture", but that it is what Heinrich Rickert calls an "intermediary domain". It means that both the objects of AI and its logic belong to fields covered simultaneously by the "Sciences of the nature" and by the "Sciences of the culture". The practical consequences of such philosophical considerations concern first the fields of application of AI: its objects cannot be reduced to the sole simulation of the nature, i.e. to a total and perfect reproduction of activities of intelligent beings. The AI influences also the culture, i.e. the medium of communication. Many of AI successes concern the way it changed – or it helped to change – the contemporaneous culture. Unfortunately, those successes have not been credited to AI. The second consequences are about the AI methods that cannot all be assimilated to logical generalizations of the diversity by general laws; for instance, the careful study of paradigmatic past cases, or more precisely of alleged past AI failures, can be valuable; it also belongs to the method of the "Sciences of the Culture", of which AI is a part of.

Apart this introduction, the paper is divided in four parts. The first one is a lesson drawn from the apparent AI failures. The second constitutes an attempt to explain why AI is supposed to have failed. The third briefly recalls the notion of "Sciences of the Artificial" as it was introduced by Herbert Simon and then describes the Neo-Kantian distinction between the "Sciences of the Nature" and the "Sciences of the Culture". The last one inventories some of the practical consequences of this distinction.

2 What Went Wrong?

Last summer a special issue of the AI Magazine [4] has published numerous of cases of alleged faulty AI systems. The goal was to understand what made them wrong. The main lesson was that, most of the time, the difficulties were not due to technical impediments, but to the social inadequacy of those AI systems to their environment. This point is crucial; it has motivated the reflexion presented here. For the sake of clarity, let us take an example about the so-called electronic elves, which are agents that act as efficient secretaries. Those "Elves" were designed to help individuals to manage their diary, to fix appointments, to find rooms for meetings, to organize travel, etc. A paper reported technical successes but difficulties with some inappropriate agent behaviors. For instance, one day, or rather one night, an elf rang his master at 3am to inform him that his 11 o'clock plane was going to be delayed. Another was unable to understand that his master was in his office for nobody, since he had to complete an important project... Many of these actions make those intelligent agents tiresome and a real nuisance, which causes their rejection by users. This example shows that social inadequacy is the main cause of AI system rejection. In the previous case, the AI programs were technically successful; the system was not accepted because it did not answer to the requirements of the social environment. The causes of inappropriateness was not in the artificial system itself, but in the adequacy of the artificial system to its environment. It is neither astonishing nor original. Many people have noticed that the failures of knowledge-based systems were mainly due to man-machine interfaces or to organizational impediments, which make them inefficient (cf. for instance [5]). This is consistent with what had said the pioneers of AI, and in particular, with Herbert Simon who has insisted on the importance of the outer environment in [2]: according to him, "*Human beings, viewed as behaving systems, are quite simple. The apparent complexity of our behavior over time is largely a reflection of the complexity of the environment in which we find ourselves.*" In other words, the difficulty would not be in reproducing intelligent behaviors, but in adapting them to the complexity of their environment.

These conclusions are so obvious and conform with the predictions that the above mentioned AI failures would have had an incentive to address both user-centered design and social studies. Nevertheless, surprisingly, since the eighties, the evolution of AI toward, for instance, the so-called "Nouvelle AI" has gone in a completely different direction: AI has been accused of oversimplifying the world. It has been said that the reproduction of high level cognitive abilities, for instance doing mathematics, reasoning or playing chess, were easier, from a computational point of view, but less valuable than the simulation of basic physiological mechanisms of perception and action. The so-called "Moravec's Paradox" [6] summarized this point; it has been frequently invoked by specialists of robotics and AI last 20 years. As a consequence, the proposed solution was to increase the complexity of the models and to build powerful machines that mimic physiological capacities [7]. This view tends to reduce AI to a simulation of the natural processes. It opens undoubtedly exciting prospects for scientists. However, this does not exhaust the project of AI, which cannot be fully assimilated to a pure reproduction of the mind, i.e. to a "naturalization".

3 "Artificiality" vs. "Culturality"

Herbert Simon has introduced the distinction between the "Sciences of the Nature" and the "Sciences of the Artificial" in a famous essay published in 1962 in the "Proceedings of the American Philosophy Society". Its original point was to introduce the notion of artificiality to describe complex artificial systems in complex environments and to make them objects of science. According to him, artificial systems have to be distinguished from natural systems, because they are produced by human beings who have in mind some goals to achieve. More precisely, artificial things are characterized by the four following points [2]:

1. They are *produced by human* (or by intelligent beings) activity
2. They imitate the nature more or less the nature, while *lacking the whole characteristics of natural things*
3. They can be characterized in terms of *functions*, *goals* and *adaptation*
4. They can be discussed both in terms of imperatives or as descriptives

Since the artificial things can be approached not only in descriptive terms of their structure, but with respect to their functions, their goals and their adaptive abilities, they cannot be reduced to natural things that have only to be objectively described from the outside, without any a priori. Their study can take into consideration the imperatives to which they are supposed to obey. As a consequence, the discipline that is in charge to study artificial things, i.e. the science of the artificial things, has to be distinguished from the sciences of the natural things. To characterize this discipline, Hebert Simon has introduced the concept of "artefact", which is defined as an interface between the "inner" environment, i.e. the internal environment of an agent, and the "outer" environment where it is plunged. As previously said, the "inner" environment is easy both to describe in terms of functions, goals and adaptation and to simulate with computers; its complexity results from the "outer" environment in which it operates. It has to be recalled that artificial things can always be studied with the methods of the "sciences of the nature", for instance a clock can be studied from a physical point of view, by analyzing the springs and the wheels it is composed of, but those "sciences of the nature" don't take into consideration the imperatives to which the artificial things are supposed to obey, their functions and their goals. Symmetrically, natural things can be investigated by the "Sciences of the artificial". More precisely, according to Herbert Simon, the "sciences of the artificial" can greatly help to improve our knowledge of the natural phenomenon. Any natural thing can be approached by building models, i.e. artificial things, that aim at simulating some of their functions. For instance, cognitive psychology has been very much improved by the use of computers that help to simulate many of our cognitive abilities.

Two critics can be addressed to the AI understood as a "science of the artificial". The first is traditional and recurrent: for more than 20 years now, scientists and philosophers criticize the oversimplified models of the so-called old-fashioned AI. In a word, they think that models have to be exact images of what they are intended to model. As a consequence, the "artefacts", taken in Herbert Simon terms, i.e. the interfaces between "inner" and "outer" environments, have no real value when the "inner" environments are too schematic. Therefore, the artificiality has to faithfully copy the reality, i.e. the nature. As a consequence, many mental and social phenomenon are viewed as natural phenomenon. For instance, the mind is reduced to physical phenomenon that result from brain activity [8] or the epistemology is identified to informational processes [9]. This tendency corresponds to the so-called "naturalization", which is very popular nowadays among philosopher [10]. Nevertheless, despite the huge amount of researches done in this area for many years now, only a few results have been obtained.

The second critic is symmetric: the notion of "artefact" does not allow to fully approach the semantical and cultural nature of all mental processes. For instance, Herbert Simon considers music as a science of the artificial, since everything that is said about the sciences of the artificial can be said about music: it requires formal structures and provokes emotions. It is partially true, however, music is not only a syntax; semantical and cultural dimensions of music exist and they are not taken into account in Simon models. Therefore, we pretend that an extension of the "science of the artificial" toward the "sciences of the culture" is required. In other words, while the first critics opens on a naturalization, i.e. on a refinement of the models, the second pursues and extends the Herbert Simon "sciences of the artificial" by reference to the Neo-Kantian "sciences of the culture" that will be presented in the next section.

4 The "Sciences of the Culture"

The notion of "Sciences of the Culture" [11] was introduced in the beginning of the 20th century by a German Neo-Kantian philosopher, Heinrich Rickert who has been very influential on many people among which were the sociologist Max Weber and the young Martin Heidegger. Its goal was to base the humanities, i.e. the disciplines like historic studies, sociology, laws, etc., on rigorous basis. He wanted to build an empirical science able to interpret human achievements as the results of mental processes. However, he thought that the scientific characterization of the mind had to be distinguished from the psychology, which approached the mental phenomenon with the methods of the physical sciences. For him, spiritual phenomenon have a specificity that cannot be reduced to physical one, even if they can be submitted to a rational and empirical inquiry. The distinction between the "sciences of the nature" and the "sciences of the culture" had to precisely establish this specificity. As we shall see in the following, according to Rickert, the underlying logic of the "sciences of the culture" totally differs from the logic of the "sciences of the nature". Let us precise that the "sciences of the culture" have nothing to see with "cultural studies": the first attempt to characterize scientifically the results of human conscious activities – politics, art, religion, education, etc.– while the second try to identify and to differentiate cultural facts from various manifestations of human activities – dances, musics, writings, sculpture, etc.–.

As previously mentioned, the "sciences of the culture" aim at understanding social phenomenon that result from human conscious activities. Obviously, physics and chemistry are out of the scope of the "sciences of the culture" because they investigate the objective properties of the world, without any interference with human activities. On the contrary, the study of religion and discrimination may participate to the "sciences of the culture". But, the distinction is not so much a difference in the objects of study than in the methods of investigation. Therefore, the *history of physics* participates to the "sciences of the culture" while some mathematical models of social phenomenon, e.g. game theory, participate to the "science of the nature". Moreover, the same discipline may simultaneously participate to the "sciences of the nature" and to the "sciences of the culture"; it is what Rickert characterizes as an intermediary domain. For instance, medicine benefits simultaneously from large empirical studies and from individual case studies; the first participate more to the logic of the "sciences of the nature" and the second to the logic of the "sciences of the culture". It even happens, in disciplines like medicine, that national traditions differ, some of them being more influenced by the "sciences of the nature", like the *evidence-based medicine*, while others participate more easily to the "sciences of the culture", like *clinical medicine* when it is based on the study of the patient history. In other words, the main distinction concerns different logics of sciences that are described hereafter.

Ernst Cassirer clearly described the different logics of sciences in many of his essays [12]. Briefly speaking, he first distinguishes the theoretical sciences like mathematics, which deal with abstract and perfect entities as numbers, figures of functions, from empirical sciences that are confronted with the material reality of the world. Then, among the empirical sciences, Ernst Cassirer differentiates the "sciences of the nature", which deal with physical perceptions, and the "sciences of the culture" that give sense to the world. According to him and to Heinrich Rickert, the "sciences of the nature" proceed by generalizing cases: they extract general properties of objects and they determine laws, i.e. constant relations between observations. As a consequence, the logic of the "sciences of the nature" is mainly inductive, even if the modalities of reasoning may be deductive or abductive. The important point is that the particular cases have to be forgotten; they have to be analyzed in general terms as composed of well defined objects that make no reference to the context of the situation. The validity of the scientific activity relies on the constance and the generality of the extracted laws. By contrast to the logic of the "sciences of the nature", the logic of the "sciences of the culture" do not proceed by generalizing multiple cases. It does not extract laws, i.e. relations between observations; it does not even work with physical perceptions, but with meaningful objects that have to be understood. In brief, the main function of the "sciences of the culture" is to give sense to the world. The general methodology is to observe particular cases and to understand them. However, they have to choose, among the particulars, individuals that are paradigmatic, i.e. which can teach general lessons that may be reused in other circumstances. In other words, the "sciences of the culture" are not interested in the singularity of cases, which has to be forgotten, but in the understandability of individuals under study. Their methods help to give sense to observations of complex individual cases.

Looking back to the "sciences of the artificial", it appears that they belong both to the "sciences of the nature", since they proceed by generalization of cases and to the "sciences of the culture", because they characterize artificial things by their functions, their goals and their adaptivity, and not only but their structure. More precisely, the artificiality, taken in the sense given by Herbert Simon, includes not only all the things that are produced by the activity of intelligent beings, but also the goals to which they are designed for. Human productions are not reducible to the material things they achieved. A clock is more than the metal it is made of. As a consequence, artificiality is also part of the culturality. The sciences that produce artefacts, i.e. the "sciences of the artificial" are undoubtedly part of the "sciences of the culture', while the culture covers a broader area, since it also includes pure interpretative activities like history .

5 Conclusion and Perspectives

The thesis developed here is that the AI weaknesses are not caused by the over-simplification of AI models, like many people pretend nowadays, but by their inadequacy to the "outer" environment. It has been shown that the notion of "science of the artificial", which was introduced by Herbert Simon, has to be extended by reference to the notion of "science of the culture" introduced by the Neo-Kantian school in the beginning of the 20th century. From a philosophical point of view, it means that AI participates to the "sciences of the culture", i.e. that it cannot be entirely reducible to a "science of the nature" or to mathematics and theoretical sciences. But it is not more reducible to the "sciences of the culture". More precisely, it is what Heinrich Rickert identifies as an "intermediary domain" that belongs simultaneously to the theoretical sciences, i.e. to formal logic and mathematics, to the empirical sciences of the nature and to the empirical sciences of the culture. The practical consequences of such philosophical considerations are twofold: they have an impact on both the methods and the objects of application of AI.

Since AI participates to the "sciences of the culture", it has to take advantage of its logic, which may enlarge the scope of its methods. Let us recall that the sciences of the culture are empirical sciences, i.e. they build knowledge from the observation of particulars. However, they don't proceed by extracting properties common to observed cases; they do not abstract knowledge from particulars. They collect data about individual cases and they attempt to understand them, i.e. to find a common cause or to give a reason for them. Let us precise that it is not to observe singularity, but to study paradigmatic cases and to explain in what respect the individual cases under study can be universalized. An excellent example of such type of studies was done by a cognitive anthropologist, Edwin Hutchins, in the book titled "Cognition in the wild" [13] where he attempted to identify the cognition in its natural habitat, in the circumstances a modern ship, and to model it. In practice, many preliminary studies should have recourse to such methods. It has to be the case with knowledge engineering and, more generally, when designing any AI concrete application. Moreover, the attentive study of past failures participates to this dimension of AI. It is not to generalize all the individual failures by extracting their common properties, as it could be in any science of the nature, but to understand the logic of the failures, to see what lessons could be drawn from these bad experiences to generalize them and to learn from it.

Lastly, the investigations of AI could focus more deliberately on cultural dimensions of the world, where there are many valuable applications. The information sciences and technologies greatly contribute to the advancement of knowledge to the point that the present age is often called a "knowledge age". However, it's a pity that AI did not participate more actively to cultural evolutions consecutive to the development of information technologies, for instance, to the Wikipedia free encyclopedia or to the social web. More generally, the knowledge quest can be greatly accelerated by the use of AI technologies. For instance, my team is working in musicology [14], in textual criticism, in social sciences, in epistemology [15] etc. But there are many other fields of applications, not only in humanities. Let us insist that such applications of AI are directly connected with cultural dimensions. So, in case of medicine, there already exist many attempts to model organs and to simulate medical diagnosis; AI had part in these successful achievements; but the new challenge now is to manage all the existing knowledge and to help researchers to find their way. This is undoubtedly the role of AI understood as a science of culture to help to achieve such tasks.

References

[1] Hutter, M.: Universal Artificial Intelligence: sequential Decisions Based on Algorithmic Probabilities. Springer, Heidelberg (2005)
[2] Simon, H.A.: The Sciences of the Artificial, 3rd edn. MIT Press, Cambridge (1996)
[3] Newell, A.: The knowledge level. Artificial Intelligence Journal 18, 87–127 (1982)
[4] Shapiro, D., Goker, M.: Advancing ai research and applications by learning from what went wrong and why. AI Magazine 29(2), 9–76 (2008)
[5] Hatchuel, A., Weil, B.: Experts in Organizations: A Knowledge-Based Perspective on Organizational Change. Walter de Gruyter, Berlin (1995)
[6] Moravec, H.: Mind Children. Harvard University Press, Cambridge (1988)
[7] Brooks, R.: Flesh and Machines: How Robots Will Change Us. Pantheon Books, New York (2002)
[8] Manzotti, R.: Towards artificial consciousness. American Philosophy Association Newsletter on Philosophy and Computers 07(1), 12–15 (2007)
[9] Chaitin, G.: Epistemology as information theory. COLLAPSE 1, 27–51 (2006)
[10] Dodig-Crnkovic, G.: Epistemology naturalized: The info-computationalist approach. American Philosophy Association Newsletter on Philosophy and Computers 06(2) (2007)
[11] Rickert, H.: Kulturwissenschaft und Naturwissenschaft, 5th edn. J.C.B. Mohr (Paul Siebeck), Tubingen (1921)
[12] Cassirer, E.: Substance and Function. Open Court, Chicago (1923)
[13] Hutchins, E.: Cognition in the Wild. MIT Press, Cambridge (1995)
[14] Rolland, P.Y., Ganascia, J.G.: Pattern detection and discovery: The case of music data mining. In: Hand, D.J., Adams, N.M., Bolton, R.J. (eds.) Pattern Detection and Discovery. LNCS (LNAI), vol. 2447, pp. 190–198. Springer, Heidelberg (2002)
[15] Ganascia, J.G., Debru, C.: Cybernard: A computational reconstruction of claude bernard's scientific discoveries. In: Studies in Computational Intelligence (SCI), pp. 497–510. Springer, Heidelberg (2007)

Beyond Public Announcement Logic: An Alternative Approach to Some AI Puzzles

Paweł Garbacz, Piotr Kulicki, Marek Lechniak, and Robert Trypuz

John Paul II Catholic University of Lublin,
al. Racławickie 14, 20-950 Lublin, Poland
{garbacz,kulicki,lechniak,trypuz}@kul.pl
http://www.l3g.pl

Abstract. In the paper we present a dynamic model of knowledge. The model is inspired by public announcement logic and an approach to a puzzle concerning knowledge and communication using that logic. The model, using notions of situation and epistemic state as foundations, generalizes structures usually used as a semantics for epistemic logics in static and dynamic aspects. A computer program automatically solving the considered puzzle, implementing the model, is built.

Keywords: knowledge representation, dynamic epistemic logic, multi-agent systems, Prolog.

1 Introduction

Knowledge representation plays an important role in AI. Moore in [7] claims that the basic application of logic in AI is an analysis of this concept. He points out two uses of formal logic in that context. Firstly, the formalism of logic can be used as a knowledge representation system; secondly, logic provides forms of deductive reasoning. The formalism of logic has two components: syntactic (language and axioms) and semantic (model theoretic structure). In the case of knowledge representation the language of logic includes epistemic operators that enable us to formulate sentences of the type: "agent a knows that φ","an agent a believes that φ", etc. Axioms set up a theory that defines valid deductions for such sentences. The semantic component of epistemic logic is usually based on structures built from sets of possible worlds. In that setting logical space of epistemically possible worlds, i.e. worlds consistent with an agent's knowledge, is considered.

McCarthy [6] argues, that from the AI point of view the semantic part of epistemic logic is more useful. The reason is that it is more intuitive and easier from the operational perspective. We share that opinion. Moreover, we think that philosophical understanding of semantic structures for epistemic logic still requires more attention. Thus, our main effort is to construct an ontologically founded model for epistemic logic. In that model, we would like to be able to represent a vast spectrum of notions about knowledge in multi-agent environments, including not only the actual knowledge of the agent himself and other agents

B. Mertsching, M. Hund, and Z. Aziz (Eds.): KI 2009, LNAI 5803, pp. 379–386, 2009.

but possible knowledge, change of knowledge and relations between knowledge and actions as well.

As a working example we will use a well-known *hat puzzle: Three people Adam, Ben and Clark sit in a row in such a way that Adam can see Ben and Clark, Ben can see Clark and Clark cannot see anybody. They are shown five hats, three of which are red and two are black. The light goes off and each of them receives one of the hats on his head. When the light is back on they are asked whether they know what the colours of their hats are. Adam answers that he doesn't know. Then Ben answers that he doesn't know either. Finally Clark says that he knows the colour of his hat. What colour is Clark's hat?*

2 Solving Puzzles in Public Announcement Logics

Public announcement logic [8,2,10] (henceforward PAL) is an extension of epistemic logic [11,4], which is a modal logic formally specifying the meaning of the formula $K_a\varphi$—"agent a knows that φ". PAL enriches epistemic logic with a dynamic operator of the form "$[\varphi]\psi$", which is read "after public and truthful announcement that φ, it is the case that ψ". That extension allows for reasoning about the change of agent's beliefs being a result of receiving new information.

The language of PAL is defined in Backus-Naur notation as follows:

$$\varphi \quad ::= \quad p \mid \neg\varphi \mid \varphi \wedge \varphi \mid K_a\varphi \mid [\varphi]\varphi$$

where p belongs to a set of atomic propositions Atm, a belongs to a set of agents $Agent$ and K is an operator of individual knowledge and $[\_]$ is an operator of public announcement.

Models of PAL are Kripke structures with valuation function and have the form $\mathcal{M} = \langle S, Agent, \sim, v\rangle$. Its components are described as below.

- S is a set of situations (or possible worlds).
- $Agent$ is a set of rational agents.
- $\sim$ is a function assigning a set of pairs of situations about which it is said that they are *indiscernible* to every element of $Agent$. Thus $s \sim_a s'$ means that for agent a, two situations, s and s', are indiscernible.
- $v : Atm \longrightarrow 2^S$ is a standard valuation function.

Satisfaction conditions for PAL in the model $\mathcal{M}$ and for $s \in S$ are defined in a standard way. For knowledge and public announcement operators they have the forms:

$$\mathcal{M}, s \models K_a\varphi \iff \forall s' \in S\ (s \sim_a s' \Longrightarrow \mathcal{M}, s' \models \varphi)$$
$$\mathcal{M}, s \models [\varphi]\psi \iff \mathcal{M}, s \models \varphi \Longrightarrow \mathcal{M}^{\varphi}, s \models \psi$$

where $\mathcal{M}^{\varphi} = \langle S', A, \sim', v'\rangle$ is characterized in such a way that: $S' = \{s \in S : \mathcal{M}, s \models \varphi\}$, for every $a \in Agent$, $\sim'_a = \sim_a \cap (S' \times S')$ and for every $p \in Atm$, $v'(p) = v(p) \cap S'$.

The epistemic logic S5 is sound and complete with respect to $\mathcal{M}$, when $\sim_a$ are equivalence relations. One can find the axiomatisation of "$[\_]$" in [10, section 4].

The concept of knowledge presupposed by PAL is such that agent knows that φ when in all situations (possible words) belonging to his epistemic possibility space, φ holds. The epistemic possibility space of agent a can be seen as a set of situations which are equally real for that agent.

By reducing the epistemic possibility space the agent minimizes his level of ignorance [5]. An omniscient agent would have in his epistemic possibility space just one single world, i.e. the real one, whereas an agent having no knowledge would consider all situations possible, or in other words, equally real.

The reduction of the level of ignorance is done in PAL by a public and truthful announcement about a real situation. For instance, an agent having no knowledge, just after receiving the information that he is wearing a red hat, will restrict his epistemic possibility space to those situations in which it is true that he is wearing a red hat.

Now let us return to the hats puzzle. The model for the hats puzzle, $\mathcal{M}^* = \langle S, Agent, \sim, v\rangle$, is such that $Agent = \{Adam, Ben, Clark\}$. Remaining elements of $\mathcal{M}^*$ are described below.

In the models of PAL a situation is usually represented by a tuple. For instance $\langle r, r, w\rangle$ is 3-tuple representing the situation in which Adam has a red hat (r), Ben has a red hat, too (r) and Clark has a white hat (w). Thus:

$$S = \{\langle r,r,r\rangle, \langle r,r,w\rangle, \langle r,w,r\rangle\ \langle r,w,w\rangle, \langle w,r,r\rangle, \langle w,r,w\rangle, \langle w,w,r\rangle\}.$$

Because the indistinguishability relation (relativized to agent a), $\sim_a$, is an equivalence relation, the equivalent classes and a quotient set (partition), $S/_{\sim_a}$, can be defined in a standard way. Intuitively, each cell of this partition is a set of the situations which are indistinguishable for agent a.

Which situations are indistinguishable for each of the three men? It is a question on which PAL can answer neither on language nor on semantic level. Although in the hats puzzle, before the men say anything, an indistinguishability relation can be precisely defined by means of *sees* relation. $sees(a, b)$ is an irreflexive relation (because it is assumed in the puzzle that nobody sees himself) expressing a fact that a sees (perceives) b. Let $\alpha^i_{a_i}$ be an expression saying that agent a_i has a hat of the colour α^i. Then:

$$\langle \alpha^1_{a_1}, \alpha^2_{a_2}, \alpha^3_{a_3}\rangle \sim_{a_i} \langle \beta^1_{a_1}, \beta^2_{a_2}, \beta^3_{a_3}\rangle \iff \forall j(sees(a_i, a_j) \rightarrow \alpha^j = \beta^j), \qquad (1)$$

where $1 \leqslant i, j \leqslant 3$ and $\alpha^1, \ldots, \beta^3$ represent red or white colour and a_1, a_2, a_3 represent Adam, Ben and Clark respectively.

Because Adam sees the (colours of the) hats of Ben and Clark, and he does not see his own hat, the epistemic possibility space for Adam is described by the set:

$$S/_{\sim_{Adam}} =$$

$$\{\{\langle r,w,r\rangle, \langle w,w,r\rangle\}, \{\langle r,r,r\rangle, \langle w,r,r\rangle\}, \{\langle r,r,w\rangle, \langle w,r,w\rangle\}, \{\langle r,w,w\rangle\}\}$$

Similarly we get the sets $S/_{\sim_{Ben}}$ and $S/_{\sim_{Clark}}$:

$$S/_{\sim_{Ben}} = \{\{\langle r,w,r\rangle, \langle w,w,r\rangle, \langle r,r,r\rangle, \langle w,r,r\rangle\}, \{\langle r,r,w\rangle, \langle w,r,w\rangle, \langle r,w,w\rangle\}\}$$

$$S/_{\sim_{Clark}} = \{\{\langle r,r,r\rangle, \langle w,r,r\rangle, \langle r,w,r\rangle, \langle w,w,r\rangle, \langle r,r,w\rangle, \langle w,r,w\rangle, \langle r,w,w\rangle\}\}$$

Now we are going to show how PAL can deal with the hats puzzle. Let us start with the assumption that set Atm has the following elements:

$$Atm = \{has(Adam,r), has(Adam,w), has(Ben,r), has(Ben,w), has(Clark,r), has(Clark,w)\} \tag{2}$$

where $has(a,r)$ and $has(a,w)$ mean that a has a red hat and a has a white hat respectively.

Because in the puzzle each man is allowed only to announce that he knows or does not know the colour of his hat, it is useful to introduce the following definition:

$$K_a Hatcolour =_{df} K_a has(a,r) \vee K_a has(a,w) \tag{3}$$

$K_a Hatcolour$ means that a knows the colour of his hat.

Now we are ready to express in PAL what has happened in the puzzle's scenario, i.e. that $Adam$ and then Ben announce that they do not know the colours of their hats, what finally leads to the situation in which $Clark$ knows that his hat is red. Formally:

$$[\neg K_{Adam} Hatcolour][\neg K_{Ben} Hatcolour] K_{Clark} has(Clark,r) \tag{4}$$

Clark gets the knowledge about the colour of his hat by reducing his set of indiscernible situations to the set in which it is true that neither Adam knows the colour of his hat nor does Ben. As a result he receives the following set of possible situations: $\{\langle r,w,r\rangle, \langle w,w,r\rangle, \langle r,r,r\rangle, \langle w,r,r\rangle\}$ and therefore formula 4 is true. Clark taking into account what $Adam$ and Ben have said is still unable to identify the real situation. However in all remained indiscernible situations $Clark$ can be sure that he has a red hat.

In the solution of the puzzle we used both PAL and its model. This points out a more important property of PAL, namely that in order to completely understand what knowledge is in this approach we need both the language of PAL and its model. In the language of PAL we could express that an agent knows something and state by axioms of S5 some of the formal properties of the agent's knowledge. The models of PAL give us the understanding of the fact that agent a knows something in situation s by reference to the indiscernibility relation.

3 Towards a General Model of Epistemic Change

Our model of epistemic change has two components: ontological and epistemological. The ontological part represents, in rather rough and ready way, the world our knowledge concerns. The epistemological part of the model represents the phenomenon of knowledge in its static and dynamic aspects. The situation at stake may have any ontic structure. Thus, there are situations "in which" certain objects possess certain properties, situations "in which" certain objects participate in certain relations or processes, etc.

A situation is *elementary* if no other situation is part of it, eg. *that Adam has a red hat* would be an elementary situation and *that both Adam and Ben have red hats* would not be an elementary situation. Let $ElemSit$ be a set of elementary ontic (possible) situations.

In the set $ElemSit$ we define the relation of copossibility ($\|$). Intuitively, $x \parallel y$ means that situation x may (ontologically) cooccur with situation y. For example, *that Adam has a red hat* is copossible with *that Ben has a red hat*, but is not copossible with *that Adam has a white hat.* The relation $\|$ is reflexive and symmetric in $ElemSit$, but is not transitive.

In what follows we will represent situations as sets of elementary situations. Let $\emptyset \notin Sit \subseteq \wp(ElemSit)$ be a set of (possible) ontic situations. Given our understanding of the relation $\|$, the following condition is accepted:

$$X \in Sit \to \forall y, z \in X y \parallel z. \tag{5}$$

We can now define the notion of a possible world:

$$X \in PossWorld \triangleq X \in Sit \wedge \forall Y (X \subset Y \to Y \notin Sit). \tag{6}$$

Let $Time = (t_1, t_2, \dots)$ be a sequence of moments. The *actual epistemic state* of an agent at a given moment will be represented by a subset of $PossWorld$: $epist(a, t_n) \subseteq PossWorld$. Any such state collectively, so to speak, represents both the agent's knowledge and his ignorance. Due to its actual epistemic state, which is represented by a set $epist(a, t_n)$, and for every $X \in Sit$, agent a may be described (at t_n) according to the following three aspects:

Definition 1. *Agent a knows at moment t_n that situation X holds (written: $K_{a,t_n}(X)$) iff $X \subseteq \bigcap epist(a, t_n)$.*

Definition 2. *Agent a knows at moment t_n that situation X does not hold (written: $\overline{K}_{a,t_n}(X)$) iff $X \cap (\bigcup epist(a, t_n)) = \emptyset$.*

Definition 3. *Agent a does not have any knowledge at moment t_n about situation X iff $\neg K_{a,t_n}(X) \wedge \neg \overline{K}_{a,t_n}(X)$.*

However, the puzzles we are dealing with do not presuppose that we know the actual epistemic state of a given agent. Thus, we extend the notion of actual epistemic state to the notion of possible epistemic state. A *possible epistemic state* of an agent represents a body of knowledge (resp. of ignorance) that the agent may exhibit given the ontic situation and epistemic capabilities of the agent. In our case, the possible epistemic states are determined by the relation of seeing, other agents' announcements and the agent's deductive capabilities.

A possible epistemic state of an agent at a given moment will be represented by the set $epist_i(a, t_n) \subseteq Sit$.

Definition 4. *Agent a knows in a possible epistemic state X that (ontic) situation Y holds (written : $K_{epist_i(a,t_n)}(Y)$) iff $Y \subseteq \bigcap X$.*

Definitions analogous to 2 and 3 can be also added.

We will use the following auxiliary notions:

- $Epist(a, t_n)$ - the set of all possible epistemic states of agent a at t_n,
- $Epist(t_n)$ - the set of sets of possible epistemic states of all agents at t_n,
- $Epist$ - the set of sets of all possible epistemic states of all agents (from $Agent$) at all moments (from $Time$).

When $a \in Agent$, then "$\sim_a$" will represent the relation of epistemological indiscernibility, which we treat as an equivalence relation. In a general case, the epistemological indiscernibility covers a number of epistemic constraints of agents.

In our case, the relation of epistemological indiscernibliity depends on the knowledge obtained thanks to the behaviour of some agent.

The relation of epistemological indiscernibility is, as the notion of knowledge itself, relative to time: $\sim_{a,t_n}$ is the relation of epistemological indiscernibility for agent a at time t_n.

It is an assumption of our approach that possible epistemic states coincide with the abstraction classes of the epistemological indiscernibliity relation:

$$Epist(a, t_n) = PossWorld / \sim_{a,t_n} \tag{7}$$

We assume that all changes of epistemic states are caused by the behaviour of agents, in particular by their utterances, by means of which they expose their (current) epistemic states and not by their inference processes.

A number of rules that govern the dynamics of epistemic states can be defined. In the hats puzzle the only rule that sets the epistemic states in motion is the following one:

If agent a (says that) he does not know that X holds, in an epistemic state $epist_i(a, t_n)$ a knows that X holds, then after the aforementioned utterance this state (i.e. $epist_i(a, t_n)$) is effectively impossible, i.e. we remove its elements from all possible epistemic states of all agents. Formally,

Rule 1. *If (a says that)* $\neg K_{a,t_n}(X)$ *and* $Y \in Epist(a, t_n)$ *and* $K_Y(X)$, *then for every* $a' \in Agent$, $Epist(a', t_{n+1}) = \delta_0(Epist(a', t_n), Y)$, *where*

Definition 5. δ_0 *maps* $Epist \times \bigcup\bigcup Epist$ *into* $Epist$ *and satisfies the following condition:*

$$\delta_0(Epist(a, t_n), X) = \begin{cases} Epist(a, t_n) \backslash \{X\}, & \text{if } X \in Epist(a, t_n), \\ (Epist(a, t_n) \backslash \{Z\}) \cup \{Z \backslash X\} & \text{if } Z \in Epist(a, t_n) \text{ and} \\ & X \cap Z \neq \emptyset, \\ Epist(a, t_n) & \text{otherwise.} \end{cases}$$

Some other possible rules are presented in [1].

It seems that the factors that trigger the process of epistemic change are of two kinds: ontological and epistemological. The ontological condition of this rule is the fact that agent a says that he does not know that a certain ontic situation holds. The epistemological condition is his epistemic state ($Y \in Epist(a, t_n)$), in which the agent knows that this situation holds ($K_Y(X)$). We may represent the epistemological conditions of rules for epistemic changes by means of the notion of epistemic state. However, in order to account for the ontological conditions, we distinguish in the set $\wp(ElemSit)$ a subset *AgentBeh* that collects types (here: sets) of ontic situations that are those conditions. An example of such type may be a set of situations in which agents say that they do not know what hat they have. In general, those conditions may be classified as agents' behaviours, which include also such "behaviours" as being silent (cf. the wisemen puzzle).

Let $a \in Agent$. A *rule for epistemic change* ρ is either

1. mapping $\rho : \bigcup Epist \times AgentBeh \times \bigcup Epist \to \bigcup Epist$ (this condition concerns rules with epistemological conditions) or
2. mapping $\rho : \bigcup Epist \times AgentBeh \to \bigcup Epist$ (this condition concerns rules without epistemological conditions).

It should be obvious that

1. if $\rho(X, Y, Z) = V$ and $X, Z \in Epist(t_n)$, then $V \in Epist(t_{n+1})$ (for rules with epistemological conditions),
2. if $\rho(X, Y) = V$ i $X \in Epist(t_n)$, to $V \in Epist(t_{n+1})$ (for rules without epistemological conditions).

The set of all such rules will be denoted by "*Rule*". For the sake of brevity, from now on we will consider only rules with epistemological conditions.

In order to obtain the solution to the puzzle at stake, one needs the following input data: set *Sit*, temporal sequence $Time = (t_n)$, set of epistemic agents *Agent*, set of sets of epistemic states of any such agent at the initial moment t_1: $Epist_1 = \{Epist(a, t_1) : a \in Agent\}$ and function $dist : Time \to \wp(ElemSit)$.

The evolution of sets of epistemic states is triggered by the ontological conditions according to the accepted rules of epistemic change. This implies that the following condition holds (For the sake of simplicity, we assume that function *dist* does not trigger more than one rule at a time.):

$$\exists \rho \in Rule\ \exists X \in AgentBeh\ [\rho(Epist(a, t_n), X, Epist(a', t_n)) = Z \wedge dist(t_n) \cap X \neq \emptyset] \to Epist(a, t_{n+1}) = Z. \tag{8}$$

We also assume that epistemological states change only when a certain rule is triggered:

$$Epist(a, t_{n+1}) \neq Epist(a, t_n) \equiv \exists \rho \in Rule\ \exists X \in AgentBeh\ \exists Y \in \bigcup Epist\ [\rho(Epist(a, t_n), X, Y) = Epist(a, t_{n+1})]. \tag{9}$$

4 Conclusion

We have presented a dynamic model of knowledge for rational agents. The model is inspired by public announcement logics and their usual semantics in a form of Kripke structures. However, the basic notions of our model, situation and possible epistemic state of an agent, are more fundamental from a philosophical point of view and enable us to define the notion of a possible world. We have also introduced rules defining dynamics of agents' knowledge in the presence of other agents' behaviour. Public announcements can be seen as special cases of such a behaviour.

The model has been successfully tested on logic puzzles concerning knowledge. Just by changing parameters of the program one can solve the hats puzzle with any number of agents and different perception relations. Two of the special cases here are known from literature puzzles: *three wise men* and *muddy children.* From the point of view of programming it is important that the formulation of the model could be easily transferred into a program in Prolog (see the prolog source at www.l3g.pl).

References

1. Garbacz, P., Kulicki, P., Lechniak, M., Trypuz, R.: A formal model for epistemic interactions. In: Nguyen, N.T., Katarzyniak, R., Janiak, A. (eds.) Challenges in Computational Collective Intelligence. Studies in Computational Intelligence. Springer, Heidelberg (2009) (forthcoming)
2. Gerbrandy, J.D., Groeneveld, W.: Reasoning about information change. Journal of Logic, Language and Information 6, 147–169 (1997)
3. Gomez-Perez, A., Corcho, O., Fernandez-Lopez, M.: Ontological Engineering. Springer, London (2001)
4. Hintikka, J.: Knowledge and Belief: An Introduction to the Logic of The Two Notions. Cornell University Press, Ithaca (1962)
5. Hintikka, J.: On the Logic of Perception. In: Models for Modalities. Reidel Publ. Comp., Dordrecht (1969)
6. McCarthy, J.: Modality, Si! Modal Logic, No! Studia Logica 59, 29–32 (1997)
7. Moore, R.C.: Logic and Representation. In: CSLI Lecture NotesNo 39. CSLI Publications, Center for the Study of Language and Information, Standford (1995)
8. Plaza, J.A.: Logic of public communications. In: Emrich, M.L., Pfeifer, M.S., Hadzikadic, M., Ras, Z.W. (eds.) Proceedings of the 4th International Symposium on Methodologies for Intelligent Systems, pp. 201–216 (1989)
9. Russell, B.: On Denoting. Mind 14, 479–493 (1905)
10. van Ditmarsch, H., van der Hoek, W., Kooi, B.: Dynamic Epistemic Logic. Synthese Library Series, vol. 337. Springer, Heidelberg (2007)
11. von Wright, G.H.: An Essay in Modal Logic. North Holland, Amsterdam (1951)

Behavioural Congruence in Turing Test-Like Human-Computer Interaction

Stefan Artmann

Friedrich-Schiller-University, Frege Centre for Structural Sciences, Jena, Germany
stefan.artmann@uni-jena.de

Abstract. Intensions of higher-order intentional predicates must be observed in a system if intelligence shall be ascribable to it. To give an operational definition of such predicates, the Turing Test is changed into the McCarthy Test. This transformation can be used to distinguish degrees of behavioural congruence of systems engaged in conversational interaction.

1 Introduction

Artificial Intelligence (AI) explores the space of possible intelligent systems. To give a demarcation of its field of research, AI must list general conditions of intelligence. Each of these conditions should imply an operationalizable criterion that, if satisfied by a system in a test, gives the tester a good reason to be more inclined to consider the system intelligent. Thus, AI ought to determine experimentally testable conditions of intelligence that are necessary, formal, and not restricted to particular species of intelligent systems. This paper regards a basic feature of intelligent systems as a necessary condition of intelligence, and operationalizes it in form of a conversational test. It is proposed that intensions of higher-order intentional predicates must be observed in a system if intelligence shall be ascribable to it. To observe these intensions, the Turing Test is changed, via the Simon-Newell Test, into the McCarthy Test, whose participants communicate also meta-information about their internal information processing. The transformation can be used to differentiate between three degrees of behavioural congruence of systems in the context of conversational interaction. In the following, these degrees will be described in an elementary manner along with the test scenarios to which they are pertinent: weak congruence and the Turing Test (*sect. 2*), structural congruence and the Simon-Newell Test (*sect. 3*), strong congruence and the McCarthy Test (*sect. 4*).

2 Black-Box Testing of Weak Congruence (Turing Test)

It is quite usual to start discussing the problem of how to define intelligence in AI, by analysing the famous Turing Test (TT), which originated in A.M. Turing's imitation game [1, 2-4].

B. Mertsching, M. Hund, and Z. Aziz (Eds.): KI 2009, LNAI 5803, pp. 387–394, 2009.

2.1 The Turing Test as a Black-Box Testing Procedure

To carry out TT one needs: two devices for sending and receiving signals; a reliable bidirectional communication channel between the devices; a human acting as an information source and destination at one of the devices; a computer acting as an information source and destination at the other device; and a sensory barrier preventing the human from perceiving the computer. A TT goes repeatedly through the following procedure: the human inputs a question into her device; the question is transmitted through the communication channel and received by the computer's device; the computer inputs an answer into its device; the answer is sent through the communication channel and received by the human's device. This sequence is to be repeated until the channel is blocked after a predetermined time interval. Then the human is asked a metaquestion: 'Did you communicate with another human or with a computer?' The computer passes TT if, and only if, the human answers that her interlocutor has been another human.

The computer acts as a black box [5] in TT: its inside is completely hidden from the human. She is forced to adopt a behaviouristic stance on her interlocutor by interpreting the received signals as externally observable linguistic behaviour. To answer the TT metaquestion, the human cannot but speculate about internal structures hidden from her, and thus about the cognitive nature of her interlocutor. A rational way out of this predicament is to compare the signals received from the unknown system, with signals that an average human might reasonably be expected to send if the latter were asked the same questions. If the real and the expected signals are, in a sense to be defined, congruent, then the human is empirically justified to suppose that her interlocutor has been another human.

2.2 Weak Congruence of Behaviour in the Turing Test

The congruence of signals, which is checked by the human in TT, must be defined precisely. R. Milner, whose Calculus of Communicating Systems is a very important realization of the idea of process algebra that broadly inspires the definitions (1)-(3) of congruence relations given below, calls a process by which a system does something, 'action'. Then the notion of interaction can be understood "in the sense that an action by one entails a complementary action by another" [6]. Since, according to Milner [6], an interaction shall also constitute a mutual observation of both systems involved, just to send s a signal and to receive r it, is insufficient for constituting a pair of complementary actions in TT. Postulating furthermore that a system can observe an action only by interacting with it, observations are interactions, and vice versa [6]. In TT, observations are composed of two communicative actions, the question $q = (s_H, r_C)$, the human H sending a signal and the computer C receiving it, and the answer $a = (s_C, r_H)$, C sending a signal and H receiving it. A pair (q_n, a_n) is the nth interaction in a TT experiment. Internal actions i_C of C that H cannot observe, are called 'reactions of C' (the reactions of other systems are defined accordingly). The imaginary system that would live up to H's expectations of how an average human might answer H's questions, is denoted by $E(H)$. By CR » CS and $E(H)R$ » $E(H)S$, let us denote sequences of reactions of C and $E(H)$, respectively, that occur between the reception of H's signal and the sending of a signal by C and $E(H)$, respectively, to H.

CR, *CS*, *E*(*H*)*R*, and *E*(*H*)*S* denote, thus, the states in which the computer and the imaginary human are after they received a signal of *H* and sent a signal to *H*, respectively. *H* cannot observe what *C*'s reactions are, or how many reactions occur, but *H* has a more or less detailed account of what *E*(*H*)'s reactions could be. A reaction of *C* corresponds to zero or any positive number of reactions of *E*(*H*), and vice versa. Now the TT-relevant type of behavioural congruence, as assessed by *H*, of the answers of *C* and *E*(*H*) can be defined.

Weak congruence. An answer $a_C = (s_C, r_H)$ of *C* and an answer $a_{E(H)} = (s_{E(H)}, r_H)$ of *E*(*H*) to a question $q = (s_H, r_{C \text{ or } E(H)})$ of *H* are *weakly congruent as to H*, written $a_C \approx_H a_{E(H)}$, if and only if two conditions are met (with *R* the reception of s_H and *S* the sending of s_C and $s_{E(H)}$, respectively): (a) if *CR* » *CS* then there exists *E*(*H*)*S* such that *E*(*H*)*R* » *E*(*H*)*S* and $s_C = s_{E(H)}$; and (b) if *E*(*H*)*R* » *E*(*H*)*S* then there exists *CS* such that *CR* » *CS* and $s_C = s_{E(H)}$. (1)

Two interlocutors of *H* cannot differ, in respect to their weakly congruent answers, internally in a way that would lead *H* to divergent observations. But weak congruence does not entail more information about the internal structure of *C* and *E*(*H*).

The human judges after every interaction whether the answer is weakly congruent to the one that she would expect from another human being. Her behaviouristic stance on the black box she talks to, allows the human to reword the TT metaquestion as follows: 'Is the black box using linguistic signs in a way which does not disappoint my expectations of how another human relates communicated signs to actually referenced objects (extensions) in a shared world of linguistic meaning?' If the human thinks that her interlocutor has answered a high enough percentage of questions like another human, and that none of its answers blatantly contradicts its human character, then the black box meets a sufficient condition for being human and, so it is presumed, for being intelligent.

Weak congruence should, in general, be the relation between a system and its specification, since the latter concerns only correctness of behaviour [6]. What is at stake in TT, is, to put it shortly, the equation *system* $C \approx_H$ *specification* *E*(*H*). Though this seems to be a very liberal condition and, thus, adequate to AI, weak congruence implies too strong a condition: a specification of the behaviour of a particular species of real intelligent systems is too restrictive for a science that studies intelligent systems in general.

3 Grey-Box Testing of Structural Congruence (Simon-Newell Test)

A first step of transforming TT into a procedure that operationalizes a necessary, formal, and species-neutral condition of intelligence, is to address the reactions of the computer during the test. Thus results the so-called 'Simon-Newell Test' (SNT) [7].

3.1 The Simon-Newell Test as a Grey-Box Testing Procedure

The analysis of TT has shown that a conversational test relying just upon how systems relate communicated signs to extensions, does not operationalize a condition of intelligence that is appropriate to AI. It is, thus, reasonable to design a test in which a system is tested whether it designates intensions of linguistic signs. R. Carnap defined the intension of a predicate for a system *X* as "the general condition which an object must fulfil for *X* to be willing to apply the predicate to it." [8] Algorithmically, the intension of a predicate *p* for *X* is a procedure within *X* that computes whether an object satisfies *p*. The intension of *p* for *X* implicitly defines the set of all those objects that *X* could denotate by using *p*. In short, intensions are "potential denotations" [9]. Now a new condition for intelligence can be proposed: only if *X* can distinguish objects by intensions, *X* should be ascribed intelligence to. Otherwise, the Shannon-McCarthy objection [10] to Turing's imitation game would become relevant; i.e., a world could then be imagined in which a machine would be considered intelligent though it just looked up, in a table, its responses to all possible linguistic inputs very fast.

Whether a system does have intensions, shall be detected by SNT [9, 7]. In SNT, a computer answers the human's questions also by truthfully giving meta-information about how it internally generated its answers. As in TT, the computer is allowed to lie – but not about the structure of its information processing. After the conversation, the human must give an answer to the same metaquestion as in TT: 'Did you communicate with another human or with a computer?' If she thinks that the procedures by which the tested system computes whether signals denote certain extensions, are those a human would follow, she is obliged to answer that she talked to another human, and the computer has then passed SNT.

In SNT, the tested system is visible to the human tester as a grey box [5]. By receiving true meta-information about the information processing of the system, the human is able to make a shadowy picture of the system's inside as if she were looking, by means of a torch, from inside the system into its mechanism. The tested system cannot, of course, be a white box: the tester would then know everything about the tested system, and a TT-like test would not make sense any more.

For carrying out SNT, a standardized language is needed that can be used to communicate meta-information in the test. Such a language must be a calculus interpretable purely in terms of information entities by abstracting details of its implementation away so that the system's answers do not contain data which would immediately reveal the non-human realization of a structure of information processing.

3.2 Structural Congruence of Behaviour in the Simon-Newell Test

The concepts and notations introduced in *sect. 2.2* are used also to define the congruence relation pertinent to SNT. In addition, by $CR \gg_i CS$ and $E(H)R \gg_i E(H)S$, let us denote the precise sequence of reactions that occurs in *C* and *E*(*H*), respectively, between the reception of *H*'s signal and the sending of a signal to *H*. In SNT, *H* does know the sequence occuring in *C* since *C* gives her information about the information processing that generated its answer to *H*'s question. Now the type of behavioural congruence of the answers of *C* and *E*(*H*), as assessed by *H*, that is needed for SNT, can be defined.

Structural congruence. An answer $a_C = (s_C, r_H)$ of C and an answer $a_{E(H)} = (s_C, r_H)$ of $E(H)$ to a question $q = (s_H, r_{C \text{ or } E(H)})$ of H are *structurally congruent as to H*, written $a_C =_H a_{E(H)}$, if and only if two conditions are met (with R the reception of s_H and S the sending of s_C and $s_{E(H)}$, respectively): (a) if $CR \gg_i CS$ then there exists $E(H)S$ such that $E(H)R \gg_i E(H)S$ and $s_C = s_{E(H)}$; and (b) if $E(H)R \gg_i E(H)S$ then there exists CS such that $CR \gg_i CS$ and $s_C = s_{E(H)}$. (2)

Two interlocutors of H cannot show, in respect to their structurally congruent answers, any differences in the structure of their information processing that are relevant either interactionally or reactionally. If two answers are structurally congruent, they are, as can easily be seen, also weakly congruent.

The human must judge after every interaction whether the answer is structurally congruent to the one that she would expect from another human being. Her generative stance on the communicated internal information processing of the grey box allows the human to put the SNT metaquestion more precisely: 'Is the grey box generating linguistic signs in a way which does not disappoint my expectations of how another human being relates communicated signs to potentially referenced objects (intensions) in a shared world of linguistic meaning?' If the human thinks that her interlocutor's mechanism of producing its anwers is such as that at work in an average human being, then the grey box meets a necessary (having intentions) and sufficient (being an average human) condition of intelligence.

The outcome of SNT depends on the qualities of information processing in a particular species of intelligent systems even in a higher degree than the outcome of TT. Compared to weak congruence, structural congruence is a much stronger relation. In SNT, the equation *system* $C =_H$ *blueprint* $E(H)$ is tested: the human's expectation of human information processing acts as a plan of the tested system's inside. This is much too strong a condition for AI: a blueprint of the information processing of a particular class of real intelligent systems is far too restrictive to be useful as a general criterion of intelligence. An important feature of SNT is, however, that the blueprint must be describable by the human in terms of a standardized information calculus.

4 Glass-Box Testing of Strong Congruence (McCarthy Test)

SNT can be transformed into a procedure that operationalizes a necessary, formal, and neutral condition of intelligence, by moderating the SNT criterion of behavioural equivalence. This is done in the so-called 'McCarthy Test' (MT) [7].

4.1 The McCarthy Test as a Glass-Box Testing Procedure

SNT raises the problem of how meta-information about a system's information processing should be represented in the messages the interlocutors exchange. J. McCarthy [11] suggested representing system states by using intentional predicates (such as 'believe', 'know', and 'being aware of') that express a system's epistemic qualification of information it processes. 'Intentional' is derived here, not from 'intention' in the sense of purpose of doing something, but from 'intentionality' in the sense of epistemic directedness on something. An intentional predicate is, hence, a predicate expressing a particular type of meta-information, namely a system's state that involves a reference of the system to information it has. Let s be a state of a system and p an information expressed by a sentence in a predefined language. Then, e.g., the intentional predicate $B(s, p)$ means that, if the system is in state s, it believes p [11]. In SNT, such definitions can be used by the computer in order to truthfully communicate meta-information about its processing of information.

What does it mean that a tested system can handle higher-order intentionality (or, in short, has higher-order intentionality)? In the simple case of beliefs about beliefs, this requires the system to have a second-order definition of an intentional predicate $B^2(T, H, B)$ at its disposal. It expresses that the system considers the human H in the test situation T to be in a state expressed by the first-order intentional predicate B (cf. [11]). This example can be generalized to a definition of nth-order intentional predicates $I^n(T, S^n, S^{n-1}, I^{n-1})$ for TT-like communication scenarios with T the current test situation, S^n the system whose state is expressed by I^n, and S^{n-1} the system to which the n-1-order intentional predicate I^{n-1} is ascribed. (If a system ascribes an intentional predicate to itself, S^n and S^{n-1} are the same.) $I^1(T, S^1, \emptyset, p)$ is the general first-order intentional predicate with p the information which the system S^1 qualifies epistemically.

A system that can handle first-order but not higher-order intentionality, is unable to represent other systems having intentionality as systems that are, in this respect, of the same kind as itself. Such a system would act in a world in which it principally could not encounter systems that qualify information epistemically; it could not even represent itself as such a system. In SNT, however, another system with intentionality, the human, is part and parcel of the situation in which the tested system communicates to suggest to the human that it is intelligent. If the human notices that the tested system does not represent her epistemic qualifications, the tested system will not pass SNT. Consequently, any intelligent system must be able to represent other systems in its TT-like test environment as having intentionality. It is, thus, reasonable to transform SNT into the more specific MT ending with the metaquestion: 'Did you communicate with another system that has intensions of higher-order intentional predicates?' If the human answers 'Yes', she has to consider the tested system, not necessarily as another human, but as satisfying a necessary condition for being intelligent. In this case, the human comes to the result that her communication with the tested system has been symmetrical in the following sense: both interlocutors must have used at least second-order intentional predicates for intensionally representing, e.g., beliefs about each other's beliefs. To respond to the MT metaquestion, the human must not compare the answers she received, to answers of an average human any more. Instead, she has to make up her mind about her expectations of intelligent answers given by any kind of communicating agent, so she should draw also on, e.g., her experiences in interacting with computers.

In MT, the human has a view of the tested system as a glass box [5]. She looks at it from the outside, observes its conversational behaviour (TT's black-box view), and receives meta-information about its internal information processing (SNT's grey-box view), but uses only a certain kind of information about the internals of the tested system for her judgement about its intelligence. This information consists of declarative knowledge about boundary conditions on information processes in the tested system, e.g., about the presence of intensions of higher-order intentional predicates.

The tested system is, thus, glassy in the sense that the human observes, from the outside, informational constraints on the processes inside. MT is a test of one of these constraints. If it is present in the tested system, and if the human can somehow reconcile her expectations of intelligent answers with those given by the system, she should be more inclined to ascribe intelligence to her interlocutor – although further aspects of its information processing might distinguish it from specimen of other classes of intelligent systems.

4.2 Strong Congruence of Behaviour in the McCarthy Test

The concepts and notations introduced in *sects. 2.2* and *3.2* are used to define the congruence relation needed for MT, too, but now the imaginary system that would live up to *H*'s expectations of how an intelligent agent could answer *H*'s questions, is denoted by *E(A)*. By $CR \gg_{IP} CS$ and $E(A)R \gg_{IP} E(A)S$, let us denote such sequences of reactions of *C* and *E(A)*, respectively, that involve the use of a set *IP* of higher-order intentional predicates between the reception of *H*'s signal and the sending of a signal by *C* and *E(A)*, respectively, to *H*. Now the MT-pertinent type of behavioural congruence of the answers of *C* and *E(A)*, as assessed by *H* in respect to *IP*, can be defined.

Strong congruence. An answer $a_C = (s_C, r_H)$ of *C* and an answer $a_{E(A)} = (s_C, r_H)$ of *E(A)* to a question $q = (s_H, r_{C \text{ or } E(A)})$ of *H* are *strongly congruent as to H in respect to IP*, written $a_C \sim_{H,IP} a_{E(A)}$, if and only if two conditions are met (with *R* the reception of s_H and *S* the sending of s_C and $s_{E(A)}$, respectively): (a) if $CR \gg_{IP} CS$ then there exists $E(A)S$ such that $E(A)R \gg_{IP} E(A)S$ and $s_C = s_{E(A)}$; and (b) if $E(A)R \gg_{IP} E(A)S$ then there exists CS such that $CR \gg_{IP} CS$ and $s_C = s_{E(A)}$. (3)

Two interlocutors of *H* cannot show, in respect to their strongly congruent answers, any difference regarding the fact that they both use intentional predicates belonging to *IP* in their information processing. Strong congruence is much weaker than structural congruence, since the detection of the former does use only a small part of the information needed for the detection of the latter. If two answers are strongly congruent, they are also weakly congruent. Strong congruence fits, thus, in between weak and structural congruence.

MT operationalizes an experimentally confirmable condition of intelligence that is necessary and as species-neutral in TT-like situations as possible. It is a test of the equation *system C* $\sim_{H,IP}$ *sketch E(A)*. If *C* shall pass MT, its answers must contain meta-information about higher-order intentional predicates $I^n(T, S^n, S^{n-1}, I^{n-1})$, with S^n the computer and S^{n-1} the human, that allow the computer to use information about the human's intentionality in its information processing. The intensions of these predicates are objects used by S^n to generate its own epistemic stance on S^{n-1}.

5 Résumé

Table 1 summarizes the three Turing Test-like human interaction procedures described above. Each test is characterized by answers to the following questions: First, how visible is the tested system to the human? Second, what stance does the human

Table 1. Overview of three Turing Test-like human-computer interaction procedures

Test Type	Visibility of Tested System	Interactional Stance of Tester	Congruence Relation
Turing	Black Box	Behaviouristic	Weak
Simon-Newell	Grey Box	Generative	Structural
McCarthy	Glass Box	Epistemic	Strong

adopt in her interaction with the tested system? Third, what formal relation is the criterion of passing the test based upon? The methodological recommendation this paper comes up with, is that AI ought to define necessary conditions of intelligence based on strongly congruent behaviour of intelligent systems. This can be done by starting out from a TT-like conversation procedure, dropping human intelligence as the sole standard up to which the answers of the unknown interlocutor should be, and focussing on single basic features of information processing in the tested system. MT is just one example of such a test of strong congruence. A series of MT-like tests could approximate a precisely defined sufficient condition of intelligence that would replace, in AI, rather vague expectations of intelligent behaviour.

References

1. Turing, A.M.: Computing Machinery and Intelligence. Mind 59, 433–460 (1950)
2. Moor, J.H. (ed.): The Turing Test: The Elusive Standard of Artificial Intelligence. Kluwer, Dordrecht (2003)
3. Shieber, S. (ed.): The Turing Test: Verbal Behavior as the Hallmark of Intelligence. MIT Press, Cambridge (2004)
4. Epstein, R., Roberts, G., Beber, G. (eds.): Parsing the Turing Test: Philosophical and Methodological Issues in the Quest for the Thinking Computer. Springer, Dordrecht (2008)
5. Battle, S.: Boxes: Black, White, Grey and Glass Box Views of Web-Services. Technical Report HPL-2003-30, Hewlett-Packard Laboratories, Bristol (2003)
6. Milner, R.: Communicating and Mobile Systems. Cambridge UP, Cambridge (1999)
7. Artmann, S.: Divergence versus Convergence of Intelligent Systems. In: Hertzberg, J., Beetz, M., Englert, R. (eds.) KI 2007. LNCS (LNAI), vol. 4667, pp. 431–434. Springer, Heidelberg (2007)
8. Carnap, R.: Meaning and Necessity. University of Chicago Press, Chicago (1956)
9. Simon, H.A., Eisenstadt, S.A.: A Chinese Room that Understands. In: Preston, J., Bishop, M. (eds.) Views into the Chinese Room, pp. 95–108. Clarendon Press, Oxford (2002)
10. McCarthy, J., Shannon, C.E.: Preface. In: Shannon, C.E., McCarthy, J. (eds.) Automata Studies, pp. v–viii. Princeton UP, Princeton (1956)
11. McCarthy, J.: Ascribing Mental Qualities to Machines. In: Ringle, M. (ed.) Philosophical Perspectives in Artificial Intelligence, pp. 161–195. Harvester Press, Brighton (1979)

Machine Learning Techniques for Selforganizing Combustion Control

Erik Schaffernicht[1], Volker Stephan[2], Klaus Debes[1], and Horst-Michael Gross[1]

[1] Ilmenau University of Technology
Neuroinformatics and Cognitive Robotics Lab
98693 Ilmenau, Germany
[2] Powitec Intelligent Technologies GmbH
45219 Essen-Kettwig, Germany
Erik.Schaffernicht@Tu-Ilmenau.de

Abstract. This paper presents the overall system of a learning, selforganizing, and adaptive controller used to optimize the combustion process in a hard-coal fired power plant. The system itself identifies relevant channels from the available measurements, classical process data and flame image information, and selects the most suited ones to learn a control strategy based on observed data. Due to the shifting nature of the process, the ability to re-adapt the whole system automatically is essential. The operation in a real power plant demonstrates the impact of this intelligent control system with its ability to increase efficiency and to reduce emissions of greenhouse gases much better then any previous control system.

1 Introduction

The combustion of coal in huge industrial furnaces for heat and power production is a particular complex process. Suboptimal control of this kind of process will result in unneccesary emissions of nitrogen oxides (NOx) and in a reduced efficiency, which equals more carbon dioxides emitted for one unit of energy.

Hence, optimal control of this process will not only reduce the costs but decrease the environmental impact of coal combustion. But finding such a control strategy is challenging. Even today, the standard approach is a human operator, which hardly qualifies as optimal. Classical control approaches are limited to simple PID controllers and expert knowledge, sometimes augmented by computational fluid dynamics (CFD) simulations in the setup stage [1].

The main reason for this is the process itself. The hazardous environment restricts sensor placements and operations. It is not possible to measure all relevant values that are known to influence the process. High temperatures, abrasive material flows, and problems like slagging (molten ash that condenses on the furnace wall) interfere with reliability of sensor readings, thus, the measurement noise and stochasticity of the observations is very high.

B. Mertsching, M. Hund, and Z. Aziz (Eds.): KI 2009, LNAI 5803, pp. 395–402, 2009.

In the light of all these facts, we approach the problem from another point of view. Our presumption is, that one has to try to extract more information from the furnace directly and to learn from the data with Machine Learning (ML) techniques to build the optimal controller.

This work is a complete overhaul of a previous system [2], but we are not aware of any other work about coal combustion and its control by ML techniques.

For several reasons we follow the basic idea of our approach to explicitly avoid the usage of expert knowledge. First, for many applications there is no expert knowledge available, especially if new information about the process comes into play. Secondly, industrial combustion processes usually are time varying due to changing plant properties or different and changing fuel-characteristics. Thus, initially available expert knowledge becomes invalid over time and needs to be updated too often. In consequence, we prefer a self-organizing, adaptive and learning control architecture that develops a valid control strategy based on historical data and past experiences with the process only.

We will give an overview of the complete control system (and the tasks it learns to solve) in Section 2. A more detailed explanation of the subsystems involved is given in the Sections 3 to 5. Finally, the adaptivity of the overall system is discussed in Section 6, and first results of online operation are shown in Section 7.

2 System Overview

The system that is presented in this paper is designed with the goal in mind to deploy it to existing power plants. Hence, it operates on top of the infrastructure provided by the plant and the main interface is formed between the Distributed Control System (DCS) of the power plant and the developed system. Besides additional computer hardware, special CMOS cameras are installed to observe the flames inside the furnace directly. The video data and the traditional sensor readings from the DCS are fed into the system, which is depicted in Fig. 1.

To handle the high dimensional pixel data generated with the cameras, a feature transformation is applied to preserve the information bearing parts while discarding large amounts of irrelevant data. Details and further references can be found in Section 3. Thereafter, a feature selection is performed on the extracted image data, the spectral data, and the sensory data from the DCS to reduce the number of considered inputs even more. A fairly simple filter approach is used for this task, which is presented in Section 4. For the adaptive controller itself, several approaches with different control paradigms were considered. We compared Nonlinear Model Predictive Control (MPC), Bayesian Process Control (BPC) and Neuroevolutionary Process Control (NEPC). Some basics about these approaches are presented in Section 5. Due to the nature of the combustion process (e.g. fuel changes), a certain life-long adaptivity of the system is urgently required. Hence all of the submethods must be able to adapt themselves given the current data, which we discuss in Section 6.

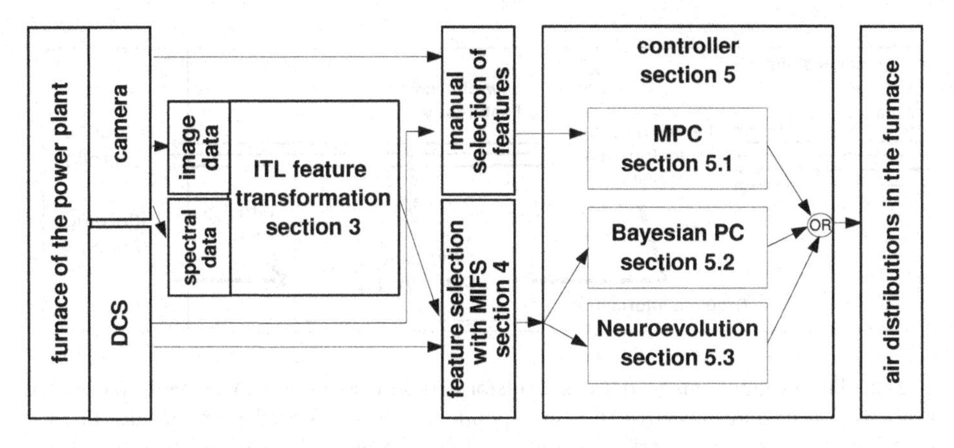

Fig. 1. This figure shows the data flow of a complete control step. The distributed control system (DCS) provides different measurements like coal feed rates or settings of the coal mill, while cameras are used to directly observe the process. Beside images from the burners, the cameras provide spectral information about the flames. Both are transformed by an Information Theoretic Learning (ITL) approach that will reduce dimensionality while maximizing the kept information about the targets. The results as well as the sensor data from the DCS are subject to a feature selection. All selected features are the inputs for two different control strategies, Bayesian Process Control (BPC) and Neuroevolutionary Process Control (NEPC). The Model Predictive Control (MPC) is fed with handselected features chosen by an human expert and is included as a reference. The selected outputs are fed back into the control system of the power plant using the DCS.

3 Image Data Processing

Using the CMOS cameras, images from the flames inside the furnace are captured. For transforming these images into information channels useful for the controller, an adaptive subspace transformation scheme based on [3] is applied. This Maximal Mutual Information approach is built upon the *Information-Theoretic Learning* (ITL) framework introduced by Principe [4]. The idea is to find a transformation that maximizes the *Quadratic Mutual Information* (QMI) between the transformed data in a certain subspace and the desired target values.

The basic adaptation loop for the optimization process is shown in Fig. 2. The original input data sample x_i is transformed by some transformation g (linear, neural network, ...) with the free parameters w into a lower dimensional space. The transformed data is denoted by y_i. The goal is to find those transformation parameters w that confer the most information into the lower dimensional space with respect to the controller's target variables.

The update rule for the parameters of the transformation is given by the following equation, where α denotes the learning rate

$$w_{t+1} = w_t + \alpha \frac{\partial I}{\partial w} = w_t + \alpha \sum_{i=1}^{N} \frac{\partial I}{\partial y_i} \frac{\partial y_i}{\partial w}. \quad (1)$$

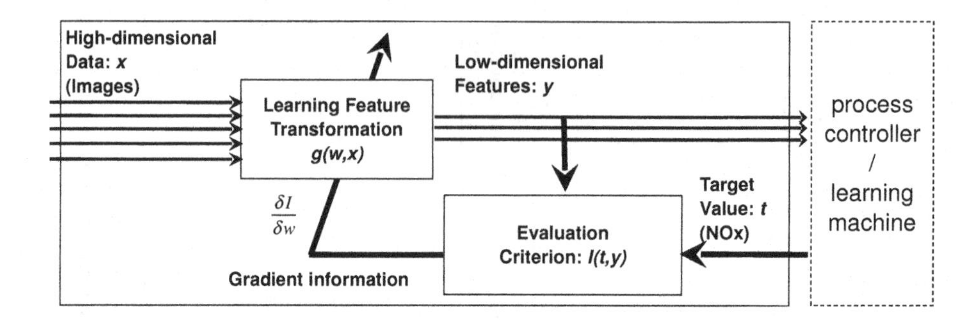

Fig. 2. The original image data is transformed into a lower-dimensional space. An evaluation criterion measures the correspondence to the desired target value (in this case NOx). From this criterion, a gradient information is derived and used to adapt the transformation parameters w.

Finding the gradient $\partial I/\partial w$ can be split into in the sample wise computation of the information forces $\partial I/\partial y_i$ and the adaption of the parameters $\partial y_i/\partial w$. The second term is dependent on the used transformation, and is quite simple for the linear case, where the transformation is computed as:

$$y_i = W^T x_i. \tag{2}$$

For the derivation of the update rule of the transformation parameters w, especially for the information forces $\partial I/\partial y_i$, we refer to details presented in [3].

Due to the intrinsic problems of learning in high dimensional input spaces like images, we compress the images to a subspace of only three dimensions, which was shown to achieve stable results. Furthermore, it was shown in [5] that this kind of transformation is superior to classic approaches like PCA for function approximation tasks in the power plant domain.

The camera system in use allows high frequency analysis of grey values, too. Image patches are observed in the frequency domain of up to 230 Hz. Using the same feature extraction technique, we were able to identify different parts of the temporal spectra that are relevant for targets like NOx.

All these new features generated from images and spectral data are sent to the feature selection module, which is described subsequently.

4 Feature and Action Selection

All features generated from images and those coming from other sensors (e.g. coal feed rates) as part of the DCS are subject to a feature selection step. The main reason for this module is the required reduction of the input space, and there are several channels that are not important or mostly redundant, which can be eliminated. A smaller input space allows an easier adaptation and generalization of the controller. For this step, we employ the MIFS algorithm [6], which is a simple approximation of the Joint Mutual Information between features f and the control targets y. This approximation uses pairwise Mutual Information only, and is hence, easily to compute.

Features f_i that maximize the following term are iterativly added to the subset of used features f_s

$$\arg\max_{f_i}(I(f_i, y) - \beta \sum_{f_s \in S} I(f_i, f_s)). \tag{3}$$

The first term describes the information between the feature and the target value and represents the relevance, while the second term represents the information between the feature and all features already in use. This represents the redundancy. The parameter β is used to balance the goals of maximizing relevance to the target value and minimize the redundancy in the subset.

With the help of this selection technique the number of inputs is trimmed down from more than 50 to approximately 20-30 features used by the controller as problem-relevant input.

The same methods were applied for action selection to reduce the number of manipulated control variables (mainly different air flows settings), but unfortunately all manipulated variables proved to be important in the 12-dimensional action space.

5 Control Strategy

In the following, we describe how to bridge the gap between the selected informative process describing features at one hand and the selected effective actions on the other hand by an adequate sensory-motor mapping realized by the controller. Practically, we tested three different control approaches.

5.1 Model Based Predictive Control

Model Based Predictive Control (MPC) is a well known approach for advanced process control of industrial processes [7]. Its basic idea is to build a mathematical model of the process to be controlled. Based on that process model, a control sequence is computed, that drives the process into the desired state within a given prediction horizon. We tested a simplified version of the MPC-family, that uses a feedforward neural network as process model. It operates with inputs defined by a human expert, rather than an automatic feature extraction.

5.2 Bayesian Process Control

Since real world processes usually are partially observable and noisy, we tested an architecture, that was designed to deal with these drawbacks. Probabilistic models [8] explicitly approximate stochastic processes. Practically, our probabilistic control approach first estimates from historical data joint probability density functions of all control, state, and target variables of the industrial process. The resulting probabilistic process model is used afterwards to build a factor graph [9]. During the so called inference process within that factor graph, a control sequence is calculated, that most probably drives the process into the desired target state, which is a minimum of NOx produced and 02 used in the combustion.

5.3 Neuroevolutionary Process Control

In addition to both control approaches described before, we also tested a neuroevolution method called Cooperative Synapse Neuroevolution (CoSyNE) [10]. CoSyNE applies evolutionary strategies to evolve a nonlinear controller based on recurrent neural networks, that drives the process as fast and accurate as possible into the desired target state. Unfortunately, evolutionary algorithms for control require extensive tests of the individual controllers with the plant in order to determine their performance, the so called fitness. Obviously, that extensive testing is not applicable for industrial processes with their real-world constraints like irreversibility and control cycle periods. Thereto, we extended the original approach proposed by [10] by a couple of alternative process models, that replace the real plant for the extensive closed-loop controller tests. Instead of a single process model, we favor a couple of different models, because that diversity ensures the development of a robust controller. On the contrary, operating on a single process model would be suboptimal, since any failure of the model may mislead the evolutionary search process for an optimal controller to a suboptimal solution or even a dead-end.

6 Adaptivity

As mentioned before, the industrial process to be controlled is time-varying for several reasons. In consequence, a suitable control system needs to be adaptive to cope with these changes automatically. Thereto, we introduced an automatic adaptivity of all components of our overall control architecture shown in Fig. 1.

Starting with the feature extraction from high dimensional image or CMOS-spectra data, we automatically retrain the underlying filters based on new data. This recalculation takes place every 6 hours and operates on image and spectral data of the last three days. This way, any change caused by either process variation, fuel change, or even camera function is taken into account automatically.

At the next stage, the feature selection is updated every 12 hours. Thereto, both process data and adaptively extracted features of the last three weeks are taken into consideration. Thus, less informative features or process measurements can be excluded from further calculation in the controller. In this way, malfunctioning sensors or even uninformative features from cameras temporarily blocked by slag will be sorted out from further observation automatically. On the other side, previously unused, but currently relevant features can be included.

Again every 12 hours or whenever the adaptive feature selection changed, also the corresponding controller is adapted to the new data by a retraining based on most recent process data of the last three weeks. In this way, the corresponding controller is adapted to new process properties of changed input-situations automatically.

7 Results of Field Trials in a Coalfired Power Plant

The adaptive and self-organizing control architectures described before, are running on a real process, namely a 200MW hard-coal fired power plant in Hamburg, Germany. The task is to minimize the nitrogen emissions (NO_x) and to increase the efficiency factor of that plant. Thereto, the controllers have to adjust the total amount of used air and its distribution between the 6 burners of the boiler. In total, 12 different actuators have to be controlled. Both amount and distribution of coal were given.

For comparison, the three control approaches Model-based Predictive Control (MPC), Bayesian Process Control (Bayes), and Neuroevolutionary Control (CoSyNE) have been tested against normal plant operation without any intelligent air optimization by an AI system (no control, only basic PID control provided by the plant control system). All controllers have been tested sequentially from April, 9th until April, 20th in 2009 during normal daily operation of the plant including many load changes.

Figure 3 shows, that the Model-based Predictive Controller (MPC), operating with manually selected inputs, can outperform the operation without any air control for both targets NO_x and O_2. The Probabilistic Controller (Bayes) operating on automatically extracted and selected features performs even better. The same holds true for the evolutionary Neuro-Controller (CoSyNE), which as well as the Bayesian controller operates on automatically extracted and selected inputs.

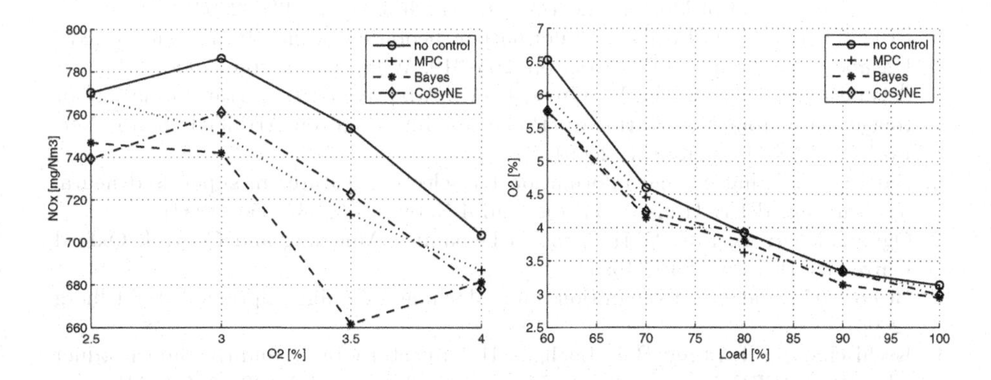

Fig. 3. Left: Comparison of NO_x emissions with different controllers plotted over O_2-concentration in the waste gas. **Right:** Comparison of efficiency factors for different controllers plotted over plant-load. The lower the O_2, the higher the efficiency factor of the plant.

Obviously, these self-organizing and adaptive control architectures, namely the Bayesian and the Neuroevolutionary Controller (at least for the more important overall O_2 value), are able to control complex industrial processes. Furthermore, their superior performance is also stable during permanent automatic adaptation of feature extraction, feature selection, and controller retraining using recent process data. In this way, these control architectures are able to cope with a large spectrum of typical real-world-problems ranging from sensor drift to time-varying processes.

8 Conclusion and Future Work

In this paper, an adaptive system for controlling a dynamic industrial combustion process is presented. Despite the potentially difficult handling of adaptive systems, we were able to show a superior performance compared to classical control approaches. An implementation of the methods presented here is successfully operating a real coal power plant.

Over time, the application of this system will result in considerable savings in coal consumption and emissions of greenhouse gases. Hence the application of Machine Learning techniques in this field have high economical and ecological impact.

References

1. Flynn, D. (ed.): Thermal power plant simulation and control. IEE London (2003)
2. Stephan, V., Debes, K., Gross, H.-M., Wintrich, F., Wintrich, H.: A New Control Scheme for Combustion Processes using Reinforcement Learning based on Neural Networks. International Journal on Computational Intelligence and Applications 1, 121–136 (2001)
3. Torkkola, K.: Feature Extraction by Non Parametric Mutual Information Maximization. Journal of Machine Learning Research 3, 1415–1438 (2003)
4. Principe, J., Xu, D., Fisher, J.: Information theoretic learning. In: Haykin, S. (ed.) Unsupervised Adaptive Filtering, pp. 265–319. Wiley, Chichester (2000)
5. Schaffernicht, E., Stephan, V., Gross, H.-M.: Adaptive Feature Transformation for Image Data from Non-stationary Processes. Int. Conf. on Artificial Neural Networks (ICANN) (to appear, 2009)
6. Battiti, R.: Using mutual information for selecting features in supervised neural net learning. IEEE Transactions on Neural Networks 5, 537–550 (1994)
7. Ogunnaike, B.A., Ray, W.H.: Process Dynamics, Modeling and Control. Oxford University Press, Oxford (1994)
8. Bishop, C.: Pattern Recognition and Machine Learning. Springer, Heidelberg (2006)
9. Kschischang, F.R., Frey, B.J., Loeliger, H.A.: Factor Graphs and the Sum-Product Algorithm. IEEE Transactions on Information Theory 47(2), 498–519 (2001)
10. Gomez, F., Schmidhuber, J., Miikkulainen, R.: Efficient non-linear control through neuroevolution. In: Fürnkranz, J., Scheffer, T., Spiliopoulou, M. (eds.) ECML 2006. LNCS (LNAI), vol. 4212, pp. 654–662. Springer, Heidelberg (2006)

Constraint-Based Integration of Plan Tracking and Prognosis for Autonomous Production*

Paul Maier[1], Martin Sachenbacher[1], Thomas Rühr[1], and Lukas Kuhn[2]

[1] Technische Universität München, Department of Informatics
Boltzmanstraße 3, 85748 Garching, Germany
{maierpa,sachenba,ruehr}@in.tum.de
[2] PARC, Palo Alto, USA
Lukas.Kuhn@parc.com

Abstract. Today's complex production systems allow to simultaneously build different products following individual production plans. Such plans may fail due to component faults or unforeseen behavior, resulting in flawed products. In this paper, we propose a method to integrate diagnosis with plan assessment to prevent plan failure, and to gain diagnostic information when needed. In our setting, plans are generated from a planner before being executed on the system. If the underlying system drifts due to component faults or unforeseen behavior, plans that are ready for execution or already being executed are uncertain to succeed or fail. Therefore, our approach tracks plan execution using probabilistic hierarchical constraint automata (PHCA) models of the system. This allows to explain past system behavior, such as observed discrepancies, while at the same time it can be used to predict a plan's remaining chance of success or failure. We propose a formulation of this combined diagnosis/assessment problem as a constraint optimization problem, and present a fast solution algorithm that estimates success or failure probabilities by considering only a limited number k of system trajectories.

1 Introduction

As markets increasingly demand for highly customized and variant-rich products, the industry struggles to develop production systems that attain the necessary flexibility, but maintain cost levels comparable to mass production. A main cost driver in production is the human workforce needed for setup steps, the design of process flows, and quality assurance systems. These high labor costs can only be amortized by very large lot sizes. Therefore, to enable individualized production with its typically small lot sizes, automation must reach levels of flexibility similar to human workers. The TUM excellence cluster "Cognition for Technical Systems" (CoTeSys) [1] aims at understanding human cognition and making its performance accessible for technical systems. The goal is to develop systems that act robustly under high uncertainty, reliably handle unexpected events, and quickly adapt to changing tasks and own capabilities.

* This work is supported by DFG and CoTeSys.

B. Mertsching, M. Hund, and Z. Aziz (Eds.): KI 2009, LNAI 5803, pp. 403–410, 2009.

A key technology for the realization of such systems is automated planning combined with self-diagnosis and self-assessment. These capabilities can allow the system to plan its own actions, and also react to failures and adapt the behavior to changing circumstances. From the point of view of planning, production systems are a relatively rigid environment, where the necessary steps to manufacture a product can be anticipated well ahead. However, from a diagnosis point of view, production systems are typically equipped with only few sensors, so it cannot be reliably observed whether an individual manufacturing step went indeed as planned; instead, this becomes only gradually more certain while the production plan is being executed. Therefore, in the presence of faults or other unforeseen behaviors, which become more likely in individualized production, the question arises whether plans that are ready for execution or already being executed will indeed succeed, and whether it is necessary to revise a plan or even switch to another plan.

To address this problem, we propose in this paper a model-based capability that estimates the success probability of production plans in execution. We assume that a planner provides plans given a system model. A plan is a sequence of action and start time pairs where each action is executed at the corresponding start time. Whenever the system produces an observation, it will be forwarded to a module that performs simultaneous situation assessment and plan prognostic using probabilistic hierarchical constraint automata (PHCA) models [2] of the system. We propose a formulation of this problem as a soft constraint optimization problem [3] over a window of N time steps that extends both into the past and the future, and present a fast but approximate solution method that enumerates a limited number k of most likely system trajectories. The resulting success or failure prognosis can then be used to autonomously react in different ways depending on the probability estimate; for instance, continue with plan execution, discard the plan, or augment the plan by adding observation-gathering actions to gain further information [4].

In the remainder of the paper, we first motivate the approach informally with an example from an automated metal machining process, and then present our algorithmic solution and experimental results.

2 Example: Metal Machining and Assembly

As part of the CoTeSys cognitive factory test-bed, we set up a customized and extended Flexible Manufacturing System (FMS) based on the iCim3000 from Festo AG (see figure 1b). The system consists of a conveyor transport and three stations: storage, machining (milling and turning), and assembly. We built a simplified model of this manufacturing system (see figure 1a) which consists only of the machining and the assembly station and allows to track system behavior over time, including unlikely component faults. In particular, the machining station can transition to a "cutter blunt" composite location, where abrasions are caused during operation due to a blunt cutter. This makes it very probable that the cutter breaks, leading to flawed products (see figure 1d). The assembly station model contains a composite location which models occasional abrasions. A vibration sensor at the assembly station can detect these abrasions, yielding binary signals "abrasion occurred" and "no abrasion occurred". However, the signal is ambiguous, since the sensor cannot differentiate between the two possible causes.

Two products are produced using a single production plan $\mathcal{P}_{prod}$: a toy maze consisting of an alloy base plate and an acrylic glass cover, and an alloy part of a robotic arm (see figure 1c). $\mathcal{P}_{prod}$ consists of these steps: (1) cut maze into base plate (one time step), (2) assemble base plate and cover (one time step), (3,4,5,6) cut robot arm part (one to four time steps). The plan takes two to six time steps (starting at $t = 0$). The plan is considered successful if both products are flawless. In our example, only a broken cutter causes the machined product to be flawed, in all other cases the production plan will succeed. Now consider the following scenario: after the second plan step (assembling the maze base plate and its cover at $t = 2$) an abrasion is observed. Due to sensor ambiguity it remains unclear whether the plan is unaffected (abrasion within assembly) or whether it might fail in the future due to a broken cutter (abrasion caused by a blunt cutter), and the question for the planner is: How likely is it that the current plan will still succeed? Our new capability allows to compute this likelihood, taking into account past observations and future plan steps.

3 Plan Assessment as Constraint Optimization over PHCA Models

Probabilistic hierarchical constraint automata (PHCA) were introduced in [2] as a compact encoding of hidden markov models (HMMs). They allow to model both probabilistic hardware behavior (such as likelihood of component failures) and complex software behavior (such as high level control programs) in a uniform way. A PHCA is a hierarchical automaton with a set of locations Σ, consisting of primitive and composite locations, a set of variables Π, and a probabilistic transition function $P_T(l_i)$. Π consists of dependent, observable and commandable variables. The state of a PHCA at time t is a set of marked locations, called a marking. The execution semantics of a PHCA is defined in terms of possible evolutions of such markings (for details see [2]).

Plan assessment requires tracking of the system's plan-induced evolution; in our case, it means tracking the evolution of PHCA markings. In previous work [5], we introduced an encoding of PHCA as soft constraints and casted the problem of tracking markings as a soft constraint optimization problem [3], whose solutions correspond to the most likely sequences of markings given the available observations. The encoding is parameterized by N, which is the number of time steps considered (for a detailed description of the encoding, see [5]).

Observations made online are encoded as hard constraints specifying assignments to observable variables at time t (e.g., $Abrasion^{(2)} = OCCURRED$), and added to the constraint optimization problem.

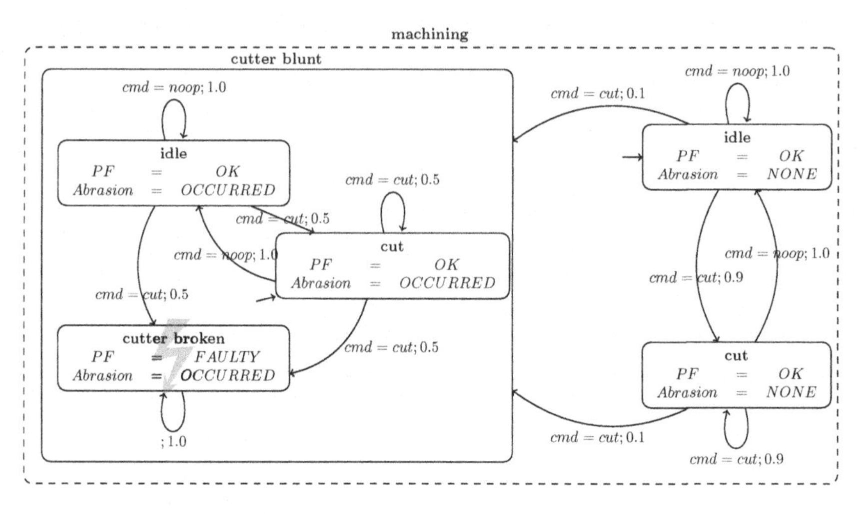

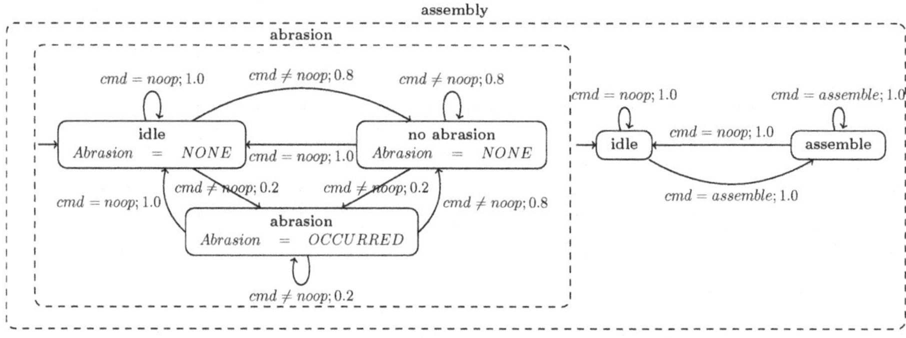

(a)

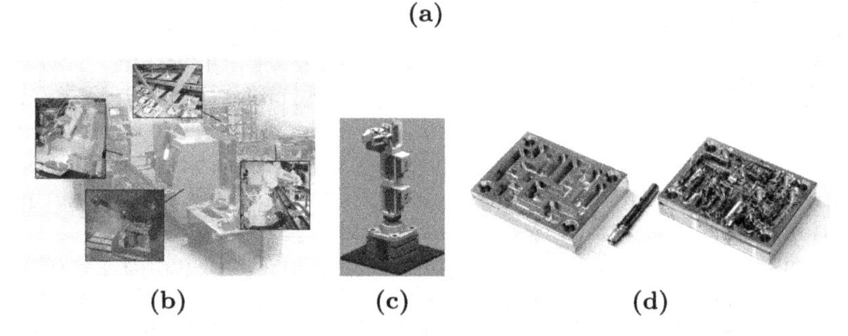

(b) (c) (d)

Fig. 1. (1a) Simplified PHCA of the manufacturing system. The machining and assembly station are modeled as parallel running composite locations (indicated by dashed borders). Variables appearing within a location are local to this location, i.e. *machining.cmd* refers globally to the command variable *cmd* within composite location *machining*. (1b) The hardware setup used for experimentation, showing storage, transport, robot and machining components. (1c) The robotic arm product. (1d) Effects of cutter deterioration until breakage in machining. Images *(c)* Prof. Shea TUM PE.

Plans are added analogously as assignments to commandable variables at time t; for example, $a^{(t)}_{cut}$ and $a^{(t)}_{assemble}$ are assignments $machining.cmd^{(t)} = cut \wedge assembly.cmd^{(t)} = noop$ and $assembly.cmd^{(t)} = assemble \wedge machining.cmd^{(t)} = noop$. Note that variables appear time independent in the PHCA notation (see, e.g., figure 1a), i.e. without superscript $^{(t)}$.

The plan's goal G is to produce a flawless product. We encode this informal description as a logical constraint $G \equiv \forall PF^{(t_{end})} \in RelevantFeatures(\mathcal{P}) : PF^{(t_{end})} = OK$ over product feature variables $PF^{(t)} \in \{OK, FAULTY\}$ at the end of the execution, t_{end}. $RelevantFeatures()$ is a function mapping a production plan to all product feature variables which define the product. Each system component is responsible for a product feature in the sense that if it fails, the product feature is not present ($PF^{(t)} = FAULTY$). In our example, there is only a single product feature PF, which is absent if the cutter is broken. The goal constraint for the above mentioned plan (three time steps long) is accordingly $PF^{(3)} = OK$.

We then enumerate the system's k most likely marking sequences, or trajectories, using a modified version of the soft constraint solver toolbar [6], which implements A* search with mini-bucket heuristics (a dynamic programming approach that produces search heuristics with variable strength depending on a parameter i, see [7]).

Combining Plan Tracking and Prognosis. To assess a plan's probability of success, we require not only to track past system behavior, but also to predict its evolution in the future. In principle, this could be accomplished in two separate steps: first, assess the system's state given the past behavior, and then predict its future behavior given this belief state and the plan. However, this two-step approach leads to a problem. Computing a belief state (complete set of diagnoses) is intractable, thus it must be replaced by some approximation (such as considering only k most likely diagnoses [8]). But if a plan uses a certain component intensely, then this component's failure probability is relevant for assessing this plan, even if it is very low and therefore would not appear in the approximation. In other words, the plan to be assessed determines which parts of the belief state (diagnoses) are relevant.

To address this dependency problem, we propose a method that performs diagnosis and plan assessment *simultaneously*, by framing it as a single optimization problem. The key idea is as follows: The optimization problem formulation is independent of where the present time point is within the N-step time window. We therefore choose it such that the time window covers the remaining future plan actions as well as the past behavior. Now solutions to the COP are trajectories which start in the past and end in the future. We then compute a plan's success probability by summing over system trajectories that achieve the goal. Again due to complexity reasons, we cannot do this exactly and approximate the success probability by generating only the k most probable trajectories. But since we have only a single optimization problem now, we don't have to prematurely cut off unlikely hypotheses and have only one source of error, compared

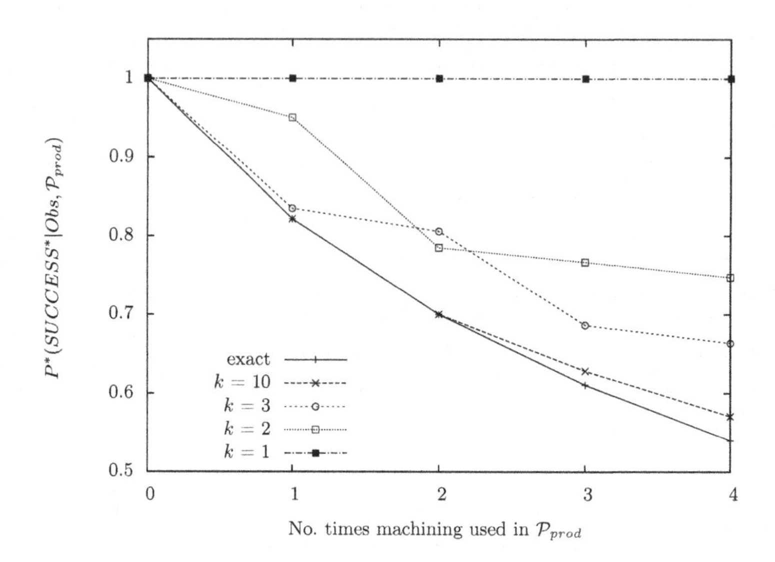

k	i	No. times machining used in $\mathcal{P}_{prod}$ (window size N, #Variables, #Constraints)				
		0 (2, 239, 242)	1 (3, 340, 349)	2 (4, 441, 456)	3 (5, 542, 563)	4 (6, 643, 670)
1	10	< 0.1 / 1.8	0.1 / 6.8	0.1 / 19.0	(mem)	(mem)
	15	0.1 / 1.9	0.3 / 4.2	0.5 / 7.8	0.5 / 16.6	0.8 / 32.0
	20	0.1 / 1.9	0.5 / 5.2	3.7 / 20.1	6.5 / 34.5	9.5 / 50.7
2	10	< 0.1 / 2.1	0.1 / 11.9	0.2 / 38.5	(mem)	(mem)
	15	0.1 / 2.2	0.3 / 5.4	0.5 / 9.7	0.6 / 28.0	0.8 / 52.0
	20	0.1 / 2.2	0.5 / 6.4	3.7 / 21.8	6.5 / 37.2	9.5 / 55.8
3	10	< 0.1 / 2.3 (e)	0.1 / 11.9	0.2 / 40.1	(mem)	(mem)
	15	0.1 / 2.4 (e)	0.3 / 5.4	0.5 / 11.4	0.6 / 29.9	0.9 / 55.5
	20	0.1 / 2.4 (e)	0.5 / 6.4	3.7 / 23.5	6.6 / 38.3	9.5 / 57.4
4	10	(e)	0.1 12.5	0.2 / 40.1	(mem)	(mem)
	15	(e)	0.3 / 5.9	0.5 / 11.4	0.6 / 30.9	0.9 / 57.2
	20	(e)	0.5 / 6.9	3.7 / 23.5	6.6 / 39.3	9.5 / 59.1
5	10	(e)	0.1 / 13.1	0.2 / 40.7	(mem)	(mem)
	15	(e)	0.3 / 6.6	0.5 / 12.0	0.6 / 33.6	0.9 / 59.5
	20	(e)	0.5 / 7.6	3.7 / 24.0	6.6 / 42.8	9.5 / 63.9
10	10	(e)	0.1 / 14.0 (e)	0.2 / 43.4 (e)	(mem)	(mem)
	15	(e)	0.3 / 6.7 (e)	0.5 / 14.7 (e)	0.6 / 36.2	0.9 / 64.8
	20	(e)	0.6 / 7.7 (e)	3.8 / 26.6 (e)	6.6 / 45.8	9.6 / 68.9

Fig. 2. Top: Approximate success probability (y-axis) of plan $\mathcal{P}_{prod}$ against varying usage of the machining station (x-axis) after the observation of an abrasion at $t = 2$. Bottom: Runtime in seconds / peak memory consumption in megabytes. (e) indicates that the exact success probability $P(SUCCESS|Obs, \mathcal{P}_{prod})$ could be computed with this configuration. (mem) indicates that A* ran out of memory (artificial cutoff at > 1 GB, experiments were run on a Linux computer with a recent dual core 2.2 Ghz CPU with 2 GB RAM).

to approximating the belief state and predicting the plan's evolution based on this estimate.

Approximating the Plan Success Probability. We denote the set of all trajectories as Θ and the set of the k-best trajectories as Θ^*. A trajectory is considered successful if it entails the plan's goal constraint. We define $SUCCESS := \{\theta \in \Theta | \forall s \in \mathcal{R}_{sol}, s \downarrow_Y = \theta : F_G(s) = \text{true}\}$, where $\mathcal{R}_{sol}$ is the set of all solutions to the probabilistic constraint optimization problem, $s \downarrow_Y$ their projection on marking variables, and $F_G(s)$ is the goal constraint. $SUCCESS^*$ is the set of successful trajectories among Θ^*. The exact success probability is computed as

$$P(SUCCESS|Obs, \mathcal{P}) = \sum_{\theta \in SUCCESS} P(\theta|Obs, \mathcal{P}) = \sum_{\theta \in SUCCESS} \frac{P(\theta, Obs, \mathcal{P})}{P(Obs, \mathcal{P})} =$$

$$= \sum_{\theta \in SUCCESS} \frac{P(\theta, Obs, \mathcal{P})}{\sum_{\theta \in \Theta} P(\theta, Obs, \mathcal{P})} = \frac{\sum_{\theta \in SUCCESS} P(\theta, Obs, \mathcal{P})}{\sum_{\theta \in \Theta} P(\theta, Obs, \mathcal{P})}$$

The approximate success probability $P^*(SUCCESS^*|Obs, \mathcal{P})$ is computed the same way, only $SUCCESS$ is replaced with $SUCCESS^*$ and Θ with Θ^*. We define the error of the approximation as $E(k) := |P(SUCCESS|Obs, \mathcal{P}) - P^*(SUCCESS^*|Obs, \mathcal{P})|$. $E(k)$ converges to zero as k goes to infinity. Also, $E(k) = 0$ if $P(SUCCESS|Obs, \mathcal{P})$ is 0 or 1. However, as the example in Figure 2 shows, $E(k)$ does in general not decrease monotonically with increasing k.

Algorithm for Plan Evaluation. Plans are generated by the planner and then advanced until they are finished or new observations are available. In the latter case, the currently executed plan $\mathcal{P}$ is evaluated using our algorithm. It first computes the k most probable trajectories by solving the optimization problem over N time steps. Then, using these trajectories, it approximates the success probability of plan $\mathcal{P}$ and compares the probability against two thresholds ω_{success} and ω_{fail}. We can distinguish three cases: (1) The probability is above ω_{success}, i.e. the plan will probably succeed, (2) the probability is below ω_{fail}, i.e. the plan will probably fail or (3) the probability is in between both thresholds, which means the case cannot be decided. In the first case we simply continue execution. In the second case we have to adapt the plan to the new situation. This is done by REPLAN($\mathcal{P}$, Θ^*), which modifies the future actions of $\mathcal{P}$ taking into account the diagnostic information contained in Θ^*. The third case indicates that not enough information about the system's current state is available. As a reaction, a procedure REPLANPERVASIVEDIAGNOSIS($\mathcal{P}$, Θ^*) implements the recently developed *pervasive diagnosis* [4], which addresses this problem by augmenting a plan with information gathering actions.

4 Experimental Results

We ran experiments for five small variations of our example scenario, where $\mathcal{P}_{prod}$ uses the machining station zero to four times. The time window size N

accordingly ranges from two to six, problem sizes range from 240 to 640 variables and 240 to 670 constraints. Figure 2 shows the success probabilities for different $\mathcal{P}_{prod}$ and k (top), and a table of the runtime in seconds and the peak memory consumption in megabytes (bottom) for computing success probabilities in the planning scenarios. In addition, the table ranges over different values for the mini-bucket parameter i. As expected, with increasing use of the machining station, $P^*(SUCCESS^*|Obs, \mathcal{P}_{prod})$ decreases. Also, runtime increases for larger time windows. With increasing k, $P^*(SUCCESS^*|Obs, \mathcal{P}_{prod})$ converges towards the exact solution. In our example, the approximation tends to be optimistic, which however cannot be assumed for the general case. Increasing k hardly seems to affect the runtime, especially if the mini-bucket search heuristic is strong (bigger i-values). For weaker heuristics the influence increases slightly. Memory consumption is affected much stronger by k. Here also, a weaker search heuristic means stronger influence of k.

5 Conclusion and Future Work

We presented a model-based method that combines diagnosis of past execution steps with prognosis of future execution steps of production plans, in order to allow the production system to autonomously react to failures and other unforeseen events. The method makes use of probabilistic constraint optimization to solve this combined diagnosis/prognosis problem, and preliminary results for a real-world machining scenario show it can indeed be used to guide the system away from plans that rely on suspect system components. Future work will concern the integration of the method into our overall planning/execution architecture, and its extension to multiple simultaneous plans. We are also interested in exploiting the plan diagnosis/prognosis results in order to update the underlying model, for instance, to automatically adapt to parameter drifts of components.

References

1. Beetz, M., Buss, M., Wollherr, D.: Cognitive technical systems — what is the role of artificial intelligence? In: Proc. KI 2007, pp. 19–42 (2007)
2. Williams, B.C., Chung, S., Gupta, V.: Mode estimation of model-based programs: monitoring systems with complex behavior. In: Proc. IJCAI 2001, pp. 579–590 (2001)
3. Schiex, T., Fargier, H., Verfaillie, G.: Valued constraint satisfaction problems: hard and easy problems. In: Proc. IJCAI 1995 (1995)
4. Kuhn, L., Price, B., Kleer, J.d., Do, M.B., Zhou, R.: Pervasive diagnosis: The integration of diagnostic goals into production plans. In: Proc. AAAI 2008 (2008)
5. Mikaelian, T., Williams, B.C., Sachenbacher, M.: Model-based Monitoring and Diagnosis of Systems with Software-Extended Behavior. In: Proc. AAAI 2005 (2005)
6. Bouveret, S., Heras, F., Givry, S., Larrosa, J., Sanchez, M., Schiex, T.: Toolbar: a state-of-the-art platform for wcsp (2004), `http://www.inra.fr/mia/T/degivry/ToolBar.pdf`
7. Kask, K., Dechter, R.: Mini-bucket heuristics for improved search. In: Proc. UAI 1999, pp. 314–332 (1999)
8. Kurien, J., Nayak, P.P.: Back to the future for consistency-based trajectory tracking. In: Proc. AAAI 2000, pp. 370–377 (2000)

Fault Detection in Discrete Event Based Distributed Systems by Forecasting Message Sequences with Neural Networks

Falk Langer, Dirk Eilers, and Rudi Knorr

Fraunhofer ESK,
Hansasstrasse 32, D-80686 Munich, Germany
{falk.langer,dirk.eilers,rudi.knorr}@esk.fraunhofer.de
http://www.esk.fraunhofer.de

Abstract. In reliable systems fault detection is essential for ensuring the correct behavior. Todays automotive electronical systems consists of 30 to 80 electronic control units which provide up to 2.500 atomic functions. Because of the growing dependencies between the different functionality, very complex interactions between the software functions are often taking place.

Within this paper the diagnosability of the behavior of distributed embedded software systems are addressed. In contrast to conventional fault detection the main target is to set up a self learning mechanism based on artificial neural networks (ANN). For reaching this goal, three basic characteristics have been identified which shall describe the observed network traffic within defined constraints. With a new extension to the reber grammar the possibility to cover the challenges on diagnosability with ANN can be shown.

Keywords: self learning fault detection, distributed embedded systems, message sequence learning, neural networks.

1 Introduction

The target of fault detection in technical systems is the detection of malfunction of some elements or the whole system. The detection of a malfunction of a single atomic element is a well known task. In large distributed systems with many interacting elements fault detection is much more complicated. Even the detection of a break down of a single element in a network is a quite difficult task [1] and [2]. Detecting failures of a distributed functionality is very complicated. Conventional fault detection systems are not applicable when the system behavior is faultily although each single network element shows the correct behavior.

Most recent approaches for fault detection in distributed systems are based on a deep knowledge of the system itself. In most cases it is an extension of the conventional fault detection for single elements. Here, the system is observed to discover the known fault symptoms. This approach is always limited to well known failures which are mostly produced by defect hardware. The next step

B. Mertsching, M. Hund, and Z. Aziz (Eds.): KI 2009, LNAI 5803, pp. 411–418, 2009.

towards an intelligent fault detection is the usage of simplified models of the system behavior. Here, an observer model is implemented and the states reported by the system are compared with the state required by the observer model. In very critical systems this technique is often used in order to ensure correct software behavior. The NASA extended this approach to the model based diagnosis (MBD) concept [3], [4]. MBD is applied for diagnosing continuous and discrete systems. For continuous systems prevalently mathematical models are used and for message based systems mostly simplified models represented by state machines are used. For the error recovery the model is feed with information about the current state of of the observed system. This can be sensor data or other information calculated or emitted by the system. If the expected system behavior did not match the calculated behavior an error is expected.

All of these approaches are based on deep knowledge of the system like known error pattern or exact model behavior. If such knowledge is not available or the system model is to complex new methods for fault detection must be found.

With the growing of technical systems not only the physical behavior measured by sensors leaks exact (mathematical) models also the behavior of the running software can not always be observed by the availability of a simplified system model [5]. This problem is extended by the usage of dynamic changeable software which leads to a state explosion in observer models [6]. In this paper a new approach for a neural network based prediction of event based discrete system behavior is introduced.

2 Fault Detection with Neural Networks

Many people are working on diagnosing faults in complex physical units. Often the behavior of this units, e.g. automotive engines, can not exactly be covered by mathematical models because there are discontinuous behaviors or the model is simply to complex to be calculated. In this case often artificial neural networks (ANN) are used to replace an exact model (e.g. [7,8,9,10,6]). The ANN is trained with the correct measured values of the system. Afterwards the ANN forecast the correct behavior which is compared with the current behavior of the system.

In difference to the previous applications of ANN, the challenge of this work is the forecast of discrete events in order to diagnose a distributed software system.

The main goal of this work is to diagnose a distributed system by observing and forecasting the network traffic. Therefore not the correctness of some physical elements (like sensors or actors) shall be observed but the internal state of the software elements. This is a new adaption of the combination of ANN and fault detection.

2.1 Basic Assumptions in Diagnosing Distributed Software

In software development, sequences are often described within sequence charts or state machines. In the most cases an external input to a software element causes an internal state change and on the other side internal state changes are causing the transmitting of some messages on the network. It is not well defined how an embedded distributed software system can be observed and diagnosed. Referring to this, basic assumptions are postulated which must be met by the distributed system to be diagnosable.

Lemma 1. *The behavior of a distributed system can be diagnosed by observing the network traffic.*

Lemma 2. *For detection faults in a distributed system the important data to observe are events which signal or causes changes in the system.*

Lemma 3. *An error occurs if the forecast of the network traffic matches not the observed behavior.*

2.2 Interpreting the Network Traffic

To detect faults in communication it is very important to make a precisely forecast of the systems network traffic. The ideal would be to forecast the complete traffic with all included payload. To learn the varied and complex properties and content of the network traffic it is important identify some basic characteristics for simplifying the learning task by ANN. The first simplifying is that only discrete information like control and status messages are interpreted. This messages can be generated by any instance and any network layer. We concentrate of learning messages generated by state changes from systems that can be described by a timed petri net.

For this kind of messages three basic characteristics for describing and forecasting an discrete message stream of a distributed functionality have been identified.

(i) Sequence. The most important characteristic is the sequence of the events. A sequence of events from a single component can be interpreted the result of the execution of a state machine where transitions generate events. Such message sequences are often described for system specification by sequence charts. Mostly single network events depends on transitions or previous states of the system. A simple sequence of transitions in a distributed embedded system would be e.g.: Start of Transmission → Data → Calculation.

(ii) Timed Messages. In embedded systems often realtime applications are running. Often specific timeouts must be considered, specific response times are required or timed cyclic messages are sent. For an accurate interpretation of a message stream the time or timeslot of the message occurrence is an important characteristic.

(iii) Parallelism. With the characteristics of (i) and (ii) a message stream between two basic functions can be described. But in a distributed system many functions are executed in parallel. In a consequence the resulting message stream on the network consists of many parallel message steams. In the case of observing the network, parallelism can be interpreted as an addition of single streams to one stream.

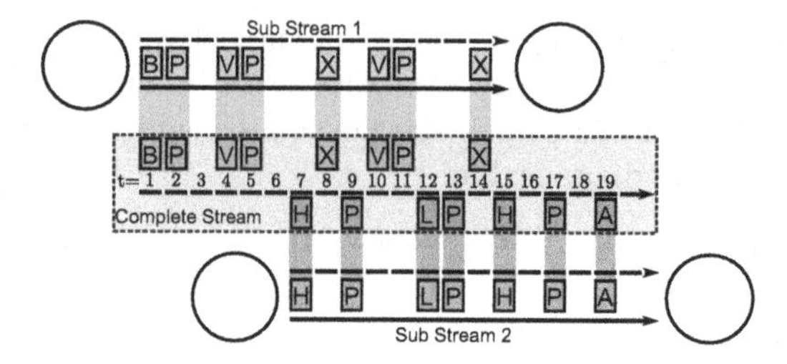

Fig. 1. Parallelism - addition of two message sequences to one mixed message stream

3 Prediction of Message Streams with ANN

In section 2.2 the composition of a message streams was shown. The reduction of three simple problems is the first step for the proposed fault detection with ANN. In this section it is shown how these problems can be learned by ANN and which types of ANN fits to the problem.

The main topic of this paper is the message sequence prediction in the context of automotive network. Classification for resolving parallelism and learning of timed message are not covered in this paper.

3.1 Models for Representing a Message Stream in Automotive Applications

When analyzing message sequences generated from embedded distributed systems like they are given in automotive networks, some basic problems are identified which have to be covered by addressing the ANN learning:

- *The message stream is continuous.* This means that sequences are repeated infinite times and one is not able to say where it begins and where it ends.
- *The sequences are not trivial.* In most cases a message in a sequence depends not only on one parent message and the dependency can be a long time in the past. Message sequences generated from the same software can have different length depending on fare away messages.

For a better reproducibility related problems which are well known to solve with ANN have to be identified. We found that reber grammars (RG) [11] (see Fig. 2(a)) with its extension to embedded RG (ERG)[12] and continuous embedded reber grammars (CERG)[13] are fitting very good to our objectives.

The CERG addresses the long term problematic and the problematic of continuous streams. Whereas RG naturally fitting to the standard sequence learning problem with the focus of short term dependencies (for example $B \rightarrow P \Rightarrow T \vee V$), ERG are an expansion to RG to address the problems of long term dependencies. (consider $B \rightarrow P \Rightarrow B$ or $B \rightarrow P \Rightarrow T \vee V$ which depends on the position in the stream).

3.2 The Proper Network

After the identification of the problems to solve the next step is to find a proper ANN which can learn an predict CERGs and some additional problems described Sec. 4 with a high accuracy. There are many different types. Traditional feed forward networks (FNN)[14] are not able to solve long term problems because they can only deal with time windows with a limited length. Recurrent neural networks (RNN) can solve a longer distance between important messages but no continuous streams where no start and endpoint is known. As an extension to RNN, long short-term memory neural networks (LSTM) [15] looking promising to solve the announced problems. Especially LSTM with forget gates introduced by [13] can explicitly deal with continuous input streams.

In [13] was shown that LSTM with forget gates can learn CERGs with a very good accuracy.

3.3 Applying the Theory

In Table 1 it has been shown that with an increasing of input streams the error in prediction of symbols decreases. The LSTM was trained with the output of a CERG. We generated in n-cycles of the CERG n streams of symbols. The net is trained until this n steams are learned successfully. Then, a new run of CERG with the output of 10^8 symbols was used to evaluate the trained net.

The challenge of the ANN is to learn and predict for every step two of the seven symbols of the RG. The representation of the symbols is done in a array with seven rows. Each row is representing its own symbol. For the learning phase this array was used as reference for the back propagation of the supervised learning. In the test phase the ANN calculates for every step the probability of each symbol and we use the two symbols with the highest probability. The prediction was correct if one of this two selected symbols are equal to the output of the parallel executed RG. There is always the choise of two symbols because of the RG allow always two possible follower symbols (see Fig. 2 (a)).

Table 1. Results from learning CERGs with LSTM. Streams are the count assembled ERG-strings, learning Parameter are learning rate (learning rate decay). Inputs is the count of symbols in the learning phase, 1st and 10th show the count of symbols until the first or tenth error occurs, error is the probability of a wrong predicted symbol.

Streams	Learn Parameter η	Inputs $[10^3]$	1st $[10^3]$	10th $[10^3]$	Error $[10^{-3}]$
100	0.5 (1.00)	387.4	3.3	39.6	37.61
1000	0.5 (1.00)	586.9	84.9	1,121.6	2.43
10000	0.5 (1.00)	2,199.5	1,316.5	5,382.8	0.28
100000	0.5 (1.00)	5,335.5	26,456.3	80,114.8	0.012

4 Extensions of CERG

For applying ANN to real world scenarios, the shown CERG did not provide enough complexity. For example CERG provide only one long term problem and does not cover problems like scalability or different distances between depending messages.

For expanding the long term problems and combining it better with short term problems an extension to the CERG the advanced reber grammar (ARG) has been introduced. (see Fig. 2).

ARG I. A composition of normal RG with an ERG. With this extension the network must deal with a long term and a short term problem depending on the position on the stream.

ARG II-n. The ARG II-n consists of at least 3 ARG I grammars which are connected as shown in Fig. 2 (c). The ARG II-n grammar can be recursively embedded whereas each ARG I grammar is replaced by an ARG II-1. The n represents the recursion depth.

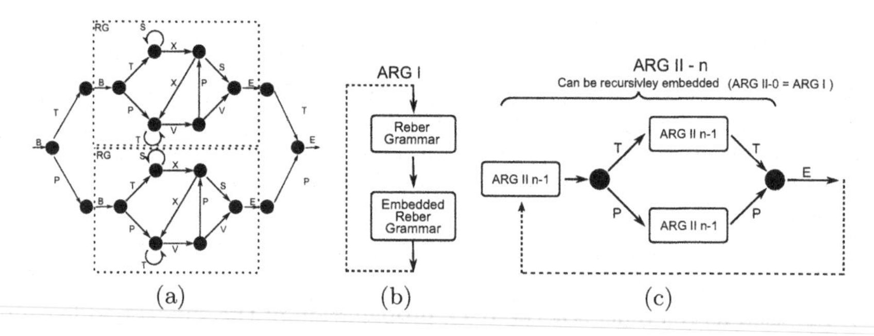

Fig. 2. The embedded reber grammar with two regular reber grammars (a), the first advanced version of reber grammar ARG I, consisting of a RG and an ERG (b) and the more complex ARG II-n grammar (for ARG II-0 = ARG I) (c)

4.1 Results

Results have been achieved on predicting continuous messages streams. In most test runs the prediction of the first symbols is difficult with tested LSTM. Very often the first error occurs within the first 100 symbols. This is because the LSTM has first to initialize its internal state which represents the position in the stream which is not known at the beginning. As in shown Table 2 shown not all training runs are completed successfully. But still every time proper LSTM configuration where found which can predict even a ARG II-3 grammar with 3^3 (21) ARG I grammars. The count of the memory blocks of the LSTM network must be adapted to the stronger ARG II-n problems.

An interesting behavior shown in Table 1 is that for larger stream lengths the error in prediction is reducing much more slowly than the required trainings

Table 2. Results of learning the ARG II-n test. The ARG II-n was learned with different recursion depths ARG II-1 to ARG II-3 and with different sizes of LSTM networks [count of memory blocks(memory cells)]. Not all epochs did learn successfully.

	Streams [10^3]	Inputs [10^3]	1st	10th [%]	Successful
ARG II-1 [4(2)]:	100,000	15,081.1	1	38,743.03	48
ARG II-2 [5(2)]:	100,000	35,552.3	1	12,141.35	10
ARG II-2 [8(2)]:	100,000	35,471.6	1	100,000.00	10
ARG II-3 [8(2)]:	100,000	80,194.4	1	13,394.69	20

stream length are growing so even zero error test steams are possible. The growing of the LSTM network for learning ARG II-n must only be linear to n (from 4 to 8 memory blocks) whereas the solved problem is growing exponential (from 3 to 21 ARG I). Even the error is still not zero. Some test runs get a zero error result (all 10^5 streams with 10^8 symbols) was predicted successfully.

5 Conclusions and Future Work

It was shown that with some restrictions it should be possible to diagnose distributed embedded software systems with ANN. Three basic problems which must be solved for an adequate prediction of message streams in embedded software systems where separated. LSTM networks have been identified to fit very good to our problem. ARG has been introduced as an extension to CERG for testing the LSTM networks with more complex problems. The results of the executed experiments show that LSTM scales very good to complex message streams. It could be shown that the error in forecasting message sequences decreases with a grater amount of training streams. Nethertheless problematic is the reliability of the forecast, especially at the beginning of observation correct message prediction was a challenge.

In the future it is necessary for a reliable fault detection to predict the exact time of an expected message and to solve the problem of parallel sequences. For this problems ANN surely can be used to overcome a generic solution. We assume that the prediction of message sequences can detect failures in communication behavior very early. It is expected that this could simplify the problem of fault propagation and even fault isolation.

Additionally the learning phase consumes a lot of computing resources and needs proper learning data which could be in real applications a problem. While the learning phase could be executed on large computers the usage of the learned LSTM networks is on embedded devices. For this purpose an evaluation of the needed resources for executing this LSTM have to be done.

However, this approach promise the recovery of failures in a distributed system which are not covered by fault detection and isolation (FDI) mechanism. This is because the most FDI mechanism are based on expert knowledge to identify and describe system behavior. But in most complex distributed system even an expert can not cover all dependencies and implicit behavior of the system. This approach can help to discover implicit dependencies and detection of failures within this dependencies.

Acknowledgments. This work is related to the project DynaSoft. This project is funded by the bavarian Ministry of Economic Affairs, Infrastructure, Transport and Technology.

References

1. Chen, W., Toueg, S., Aguilera, M.: On the quality of service of failure detectors. IEEE Transactions on Computers 51(5), 561–580 (2002)
2. Satzger, B., Pietzowski, A., Trumler, W., Ungerer, T.: A new adaptive accrual failure detector for dependable distributed systems. In: Proceedings of the 2007 ACM symposium on Applied computing, Seoul, Korea, pp. 551–555. ACM, New York (2007)
3. Williams, B.C., Nayak, P.P., Nayak, U.: A model-based approach to reactive self-configuring systems. In: Proceedings of AAAI 1996, pp. 971–978 (1996)
4. Cimatti, A., Pecheur, C., Cavada, R.: Formal verification of diagnosability via symbolic model checking. In: Proceedings of the 18th International Joint Conference on Artificial Intelligence IJCAI 2003, pp. 363–369 (2003)
5. Ermagan, V., Krueger, I., Menarini, M., Mizutani, J.-i., Oguchi, K., Weir, D.: Towards Model-Based Failure-Management for automotive software. In: Fourth International Workshop on Software Engineering for Automotive Systems, 2007. ICSE Workshops SEAS 2007, p. 8 (2007)
6. Djurdjanovic, D., Liu, J., Marko, K., Ni, J.: Immune systems inspired approach to anomaly detection and fault diagnosis for engines. In: International Joint Conference on Neural Networks, IJCNN 2007, pp. 1375–1382 (2007)
7. Vemuri, A., Polycarpou, M.: Neural-network-based robust fault diagnosis in robotic systems. IEEE Transactions on Neural Networks 8(6), 1410–1420 (1997)
8. Atkinson, C.M., Long, T.W., Hanzevack, E.L.: Virtual sensing: a neural network-based intelligent performance and emissions prediction system for on-board diagnostics and engine control. In: The 1998 SAE International Congress & Exposition, pp. 39–51 (1998)
9. Chen, Y.M., Lee, M.L.: Neural networks-based scheme for system failure detection and diagnosis. Mathematics and Computers in Simulation 58(2), 101–109 (2002)
10. Isermann, R.: Model-based fault-detection and diagnosis-status and applications. Annual Reviews in control 29(1), 71–85 (2005)
11. Reber, A.S.: Implicit learning of artificial grammars. Journal of Verbal Learning & Verbal Behavior 6(6), 855–863 (1967)
12. Cleeremans, A., Servan-Schreiber, D., McClelland, J.L.: Finite state automata and simple recurrent networks. Neural Comput. 1(3), 372–381 (1989)
13. Gers, F.A., Schmidhuber, J.A., Cummins, F.A.: Learning to forget: Continual prediction with LSTM. Neural Comput. 12(10), 2451–2471 (2000)
14. Rosenblatt, F.: The perceptron: A probabilistic model for information storage and organization in the brain. Psychological review 65(6), 386–408 (1958)
15. Hochreiter, S., Informatik, F.F., Schmidhuber, J.: Long short-term memory. Technical Report FKI-207-95, Fakultät für Informatik, Technische Universität München (1995)

Fuzzy Numerical Schemes for Hyperbolic Differential Equations

Michael Breuss[1] and Dominik Dietrich[2]

[1] Saarland University, Germany
breuss@mia.uni-saarland.de
[2] German Research Centre for Artificial Intelligence (DFKI GmbH), Bremen, Germany
dietrich@dfki.de

Abstract. The numerical solution of hyperbolic partial differential equations (PDEs) is an important topic in natural sciences and engineering. One of the main difficulties in the task stems from the need to employ several basic types of approximations that are blended in a nonlinear way. In this paper we show that fuzzy logic can be used to construct novel nonlinear blending functions. After introducing the set-up, we show by numerical experiments that the fuzzy-based schemes outperform methods based on conventional blending functions. To the knowledge of the authors, this paper represents the first work where fuzzy logic is applied for the construction of simulation schemes for PDEs.

1 Introduction

Fuzzy logic (FL) is an important tool in computer science [9,6,19] and engineering [10,18,16,7]. It allows to control complex processes based on a small number of expert rules where an explicit knowledge of the behaviour of the considered system is given. In this paper, we consider an entirely new field of application for fuzzy logic: the construction of numerical schemes for hyperbolic partial differential equations (PDEs). Such equations arise in many disciplines, e.g., in gas dynamics, acoustics, geophysics, or bio-mechanics, see e.g. [11] for an overview. They describe a time-dependent transport of a given quantity without an inherent tendency to an equilibrium.

Hyperbolic numerics. The difficulty to handle hyperbolic PDEs stems from the fact that their solutions usually involve the formation of discontinuities. In order to cope with these, modern high-resolution schemes employ a nonlinear mixture of approximation schemes. For components of such a blending, two classic candidates to which we stick in this work are the monotone upwind method and the Lax-Wendroff (LW) method that is second order accurate. Second order accurate methods such as the LW scheme give a reasonable accuracy on smooth solutions, but they fail near discontinuities. Monotone schemes such as the upwind method are always only first order accurate, yet they offer the advantage that they can deal with discontinuities. The idea behind high-resolution (HR) schemes is to take the best of these two worlds, performing a blending in such

B. Mertsching, M. Hund, and Z. Aziz (Eds.): KI 2009, LNAI 5803, pp. 419–426, 2009.

a way that a monotone scheme is used at discontinuities while a higher order method is taken at smooth solution parts. For a more detailed exposition of these topics see e.g. [11].

Fuzzy logic for high-resolution schemes. The desired blending of first and second order approximations is performed by a so-called *limiter function* completely defining the HR scheme and its approximation properties. In the design of such a limiter only a few restrictions are imposed in the hyperbolic theory. Consequently, a wide variety of limiters has been proposed in the literature, motivated by experiences with computational examples; see [17] for an overview. However, the main *expert rules* in the design of a limiter are due to the already mentioned basic idea: *(i)* to use a monotone scheme at discontinuities, and *(ii)* to use a higher order scheme in smooth solution parts. It seems natural to ask if the blending process can be formalised using FL.

Our contribution. To our best knowledge, we employ FL for the first time within the literature for the construction of numerical schemes for PDEs here by applying the concept of FL at HR schemes for hyperbolic PDEs. Moreover, our work goes much beyond a proof-of-concept: we show that the use of FL yields significant improvements in the approximation quality.

Related work. As we have clarified above, to our knowledge no work has dealt before with the area of application we consider in this paper. Thus, we cannot rely on previous work in this field.

Structure of the paper. In Paragraph 2, we briefly review the fuzzy tools we employ in this work. Paragraph 3 introduces the mathematical background on hyperbolic PDEs and flux limiter. This is followed by an exposition on the construction of fuzzy limiters in Section 4. In Paragraph 5 we show results of numerical simulations. The paper is concluded by Paragraph 6.

2 Fuzzy Systems

Fuzzy logic is derived from *fuzzy set theory* [20], where the general idea is to allow not only for full membership or no membership, represented by the membership values 1 and 0, respectively, but also for *partial membership*, represented by a membership value in the interval $[0, 1]$. Consequently, a fuzzy set F over a domain G is represented by a membership function $\mu_F : G \to [0, 1]$. Similarly, in fuzzy logic the degree of truth is not restricted to true (1) and false (0), but can take any value of the interval $[0, 1]$, which builds the foundation for approximate reasoning and fuzzy control.

A *fuzzy system* is an expert system based on fuzzy logic. It consists of four components: fuzzy rules, fuzzifier, inference engine, and defuzzifier. *Fuzzy rules* are of the form "IF A then B" where A and B are fuzzy sets. A convenient way is to represent fuzzy sets by a (set of) parameterised function. Typical examples are monotonic decreasing functions, piecewise linear functions, exponential functions, or symmetric triangular functions. Moreover, it is common to attach a name to each fuzzy set and to identify the set by its name, resulting in more natural rules. The *fuzzifier* maps discrete input values to fuzzy sets, thus activating the rules. The *inference engine* maps fuzzy input sets to fuzzy output sets and handles the way in which rules are combined. Finally, the *defuzzifier* transforms the fuzzy output set to a discrete output value.

Fuzzy rules can either be formulated by an expert or be extracted from given data. Once all parameters have been fixed, a fuzzy system defines a nonlinear mapping, called control function, from the input domain to the output domain. For further details we refer to [7,16].

An advantage of fuzzy expert systems over other approaches is that often a small number of expert rules are sufficient to encode explicit knowledge about the problem to be controlled. This is because fuzzy reasoning adapts a rule to a given situation by modifying the conclusion of the rule, based on a difference analysis between the situation encoded in the premise and the given situation. In this sense, we can see fuzzy reasoning as a kind of knowledge based interpolation.

Additionally, the control function can easily be adapted by changing the fuzzy rules, the inference engine or the parameters of the underlying fuzzy sets, either manually, or by applying learning and optimisation techniques [1,2,12]. One particularly simple but powerful adaptation consists of applying *hedges* or *modifiers*, which are functions on fuzzy sets which change their shape while preserving their main characteristics. Within this paper we consider the standard hedges given by the *contrast operator*, the *dilation operator*, and the *concentration operator* (see [7,16] for details).

For our application, fuzzy systems seem to offer a natural framework to represent so-called flux limiter schemes, see next paragraph. Modification of the parameter results in a new control function, which corresponds to a new numerical scheme within our application. In particular it allows the use of learning and optimisation techniques in the context of hyperbolic PDEs.

3 Hyperbolic PDEs and Flux Limiter

Hyperbolic PDEs. Let us consider the usual format of a hyperbolic PDE

$$\frac{\partial}{\partial t}u(x,t) + \frac{\partial}{\partial x}f\left(u(x,t)\right) = 0\,; \qquad (1)$$

- $x \in \mathbb{R}$ and $t \in \mathbb{R}^+$ are typically interpreted as space and time variables, respectively, whereas $\partial/\partial t$ and $\partial/\partial x$ are the corresponding spatial and temporal differential operators,
- $u(x,t)$ is the unknown density function of a quantity of interest,
- f denotes the flux function that describes the flow behaviour.

The PDE (1) needs to be supplemented by an initial condition $u_0(x)$. Then, the differential equation may be understood to encode the temporal change of the beginning state u_0. Note that we only consider here the one-dimensional spatial setting, as our developments are readily extendable to higher dimensions.

Without going into details, let us remark that solving (1) is a non-trivial task, as the solution involves the formation of discontinuities even if the initial function u_0 is arbitrarily smooth [5].

The basic numerical set-up. To make up a numerical scheme, one needs to give a discrete representation of the computational domain in space and time, as well as a discrete representation of the differential operators in (1). Thus, we define a grid in space and time featuring the widths Δx and Δt, respectively. For a discrete representation of derivatives we need the approximation of $u(x,t)$ at the grid points. Indicating discrete data by large letters, we write U_j^n for the approximation of $u(x,t)$ at the point $(j\Delta x, n\Delta t) \in \mathbb{R} \times \mathbb{R}^+$.

Most numerical schemes for hyperbolic PDEs mimic the format of (1), as this proceeding ensures the validity of useful properties. For the discretisation point with indices (j,n) in space and time, they can be read as

$$\frac{U_j^{n+1} - U_j^n}{\Delta t} + \frac{F_{j+1/2}^n - F_{j-1/2}^n}{\Delta x} = 0\,, \tag{2}$$

The indices $j \pm 1/2$ indicate a *numerical flux* between the points j and $j+1$, or $j-1$ and j, respectively. The formula (2) is then evaluated for each spatial point j of the computational domain; at the end points of the necessarily finite interval in space one needs numerical boundary conditions. The process begins at the time level $n=0$, building up a solution iterating from time level to time level until a prescribed stopping time.

The key to success in the definition of a suitable scheme is the construction of the so-called *numerical flux function* F. In our paper, we exactly elaborate on this topic, constructing via F high-resolution schemes.

Flux limiters. Generally speaking, there exist two powerful frameworks to construct HR schemes for hyperbolic PDEs, using so-called slope limiters and flux limiters, respectively; c.f. [11,17] for detailed expositions. These approaches are only roughly equivalent. The slope limiter construction is based on geometric insight making use of the slopes of higher order interpolation polynomials incorporated in such schemes. In our work, we use the flux limiter framework that relies on the physical interpretation of scheme components as fluxes of the quantity considered in the underlying PDE.

As indicated, the basic idea is to use a first order monotone scheme at discontinuities and a high order scheme in smooth solution parts. Both schemes are usually cast in the format (2), and we identify them via a monotone numerical flux $F_{j\pm 1/2,M}^n$ and a high order numerical flux $F_{j\pm 1/2,H}^n$, respectively.

In order to distinguish discontinuities from smooth solution parts, we also need at $j \pm 1/2$ and time level n a *smoothness measure* we denote by $\Theta_{j\pm 1/2}^n$. This is usually computed making use of the ratio of consecutive slopes depending on the flow direction. In the usual basic set-up the flow is given from left to right, and the smoothness measure is computed as

$$\Theta_{j+1/2}^n := \frac{U_{j+1}^n - U_j^n}{U_j^n - U_{j-1}^n}\,. \tag{3}$$

A *limiter function* Φ is then a function evaluating Θ and assigning it a reasonable output, so that the construction idea of HR schemes is met and we can write the high-resolution numerical flux $F_{j+1/2,HR}^n$ as

$$F_{j+1/2,HR}^n = F_{j+1/2,M}^n + \Phi\left(\Theta_{j+1/2}^n\right) F_{j+1/2,H}^n\,. \tag{4}$$

Thus, HR methods with numerical fluxes as in (4) can be viewed as a monotone scheme plus a correction. Note that for $\Phi = 0$ the monotone numerical flux is obtained, whereas $\Phi = 1$ yields the high-resolution numerical flux.

One should note that there are only a few theoretical restrictions on the definition of the limiter function Φ. The most important one is concerned with ensuring the so-called *total variation (TV) stability* of the numerical scheme, which means in practice that no oscillations are generated numerically if the scheme is TV stable, c.f. [11]. For the proceeding in our work, it is important that the limiter function is in the corresponding so-called *TV region*, see next paragraph.

4 Fuzzy Limiter Construction

From an abstract point of view the control problem we consider here is to determine the flux limiter value $\Phi \in [0, 2]$, given a smoothness measure Θ. To get a TV stable method, the control function must lie in the TV-region, which is sketched in Fig. 1 as the shaded area. Modelling a solution to this control problem using fuzzy logic requires the following steps: *(i)* Defining fuzzy sets on the input and the output domain *(ii)* Choosing a suitable inference mechanism and defuzzification method, and *(iii)* Defining a suitable rule base. Rather than starting from scratch, it is a good idea to look at a simple, already existing HR scheme. The probably simplest is the *Minmod* scheme, which linearly blends from the upwind scheme to the LW scheme, as shown in Fig. 1.

One can clearly see how the expert knowledge, (i) to use LW if the solution is smooth, and (ii) to use the Upwind method if the solution is not smooth, has been translated into the behaviour of the numerical scheme. Moreover, we can also see how the scheme fits smoothly into the framework of FL: We can think of a region in which the flux limiter is constant to correspond to a situation in which a single rule fires with full activation degree; the remaining parts correspond to situations in which several rules are active (to some degree) and their result combined by means of FL.

As we have identified two key situations in the case of the Minmod scheme, we define two fuzzy sets "smooth" and "extremum" on the input parameter, which we call "smoothness". Similarly, we define two (singleton) fuzzy sets "UP" (corresponding to a flux limiter value of 0) and "LW" (corresponding to a flux limiter value 1) on the output parameter, which we call "limiter". Our modelling

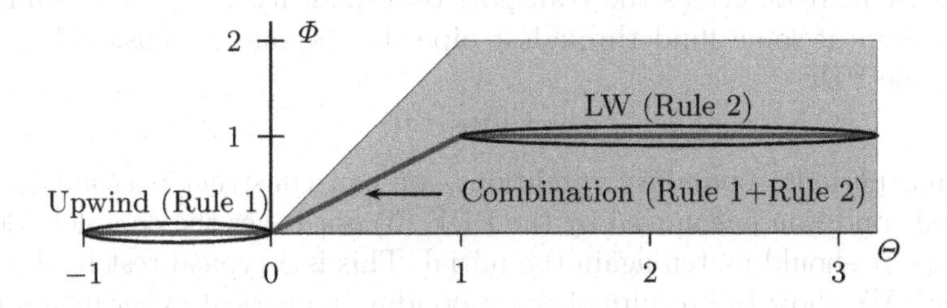

Fig. 1. TV-region and control function of the Minmod scheme

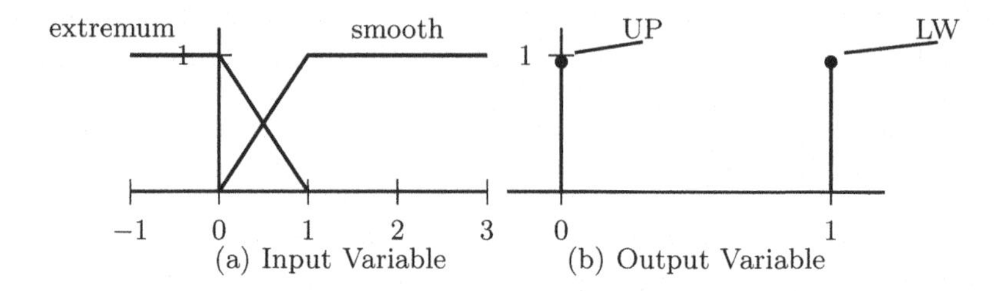

Fig. 2. Fuzzy sets for the Minmod limiter

is shown in Fig. 2. This allows us to explicitly express the construction idea of the Minmod numerical scheme in form of the following rule base:

If **smoothness** is **extremum** then **limiter** is UP
If **smoothness** is **smooth** then **limiter** is LW

Choosing an inference engine consisting of the min operator as implication operator, the max operator as aggregation operator, and the centroid method for defuzzification, we exactly obtain the control function of the Minmod scheme.

Similarly, other prominent TV-scheme such as the Superbee scheme or the MC scheme (see [11] for details) can easily be modelled within the FL framework.

5 Experimental Study

In addition to providing a common framework and embedding existing schemes into it, new numerical schemes can be constructed by modifying the parameters within of the FL setting. This is a common step in FL design, often called *parameter tuning*. As this is usually a time consuming task, complex learning techniques have been developed to automate it [1,2,12]. Instead of relying on these we will follow the more pragmatic approach of modifying a numerical scheme only by applying the fuzzy modifiers concentration, dilation, contrast to the input fuzzy sets. The benefits are (i) simplicity and (ii) the guarantee that only the blending between the specified key situations is modified.

To evaluate our approach we consider the *linear advection equation*, which is probably the simplest hyperbolic PDE and which is commonly used as benchmark problem. It describes the transport of a quantity within a medium, e.g., the advection of some fluid through a pipe. In the one dimensional case it is given by the PDE

$$u_t + \overline{a} u_x = 0. \tag{5}$$

Employing periodic boundary conditions, we can construct a situation where our initial condition has moved by the PDE (5) once over the complete domain, i.e., ideally it should match again the initial. This is a typical test in the hyperbolic field. We show the results of corresponding numerical experiments for the parameters $\Delta t = 0.0025$, $\Delta x = 0.01$.

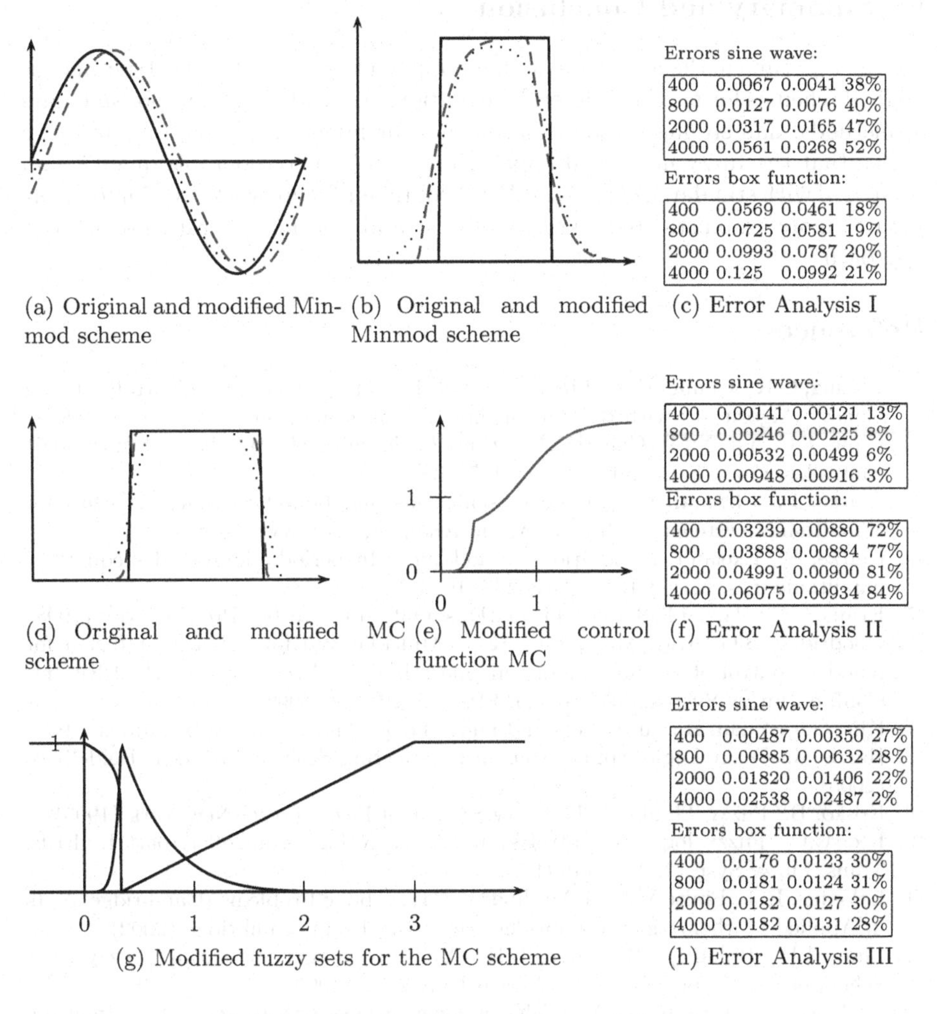

Errors sine wave:

400	0.0067	0.0041	38%
800	0.0127	0.0076	40%
2000	0.0317	0.0165	47%
4000	0.0561	0.0268	52%

Errors box function:

400	0.0569	0.0461	18%
800	0.0725	0.0581	19%
2000	0.0993	0.0787	20%
4000	0.125	0.0992	21%

(a) Original and modified Minmod scheme (b) Original and modified Minmod scheme (c) Error Analysis I

Errors sine wave:

400	0.00141	0.00121	13%
800	0.00246	0.00225	8%
2000	0.00532	0.00499	6%
4000	0.00948	0.00916	3%

Errors box function:

400	0.03239	0.00880	72%
800	0.03888	0.00884	77%
2000	0.04991	0.00900	81%
4000	0.06075	0.00934	84%

(d) Original and modified MC scheme (e) Modified control function MC (f) Error Analysis II

Errors sine wave:

400	0.00487	0.00350	27%
800	0.00885	0.00632	28%
2000	0.01820	0.01406	22%
4000	0.02538	0.02487	2%

Errors box function:

400	0.0176	0.0123	30%
800	0.0181	0.0124	31%
2000	0.0182	0.0127	30%
4000	0.0182	0.0131	28%

(g) Modified fuzzy sets for the MC scheme (h) Error Analysis III

Fig. 3. Results of the experimental study

For an error analysis, we give in Fig. 3(c,f,h), from left to right, *(i)* the number of iterations, *(ii)* the L_1-error of (c) the Minmod, (f) the MC, and (h) the Superbee scheme, *(iii)* the L_1-error of the corresponding Fuzzy scheme, and *(iv)* the error improvement by our approach. Let us stress, that the algorithms constructed by us perform significantly better than the original schemes.

Fig. 3(a,b,d) graphically show the difference between the original scheme (dotted), the modified scheme (red dashed line), and the reference solution (black line). Fig. 3(g) shows an example of a modified fuzzy set resulting in the control function Fig. 3(e). For more complex PDEs the improvements are similar.

6 Summary and Conclusion

For the first time in the literature, we have applied fuzzy logic for constructing numerical schemes for partial differential equations. Considering an important class of methods, namely high-resolution schemes for hyperbolic equations, we have shown that the fuzzy numerical approach results in a considerable quality gain compared with standard schemes in that field. In our future work, we aim to incorporate more sophisticated data analysis and learning strategies in our approach.

References

1. Chiang, C.K., Chung, H.Y., Lin, J.J.: A Self-Learning Fuzzy Logic Controller Using Genetic Algorithms with Reinforcements. Transaction on Fuzzy Systems 5 (1997)
2. Chin, T.C., Qi, X.M.: Genetic algorithms for learning the rule base of fuzzy logic controller. Fuzzy Sets and Systems, 1–7 (1998)
3. Clarke, F., Ekeland, I.: Nonlinear oscillations and boundary-value problems for Hamiltonian systems. Arch. Rat. Mech. Anal. 78, 315–333 (1982)
4. Clarke, F., Ekeland, I.: Solutions périodiques, du période donnée, des équations hamiltoniennes. Note CRAS Paris 287, 1013–1015 (1978)
5. Evans, L.C.: Partial Differential Equations. Oxford University Press, Oxford (1998)
6. Goodridge, S.G., Kay, M.G., Luo, R.C.: Multi-layered fuzzy behavior fusion for reactive control of an autonomous mobile robot. In: Procs. of the 6th IEEE Int. Conf. on Fuzzy Systems, Barcelona, SP, pp. 579–584 (1996)
7. Klir, G.J., Yuan, B.: Fuzzy Sets and Fuzzy Logic, Theory and Applications (1995)
8. Harris, J.: Fuzzy Logic Applications in Engineering Science. Springer, Heidelberg (2005)
9. Kosko, B.: Fuzzy Thinking: The New Science of Fuzzy Logic, New York (1993)
10. Lee, C.C.: Fuzzy logic in control systems: Fuzzy logic controller, part I. IEEE, Trans. Fuzzy Syst., 2, 4–15 (1994)
11. LeVeque, R.J.: Finite Volume Methods for Hyperbolic Problems (Cambridge Texts in Applied Mathematics). Cambridge University Press, Cambridge (2002)
12. Lim, M.H., Rahardja, S., Gwee, B.H.: A GA paradigm for learning fuzzy rules School of EEE, Nanyang Technological University (1996)
13. Michalek, R., Tarantello, G.: Subharmonic solutions with prescribed minimal period for nonautonomous Hamiltonian systems. J. Diff. Eq. 72, 28–55 (1988)
14. Rabinowitz, P.: On subharmonic solutions of a Hamiltonian system. Comm. Pure Appl. Math. 33, 609–633 (1980)
15. Tarantello, G.: Subharmonic solutions for Hamiltonian systems via a $\mathbb{Z}_p$ pseudoindex theory. Annali di Matematica Pura (to appear)
16. Terano, T., Asai, K., Sugeno, M.: Fuzzy Systems Theory and Its Applications. Academic, New York (1992)
17. Toro, E.F.: Riemann Solvers and Numerical Methods for Fluid Dynamics, 2nd edn. Springer, Heidelberg (1999)
18. Von Altrock, C.: Recent Successful Fuzzy Logic Applications in Industrial Automation IEEE. In: Fifth International Conference on Fuzzy Systems (1996)
19. Xu, D., Keller, J.M., Popescu, M.: Applications of Fuzzy Logic in Bioinformatics (Advances in Bioinformatics and Computational Biology) (2008)
20. Zadeh, L.A.: Fuzzy Sets. Information and Control (1965)

Model-Based Test Prioritizing – A Comparative Soft-Computing Approach and Case Studies

Fevzi Belli[1], Mubariz Eminov[2], and Nida Gokce[3]

[1] Department of Computer Science, Electrical Engineering and Mathematics, University of Paderborn, Germany
belli@upb.de

[2] Faculty of Engineering, Department of Computer Engineering, Halic University, Istanbul, Turkey
mubarizeminli@halic.edu.tr

[3] Faculty of Arts and Sciences, Department of Statistics, Mugla University, Turkey
nidagokce@yahoo.com

Abstract. Man-machine systems have many features that are to be considered simultaneously. Their validation often leads to a large number of tests; due to time and cost constraints they cannot exhaustively be run. It is then essential to prioritize the test subsets in accordance with their importance for relevant features. This paper applies soft-computing techniques to the prioritizing problem and proposes a graph model-based approach where preference degrees are indirectly determined. Events, which imply the relevant system behavior, are classified, and test cases are clustered using (i) unsupervised neural network clustering, and (ii) Fuzzy c-Means clustering algorithm. Two industrial case studies validate the approach and compare the applied techniques.

Keywords: model-based test prioritizing, adaptive competitive learning, fuzzy c-means, clustering, neural networks.

1 Introduction and Related Work

Testing is one of the important, traditional, analytical techniques of quality assurance widely accepted in the software industry. There is no justification, however, for any assessment of the correctness of software under test (SUT) based on the success (or failure) of a single test, because potentially there can be an infinite number of test cases. To overcome this shortcoming of testing, formal methods model and visualize the relevant, desirable features of the SUT. The modeled features are either functional or structural issues, leading to *specification-* or *implementation-oriented* testing, respectively. Once the model is established, it "guides" the test process to generate and select test cases, which form sets of test cases (*test suites*). The test selection is ruled by an *adequacy criterion* for measuring the effectiveness of test suites [8]. Most of the existing adequacy criteria are *coverage*-oriented [18].

The test approach introduced in this paper is specification- and coverage-oriented. *Event sequence graphs* (*ESG*, [2,3]) are favored for modeling which view the

B. Mertsching, M. Hund, and Z. Aziz (Eds.): KI 2009, LNAI 5803, pp. 427–434, 2009.

behavior of SUT and its interaction with user as events. Because of the large amount of features to be considered, the number of test cases and thus the costs of testing often tend to run out of the test budget. Therefore, it is important to test the most important items first which leads to the *test prioritization problem* [12]:

Given: A test suite T; the set of permutations of T (symbolized as PT); a function f from PT to the real numbers representing the preference of the tester while testing.

Problem: Find $T' \in PT$ such that $(\forall T'')\ (T'' \neq T')\ [f(T') \geq f(T'')]$

The ESG approach generates test suites through a finite sequence of discrete events. The underlying optimization problem is a generalization of the *Chinese Postman Problem* (*CPP*) [13] and algorithms given in [2,3] differ from the well-known ones in that they satisfy not only the constraint that a minimum total length of test sequences is required, but also fulfill the coverage criterion with respect to converging of all event pairs represented graphically. To overcome the problem that an exhaustive testing might be infeasible, the paper develops a *prioritized* version of the mentioned test generation and optimization algorithms, in the sense of "divide and conquer" principle. This is the primary objective and the kernel of this paper which is novel and thus, to our knowledge, has not yet been worked out in previous works, including ours [2,3].

The approaches assign to each test generated a *degree of its preference* which is estimated for all including events and qualified by several attributes that depend on the features of the project, and their values are justified by their significance to the user. For clustering, unsupervised neural networks (NN) [14] and Fuzzy c-Means (FCM) analysis [9,10] are used.

Section 2 explains the background of the approach, summarizing ESG notation and NN and FCM clustering algorithms. Section 3 describes the proposed prioritized graph-based testing approach. Two case studies empirically validate the approach in Section 4 which also identifies and analyzes its characteristic issues. Section 5 summarizes the results, gives hints to further researches and concludes the paper.

2 Background and Approach

2.1 Event Sequence Graphs for Test Generation

Basically, an *event* is an externally observable phenomenon, such as an environmental or a user stimulus, or a system response, punctuating different stages of the system activity. A simple example of an ESG is given in Fig.1.

Mathematically speaking, an ESG is a directed, labeled graph and may be thought of as an ordered pair $ESG=(\alpha, E)$, where α is a finite set of nodes (vertices) uniquely labeled by some input symbols of the alphabet Σ, denoting events, and E: $\alpha \rightarrow \alpha$, a

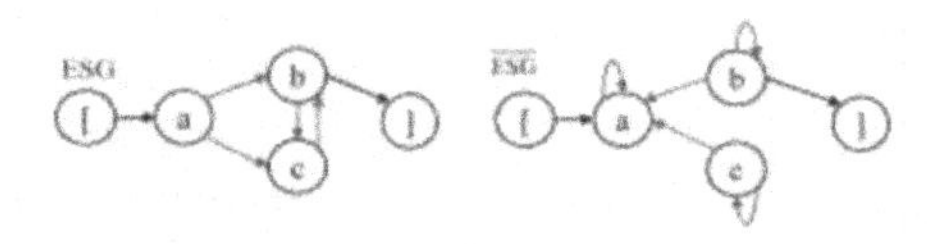

Fig. 1. An event sequence graph ESG and $\overline{ESG}$ as the complement of the given ESG

precedence relation, possibly empty, on α. The elements of E represent directed arcs (edges) between the nodes in α. Given two nodes a and b in α, a directed arc ab from a to b signifies that event b can follow event a, defining an *event pair* (*EP*) ab (Fig. 1). The remaining pairs given by the alphabet Σ, but not in the ESG, form the set of *faulty event pairs* (*FEP*), e.g., ba for ESG of Fig.1. As a convention, a dedicated, start vertex e.g., *[*, is the *entry* of the ESG whereas a final vertex e.g., *]* represents the *exit*. Note that *[* and *]* are not included in Σ. The set of FEPs constitutes the *complement* of the given ESG ($\overline{ESG}$ in Fig.1).

A sequence of $n+1$ consecutive events that represents the sequence of n arcs is called an *event sequence (ES) of the length n+1*, e.g., an *EP* (*event pair*) is an ES of length 2. An ES is *complete* if it starts at the initial state of the ESG and ends at the final event; in this case it is called a *complete ES* (*CES*). Occasionally, we call CES also as *walks* (or *paths*) through the ESG given [2,3].

Completeness Ratio (CR) is a metric which explains density of edges in the ESG and is defined as follows:

$$CR = |E|/|V|^2 \tag{1}$$

where $|E|$ is the number of edges in the ESG and $|V| = n$ is the number of nodes (vertex) in the ESG. CR takes the values between 0 and 1.Value 1 shows that ESG is completed graph and Value 0 means null graph. As the values are getting closer to 1, the density of the graph gets bigger. ESG concept and definitions informally introduced above are sufficient to understand the test prioritization approach represented in this paper. For more information on ESG refer to ([2,3,7,16]).

2.2 Neural Network-Based Clustering

In this study, we have chosen unsupervised NN where competitive learning CL algorithm can adaptively cluster instances into clusters. For clustering of an unstructured data set dealing especially with vector quantization, unsupervised learning based on clustering in a NN framework is fairly used. In clustering a data set $X = \{x_1, \dots, x_i, \dots x_n,\} \subset \mathbb{R}^p$, $\in \mathbb{R}^p$ *containing events* (nodes) is portioned into c *number of clusters* each of which contains a subset S_k defined as:

$$X = \bigcup_{k=1}^{c} S_k \text{ with } S_k \cap S_j = 0 \quad \forall k \neq j \tag{2}$$

Each S_k cluster is represented by a cluster center (*prototype*) that corresponds to a weight vector $\widetilde{w}_k = (\widetilde{w}_{k1}, \dots, \widetilde{w}_{kj}, \dots, \widetilde{w}_{kp}) \in \mathbb{R}^p$ and after finding a trained value of all weight vectors $\widetilde{W} = \{\widetilde{w}_1, \dots, \widetilde{w}_k, \dots, \widetilde{w}_c\} \subset \mathbb{R}^p$ the data set $X \epsilon \mathbb{R}^p$ is divided into k^{th} cluster by the condition:

$$S_k = \left\{ x \in \mathbb{R}^p \,\middle|\, \sum_{j=1}^{p} \tilde{x}_{ij}\tilde{w}_{kj} \geq \sum_{j=1}^{p} \tilde{x}_{ij}\tilde{w}_{gj} \;\; \forall k \neq g \right\} \quad i = 1,\dots,n \quad j = 1,\dots,p \quad k = 1,\dots,c \quad g = 1,\dots,c \in \mathbb{N} \tag{3}$$

Training: The initial values of the weight vectors are randomly allotted. It negatively influences the clustering performance of standard CL algorithm that is explained in literature [5,11,14]. In order to get a better clustering performance, in this paper we use the *adaptive* CL algorithm by deleting over specified weight vectors [15,16]. The main distinguishing properties of this algorithm from standard CL are: Both data points $\tilde{x}_i$ and weight vectors $\tilde{w}_k$ are normalized to a unit length, i.e., they are presented as the unit vector the length of which is 1 (one). In adaptive CL algorithm, the winner weight vector $\tilde{w}_w$ is determined by a dot product of data point $\tilde{x}_i$ and weight vector $\tilde{w}_k$ as in (4), and the updating rule of a winner weight vector [5] is based on the adjusting equation expressed as in (5)

$$\tilde{w}_w = arg\ max_k\{\textstyle\sum_{j=1}^{p} \tilde{x}_{ij}\, \tilde{w}_{kj}\}\quad i = 1, \ldots, n \quad k = 1, \ldots, c \tag{4}$$

$$\Delta\tilde{w}_w(t) = \eta(t)(\tilde{x}_i/p - \tilde{w}_w) \tag{5}$$

where η is a learning rate. There is a deletion mechanism [7,16] that eliminates one weight vector, w_s that corresponds to cluster which has a minimum intra-cluster partition error, i.e., $D_k \geq D_s$, for all k, and it proceeds until the number of weight vectors is equal to the predetermined one. D_k is determined as in (6):

$$D_k = \frac{1}{p}(\textstyle\sum_{\tilde{x}\in S_k} \tilde{x}\, \tilde{w}_w) \qquad k = 1, \ldots, c \tag{6}$$

2.3 FCM-Based Clustering

Fuzzy c-means (FCM) algorithms as a fuzzy version of hard c-means (HCM), crisp partition, as introduced in [9] and improved by "fuzzifier m" in [1]. In real applications there are very often no sharp boundaries among clusters so that fuzzy clustering is often better suited for the data. FCM (probabilistic) clustering assigns the data into c fuzzy clusters each of which is represented by its center, called a *prototype*, as a representative of data. There every datum may belong simultaneously to several clusters with corresponding *membership degree (MD) that* has value between zero and one.In FCM, for optimal partitioning the alternating optimization algorithm for minimizing the objective function [6, 7, 10] presented in (7) is used:

$$J(X, U, V) = \textstyle\sum_{k=1}^{c}\sum_{i=1}^{n}(u_{ki})^m d^2(v_k, x_i) \tag{7}$$

2.4 Prioritized ESG-Based Testing

We consider the testing process is based on the generation of a test suite from ESG that is a discrete model of a SUT. To generate tests, a set of ESGs is derived which are input to the generation algorithm to be applied. Our prioritized testing approach is based on the ESG-based testing algorithms as introduced in [2,3]. The constraints on total lengths of the tests [2,3] generated are enable a considerable reduction in the cost of the test execution. To solve the test prioritizing problem, several algorithms have been introduced [4,8]. However, this kind of prioritized testing is computationally expensive and hence restricted to deal with short test cases only.

The ordering of the CESs is in accordance with their importance degree which is defined indirectly, i.e., by classification of events that are the nodes of ESG and represent objects (modules, components) of SUT. For this aim, firstly events are presented as a *multidimensional data vector* $x_i = (x_{i1}, \ldots, x_{ip})$ then; a data set $X = \{x_1, \ldots, x_n\} \subset \mathbb{R}^p$ is constructed which divided into c groups. The groups are constructed by using both Adaptive CL algorithm and FCM clustering algorithm and then classification procedure as explained in detail in our previous works [7,16]. Afterwards, the importance degree of these groups has been assigned according to length of their corresponding weight vector.

Importance index $(Imp(x_i))$ of ith event belonging to k^{th} group is defined as follows:

$$Imp(x_i) = c - ImpD(S_k) + 1 \tag{8}$$

where c is the optimal number of the groups; $ImpD(S_k)$ is the importance degree of the group S_k where the ith event belongs to and this importance degree is determined by comparing the length of obtained groups weight vectors. Finally, the assignment of preference degrees to CESs is based on the rule that is given in [7,16].

Therefore, the preference degree of CES can be defined by taking into account both the importance of events and the frequency of occurrence of event(s) within them, and it can be formulated as follows:

$$PrefD(CES_q) = \sum_{i=1}^{n} Imp(x_i)\mu_{S_k}(x_i)f_q(x_i) \quad q = 1, \ldots r \in \mathbb{N} \tag{9}$$

where r is the *number of CESs*, $Imp(x_i)$ is importance index of the ith event (8), $\mu_{S_k}(x_i)$ is MD of the ith event belonging to the group S_k (it is 0 or 1 in NN-based clustering, and it takes any value between 0 and 1 in FCM-based clustering), and $f_q(x_i)$ is frequency of occurrence of event i^{th} within CES_q. Finally, the CESs are ordered, scaling of their preference degrees (9) based on the events which incorporate the importance group(s). We propose *indirect determination of preference algorithm* for prioritized ESG-based testing algorithm in our previous works [7,16].

3 Case Studies

Data borrowed from two industrial projects are used in the following case studies which are mentioned in our previous papers [7,16]: A marginal strip mower (Fig.2.(a)), a web-based tourist portal (Fig.2.(b)) and an example ESG (Fig.2.(c)). The construction of ESG and the generation of test cases from those ESGs have been explained in the previous papers of the first author [2,3]. Classification of the events using the results of Adaptive CL algorithm is accomplished according to equality (3) where each event is assigned crisply to one group only. After FCM clustering, each event belongs to only one group according to its maximum MD. For two clustering approach, importance degrees ($ImpD(S_k)$) of groups are determined by comparing the length of corresponding center vectors (ℓ), and their values that are given in Table 1.

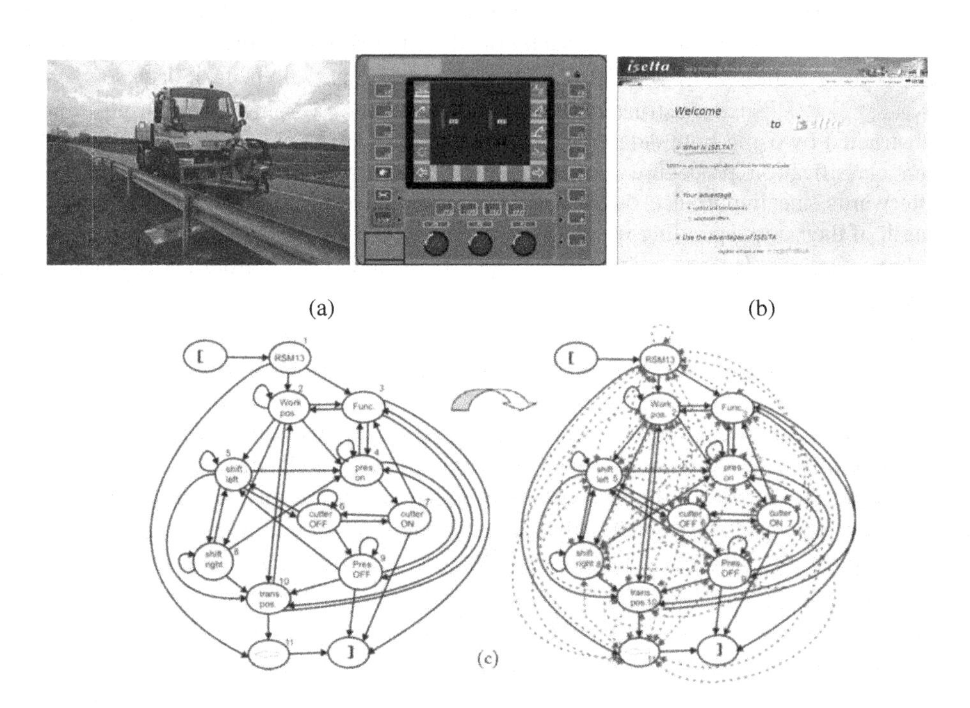

Fig. 2. (a) The marginal strip mower mounted on a vehicle and its display unit as a control desk. (b) The web-based system ISELTA. (c) Example ESG.

Table 1. Ranking of CESs of the example ESG using algorithms selected

Adaptive CL					*FCM*				
Groups	*Events*	*MD*	ℓ	$ImpD(S_k)$	*Groups*	*Events*	*MD*	ℓ	$ImpD(S_k)$
$S_1^{(A)}$	3	1	2,094	1	$S_4^{(F)}$	5	0,934	1,8208	4
	4	1				4	0,445		
	10	1				9	0,272		
$S_2^{(A)}$	1	1	1,344	6	$S_6^{(F)}$	11	0,999	1,9780	3
$S_3^{(A)}$	2	1	1,589	3	$S_5^{(F)}$	8	0,890	1,7341	5
	5	1				10	0,516		
$S_4^{(A)}$	9	1	1,788	2	$S_1^{(F)}$	3	0,996	2,2471	1
$S_5^{(A)}$	7	1	1,544	5	$S_2^{(F)}$	7	0,811	1,4073	6
	8	1				6	0,645		
	11	1				1	0,362		
$S_6^{(A)}$	6	1	1,550	4	$S_3^{(F)}$	2	0,985	2,0235	2

Table 2. Ordering of CESs of the example ESG using methods selected

CESs		*Adaptive CL*		*FCM*	
No	*Walks*	*PrefD(CES$_q$)*	*Order No*	*PrefD(CES$_q$)*	*Order No*
CES$_1$	[1 3 4 7 3 10 2 2 4 4 9 4 3]	64	**2**	36,147	**1**
CES$_2$	[1 2 2 5 4 7 6 6 7]	29	**5**	17,262	**6**
CES$_3$	[1 2 5 6 5 5 8 8 5 10 11]	36	**3**	25,734	**3**
CES$_4$	[1 2 3 2 8 10 3]	29	**4**	24,979	**4**
CES$_5$	[1 11]	3	**7**	4,359	**7**
CES$_6$	[1 2 10 2 5 6 9 9 8 4 9 10 2 4 9]	66	**1**	28,360	**2**
CES$_7$	[1 2 4 9 5 10 11]	28	**6**	15,270	**5**

The preference degrees of the CESs are determined by (9), and the ordering of the CESs is represented in Table 2. Consequently, we have a ranking of test cases to make the decision of which test cases are to be primarily tested. For all four considered ESGs examples, to compare the performance of clustering approaches the mean square error (MSE) (10) in accordance with corresponding level of CR are determined and compared in Table 3.

$$MSE = \frac{1}{np} \sum_{k=1}^{c} \sum_{x \in S_k} d(x, \bar{x}_k) \qquad (10)$$

Table 3. Completeness Ratio (CR) and clustering error for each examples

EXAMPLES		ESG1	ESG2	ESG3	ESG4
	CR	0,31	0,17	0,10	0,83
MSE (10)	FCM	0,022	0,037	0,032	0,014
	Adaptive CL	0,030	0,036	0,037	0,027

4 Conclusions

The model-based, coverage-and specification-oriented approach described in this paper provides a novel and effective algorithms for ordering test cases according to their degree of preference. Such degrees are determined indirectly through the use of the events specified by several attributes, and no prior knowledge about the tests carried out before is needed. Those are important and consequently, the approach introduced radically differs from the existing ones. This approach is useful when an ordering of the tests due to restricted budget and time is required.

In this paper, the events (nodes of ESG) are classified by using both unsupervised NN based adaptive CL and FCM clustering algorithm, and their classification results are investigated comparatively. After carrying out corresponding clustering, in order to classify the events, the nearest centroid (weight) rule and the maximum MD based deffuzzification procedure is applied, respectively. So the crisply and fuzzily assigning groups of events are produced. FCM clustering based prioritization of CESs is more plausible and convenient than the former due to construction of the fuzzy qualified groups of events that are important when groups to be obtained are interloped. However, computational complexity of FCM clustering is higher than the other and hence choice of the appropriate clustering algorithm in dependence of data structure is needed. For this aim, usage of CR metric defined in (1) and evaluating its value is proposed. As seen from Table 5, for little CR value (less than approximately 0.18), the adaptive CL algorithm should be preferred as the simple method (no difference between MSE). But for greater CR value (more than nearly 0.31), FCM clustering is more suitable due to the little value of MSE.

References

1. Dunn, J.C.: A Fuzzy Relative of the ISODATA Process and Its Use in Detecting Compact Well-Separated Clusters. Journal of Cybernetics 3, 32–57 (1973)
2. Belli, F., Budnik, C.J., White, L.: Event-Based Modeling, Analysis and Testing of User Interactions – Approach and Case Study. J. Software Testing, Verification & Reliability 16(1), 3–32 (2006)
3. Belli, F., Budnik, C.J.: Test Minimization for Human-Computer Interaction. J. Applied Intelligence 7(2) (2007) (to appear)
4. Bryce, R.C., Colbourn, C.C.: Prioritized Interaction Testing for Pair-wise Coverage with Seeding and Constraints. Information and Software Technology 48, 960–970 (2006)
5. Rummelhart, D.E., Zipser, D.: Competitive Learning. J. Cognitive Science 9, 75–112 (1985)
6. Eminov, M.E.: Fuzzy c-Means Based Adaptive Neural Network Clustering. In: Proc. TAINN 2003, Int. J. Computational Intelligence, pp. 338–343 (2003)
7. Belli, F., Eminov, M., Gökçe, N.: Coverage-Oriented, Prioritized Testing – A Fuzzy Clustering Approach and Case Study. In: Bondavalli, A., Brasileiro, F., Rajsbaum, S. (eds.) LADC 2007. LNCS, vol. 4746, pp. 95–110. Springer, Heidelberg (2007)
8. Binder, R.V.: Testing Object-Oriented Systems. Addison-Wesley, Reading (2000)
9. Bezdek, J.C.: Pattern Recognition with Fuzzy Objective Function Algorithms. Plenum Press, New York (1981)
10. Hoppner, F., Klawonn, F., Kruse, R., Runkler, T.: Fuzzy Cluster Analysis. John Wiley, Chichester (1999)
11. Fu, L.: Neural Networks in computer Intelligence. McGraw Hill, Inc., New York (1994)
12. Elbaum, S., Malishevsky, A., Rothermel, G.: Test Case Prioritization: A Family of Empirical Studies. IEEE Transactions on Software Engineering 28(2), 182–191 (2002)
13. Edmonds, J., Johnson, E.L.: Matching: Euler Tours and the Chinese Postman. Math. Programming, 88–124 (1973)
14. Eminov, M., Gokce, N.: Neural Network Clustering Using Competitive Learning Algorithm. In: Proc. TAINN 2005, pp. 161–168 (2005)
15. Kim, D.J., Park, Y.W., Park, D.J.: A Novel Validity Index for Clusters. IEICE Trans. Inf & System, 282–285 (2001)
16. Belli, F., Eminov, M., Gökçe, N.: Prioritizing Coverage-Oriented Testing Process- An Adaptive-Learning-Based Approach and Case Study. In: The Fourth IEEE International Workshop on Software Cybernetics, IWSC 2007, Beijing, China, July 24 (2007)
17. Memon, A.M., Pollack, M.E., Soffa, M.L.: Hierarchical GUI Test Case Generation Using Automated Planning. IEEE Trans. Softw. Eng. 27/2, 144–155 (2001)
18. Zhu, H., Hall, P.A.V., May, J.H.R.: Software Unit Test Coverage And Adequacy. ACM Computing Surveys 29(4) (December 1997)

Comparing Unification Algorithms in First-Order Theorem Proving

Kryštof Hoder and Andrei Voronkov

University of Manchester, Manchester, UK

Abstract. Unification is one of the key procedures in first-order theorem provers. Most first-order theorem provers use the Robinson unification algorithm. Although its complexity is in the worst case exponential, the algorithm is easy to implement and examples on which it may show exponential behaviour are believed to be atypical. More sophisticated algorithms, such as the Martelli and Montanari algorithm, offer polynomial complexity but are harder to implement.

Very little is known about the practical perfomance of unification algorithms in theorem provers: previous case studies have been conducted on small numbers of artificially chosen problem and compared term-to-term unification while the best theorem provers perform set-of-terms-to-term unification using term indexing.

To evaluate the performance of unification in the context of term indexing, we made large-scale experiments over the TPTP library containing thousands of problems using the COMPIT methodology. Our results confirm that the Robinson algorithm is the most efficient one in practice. They also reveal main sources of inefficiency in other algorithms. We present these results and discuss various modification of unification algorithms.

1 Introduction

Unification is one of the key algorithms used in implementing theorem provers. It is used on atoms in the resolution and factoring inference rules and on terms in the equality resolution, equality factoring and superposition inference rules. The performance of a theorem prover crucially depends on the efficient implementation of several key algorithms, including unification.

To achieve efficiency, theorem provers normally implement unification and other important operations using *term indexing*, see [11,9]. Given a set L of *indexed terms*, and a term t (called the *query term*), we have to retrieve the subset M of L that consists of the terms l unifiable with t. The retrieval of terms is interleaved with insertion of terms to L and deletion of them from L. Indexes in theorem provers frequently store 10^5–10^6 complex terms and are highly dynamic since insertion and deletion of terms occur frequently. Our paper is the first ever study of unification algorithms in the context of term indexing.

The structure of this paper is the following. Section 2 introduces the unification problem, the notion of inline and post occurs checks and several unification algorithms. Section 3 presents implementation details of terms and relevant algorithms in the theorem prover Vampire [12], explains the methodology we used to measure the performance of the unification retrieval, and presents and analyses our results. To this end,

B. Mertsching, M. Hund, and Z. Aziz (Eds.): KI 2009, LNAI 5803, pp. 435–443, 2009.

we measure the performance of four unification algorithms on hundreds of millions of term pairs obtained by running Vampire on the TPTP problem library.

Section 4 discusses related work and contains the summary of this work.

Due to the page limit, we omit many technical details. They can be found in the full version of this paper available at `http://www.cs.man.ac.uk/~hoderk/ubench/unification_full.pdf`

2 Unification Algorithms

A *unifier* of terms s and t is a substitution σ such that $s\sigma = t\sigma$. *A most general unifier* of two terms is their unifier σ such that for any other unifier τ of these two terms there exist a substitution ρ such that $\tau = \rho\sigma$. If two terms have a unifier, they also have a *most general unifier*, or simply *mgu*, which is unique modulo variable renaming. The *unification problem* is the task of finding a most general unifier of two terms.

For all existing unification algorithms, there are three possible outcomes of unification of terms s and t. It can either succeed, so that the terms are unifiable. It can fail due to a *symbol mismatch*, which means that at some point we have to unify two terms $s' = f(s_1, \ldots, s_m)$ and $t' = g(t_1, \ldots, t_n)$ such that f and g are two different function symbols. Lastly, it can fail on the *occurs check*, when we have to unify a variable x with a non-variable term containing x.

Unification algorithms can either perform occurs checks as soon as a variable has to be unified with a non-variable term or postpone all occurs checks to the end. We call occurs checks of the first kind *inline* and of the second kind *post* occurs checks.

When we perform unification term-to-term, the post occurs check seems to perform well, also somehow confirmed by experimental results in [4]. However, when we retrieve unifiers from an index, we do not build them at once. Instead, we build them incrementally as we descend down the tree perfoming *incremental unification*. In this case, we still have to ensure that there is no occurs check failure. It brings no additional cost to algorithms performing inline occurs check, but for post occurs check algorithms it means that the same occurs check labour may have to be performed unnecessarily over and over again. On the other hand, postponing occurs check may result in a (cheap) failure on comparing function symbols. Our results in Section 3 confirm that algorithms using the inline occurs check outperform those based on the post occurs check.

In the rest of this paper, x, y will denote variables, f, g different function symbols, and s, t, u, v terms. We consider constants as function symbols of arity 0. All our algorithm will compute *triangle form* of a unifier. This means a substitution σ such that some power $\theta = \sigma^n$ of σ is a unifier and $\sigma^{n+1} = \sigma^n$. We will denote such θ as σ^*. When two terms have an exponential size mgu, it has a polynomial-size triangle form.

The Robinson Algorithm ROB. This is a is a simple unification algorithm [13] with the worst-case exponential time complexity. It starts with an empty substitution σ and a stack of term pairs S that is initialized to contain a single pair (s_0, t_0) of terms s_0 and t_0 that are to be unified. At each step we remove a term pair (s, t) from the stack and do the following.

1. If s is a variable and $s\sigma \neq s$, the pair $(s\sigma, t)$ is put on the stack S; and similar for t instead of s.

2. Otherwise, if s is a variable and $s = t$, do nothing.
3. Otherwise, if s is a variable, an occurs check is performed to see if s is a proper subterm of $t\sigma^*$. If so, the unification fails, otherwise we extend σ by $\sigma(s) = t$.
4. Otherwise, if $s = f(s_1, \ldots, s_n)$ and $t = f(t_1, \ldots, t_n)$ for some function symbol f, the pairs $(s_1, t_1), \ldots, (s_n, t_n)$ are put on the stack S.
5. Otherwise the unification fails.

When there is no pair left on the stack, σ^* is an mgu of s_0 and t_0. The occurs check is performed before each variable binding, which makes ROB an inline occurs check algorithm.

The Martelli-Montanari Algorithm MM. We call a set of terms a *multi-equation.* We say that two multi-equations M_1 and M_2 can be *merged*, if there is a variable $x \in M_1 \cap M_2$. The merge operation then replaces M_1 and M_2 by $M = M_1 \cup M_2$. A set of multi-equations is said to be in *solved form*, if every multi-equation in this set contains at most one non-variable term and no multi-equations in the set can be merged.

Let us inductively define the notion of *weakly unifiable terms s, t with the disagreement set E*, where E is a set of multi-equations.

1. If $s = t$ then s and t are weakly unifiable with the empty disagreement set.
2. Otherwise, if either s or t is a variable, then s and t are weakly unifiable with the disagreement set $\{\{s, t\}\}$.
3. Otherwise, if $s = f(s_1, \ldots, s_n)$, $t = f(t_1, \ldots, t_n)$ and for all $i = 1, \ldots, n$ the terms s_i and t_i are weakly unifiable with the disagreement set E_i, then s and t are weakly unifiable with the disagreement set $E_1 \cup \ldots \cup E_n$.
4. In all other cases s and t are not weakly unifiable.

It is not hard to argue that weak unifiability is a necessary condition for unifiability.

The Martelli-Montanari algorithm unifying terms s_0 and t_0 maintains a set of multi-equations $\mathcal{M}$, initially equal to $\{\{s_0, t_0\}\}$. Until $\mathcal{M}$ is in solved form, it merges all multi-equations in $\mathcal{M}$ that can be merged, and for each multi-equation $M \in \mathcal{M}$ containing two non-variable terms s and t, if s and t are weakly unifiable with the disagreement set E, we set $\mathcal{M} := (\mathcal{M} \setminus M) \cup \{M \setminus \{t\}\} \cup E$. If they are not weakly unifiable, the unification fails.

When $\mathcal{M}$ is in solved form, an occurs check is performed. A directed graph G is built, so that its vertices correspond to multi-equations in $\mathcal{M}$ containing non-variable terms, and an edge is going from vertex M_1 to M_2 iff the non-variable term in M_1 contains a variable in M_2. The occurs check is succesful if the graph G is acyclic. When the occurs check succeeds, one can extract the triangle form of an mgu from $\mathcal{M}$. For a proof of correctness and termination in almost linear time, see [8].

In our implementation of the algorithm, we maintain and merge the variable parts of multi-equations using the union-find algorithm [16], and check that the graph G is acyclic using the topological sort algorithm [7]. The occurs check is performed as the last step of the algorithm, which makes MM a post occurs check algorithm.

The Escalada-Ghallab Algorithm EG. In order to examine a post occurs check algorithm that aims to be practically efficient, we implemented this algorithm presented in

[4]. To make the algorithm competitive with inline occurs check algorithms on the incremental unification problem, we made the EG occurs check incremental. The details can be found in the full version of this paper.

PROB. Inspired by our experiments described below we implemented a modification PROB of the Robinson algorithm having polynomial worst-case time complexity. It provides an easy-to-implement polynomial-time alternative to ROB. In PROB, we keep track of term pairs that previously occurred in the stack S. When we encounter such a pair again, we simply skip it. In the occurs-check routine, we similarly maintain a set of bound variables that have already been checked or are scheduled for checking. Such variables are then skipped. In the implementation we do not keep track of pairs that contain an unbound variable at the top. Practical results have shown that this happens frequently and that the cost of keeping track of such pairs does not pay off.

We modified all the above mentioned algorithms to work on the *substitution tree index*. Due to a lack of space we simply refer the reader to [11] for their description. We also had to modify algorithms to make them *incremental* so that the computed unifier can be refined to unify also a new pair of terms. These incremental algorithms can be implemented to retrieve unifiable terms from a substitution tree as follows. We traverse the tree depth-first, left-to-right. When we move down the tree to a node containing a substitution $x = t$, we extend the currently computed substitution to be also a unifier of (x, t). When we return to a previously visited node, we restore the previous substitution and, in the case of MM, the previous value of $\mathcal{M}$.

3 Implementation and Experiments

We implemented four algorithms for retrieval of unifiers, corresponding to the unification algorithms of Section 2. In this section we describe the data structures and algorithms used in the new version of Vampire [12].

We use shared Prolog terms to implement terms and literals. In Prolog, non-variable terms are normally implemented as a contiguous piece of memory consisting of some representation of the top symbol followed by a sequence of pointers to its subterms (actually, in the reverse order). We add to this representation *sharing* so that the same term is never stored twice. Besides conserving memory, this representation allows for constant-time equality checking. Another difference with Prolog terms is that, when an argument is a variable, Prolog stores a pointer pointing to itself, while we store the variable number.

When performing an inference on two different clauses (and in some cases even on two copies of the same clause), we must consider their variables as disjoint, although some variables may be the same, that is, have the same number. To deal with this, we use the idea of *variable banks* used in several theorem provers, including Waldmeister [6], E [15] and Vampire [12].

Terms whose variables should be disjoint are assigned different *bank indexes*. One can view it as adding a subscript to all variables in a term — instead of terms $f(x, y)$ and $f(y, a)$, we will work with terms $f(x_0, y_0)$ and $f(y_1, a)$. In practice it means that when it is unclear from which clause a term origins, we store a pair of the term and a bank index instead of just the term.

Substitutions that store unifiers are stored as maps from pairs (variable number, bank index) to pairs (term pointer, bank index). Those maps are implemented as double hash tables[5] with fill-up coefficient 0.7 using two hash functions. The first one is a trivial function that just returns the variable number increased by a multiple of the bank index. This function does not give randomly distributed results (which is usually a requirement for a hash function), but is very cheap to evaluate. The second hash function is a variant of FNV. It gives uniformly distributed outputs, but is also more expensive to evaluate. It is, however, evaluated only if there is a collision of results of the first function.

The union-find data structures of EG and MM are implemented on top of these maps. In EG, we use path compression as described in [4]. In MM, it turned out that the path compression led to lower performance, so it was omitted.

Benchmarking Methodology. Our benchmarking method is COMPIT [9]. First, we log all index-related operations (insertion, deletion and retrieval) performed by a first-order theorem prover. This way we obtain a description of all interactions of the prover with the index and it is possible to reproduce the indexing process without having to run the prover itself. Moreover, benchmarks generated this way can be used by other implementations, including those not based on substitution trees, and we welcome comparing our implementation of unification with other implementations.

The main difference of our benchmarking from the one presented in [9] is that instead of just success/failure, we record the number of terms unifiable with the query term. This reflects the use of unification in theorem provers, since it is used for generating inferences, so that *all* such inferences with a given clause have to be performed.

We created two different instrumentations of the development version of the Vampire prover, which used the DISCOUNT [2] saturation algorithm. The first instrumentation recorded operations on the unification index of selected literals of active clauses (the *resolution index*). The second one recorded operations on the unification index of all non-variable subterms of selected literals of active clauses (the *superposition index*).

Both of these instrumentations were run on several hundred randomly selected TPTP problems with the time limit of 300s to produce benchmark data[1]. In the end we evaluated indexing algorithms on all of these benchmarks and then removed those that ran in less than 50ms, as such data can be overly affected by noise and are hardly interesting in general. This left us with about 40 percent of the original number of benchmarks[2], namely 377 resolution and 388 superposition index benchmarks.

Results and Analysis. We compared the algorithms described above. Our original conjecture was that MM would perform comparably to ROB on most problems and be significantly better on some problems, due to its linear complexity. When this conjecture showed to be false, we added the PROB and EG algorithms, in order to find a well-performing polynomial algorithm.

[1] Recording could terminate earlier in the case the problem was proved. We did not make any distinction between benchmarks from successful and unsuccessful runs.

[2] This number does not seem to be that small, when we realise that many problems are proved in no more than a few seconds. Also note that in problems without equality there are no queries to the superposition index at all.

On a small number of problems (about 15% of the superposition benchmarks and none of the resolution ones) the performance of ROB and MM was approximately the same ($\pm 10\%$), but on most of the problems MM was significantly slower. On the average, it was almost 6 times slower on the superposition benchmarks and about 7 times slower on the resolution benchmarks. On 3% of the superposition benchmarks and 5% of the resolution benchmarks, MM was more than 20 times slower.

The only case where MM was superior was in a handcrafted problem designed to make ROB behave exponentially containing the following two clauses:

$$p(x_0, f(x_1, x_1), x_1, f(x_2, x_2), x_2, \ldots, x_9, f(x_{10}, x_{10}));$$
$$\neg p(f(y_0, y_0), y_0, f(y_1, y_1), y_1, \ldots, y_9, f(y_{10}, y_{10}), y_{11}).$$

This problem was solved in no time by MM and PROB and in about 15 seconds by ROB.

In general, PROB has shown about the same performance as ROB. It was only about 1% slower, so it can provide a good alternative to ROB if we want to avoid the exponential worst-case complexity of the ROB. EG did not perform as bad as MM, but it was still on the average over 30% slower than ROB.

Table 1 summarises the performance of the algorithms on resolution and superposition benchmarks. The first two benchmarks in each group are those on which MM performed best and worst relatively to ROB, others are benchmarks from randomly selected problems. In the table, *term size* means the number of symbols in the term; *average result count* is the average number of results retrieved by a query, and *query fail rate* is the ratio of queries that retrieved no results. The last three numbers show the use of substitutions in the index—the number of successful unification attempts, failures due to symbol mismatch, and failures due to an inline occurs check.

To determine the reason for the poor performance of MM, we used a code profiler on benchmarks displaying its worst performance. It turned out that over 90% of the measured time was being spent on performing the occurs checks, most of it actually on checking graph acyclicity. It also showed that the vast majority of unification requests were just unifying an unbound variable with a term. Based on this, we tested an algorithm performing the PROB occurs checks instead of the MM ones after such unifications. This caused the worst-case complexity to be $O(n^2)$ but improved the average performance of MM from about 600% worse than ROB to just about 30% worse.

Our results also show empirically that the source of terms matters. For example, the MM algorithm gives relatively better performance on the superposition index, than it does on the resolution index. Other papers do not make such a distinction, for example [4] just states "theorem proving" as the source of the benchmarked term pairs.

4 Related Work and Summary

There is another comparison of ROB and MM in [1], which presents a proof that on a certain random distribution of terms the expected average (according to some measure) number of steps of ROB is constant, while the expected number of MM steps is linear in the size of terms. However, the use of random terms is essential for their results. A practical comparison of ROB, MM and EG is undertaken in [4], but this comparison is

Table 1. ROB and MM comparison on selected benchmarks

Problem	Time [ms]				Relative		All			Maximal	Avg. term size		Avg	Query	Unification outcomes		
	MM	EG	ROB	PROB	MM	EG	ops	Ins	Dels	index size	indexed	query	res cnt	fail rate	success	mism.	o.c. fail
Resolution index benchmarks																	
AGT022+2	2921	2831	2285	2303	1.3	1.2	175346	87673	0	87673	3.2	3.2	134.7	0.2	1275420	16392	0
SET317-6	51997	2600	1958	1915	26.6	1.3	68338	33440	1458	31982	10.6	10.6	52.2	0.0	2025401	5079	63
ALG229+1	1853	720	474	497	3.9	1.5	23861	11447	967	10480	7.8	7.8	54.8	0.5	420047	461	15128
ALG230+3	1490	1046	752	711	2.0	1.4	48025	23912	201	23711	2.9	2.9	40.3	0.4	768620	380	1569
CAT028+2	675	399	295	306	2.3	1.4	17989	8752	485	8267	3.6	3.6	47.5	0.2	302026	185	899
CAT029+1	3065	520	400	417	7.7	1.3	12897	6426	45	6381	11.6	11.6	114.1	0.3	498626	3	155
FLD003-1	6058	1210	941	949	6.4	1.3	14384	7187	9	7178	7.2	7.2	312.2	0.0	1247011	187	0
FLD091-3	4626	890	690	736	6.7	1.3	23037	11505	26	11479	7.7	7.7	239.4	0.0	798127	97	0
LAT289+2	1331	850	625	629	2.1	1.4	32447	16076	295	15781	3.2	3.2	36.2	0.3	608904	2678	1585
LAT335+3	1482	1002	730	742	2.0	1.4	42330	21044	242	20802	3.0	3.0	44.1	0.3	756292	252	1930
LCL563+1	4972	564	445	431	11.2	1.3	5899	2868	163	2705	14.3	14.3	135.2	0.2	441681	470	7
NUM060-1	9658	1480	1098	1104	8.8	1.3	101608	49138	3331	45807	9.8	9.8	38.4	0.0	1001317	16940	242
SET170-6	48152	2495	1864	1830	25.8	1.3	71396	35332	731	34601	10.6	10.6	49.8	0.0	1915322	5694	63
SET254-6	13807	1759	1259	1260	11.0	1.4	78914	38729	1455	37274	10.6	10.6	44.0	0.0	1244861	9979	63
SET273-6	13833	1752	1261	1268	11.0	1.4	78680	38643	1393	37250	10.7	10.7	44.0	0.0	1243272	9605	63
SET288-6	51151	2606	1946	1924	26.3	1.4	68348	33445	1458	31987	10.6	10.6	52.2	0.0	2025514	5079	63
SEU388+1	3641	821	610	603	6.0	1.3	27895	13911	73	13838	6.4	6.4	103.9	0.1	683318	10	2792
TOP031+3	1664	1089	821	831	2.0	1.3	42771	21273	225	21048	3.0	3.0	43.2	0.3	823743	3809	1808
Superposition index benchmarks																	
SEU388+1	55	54	57	53	0.96	0.9	63410	63194	200	62994	2.7	4.2	2.9	0.4	38	0	0
SET288-6	48717	2484	1808	1824	26.95	1.4	71228	35279	669	34610	10.6	10.6	49.8	0.0	1913644	5336	63
ALG229+1	80	75	71	72	1.13	1.1	63466	56466	6673	49819	4.2	10.0	4.1	0.7	1916	127	0
ALG230+3	1489	1058	764	765	1.95	1.4	49787	24780	227	24553	2.9	2.9	37.0	0.4	744009	391	1639
CAT028+2	719	432	314	321	2.29	1.4	18885	9368	149	9219	3.6	3.6	47.2	0.2	322684	238	872
CAT029+1	3073	523	387	419	7.94	1.4	12917	6436	45	6391	11.6	11.6	114.3	0.3	498651	3	155
FLD003-1	6157	1181	1010	983	6.10	1.2	14384	7187	9	7178	7.2	7.2	312.2	0.0	1246881	187	0
FLD091-3	4655	890	733	718	6.35	1.2	23003	11488	26	11462	7.7	7.7	239.7	0.0	798205	105	0
LAT289+2	1334	851	619	642	2.16	1.4	32352	16106	140	15966	3.2	3.2	36.2	0.3	605700	2753	1583
LAT335+3	1728	1150	855	862	2.02	1.3	43904	21829	246	21583	3.0	3.0	42.3	0.3	842139	4522	1797
LCL563+1	5711	636	489	503	11.68	1.3	6158	3005	148	2857	14.5	14.5	142.5	0.2	496902	497	7
NUM060-1	8820	1457	1098	1077	8.03	1.3	118475	57574	3326	54248	9.6	9.6	31.1	0.0	950386	23750	244
SET170-6	13700	1709	1255	1256	10.92	1.4	78669	38642	1385	37257	10.7	10.7	44.0	0.0	1243409	9604	63
SET254-6	13809	1725	1279	1262	10.80	1.3	78904	38724	1455	37269	10.6	10.6	44.0	0.0	1245069	9980	63
SET273-6	51081	2569	1909	1932	26.76	1.3	68156	33380	1396	31984	10.6	10.6	52.3	0.0	2023900	4718	63

not of much use for us since it is only done on a small number of examples (less than 100 term pairs all together), many of them hand-crafted, and uses no term indexing.

There are several unification algorithms not considered here, which we did not evaluate for the reasons explained below. The Paterson algorithm [10] has linear asymptotic time complexity, but according to [4], this benefit is redeemed by the use of complex data structures to the extent that it is mainly of theoretical interest. The Corbin-Bidoit algorithm [3] might look promising, as it uses an inline occurs check, but it requires input terms to be dags modified during the run of the algorithm which we cannot do because of term sharing. The Ruzicka-Privara algorithm [14], which is an improvement of the Corbin-Bidoit one, suffers from the same problem, and moreover uses a post occurs check.

Summary. We studied the behaviour, in the framework of term indexing, of four different unification algorithms: the exponential time Robinson algorithm, the almost linear time Martelli-Montanari and Escalada-Ghallab algorithms, and a polynomial-time modification of the Robinson algorithm. To this end, we used the appropriately modified COMPIT method [9] on a substitution tree index. The evaluation has shown that the Martelli-Montanari and Escalada-Ghallab algorithms, although asymptotically superior in the worst case, in practice behave significantly worse than the other two. The main cause of this behaviour was the occurs-check that verified acyclicity of the substitution. On the other hand, the PROB algorithm turned out to perform comparably to the Robinson one, while having the advantage of being polynomial in the worst case.

The benchmarks are available at `http://www.cs.man.ac.uk/~hoderk/`

References

1. Albert, L., Casas, R., Fages, F., Torrecillas, A., Zimmermann, P.: Average case analysis of unification algorithms. In: Jantzen, M., Choffrut, C. (eds.) STACS 1991. LNCS, vol. 480, pp. 196–213. Springer, Heidelberg (1991)
2. Avenhaus, J., Denzinger, J., Fuchs, M.: Discount: A system for distributed equational deduction. In: Hsiang, J. (ed.) RTA 1995. LNCS, vol. 914, pp. 397–402. Springer, Heidelberg (1995)
3. Corbin, J., Bidoit, M.: A rehabilitation of Robinson's unification algorithm. In: IFIP Congress, pp. 909–914 (1983)
4. Escalada-Imaz, G., Ghallab, M.: A practically efficient and almost linear unification algorithm. Artif. Intell. 36(2), 249–263 (1988)
5. Guibas, L.J., Szemerédi, E.: The analysis of double hashing. J. Comput. Syst. Sci. 16(2), 226–274 (1978)
6. Hillenbrand, T., Buch, A., Vogt, R., Löchner, B.: Waldmeister: High-performance equational deduction. Journal of Automated Reasoning 18(2), 265–270 (1997)
7. Kahn, A.B.: Topological sorting of large networks. Commun. ACM 5(11), 558–562 (1962)
8. Martelli, A., Montanari, U.: An efficient unification algorithm. ACM Trans. Program. Lang. Syst. 4(2), 258–282 (1982)
9. Nieuwenhuis, R., Hillenbrand, T., Riazanov, A., Voronkov, A.: On the evaluation of indexing techniques for theorem proving. In: Goré, R.P., Leitsch, A., Nipkow, T. (eds.) IJCAR 2001. LNCS (LNAI), vol. 2083, pp. 257–271. Springer, Heidelberg (2001)

10. Paterson, M.S., Wegman, M.N.: Linear unification. In: STOC 1976, pp. 181–186. ACM, New York (1976)
11. Ramakrishnan, I.V., Sekar, R.C., Voronkov, A.: Term indexing. In: Robinson, A., Voronkov, A. (eds.) Handbook of Automated Reasoning, pp. 1853–1964 (2001)
12. Riazanov, A., Voronkov, A.: The design and implementation of Vampire. AI Commun. 15(2,3), 91–110 (2002)
13. Robinson, J.A.: A machine-oriented logic based on the resolution principle. J. ACM 12(1), 23–41 (1965)
14. Ruzicka, P., Prívara, I.: An almost linear robinson unification algorithm. In: Koubek, V., Janiga, L., Chytil, M.P. (eds.) MFCS 1988. LNCS, vol. 324, pp. 501–511. Springer, Heidelberg (1988)
15. Schulz, S.: E — a brainiac theorem prover 15(2-3), 111–126 (2002)
16. Tarjan, R.E.: Efficiency of a good but not linear set union algorithm. J. ACM 22(2), 215–225 (1975)

Atomic Metadeduction

Serge Autexier and Dominik Dietrich

German Research Centre for Artificial Intelligence (DFKI GmbH), Bremen, Germany

Abstract. We present an extension of the first-order logic sequent calculus SK that allows us to systematically add inference rules derived from arbitrary axioms, definitions, theorems, as well as local hypotheses – collectively called assertions. Each derived deduction rule represents a pattern of larger SK-derivations corresponding to the use of that assertion. The idea of metadeduction is to get shorter and more concise formal proofs by allowing the replacement of any assertion in the antecedent of a sequent by derived deduction rules that are available locally for proving that sequent. We prove the soundness and completeness for atomic metadeduction, which builds upon a permutability property for the underlying sequent calculus SK with liberalized δ^{++}-rule.

1 Introduction

In spite of almost four decades of research on automated theorem proving, mainly theorems considered easy by human standards can be proved fully automatically without human assistance. Many theorems still require a considerable amount of user interaction, and will require it for the foreseeable future. Hence, there is a need that proofs are presented and ideally constructed in a form that suits human users in order to provide an effective guidance.

To come close to the style of proofs as done by humans, Huang [9] introduced the so-called *assertion-level*, where individual proof steps are justified by axioms, definitions, or theorems, or even above at the so-called *proof level*, such as "by analogy". The idea of the assertion-level is, for instance, that given the facts $U \subset V$ and $V \subset W$ we can prove $U \subset W$ directly using the assertion:

$$\subset_{Trans}\colon \forall U.\forall V.\forall W.U \subset V \wedge V \subset W \Rightarrow U \subset W$$

An assertion level step usually subsumes several deduction steps in standard calculi, say the classical sequent calculus [8]. To use an assertion in the classical sequent calculus it must be present in the antecedent of the sequent and be processed by means of decomposition rules, usually leading to new branches in the derivation tree. Some of these branches can be closed by means of the axiom rule which correspond to "using" that assertion on known facts or goals.

Huang followed the approach of a human oriented proof style by hiding decomposition steps once detected by abstracting them to an assertion application. Since he was mainly concerned with using the abstract representation for proof presentation in natural language [9,7] there was no proof theoretic foundation for the assertion level. Hence, assertion level proofs could only be checked once

B. Mertsching, M. Hund, and Z. Aziz (Eds.): KI 2009, LNAI 5803, pp. 444–451, 2009.

expanded to the underlying calculus, and the actual proof had still to be found at the calculus level and only proof parts of a specific form could be abstracted.

More recently, work was devoted to analyze the assertion-level proof theoretically: [11] defined *supernatural deduction* that extends the natural deduction calculus by inference rules derived from assertions and showed its soundness and completeness. This work was extended to the classical sequent calculus in [5] to obtain the *superdeduction calculus.* However, both approaches are restricted to closed, universally quantified equations or equivalences and the premises and conclusions of the derived inference rules are restricted to atomic formulas. In this paper we extend that work to derive and use inference rules from arbitrary formulas, including non-closed formulas such as local hypotheses, but still allow only for atomic premises and conclusions. Hence the name *atomic metadeduction.* Compared to [5] we use a different meta-theory based on a sequent calculus with a liberalized δ-rule (δ^{+^+}, [4]) which enables the necessary proof transformations to establish soundness and completeness.

The paper is organized as follows: In Sec. 2 we present a minimal sequent calculus for first-order logic with liberalized δ^{+^+}-rule and give two permutability results due to the use of that rule. In Sec. 3 we present the technique to compute derived inference rules from arbitrary assertions. In Sec. 4 we prove the soundness and completeness of the calculus using derived inference rules, define the metadeduction calculus and prove that the rule that allows us to move assertions from sequents to the inference level is invertible, i.e. we do not lose provability by applying it. We conclude the paper by summarizing the main results and comparing it to related work in Sec. 5.

2 Sequent Calculus with Liberalized Delta Rule

The context of this work is first-order logic. First-order terms and atomic formulas are build as usual inductively over from functions $\mathcal{F}$, predicates $\mathcal{P}$ and variables $\mathcal{V}$. The *formulas* are then build inductively from falsity $\bot$, atomic formulas, the connective $\Rightarrow$ and universal quantification $\forall$. For the formal parts of this paper we use the restricted set of connectives, but also the other connectives for sake of readability. Finally, *syntactic equality* on formulas is modulo renaming of bound variables (α-renaming) and denoted by $=$. Our notion of *substitution* is standard: A substitution is a function $\sigma : \mathcal{V} \rightarrow T(\Sigma, \mathcal{V})$ which is the identity but for finitely many $x \in \mathcal{V}$ and whose homomorphic extension to terms and formulas is idempotent. We use $t\sigma$ to denote the application of σ to t.

The sequent calculus for first-order logic is given in Fig. 1 is mostly standard. The specificities are: (i) The axiom rule AXIOM is restricted to atomic formulas; (ii) the $\forall_L$-rule allows us to substitute terms with free variables to postpone the choice of instances; (iii) there is a substitution rule SUBST to substitute free variables globally in the derivation tree; the idempotency of substitutions ensures the admissibility of the substitution; (iv) the $\forall_R$-rule uses Skolemization with an optimization regarding the used Skokem-function: Standard Skolemization requires that the Skolem-function f is new wrt. the *whole* sequent and takes

$$\textsc{Axiom}\ \frac{}{\Gamma, A \vdash A, \Delta}\ A \text{ atomic} \qquad \bot_L\ \frac{}{\Gamma, \bot \vdash \Delta} \qquad \textsc{Contr}_L\ \frac{\Gamma, F, F \vdash \Delta}{\Gamma, F \vdash \Delta}$$

$$\Rightarrow_L \frac{\Gamma \vdash F, \Delta \quad \Gamma, G \vdash \Delta}{\Gamma, F \Rightarrow G \vdash \Delta} \qquad \Rightarrow_R \frac{\Gamma, F \vdash G, \Delta}{\Gamma \vdash F \Rightarrow G, \Delta} \qquad \forall_L \frac{\Gamma, F[t/x] \vdash \Delta}{\Gamma, \forall x.F \vdash \Delta}\ \begin{array}{l} t \text{ term that} \\ \text{may contain} \\ \text{free variables} \end{array}$$

$$\forall_R \frac{\Gamma \vdash F[f_{[\forall x.F]}(\overrightarrow{Z})/x], \Delta}{\Gamma \vdash \forall x.F, \Delta}\ \overrightarrow{Z} = \mathcal{FV}(\forall x.F) \qquad \textsc{Subst}\ \nabla_x \overset{[t/x]}{\longrightarrow} \nabla_t \text{ if } [t/x] \text{ substitution}$$

Fig. 1. The Sequent Calculus SK with liberalized δ^{++}-rule and Skolemization

as arguments *all* variables that occur free in the sequent. Contrary to that we use the even more liberalized δ^{++} approach [4]: when eliminating the universal quantification for some succedent formula $\forall x.F$, first it allows us to take the *same* Skolem function for all formulas that are equal modulo α-renaming to $\forall x.F$; such Skolem-functions are denoted by $f_{[\forall x.F]}$ where $[\forall x.F]$ denotes the set of all formulas equal to $\forall x.F$. Secondly, the arguments to the Skolem function are only those variables that actually occur freely in $\forall x.F$. The result of using the δ^{++}-approach is that $\forall_R$ for some sequent $\Gamma, \forall x.F \vdash \Delta$ is invariant with respect to different Γ and Δ, and that it allows for shorter and also more natural proofs (see [12] for a survey and [13] for soundness and completeness proofs[1]). The CUT-rule $\frac{\Gamma \vdash F, \Delta \quad \Gamma, F \vdash \Delta}{\Gamma \vdash \Delta}$ is admissible for this sequent calculus, that is every proof using CUT can be transformed into a proof without CUT.

For every rule, any formula occurring in the conclusion but not in any of the premises is a *principal formula* of that rule[2], while any formula occurring in some premise but not in the conclusion is a *side formula*. All rules have the subformula property, that is all side formulas are subformulas of the principal formulas (or instances thereof).

Permutability. The use of the liberalized δ^{++}-rule ($\forall_R$) allows for a specific permutability result on proof steps. The observation is that applying it to $\Gamma \vdash \forall x.F, \Delta$ the used Skolem function and variables used as arguments only depend on the formula $\forall x.F$ and variables that occur free in it. Hence, if we always introduce new free variables in $\forall_L$-applications and postpone the application of SUBST to the end of any proof attempt (only followed by AXIOM-rule applications), the chosen Skolem-functions and their arguments only depend on F and any previous $\forall_L$-applications to a formula of which $\forall x.F$ is a subformula. As a result the $\forall_R$-step with principal formula $\forall x.F$ can be permuted with any proof step having a principal formula that is "independent" from $\forall x.F$. More generally, any two successive rule applications with principal formulas that are "independent" of each other can be permuted (cf. Lemma 2.3 in [3])).

Given two successive sequent rule applications of respective principal formulas F and G. F and G are *independent* from each other, if G is not a side formula for F, i.e., not a subformula of F. Otherwise we say that G is a subformula of F.

[1] [13] who uses explicit variable conditions instead of Skolemization and substitution, but the same Eigenvariable for all syntactically equal formulas.

[2] For CONTR$_L$, F is the principal formula and its copy in the premise a side formula; for AXIOM both A are principal formulas and so is $\bot$ in the rule $\bot_L$.

The notion of independent formulas serves to define rule applications that are irrelevant in a proof. A rule application R in some proof is *irrelevant*, iff none of the subformulas of R's principal formula is an active partner in an AXIOM- or $\bot_L$-rule application in that proof. A proof without irrelevant rule applications is called *concise* and it is folklore to show that any proof can be turned into a concise proof by removing all irrelevant proof steps. Throughout the rest of this paper we will assume concise proofs.

A consequence of being able to permute proof steps working on independent principal formulas is that we can group together in any SK* derivation all rules working on a principal formula and all its subformulas, where SK* denotes SK without the SUBST rule and the restriction of always introducing new free variables in $\forall_L$-steps. To formalize that observation, we introduce the concept of A-active derivations to denote those derivations where only rule applications with principal formula A or one of its subformulas are applied.

Definition 1 (A-Active/Passive Derivations). *Let L, L' be multisets of formulas and D a derivation (possibly with open goals) for the sequent $\Gamma, L \vdash L', \Delta$. The derivation D is* (L, L')-active, *if it contains only calculus rules having a formula from L or L' or one of their subformulas as principal formula. Conversely, we say D is* (L, L')-passive *if it contains no calculus rule that has some formula from L or L' or one of their subformulas as principal formula. If $L = \{A\}$ and $L' = \emptyset$ (respectively if $L' = \{A\}$ and $L = \emptyset$) then we agree to say D is A-active if it is (L, L')-active and A-passive if it is (L, L')-passive.*

It holds that every A-active rule followed by A-passive derivations can be permuted to first have an A-passive derivation followed by applications of the A-active rule on the different open sequents (cf. Corollary 2.7 in [3]). Using that we can transform any SK* derivation into one composed of A-active derivations (cf. Lemma 2.8 in [3]). However, we can do even better and move in any A-active derivation all applications of CONTR$_L$ on A or one of its subformulas downwards to be applied on A already. To formalize this observation, we introduce the notion of *contraction-free* derivations which are derivations without applications of the contraction rule (cf. Lemma 2.9 in [3]).

3 Derived Sequent Rules

In this section we present the technique to obtain derived inference rules for an arbitrary formula. As motivation consider the proof of the following simple statement about natural numbers $\forall n, m.n < m \Rightarrow \exists u.n + u = m$ under the assertions $\forall x.x = 0 \vee \exists y.x = s(y)$, $\forall n, m.s(n) + m = s(n + m)$, $\forall n, m.n < m \Rightarrow s(n) < s(m)$, $\forall n.n < 0 \Rightarrow \bot$, and $\forall n, m.s(n) = s(m) \Rightarrow n = m$. During the proof, which is done by induction over n outside of the logic, many steps consists of the application of one of these assertions by decomposing and instantiating it in an appropriate way, thereby yielding several branches in the derivation tree, some of which can be closed using available facts and the axiom rule. Consider for

$$
\dfrac{\dfrac{\dfrac{\text{AXIOM}\ \dfrac{①}{\Gamma \vdash u < v \vdash \Delta} \qquad \dfrac{②}{\Gamma, s(u) < s(v) \vdash \Delta}\ \text{AXIOM}}{\Gamma, u < v \Rightarrow s(u) < s(v), u < v \vdash \Delta}\ \Rightarrow_L}{\Gamma, \forall m.u < m \Rightarrow < m \vdash \Delta}\ \forall_L}{\Gamma, \forall n, m.n < m \Rightarrow s(n) < s(m) \vdash \Delta}\ \forall_L
$$

Fig. 2. Example derivation

example the derivation in Fig. 2 derivation. It shows given the facts $\Gamma = \{u < v\}$ we can show $\Delta = \{s(u) < s(v)\}$.

The idea of atomic metadeduction is to allow the lifting of assertions, such as $\forall n, m.n < m \Rightarrow s(n) < s(m)$ and $u < v$ in the example above, to the level of inference rules at any time during the search. In general, there will be several possibilities to apply an assertion, and the number of the new branches created by the application of the assertion will depend on the available facts Γ and goals Δ. In the example above, we obtain the following possibilities:

$$
<① \ \frac{\Gamma, u < v, s(u) < s(v) \vdash \Delta}{\Gamma, u < v \vdash \Delta} \qquad <② \ \frac{\Gamma \vdash n < m, \Delta}{\Gamma \vdash s(n) < s(m), \Delta} \qquad <①② \ \frac{}{\Gamma, n < m \vdash s(n) < s(m), \Delta}
$$

For instance, if $\Delta = \{\}$ and $\Gamma = \{u < v\}$, then the axiom rule is no longer applicable in ① which gets a new open sequent. Note that this remains a valid derivation, if we add arbitrary formulas Γ' to the antecedent or Δ' to the succedent. We get a variety of these inferences depending on which application of axiom rules are enabled by filling the Γ and Δ; these rules all represent one possible application of the assertion. However, if there is not at least one axiom rule application, then we do not consider this as an application of the assertion (otherwise it would always be applicable); moreover, this derivation is somehow superfluous if none of the subformulas of the assertions is used in the proof. Skolem functions introduced by $\forall_R$-rules are always the same, which results from the use of the δ^{++} rule, where we use the same Skolem function for the same formulas. In the case of derived rules, these are always the subformulas of the assertion which are always the same.

$$
\dfrac{\dfrac{\dfrac{\vdots \qquad \dfrac{\dfrac{\dfrac{\dfrac{\dfrac{}{n < f_1(m), m = s(f_1(m)) \vdash_{H_1,H_2} n < f_1(m)}\ \text{AXIOM}}{n < f_1(m), m = s(f_1(m)) \vdash_{H_1,H_2} n + u = f_1(m)}\ \text{HYP1}}{n < f_1(m), m = s(f_1(m)) \vdash_{H_1,H_2} s(n+u) = s(f_1(m))}\ =_s}{s(n) < s(f_1(m)), m = s(f_1(m)) \vdash_{H_1,H_2} s(n+u) = s(f_1(m))}\ <_s}{s(n) < m, m = s(f_1(m)) \vdash_{H_1,H_2} \exists u.s(n) + u = m}\ \forall_R,=^*}{s(n) < m \vdash_{H_1,H_2} \exists u.s(n) + u = m}\ \text{NAT}}{\vdash_{H_1,H_2} s(n) < m \Rightarrow \exists u.s(n) + u = m}\ \Rightarrow_R}{\vdash_{H_1,H_2} \forall m.s(n) < m \Rightarrow \exists u.s(n) + u = m}\ \forall_R
$$

Fig. 3. Induction Step of the example

Fig. 3 shows the induction step of the example statement after lifting the induction hypothesis (which contains free variables), resulting in two new inferences H_1 and H_2, where f_1 stands for the Skolem constant $f_{\exists y.X=s(y)}$ and

$$\frac{\Gamma_1, H_1 \vdash H_2, \Delta_1 \quad \dots \quad \Gamma_n, H_1 \vdash H_2, \Delta_n}{H_1, F \vdash H_2} \qquad \frac{\Gamma_{i_1}, \sigma(H_1) \vdash \sigma(H_2), \Delta_{i_1} \quad \dots \quad \Gamma_{i_l}, \sigma(H_1) \vdash \sigma(H_2), \Delta_{i_l}}{\sigma(H_1), F \vdash \sigma(H_2)}$$

(a) **(b)**

Fig. 4. Intermediate Stages of the Computation of Derived Rules

$\Gamma = s(n) < m$. Systematizing the derivation of rules results in the following rule synthesis procedure:

Definition 2 (Derived Inference Rules). *Let SK0 denote the subset of SK* without the rule* CONTR$_L$. *Given a not necessarily closed formula F, we compute the derived rules from F as follows: Take the sequent $H_1, F \vdash H_2$, where H_1 and H_2 are place-holders for lists of formulas and apply exhaustively all rules from SK0. All obtained derivation trees are of the form in Fig. 4a Then enable one or more application of the* AXIOM *rule by instantiating H_1 and H_2 with atoms respectively from some Δ_i or some Γ_i which results in (multiple) derivations of the form in Fig. 4b, where σ is the respective instantiation of H_1 and H_2. For each of these trees we introduce the derived rule*

$$\text{BY F}\ \frac{\Gamma, \Gamma_{i_1}, \sigma(H_1) \vdash \sigma(H_2), \Delta_{i_1}, \Delta \qquad \dots \qquad \Gamma, \Gamma_{i_l}, \sigma(H_1) \vdash \sigma(H_2), \Delta_{i_l}, \Delta}{\Gamma, \sigma(H_1) \vdash \sigma(H_2), \Delta}$$

Using that we can derive for any, not necessarily closed formula F a set of derived so-called F-rules. Note that the derived rules are strongly interrelated; they can be divided in forward rules, working on antecedent formulas, and backward rules, working on the succedent and possibly on the antecedent. While the rules are in the spirit of Huang's assertion level and reflect all possibilities to apply an assertion – hence well suited for an interactive setting – they introduce redundancy in the search space if used without care in an automated setting.

4 The Metadeduction Calculus

We now formally define the metadeduction calculus: First, we define *theory sequent calculi* and prove their soundness and weak completeness. We then define the metadeduction calculus by adding a lifting rule that enables to replace assertions from the antecedent of some open goal sequent by corresponding inference rules in the meta-level theory of the sequent and prove, besides its soundness, its *inversion property*, that is we do not lose provability by this operation.

Definition 3 (Theory Sequent Calculus). *Let Th be a set of not necessarily closed formulas: Then we denote by $\Gamma \vdash_{Th} \Delta$ a* theory sequent wrt. *Th and we allow to write $\Gamma \vdash_{Th,F} \Delta$ to denote the sequent wrt. the theory Th augmented by the formula F. The theory sequent calculus consists of the sequent calculus rules of SK and for each theory sequent $\Gamma \vdash_{Th} \Delta$ of all rules derived from all formulas in Th. The* SUBST*-rule now affects the antecedent and succedent of sequents and the formulas in the attached theory Th (resp. the derived rules).*

We prove soundness of the theory sequent calculus by constructing from any proof for some sequent $\Gamma \vdash_{Th,F} \Delta$ using F-rules a proof for $\Gamma, F \vdash_{Th} \Delta$ not using F-rules. This allows us to eliminate step by step all theory formulas and end up in the classical sequent calculus (that is, $Th = \emptyset$).

Theorem 1 (Soundness). *For all sets of formulas Th, all formulas F and all proofs of $\Gamma \vdash_{Th,F} \Delta$ there exists a proof for $\Gamma, F \vdash_{Th} \Delta$ without F-rules.*

Conversely, we prove completeness by constructing from any proof of some sequent $\Gamma, Th \vdash \Delta$ a proof for $\Gamma \vdash_{Th} \Delta$. The completeness proof relies on an *self-derivability property* for derived rules, that is, if we lift an assertion F from the antecedent of our current open goal sequent to the calculus level then F is still derivable using F-rules. This allows us to reuse the proof of $\Gamma, Th \vdash \Delta$ using CUT (cf. [3], Sec. 4 for more details). Because of the use of CUT, we call the obtained result weak completeness. Future work is devoted to transform a given proof to an assertion proof without CUT using our permutability results.

Lemma 1 (Selfderivability). *For every formula F there is a proof of $\vdash_F F$.*

Theorem 2 (Weak Completeness). *For all sets of formulas Th and proofs Π for $\Gamma, Th \vdash \Delta$ there exists a proof Π_{Th} for $\Gamma \vdash_{Th} \Delta$ possibly using* CUT.

We further extend the theory sequent calculus by a rule to lift arbitrary assertions F from the antecedent of sequents to the calculus level at any stage of the proof and to apply the henceforth derived F-rules at any time: LIFT $\frac{\Gamma \vdash_{Th,F} \Delta}{\Gamma, F \vdash_{Th} \Delta}$. Due to Lemma 1 the LIFT-rule has the inversion property. We call the resulting calculus the *metadeduction*-calculus which soundness and completeness directly follows from the Theorems 1 and 2.

5 Conclusion and Related Work

In this paper we have presented atomic metadeduction as an extension to a first-order sequent calculus to systematically synthesize new derived inference rules from assertions at any time during the proof search. Using metadeduction, proofs can directly be constructed, checked, and presented at the assertion level. We have shown soundness, completeness of the theory sequent calculi and proved the inversion property of the LIFT-rule.

The idea of extending the natural deduction calculus or the sequent calculus by new deduction rules is not new [10,11,5] and was discussed in Sec. 1. Compared to that, we allow for derived rules from arbitrary, even non-closed assertions, which allows us to use intermediate facts as derived rules, for instance the induction hypthesis of an inductive proof or case conditions in a case analysis. However, in contrast to these works we have no cut admissibility result so far, which is a topic for future work. We expect to obtain similar results at least for the restricted fragment of closed equivalences of the form $P \Leftrightarrow Q$ where P is atomic – if not even for a larger fragment of formulas – by following ideas of [6].

Closely related are also focusing derivations [1] to eliminate inessential nondeterminism by alternating phases of asynchronous (invertible) and synchronous (non-invertible) steps. Focusing derivations decompose a chosen formula and, if the formula was an antecedent formula, it corresponds roughly to the synthesis of derived rules from it; the difference is that we apply both synchronous and asynchronous rules and at least one AXIOM-rule, which excludes applications of derived rules of definitely irrelevant formulas. Moreover, derived F-rules remain available, while focusing consumes F. Finally, the use of derived rules allows to study in future work how to adapt proof search techniques known from other calculi, such as, for instance, the use of term indexing techniques on the level of derived inference rules, or proof strategies based on term orderings as in *superposition calculi*.

Further future work will also be concerned with relaxing the atomicity restriction (requiring more complex proof transformations) as well as investigating how to adapt metadeduction to deep inference[3] (following ideas from [2]) to eventually support the application of derived rules on subformulas.

References

1. Andreoli, J.-M.: Logic programming with focusing proofs in linear logic. J. Log. Comput. 2(3), 297–347 (1992)
2. Autexier, S.: The CoRE calculus. In: Nieuwenhuis, R. (ed.) CADE 2005. LNCS (LNAI), vol. 3632, pp. 84–98. Springer, Heidelberg (2005)
3. Autexier, S., Dietrich, D.: Soundness and weak completeness of proof calculi augmented with derived inference rules restricted to atomic formulas. Seki-Report SR-09-03, DFKI Bremen (2009)
4. Beckert, B., Hähnle, R., Schmitt, P.H.: The even more liberalized δ-rule in free variable semantic tableaux. In: Mundici, D., Gottlob, G., Leitsch, A. (eds.) KGC 1993. LNCS, vol. 713, pp. 108–119. Springer, Heidelberg (1993)
5. Brauner, P., Houtmann, C., Kirchner, C.: Principle of superdeduction. In: Ong, L. (ed.) Proceedings of LICS, July 2007, pp. 41–50 (2007)
6. Burel, G., Kirchner, C.: Cut elimination in deduction modulo by abstract completion. In: Artemov, S., Nerode, A. (eds.) LFCS 2007. LNCS, vol. 4514, pp. 115–131. Springer, Heidelberg (2007)
7. Fiedler, A.: P.rex: An interactive proof explainer. In: Goré, R.P., Leitsch, A., Nipkow, T. (eds.) IJCAR 2001. LNCS (LNAI), vol. 2083, pp. 416–420. Springer, Heidelberg (2001)
8. Gentzen, G.: The Collected Papers of Gerhard Gentzen (1934-1938). In: Szabo, M.E. (ed.) North Holland, Amsterdam (1969)
9. Huang, X.: Human Oriented Proof Presentation: A Reconstructive Approach, Sankt Augustin, Germany. DISKI. Infix, vol. 112 (1996)
10. Prawitz, D.: Natural deduction; a proof-theoretical study. Stockholm Studies in Philosophy, vol. 3. Almqvist and Wiksells (1965)
11. Wack, B.: Typage et déduction dans le calcul de réécriture. Thèse de doctorat, Université Henri Poincaré (Nancy 1) (October 2005)
12. Wirth, C.-P.: Descente infinie + Deduction. Logic J. of the IGPL 12(1), 1–96 (2004)
13. Wirth, C.-P.: Hilbert's epsilon as an operator of indefinite committed choice. J. of Applied Logic (2007)

[3] `alessio.guglielmi.name/res/cos/`

Toward Heterogeneous Cardinal Direction Calculus

Yohei Kurata and Hui Shi

SFB/TR 8 Spatial Cognition, Universität Bremen
Postfach 330 440, 28334 Bremen, Germany
{ykurata,shi}@informatik.uni-bremen.de

Abstract. Cardinal direction relations are binary spatial relations determined under an extrinsically-defined direction system (e.g., *north of*). We already have point-based and region-based cardinal direction calculi, but for the relations between other combinations of objects we have only a model. We are, therefore, developing *heterogeneous cardinal direction calculus*, which allows reasoning on cardinal direction relations without regard to object types. In this initial report, we reformulate the definition of cardinal direction relations, identify the sets of relations between various pairs of objects, and develop the methods for deriving upper approximation of converse and composition.

Keywords: cardinal direction relations, heterogeneous spatial calculi, converse, composition.

1 Introduction

Imagine that you are at the peak of *Mt. Fuji*. You can see *Fuji-gawa river* flowing from the northwest to the southwest, and also the city of *Tokyo* spreading out in the north-east. Then, what can you say about the relation between *Fuji-gawa* and *Tokyo*? This question concerns *cardinal direction relations* [1], which are determined under an extrinsically-defined direction system. In addition to compass directions like *north* and *south*, these relations are applicable to capture such relations as *above* and *left-of* in shelves and storages. *Cardinal direction calculi* are the mechanisms that realize spatial reasoning on such cardinal direction relations. So far cardinal direction calculi have been developed for the relations between points [1, 2] and those between regions [3-6]. In our example question, however, *Mt. Fuji*'s peak, *Fuji-gawa*, and *Tokyo* are modeled as a point, a line, and a region, respectively. Accordingly, their arrangement may not be deduced *properly* in the previous calculi, even though we can treat them as regions and conduct spatial reasoning at the cost of reliability.

Actually, most spatial calculi have targeted spatial relations between single-type objects. Kurata [7], therefore, insisted on the necessity of *heterogeneous spatial calculi*, which allow spatial reasoning beyond the difference of object types. As a first step, he developed *9-intersection calculi* that realize topology-based reasoning on multi-type objects [7]. Following this effort, we are currently developing *heterogeneous cardinal relation calculus*, since cardinal direction relations are one of the most basic sets of spatial relations. This article is the first report of this work.

B. Mertsching, M. Hund, and Z. Aziz (Eds.): KI 2009, LNAI 5803, pp. 452–459, 2009.

The remainder of this paper is structured as follows: Section 2 reviews existing work on cardinal direction relations. Section 3 reformulates the model of cardinal direction relations, under which Section 4 identifies the sets of cardinal direction relations between various pairs of objects. Section 5 introduces the methods for deriving upper approximation of converse and composition of these directional relations. Finally, Section 6 concludes the discussion.

In this paper, we consider a 2D Euclidean space $\mathbb{R}^2$ with a coordinate system in which the x- and y-axes are aligned with the north and east, respectively. Simple regions and non-branching line segments are called *regions* and *lines* for short, and lines parallel to x-axis, those parallel to y-axis, and other lines are called *HLines*, *VLines*, and *GLines* (or *horizontal, vertical, and generic lines*), respectively.

2 Related Work

In general, directional relations concern the orientation of the target object (*referent*) with respect to the reference object (*relatum*). Frank [1] introduced *cone-based* and *projection-based* frames for distinguishing directional relations between two points. The frames are placed such that the relatum comes to the center of the frames. In a spatio-linguistic viewpoint, these frames are sorts of *extrinsic frames of spatial reference* whose direction is fixed in the environment [8]. Frank [1] assessed the converse and composition of his directional relations and identified the presence of some indeterminate compositions (for instance, the composition of *north-of* and *south-of* yields *north-of*, *south-of*, and *equal*). Together with the converse and composition operations, Frank's model forms (*Point-Based*) *Cardinal Direction Calculus*. Ligozat [2] applied this calculus to constraint-based reasoning, identifying maximum tractable subsets of the relations. *Star Calculus* [9] generalizes cardinal direction calculus, such that it captures directional relations in arbitrary granularity.

Papadias and Sellis [3] studied the cardinal direction relations between two regions making use of their minimum bounding rectangles (*MBRs*). Regions are mapped to intervals on each axis. Thus, by applying MBRs, 13×13 directional relations were distinguished, as well as the converse and composition of those relations were determined based on Allen's [10] interval calculus,.

Goyal and Egenhofer [4] introduced a *directional-relation matrix*, whose 3×3 elements correspond to the intersections between the referent and the 3×3 *tiles*. The tiles are determined by four lines l_1: $x = \inf_x(B)$, l_2: $x = \sup_x(B)$, l_3: $y = \inf_y(B)$, and l_4: $y = \sup_y(B)$, where $\inf_{x/y}(B)$ and $\sup_{x/y}(B)$ are the greatest lower bound and the least upper bound of the projection of the relatum B on x/y-axis. This approach is applicable to any pair of objects, but depending on the relatum's shape some tiles may be collapsed to a line or a point. In the simplest approach, directional relations are distinguished by the presence or absence of the 3×3 referent-tile intersections. Based on this framework, calculus aspects of region-based cardinal directions were studied in [5, 6]. Cicerone and Felice [6] identified all realizable pairs of the cardinal directional relation between A and B and that between B and A. Skiadopoulos and Koubrakis [5] identified a method for deriving (weak) compositions of cardinal direction relations between two regions.

3 Our Model of Cardinal Direction Relations

We introduce four iconic representations of spatial arrangements of two objects, namely *CD pattern*, *CD^+ pattern*, *PCD pattern*, and *PCD^+ pattern* (Fig. 1a). The CD pattern captures how the referent A intersects with the 3×3 tiles determined by the relatum B, namely NW_B, NC_B, NE_B, MW_B, MC_B, ME_B, SW_B, SC_B, and SE_B. The CD pattern has 3×3 black-and-white cells, each of which is marked if the corresponding element in $\begin{bmatrix} A \cap NW_B & A \cap NC_B & A \cap NE_B \\ A \cap MW_B & A \cap MC_B & A \cap ME_B \\ A \cap SW_B & A \cap SC_B & A \cap SE_B \end{bmatrix}$ is non-empty. This follows the direction matrix in [4], but in our model the tiles are defined as a set of jointly-exclusive and pairwise-disjoint partitions of the space (by this we avoid strange situations where a point-like referent intersects with two tiles at the same time). First, two sets of JEPD partitions, $\{W_B, C_B, E_B\}$ and $\{S_B, M_B, N_B\}$, are defined as Eqs. 1-2. Then, the tiles are defined as the intersections of these two partition sets, like $NW_B = N_B \cap W_B$. Note that NC_B, MW_B, MC_B, ME_B, and SC_B may be collapsed depending on the type of the relatum (Fig. 2).

$$S_B = \{(x,y) \in \mathbb{R}^2 | y < \inf_y(B)\} \;\; N_B = \{(x,y) \in \mathbb{R}^2 | y > \sup_y(B)\} \;\; M_B = \mathbb{R}^2 \setminus (S_B \cup N_B) \quad (1)$$

$$W_B = \{(x,y) \in \mathbb{R}^2 | y < \inf_y(B)\} \;\; E_B = \{(x,y) \in \mathbb{R}^2 | x > \sup_x(B)\} \;\; C_B = \mathbb{R}^2 \setminus (W \cup E_B) \quad (2)$$

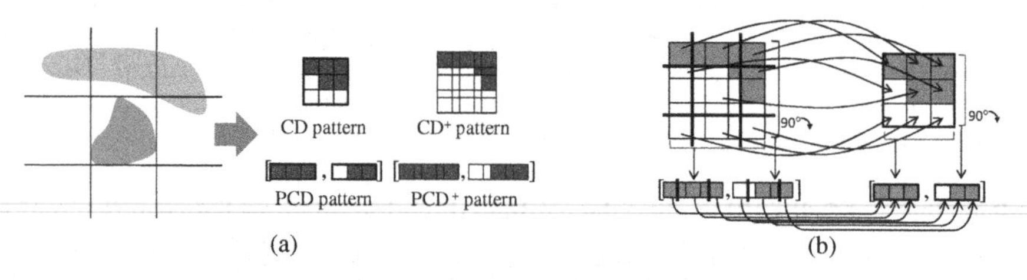

Fig. 1. (a) Four types of iconic representations of a spatial arrangement featuring cardinal directions and (b) converse between these four representations

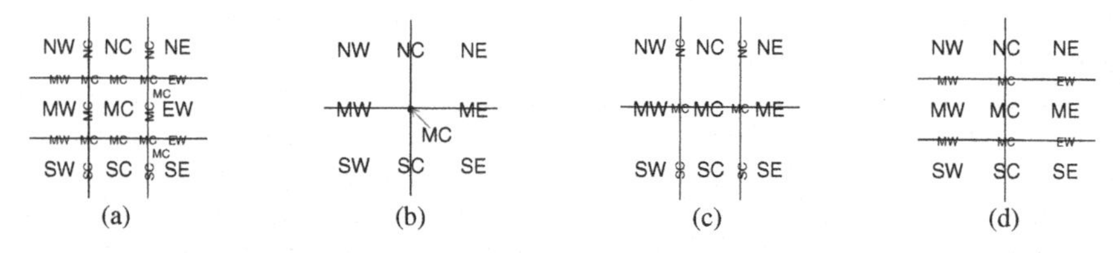

Fig. 2. Tiles determined by the relatum, which is (a) a simple region or a GLine, (b) a point, (c) a HLine, and (d) a VLine, respectively

The CD^+ pattern is a refinement of the CD pattern, which captures how the referent intersects with *n×m fields*. The CD^+ pattern has $n \times m$ black-and-white cells, among which the ith left and j^{th} bottom cell is marked if A intersects with the ith left and j^{th} bottom field. The fields are determined by four lines l_1: $x = \inf_x(B)$, l_2: $x = \sup_x(B)$,

l_3: $y = \inf_y(B)$, and l_4: $y = \sup_y(B)$, such that not only 2D blocks separated by these lines, but also 1D boundaries between two blocks and 0D boundary points between four blocks are all considered independent fields. Naturally, 5×5, 3×3, 5×3, and 3×5 fields are identified when the relatum is a region or a GLine, a point, a HLine, and a VLine, respectively.

The PCD pattern consists of two sub-patterns, called *PCD_x* and *PCD_y patterns*. They have three black-and-white cells, each marked if the corresponding element in $[A\cap W_B \; A\cap C_B \; A\cap E_B]$ and $[A\cap S_B \; A\cap M_B \; A\cap N_B]$ is non-empty, respectively. Essentially, $PCD_{x/y}$ patterns are the projection of the CD pattern onto *x*-/*y*-axis. Finally, the PCD$^+$ pattern is a counterpart of PCD pattern for the CD$^+$ pattern.

In this paper, *directional relations* refer to the spatial arrangements distinguished by CD patterns, while *fine-grained directional relations* refer to the spatial arrangements distinguished by the CD$^+$ patterns. We primarily use CD patterns because they allow the unified representation of directional relations by 3×3 binary patterns without regard to object types. Another practical reason is that CD$^+$ patterns may distinguish an overwhelming number of spatial arrangements (up to $2^{5\times5}$).

4 Identification of Directional Relations

Even though the CD pattern icon distinguishes $2^{3\times3}$ patterns, not all of them are effective as the representation of directional relations (i.e., some patterns have no geometric instance). We, therefore, identify the sets of all effective CD patterns when the referent and relatum are points, H/V/GLines, simple regions, and their combinations (in total 5×5 cases). Suppose *A* be the referent and *B* be the relatum that defines $n{\times}m$ fields. First, we consider $2^{n\times m}$ CD$^+$ patterns, from which we remove geometrically-impossible CD$^+$ patterns by the following constraints:

- if *A* is a point, it intersects with exactly one field;
- if *A* is a H/V/GLine, it intersects with at least one non-0D field, and all fields with which *A* intersects must be connected; in addition,
  - if *A* is an HLine/VLine, all fields with which *A* intersects must be horizontally/vertically aligned;
  - if *A* is a GLine, letting ***C*** be the set of connected components formed by non-0D fields with which *A* intersects, and ***P*** be the set of 0D fields with which *A* intersects and which connect to more than one connected component in ***C***, there are at most two connected components in ***C*** that are connected to exactly one field in ***P*** (otherwise *A* is branching; compare Figs. 3a-3a'); and
- if *A* is a simple region, it intersects with at least one 2D field, and all fields with which *A* intersects must be connected, even if all 0D fields are excluded (otherwise *A* has a spike or disconnected interior; compare Figs. 3b-3b').

Then, the remaining CD$^+$ patterns are converted to CD patterns as shown in Fig. 1b. We confirmed that these CD patterns are all effective (i.e., we succeeded to draw an instance for every CD pattern). Thus, these CD patterns represent the complete set of directional relations. For instance, we identified 222 directional region-region relations (four more than the relations listed in [6], due to the slightly different definition of relations). Table 1 shows the number of the directional relations we have identified for 5×5 cases, while Table 2 summarizes their patterns.

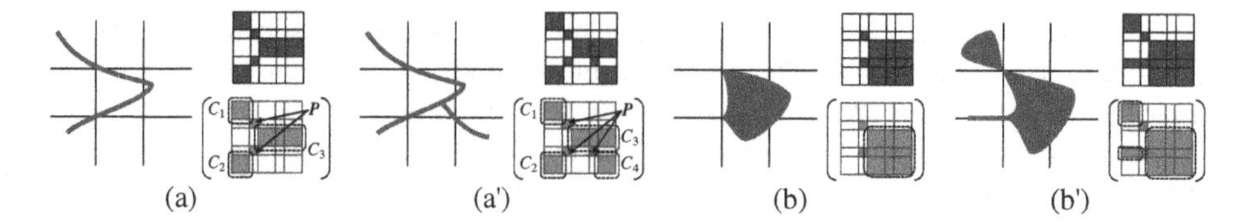

Fig. 3. Examples of CD⁺ patterns when the referent is a simple or non-simple object

Table 1. Numbers of cardinal direction relations distinguished by our framework

		Relatum				
		Point	HLine	VLine	GLine	Region
Referent	Point	9	9	9	9	9
	HLine	15	18	15	18	18
	VLine	15	15	18	18	18
	GLine	254	300	300	308	308
	Region	106	146	146	222	222

Table 2. Cardinal direction relations distinguished by our framework (icons with "×4"/"×2" also represent the CD patterns derived by rotating {90,180,270}/180 degree)

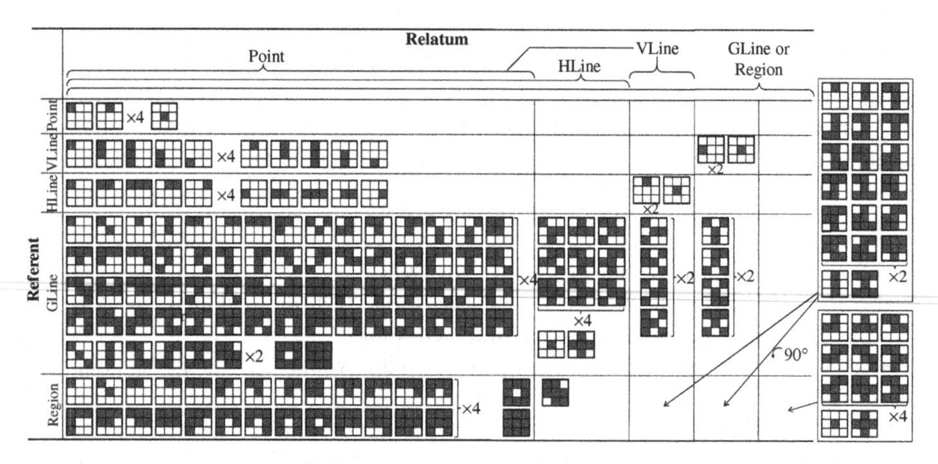

5 Projection-Based Spatial Inference

Given the directional relation between two regions A and B as , what are possible relations between B and A? Similarly, given the relation between a GLine C and a point D and that between D and a region E as and , respectively, what are possible relations between C and E? Figs. 4a-b show one possible solution for each question, but what else? The answers to these questions are derived by *converse* and *composition* operations. In general, given the relation r_1 between the objects O_1 and O_2 and r_2 between O_2 and O_3, the converse of r_1 returns the relation(s) that may hold between O_2 and O_1, while the (weak) composition of r_1 and r_2 returns the relation(s) that may hold between O_1 and O_3. The converse and composition operations are fundamentals of qualitative spatial calculi [11].

We introduce methods for deriving upper approximations of converse and composition of cardinal directional relations. These methods project spatial objects onto x- and y-axes, conduct converse/composition on each axis, and synthesize the results to derive the candidates for the converse/composition. The basic idea comes from MBR-based spatial inference in [3], but our target is not limited to regions.

By projection onto an axis each object is mapped to an interval or a point. First, suppose that the referent A and the relatum B correspond to two intervals i_A and i_B on the x-axis. Then, the PCD^+_x-pattern for A and B corresponds to the interval relation between i_A and i_B. Naturally, the converse of this PCD^+_x-pattern is determined by the converse of this interval relation (Fig. 5a). Similarly, if A and B correspond to two intervals on the y-axis, the converse of their PCD^+_y-pattern is determined by the converse of the corresponding interval relation. This mapping from $PCD^+_{x/y}$ patterns to their converse is coarsened based on the correspondence between $PCD^+_{x/y}$ and $PCD_{x/y}$ patterns in Fig. 6. Fig. 7a summarizes the result. By combining the converse of a PCD_x pattern and that of a PCD_y pattern determined by Fig. 7a, we can determine the converse of a PCD pattern. Making use of this projection-based converse, the candidates for the converse of the given directional relation can be derived. For instance, imagine that A and B are regions whose directional relation is ▦ (Fig. 4a). The PCD pattern for A and B is ▬,▬. The converse of this PCD pattern is {▬,▬, ▬,▬, ▬,▬, ▬,▬}, since ▬'s converse is {▬, ▬} and ▬'s converse is {▬, ▬} (Fig. 7a). From these four PCD patterns, we can derive ten CD patterns {▦, ▦, ▦, ▦, ▦, ▦, ▦, ▦, ▦, ▦}, but ▦ is removed because it does not represent any directional region-region relation (Table 2). As a result, we obtain nine directional region-region relations as the candidates for the converse of the given relation.

When both the referent and relatum correspond to a point on x/y axis, we can derive the mapping from $PCD_{x/y}$ patterns to their converse (Fig. 7b), making use of the correspondence between $PCD^+_{x/y}$ patterns and point-order relations (Fig. 5b). Similarly, when the referent and relatum correspond to a point and an interval, we can derive the mapping from $PCD_{x/y}$ patterns to their converse (Figs. 7c-d), making use of the correspondence between $PCD^+_{x/y}$ patterns and point-interval relations (Fig. 5c).

In a similar way, we consider projection-based compositions. Suppose that three objects A, B, and C correspond to the intervals i_A, i_B, and i_C on the x-axis. Then, the PCD^+_x pattern representing A-B relation and that representing B-C relation correspond to the interval relation between i_A and i_B and that between i_B and i_C, respectively. Naturally, the composition of these two PCD^+_x patterns is determined by the composition of these interval relations. Thus, from Allen's composition table [10], we can derive the composition table of $PCD^+_{x/y}$ patterns. By coarsening this table based on the correspondence between $PCD^+_{x/y}$ and $PCD_{x/y}$ patterns (Fig. 6), the composition table of $PCD_{x/y}$ patterns was developed (Table 3). We skip the detail, but this table can be used even when the referent, relatum, or both correspond to points (instead of regions), taking the realizability of $PCD_{x/y}$ patterns into account (see * in Table 3). Consequently, using Table 3, we can determine the compositions of two PCD patterns, from which the candidates for the composition of the given directional relations can be derived. For instance, the projection-based composition of a GLine-point relation ▦ and a point-region relation ▦ (Fig. 4b) is derived from the composition of ▬ and ▬ and that of ▬ and ▬—the result is {▦, ▦, ▦} (this corresponds to the *Fuji-gawa–Tokyo* relation in Section 1).

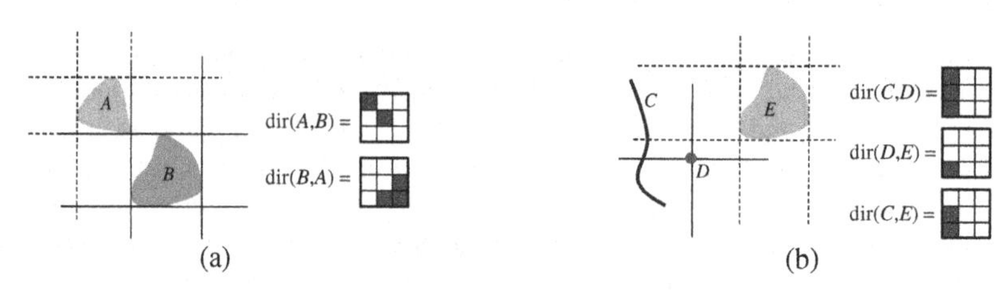

Fig. 4. Examples of projection-based converse and composition

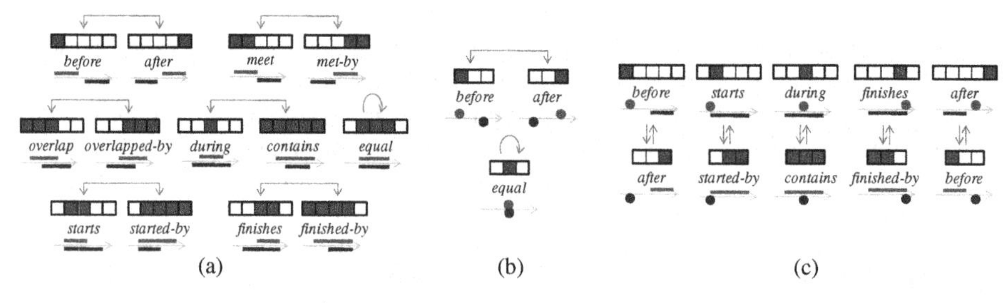

Fig. 5. Mapping from $PCD^{+}_{x/y}$ patterns to their converse, when the referent and relatum correspond to (a) two intervals, (b) two points, and (c) a point and an interval, respectively

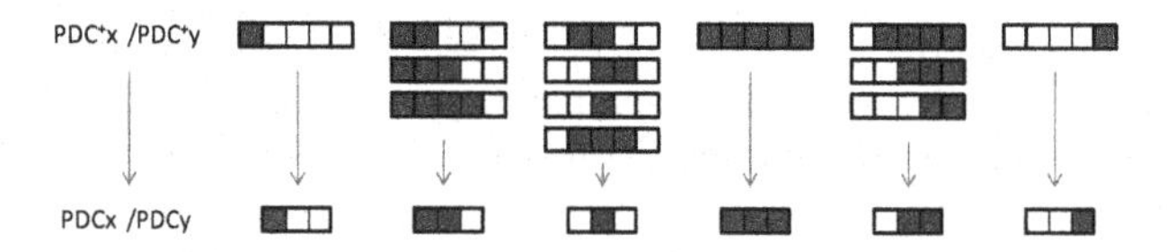

Fig. 6. Correspondence between $PCD^{+}_{x/y}$ and $PCD_{x/y}$ patterns when the $PCD^{+}_{x/y}$ patterns have higher granularity (otherwise $PCD^{+}_{x/y}$ and $PCD_{x/y}$ patterns are equivalent)

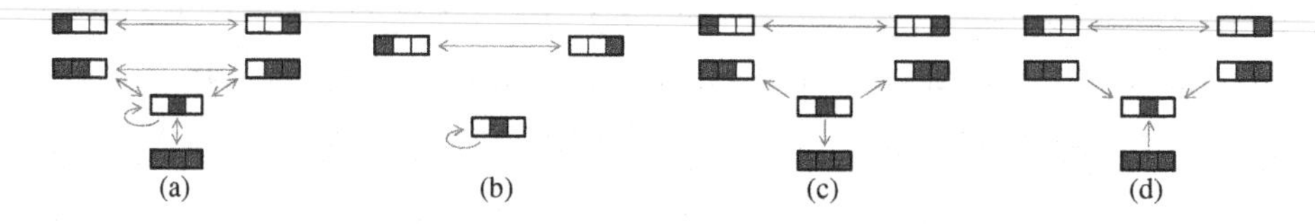

Fig. 7. Converse of $PCD_{x/y}$ patterns when the referent and relatum correspond to (a) two intervals, (b) two points (c) a point and an interval, and (d) an interval and a point on an axis

Table 3. Composition table of $PCD_{x/y}$ patterns

* , , and in the composition results are removed when they correspond to point-point or point-interval relations, and is removed when they correspond to interval-point relations

6 Conclusions and Future Work

In this paper, we developed the basis of *heterogeneous cardinal relation calculus*, which aims at the support of direction-featured spatial reasoning without regard to object types. We captured cardinal direction relations with bitmap-like icons with 3×3 cells and identified sets of relations between points, lines, regions, and their combinations. Then, we developed methods for deriving upper approximation of converse and composition of these relations. Actually, these approximations agree with the precise converse and composition when the source relations are *convex* (i.e., the icon's black cells form a single rectangle); otherwise, the results of projection-based converse or composition may include unnecessary patterns and, therefore, other methods are necessary for deriving more precise converse and composition. These methods, as well as the proof of the above fact, will be reported in our next report.

Acknowledgment

This work is supported by DFG (Deutsche Forschungsgemeinschaft) through the Collaborative Research Center SFB/TR 8 Spatial Cognition—Strategic Project "Spatial Calculi for Heterogeneous Objects" and Project I3-[SharC].

References

1. Frank, A.: Qualitative Spatial Reasoning about Cardinal Directions. In: Mark, D., White, D. (eds.) AutoCarto 10, ACSM/ASPRS (1990)
2. Ligozat, G.: Reasoning about Cardinal Directions. Journal of Visual Language and Computing 9(1), 23–44 (1998)
3. Papadias, D., Sellis, T.: Qualitative Representation of Spatial Knowledge in Two-Dimensional Space. VLDB Journal 3(4), 479–516 (1994)
4. Goyal, R., Egenhofer, M.: Consistent Queries over Cardinal Directions across Different Levels of Detail. In: Tjoa, A.M., Wagner, R., Al-Zobaidie, A. (eds.) 11th International Workshop on Database and Expert Systems Applications, pp. 876–880 (2000)
5. Skiadopoulos, S., Koubarakis, M.: Composing Cardinal Direction Relations. Artificial Intelligence 152(2), 143–171 (2004)
6. Cicerone, S., Felice, P.: Cardinal Directions between Spatial Objects: The Pairwise-Consistency Problem. Information Science 164(1-4), 165–188 (2004)
7. Kurata, Y.: 9-Intersection Calculi for Spatial Reasoning on the Topological Relations between Multi-Domain Objects. In: Guesgen, H., Bhatt, M. (eds.) IJCAI Workshop on Spatial and Temporal Reasoning (2009)
8. Levinson, S.: Frame of Reference. In: Space in Language and Cognition: Explorations in Cognitive Diversity, pp. 24–61. Cambridge University Press, Cambridge (2003)
9. Renz, J., Mitra, D.: Qualitative Direction Calculi with Arbitrary Granularity. In: Zhang, C., Guesgen, H.W., Yeap, W.-K. (eds.) PRICAI 2004. LNCS (LNAI), vol. 3157, pp. 65–74. Springer, Heidelberg (2004)
10. Allen, J.: Maintaining Knowledge about Temporal Intervals. Communications of the ACM 26(11), 832–843 (1983)
11. Ligozat, G., Renz, J.: What Is a Qualitative Calculus? A General Framework. In: Zhang, C., Guesgen, H.W., Yeap, W.-K. (eds.) PRICAI 2004. LNCS (LNAI), vol. 3157, pp. 53–64. Springer, Heidelberg (2004)

The Scared Robot: Motivations in a Simulated Robot Arm

Martin V. Butz and Gerulf K.M. Pedersen

University of Würzburg
Röntgenring 11
97070 Würzburg, Germany
{mbutz,gerulf}@psychologie.uni-wuerzburg.de

Abstract. This paper investigates potential effects of a motivational module on a robotic arm, which is controlled based on the biological-inspired SURE_REACH system. The motivational module implements two conflicting drives: a goal-location drive and a characteristic-based drive. We investigate the interactions and scaling of these partially competing drives and show how they can be properly integrated into the SURE_REACH system. The aim of this paper is two-fold. From a biological perspective, this paper studies how motivation-like mechanisms may be involved in behavioral decision making and control. From an engineering perspective, the paper strives for the generation of integrated, self-motivated, live-like artificial creatures, which can generate self-induced, goal-oriented behaviors while safely and smartly interacting with humans.

1 Introduction

Artificial intelligence systems are gradually becoming more complex as increasingly more autonomy is added into robotic systems. Despite these facets of intelligence and autonomy, behavior that is generally similar to what can be observed in animals is still only partially achieved at most. In this work, we move one step in the direction of biologically-inspired autonomous artificial intelligence by introducing motivations to a simulated robot arm system.

Motivations can be included by defining multiple goals for the system and the interaction of the system with these goals will result in certain behavioral autonomy and particular self-induced goal-directedness. It has previously been investigated how multiple goals can be included in a reinforcement learning system [1]. Elsewhere [2], the goals were gathered into a motivational module for a robotic system. The principles of these works were further extended in [3], where motivations were added to a neural reinforcement learning system to obtain behaviors resembling that of rodents.

In this paper, we investigate how similar types of motivations can change the behavior of a biologically-inspired, neurally-controlled robotic arm control system—the SURE_REACH model [4]. Adding multiple motivational drives, it is possible to obtain interesting autonomous behaviors. We investigate the interplay

B. Mertsching, M. Hund, and Z. Aziz (Eds.): KI 2009, LNAI 5803, pp. 460–467, 2009.

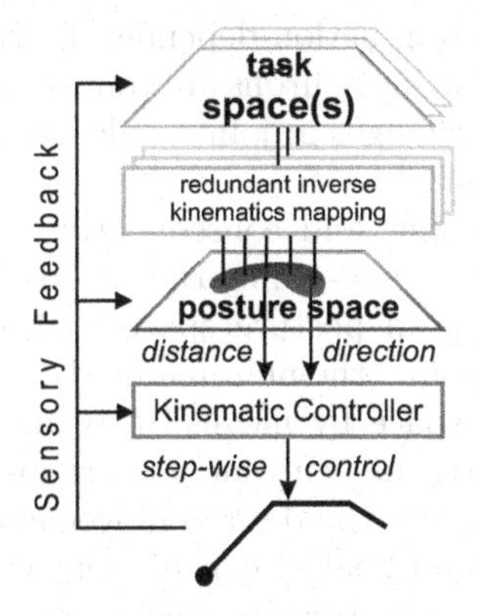

Fig. 1. SURE_REACH model with kinematic controller

of two drives in this paper: a location-based drive, which corresponds to a desired goal, and a characteristics-based drive, which corresponds to undesirable robot arm configurations. Since these drives may be conflicting, the system needs to change its behavior according to the settings of the drives.

The paper first gives an overview of the used robotic system. After the system description, a more detailed description of the drives and how they are used is given in Section 3. Next, in Section 4 a series of experiments is conducted to investigate the influence of the motivational drives on the system. Finally, the implications of the results are discussed.

2 The SURE_REACH Arm Control System

The system that is used for this investigation of behavior is the SURE_REACH control architecture applied to a kinematic model of a robotic arm with three joints [4]. The original SURE_REACH system learned the inverse kinematics and the sensorimotor correlations within its architecture. However, even though the SURE_REACH version we are using for this work uses the same knowledge representation, it is not learned but provided to the system, seeing that the inverse kinematics of the arm system are fully known. The general architectural framework is shown in Figure 1.

In particular, SURE_REACH represents its knowledge in two spatial layers, which are encoded by an overlapping population of neurons with local receptive fields. We use a population of neurons that are distributed in a uniformly spaced grid. The posture space (also known as configuration space) encodes arm postures, that is, the joint angle space of the arm. The task space (also known as location space) encodes joint and end-point locations. These two spaces are associated by inverse kinematics mappings, which associate the local task space neurons with the corresponding, coinciding posture space neurons. One particular end-point location or posture configuration is represented by the overlapping, neighboring neurons in the space. In addition, a sensorimotor mapping associates posture space neurons action-dependently with each other. In the preprogrammed version, neighboring neurons are connected and the connections are linked to the motor vector (changes in joint angles) necessary to move from one posture to a neighboring one.

Given a particular end-point, and consequently the overlapping neurons in task space that represent this end-point, the inverse kinematics mapping can co-activate all those neurons in posture space that yield postures that overlap with the activated end-point (the null-manifold). The sensorimotor mapping allows planning in posture space by means of dynamic programming, that is, value iteration. Given a particular current posture and the corresponding neurons that represent this posture, the desired movement direction and estimated distance towards the activated goal(s) can be determined. Overall, the arm is controlled by means of closed-loop control using its representations to update desired movement directions given current posture perceptions.

3 Motivational Drives

SURE_REACH is a psychological model of arm reaching behavior [4] but it is also a useful algorithm to resolve redundancies and reach goal locations or goal postures with the highest flexibility possible for a given arm. However, so-far, SURE_REACH cannot activate goals nor constraints internally. Thus, we investigate how we can add a motivational module to the SURE_REACH architecture in order to generate a self-motivated system that pursues its goals due to internal drives. In particular, we introduce a goal location drive and a security drive.

The location-based drive is implemented as a location goal, which is identical to the original SURE_REACH formulation. However, the inclusion of the security drive requires the introduction of an additional parameter in the neuronal representation, a fear factor, f, which is inversely proportional to the level of security in the system. This factor indicates the level of fear the system has for achieving the position where the corresponding neuron is activated. The values for the fear factor should be in the range $[0, 1]$, where 0 indicates no fear and 1 indicates a maximum level of fear.

Weighting the level of fear relative to the desire of the system to reach a goal location, whose importance is kept constant, should thus result in a variety of behaviors.

3.1 Setting the Level of Fear

The fear factor is incorporated directly into the propagation of activation in the neuronal representation. In many situations it is desired for a robot to avoid obstacles as well as extreme postures that might limit further movement, and for this purpose such a security drive would be beneficial. Investigating the behavior of the system due to the security drive without using obstacles would be to set a higher fear factor as the robotic arm comes close to the limits of the movement range.

A simple way of setting the level of fear for avoiding extreme angles is to use a piecewise-linear function, $g(x_i)$, where the level of fear for each degree of freedom can be set individually. The function is given by

$$g(x_i) = \begin{cases} \frac{1}{n}\left(1 - \frac{x_i}{b_i}\right) & \texttt{given } x_i < b_i \\ \frac{1}{n}\left(1 - \frac{1-x_i}{b_i}\right) & \texttt{given } x_i > 1 - b_i \\ 0 & \text{otherwise} \end{cases}, \quad (1)$$

where n is the number of dimensions, x_i is a fraction in the range $[0, 1]$ that indicates the location of the neuron in the movement range along the ith dimension, and $b_i < 0.5$ is a fraction that indicates the size of the border region where fear should be non-zero. Considering the dimensions as independent, the overall fear level can simply be determined by summing the individual fear levels: $f(\boldsymbol{x}) = \sum_{i=1}^{n} g(x_i)$, where $\boldsymbol{x}$ is a vector containing the fractional positions, x_i, of the posture neuron in each dimension.

Balancing fear and goal motivations is not straight-forward since the current importance and the influence of the property-based fear motivation can be scaled somewhat arbitrarily. Elsewhere [3] it was shown that location-based and property-based motivations may be handled differently in that property-based motivations should be included in the propagation mechanism (value iteration) of the location-based motivations. Thus, the propagated activation level of a given neuron is a combination of goal-based activation and fear impact on the propagation:

$$a_i^p = \max\left\{a_i^g, \max_j \left\{a_j^p - c_{ji} - w_f f_j\right\}\right\}, \quad (2)$$

where a_i^g is the level of activation indicated by a goal located within the range of the neuron, c_{ji} is the movement cost of moving from the location of neuron j to neuron i, f_j is the level of fear for neuron j, and w_f is a weight that scales the fear impact. Thus, goal-values are propagated using a loss function that is a combination of the actual movement cost plus the "fear-cost" of moving through a particular region. The propagation mechanism above does, however, have one weakness: The goal activation is not affected by the fear level. Thus, for a movement to a goal location, where many redundant goal postures are possible, the level of activation will be dominated by these activation goals. To compensate for that, the final activation of a neuron a_i is realized as $a_i = (1 - w_f f_i)\, a_i^p$.

4 Experiments

A series of experiments were designed to investigate how the arm would react to the introduced fear. The experiments focus first on movements with moderate fear given a particular posture goal. Next, we investigate the impact of location goals and additionally vary the fear factor w_f showing that with increasing fear robot arm behavior differs significantly on a continuous spectrum. The considered system consists of three joints with lengths 0.6, 0.5, and 0.4, respectively.

The corresponding movement ranges of the joints are $[-4.5; 1.5]$, $[-3, 3]$, and $[-3, 3]$. The number of neurons per dimension in both posture and task spaces are 15, and the movement cost between neurons is set to 0.01 for all transitions. All experiments use a boundary fraction $b_i = 0.3$ for the fear in each dimension.

4.1 Posture Goal

To investigate the reaching capability of one exact posture under the influence of the fear drive, the arm was placed in the initial posture: 0.87, −0.70, −1.92. The goal posture was set to −3.49, −0.70, −1.92. In a setting without fear, the smallest change in arm angles corresponds to the optimal path. In the experiment, this corresponds to moving joint one to the desired angle while keeping the other joint angles constant.

Figure 2 shows the comparison between the movements performed by the arm with and without the proposed behavior module. Without fear, the arm primarily moves joint one as expected. The additional slight movements of the other joints is due to the discrete distribution of the neurons in the posture space. When fear is activated with a weight, w_f, of 1, (Figure 2(b)), the primary movement is performed by joint one, similar to the case without fear. When it comes to joint two, there is again some slight movement. Compared to the original system the fear did hardly affect this joint since the start and goal joint angles lie outside of the fear border region of 0.3. However, for the third joint there is a clear effect. During the arm movement, the joint moves initially toward a less extreme angle before it returns to the desired angle. This confirms that the included fear motivation had a significant effect on the movement of the system for the joint located in the region with non-zero fear level.

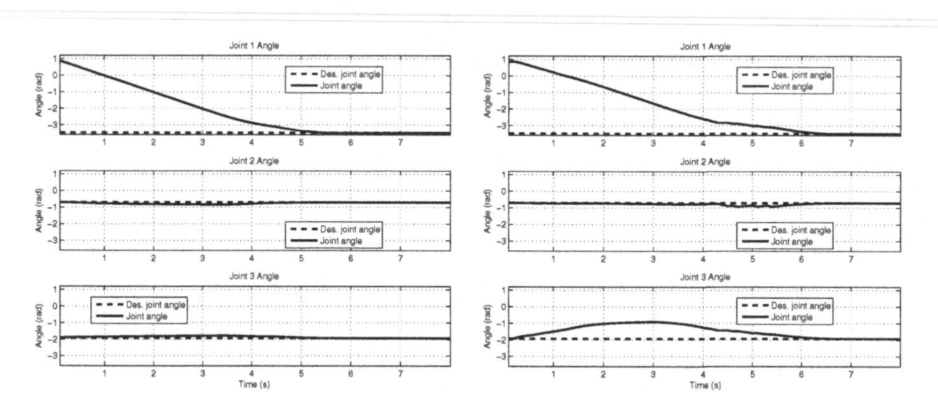

(a) Original SURE_REACH with posture goal

(b) SURE_REACH with posture goal and behavior included

Fig. 2. When moving towards a desired *goal posture*, the movement of the arm is affected by the security drive 2(b) when compared to the original SURE_REACH approach 2(a)

4.2 Location Goal

The strength of SURE_REACH, however, does not only lie in flexible path planning within the arm configuration space, but lies also in the flexible redundancy resolution of alternative end-postures (plus the resolution of alternative paths to a posture) given a hand location goal. Thus, we now use a desired hand goal location in task space in order to see how behavior for movements towards the resulting ambiguous set of postures (the null space for the particular goal location) would be affected by the fear motivation.

Figure 3 shows that the movement path to the goal strongly differs when the influence of the fear motivation is increased. Without fear, the majority of the

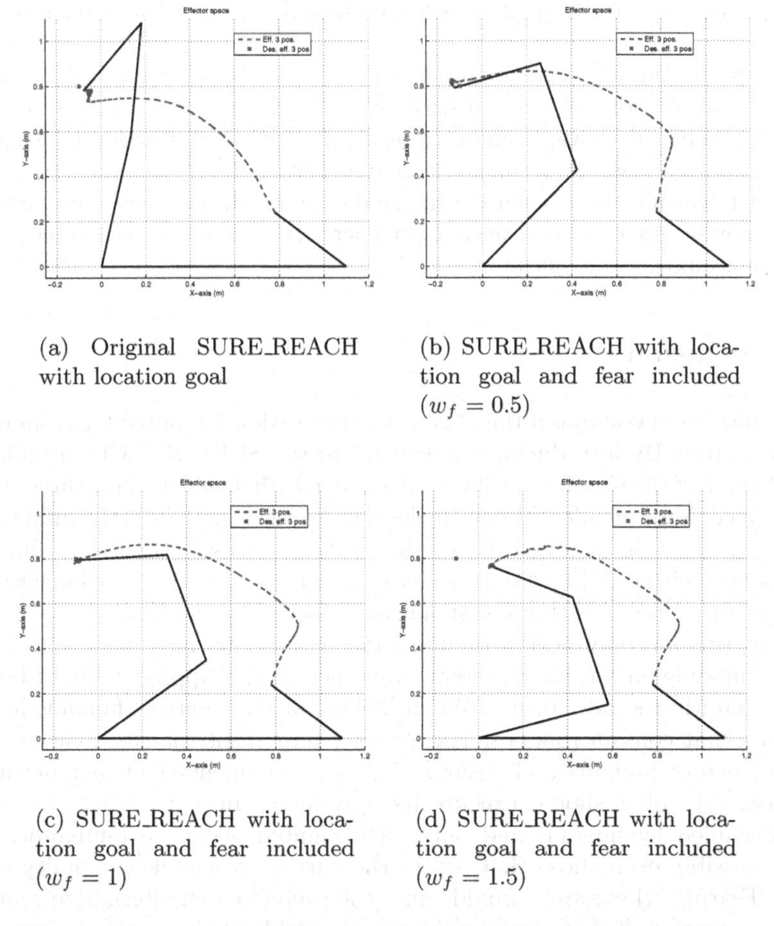

(a) Original SURE_REACH with location goal

(b) SURE_REACH with location goal and fear included ($w_f = 0.5$)

(c) SURE_REACH with location goal and fear included ($w_f = 1$)

(d) SURE_REACH with location goal and fear included ($w_f = 1.5$)

Fig. 3. When moving towards a desired *goal location*, the movement of the arm is affected by the security drive, 3(b) through 3(d), compared to the original SURE_REACH approach 3(a). In 3(d) the goal is not reached because the very strong fear drive prevents that the goal activity reaches the start location.

movement was performed by the innermost joint and the other joints were moved only marginally—effectively minimizing the changes in posture angles necessary to reach the goal location. However, since the resulting posture is located in a region of non-zero fear, the movement and end-posture of the arm strongly differ when the fear motivation is activated at varying levels. The path immediately moves towards the center region of the arm control space and for the lower fear levels, w_f values of 0.5 and 1, the goal location is reached with a "comfortable" posture. For the high fear level $w_f = 1.5$, the arm does not reach the desired goal, but settles in a posture short of the goal. This effect was expected, as an increased weight will have a greater influence on the activation propagation in formula (2), and thus also on the resulting movement. The effect is that the gradient towards the goal is weaker than the gradient due to the fear drive when entering the increasing fear region on the goal side. Thus, the arm moves towards that region as close to the goal as possible but did not "dare" to enter the feared region.

For experiments with even higher fear levels, $w_f = 4$ and higher, the propagation of the activation becomes so strongly affected that the arm does not move at all. A similar effect was seen on experiments with high fear levels using goal postures. In this case, the propagated goal activity quickly decreases to zero due to the high fear impact, preventing the goal activities from exiting the surrounding fear region. As a consequence, no goal activity reaches the arm start posture and thus no movement takes place.

5 Conclusion

This paper has investigated the effect of a motivational module for a simulated robotic system. By introducing a fear drive to the SURE_REACH formulation, the behavior of the simulated robot arm changed when extreme postures (feared postures) could be avoided. When the level of fear was increased, the influence on the movement of the arm varied accordingly. With excessively high fear impacts, the arm even either did to "dare" to move into the fear region any longer or did not move at all because the goal signal was blocked by the fear.

So far, this has only been a preliminary investigation into the effect of a behavioral module on a robotic system. Nonetheless, this approach might become highly useful in the near future. When looking at the safety of humans in environments that contain robots it is extremely important that a robot considers the presence of humans and assures that no (detrimental) impact occurs. In this case, the robot should execute its movements in a way that circumvent areas occupied by humans and only allow limited access to human-occupied areas—possibly even those that are in the current reachable proximity of humans. "Fearing" these areas would cause the robot to seek alternative routes or even stop moving, if there is no such route available at this point.

Besides this potential impact on even actual industrial robotic systems, there is still more work to be done when modeling animal behavior. Clearly, in animals and humans motivational drives are not fixed to a particular weight level but are fluent dependent on the current internal state of the animal and also the environmental circumstances. Thus, the impact of each motivation needs to be adaptive. A first step in this direction has been done by adding homeostatic reservoirs, which control motivation weights [5,3]. Further work utilizing other behavioral modifiers, such as curiosity mechanisms [6], has the potential of producing even more autonomous behavior, as illustrated elsewhere for the case of a simulated mobile robot [3].

In closing, we would like to emphasize that we are intentionally using a rather loose terminology. "Fear" may be considered a general place-holder for a decision-theoretic influence that increases the movement cost—possibly in the form of a loss function. If the movement cost increases we get different, non-continuous behavioral effects, which yield emergent behavioral properties. Moreover, using motivations, multiple task priorities can be flexibly combined and even propagated through various, concurrent spatial representations—such as task and configuration spaces. Thus, it is hoped that the concept of motivations integrated into multiple, concurrent spatial representations is perceived as an intuitive tool for the design of flexible and highly adaptive robot behavior.

Acknowledgments

The authors acknowledge funding from the Emmy Noether program (German Research Foundation, DFG, BU1335/3-1) and thank the COBOSLAB team.

References

1. Sprague, N., Ballard, D.H.: Multiple-goal reinforcement learning with modular sarsa(o). In: Proceedings of the Eighteenth International Joint Conference on Artificial Intelligence, IJCAI 2003 (2003)
2. Konidaris, G., Barto, A.: An adaptive robot motivational system. In: Nolfi, S., Baldassarre, G., Calabretta, R., Hallam, J.C.T., Marocco, D., Meyer, J.-A., Miglino, O., Parisi, D. (eds.) SAB 2006. LNCS (LNAI), vol. 4095, pp. 346–356. Springer, Heidelberg (2006)
3. Shirinov, E., Butz, M.V.: Distinction between types of motivations: Emergent behavior with a neural, model-based reinforcement learning system. In: 2009 IEEE Symposium Series on Artificial Life (ALIFE 2009) Proceedings, pp. 69–76 (2009)
4. Butz, M.V., Herbort, O., Hoffmann, J.: Exploiting redundancy for flexible behavior: Unsupervised learning in a modular sensorimotor control architecture. Psychological Review 114(4), 1015–1046 (2007)
5. yuh Hsiao, K., Roy, D.: A habit system for an interactive robot. In: Castelfranchi, C., Balkenius, C., Butz, M.V., Ortony, A. (eds.) From Reactive to Anticipatory Cognitive Embodied Systems: Papers from the AAAI Fall Symposium, pp. 83–90. AAAI Press, Menlo Park (2005)
6. Schmidhuber, J.: A possibility for implementing curiosity and boredom in model-building neural controllers. In: Meyer, J.A., Wilson, S.W. (eds.) From Animals to Animats, pp. 222–227. MIT Press, Cambridge (1991)

Right-of-Way Rules as Use Case for Integrating GOLOG and Qualitative Reasoning*

Florian Pommerening, Stefan Wölfl, and Matthias Westphal

Department of Computer Science, University of Freiburg,
Georges-Köhler-Allee, 79110 Freiburg, Germany
{pommeren,woelfl,westpham}@informatik.uni-freiburg.de

Abstract. Agents interacting in a dynamically changing spatial environment often need to access the same spatial resources. A typical example is given by moving vehicles that meet at an intersection in a street network. In such situations right-of-way rules regulate the actions the vehicles involved may perform. For this application scenario we show how the Golog framework for reasoning about action and change can be enhanced by external reasoning services that implement techniques known from the domain of Qualitative Spatial Reasoning.

1 Introduction

Agents interacting in a dynamically changing spatial environment often need access to the same spatial resources. A typical example is given by vehicles meeting at an intersection in a street network. In such situations right-of-way rules regulate the actions the vehicles involved may perform. For instance, the German right-of-way regulations (StVO § 8 and § 9) present a detailed set of rules on how the agents controlling these vehicles should proceed if the intersection is not regulated by special traffic signs (stop sign, traffic lights, etc.).

In this paper we consider this particular application scenario and present an approach that allows for determining rule-compliant actions for the involved vehicles in a real-time environment. Action choices are represented within the Golog framework for reasoning about action and change [1], which is based on the Situation Calculus [2]. To detect actions that may lead to collisions, we use methods known from the field of Qualitative Spatial Reasoning (QSR), that is, we use constraint-based and neighborhood-based reasoning to detect dangerous actions[1]. Qualitative spatial reasoning is integrated in the IndiGolog-framework [4] by calling an external reasoner.

Qualitative reasoning and neighborhood graphs have been previously used to determine rule-compliant actions in the SailAway demonstrator (see, e.g., [5]

* This work was partially supported by Deutsche Forschungsgemeinschaft as part of the Transregional Collaborative Research Center *SFB/TR 8 Spatial Cognition*, project R4-[LogoSpace].

[1] A quite different approach to reasoning about moving objects in a traffic scenario is to directly represent trajectories of vehicles [3].

B. Mertsching, M. Hund, and Z. Aziz (Eds.): KI 2009, LNAI 5803, pp. 468–475, 2009.

and [6]). The main difference to the approach used there is that we use a generic formalism to represent actions and change, namely the Golog language. A Golog representation of actions and space has also been discussed in [7]. However, the reasoning process was not externalized in the method discussed there.

In the next section we present the general idea as well as the main components of our spatial agent framework, which is based on the Golog interpreter IndiGolog and uses qualitative reasoning methods as an external method in order to trigger rule-compliant actions. More details regarding our implementation are then presented in section 3. We show how traffic rules can be represented within Golog, how spatial neighborhood graphs can be used to formulate reasoning tasks for qualitative reasoning, and how individual actions are mapped to these reasoning results. In particular, we present a reusable method that allows for integrating external reasoning in IndiGolog. In section 4 we provide a short evaluation of the system. Finally, in section 5 we give a short summary and an outlook on research problems to be addressed in future work.

2 Integrating Qualitative Reasoning in Golog

For our problem we consider a number of cars meeting at an intersection. Each car travels at a given speed on a planned route through the intersection. Further, each car (driver) aims at crossing the intersection as fast as possible. Since this is not always possible, each car has the ability to accelerate until it reaches its full speed or to decelerate until it stops. The right-of-way rules state which car has to stop in order to avoid a collision. We assume that the intersection has no specific signs regulating the right of way. Further, we also assume right-hand traffic, i.e., left-hand driving, and the right-of-way rule common in continental Europe, i.e., priority is given to the right vehicle.

To represent the essential aspects of this problem, we consider a limited number of cars (at most four) that try to cross an intersection, which consists of four streets entering from the four cardinal directions. For each car, we model the planned route, the current position on this route, maximal speed, and whether it is currently accelerating or decelerating. The route specifies the complete continuous path from the current position of the car to the final destination after crossing the intersection. We try to find crash-free behaviors for all cars. The term *behavior* just refers to a sequence of actions with timestamps, where possible actions are *accelerate*, *hold*, or *decelerate*. In particular, we assume that the given routes are never altered.

The idea, which is presented in more detail in the following, is to represent the action choices of the agents within IndiGolog whereas spatial reasoning tasks are delegated to a specialized reasoner. If two or more vehicles are approaching the intersection, we first check whether the planned routes could result in a collision. If this is not the case, all vehicles proceed as planned; otherwise we determine those cars that need to decelerate in order to avoid the collision.

IndiGolog. Golog describes a family of programming languages based on the situation calculus, a multi-sorted second-order logic. The situation calculus defines three sorts, distinguishing the concepts of *action*, *situation*, and *object*. A situation consists of a sequence of actions concatenated with the distinguished function $do(a, s)$ that returns the situation resulting from executing action a in situation s. These sequences are grounded in some initial situation S_0. Changing aspects of the domain are modeled by functional and relational *fluents*. Fluents are domain-specific functions or predicates with one situation parameter. To define an action and its effects on fluents, three types of axioms are distinguished: *Action precondition axioms* define whether an action is applicable in a situation. *Initial state axioms* describe the value of fluents in the initial situation S_0. Finally, *successor state axioms* are used to express how actions influence fluents.

Golog [1] adds a notion of executability to the situation calculus by introducing an interpreter macro, which transforms a given Golog program into a situation calculus formula. Golog also defines elements typical for programming languages like sequences, conditional executions, loops, and procedures.

In our implementation the Golog variant IndiGolog [4] was used, since it offers two important features. The first one is the `search`-operator that tries to find a possible execution of a non-deterministic Golog program instead of executing it online. The second feature is the environment manager that communicates with multiple device managers and allows for sending *sensing actions* to them. When a sensing action is executed, the execution of the Golog program pauses until the action is finished. The return code of the action can influence (i.e. sense) the values of several fluents. This way of injecting data into a running Golog program allows an easier integration of external programs.

Qualitative Reasoning. To solve problems containing rules like the right of way, it is often helpful to abstract from the absolute numerical values early on and consider only qualitative descriptions of the situations. A variety of different constraint-based qualitative calculi has been used in other studies in order to reduce the size of the state space by abstraction in such cases (e.g. [5]).

In this paper, we qualitatively describe the positions of cars relative to the center of the intersection. For this, a STAR calculus [8] seems appropriate. It describes positional information between points via a partition of the plane into a given number of sectors, which are obtained by m distinct, intersecting lines, thus giving the form of a star (see Fig. 1a). One distinguishes positions on the

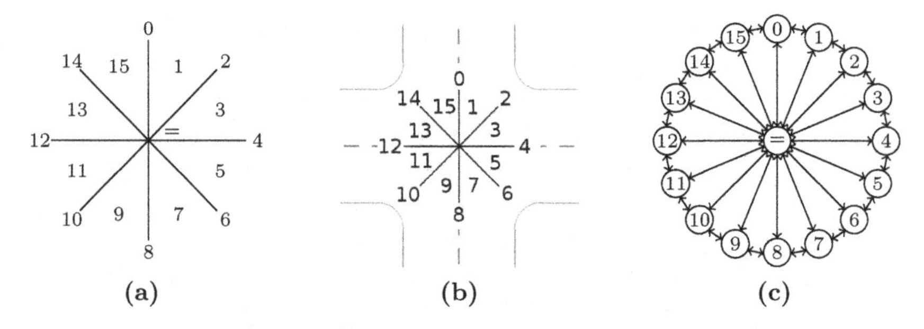

Fig. 1. $\mathrm{STAR}_4(0)$ and its conceptual neighborhood graph

lines and positions in the sectors between them, thereby obtaining $4 \cdot m + 1$ distinct cases including equality of points. $\text{STAR}_4(0)$ is expressive enough for our problem, since we can describe the necessary positions that influence the behavior selection for the cars (see Fig. 1b). To reason about changes in the positions of cars over time, we use the conceptual neighborhood graph (CNG) [9] that makes the possible continuous changes of these relations explicit (Fig. 1c).

3 Implementation Details

The implementation is split up into five different components that communicate via sockets (cf. Fig. 2). First, a Java program contains the full simulation and represents the physical world, in which cars move based on natural laws and crashes can be detected, but not predicted. Cars in the simulation can only change their behavior after receiving `accelerate` or `brake` messages from the Golog program. Thus a car that receives a `brake` message will slow down until the Golog program signals a clear path. Cars can also be set to the state `ignore`, which is done by the Java program if a car leaves the intersection and by the Golog program if a car's trajectory has no conflicts with any other trajectory. In this state, cars accelerate until they leave the area and are neglected in the reasoning process.

The Golog program is called by the Java program with a qualitative representation of the world, which can be easily generated from the complete simulation. For each car it contains the starting position, the current position and the goal position, all as STAR-relations relative to the center of the intersection. The drivers can get this information from watching the other cars and their turning signals. Whenever the Golog program detects a possible collision, it tries to avoid a crash by sending the correct movement messages back to the Java program using a device manager also written in Java. To detect collisions in the trajectories, the Golog program integrates the generic qualitative reasoner

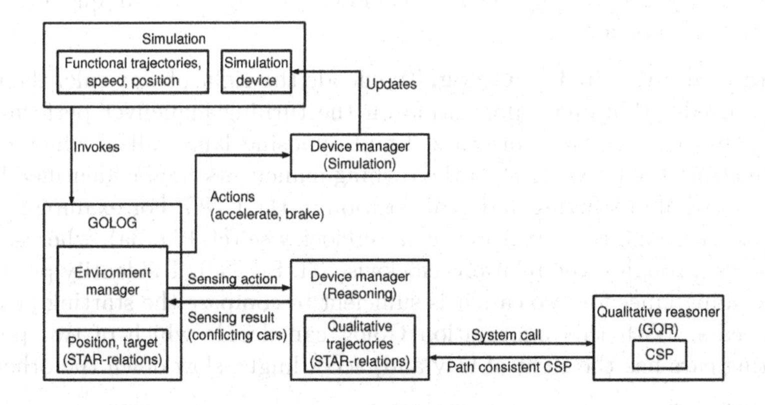

Fig. 2. Communication of components

GQR [10] with a second device manager, which transforms the trajectories into a qualitative reasoning task, starts GQR, and sends the result back as a sensing result.

Encoding of Trajectories. To reason with the trajectory of a car, the Golog program first needs to recreate it from the current and goal position. The full (functional) trajectory from the simulation is not used here since it would be implausible for a driver to know the absolute coordinates of each car at every moment. Instead the trajectories are expressed qualitatively as a sequence of relations to the center of the intersection. The sequence describes the relations the car will pass through on its way from the current to the goal position.

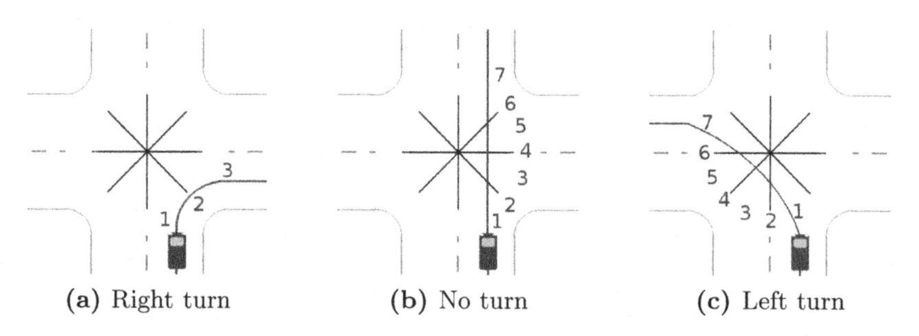

(a) Right turn (b) No turn (c) Left turn

Fig. 3. Standard turns

For a standard 4-way intersection all possible turning maneuvers (a right or left turn or driving straight through) go through at most seven neighboring relations in the $\text{STAR}_4(0)$ calculus (cf. Fig. 3). Since we consider only tangential turning, no car travels through the center of the intersection. The qualitative trajectory can thus be reconstructed by finding a path inside the CNG with at most seven nodes that transforms the current position to the goal position. For standard 4-way intersections such a path always exists and is unique for a given start and goal position.

Encoding of rules in IndiGolog. To encode the right-of-way rule, the Golog program needs additional information on the turning maneuver performed by each car and the notion of prioritized and opposing lanes, all of which can be extracted from the CNG. Each of the turning maneuvers can be identified by the relative position of starting and goal position in the CNG. For example, a right turn moves through three relations counterclockwise (cf. Fig. 3a), whereas a left turn moves through seven relations clockwise (cf. Fig. 3c). To identify prioritized and oncoming lanes for two cars it is sufficient to compare the starting positions of both cars. With this information Golog can decide which of two possibly conflicting cars has the right of way and, accordingly, slow down the other car.

Qualitative Reasoning with GQR. In order to apply GQR, the reasoning problem has to be transformed into a constraint satisfaction problem in the

STAR calculus. This is done by introducing one variable to represent the center of the intersection and one for each entry in the qualitative trajectory of each car. Since the trajectories of the cars are generated from a CNG, we can add an inequality constraint between each two consecutive elements from a car's trajectory. All other added constraints are constraints between a trajectory variable and the center of the intersection determined by the entry in the trajectory.

GQR is called to calculate a path-consistent equivalent CSP, which contains the possible relations between each two trajectory variables. If all constraints between the trajectory variables of two cars do not mention equality, the reasoner can guarantee that these two cars will never be in the same relation to the center of the intersection.

The device manager returns a list of car tuples to the Golog program, containing the pairs of cars where the reasoner cannot guarantee a conflict-free trajectory. The reasoning process done here basically amounts to checking if two trajectories share a relation and only demonstrates the integration of GQR as a high-performance tool for qualitative reasoning within the situation calculus.

4 Evaluation

Whether the behavior found by the program is crash-free and rule-compliant crucially depends on the speed of the cars and how often the Golog program can be started to determine an action for the cars. Obviously, crashes might happen if the first call to the reasoner returns an action too late for a car to brake in time. The timing currently implemented allows for approximately six calls to the reasoner in the time it takes one car to approach an intersection, cross it and reach a safe distance (all numbers with respect to a 2 GHz PC with 3 GB RAM). With this timing most of the combinations of up to four cars can be resolved in a crash-free and rule-compliant way[2].

For the runtime measurement seven different test cases were executed until all cars reached their goal. The test cases are listed in Fig. 4b and ordered by the number of cars involved and the total number of possible crash situations, i.e., the sum of common STAR-relations for each pair of cars. Each test case called the Golog program between 11 and 23 times. For the evaluation we measured the execution times averaged over all calls. Fig. 4a shows the time it took to consult the prolog files through JPL (CONSULT), the time to initialize IndiGolog and set up all device managers (INIT), the time spent in the main IndiGolog procedure (GOLOG; without external reasoning) as well as the time needed by the external reasoner GQR (REASONER). The time needed to shut down IndiGolog and close all device managers is not depicted.

The interesting components are *GOLOG* and *REASONER*. *GOLOG* strongly reacts to the number of cars in the situation and is mainly responsible for the increase in overall running time of a reasoning step. *REASONER* on the other

[2] We currently do not handle deadlock situations in which no car has priority, e.g., four cars arriving at an intersection simultaneously from the four cardinal directions, where each car wants to cross the intersection in a straight line.

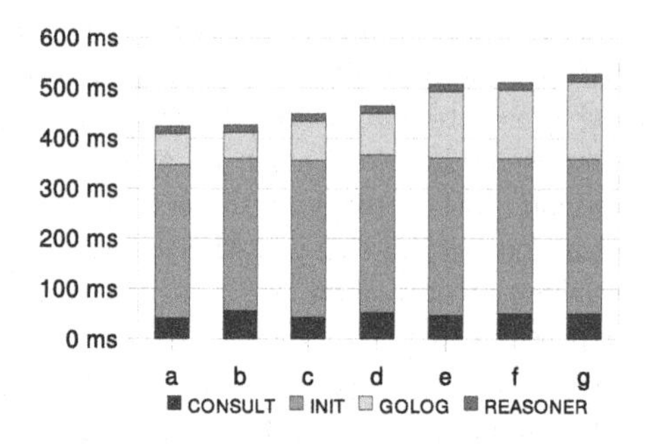

(a) Average runtime of components

Test	# Cars	# Conflicts
a	1	0
b	2	0
c	2	1
d	2	3
e	3	4
f	3	6
g	3	9

(b) Test case scenarios

Fig. 4. Runtime evaluation

hand only shows slight changes despite the quadratic growth of the CSP. This could be the effect of the small problem sizes considered here. Seeing how the runtime of the Golog program scales with the problem size explains the demand for outsourcing parts of Golog programs to external reasoners. The Golog program presented here is only responsible for a very limited part of the calculation, but still makes up a significant amount of the overall runtime. By outsourcing problems to external reasoners, Golog can use the additional computation time to work on a high-level progam that uses the results of those reasoning processes to reach its goal.

5 Conclusion

The two main results presented here are a working example that uses qualitative reasoning to find rule-compliant crash-free behavior for a number of cars, and a reusable method to integrate external reasoning in IndiGolog. The main limitations of the current implementation is that actions for the individual vehicles are selected from the point of a traffic controller that has complete knowledge about the approaching vehicles. Therefore the current framework cannot be considered a multi-agent framework for reasoning about action and change in spatial environments. It would be interesting to integrate external reasoning into other Golog variants such as MIndiGolog [11] and combine their features with qualitative reasoning. To that end one would have to add features like sensing actions that allow for an execution of actions outside of the Golog context.

Moreover, the current implementation could be extended by adding further sensing actions to request information that the device manager can calculate concurrently to the execution of the Golog program. An interesting example of this would be to ignore conflicts that are rendered invalid by other conflicts because the car causing the conflict will brake anyway. This could also be used to act on incomplete knowledge. If, for example, the intersection is not completely visible from all streets, one could add relations between cars as soon as they are visible and restart the reasoning process to react to the new situation.

Finally, the method for integrating external reasoners in Golog could be simplified by an extension of Golog implementations. The method presented in the paper works well on a per-application basis, but has to be reimplemented in each application. Some of the basic features of this implementation (especially the base class of the Java device manager) can be used to develop a language extension of the IndiGolog syntax that provides an easy-to-use interface to external reasoning services as provided by constraint solvers.

References

1. Levesque, H.J., Pirri, F., Reiter, R.: Foundations for the situation calculus. Electronic Transactions on Artificial Intelligence 2(18), 159–178 (1998)
2. McCarthy, J.: Situations, actions and causal laws. Technical Report Memo 2, Stanford University Artificial Intelligence Laboratory, Stanford, CA (1963)
3. de Weghe, N.V., Cohn, A.G., Maeyer, P.D., Witlox, F.: Representing moving objects in computer-based expert systems: the overtake event example. Expert Syst. Appl. 29(4), 977–983 (2005)
4. Giacomo, G.D., Levesque, H.J.: An incremental interpreter for high-level programs with sensing. In: Levesque, H.J., Pirri, F. (eds.) Logical Foundations for Cognitive Agents: Contributions in Honor of Ray Reiter, pp. 86–102. Springer, Heidelberg (1999)
5. Dylla, F., Frommberger, L., Wallgrün, J.O., Wolter, D., Nebel, B., Wölfl, S.: SailAway: Formalizing navigation rules. In: Proceedings of the Artificial and Ambient Intelligence Symposium on Spatial Reasoning and Communication, AISB 2007 (2007)
6. Dylla, F., Wallgrün, J.O.: Qualitative spatial reasoning with conceptual neighborhoods for agent control. Journal of Intelligent and Robotic Systems 48(1), 55–78 (2007)
7. Dylla, F., Moratz, R.: Exploiting qualitative spatial neighborhoods in the situation calculus. In: Freksa, C., Knauff, M., Krieg-Brückner, B., Nebel, B., Barkowsky, T. (eds.) Spatial Cognition IV. LNCS (LNAI), vol. 3343, pp. 304–322. Springer, Heidelberg (2005)
8. Renz, J., Mitra, D.: Qualitative direction calculi with arbitrary granularity. In: Zhang, C., Guesgen, H.W., Yeap, W.-K. (eds.) PRICAI 2004. LNCS (LNAI), vol. 3157, pp. 65–74. Springer, Heidelberg (2004)
9. Freksa, C.: Conceptual neighborhood and its role in temporal and spatial reasoning. In: Singh, M., Travé-Massuyès, L. (eds.) Decision Support Systems and Qualitative Reasoning, pp. 181–187. North-Holland, Amsterdam (1991)
10. Gantner, Z., Westphal, M., Wölfl, S.: GQR - A fast reasoner for binary qualitative constraint calculi. In: Proceedings of the AAAI 2008 Workshop on Spatial and Temporal Reasoning. AAAI Press, Menlo Park (2008)
11. Kelly, R.F., Pearce, A.R.: Towards high-level programming for distributed problem solving. In: IAT 2006: Proceedings of the IEEE/WIC/ACM international conference on Intelligent Agent Technology, Washington, DC, USA, pp. 490–497. IEEE Computer Society, Los Alamitos (2006)

Assessing the Strength of Structural Changes in Cooccurrence Graphs

Matthias Steinbrecher and Rudolf Kruse

Department of Knowledge Processing and Language Engineering
Otto-von-Guericke University of Magdeburg
Universitätsplatz 2
39106 Magdeburg, Germany
{msteinbr,kruse}@ovgu.de

Abstract. We propose a heuristic for assessing the strength of changes that can be observed in a sequence of cooccurrence graphs from one graph to the next one. We represent every graph by its bandwidth-minimized adjacency matrix. The permutation that describes this minimization is applied to the matrices of the respective following graph. We use a repair count measure to assess the quality of the approximation that is then used to determine whether time frames shall be merged.

1 Introduction

The identification of patterns inside large databases comprises a core objective of any data mining procedure [1,2]. In this paper we address the analysis of so-called cooccurrence graphs. These graphs arise quite naturally wherever one is interested in the fact that two entities share some property with respect to a so-called location (which not necessarily has to be a spatial artifact but often is). Examples for this may be internet users visiting the same web pages or online shoppers purchasing the same set of products. A visual representation would be an undirected graph were the nodes represent the web pages (or products) and the edges model the (set of) users that accessed (or purchased) the respective entity. The number of users is normally tracked by introducing a weight for every edge.

Our experience from multiple industrial cooperations in the field of data mining in the past years has shown that time is a dimension of growing interest. Financial partners not only asked for the identification of malicious credit contracts but mainly how their respective models changed during the last months. Experts from the automobile manufacturing industry are interested in failure patters among the sold cars [3], however, the recognition of slow evolvements of such patterns can be much more helpful since it may allow for addressing a (potential) problem before it actually becomes one.

We addressed this problem for several types of patterns [4,5] of which cooccurrence graphs are the most recent ones. As stated above, users are merely interested in slight model changes rather than large abrupt variations. This has

B. Mertsching, M. Hund, and Z. Aziz (Eds.): KI 2009, LNAI 5803, pp. 476–483, 2009.

mainly two reasons which depend on the cause of change: If the change stems from a systematic user-controlled process (e. g. rolling out the new car model, changing the website layout of a shopping portal) they are known and do not convey valuable information. It rather deteriorates the analysis since subtle model shifts which are more interesting may be hidden. The second reason why an abrupt (unintended) change is of less user interest amounts to the fact that it might be already too late, that is the ongoing changes are already visible and subject to user countermeasures. The above motivation, however, stressed that we intend to find evolving pattern changes before they become obvious.

Our claim therefore is to find a way of separating phases with a high amount of model perturbation from phases with much less variations and then apply our methods [4] to these subsets of data to arrive at a more reliable result.

The remainder of this paper is organized as follows: Section 2 introduces the notation used throughout the paper. Section 3 motivates and presents the model change detection procedure whereas section 4 presents a real-world example to show the applicability of our method. We conclude the paper in section 5.

2 Graph Representation

In this work we deal exclusively with undirected graphs for which we use two well-known equivalent representations: layouting the graph and the corresponding adjacency matrix. A graph is a tuple $G = (V, E)$ with vertices V and edge set E with $E \subseteq V \times V \setminus \{(v, v) \mid v \in V\}$, and the constraint $(u, v) \in E \Rightarrow (v, u) \in E$ to emphasize the undirected character. We will interpret the graphs as cooccurrence graphs where edges determine the number of cooccurrences (of whatever kind). This is taken into account with an edge weight function for every edge $e = (u, v)$:

$$w : E \rightarrow \mathbb{N}_0 \;\text{ with }\; w(e) = w(u, v) = w(v, u).$$

This weight is represented as the edge width, thus we use the notion *width* and *weight* interchangeably. A zero width indicates that the respective edge is not contained in the graph.

Another representation of a graph $G = (V, E)$ is by its *adjacency matrix* $\text{adj}(G) = (w_{ij})_{|V| \times |V|}$. This quadratic (and for undirected graphs also symmetric) matrix contains the edge weights: $w_{ij} = w(i, j)$ with i and j being vertices. Figure 1 shows an example graph with its corresponding adjacency matrix. We will depict large matrices as so-called matrix plots, that is as a two-dimensional pixel array where every matrix entry is represented by a shaded pixel. The darker the color, the larger the respective matrix cell entry (the respective ranges are given for every figure).

3 Model Change Detection

As motivated in the introduction, our contribution comprises a heuristic to assess the quantity of structural change from time frame to time frame in a series of

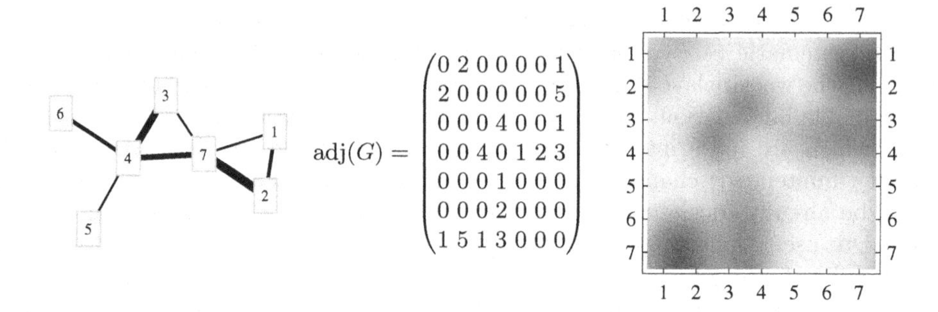

Fig. 1. Left: An undirected graph G width edge weights indicated as edge widths. Middle: Corresponding adjacency matrix adj(G). Right: Matrix plot of the same matrix. Light gray as in the background indicates 0 while black equals 5. The axes labels correspond to the vertices.

cooccurrence graphs. The evaluation section will explain the underlying data in more detail, however, the reader might find it helpful to know where our ideas came from: A large 3D online playing community is tracking which players pass by certain locations of interest. Every log entry contains player ID, time and location ID. One of the questions was: Are there temporal changes within the visiting patterns of the players? Since everybody can contribute buildings and other landmarks, it is important to focus on those periods where slight changes in visiting patterns are not superimposed by large modifications.

3.1 Objective

The question we intend to answer is: Given a series of cooccurrence graphs (one for each time frame), which frames should remain and which ones can be merged to create a data set that is then subject of a search for subtle changes? And is it possible to deliver visual cues that allow the user to follow the suggestions without inspecting the numerical data in full depth?

3.2 Representing a Graph of a Single Time Frame

The visual assessment whether a large or small structural change (that is a change of the values of the adjacency matrix) did take place can be hard if the graphs are depicted layouted as in the left part of figure 1. We do not address this problem by investigating different layout methods since the graphs will be very dense in real-world applications and thus not very instructive for the user anyway. We suggest to use the adjacency matrix instead. In its original form (that is with a predefined vertex order which corresponds to the row/column order) it will, however, be rather scattered since empty rows/columns refer to vertices that are isolated. Therefore, we bring the matrix in a minimal bandwidth form [6] and use the corresponding permutation as a representation of the corresponding time frame.[1]

We thus arrive at a series of adjacency matrices $\text{adj}(G_t)$ with corresponding permutations π_t where t indicates the respective time frame. Since the process of bandwidth minimization arranges nonzero matrix entries (that is, edges of the graph) close to the main diagonal, it semantically corresponds to relabeling the graph vertices such that the absolute difference of vertex indices of vertices connected by an edge is smaller than the difference of indices of vertices without an edge in between.

3.3 Assessing Model Change

The main idea to assess the similarity of the graphs of two adjacent time frames is to apply the permutation π_{t-1} on the matrix $\text{adj}(G_t)$ and check how good the result $\pi_{t-1} \circ \text{adj}(G_t)$ is compared to $\pi_t \circ \text{adj}(G_t)$. Therefore, an appropriate evaluation measure is needed to assess how good the last month's minimization permutation was able to approximate this month's minimized matrix.

Figure 2 shows a real-world example that motivates our so-called *repair count* measure. The left matrix shows the original minimal bandwidth adjacency matrix $\pi_t \circ \text{adj}(G_t)$ of the graph of a particular month t. The matrix in the middle shows the approximation after having applied the previous month minimizing permutation, that is the shown matrix is $\pi_{t-1} \circ \text{adj}(G_t)$. The right matrix shows the result after the permutation π_{t-2} has been applied to matrix $\text{adj}(G_t)$. To quantify what is visually and intuitively obvious, namely the better approximation by the middle rather than by the right matrix, we consider the "repair" of a matrix to the permutation that brings it again to a minimal bandwidth form (not necessarily the original one). Therefore, the respective permutation that achieves this task is evaluated with respect to the number and distance of repositionings. That is, we ask for the sum of the absolute differences between the original position of a column/row and the new one. We define the repair count of a permutation (with respect to an implicit neutral permutation π_0 which does not permute anything) as

$$\text{rc}(\pi) = \sum_{i=1}^{n} |i - \pi(i)| .$$

The procedure for determining which time frames to retain and which may be subject to merging is as follows:

1. Cast the original data into a series of (cooccurrence) graphs $G_1, \ldots, G_r$.
2. For every graph G_t, $1 \leq t \leq r$, calculate a repair permutation π_t that minimizes the corresponding adjacency matrix $\text{adj}(G_t)$ with respect to the bandwidth.

[1] The used algorithm actually cannot guarantee that the resulting matrix has minimal bandwidth since this optimization problem is NP-hard.

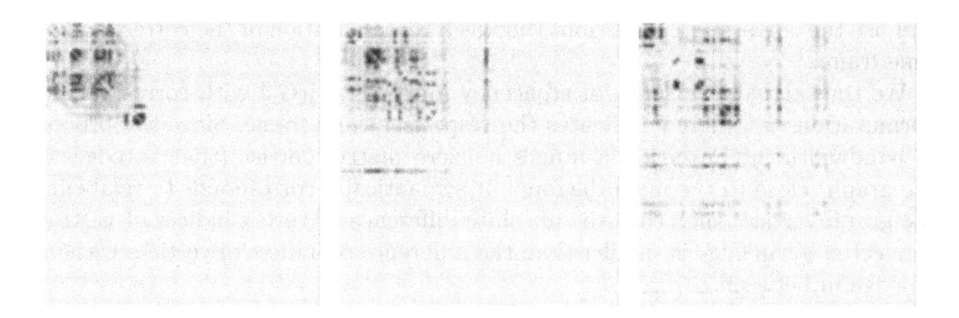

Fig. 2. Left: A minimized bandwidth adjacency matrix of a graph with 101 vertices. Middle and right: Two different permutations applied to the left matrix: The repair count measure quantifies the number and size of steps that are necessary to "repair" the matrices and result in the left one. The repair count of the matrix in the middle is smaller than the right one which meets intuitive user observations. Background color indicates 0, black represents values ≥ 100.

3. Let ρ_t, $2 \leq t \leq r$, be the repair permutation that brings back the matrix $\pi_{t-1} \circ \mathrm{adj}(G_t)$ to minimal bandwidth form.
4. Calculate for every border between time frames $t-1$ and t the repair count $\mathrm{rc}(\rho_t)$.
5. Let the user choose a (data-dependent) threshold θ that states the maximum repair count beyond which we will establish a new time frame.
6. Test all time frame borders and remove those with a repair count smaller than θ.

4 Evaluation

We will now apply the proposed method to a real-world dataset from which the main ideas originate. As sketched in section 3, the underlying data set represents online player contacts within a 3D environment. The logs span a period of 8 months (April to November) and contains over 1 million contacts for 101 locations (which comprise the vertices of the graphs).

The objective is to come up with a temporal discretization of the data that yields better results when looking for subtle changes in the visiting patterns than it would when using the given fixed subdivision by months.

The actual results we refer to are obtained with the technique described in [4]: We intend to find subgraphs (of the large graph containing 101 vertices) which have some user-specified temporal behavior such as becoming more dense or less balanced over time. We suggested various graph measures that assess certain structural properties and let the user fuzzify the respective change rate domains, that is come up with a set of fuzzy sets that quantify whether a series of graph measures is fast increasing, slowly increasing or stable. We then can compute for every series of subgraphs (of the initial cooccurrence graphs) to which degree it satisfies the user's concept.

For this experiment we tracked a subgraph induced by a vertex set of 9 nodes. These nodes have been identified by a simple heuristic given in [4] which considers only vertices of edges with a weight exceeding some predefined threshold. However, every other way of selecting this vertex subset would do (e. g. by using certain definitions of frequent subgraphs [7,8]). The intention shall be to assess how the edge weights balance is behaving. The standard deviation is used to assess the balance of the edges. The left chart of figure 3 shows the standard deviation of the subgraph in all eight months. If we assess the growth (or decline) of this time series with the slope of a regression line (as it is used in [4]), we would conjecture an increasing trend as indicated by the dashed line. This is formally certainly correct, however, the chart suggests that a linear approximation with only one piece is inappropriate since the strong increase is followed by relatively stable values.

We therefore apply the technique suggested in the previous section to get hints for the new time frame borders. The right chart of figure 3 shows the seven repair counts for every adjacency matrix (except the first, of course). The threshold was chosen to be $\theta = 500$. All borders that have a repair count below this value are dropped, all others are kept.

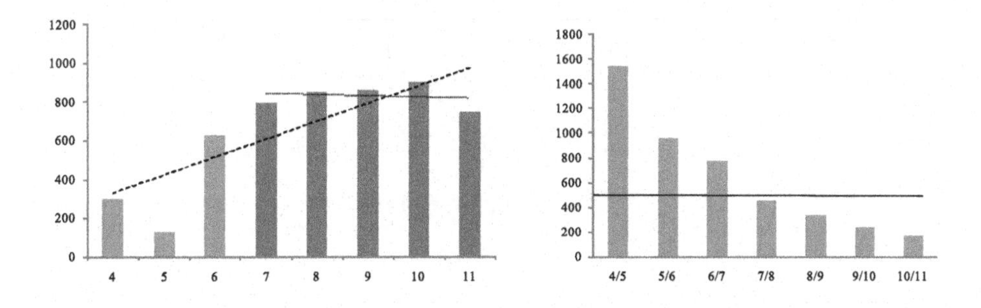

Fig. 3. Left: Standard deviations of edge weights of a designated subgraph. Right: Repair counts for all 7 borders between the 8 time frames of one month length of the data set. The threshold was chosen to be $\theta = 500$, that is we arrived at 4 new time frames.

Therefore, we combine the months July to November and keep May, June and July (and April anyway since it has no preceeding month in the data set). If we now assess the standard deviations in the new time frames, one can clearly assess that the large time frame of the combined months (indicated by the dark gray columns in the left of figure 3) now shows a considerably different development of the edge deviations as depicted by the dotted line.[2] Figure 4 shows all seven original adjacency matrices with their approximation by the previous months model below.

[2] Note that this line was computed by again using a monthwise subdivision but now according the new borders.

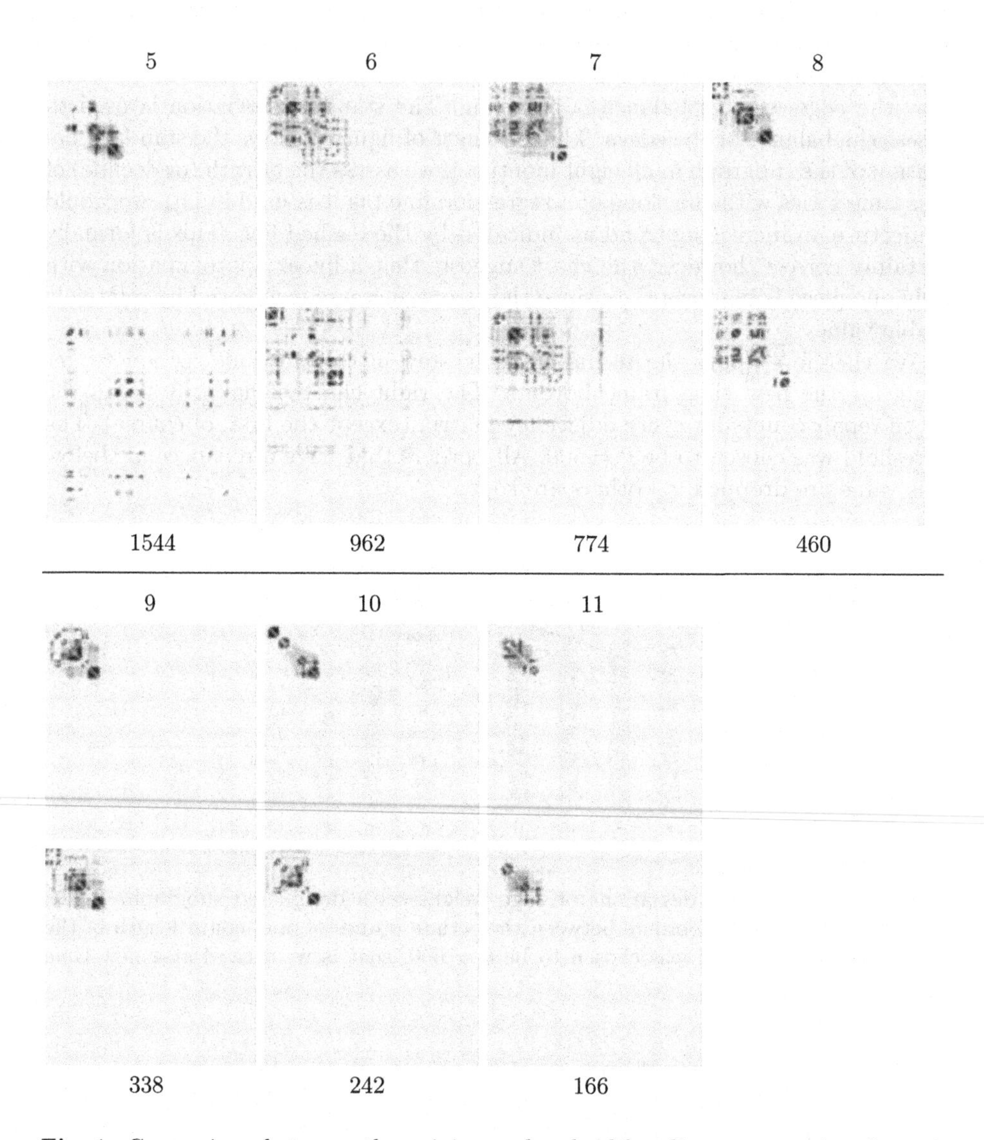

Fig. 4. Comparison between the minimum bandwidth adjacency matrix of every month's graph (respective upper row) and the matrix approximated by the preceeding month (respective lower row). The values below the matrices give the repair counts to create a minimized bandwidth matrix back from the approximations. Background color indicates 0, black represents values ≥ 100.

5 Conclusion

We motivated a method to assess the strength of model changes from one time frame to the next one within a series of cooccurrence graphs. The main objective was to suggest which borders may be dropped (or equivalently which frames may be combined). To solve this task, we used bandwidth-minimized adjacency matrices to represent the graphs. The scatteredness of these matrices provides a visual cue to the user for the strength of model change, while the repair count underpins it with specific numerical values.

We provided empirical evidence that our method can help suggest new potential borders for a given data set. The original data set was subdivided into 8 equidistant frames, while the new one contained 4 frames with the last 5 months combined into a single frame. This allowed to better assess the temporal change of a dedicated subgraph as the values were not subject to a considerable change within the last time frame.

However, since the proposed method is a heuristic, it is possible to handcraft example data sets where the proposed method fails: A graph series with alternating egdes from one month to the next will have a sequence of equally compact bandwidth-minimized adjacency matrices. That is, no large repair count is to be expected while the graphs did change entirely from month to month.

References

1. Agrawal, R., Imielinski, T., Swami, A.N.: Mining Association Rules between Sets of Items in Large Databases. In: Buneman, P., Jajodia, S. (eds.) Proc. ACM SIGMOD Int. Conf. on Management of Data, Washington, D.C, May 26-28, pp. 207–216. ACM Press, New York (1993)
2. Borgelt, C., Steinbrecher, M., Kruse, R.: Graphical Models — Methods for Data Analysis and Mining, 2nd edn. Wiley Series in Computational Statistics. John Wiley & Sons, United Kingdom (2009)
3. Steinbrecher, M., Rügheimer, F., Kruse, R.: Application of Graphical Models in the Automotive Industry. In: Prokhorov, D. (ed.) Computational Intelligence in Automotive Applications. Studies in Computational Intelligence, vol. 132, pp. 79–88. Springer, Berlin (2008)
4. Steinbrecher, M., Kruse, R.: Fuzzy Descriptions to Identify Temporal Substructure Changes of Cooccurrence Graphs. In: Conf. International Fuzzy Systems Association (IFSA), Lisbon, Portugal (to appear, 2009)
5. Steinbrecher, M., Kruse, R.: Identifying Temporal Trajectories of Association Rules with Fuzzy Descriptions. In: Proc. Conf. North American Fuzzy Information Processing Society (NAFIPS 2008), New York, USA, May 2008, pp. 1–6 (2008)
6. Cuthill, E., McKee, J.: Reducing the bandwidth of sparse symmetric matrices. In: Proc. of the 24th National Conference of the Association for Computing Machinery (ACM), 1122 Ave. of the Americas, New York, N.Y, pp. 147–172 (1969)
7. Bringmann, B., Nijssen, S.: What Is Frequent in a Single Graph? In: Washio, T., Suzuki, E., Ting, K.M., Inokuchi, A. (eds.) PAKDD 2008. LNCS (LNAI), vol. 5012, pp. 858–863. Springer, Heidelberg (2008)
8. Bollobás, B.: The Evolution of Random Graphs — the Giant Component. In: Random Graphs, 2nd edn., pp. 130–159. Cambridge Univ. Press, Cambridge (2001)

Maximum a Posteriori Estimation of Dynamically Changing Distributions

Michael Volkhardt, Sören Kalesse, Steffen Müller, and Horst-Michael Gross*

Neuroinformatics and Cognitive Robotics Lab,
Ilmenau University of Technology, Germany
{michael.volkhardt,steffen.mueller,
horst-michael.gross}@tu-ilmenau.de
http://www.tu-ilmenau.de/neurob

Abstract. This paper presents a sequential state estimation method with arbitrary probabilistic models expressing the system's belief. Probabilistic models can be estimated by Maximum a posteriori estimators (MAP), which fail, if the state is dynamic or the model contains hidden variables. The last typically requires iterative methods like expectation maximization (EM). The proposed approximative technique extends message passing algorithms in factor graphs to realize online state estimation despite of hidden parameters. In addition no conjugate priors or hyperparameter transition models have to be specified. For evaluation, we show the relation to EM and discuss the transition model in detail.

Keywords: state estimation, Bayesian filter, MAP, EM, factor graph, conjugate prior.

1 Introduction

Probabilistic modeling techniques provide an appropriate tool set, when dealing with uncertainties in arbitrary systems. When tracking system states, probability theory models the system's belief and defines the required base operations for state estimation [1]. Recently, graphical models have been established as a powerful tool to visualize probabilistic models, whereat the influence of graph theory allows efficient algorithms for probabilistic inference [2], [3], [4], [5]. As described subsequently, this work extends factor graphs, which provide a powerful representation of graphical models for inference by explicitly modeling the model variables and their dependencies [6]. The exchange of messages in acyclic factor graphs is made possible by sequential message passing [7], while loopy belief propagation provides a message passing scheme for cyclic factor graphs [8].

Our aim is the tracking of uncertain system states modeled by arbitrary complex probability distributions (discrete, continuous or mixed). These can be expressed by graphical models or factor graphs, respectively. The parameters of

* This work is partially supported by EU-FP7-ICT Grant #21647 to H.-M. Gross.
M. Volkhardt, S. Müller, H.-M. Gross are with Neuroinformatics and Cognitive Robotics Lab, Ilmenau University of Technology, 98684 Ilmenau, Germany michael.volkhardt@tu-ilmenau.de

B. Mertsching, M. Hund, and Z. Aziz (Eds.): KI 2009, LNAI 5803, pp. 484–491, 2009.

stationary probability distributions can be learned from data samples using maximum likelihood estimation (ML), maximum a posteriori estimation (MAP) or expectation-maximization algorithm (EM). ML fits a parameter set of a probabilistic model to a given data set by solving an optimization problem. MAP augments ML with a conjugate prior distribution on the unknown parameters. By using Bayes' theorem, this distribution can be adapted sequentially with new observations [9]. If the model depends on unobserved latent variables, a sequential estimation is not possible and MAP estimation fails. EM overcomes this problem by iteratively calculating the hidden parameters in a first step and selecting the parameters of interest with ML in a second step in order to maximize the likelihood of the data [10]. Unfortunately, EM is not a sequential method and therefore requires the complete data set of observations.

MAP as well as EM fail, if the probability distribution changes permanently. When using Bayesian filtering for tracking states expressed by complex distributions, it is very hard to model the transition model and the conjugate prior of the system's state. This paper addresses these problems and implements an online message passing algorithm in extended factor graphs that allows approximative state tracking.

The remainder of this paper is organized as follows. Section 2 describes problems that occur while estimating a system's state represented by a complex probability distribution. To solve these problems, dynamic MAP estimation in extended factor graphs is presented in Sect. 3. For evaluation we show the relation of the developed method to EM algorithm. The paper concludes with a discussion on the assumptions and limitations of the algorithm and a summary.

2 Bayesian Filtering of Complex Probability Distributions

We assume an exemplary system state modeled by a factor graph with random variable X that is dependent on an unobserved variable Y. The belief on the system's state hence is defined by factor potential $p(\mathbf{Z}) = p(X|Y)$. This conditional probability distribution is defined by parameters $\mathbf{\Theta}$. To estimate the unknown system's state, we search for probability distribution $p(\mathbf{\Theta})$.

MAP estimation can be applied to sequentially update the unknown probability distribution $p(\mathbf{\Theta})$ if distribution $p(\mathbf{Z})$ is constant and all variables are observed. In the context of state estimation, the problem of a permanently changing probability arises. By introducing a transition model the Bayesian filter is able to track the system's belief over time. The recursive Bayesian filter equation adapted to parameter estimation is defined by:

$$Bel(\mathbf{\Theta}_N) = \eta\, p(\mathbf{Z} = \mathbf{z}_N|\mathbf{\Theta}_N) \int_{\mathbf{\Theta}_{N-1}} p(\mathbf{\Theta}_N|\mathbf{\Theta}_{N-1}) Bel(\mathbf{\Theta}_{N-1})\, d\mathbf{\Theta}_{N-1} \ , \quad (1)$$

where $p(\mathbf{Z} = \mathbf{z}_N|\mathbf{\Theta}_N)$ denotes the observation model, $p(\mathbf{\Theta}_N|\mathbf{\Theta}_{N-1})$ denotes the transition model and $Bel(\mathbf{\Theta}_{N-1})$ is the prior belief on the unknown parameters. The left term of the equation describes the posterior probability and

$\eta = p(\mathbf{Z} = \mathbf{z}_N)$ is a normalization term. Figure 1 shows the parameter estimation on the basis of the Bayesian filter in a factor graph, whereby arrows indicate the messages sent to estimate the posterior belief.

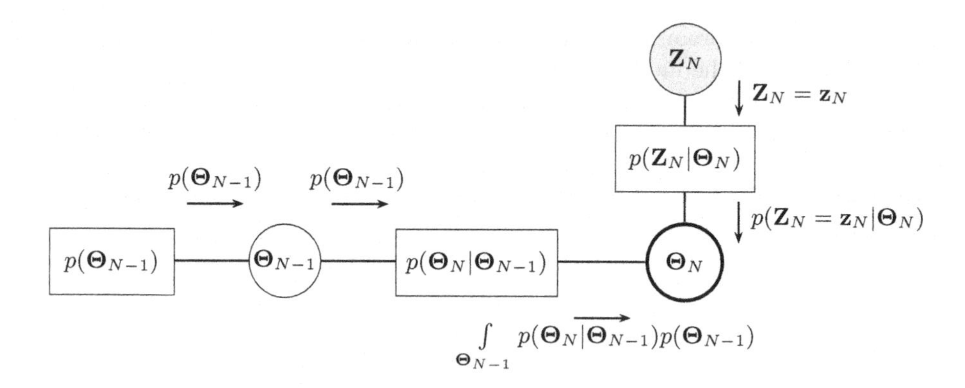

Fig. 1. Parameter estimation with Bayesian filter. The system state $\mathbf{\Theta}$ can be updated by new incoming observations $\mathbf{z}_N$.

When using arbitrary distributions for state representation several problems arise. First, when sending a message from variable node $\mathbf{\Theta}_{N-1}$ to factor node $p(\mathbf{\Theta}_N|\mathbf{\Theta}_{N-1})$ the marginal of the probability distribution $p(\mathbf{\Theta}_{N-1})$ is required. This corresponds to the conjugate prior of the distribution $p(X|Y)$. For arbitrary distributions, modeling the conjugate prior is very hard [9]. Another problem is the design of the conditional probability distribution $p(\mathbf{\Theta}_N|\mathbf{\Theta}_{N-1})$ which has to transform the hyperparameters into a new time step. Additionally, some variables of the state representation can depend on unobserved variables.

3 MAP Estimation of Dynamically Changing Distributions

This section presents an extended factor graph structure and a loopy belief propagation algorithm with augmented message types, that addresses the aforementioned problems. Without loss of generality, we show the state estimation of probability distribution $p(\mathbf{Z}) = p(X|Y)$, where variable Y is hidden.

Figure 2 shows the proposed architecture. We assume a two layered dynamic factor graph, that combines the inference on latent variables in each time step with the estimation of factor potentials over time. The lower layer represents the system's state in its factor potentials and corresponds to a factor graph with conventional algorithms for message passing and inference. The upper layer replaces the transition model and the conjugate prior of the state's probability distribution defined by the factor potential, which is going to develop over time (compare Fig. 1). Thereto, every factor node of the lower layer is linked to a

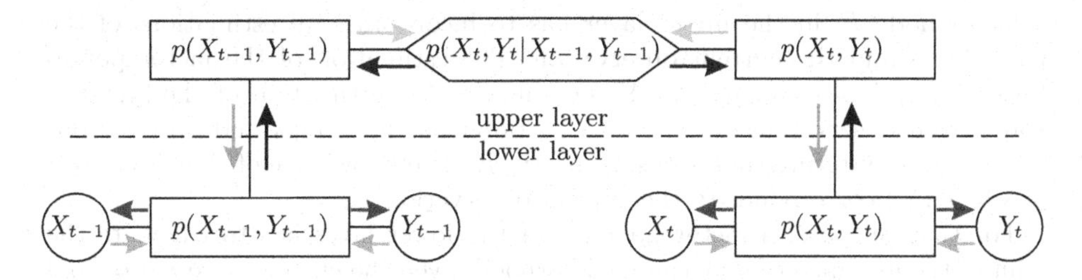

Fig. 2. Extended dynamic factor graph. The figure shows a two layered factor graph, which avoids modeling of conjugate priors and the transition model of the hyper-parameters by estimating factor potentials over time. The messages in the proposed architecture are coded in terms of color.

counterpart in the upper layer, where it acts similar to a variable node in conventional factor graphs. We introduce a new node type denoted by the diamond shape: the hyperfactor node. It acts exactly like a factor node in conventional factor graphs and represents the transition model. Note that we integrated the factor potential $p(Y)$ in distribution $p(X|Y)$ to receive a factor potential $p(X,Y)$ to simplify the example.

3.1 Message Types

We apply loopy belief propagation for message passing, where each node iteratively calculates and updates all outgoing messages. Not yet calculated messages are supposed to be uniformly distributed.

A message from a variable node V_i to a factor node F_j (yellow arrows in Fig. 2) is defined by common sum product algorithm:

$$\mu_{V_i \to F_j}(\mathbf{X}_{V_i}) = \prod_{F_k \in ne(V_i) \setminus F_j} \mu_{F_k \to V_i}(\mathbf{X}_{V_i}) \ . \tag{2}$$

The variable node V_i multiplies all incoming messages from connected factor nodes $F_k \in ne(V_i)$, except destination node F_j.

A message from a factor node F_i to a variable node V_j (red arrows in Fig. 2) is also defined by common sum product algorithm:

$$\mu_{F_i \to V_j}(\mathbf{X}_{V_j}) = \int_{V_k \in ne(F_i) \setminus V_j} \left(\mu_{F_i^u \to F_i^l}(\mathbf{X}_{F_i}) \prod_{V_k \in ne(F_i) \setminus V_j} \mu_{V_k \to F_i}(\mathbf{X}_{V_k}) \right) dV_k \ . \tag{3}$$

The factor node F_i multiplies all incoming messages from connected variable nodes $V_k \in ne(F_i)$, except destination node V_j with its factor potential. The factor potential complies with the message from the upper layer $\mu_{F_i^u \to F_i^l}(\mathbf{X}_{F_i})$ defined in (5). The result is marginalized on the correct type of the destination node by integrating over all other variables.

The factor potentials should be transferred in the new state by applying the transition model in the hyperfactor nodes. Therefore, the factor potential of a factor node F_i in the upper layer has to incorporate all estimations of the parameters and the current observation. The estimation of the factor potentials $p(X_{i-1}, Y_{i-1})$ and $p(X_{i+1}, Y_{i+1})$ comprise the estimation of the system's state given observations $x_1, \ldots, x_{i-1}$ of previous time steps and observations $x_{i+1}, \ldots, x_N$ of future time steps, respectively. The knowledge of the lower layer $p(X = x_i, Y)$ corresponds to the current observation x_i.

For that purpose, a factor node F_i^l of the lower layer has to calculate the joint of the marginal of its local neighborhood given the current observation and sends it to the factor node F_i^u in the upper layer (dark blue arrows in Fig. 2):

$$\mu_{F_i^l \to F_i^u}(\mathbf{X}_{F_i}) = \prod_{V_k \in ne(F_i^l)} \mu_{V_k \to F_i^l}(\mathbf{X}_{V_k}) \prod_{V_k \in ne(F_i^l)} \mu_{F_i^l \to V_k}(\mathbf{X}_{V_k}) \ . \tag{4}$$

The factor node multiplies all incoming messages $\mu_{V_k \to F_i^l}(\mathbf{X}_{V_k})$ from connected variable nodes $V_k \in ne(F_i^l)$ and all outgoing messages $\mu_{F_i^l \to V_k}(\mathbf{X}_{V_k})$ to these nodes. The incoming messages from connected variable nodes correspond to messages, that the variable nodes received themself from possible existing other factor nodes in the lower layer and possible observations. To receive the correct marginals of the variable nodes in the local neighborhood, the messages from the factor node F_i^l to the variable nodes are multiplied.

The factor node of the lower layer should use the estimation from all time steps, except the current, to estimate its potential. Therefore, a message from the upper layer to the lower layer (light blue arrows in Fig. 2) results in:

$$\mu_{F_i^u \to F_i^l}(\mathbf{X}_{F_i}) = \prod_{H_k \in ne(F_i^u)} \mu_{H_k \to F_i^u}(\mathbf{X}_{F_i}) \ . \tag{5}$$

The factor node multiplies all incoming messages from all connected hyperfactor nodes $H_k \in ne(F_i^u)$.

The algorithm to calculate a message from a hyperfactor node H_i to a factor node F_j (dark green arrows in Fig. 2) is exactly the same as a message of a factor node in traditional sum product algorithm:

$$\mu_{H_i \to F_j}(\mathbf{X}_{F_j}) = \int_{F_k \in ne(H_i) \setminus F_j} \left(p\left(ne\left(H_i\right)\right) \prod_{F_k \in ne(H_i) \setminus F_j} \mu_{F_k \to H_i}(\mathbf{X}_{F_k}) \right) dF_k \ . \tag{6}$$

The hyperfactor node multiplies all incoming messages from connected factor nodes $F_k \in ne(H_i)$, except destination node F_j with its potential $p(ne(H_i))$. The result is marginalized on the correct domain of the destination node.

Finally, we have to define a message from a factor node F_i^u in the upper layer to a hyperfactor node H_j (light green arrows in Fig. 2):

$$\mu_{F_i^u \to H_j}(\mathbf{X}_{F_i}) = \left(\prod_{H_k \in ne(F_i^u) \setminus H_j} \mu_{H_k \to F_i^u}(\mathbf{X}_{F_i}) \right) \oplus \mu_{F_i^l \to F_i^u}(\mathbf{X}_{F_i}) \ . \tag{7}$$

The factor node F_i^u multiplies all messages from connected hyperfactor nodes $H_k \in ne(F_i^u)$, except the destination H_j. This corresponds to the previous and future estimations. The message from the lower layer $\mu_{F_i^l \to F_i^u}(\mathbf{X}_{F_i})$ corresponds to the current observation. It is added by a MAP estimate step denoted by operator $\oplus$. The realization of the MAP estimation function depends on the type of distribution used within the factor potentials. The estimate step is possible even with hidden parameters, because the algorithm provides an estimation for these parameters. This concept is closely related to EM.

3.2 Relation to Expectation Maximization

EM finds maximum likelihood solutions for models having latent variables. Given a joint distribution $p(X, Y|\mathbf{\Theta})$, where variable Y is latent, the goal is to maximize $p(X|\mathbf{\Theta})$ with respect to $\mathbf{\Theta}$ [2]. Figure 3 shows the E step of EM interpreted as an inference problem in a factor graph. If one assumes a prior on the parameters $\mathbf{\Theta}$, the marginal $p(Y|X, \mathbf{\Theta})$ can be inferred given all observations $X = \{x_1 \dots x_N\}$ by message passing. The factor graph in Fig. 3 matches the lower layer of the extended factor graph presented in Fig. 2. Therefore, the calculation of the marginals and the message from the lower to the upper layer defined in (4) correspond to the expectation of an observation.

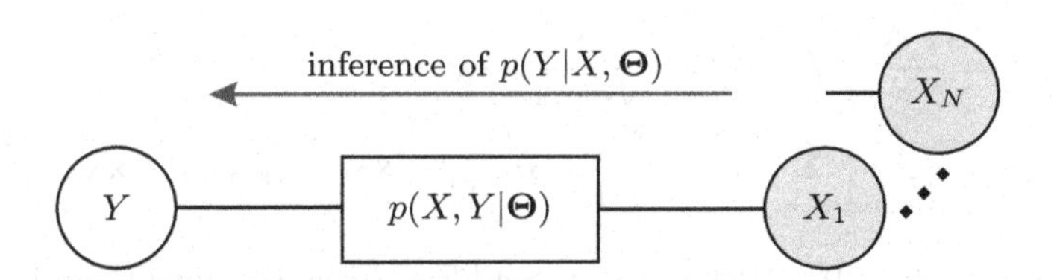

Fig. 3. Expectation step in a factor graph. The inference of marginal $p(Y|X, \mathbf{\Theta})$ corresponds to the E step of EM algorithm.

With the inferred marginal $p(Y|X, \mathbf{\Theta})$, all parameters of the probability distribution can be treated as being observed, and the maximization step follows. The M step evaluates $\mathbf{\Theta}'$ with:

$$\mathbf{\Theta}' = \arg\max_{\mathbf{\Theta}} \left(\sum_Y p(Y|X, \mathbf{\Theta}) \ln p(X, Y|\mathbf{\Theta}) \right) \tag{8}$$

The two steps are iterated successively to update the estimation of the model parameters. In the proposed architecture the M step is replaced by a MAP estimation step, which integrates the message of the lower layer into the factor potential in each time step in (7). Therefore, the belief of the state – represented by the parameters of the factor potential – is updated sequentially by splitting up the sums of the arg max operation in (8) into local operations in each time step of a dynamic factor graph. The MAP estimation step of discrete distributions using conjugate priors is equivalent to a weighted sum of the distributions [2]. For $\lim_{\Delta x \to 0}$ we can transfer this observation to arbitrary distributions. Therefore, in case of hardly manageable conjugate priors of arbitrary distributions we suggest to approximate the MAP estimation step (operation $\oplus$ in (7)) by a weighted sum. In case of manageable conjugate prior distributions on hyperparameters a regular MAP estimation step can be applied instead. The iterations of the loopy belief propagation correspond to the iterations of EM algorithm. Additionally, the hyperfactor node in the upper layer introduces a transition model, that has to be considered in the estimation process.

3.3 Transition Model

Generating the transition model is very intuitive, because it acts directly on the variables of the factor potentials instead of the hyperparameters of the model. For convenience, we assume Gaussians for variable $\mathbf{X}$ with $p(\mathbf{x}|\boldsymbol{\mu}, \boldsymbol{\Sigma})$. One receives similar results by using discrete or mixed distributions [2]. When using an identical transition model the common MAP estimation can be utilized to add the current observation to the estimation in the upper layer like in (7).

After N observations, there should exist estimations for the parameters $\boldsymbol{\mu}$ and $\boldsymbol{\Sigma}$ with $\hat{\boldsymbol{\mu}}_N$ and $\hat{\boldsymbol{\Sigma}}_\mathbf{N}$. The parameter of the Gaussian after the integration of the new observation $\mathbf{X} = \mathbf{x}_{N+1}$ result in:

$$\begin{aligned} \hat{\boldsymbol{\mu}}_{N+1} &= \frac{N}{N+1}\,\hat{\boldsymbol{\mu}}_N + (1 - \frac{N}{N+1})\,\mathbf{x}_{N+1} \\ \hat{\boldsymbol{\Sigma}}_{N+1} &= \frac{N}{N+1}\,\hat{\boldsymbol{\Sigma}}_N + \frac{N}{(N+1)^2}(\hat{\boldsymbol{\mu}}_{N+1} - \mathbf{x}_{N+1})(\hat{\boldsymbol{\mu}}_{N+1} - \mathbf{x}_{N+1})^T \;. \end{aligned} \tag{9}$$

Intuitively, the new observation $\mathbf{x}_{N+1}$ is added to the estimation and weighted. The weight accounts for the prior knowledge about the location and form of the distribution given the previous observations. We introduce an additional weighting factor $\alpha = [0, 1]$ to weaken the influence of the former observations of the MAP estimation step in the upper layer in (7). Thus the term $\frac{N}{N+1}$ of (9) is weighted as follows:

$$\frac{\alpha N}{\alpha N + 1} \;. \tag{10}$$

The weighting factor α is set to 1 if the transition model is identical:

$$p(\boldsymbol{\Theta}_N|\boldsymbol{\Theta}_{N-1}) = \begin{cases} 1 & \text{, if } \boldsymbol{\Theta}_N = \boldsymbol{\Theta}_{N-1} \\ 0 & \text{, else} \end{cases} \tag{11}$$

Once the transition model is non identical, it blurs the unknown parameters of the distribution in the hyperfactor nodes of every time step. Hence, the weight of former observations has to be adopted by decreasing α. The weighting factor is calculated empirically depending on the uncertainty of the transition model. Therefore, α constitutes the difference between keeping previous learned knownledge or integrate new knowledge into the model given by an observation.

4 Discussion and Conclusion

The used loopy belief propagation only provides an approximation of the true marginals. Additionally, the convergence of loopy belief propagation is not proven for arbitrary graph structures. Because the concept is closely related to EM, it is not guaranteed that the algorithm converges to the global maximum. The weighting of the current observation in (10) when using non identical transition models is hardly possible if the upper layer of the extended factor graph contains loops. This is primarily induced by circulating messages in the upper layer, which make it nearly impossible to find an adequate value of α. The algorithm works for arbitrary distributions as long as the MAP estimation operation of one observation to the distribution is defined or a weighted sum can be approximated. Last but not least, the algorithm is only real time capable for small graphs with few nodes. Otherwise, loopy belief propagation algorithm needs many computational expensive iterations to converge.

This paper presented a concept to allow handling of online state estimation represented by dynamic probability distributions without the need for conjugate priors or hyperparameter transition models. The proposed approximative method extends loopy belief propagation in factor graphs. The key idea is to substitute the ML in the M step of EM by a MAP estimation, which can be applied recursively. The E step is done locally and the iterations of the EM algorithm are shifted into the iterations of loopy belief propagation. Despite of MAP estimators that do not work with hidden parameters or EM algorithm that needs the complete data set of observations, our algorithm offers the possibility to track complex non-stationary system states over time, even if they depend on hidden parameters.

References

1. Thrun, S., Burgard, W., Fox, D.: Probabilistic Robotics. MIT Press, Cambridge (2005)
2. Bishop, C.M.: Pattern Recognition and Machine Learning. Springer Science+ Business Media, Secaucus, NJ, USA (2006)
3. Lauritzen, S.L.: Graphical Models. Clarendon Press, Oxford (1996)
4. Jordan, M.I., Sejnowski, T.J.: Graphical Models: Foundations of Neural Computation. MIT Press, Cambridge (2001)
5. Murphy, K.P.: Dynamic Bayesian Networks: Representation, Inference and Learning. UC Berkeley (2002)
6. Kschischang, F.R., Frey, B.J., Loeliger, H.A.: Factor graphs and the sum-product algorithm. IEEE Transactions on Information Theory 47, 498–519 (2001)
7. Pearl, J.: Probabilistic Reasoning in Intelligent Systems: Networks of Plausible Inference. Morgan Kaufmann Publishers, San Mateo (1988)
8. Frey, B.J., Kschischang, F.R., Loeliger, H.A., Wiberg, N.: Factor graphs and algorithms. In: Proceedings of the 35th ACOCCC 1998, pp. 666–680 (1997)
9. Diaconis, P., Ylvisaker, D.: Conjugate priors for exponential families. Annals of Statistics 7, 269–281 (1979)
10. Dempster, A.P., Laird, N.M., Rubin, D.B., et al.: Maximum likelihood from incomplete data via the EM algorithm. JRSS 39, 1–38 (1977)

Kinesthetic Bootstrapping: Teaching Motor Skills to Humanoid Robots through Physical Interaction

Heni Ben Amor, Erik Berger, David Vogt, and Bernhard Jung

VR and Multimedia Group, TU Bergakademie Freiberg, Freiberg, Germany
{amor,bergere,vogt3,jung}@mailserver.tu-freiberg.de

Abstract. Programming of complex motor skills for humanoid robots can be a time intensive task, particularly within conventional textual or GUI-driven programming paradigms. Addressing this drawback, we propose a new programming-by-demonstration method called *Kinesthetic Bootstrapping* for teaching motor skills to humanoid robots by means of intuitive physical interactions. Here, "programming" simply consists of manually moving the robot's joints so as to demonstrate the skill in mind. The bootstrapping algorithm then generates a low-dimensional model of the demonstrated postures. To find a trajectory through this posture space that corresponds to a robust robot motion, a learning phase takes place in a physics-based virtual environment. The virtual robot's motion is optimized via a genetic algorithm and the result is transferred back to the physical robot. The method has been successfully applied to the learning of various complex motor skills such as walking and standing up.

1 Introduction

Research in robotics and AI has lead to the emergence of increasingly complex anthropomorphic robots, such as humanoids and androids. In order to be meaningfully applied in human inhabited invironments, anthropomorphic robots need to possess a variety of physical abilities and skills. However, programming such skills is a labour and time intensive task which requires a large amount of expert knowledge. In particular, it often involves transforming intuitive concepts of motions and actions into formal mathematical descriptions and algorithms. Even in GUI-based programming environments where complex robot movements are specified as sequences of robot postures defined in the graphical user interface, much time is usually spent for parameter tweaking. Due to their relatively large number of degrees of freedom, this process becomes particularly cumbersome for the case of humanoid robots. To reduce the complexity of this task, more natural and intuitive approaches to programming robot skills are called for.

This paper presents a new programming-by-demonstration method for bootstrapping robotic motor skills through kinesthetic interactions. A human teacher instructs the robot by manually moving the robot's joints and body to postures that approximate the intended movement. Then, an automatic optimization

B. Mertsching, M. Hund, and Z. Aziz (Eds.): KI 2009, LNAI 5803, pp. 492–499, 2009.

phase takes place during which the robot learns a motor skill that still resembles but also compensates for likely imperfections of the demonstrated movement. This learning phase makes use of a physics-based virtual enviroment, where a large amount of movement variations can be tried out very quickly without the need for human intervention. In this way, the Kinesthetic Bootstrapping method introduced in this paper both simplifies and reduces the time for programming of robotic motor skills.

2 Related Work

The work presented in this paper can be regarded as a variant of imitation learning. In the context of robotics, the goal of imitation learning is to allow human teachers to program robotic agents by conveying a demonstration of the desired behavior. This is often realized using expensive motion capture or virtual reality techniques in order to record the example motions. In [1] the walking gait of a human demonstrator is first recorded through motion capture and then adapted for imitation by a small humanoid robot. However, such approaches need to tackle the *correspondence problem*: the question of how to map the joint information of the user onto the robot's body [2]. A variation of this approach can be found in [3]. Here, actions are recorded while the human demonstrator is interacting with a virtual reality environment using a data glove. Recently, more intuitive approaches to programming by demonstration have been pursued. In [4], direct kinesthetic interaction with humanoid robots has been used for teaching manipulation skills. In this approach, the instructor has to repeatedly convey demonstrations and provide corrective feedback to the learning robot. A similar approach has been used by Tani et al. [5] to encode demonstrated behaviors using recurrent neural networks. In both papers the imitated behaviors where limited to the upper body of the robot and did not involve complex, dynamic motions. In contrast to these works, the approach introduced in this paper works even with a *single* example demonstration but can also deal with several examples. The robot can adapt and improve the provided example autonomously, without relying on any further interaction with the instructor. Further, providing the example motion through a kinesthetic modality, allows us to increase the naturalness of the interaction. This is closely related to "physical programming languages" [6] found in human-computer interaction research.

3 Kinesthetic Bootstrapping

When learning a new physical skill, children are often supported by their parents. This allows to transmit knowledge on how to solve the task at hand and, thus, overcome learning barriers. In these situations, kinesthetic interactions serve as a communication channel between the parent and the learning child. The bodily experience resulting from these interactions helps to reduce the amount of time needed for acquiring the skill. Still, the child has to go through an unassisted learning phase in order to fully master the skill. Kinesthetic Bootstrapping

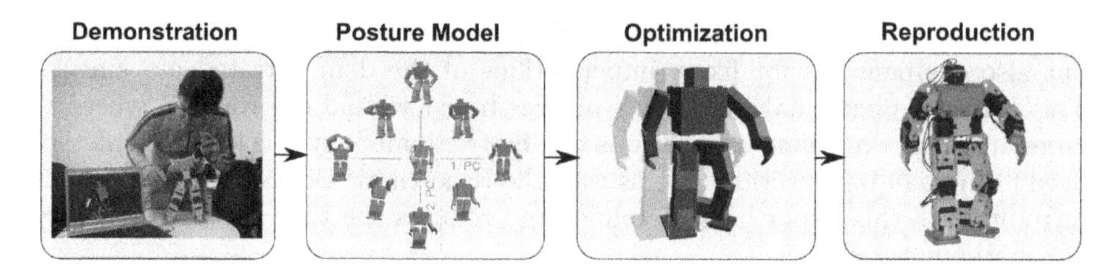

Fig. 1. Overview of the Kinesthetic Bootstrapping approach. After kinesthetic interaction, a posture model is created. A simulator is used to optimize the demonstrated skill. The result is then applied on the real robot.

applies the same principle to the programming of humanoid robots. A human conveys a demonstration of the task at hand through kinesthetic interactions. The bodily experience allows the robot to draw important information on the task at hand. This information is used to "bootstrap" the robot's knowledge, which is then used in a learning phase to reproduce the skill without any assistance. Figure 1 shows an overview of the learning approach used in Kinesthetic Bootstrapping.

First, the teacher moves the joints of the robot in order to convey a demonstration of the intended motion or behavior. This is done continuously without relying on keyframes or another kind of discretization. During the demonstration, the motor configurations of the robot are recorded with a frequency of 20 Hz. The robot used in this study is a *Bioloid* robot with 18 servo-motors (i.e. 18 degrees-of-fredom). In each step, the state of each of all servo-motors is recorded, resulting in an 18-dimensional posture vector p. Once the user finishes the demonstration, all posture vectors are collected in order to compute a low-dimensional posture model of the skill. Next, using the extracted posture model, different variations of the skill are evaluated. This is done in a physics-based virtual reality simulation of the robot. The simulator allows us to optimize the motion without harming the robot hardware and without reyling on human assistance. In particular, when the demonstrated motion is very dynamic, such as a standing up motion, it is important for the robot to learn how to account for the external (stabilizing) forces previously applied by the human teacher. Once the optimization phase is finished, the learned motion is transfered to the physical robot and replayed outside of the simulation. In the remainder of this section, we will explain each phase of the learning process in more detail.

3.1 Simulator

As mentioned above, learning and adaptation of the demonstrated skill is performed in a physics-based virtual reality simulator. The simulator is based on the Open Dynamics Engine (ODE) and contains a precise model of the Bioloid humanoid robot. For calibrating the model, each motor is automatically moved by the calibration software and the time needed to reach a given configuration by the real and simulated robot is measured. The difference between the two time values, i.e. the discrepancy between the real and simulated world, is used to adapt the values of the low-level PID controller in simulation, so as to better fit the movements of the real robot. An important feature of this simulator is an abstraction layer for the control of the robot. This layer allows it to control the real or simulated robot or both using the same interface. Thus, the user can always decide whether to apply the current program in reality or simulation. During the learning phase, the simulator is used for reproduction of different variations of the originally demonstrated motion. In a trial-and-error fashion each variation is executed by the virtual robot, evaluated and the result used for further optimization.

3.2 Low-Dimensional Posture Models

The kinesthetically recorded demonstration can be regarded as a template from which important information about the skill in mind is inferred. More precisely, the demonstration is used to compute a model, from which new variations of the skill can be synthesized. In the following, such models will be referred to as *low-dimensional posture models*. They are extracted by applying dimensionality reduction techniques on the dataset P of recorded postures p_i. The resulting low-dimensional space of postures can have arbitrary dimensions d, with $d << 18$. Without loss of generality, in the following explanation, we will use a two-dimensional posture space ($d = 2$).

In Figure 2 (left) we see an example of a low-dimensional posture model. The model was computed based on demonstrations of two-handed grabbing or grasping. Different techniques such as PCA, LLE or Isomap can be used for this purpose. Figure 2 (right) depicts the reprojection error of applying such techniques to the robot postures. We found that for most skills, even with a

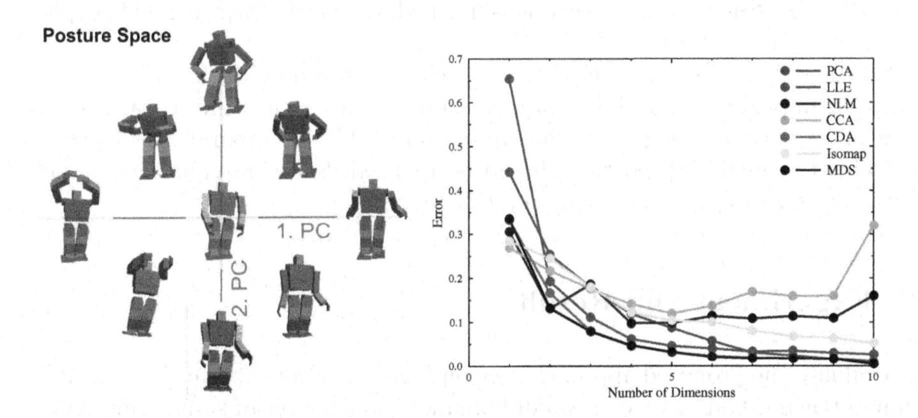

Fig. 2. Left: A representation of the low-dimensional space corresponding to the posture model for grasping with both hands. Right: The projection error resulting from applying different dimensionality reduction techniques to the recorded robot postures.

simple PCA, 95% of the original information can be retained using only four principal components.

Each position in the posture space corresponds to a posture of the robot. In the figure we see the postures resulting from projecting some points back into the space of 18 joint values. Because we get a continuous space, we can compute interpolations and extrapolations of recorded postures. Motions can be synthesized by simply specifying a trajectory in this space. Each point along the trajectory reflects a posture of the robot at a particular time step of the motion.

3.3 Learning

After dimensionality reduction, machine learning techniques are used to optimize the demonstrated skill. As a starting point for the learning algorithm, we use the low-dimensional trajectory of the demonstration. For this, the set P is projected into the posture space, yielding a new set P' of points specifying a d-dimensional trajectory. Next, P' is approximated using n control points $C = \{c_1, .., c_n\}$ specifying a spline curve. The spline can be regarded as a highly compressed representation of the demonstrated motion. Instead of using all points of the original trajectory P' only a limited number n of control points is used for learning.

More precisely, the control points C are used as an initializing individual of a real-coded Genetic Algorithm (GA). A set of slightly perturbed variants of C are created in the initial population of the GA. Each individual is then processed, and the corresponding motion executed by the simulated robot. This is done, by reprojecting each point along the encoded trajectory back to the original space of joint values. Using a user-provided fitness function, each individual is then evaluated and assigned a fitness value. Once all fitness values are determined, the best chromosomes are selected, mated and mutated according to the typical rules of a GA. Finally, when learning is finished, the newly learned skill is applied on the real physical robot.

When performing dynamic motions, such as a standing up behavior, timing plays an important role. Therefore, we add add a special time parameter for each of the control points into the chromosome. The time parameter indicates at which timestep each posture should be realized. Each individual in the GA, thus, consists of the set of values $\{c_1, t_1, .., c_n, t_n\}$.

4 Experiment and Results

To evaluate the proposed approach, we conducted a set of experiment in which a human teacher had to teach a small humanoid robot a set of skills using Kinesthetic Bootstrapping. Among others, the robot learned to perform a headstand, stand up by itself, and walk. In the following we will focus on the standing up and walking skills. In all experiments, PCA was used as a dimensionality reduction technique. The number of dimensions d was set to 4.

The teacher was given about 15 minutes time to kinesthetically demonstrate the respective skill. In Figure 3 we see the result of directly replaying the demonstrated skill. Because of the missing support of the teacher, the robot failed to stand up by itself. Next, we run the optimization as described in section 3.3. The number of control points n was 25. For the 'standing up' skill, the fitness values were determined based on the sum of the z-values (=height) of the robot's head position. The trial was aborted, if the robot's head was below a given threshold, i.e. the robot fell down during the simulation. Figure 4 shows the result of the optimized skill in simulation, and after application on the real robot. As can be seen, the robot learned to stand up by modifying the original motion. In particular, the hip motion was changed such that the robot can lift the torso up, without losing balance (2. picture from right). By moving the hip backwards to an extreme position, the zero moment point of the robot remains between the legs. The result is an elegant solution to the problem of standing up.

For the walking skill, the fitness value of each individual was determined, using the distance traveled from the starting position without falling down. The number n of control points was set to 12. In Figure 5(left) we see the low-dimensional trajectories resulting from the control points of the walking skill before and

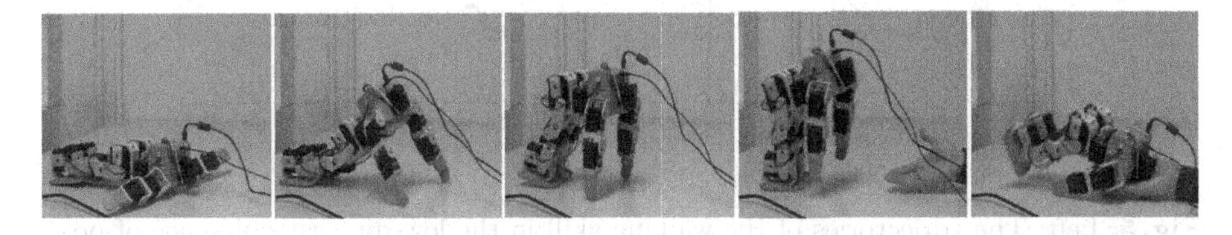

Fig. 3. Direct replay of a demonstrated standing up skill by the small humanoid robot. The robot fails to stand up, because of the missing support forces of the human teacher.

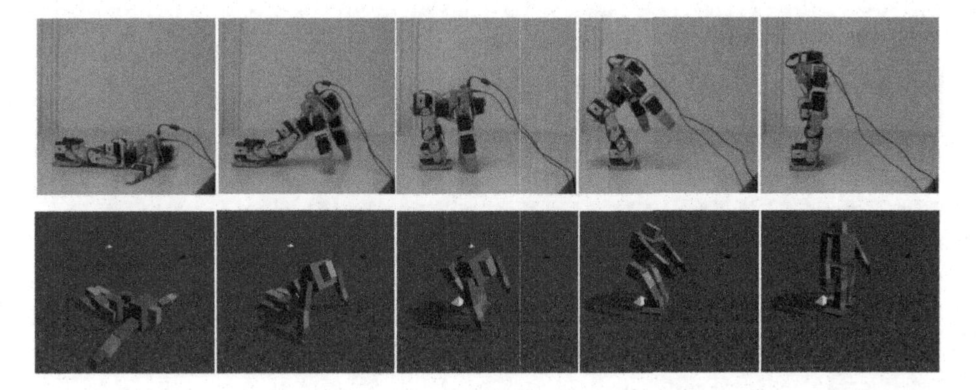

Fig. 4. Results of applying the evolved standing up behavior in simulation and on the real robot. The robot learned to move the hip backwards to an extreme position, so as to pull up the torso without falling forwards.

after optimization. The trajectories show the values of the control points in the first and second principal component. The lower-order components, in this case the first and second component, contain the "most important" aspects of the data. Thus, by visualizing the first two components, we can see most important changes to the robot motion. Figure 5(right) shows the evolution of the fitness values during optimization. With each generation of the GA, the robot managed to travel larger and larger distances. However, after the GA finished, we found that the best individual in the simulation did not lead to a stable walk in reality. This unveils a common pitfall of using a simulator: even the best simulation is only an approximation of the real world. Fortunately, GAs allow for a simple solution to this problem. By testing the best individuals of earlier generations, we can search for solutions that are transferable to the real world. In Figure 6 we see

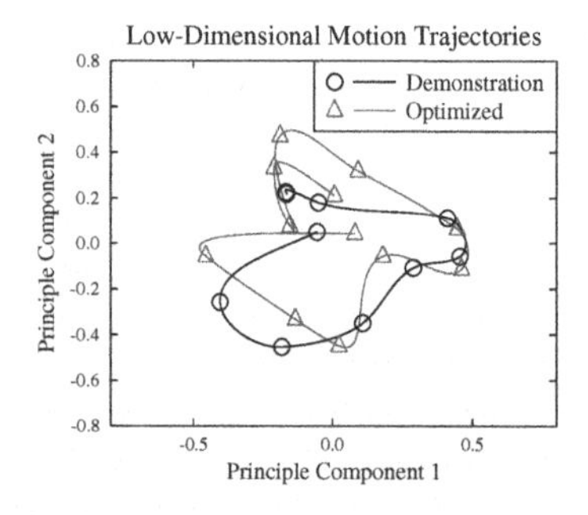

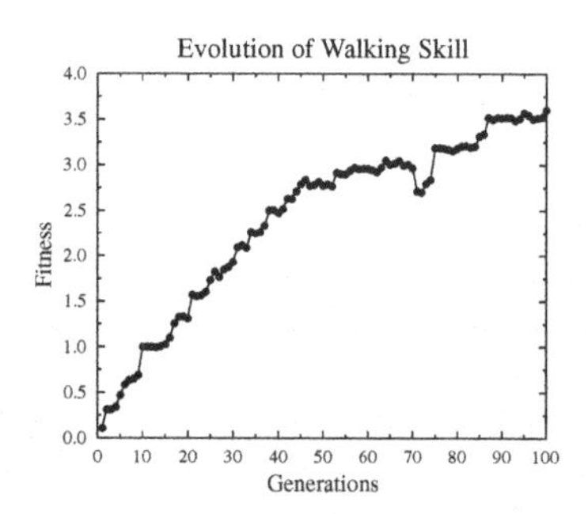

Fig. 5. Left: The trajectories of the walking skill in the low-dimensional space of postures. Right: The evolution of the fitness value during the learning of the skill. The fitness was determined using the distance traveled by the robot.

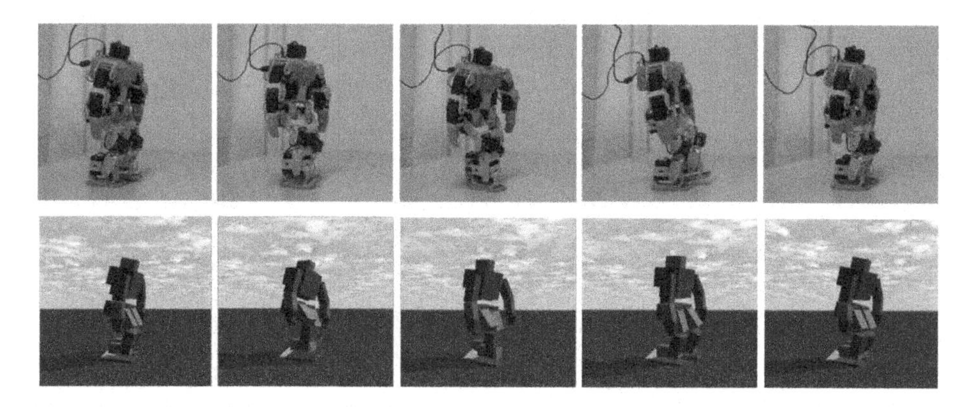

Fig. 6. The result of applying the evolved walking behavior in simulation and on the real robot. Learning this skill only involved a 5 minute kinesthetic demonstration, in which the legs where moved by the human, and the specification of the fitness function.

an evolved stable walking pattern. It corresponds to the fittest individual from generation 50. In later generation, the GA exploited the characteristics of the simulator too much and, thus, generated an individual that was not applicable on the real environment.

5 Conclusion

Up to now, the standard method for creating new motions and behaviors is through low-level programming or through the use of graphical user interfaces. Both approaches are labour intensive and do not support the automatic optimization of the specified behavior. In this paper we presented a new approach to programming humanoid robots, which relies on physical interaction between a human teacher and a learning robot. By optimizing the demonstrated behavior in a virtual environment we can speed up learning times and reduce the need for human intervention. Further, by introducing *low-dimensional posture models*, we were able to integrate human knowledge into the learning process. In the future we will investigate the use of low-dimensional posture models in conjunction with other learning techniques such as Reinforcement Learning or Neural Networks. We also plan to use the described technique on more sophisticated android robots with more degrees of freedom.

References

1. Chalodhorn, R., Grimes, D.B., Grochow, K., Rao, R.P.N.: Learning to walk through imitation. In: IJCAI, pp. 2084–2090 (2007)
2. Nehaniv, C.L., Dautenhahn, K.: The correspondence problem. In: Imitation in animals and artifacts, pp. 41–61. MIT Press, Cambridge (2002)
3. Aleotti, J., Caselli, S., Reggiani, M.: Leveraging on a Virtual Environment for Robot Programming by Demonstration. In: IEEE/RSJ Intl. Conf. on Intelligent Robots and Systems IROS 2003, Workshop on Robot Programming by Demonstration, Las Vegas, USA (2003)
4. Hersch, M., Guenter, F., Calinon, S., Billard, A.: Dynamical system modulation for robot learning via kinesthetic demonstrations. IEEE Trans. on Robotics (2008)
5. Tani, J., Nishimoto, R., Namikawa, J., Ito, M.: Codevelopmental learning between human and humanoid robot using a dynamic neural-network model. IEEE Transactions on Systems, Man, and Cybernetics, Part B 38(1), 43–59 (2008)
6. McNerney, T.S.: From turtles to tangible programming bricks: explorations in physical language design. Personal Ubiquitous Comput. 8(5), 326–337 (2004)
7. Frei, P., Su, V., Mikhak, B., Ishii, H.: Curlybot: designing a new class of computational toys. In: CHI 2000: Proceedings of the SIGCHI conference on Human factors in computing systems, pp. 129–136. ACM, New York (2000)

To See and to Be Seen in the Virtual Beer Garden - A Gaze Behavior System for Intelligent Virtual Agents in a 3D Environment

Michael Wissner, Nikolaus Bee, Julian Kienberger, and Elisabeth André

Institute of Computer Science, Augsburg Univeristy,
86135 Augsburg, Germany
{wissner,bee,andre}@informatik.uni-augsburg.de

Abstract. Aiming to increase the believability of intelligent virtual agents, this paper describes the implementation of a parameterizable gaze behavior system based on psychological notions of human gaze. The resulting gaze patterns and agent behaviors cover a wide range of possible uses due to this parametrization. We also show how we integrated and used the system within a virtual environment.

1 Introduction

The Virtual Beer Garden is a 3D environment implemented with the Horde3D GameEngine [6]. It serves as a meeting place for virtual agents which can engage in social interaction with each other. Agents can stroll through the beer garden, meet other agents by moving towards them, and interact with them using gestures, facial expression and synthesized speech. Furthermore, agents can interact with the environment itself, such as sitting down on a bench or drinking from a mug. Both kinds of interactions are driven by desires previously assigned to the agents, such as hunger, thirst or sociability.

In previous work, we investigated how to successfully integrate a user's avatar into this environment, letting them partake in the agents' social group dynamics (see [16]). However, the agents were lacking natural gaze behaviors. When a user navigated his avatar through the environment, other agents did not seem to take notice of the avatar and vice versa. We felt that this greatly reduced believability, both of the agents themselves and the virtual environment in general.

As a consequence we present a gaze behavior system that generates natural gaze behaviors for agents in the Virtual Beer Garden. The main challenges in this endeavor are owed to the fact that the Virtual Beer Garden is a detailed and dynamic environment. It contains an abundance of perceivable entities, many of which can also be moving. Also, the perceiving agents might be moving as well.

With this work, we try to achieve two main goals: The first is to counter the above-mentioned lack in believability, while the second is to move towards autonomous behavior generation for user-controlled avatars in virtual environments.

B. Mertsching, M. Hund, and Z. Aziz (Eds.): KI 2009, LNAI 5803, pp. 500–507, 2009.

2 Related Work

There are several psychological notions of human gaze and attention which we should take into account when creating gaze behaviors for virtual characters. One is the control of gaze which can be either top-down or bottom up. Top-down or endogenous control comes from within, is based on an active decision and serves a task- or goal-oriented purpose. Bottom-up or exogenous control is created by an external stimulus, such as a sudden motion (see Posner [15], Wolfe [19], Neisser [9]).

The second is the orientation of attention and how attention can be shifted. Overt attention is the process of actually directing the eyes towards a target, thus shifting the entire field of view. Covert attention, on the other hand, is the mental process of just focusing on a target without moving the eyes (see [15]).

Finally, Wolfe [19] states that a human's visual field contains far more information than the perceptual system can handle. He describes two ways to solve this: One is to limit the amount of information that is sensed, the other is to limit the amount of sensed information that is actually perceived.

Plenty of previous work exists on gaze behavior for virtual characters. Table 1 shows the related work as well as our approach, with the behavior of the respective characters characterized by the following criteria: Embodiment (Talking Head, Torso or Full Body), Number of agents with gaze behavior, Autonomy (completely autonomous agents or semi-autonomous user avatars), Kind of Interaction (towards the user or other characters, with "User (*)" indicating

Table 1. Related Work on Virtual Characters

	Embodiment	Agents	Autonomy	Interaction	Locomotion
Bee et al. [1]	Head	1	Comp.	User	No
Cohen et al. [2]	Head	2 or Many	Semi	User and Character	No
Gillies and Ballin [3]	Full	2	Semi	Character	No
Gillies and Dodgson [4]	Full	1	Comp.	–	Yes
Hill [5]	–	1	Comp.	–	–
Khullar and Badler [7]	Full	1	Comp.	–	Yes
Pedica and Vilhjálmsson [10]	Full	Many	Semi	Character	Yes
Pelachaud and Bilvi [11]	–	2	Comp.	Character	–
Peters and O'Sullivan [12]	Full	1	Comp.	Character	Yes
Peters et al. [13]	–	2	Comp.	Character	–
Poggi et al. [14]	Head	1	Comp.	User (*)	No
Vilhjálmsson and Cassell [17]	Torso	2	Semi	User	Yes
Vinayagamoorthy et al. [18]	Full	2	Semi	User	Yes
Our Work	Full	Many	Comp. or Semi	User or Character	Yes

one-way-communication without input from the user) and Locomotion (whether the characters in question move around). An entry of '–' means that a criterion is not applicable or not explicitly specified. Our system features many fully embodied agents that move around the environment. Their behavior can be either completely autonomous or semi-autonomous. Comparing this to the other entries in the table, we see that our approach covers a wide range of features (such as the possibility to switch the interaction target and autonomy) which makes the resulting agents and their behavior suitable for an equal wide range of applications and scenarios for experiments.

3 Implemented Gaze Behavior System

3.1 Field of View

While a human's visual field (i.e. the area in which the existence of other entities can be perceived) extends to about 95 degrees on either side (see [8]), we chose to limit our agents' field of view to an area in which entities can actually be identified. The corresponding parameter in our model is an angle that extends to either side of a centered and straight "focus line" affixed to the agent's forehead. Thus, as the agent's head turns, its field of view will shift as well.

Besides this horizontal limitation, the agent's field of view is also limited in regard to the distance at which entities can be identified: Each agent can have a certain number of perception areas, each defined by a radius around the agent and an identification probability. The closer an area (and any entities in it) is located to the agent, the higher the probability of successfully identifying an entity. Usually, this probability should be 100 percent for the area closest to the agent, and linearly decreasing in every further area. Outside of the last specified area the probability is zero and thus proper identification is no longer possible.

Figure 1 shows an agent along with its cone-shaped field of view, divided into four perception areas.

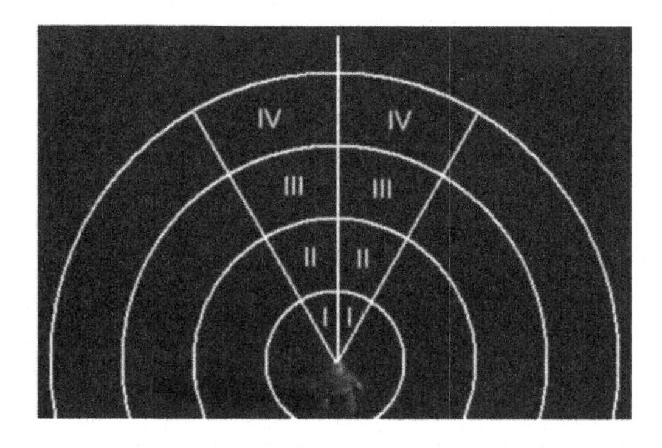

Fig. 1. An Agent's Field of View

3.2 Perception and Identification

In our gaze behavior system, the perception and identification process is distributed across three different modules: The monitoring module surveils an agent's field of view, checking whether other entities enter it, either because they moved on their own or because the agent itself moved or turned its head. The identification module tries to identify such entities, using the identification probability of the respective perception area within the field of view. Finally, the memory module keeps track of perceived entities and the area they were perceived in by putting them in one of two lists, depending on whether the identification was successful. These lists are used to ensure two things: First, there will be no further identification attempts for an already identified entity. Second, the next identification attempt for an unidentified entity will only happen if the entity appears in a nearer perception area. The reason behind this is that it would be unrealistic if the agent, after not being able to identify a target, could suddenly identify it from the same or a greater distance.

3.3 Implemented Gaze Patterns

We implemented four different kinds of gaze patterns for our virtual agents:

- Looking straight ahead: This is an agent's basic gaze behavior, it will just gaze straight ahead without shifting its field of view.
- Idle-viewing: After standing still for a certain, specified amount of time, an agent's gaze will start wandering around, performing what we call idle-viewing. This is done by selecting random points in space and having the agent look at each of them one by one, slowly turning the head in between. These points are not entirely random, however, as they are only generated within the boundaries of the agent's field of view, both on the horizontal and the vertical axis. Other names for this pattern include "undirected attention" [4] or "spontaneous looking" [12].
- Visual search: To improve an agent's chance of spotting entities of interest, the visual search pattern can be employed. Here, the agent will slowly turn its head left and right, as far as the edges of its field of view. Thus, by alternately turning until the "focus line" is aligned with the respective former edge of the field of view, the area covered by the agent's gaze is essentially doubled. As an example, if the agent's field of view was 60 degrees, a search would be started by turning its head 30 degrees to the right. The field of view is still 60 degrees but now it covers the area from zero to 60 degrees, as opposed to the range from -30 to 30 degrees when the agent was looking straight ahead. After that, the agent would turn its head all the way to -30 degrees. Altogether, the agent's gaze now covers the area from -60 to 60 degrees, though not at the same time.
- Fixating: When an agent perceives (but not necessarily identifies) an entity of interest, the agent will stare at it for a short time, fixing its "focus line" on this target. Afterwards, the agent will continue with whichever behavior it was performing before (such as idle-viewing).

The Horde3D GameEngine contains a module for Inverse Kinematics, allowing for agents to point to arbitrary points in space with any of their limbs. This allowed us to very easily implement the above-mentioned patterns by simply turning an agent's head, along with its "focus line", towards a specific point in space.

3.4 Overall Agent Gaze Behaviors

By applying and combining the four gaze patterns mentioned previously, we created three different overall gaze behaviors for our agents. Whenever an agent is added to the virtual environment, it is given one of these behaviors which will govern its actions. However, to create more diversity among the agents (even those employing the same behavior), each behavior can be further customized by adjusting the parameters explained throughout this chapter.

- Stationary: A stationary agent is just perceiving its surroundings, it is standing still and gazes straight ahead, without moving around or turning its head. However, after having identified another entity, it will first stare at them and then react in a previously specified manner appropriate to said entity.
- User-controlled: This agent can be directly controlled by a user and becomes their avatar within the virtual environment. The user can navigate the agent through the environment and the agent will gaze straight ahead while walking. As soon as the user no longer provides any input, the agent will stop and start idle-viewing. Regardless of moving or not, the agent will always react to other entities (as described above for a stationary agent).
- Autonomous: An autonomous agent moves around the environment on its own by following a set of previously defined waypoints, stopping now and then. When doing this, its gaze behavior is similar to that of a user-controlled avatar. However, an autonomous agent can decide to enter a special search mode. While in this mode, the agent will still move between waypoints but it will stop and perform a visual search at some points for a specified amount of time.

Comparing our approach with the psychological notions described in chapter 2, we find that our system contains both endogenous (e.g. visual search) and exogenous (fixating perceived entities) causes of gaze behavior. In case of a conflict between these, exogenous control takes precedence by interrupting other behaviors. Furthermore, our model includes both forms of attention, covert (agents can perceive entities anywhere within their field of view) and overt (agents fix their gaze on entities of interest). Also, we prevent perceptual overload by limiting the amount of sensible entities to a predefined set.

4 Sample Application

As a show case and testing environment for our gaze behavior system, we created a sample application within the Virtual Beer Garden.

This sample application contains eleven stationary agents and one "protagonist" agent that can either be autonomous or controlled by the user. Each of the stationary agents belongs to one of three categories in regard to its relationship to the protagonist: The protagonist can either have positive, negative or neutral feelings towards each of them (similar to [3], where agents' affiliations influence their behavior). Each agent also has a unique name that serves as an identifier. In this scenario there are no other entities of interest besides the other agents.

The application displays additional information for the user, including which behavior the protagonist is currently employing, its current status (moving, idle-viewing, searching), other agents currently within its field of view along with the respective perception areas and the last successfully identified agent. Figure 2 shows this additional information as the protagonist approaches and idly gazes upon a group of other agents. Also note the representation of the "focus line". The user can either start to steer the protagonist into the beer garden and towards the other agents or switch it to autonomous behavior. As soon as the protagonist successfully identifies another agent, it is able to read that agent's name from the environmental annotation. Matching this name against an

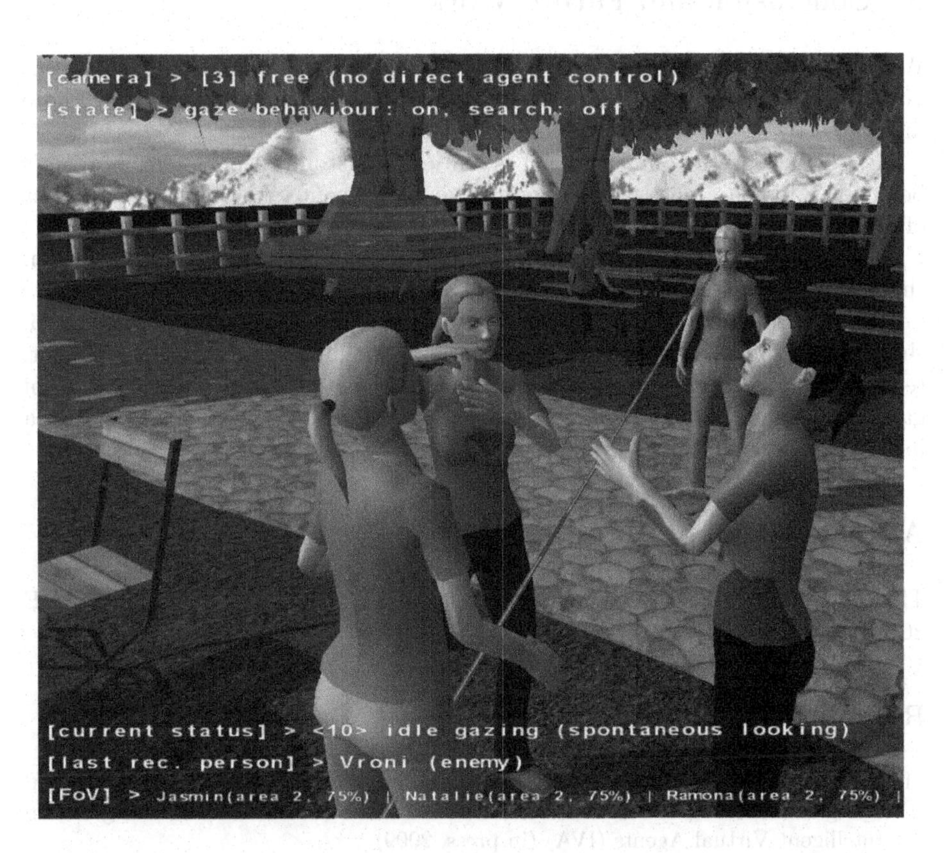

Fig. 2. Protagonist perceiving (but not yet identifying) a Group of other Agents

internal representation, the protagonist can then determine its feelings towards that agent and react accordingly. This comprises waving for positive feelings, no action for neutral feelings and suddenly averting gaze for negative feelings.

We used the sample application to conduct an informal study which showed that the integration of our gaze system into the beer garden resulted in a higher believability of the agent behavior: The impression, that agents can really see other agents and can be seen by them while walking around the environment removes of one the main impairments of believability we had with the previous version of the beer garden. Also, although the stationary behavior might not seem very sophisticated, we were satisfied with the results. By having the respective agents play animations that suggested preoccupation (e.g. eating, chatting or standing behind a counter), the fact that they were simply gazing straight ahead did not seem to impair believability. Finally, the visual search seemed a bit too mechanical in its current form with the agent coming to a full stop, then looking around and then moving again.

5 Conclusion and Future Work

We implemented a gaze behavior system for virtual agents in a virtual 3D environment, covering different kinds of gaze behaviors such as idling or those motivated by extrogenous or endogenous factors.

A main feature of our approach is the easy customization for each agent's behavior. This parametrization allows us to experiment with different variations of gaze behaviors that can be used within any virtual environment.

We integrated the implemented gaze behavior system into a test application, the Virtual Beer Garden, and conducted an informal study.

Ideas for future work include an improvement in the visual search pattern, the addition of objects as entities of interest within the virtual environment, testing the gaze behavior system with multiple moving agents, an addition of gaze behavior for communicative purposes as well as a user studies to obtain a formal evaluation of our system and substantiate our findings.

Acknowledgments

This work has been funded in part by the DFG under grant agreement RE 2619/2-1 (CUBE-G).

References

1. Bee, N., André, E., Tober, S.: Breaking the ice in human-agent communication: Eye-gaze based initiation of contact with an embodied conversational agent. In: Intelligent Virtual Agents (IVA) (in press, 2009)
2. Cohen, M., Colburn, A., Drucker, S.: The role of eye gaze in avatar mediated conversational interfaces (2000)

3. Gillies, M., Ballin, D.: Integrating autonomous behavior and user control for believable agents. In: AAMAS 2004: Proceedings of the Third International Joint Conference on Autonomous Agents and Multiagent Systems, pp. 336–343 (2004)
4. Gillies, M., Dodgson, N.: Eye movements and attention for behavioural animation. The Journal of Visualization and Computer Animation 13(5), 287–300 (2002)
5. Hill, R.: Perceptual attention in virtual humans: Towards realistic and believable gaze behaviors. In: Proceedings of the AAAI Fall Symposium on Simulating Human Agents (2000)
6. Horde3D GameEngine. University of Augsburg (2008), `http://mm-werkstatt.informatik.uni-augsburg.de/projects/GameEngine/`
7. Khullar, S.C., Badler, N.: Where to look? automating attending behaviors of virtual human characters. Autonomous Agents and Multi-Agent Systems 4(1-2), 9–23 (2001)
8. Kim, Y., van Velsen, M., Hill, R.: Modeling dynamic perceptual attention in complex virtual environments. In: Panayiotopoulos, T., Gratch, J., Aylett, R.S., Ballin, D., Olivier, P., Rist, T. (eds.) IVA 2005. LNCS (LNAI), vol. 3661, pp. 266–277. Springer, Heidelberg (2005)
9. Neisser, U.: Cognitive psychology. Appleton-Century-Crofts, New York (1967)
10. Pedica, C., Vilhjálmsson, H.: Social perception and steering for online avatars. In: Prendinger, H., Lester, J.C., Ishizuka, M. (eds.) IVA 2008. LNCS (LNAI), vol. 5208, pp. 104–116. Springer, Heidelberg (2008)
11. Pelachaud, C., Bilvi, M.: Modelling gaze behavior for conversational agents. In: Rist, T., Aylett, R.S., Ballin, D., Rickel, J. (eds.) IVA 2003. LNCS (LNAI), vol. 2792, pp. 93–100. Springer, Heidelberg (2003)
12. Peters, C., O'Sullivan, C.: Bottom-up visual attention for virtual human animation. In: Computer Animation for Social Agents, pp. 111–117 (2003)
13. Peters, C., Pelachaud, C., Bevacqua, E., Mancini, M., Poggi, I.: A model of attention and interest using gaze behavior. In: Panayiotopoulos, T., Gratch, J., Aylett, R.S., Ballin, D., Olivier, P., Rist, T. (eds.) IVA 2005. LNCS (LNAI), vol. 3661, pp. 229–240. Springer, Heidelberg (2005)
14. Poggi, I., Pelachaud, C., de Rosis, F.: Eye communication in a conversational 3d synthetic agent. AI Communications 13(3), 169–182 (2000)
15. Posner, M.I.: Orienting of attention. The Quarterly Journal of Experimental Psychology 32(1), 3–25 (1980)
16. Rehm, M., Endrass, B., Wissner, M.: Integrating the user in the social group dynamics of agents. In: Workshop on Social Intelligence Design, SID (2007)
17. Vilhjálmsson, H., Cassell, J.: Bodychat: Autonomous communicative behaviors in avatars. In: AGENTS 1998, pp. 269–276 (1998)
18. Vinayagamoorthy, V., Garau, M., Steed, A., Slater, M.: An eye gaze model for dyadic interaction in an immersive virtual environment: Practice and experience. Computer Graphics Forum 23(1), 1–12 (2004)
19. Wolfe, J.M.: Guided search 2.0: A revised model of visual search. Psychonomic Bulletin and Review 1(2), 202–238 (1994)

Requirements and Building Blocks for Sociable Embodied Agents

Stefan Kopp, Kirsten Bergmann, Hendrik Buschmeier, and Amir Sadeghipour

Sociable Agents Group, Cognitive Interaction Technology (CITEC), Bielefeld University
P.O. Box 10 01 31, 33501 Bielefeld, Germany

Abstract. To be sociable, embodied interactive agents like virtual characters or humanoid robots need to be able to engage in mutual coordination of behaviors, beliefs, and relationships with their human interlocutors. We argue that this requires them to be capable of flexible multimodal expressiveness, incremental perception of other's behaviors, and the integration and interaction of these models in unified sensorimotor structures. We present work on probabilistic models for these three requirements with a focus on gestural behavior.

1 Introduction

Intelligent agents are nowadays employed as assistants to desktop interfaces, as chatbots on webpages, as instructors in entertainment systems, or as humanoid robots that shall assist household tasks. In in all of these contexts they embody (part of) the user interface with the goal to elevate the interaction between the human and the machine toward levels of natural conversation. However, embodied agents are yet to master a number of capabilities, the most crucial of which are (1) being conversational, i.e., capable of multimodal face-to-face dialogue, (2) being cooperative in reciprocal interaction and joint tasks; (3) being convergent, i.e., able to mutually adapt to and coordinate with a user on a short timescale as well as over longer periods of time, and (4) being companionable, i.e. meet the social dimensions of the former three. All four requirements are inter-connected and must be considerd equally important for agents to become sociable. The first one has been tackled in particular in the field of embodied conversational agents (ECAs [1]), the second one in the realm of collaborative systems [2]. In this paper we focus on the third requirement, being convergent.

Natural interaction is characterized by many inter-personal coordinations when individuals feel connected and communicate with ease. For example, behavior congruence, linguistic alignment, interactional synchrony, or fluent back-channeling have been reported (cf. [3]). These mechanisms help to enhance coordination between interacting individuals and significantly eases their joint task of exchanging meaning with signals [4]. We refer to this state as one of "social resonance" to underline the importance of real-time coordination and mutual contingency in the behaviors and mental states of the participants, as well as the dynamics of this interplay. Now, the research question is can we achieve and leverage on such qualities for embodied human-agent interaction? In Sect. 2 we start by analysing which coordinations mechanisms embodied agents would need to be endowed with to that end. We argue that this cannot pertain to a single

B. Mertsching, M. Hund, and Z. Aziz (Eds.): KI 2009, LNAI 5803, pp. 508–515, 2009.

level of reciprocating to a particular behavior, but ultimately implies global design criteria for the construction of conversational agents or robots. In Section 3, we focus on three of them with an application to communicative gestures: flexible multimodal expressiveness, incremental understanding through mirroring, and the integration of these models such that the production and the perception of conversational behavior ground and coalesce in the same sensorimotor structures. We present current work that employs state-of-the-art AI techniques to bring these principles to bear in building blocks for interactive agents.

2 Requirement Analysis

Starting from the concept of "social resonance", we note that it embraces many phenomena of face-to-face interaction that have three things in common: (1) they are *interactively contingent*, i.e. their occurrence is causally linked to the interaction context including the partner, (2) they act in a *coordinative* fashion between the interactants, and (3) their occurrence correlates with both the *communicative success*, e.g., fewer misunderstandings, faster goal attainment, less effort in relation to gain, as well as the *social success* of the interaction, e.g., affiliation, prosocial behavior, likelihood of interacting again. The behavioral patterns we are referring to have been described in literature with a lot of of different, often overlapping terms (see [4] for a detailed review). A closer inspection suggests a grouping of different components. In the perspective of conversation as joint action [5] for example, *cooperation and collaboration* are essential for continued dialogic exchange. There, one important coordination device to build a shared basis of common ground, increase familiarity, or lend support is appropriate linguistic and embodied feedback (e.g., backchannels, head-nods). With it, interlocutors continuously show whether they hear and understand each other (or not) and what their attitude towards the speakers current utterance is. Listeners give feedback incrementally (while an utterance is still ongoing) and on different levels of perception and understanding, in order to collaboratively provide closure for single communicative acts and to support the speakers in communicating their thoughts as best as they can – given the situation. Another way of coordinating crops out as *convergence and synchrony* of numerous aspects of the behavior of interaction partners (e.g. lexical choice, phonologic features, duration of pauses, body posture, mimicry). Such phenomena occur fast and lead to behaviors that resemble those of another individuals when we evaluate them positively and when we want to be evaluated positively by them.

Altogether, three kinds of mechanisms can be differentiated, acting on different time scales and serving different coordinative functions: (1) *Behavior coordination* (BHC) lets interactants assimilate their behavior in form, content or timing; (2) *Belief coordination* (BLC) leads to compatible assumptions and convictions, about each other and about specific domains or tasks; (3) *Relationship coordination* (RC) regulates the attitudes and feelings individuals have toward each other. These three kinds of mechanisms bring about inter-personal coordination implicitly and work in parrallel (and jointly) with the commonly conceived exchange of dialog acts. Notably, they are not independent, but connected and inter-related (cf. [4]): BLC is required for RC, as feedback and common ground are prerequisites for establishing familiarity, trust, and rapport. The other way around, a positive relationship (RC) eases belief coordination and fosters task collaboration. BHC correlates with RC, as mimicry and synchrony are selective and correlate with rapport or affiliation. BHC and BLC are connected, as aligned communicative behavior facilitates person understanding and reflects shared mental representations and common ground. In sum, it is the triad of the coordination mechanisms that creates a state of closeness between interactants that makes the subjective process of constructing meaning-bearing singals better comprehensible and predictable for each other.

Evidence suggests that humans assume, up-front, such qualities also in interactions with artificial interlocutors. It is commonly acknowledged that computers are social actors and this holds in particular for embodied agents, which are known to induce numerous social effects comparable to those when interacting with real humans [6]. Indeed, a growing body of work on embodied agents that can live up to these expectations has started. The most sophisticated technical accounts have been proposed in the realm of "relational agents" [7] and collaborative systems, targeting longer-term qualities like solidarity and companionship. This work pushes standard models of communication by augmenting messages with "social meaning". This allows for deliberate relationship coordination (RC), e.g., by choosing to avoid face threads or to employ social dialogue moves. Others have built agents that recognize or express affective states and show emotional empathy [8,9], but this is yet to be tied up with the bigger picture of coordinations in dialogic exchange. With regard to short-term behavior coordination agents that mimic a user's head movements are indeed rated more persuasive and positive [10]. Gratch et al. [11] developed a story-listening agent that performs head nods and postural mirroring as rule-based contingent feedback, which was found to increase instant user rapport and to comfort users with social anxiety.

Unlike others [7], we opt for starting with the short-term components of social resonance, i.e., the behavior and knowledge coordinations that facilitate communication right from the start. There is only little modeling work on this topic, and existing systems basically employ simple mapping rules, mainly for the purpose of enabling evaluation studies (e.g. [11]). Like others in social robotics [12], we want to build embodied agents capable of rich social interactions and we explore how we can, to that end, benefit from adopting "design principles" as suggested by recent research on embodied communication [13,14]. One central principle is to model an unified sensorimotor basis of socio-communicative behavior, and to employ this basis for an incremental behavior perception and understanding, a flexible production of meaningful social actions, and the simulation of coordination mechanisms that are likely to be mediated by this basis. In the following section, we describe approaches that try to bring this principle to bear in work on actual building blocks of sociable embodied agents.

3 Social Behavior Perception-Production Integration

Modeling social resonance requires, on the one hand, to flexibly generate conversational behavior from communicative intentions (top-down). On the other hand, perceiving other agent's social behavior has to be grounded (bottom-up) in the same sensorimotor structures, to connect first-person with third-person knowledge and hypothesize likely interpretations and appropriate responses [14]. We focus on gestural behavior as object of investigation and briefly present in the following probabilistic models for these processes as well as their fusion.

3.1 Behavior Generation with Bayesian Decision Networks

To endow virtual agents with flexible expressiveness we developed a generation model that turns communicative intent into speech and accompanying iconic gestures. Our architecture simulates the interplay between these two modes of expressiveness by interactions between modality-specific modules at each of three stages (Fig. 1, right; for details see [15]): First, *Image Generator* and *Preverbal Message Generator* are concerned with content planning, i.e., they select and activate knowledge from two kinds of knowledge representation. For speech this is a propositional representation of conceptual spatial knowledge. For gesture, we employ Imagistic Description Trees (IDTs; [16]), a computational representation for modeling visuo-spatial imagery of objects shapes. Second, specific planners are integrated to carry out the formation of concrete verbal and gestural behavior. The *Speech Formulator* employs a microplanner using a Lexicalized Tree Adjoining Grammar (LTAG) to generate natural language sentences. The *Gesture Formulator* composes and specifies, on-the-fly, the morphology of a gesture as a typed attribute-value matrix. Finally, *Motor Control* and *Phonation* are concerned with the realization of synchronized speech and gesture animations.

Especially challenging is the task of gesture formulation as, in contrast to language or other gesture types such as emblems, iconic gestures have no conventionalized form-meaning mapping. Rather, recent findings indicate that this mapping is determined not only by the visuo-spatial features of the referent, but also by the overall discourse context as well as concomitant speech, and its outcome varies considerably across different speakers [15]. As illustrated in the left of Fig. 1, we employ Bayesian decision networks (BDNs) whose structure and probability distributions is learned from empirical corpus data (25 dyads, ~5000 gestures) and then supplemented with decision nodes (red boxes). Influences of three types of variables manifest themselves in dependencies (edges) between the respective chance nodes: (1) referent features, (2) discourse context, and (3) the previously performed gesture (for details see [17]). BDNs are suitable

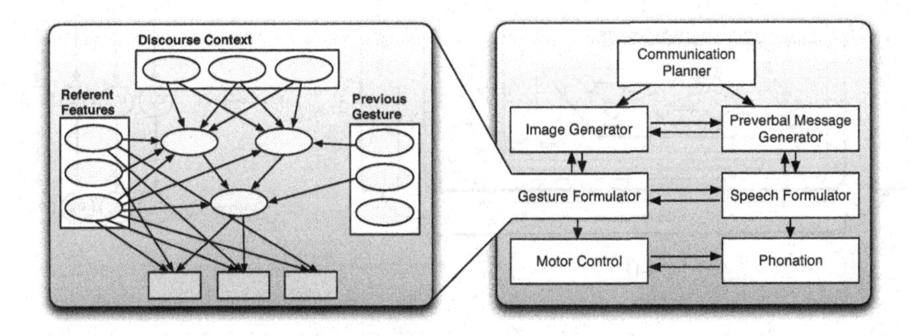

Fig. 1. Overview of the speech and gesture generation model (right), and a zoom in onto the Bayesian decision network for gesture formulation (left)

for gesture formation since they provide a way to combine probabilistic (data-driven) and model-based decision-making. Another rationale of using this method, as discussed shortly, is to prepare the design principle of model integration in order to ground the top-down generation process modeled here in sensorimotor structures that are also the basis of bottom-up gesture perception. Such a model is described next.

3.2 Probabilistic Resonances for Embodied Behavior Perception

Embodied behavior perception entails to recruit and actively involve one's own motor structures into the observation of other's behaviors, such that the motor representations that correspond to an observation immediately start to"resonate". We propose a hierarchical sensorimotor system whose layers span from kinematic movement features towards the goal and meaning of a behavior (see Fig. 2(a)). We differentiate between three major levels of a unified sensorimotor representation, modeled as hierarchically connected graphs: (1) The motor command graphs represent motor primitives (controlling small segments of a gestural movement) as edges and the intermediate states as nodes, separately for each body part; (2) a motor program captures the whole movement of a body part performing a gesture and equals a path in the corresponding motor command graph; (3) the motor schema level groups different allowed variants (motor programs) of a gesture into a single cluster. Such a generalization allows to forward the problem of interpreting a gesture from a pure feature analysis to a concurrent, incremental mapping of different aspects of an observation onto own experiences.

Bayesian inference is applied in utilizing these hierarchical levels for movement perception [18]. A hierarchical Bayesian network (Fig. 2(b)) models the causal influences in-between the different levels. At each level, forward models make probabilistic predictions of the continuation of the movement if it were an instance of a particular motor command, program, or schema, given the evidence at hand and current a priori

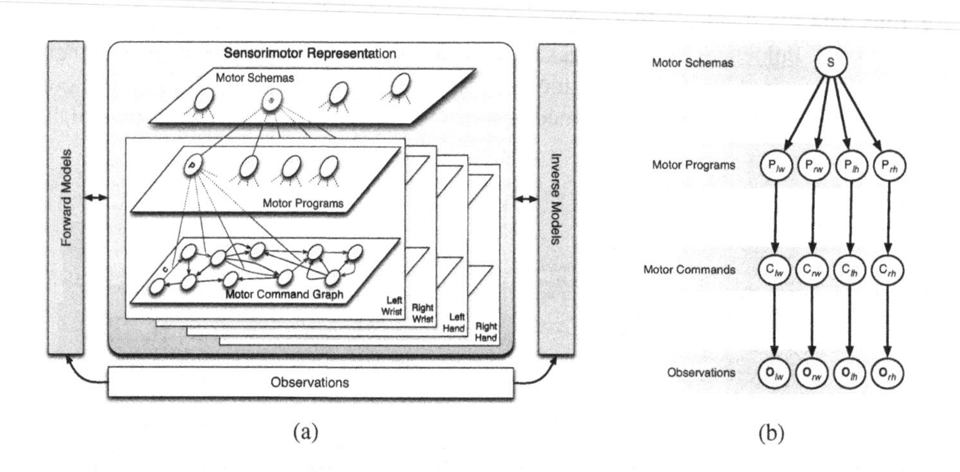

Fig. 2. (a) Hierarchical sensorimotor system for bottom-up grounding in different motor levels for hand-arm gesture perception, (b) the Bayesian network modeling relations between different levels of the sensorimotor representation.

probabilities. The resulting conditional probabilities refer to the agent's certainty in observing a known movement (see [18] for results when applied to real gesture data). In the case of observing a novel gesture (i.e. probabilities fall below a threshold), inverse models are employed to analyze and then acquire the observed movement into the three representation levels of motor knowledge.

3.3 Towards Integrating Top-Down and Bottom-Up Processing

The integration of the two models for top-down behavior generation and bottom-up perception is subject of ongoing work, confined to an application domain of object and route descriptions. Here we discuss how using a single representational system for both sensory and motor aspects of gestural behavior provides a substrate for perception-action links and, as a result, an architecture for embodied gesture understanding, imitation, and inter-personal coordination of gesture production.

In the latter case, an utterance is planned starting from a communicative goal such as "describe-construction entity-1 subpart-3". As described in Sect. 3.1, the activated parts of the imagistic and propositional content represenations build the basis for speech- and gesture formulation processes (see Fig. 1). The BDN is employed to derive a suitable gesture by specifying a representation technique with certain morphological features, such as handshape, hand orientation or movement trajectory. Now, these values are not handed on to a motor planner, but assigned to the corresponding motor levels in the hierarchical sensorimotor system (see Fig. 2). As a first approach we conceive of assigning these values directly to motor programs in the agent's repertoire (e.g. for drawing a circular trajectory or forming a fist). These activiations percolate probabilistically top-down to the motor command level, which details the planned movement trajectories and postures to be performed. This approach calls for a detailed motor repository of the agent, which becomes mitigated when starting to exploit the motor schema level, e.g., by mapping a general representation technique to a certain motor schema and specifying only the remaining, context-dependent aspects.

This architecture also models an embodied approach to gesture understanding. As described above, perceived movements of the relevant body parts (wrist positions and postures of both hands) activate the most likely motor commands via the forward models. These activations percolate in the Bayesian motor network up to the motor schema level, and only decrease gradually over time. As in the generation case, the *winner* schema and its related motor programs are associated with morphological features which are now associated with the corresponding leaf nodes in the gesture generation BDN. Bayesian networks naturally allow for bi-directional inference, to calculate the likely consequences of causal node states (causal inference) or to diagnose the likely causes of dependent node values (diagnostic inference). Doing the latter the agent can derive likelihoods for a number of contextual parameters that shape gesture use, notably, the visuo-spatial features of the referent object, the discourse context (information structure, information state), and the communicative goal of the speaker. This inference can be further supported by simply inserting the evidence about the user's previous gesture.

Note that some of the geometrical features of the referent are not encoded in the BDN as chance nodes, but enter gesture generation via rule-based decision nodes. Problematic here is that the use of diagnostic inferences reveals the fact that some nodes feed into both conditional chance nodes and rule-based decision nodes. Consequently, when used in the inverse direction the problem of arbitrating between two value hypothesis arises, one of which is not quantified in terms of certainty. Here, we will employ the form-meaning mapping rules described in [16] to transform morphological features into imagistic representations (IDTs) that will help to disambiguate the determination of the respective features. In the medium term, the solution must be to replace the decision nodes in the generation BDN with chance nodes and to acquire the required larger set of training data in a social learning scenario between the agent and humans. We also note that the resulting imagistic representation will be underspecified in accord with the iconic gesture's vague, evanescent form and characteristically underspecified meaning.

In sum, connecting the generation BDN with the probabilistic sensorimotor structures leads to an agent architecture that grounds social behavior in resonant sensorimotor structures: top-down generation of coverbal gestures results in activations at the motor level, and motor resonances induced by observing a gesture yield its likely interpretations bottom-up. Either way, activation values (probabilities) are not reset directly after perception/generation. Rather we let them decline following a sigmoidal descent towards default values. In this way, the agent's behavior production is biased towards previously perceived behavior (accounting for the so-called *perception-behavior expressway* [19]) and thus allows an embodied agent to engage in gestural alignment and mimicry phenomena between interlocutors (cf. [20]). The other way around, gesture perception is biased towards previously self-generated gestures, which amounts to *perceptual resonance* [21], another suggested mechanism of coordination in social interaction.

4 Conclusion

In this paper, we have analysed which inter-personal coordination mechanisms embodied agents may need to be able to engage in to become sociable. We have pointed out that, as suggested by the sensorimotor grounding of social behavior and intersubjectivity, such agents utlimately need to be based on close integration of models for behavior perception and generation, and an incremental processing on various levels of modularity in a cognitively plausible agent architecture. One stepstone in this direction is the development of a sensorimotor basis in which flexible generation and incremental perception of socio-communicative behavior are grounded in. The work we have outlined here has yielded promising first results and is underway to bring these principles to bear in the development of important building blocks for sociable embodied agents.

Acknowledgements. This research is supported by the Deutsche Forschungsgemeinschaft (DFG) in the Collaborative Research Center 673 "Alignment in Communication" and the Center of Excellence in "Cognitive Interaction Technology" (CITEC).

References

1. Cassell, J., Sullivan, J., Prevost, S., Churchill, E. (eds.): Embodied Conversational Agents. The MIT Press, Cambridge (2000)
2. Grosz, B.: Collaborative systems. A.I. Magazine 17(2), 67–85 (1994)

3. Wallbott, H.G.: Congruence, contagion, and motor mimicry: mutualities in nonverbal exchange. In: Markova, C.F.I., Graumann, K.F. (eds.) Mutualities in Dialogue, pp. 82–98. Cambridge University Press, Cambridge (1995)
4. Kopp, S.: From communicators to resonators – coordination through social resonance in face-to-face communication with embodied agents. In: Speech Communication (accepted)
5. Clark, H.H.: Using Language. Cambridge University Press, Cambridge (1996)
6. Krämer, N.C.: Social effects of virtual assistants. a review of empirical results with regard to communication. In: Prendinger, H., Lester, J.C., Ishizuka, M. (eds.) IVA 2008. LNCS (LNAI), vol. 5208, pp. 507–508. Springer, Heidelberg (2008)
7. Bickmore, T.: Relational Agents: Effecting Change through Human-Computer Relationships. PhD thesis, Massachusetts Institute of Technology (2003)
8. Becker, C., Kopp, S., Pfeiffer-Lessmann, N., Wachsmuth, I.: Virtual humans growing up: From primary toward secondary emotions. Künstliche Intelligenz 1, 23–27 (2008)
9. Ochs, M., Pelachaud, C., Sadek, D.: An empathic virtual dialog agent to improve human-machine interaction. In: Seventh International Joint Conference on Autonomous Agents and Multi-Agent Systems, AAMAS 2008 (2008)
10. Bailenson, J.N., Yee, N.: Digital chameleons - automatic assimilation of nonverbal gestures in immersive virtual environments. Psychological Science 16(10), 814–819 (2005)
11. Gratch, J., Okhmatovskaia, A., Lamothe, F., Marsella, S., Morales, M., van der Werf, R., Morency, L.P.: Virtual rapport. In: Gratch, J., Young, M., Aylett, R.S., Ballin, D., Olivier, P. (eds.) IVA 2006. LNCS (LNAI), vol. 4133, pp. 14–27. Springer, Heidelberg (2006)
12. Breazeal, C., Buchsbaum, D., Gray, J., Gatenby, D., Blumberg, B.: Learning from and about others: Towards using imitation to bootstrap the social understanding of others by robots. In: Artificial Life (2004)
13. Wachsmuth, I., Lenzen, M., Knoblich, G. (eds.): Embodied communication in humans and machines. Oxford University Press, Oxford (2008)
14. Gallese, V., Keysers, C., Rizzolatti, G.: A unifying view of the basis of social cognition. Trends in Cognitive Science 8, 396–403 (2004)
15. Bergmann, K., Kopp, S.: Increasing expressiveness for virtual agents–Autonomous generation of speech and gesture. In: Proceedings of the 8th International Conference on Autonomous Agents and Multiagent Systems (2009)
16. Sowa, T., Wachsmuth, I.: A model for the representation and processing of shape in coverbal iconic gestures. In: Proc. KogWi 2005, pp. 183–188 (2005)
17. Bergmann, K., Kopp, S.: Gnetic–Using bayesian decision networks for iconic gesture generation. In: Proceedings of the 9th Conference on Intelligent Virtual Agents (2009)
18. Sadeghipour, A., Kopp, S.: A probabilistic model of motor resonance for embodied gesture perception. In: Proceedings of the 9th Conference on Intelligent Virtual Agents (2009)
19. Dijksterhuis, A., Bargh, J.: The perception-behavior expressway: Automatic effects of social perception on social behavior. Advances in Experimental Social Psychology 33, 1–40 (2001)
20. Kimbara, I.: On gestural mimicry. Gesture 6(1), 39–61 (2006)
21. Schutz-Bosbach, S., Prinz, W.: Perceptual resonance: action-induced modulation of perception. Journal of Trends in Cognitive Sciences 11(8), 349–355 (2007)

Modeling Peripersonal Action Space for Virtual Humans by Learning a Tactile Body Schema

Nhung Nguyen and Ipke Wachsmuth

Artificial Intelligence Group
Faculty of Technology
Bielefeld University
33594 Bielefeld, Germany
{nnguyen,ipke}@techfak.uni-bielefeld.de

Abstract. We propose a computational model for building a tactile body schema for a virtual human. The learned body structure of the agent can enable it to acquire a perception of the space surrounding its body, namely its peripersonal space. The model uses tactile and proprioceptive informations and relies on an algorithm which was originally applied with visual and proprioceptive sensor data. As there is not only a technical motivation for devising such a model but also an application of peripersonal action space, an interaction example with a virtual agent is described and the idea of extending the reaching space to a lean-forward space is presented.

Keywords: virtual agents, virtual environment, body schema, peripersonal space.

1 Introduction and Related Work

In order to carry out sophisticated interaction tasks in a spatial environment like a virtual world, one requisite is to perceive how far away objects in the peripersonal space are in relation to the protagonist's own body. The peripersonal action space is the space which immediately surrounds our body, in which we can reach, grasp and manipulate objects with our limbs without leaning forward. It is thus to be conceived of as a sensory space to be delimitated from social perception of space as in social proxemics. The ability of virtual humans to perceive and adapt to their peripersonal space enables them to manipulate and also to avoid objects while moving their limbs through this space. Additionally, it raises more interpersonal interaction possibilities with other agents or with human partners.

In humans the representation of peripersonal space is intimately connected to the representation of the body structure, namely the body schema [6]. The most comprehensive definition of the body schema, as a neural representation, which integrates sensor modalities, such as touch, vision and proprioception, was provided by Gallagher [3]. This integration or mapping across the different modalities is adaptive and explains phenomena like tool use as an integration of tools into the body schema [9]. Learning of body schema is very versatile. We can not only learn configurations of a body structure, but according to Holmes and Spence [6] it also supports learning of the space surrounding the body.

B. Mertsching, M. Hund, and Z. Aziz (Eds.): KI 2009, LNAI 5803, pp. 516–523, 2009.

To our knowledge, work on reaching space for embodied agents has yet been done isolated from body schema acquisition. In work by Goerick et al. [4] the concept of peripersonal space is used in order to structure the visual field of a robot. Work of Zhao et al. [13] and Huang et al. [7] aim at enabling a virtual agent to carry out reaching movements in their virtual workspace. Both approaches neither regard reaching space as represented in body-centered coordinates nor do they consider a body schema as basis for reaching or peripersonal action space, respectively. Although the topic of body schema acqusition is mainly treated by roboticists (e.g. [2]) and has yet not been applied to virtual agents, we want to point out how learning a body schema can also further the design of virtual humans and characters.

In this paper we will show how to model a tactile body schema for a virtual agent and how this can be used to build a representation of its peripersonal action space. Preconditions for the tactile body schema are our work on building touch sensors and motor abilities for a virtual agent. For learning a body schema, we base our computational model on the algorithm proposed by [5]. Unlike their approach, we will not use vision but will feed touch and joint information into the algorithm, in order to learn a tactile body schema, which therefore gets along without any visual information. Combining it with motor abilities, the virtual human is able to perceive its peripersonal space. This can also be regarded as a proof of concept which shows that the spatial representation of the body and peripersonal space, respectively, are not bound to visual information, since congenitally blind people are also able to perceive their peripersonal space.

For a fuller description of these ideas see [11]. Beyond this, the present paper describes how a virtual human's peripersonal space is related to reaching space and how it extends, by bending the torso, to a "lean-forward space".

The remainder of this paper is organized as follows. In the next section, we describe how virtual sensors were realized and prepared in order to feed our model of tactile body schema, described in Section 3. In Section 4 we present a demonstration scenario in which the tactile body schema can make an impact on peripersonal space. Finally, in Section 5 we give a brief conclusion and an outlook on future work concerning the interaction abilities of our virtual human Max.

2 Touch Perception and Proprioception for a Virtual Human

In this section we will first describe how a virtual sense of touch was realized for the virtual human Max [12]. In order to feed our computational model which we present in Section 3, we had to prepare the sensory data from the touch modality and complement it with sensory data from the motor modality.

The touch receptors were developed and technically realized for Max's whole virtual body. These receptors allow for differentiating between different qualities of tactile stimulation. Findings from studies on the human tactile systems were incorporated to build an artificial sense of touch for Max. Max has a segmented body, i.e. his virtual graphical embodiment consists of several geometry parts. Around every geometry representing a limb of Max's body, 17 proximity geometries were added forming a "proximity aura". Below the proximity aura, the surface of Max's body is covered with a virtual "skin". The virtual skin consists of flat quadrangle geometries varying in size,

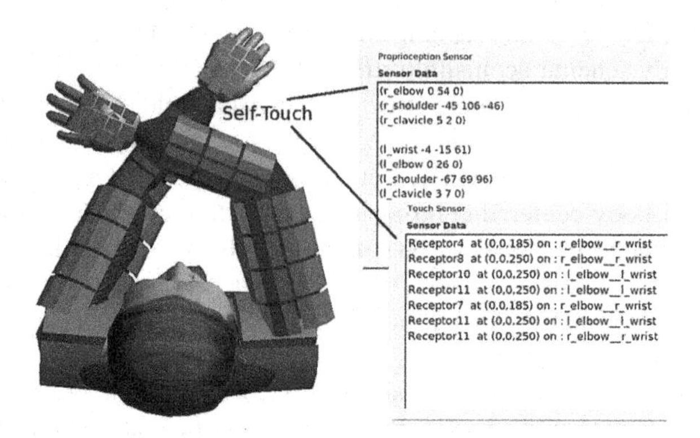

Fig. 1. Tactile body schema learning: For each random posture, sensory consequences are output by the sensory systems. The touch sensor provides an ID of the receptor the limb it is attached to, and the position in the frame of reference (FOR) of the corresponding limb. Angle data for the involved joints are output by the motor system, representing the proprioceptive information.

each representing a single skin receptor (see Figure 1). In humans, the somatosensory modality is represented in body-part-centered reference frames [6]. This aspect is also modeled by the virtual proximity auras. Each skin receptor is assigned to a unique body limb, that means, the receptors' locations and distances are not centrally encoded. Any geometry's collision with a skin receptor is regarded as tactile stimulus. This also includes skin receptors colliding with each other which is crucial for identifying self-touch. In the computational model described in Section 3, for each triggered skin receptor, the touch sensor provides the assignment to the unique body limb and its position in the frame of reference (FOR) of that corresponding limb.

In addition, we need proprioceptive information about Max's body, i.e. his sense of the orientations and positions of his limbs in space. We will refer to it as the angle configuration of the joints in Max's body skeleton. The virtual agent's body has an underlying anthropomorphic kinematic skeleton which consists of 57 joints with 103 Degrees of Freedom (DOF) altogether [8]. Everytime Max is executing a movement, the joint angle informations of the involved joints are output. Synchronously with the tactile informations, the proprioceptive informations can be observed. In Figure 1 we can see the data for a sample posture, where Max is touching his own arm. In the next section we will explain how these input data can be integrated to form a body schema.

3 A Computational Model of Peripersonal Space

For the purpose of perceiving and acting in peripersonal space, a tactile body schema is sufficient. We do not need a precise representation of the physical properties of the body, rather we need the kinematic structure and functions of the body for controlling and predicting the sensory consequences and movements with regard to tactile stimulations coming from objects located within the reaching space. In this section we present our model on how to learn a tactile body schema for our virtual human Max. The idea is to integrate tactile and proprioceptive information from his virtual body. In a first step,

Max executes random motor actions resulting in random body postures. For each posture he perceives proprioceptive data from his joints and tactile stimuli when touching himself (see Fig. 1).

Following Hersch et al. [5] we consider the body schema as a tree of rigid transformations. The kinematic tree is prescribed by the skeleton of the virtual human Max with the hip joint as the root node. In this tree each node corresponds to a joint in Max's skeleton and each edge corresponds to a limb between two joints (for more details see [11]). In our model the touch receptors are attached to the limbs (see Section 2) and their position is represented in the limb's FOR. In the kinematic tree representation, the touch receptors can therefore be represented as located along the edges. Following any path linking one joint to another represents a kinematic chain which transforms the FOR centered on one joint to the FOR centered on the other joint. Max's skeleton prescribes the hierarchy of the FOR transformations. We can transform the position for one touch receptor, given in the FOR of the corresponding limb, into any other touch receptor position also given in the FOR of its corresponding limb. Following an edge in direction to the root node a FOR translation T_i and a rotation R_i associated to the respective joint i (numbers are free chosen) have to be carried out, in the other direction we use the inverse FOR transformation (see Figure 2).

So far, we use the number of joints and the hierarchy of Max's skeleton as prior knowledge about his body structure. However, what is not yet known is the position and orientation of these joints which also determine the limb lengths. This is where the algorithm proposed by Hersch et al. (see Eq. (14) and (15) in [5]) comes in. We can use the algorithm straightforward, since it provides a new and general approach in online adapting joint orientations and positions in joint manipulator transformations. Our challenge in using this algorithm is the adaptation to a case different from the one it was

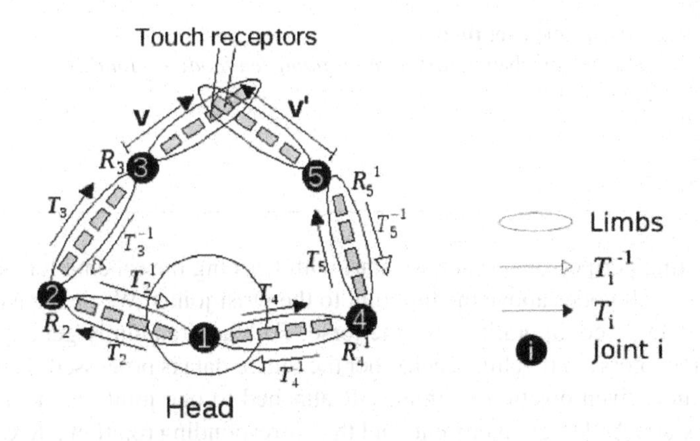

Fig. 2. Kinematic schema of Max touching himself. The following composition transforms the position **v** (given in the FOR centered on joint 3) of a touch receptor into the FOR centered on joint 5: $\mathbf{R}_5^{-1} \circ \mathbf{T}_5^{-1} \circ \mathbf{R}_4^{-1} \circ \mathbf{T}_4^{-1} \circ \mathbf{T}_2 \circ \mathbf{R}_2 \circ \mathbf{T}_3 \circ \mathbf{R}_3$. Note that retracing the same chain in the opposite direction transforms the position of the other touch receptor **v'** (given in the FOR centered on joint 5) into the FOR centered on joint 3.

originally applied to. In our case we do not use visual and joint angle data but instead, replace all visual by tactile information in order to update all the rigid transformations along the generated kinematic chains. In order to use the algorithm, we have to start with an onset body schema which is an initial guess of Max's target body schema. It is described on the one hand by known parameters and on the other hand by initially guessed parameters. The parameters which are not known yet are the joint orientations and their positions ($\mathbf{a}_i$ and $\mathbf{l}_i$ for joint i), determining the body segment lengths. Thus we choose the orientations randomly and assign the segment lengths with small values. The randomly assigned parameters can then be adapted and updated by the algorithm.

Algorithm 1. Pseudo code: Tactile learning process

1: **repeat**
2:     **for all** *torsojoints* **do**
3:         choose random angle θ
4:         set torsojoint of current body schema to θ
5:     **end for**
6:     **if** two touch receptors trigger **then**
7:         $pos_i \leftarrow$ position of touch receptor with ID i
8:         $pos_j \leftarrow$ position of touch receptor with ID j
9:         $joint_n \leftarrow$ joint of limb n where pos_i is attached to
10:         $joint_m \leftarrow$ joint of limb m where pos_j is attached to
11:     **end if**
12:     Set Transformation $T \leftarrow$ kinematic chain ($startnode \leftarrow joint_m$, $endnode \leftarrow joint_n$)
13:     $pos_j = T\ (\ pos_i\ )$
14:     **for** k = *startnode* to *endnode* **do**
15:         update Δl_i
16:         update Δa_i
17:     **end for**
18:     **if** pos_j not transformed yet **then**
19:         Set $T \leftarrow$ kinematic chain ($startnode \leftarrow joint_n$, $endnode \leftarrow joint_m$)
20:         $pos_i = T\ (\ pos_j\ )$
21:         GOTO 14
22:     **end if**
23: **until** $(pos_j$ - T$(pos_i)) = 0$

For modeling peripersonal space we start with learning the schema for Max's torso, which includes all nodes above the hip joint to the wrist joints. We then have to choose random joint angle configurations for the torso. For each randomly generated posture where skin receptors are touching each other the sensor data is processed. The algorithm takes as input a given position $\mathbf{v}_n$ in a FOR attached to one joint, its given transform $\mathbf{v'}_n$ in a FOR attached to another joint, and the corresponding rotation angles θ_i at joint i. In our case the input data are the positions of two touch receptors touching each other in the FOR of their corresponding limbs, both provided by the touch sensor (see Figure 2). Interestingly, both positions can take over the role of the input vectors $\mathbf{v}_n$ and $\mathbf{v'}_n$. This is also illustrated in the pseudo code for the tactile learning process in Algorithm 1. Additionally, the angle values of the joints involved in the current posture are input to the algorithm. It then takes the sensor data for updating its guesses of the joint orientations ($\Delta \mathbf{a}_i$) and positions ($\Delta \mathbf{l}_i$) of the involved kinematic chain. In the adaptation process the idea is to use the update algorithm from Hersch et al. two times for each posture (see Algorithm 1, Line 18-22). In a first process the transformation of the position $\mathbf{v}_n$ of one touch receptor is transformed into the FOR of the other touch receptor (Line 13). This is used to update the current body schema (Line 14-16), in a second pass the angles of the postures stay the same, but the kinematic chain linking the two touch receptors is retraced to transform the position $\mathbf{v'}_n$ of the other touch receptor. Note that this "double-use" is only possible in the case of learning a tactile body schema. After completion the learned body schema expectedly contains the kinematic functions derived from the sensory input. This can be used to control Max's movements with regard to tactile stimuli.

4 Peripersonal Space in Interaction

Based on the work presented in Section 2, we devised the computational model in Section 3 for building a body-representation for the virtual humanoid Max. This model can enable him to acquire a perception of his peripersonal space. In an interaction scenario Max is to interact in a CAVE-like environment with a human partner as shown in Figure 3. In our test scenario both partners are standing at a table with several objects located on it. Let us assume that Max is (technically) "blindfolded". The interaction partner, aware of Max's inability to see, asks him to reach out for an object near to his body. Max then explores his peripersonal space with one hand. As soon as he touches it, the partner could ask him to carry out tasks, such as touching the object with the other hand or putting it as far away from him as possible. The first task is supported by the tactile body schema which contains the kinematic transformations relating two touch receptors. This can be used to compute a movement to the respective position.

The task of putting or reaching an object as far away as possible is an interesting aspect relating to peripersonal action space. McKenzie et al. [10] showed that at the age

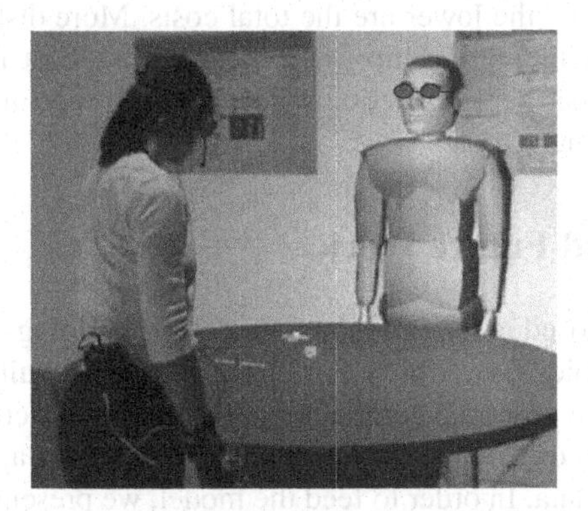

Fig. 3. Virtual agent Max with a human interaction partner standing around a table in a CAVE-like Virtual Reality environment. By means of his peripersonal space Max may perceive objects located on the table in front of him as near or far away from his body.

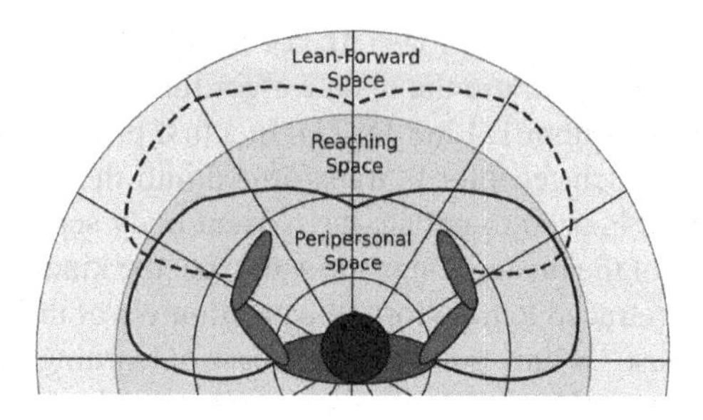

Fig. 4. Peripersonal space as subspace of Reaching space (spanned by body rotation) which extends to Lean-forward space (dashed line) by employing the hip joint

of 8 months, human infants perceive that leaning forward extends their effective reaching space in order to grasp objects, moreover at the age of 10 months they additionally perceive the effective limits of leaning and reaching. Applied to Max this means he has to learn the kinematic function of leaning forward in order to shift his peripersonal space. We can distinguish two cases similar to the work of Huang et al. [7]: Reaching objects within the peripersonal space, only using the arms and reaching objects outside of the reaching space, moving the whole torso. We refer to the latter case as "lean-forward space" shown in Figure 4. We can model the cases by relating the joint movements to human movement behavior; humans tend to adopt joint angle configurations which are comfortable [13]. The cases differ in the amount of joints included: In the first case shoulder, elbow and wrist joints are needed, whereas in the second case the hip joint supplements the movement. It is for example unlikely to lean forward when a target is near to the body and easy to reach, since we need more effort for bending the whole torso.

Our approach is to model this effort by using cost functions assigned to joints, similar to work of Cruse et al. [1]. Unlike them, we want to describe locations in peripersonal space depending on the distances in relation to certain body parts. The summed cost values depend on all involved joints of a whole posture. The more proximate an object is in relation to the body, the lower are the total costs. More distant locations can e.g. only be reached by including the hip joint, therefore the cost for moving it is high. Associating cost with peripersonal action space, hence, brings in a "feel" for the effort involved to reach an object.

5 Conclusion and Future Work

In this paper, we proposed a computational model for building a tactile body schema for the virtual humanoid Max, which can enable him to acquire a perception of his peripersonal space. The proposed computational model uses tactile and proprioceptive informations and relies on an algorithm, which was originally applied with visual and proprioceptive sensor data. In order to feed the model, we presented work on obtaining the nessessary sensory data from touch sensors and the motor system. Based on this, we described the learning process for a tactile body schema. The proposed approach of learning the body structure can not only be applied to other virtual agents but also to robots, provided that they have tactile sensors. The next step in our work will be to test the proposed model for its online learning features. This is especially very relevant for sophisticated computer games where players can design and predefine creatures even with more unusual kinematic structures, not comparable to humanoid ones. Therefore methods which take this pre-knowledge for learning body structures lend themselves for an immediate use in character animation. Based on the motivation to gain an understanding on how humans develop a sensation for the space surrounding their body, in future work we will investigate how spatial perspective models of two agents can be aligned. The aspect of computer games and the planned work on spatial perspective models are discussed in some more detail in [11].

Acknowledgment

This research is carried out in the context of the Collaborative Research Center 673 "Alignment in Communication" granted by the Deutsche Forschungsgemeinschaft.

References

1. Cruse, H., Bruewer, M., Dean, J.: Control of three- and four-joint arm movement: Strategies for a manipulator with redundant degrees of freedom. Journal of motor behavior 25(3), 131–139 (1993)
2. Fuke, S., Ogion, M., Asada, M.: Body image constructed from motor and tactile image constructed from motor and tactile images with visual information. International Journal of Humanoid Robotics (IJHR) 4(2), 347–364 (2007)
3. Gallagher, S.: How the body shapes the mind. Clarendon Press, Oxford (2005)
4. Goerick, C., Wersing, H., Mikhailova, I., Dunn, M.: Peripersonal space and object recognition for humanoids. In: Proceedings of the IEEE/RSJ International Conference on Humanoid Robots (Humanoids 2005), Tsukuba, Japan, pp. 387–392. IEEE Press, Los Alamitos (2005)
5. Hersch, M., Sauser, E., Billard, A.: Online learning of the body schema. International Journal of Humanoid Robotics 5(2), 161–181 (2008)
6. Holmes, N., Spence, C.: The body schema and multisensory representation(s) of peripersonal space. Cognitive Processing 5(2), 94–105 (2004)
7. Huang, Z., Eliëns, A., Visser, C.T.: Is it within my reach? - an agents perspective. In: Rist, T., Aylett, R.S., Ballin, D., Rickel, J. (eds.) IVA 2003. LNCS (LNAI), vol. 2792, pp. 150–158. Springer, Heidelberg (2003)
8. Kopp, S., Wachsmuth, I.: Synthesizing multimodal utterances for conversational agents. Comput. Animat. Virtual Worlds 15(1), 39–52 (2004)
9. Maravita, A., Iriki, A.: Tools for the body (schema). Trends in Cognitive Sciences 8(2), 79–86 (2004)
10. McKenzie, B.E., Skouteris, H., Day, R., Hartman, B., Yonas, A.: Effective action by infants to contact objects by reaching and leaning. Child Development (64), 415–429 (1993)
11. Nguyen, N., Wachsmuth, I.: Modeling peripersonal action space for virtual humans using touch and proprioception. In: Proceedings of 9th International Conference on Intelligent Virtual Agents (to appear, 2009)
12. Nguyen, N., Wachsmuth, I., Kopp, S.: Touch perception and emotional appraisal for a virtual agent. In: Proceedings Workshop Emotion and Computing - Current Research and Future Impact, KI 2007, Osnabrueck, pp. 17–22 (2007)
13. Zhao, L., Liu, Y., Badler, N.I.: Applying empirical data on upper torso movement to real-time collision-free reach tasks. In: Proceedings of the 2005 SAE Digital Human Modeling for Design and Engineering Conference and Exhibition (2005)

Hybrid Control for Embodied Agents Applications

Jan Miksatko and Michael Kipp

DFKI, Embodied Agents Research Group, Saarbrücken, Germany
{jan.miksatko,michael.kipp}@dfki.de

Abstract. Embodied agents can be a powerful interface for natural human-computer interaction. While graphical realism is steadily increasing, the complexity of believable behavior is still hard to create and maintain. We propose a hybrid and modular approach to modeling the agent's control, combining state charts and rule processing. This allows us to choose the most appropriate method for each of the various behavioral processes, e.g. state charts for deliberative processes and rules for reactive behaviors. Our long-term goal is to architect a framework where the overall control is split into modules and submodules employing appropriate control methods, such as state-based or rule-based technology, so that complex yet maintainable behavior can be modeled.

1 Introduction

Embodied agents (or anthropomorphic agents or avatars) can be a powerful user interface. The characters communicate with the human user using verbal and nonverbal channels (such as gaze, gestures or facial expressions), they efficiently handle turn-taking and can express their emotions and personality [1]. Virtual characters have already been applied to education, computer games, training environments, sign language communication, interactive drama, and therapy. With the rise of the 3D internet and social online platforms and games they may become a universally present device in the near future.

Virtual character applications must control all aspects of agent behavior, from low-level reactions to higher level reasoning. Low-level behaviors include gesture, gaze behavior, distance regulation or avoiding obstacles. Higher-level reasoning includes path planning, dialogue management or managing emotions. This modeling of the agent's mind has been approached in different fields like virtual characters, multi-agent systems, robotics and cognitive sciences. The approaches differ in how far they integrate a theory (cognitive architectures) or remain completely generic (multi-agent systems). Highly generic approaches leave the whole development work to developers and are unsuitable for non-experts, and theory-driven approaches may necessitate cumbersome workarounds for situations not covered in the theory. Since in our previous work we have created applications with heterogeneous control paradigms – using plan-based, rule-based and state-based approaches – we are now aiming at creating a unified framework where various technologies can be used in a complementary yet integrated fashion to

B. Mertsching, M. Hund, and Z. Aziz (Eds.): KI 2009, LNAI 5803, pp. 524–531, 2009.

allow intuitive and maintainable authoring of complex, interactive behavior. The modules we suggest for implementing different aspects of control are: extended state charts (XSC), rule-based modules and a hybrid module that combines the two former ones. Since our approach is highly motivated by pragmatic concerns, we plan to iterate through a number of interactive applications that will serve as a testbed for our framework. The first of these applications is an e-learning system since pedagogical scenarios have been shown to benefit greatly from the presence of virtual characters [2,3]. ITeach, a virtual vocabulary trainer, which we will introduce in this paper. While ITeach has specific requirements (e.g. managing linguistically rich representations and a pedagogical module) we derive a first sketch of a general architecture from it.

2 Related Work

In robotics, the dichotomy between reactive and deliberative behavior has become much more obvious early on. While Brooks' subsumption architecture [4] made a radical shift toward reactive behaviors, Gat [5] suggested a three-layered architecture to unify reactive and deliberative tasks. The three layers are the *controller* (reactive behaviors), the *deliberator* (planning), and the *sequencer* (managing the realization of actions). These architectural entities find their counterparts in our approach, however we do not restrict the developer in how many modules (e.g. multiple reactive modules) to implement and in which programming paradigm (rules, finite state, scripting).

To clarify what we mean by "different paradigms" we will briefly review our previous work in this respect. The CrossTalk [6] system was a multi-party interaction between a human user and three virtual characters. Part of the interaction was modelled with finite state machines, another part was natively plan-driven, but both parts ran seamlessly in a final plan-based system. It was an early attempt to integrate plan-based processing with FSMs. COHIBIT [7] is an edutainment exhibit for theme parks in an ambient intelligence environment. Visitors interact with two virtual characters whose dialogue is controlled by a large hierarchical finite state machine (HFSM). ERIC is an affective embodied agent for realtime commentary on a horse race [8], based on parallel rule engines for reasoning about dynamically changing events, generating natural language and emotional behavior. In IGaze [9], a semi-immersive human-avatar interaction system of an interview scenario, it was shown that finite state machines can nicely model reactive gaze behavior.

The Scenemaker authoring tool facilitates the development of embodied agents systems [10]. In Scenemaker, the control flow of the interaction is separated from content, such as speech utterances, gestures, camera motion etc. The content is organized into indexed scenes that are controlled by a scene flow, a hierarchical finite state machine, similar to state charts [11]. Parts of our approach (see Sec. 4.1) can be seen as a direct continuation and extension of the Scenemaker.

3 Use Case: The ITeach Application

Our control framework is motivated from an example e-learning application, ITeach, where an interactive vocabulary trainer presents *flash cards* with vocabulary using speech, gesture and images (Figure 1). The user interacts face-to-face with the life-size agent.

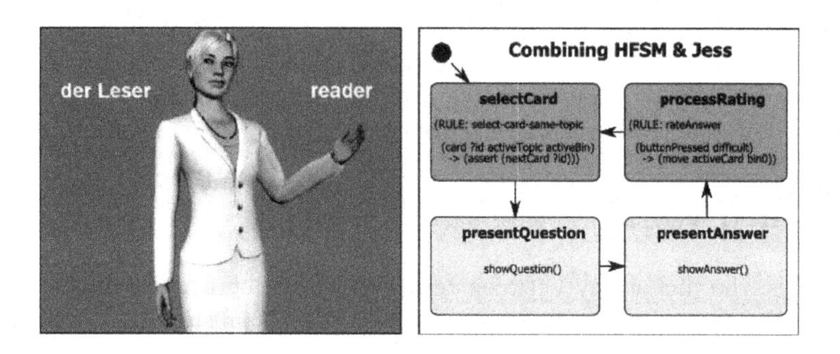

Fig. 1. The ITeach agent "Gloria" (left) and a simplified hybrid module (right)

The flash card learning system is a question/answer game where a card with an expression in a foreign language is presented by the agent and the user answers with an equivalent in his/her native language (or vice versa). The user then rates the answer with respect to how well the word was known. A simple yet effective pedagogical model now moves the cards between multiple *bins* depending on this rating and regulates the time interval between repetitions.

The ontological vocabulary representation connects linguistic information (e.g. grammar, categorical relations) with meaning. Such knowledge allows the agent to generate gestures from semantic components of the word and to answer additional user questions about, e.g., the past tense or synonyms. The agent's believability is increased by employing gaze strategies [9] and regulating the agent-user distance based on the user's location. The pedagogical module monitors the user's progress and gives feedback using verbal comments, facial expressions and gestures. The agent is equipped with an emotion model that allows to generate surprise, joy, disappointment etc. depending on user performance and the agent's personality. Furthermore, we are planning to increase the setup's instrumentation by giving the user pen and tablet for writing the words. Pen and tablet are being tracked for automatic recognition of user actions (start/finish writing) and steering the agent's gaze (looking at tablet).

In this scenario, we identified the following requirements for the control task:

- **Dialogue Management:** How to specify the overall user-agent interaction, e.g. QA dialogues, taking the past interaction history into account.
- **Interactivity:** The "regular" interaction can be interrupted at any time, (e.g. user questions), must be met by an appropriate reaction and then return to the previously interrupted sequence.

- **Domain Intelligence:** The agent has background knowledge about the topic, maintains knowledge about the world and is aware of the past actions. For instance, the agent knows that the currently presented word, an apple, is a fruit and can enumerate not yet presented fruits with similar shape.
- **Reactive Behavior:** Behaviors, that respond immediately to changes in the environment (e.g. user's position), make the user-avatar interaction a tightly coupled feedback loop and ultimately increase believability.
- **Varied and non-repetitive behavior:** The agent's behavior should be varied and non-repetitive using composition and randomization.
- **Multimodal Input Processing:** Continuously monitor several input channels such as user position, button events or pen/tablet writing, fusing the input information and reacting to it.
- **Multimodal Output Generation:** Based on the current state of the system, the agent's emotional state and user's input, generate character's gestures, facial expression, speech, camera position, gaze plus graphics and text relevant to the presented word.
- **Visual Authoring Tool:** In order to make the development accessible to non-experts (i.e. pedagogical advisors, professional authors), the authoring tool should have an intuitive visual interface.

Each requirement implies a different, best suited modeling technique. For instance, basic reactive behaviors (e.g. looking at user if user moves) are best implemented as a set of rules, whereas the overall flow of the interaction (welcome – learning – repetition – ...) is best modeled as a state chart. Our framework allows to combine different techniques, encapsulated in modules.

4 System Framework

The ITeach system is a processing pipeline as depicted in Fig. 2. Inputs like user location, button events, hand/pen position are pre-processed by the Input Interpretation modules (e.g., translate to system coordinates) and passed through the Input Fusion Module (e.g., recognize writing begin/end events) to the relevant Control module(s). The Control Modules are divided into a deliberative and a reactive layer. The reactive layer contains gaze control, implemented as an extended state chart (XSC, cf. Sec. 4.1), and distance regulation, implemented using rules (Sec. 4.2). The deliberative layer contains the main control module, a hybrid between XSC and rules (Sec. 4.3), which models the dialogue between user and agent, manages the vocabulary ontology and the pedagogical model, and handles interrupt events such as *user asked a question* or *user pressed rating high button*. Another deliberative module is an OCC emotion model implemented in JESS [12]. It receives emotion eliciting conditions (e.g. user knows right word → good event) and outputs an emotion state (joy, disappointment, satisfaction etc.) that is used for behavior generation (smile, frown, body posture etc.). Both reactive and emotion modules can be easily reused in another application.

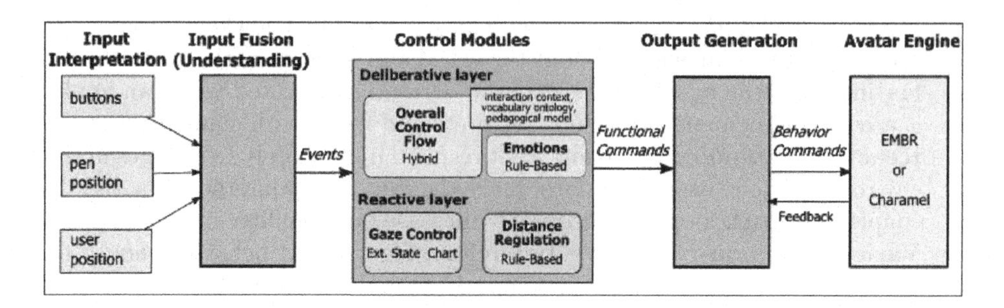

Fig. 2. Overall architecture of the ITeach system

The Output Generation module receives functional commands from the Control modules, such as *present a question*, *play a scene script* or *avert gaze* and translates them to low-level behavior commands for the Avatar engine. The *play a scene script* command executes a script [10] that resembles a movie script with dialogue utterances and stage directions for controlling, for instance, gestures or facial expressions. Scenes can be clustered into groups, where one is randomly chosen each time the group is called, to increase variety and avoid repetitive behavior. Note that the functional commands are independent from the Avatar engine so that different avatar engines can be connected. Currently, we use both a commercial engine from Charamel[1] and our own research prototype engine called EMBR (Embodied Agents Behavior Realizer) [13].

In general, the underlying framework of the ITeach application is a system that loosely integrates modules of different technologies. The modules run in parallel and communicate via messages. Each module can be implemented using a control technology described in the following sections and the framework transparently handles data conversions among them.

4.1 Extended State Chart (XSC) Module

Our *extended state charts* (XSC) extend traditional state charts [11] with transition types from the SceneMaker system[10]: probabilistic, timeout and interrupt transitions. Our extensions are fully embedded in the SCXML standard[2], an emerging W3C standard for describing Harel's state charts.

Behavior is represented by states and transitions. Actions are attached to either state or transition and executed as the graph is traversed. A state can be a superstate containing another state chart. The transitions may have an event and/or a condition attached. An incoming message triggers a transition with a satisfied condition, a matching event or with both satisfied condition and matching event. If no transition can be selected for the current state, transitions of the parent superstate will be checked. The state chart maintains a context memory containing variables scoped by the state hierarchy. These variables are

[1] http://www.charamel.de
[2] http://www.w3.org/TR/scxml/

used in the conditions and the actions. Inspired by the SceneMaker [10] we added three new transition types: *Timeout edges* model wait behavior. If mixed with standard edges, either an event arrives and/or a condition is satisfied until the timeout expires, or the timeout edge is taken. *Probabilistic edges* model random branching by randomly selecting an outgoing transition; they cannot be mixed with standard edges. *Interrupt edges* are attached to supernodes and handle interruptive events. The execution of the supernode is terminated if an event matching the interrupt edge arrives and a conditional constraint is satisfied.

In the ITeach application, the main deliberative module is modeled using an XSC. The visual nature of the XSC intuitively models the dialogue. Interrupt edges can handle unexpected user events for increased interactivity. Probabilistic edges are used to add variability. Additionally, the gaze module of ITeach is also modeled with an XSC according to [9]. Advantages of the XSC are that they are fast and easy to debug. The drawbacks are the need to model all possible variations as states, although this is alleviated with the help of superstates that wrap complex functionality and interruptive edges that are inherited by all substates.

4.2 Rule-Based Module

In a rule-based system the behavior is declaratively encoded in a knowledge base containing initial facts and condition-action rules. Our framework uses JESS (Java Expert System Shell) [14] as a rule processor. An incoming message inserted into the working memory as a fact and processed by JESS. Triggered rules can dispatch commands to the Output Processor. Rule-based systems are particularly useful for two task types. First, for encoding and using expert knowledge. The ITeach system uses JESS along with an ontology in Protégé [3] to reason about vocabulary (e.g. selecting the next word depending on the topic of the current word). Second, for behavior systems that are based on intensive if-then branching, it is easier to understand and maintain them if written in a rule-based language. This is the case for some kinds of low-level behavior such as Distance regulation module of the ITeach system. However, if the knowledge base becomes large rule-based technologies can be difficult to maintain and debug. For such cases it is better to use state charts (Section 4.1). However, in some cases one may want to use both state charts and rules in an integrated fashion (Section 4.3).

4.3 Hybrid Module

A hybrid module is an XSC module (Section 4.1) with rules written in JESS (see Fig. 1). The framework transparently updates the JESS knowledge base (KB) from/to the XSC context memory before and after executing a JESS action. The synchronization is designed in such a way that the JESS rules can reason about XSC variables and, in the other direction, the conditions and actions of the XSC can use variables that mirror JESS facts. The mapping is done as follows:

---
[3] http://protege.stanford.edu

XSC variable type	*JESS type*
primitive type (integer, float...)	ordered fact: `(varName value)`
boolean	simple fact: `(varName)`
struct	unordered fact with struct fields as slots: `(varName (field1 val1) (field2 val2) ..)`
set (struct type)	set of unordered facts: `(varName (field1 valA1) ..)` `(varName (field1 valB1) ..) ..`

This data synchronization preserves the state hierarchy of the XSC: the JESS facts are scoped in the same way as the XSC variables and the JESS facts are organized into modules corresponding to (i.e. named after) the (super)states of the XSC. In the ITeach system, various actions in the main deliberative module are implemented in JESS, for instance, card selection (reasoning based on card history and ontological representation of vocabulary) or user rating of cards (intensive if-then branching). The hybrid approach offers two advantages: (1) an appropriate technology for actions that involve reasoning or intensive if-then branching, and (2) runtime modifications of the actions since JESS is a scripted language.

One of the challenges is maintaining the knowledge of the hybrid model (partly represented as JESS facts, partly as XSC variables, partly synchronized) and keeping track of activated rules when a rule in state A actives rules in state B of the XSC. Making this more robust and intuitive (e.g. by visual feedback) is part of ongoing work.

5 Conclusion and Future Work

Controlling interactive embodied agents is a challenging problem requiring dialogue management, interactivity, domain intelligence and reactive behavior. Different programming paradigms are appropriate for implementing low-level reactive behavior on the one hand and high-level reasoning on the other hand. We try to console these different requirements in a hybrid and modular architecture. This means splitting the control into independent modules, each implemented using a control technology appropriate for the given task: Extended state charts (XSC) allow intuitive and visual coding of low-level reactive behaviors such as gaze. Rule-based systems can elegantly process knowledge items and represent behaviors with intensive if-then branching, in our case reactive behaviors, or input fusion/output fission procedures. We propose a third hybrid module, an XSC with rule processing useful for implementing the deliberative module, i.e. modeling the overall flow of the application, including the user-agent dialogue, along with simple reasoning and knowledge management at certain states of the interaction. The architecture and application presented in this paper is the first step in a long-term iterative development and research towards an authoring framework for embodied agents. In future work, we want to extend our control approaches by a planning module, e.g. for generating dialogue. Additionally, we will design a visual interface, that simplifies the development and debugging, and enables non-experts users to author embodied agents applications.

Acknowledgements. This research has been carried out within the framework of the Excellence Cluster Multimodal Computing and Interaction (MMCI), sponsored by the German Research Foundation (DFG). We would also like to thank to Charamel GmbH for providing us with their Avatar engine.

References

1. Vinayagamoorthy, V., Gillies, M., Steed, A., Tanguy, E., Pan, X., Loscos, C., Slater, M.: Building Expression into Virtual Characters. In: Eurographics Conference State of the Art Report, Vienna (2006)
2. Lester, J.C., Converse, S.A., Kahler, S.E., Barlow, S.T., Stone, B.A., Bhogal, R.: The persona effect: Affective impact of animated pedagogical agents. In: Proceedings of CHI 1997 Human Factors in Computing Systems, pp. 359–366. ACM Press, New York (1997)
3. Bailenson, J., Yee, N., Blascovich, J., Beall, A., Lundblad, N., Jin, M.: The use of immersive virtual reality in the learning sciences: Digital transformations of teachers, students, and social context. The Journal of the Learning Sciences 17, 102–141 (2008)
4. Brooks, R.A.: Intelligence without representation. Artificial Intelligence (47), 139–159 (1991)
5. Gat, E.: Integrating reaction and planning in a heterogeneous asynchronous architecture for mobile robot navigation. In: Proceedings of the National Conference on Artificial Intelligence (AAAI), pp. 809–815 (1992)
6. Klesen, M., Kipp, M., Gebhard, P., Rist, T.: Staging exhibitions: Methods and tools for modeling narrative structure to produce interactive performances with virtual actors. Virtual Reality. Special Issue on Storytelling in Virtual Environments 7(1), 17–29 (2003)
7. Ndiaye, A., Gebhard, P., Kipp, M., Klesen, M., Schneider, M., Wahlster, W.: Ambient intelligence in edutainment: Tangible interaction with life-likeexhibit guides. In: Maybury, M., Stock, O., Wahlster, W. (eds.) INTETAIN 2005. LNCS (LNAI), vol. 3814, pp. 104–113. Springer, Heidelberg (2005)
8. Strauss, M., Kipp, M.: Eric: A generic rule-based framework for an affective embodied commentary agent. In: Proceedings of the 7th International Conference on Autonomous Agents and Multiagent Systems (2008)
9. Kipp, M., Gebhard, P.: Igaze: Studying reactive gaze behavior in semi-immersive human-avatar interactions. In: Prendinger, H., Lester, J.C., Ishizuka, M. (eds.) IVA 2008. LNCS (LNAI), vol. 5208, pp. 191–199. Springer, Heidelberg (2008)
10. Gebhard, P., Kipp, M., Klesen, M., Rist, T.: Authoring scenes for adaptive, interactive performances. In: Proceedings of the Second International Joint Conference on Autonomous Agents and Multiagent Systems, pp. 725–732 (2003)
11. Harel, D.: Statecharts: A visual formalism for complex systems. Sci. Comput. Program. 8(3), 231–274 (1987)
12. Ortony, A., Clore, G.L., Collins, A.: The Cognitive Structure of Emotions. Cambridge University Press, Cambridge (1988)
13. Heloir, A., Kipp, M.: Embr - a realtime animation engine for interactive embodied agents. In: Proceedings of the 9th International Conference on Intelligent Virtual Agents, IVA 2009 (2009)
14. Friedman-Hill, E.J.: Jess, the java expert system shell, Livermore, CA. Distributed Computing Systems, Sandia National Laboratories (2000)

Towards System Optimum: Finding Optimal Routing Strategies in Time-Dependent Networks for Large-Scale Evacuation Problems

Gregor Lämmel[1] and Gunnar Flötteröd[2]

[1] Berlin Institute of Technology (TU Berlin), Transport Systems Planning and Transport Telematics, Berlin, Germany
laemmel@vsp.tu-berlin.de
[2] Ecole Polytechnique Fédérale de Lausanne (EPFL), Transport and Mobility Laboratory, Lausanne, Switzerland
gunnar.floetteroed@epfl.ch

Abstract. Evacuation planning crucially depends on good routing strategies. This article compares two different routing strategies in a multi-agent simulation of a large real-world evacuation scenario. The first approach approximates a Nash equilibrium, where every evacuee adopts an individually optimal routing strategy regardless of what this solution imposes on others. The second approach approximately minimizes the total travel time in the system, which requires to enforce cooperative behavior of the evacuees. Both approaches are analyzed in terms of the global evacuation dynamics and on a detailed geographic level.

1 Introduction

The evacuation of whole cities or even regions is a problem of substantial practical relevance, which is demonstrated by recent events such as the evacuation of Houston because of Hurricane Rita or the evacuation of coastal cities in the case of tsunamis.

The development of evacuation simulations relies strongly on results obtained in the field of transportation modeling. Like in transportation, one can distinguish static approaches, e.g., [18], and dynamic approaches, e.g., [16]. A typical static evacuation simulation is MASSVAC [7]. The obvious shortcoming of static models is that they do not capture dynamic effects, which are highly relevant in evacuation situations. Consequently, many dynamic traffic assignment (DTA) models have been applied for evacuation simulations, e.g., MITSIM [8], DYNASMART [10], and PARAMICS [3].

Another aspect according to which transportation models may be classified is their granularity: Microscopic models represent every trip-maker individually, whereas macroscopic models aggregate traffic into continuous streams. All of the above DTA packages rely on microscopic traffic models. Further microscopic approaches that have been applied to the simulation of evacuation dynamics

B. Mertsching, M. Hund, and Z. Aziz (Eds.): KI 2009, LNAI 5803, pp. 532–539, 2009.

are cellular automata [9] and the social force model [6]. Random utility models are also applicable to the microscopic modeling of pedestrian dynamics, however, they are yet to be applied in evacuation scenarios [2]. Examples of software packages based on macroscopic models are ASERI [17] and Simulex (www.iesve.com).

This paper evaluates the following two routing strategies with a learning-based multi-agent (micro)simulation in a real-world evacuation scenario: 1. A strategy where every agent learns an evacuation route of minimal travel time, regardless of the consequences for others. This selfish learning behavior leads towards a Nash equilibrium, where nobody can gain by switching to a different route. This strategy is called "user optimal" in transportation. 2. A "system optimal" strategy, where the total travel time of all agents is minimized. Here, learning agents are no longer optimizing their individual travel times only but in some way also care about others.

The added value of the agent-based approach is its natural representation of individual travelers as software agents that interact in a simulated version of the real world (a virtual environment). The agent-based approach has an edge over macroscopic models in that it allows (at least technically) for a much higher model resolution. However, this comes at the price of greater difficulties in its mathematical treatment. The agent-based routings presented in this article are therefore only of an approximate nature, and they are enforced exclusively by modifying the information provided to replanning agents.

The remainder of this article is organized as follows. Section 2 outlines the simulation framework. Section 3 describes the investigated routing strategies in detail. Section 4 presents simulation results, and Section 5 concludes the article.

2 Simulation Framework

We implement our experiments in the MATSim simulation framework. Since the details of this system are described elsewhere, e.g., [11] and www.matsim.org, only a brief description is given here.

MATSim always starts with a synthetic population, which is based as much as possible on existing information such as census data. Every synthetic individual possesses one or several plans. These plans represent the different traveling intentions of that individual. In an evacuation context, a plan corresponds to a route from an individual's current location to a safe place. Plans are generated by an iterative learning mechanism. In every iteration, one plan is selected by every agent for execution in the virtual environment. The learning logic tests different plans, eventually discards inferior plans, and sometimes generates new plans [4].

The virtual environment is a pedestrian traffic flow simulation, where each street (link) is represented by a first-in/first-out queue with three parameters [5]: minimum link traversal time, maximum link outflow rate (in evacuees per time unit), and link space capacity (in evacuees). The link space capacity limits the number of agents on the link and generates spillback if the link is filled up. In the context of a tsunami evacuation, an additional difficulty results from the

Algorithm 1. Nash equilibrium routing

1. initialize $\tau_a(k)$ with the free-flow travel time for all links a and time steps k
2. repeat for many iterations:
   (a) recalculate routes based on link costs $\tau_a(k)$
   (b) load vehicles on network, obtain new $\tau_a(k)$ for all a and k

fact that a flooded link becomes unavailable. Reference [11] describes in detail how this issue is resolved.

3 Routing Solutions

Each agent iteratively adjusts its evacuation plan during the simulation. After each iteration, every agent calculates the cost of the most recently executed plan. Based on this cost, the agent revises its plans. Some agents generate new plans using a time-dependent Dijkstra algorithm. The other agents select an existing plan, which they have previously used. This selection is realized through a Multinomial Logit model, e.g., [1], that stabilizes the simulation dynamics by allowing somewhat inferior plans to be considered for execution as well.

In the following, we discuss two different cost functions that approximately lead either to user optimal or to system optimal routing solutions. Note that we modify the agents' routing behavior only by adjusting the costs based on which the routing and the plan choice are conducted, but we do not change the replanning logic itself. For simplicity, we subsequently omit the attribute "approximate" in conjunction with either strategy.

3.1 Nash Equilibrium Approach

In a Nash equilibrium, no agent can gain by unilateral deviation from its current plan [14]. The cost function provided to replanning agents in the Nash equilibrium approach only comprises travel times. Formally, the real-valued time is discretized into K segments ("bins") of length T, which are indexed by $k = 0...K-1$. The time-dependent link travel time when entering link a in time step k is denoted by $\tau_a(k)$. Alg. 1 drafts the Nash-equilibrium routing logic.

3.2 System Optimal Approach

A system optimal routing solution minimizes the total travel time in the system. Although a system optimum is a cooperative routing strategy, it can be obtained by the same self-serving routing logic that is employed to calculate a Nash equilibrium. The only difference is that for a system optimum, the travel time based on which agents evaluate their routes needs to be replaced by the marginal travel time [15]. The marginal travel time of a route is the amount by which the total system travel time changes if one additional vehicle drives along that route. It is the sum of the cost experienced by the added vehicle and the cost imposed on other vehicles. The latter is denoted here as the "social cost". The subsequently developed approximation of the social cost term is based on continuous quantities and ignores for simplicity the interrelations of different links in the network. A discretized version is given at the very end.

Assume that the "causative" agent (unit) for which we would like to calculate the social cost it generates is of mass (size) dn and enters the considered isolated link at time t_0. If there is no congestion on the link, the agent can leave the link unhindered after the free-flow travel time τ^{free} and does not incur any cost on other agents further upstream. If there is congestion, however, there also is a positive social cost, which can be calculated in the following way.

The effect of the causative agent persists only as long as the queue it went through persists – the only trace it can possibly leave in the system is a changed state of this queue. Assume that the queue encountered when entering the link at t_0 dissolves at $t^{\text{e}}(t_0)$. Now, consider another "affected" agent that enters the link at $t_1 > t_0$, and assume that this agent leaves the link before $t^{\text{e}}(t_0)$. Denote by $n(t_1)$ the occupancy (in agent units) of the link at the affected agent's entry time t_1 and by $Q^{\text{out}}(t)$ the accumulated outflow (in agent units) of the link until time t. The exit time t_2 of the affected agent solves

$$\begin{aligned} &Q^{\text{out}}(t_2) - Q^{\text{out}}(t_1) = n(t_1) \\ &\qquad\qquad \Rightarrow t_2 = (Q^{\text{out}})^{-1}(n(t_1) + Q^{\text{out}}(t_1)). \end{aligned} \tag{1}$$

Denote by $d\tau(t_1)$ the additional travel time experienced by the affected agent because of the causative agent. If the latter had not entered the link, the following would hold:

$$\begin{aligned} &Q^{\text{out}}(t_2 - d\tau(t_1)) - Q^{\text{out}}(t_1) = n(t_1) - dn \\ &\qquad\qquad \Rightarrow t_2 = d\tau(t_1) + (Q^{\text{out}})^{-1}(n(t_1) - dn + Q^{\text{out}}(t_1)). \end{aligned} \tag{2}$$

A combination of (1) and (2) yields

$$d\tau(t_1) = (Q^{\text{out}})^{-1}(n(t_1) + Q^{\text{out}}(t_1)) - (Q^{\text{out}})^{-1}(n(t_1) - dn + Q^{\text{out}}(t_1)). \tag{3}$$

In order to calculate the social cost $C(t_0)$ generated by the causative agent, these terms are integrated over the entire span of entry times during which the queue at the downstream end of the link is encountered:

$$C(t_o) = \int_{t_1=t_o}^{t^{\text{e}}(t_0)-\tau^{\text{free}}} d\tau(t_1) q^{\text{in}}(t_1) dt_1 \tag{4}$$

where $q^{\text{in}}(t_1)$ is the entry flow rate at t_1 such that $q^{\text{in}}(t_1)dt_1$ is the affected agent mass entering at t_1.

In the following, a simplification of (4) is presented. Stationary flow conditions are assumed in that $q^{\text{in}}(t) \equiv q^{\text{out}}(t) \equiv \bar{q}$, which implies $Q^{\text{out}}(t) \equiv \bar{q}t$ and $Q^{\text{in}}(t) \equiv \bar{q}t + n(0)$, where $n(0)$ is the occupancy of the link at time $t = 0$. A substitution of this in (3) yields $d\tau(t_1) \approx dn/\bar{q}$ and, when substituted in (4),

$$C(t_0) \approx dn/\bar{q} \cdot (Q^{\text{in}}(t^{\text{e}}(t_0) - \tau^{\text{free}}) - Q^{\text{in}}(t_0)). \tag{5}$$

Algorithm 2. System optimum approach

1. initialize $C_a(k) \equiv 0$ and $\tau_a(k) \equiv \tau_a^{\text{free}}$ for all links a and time steps k
2. repeat for many iterations:
   (a) recalculate routes based on link costs $\tau_a(k) + C_a(k)$
   (b) load vehicles on network, obtain new $\tau_a(k)$ for all a and k
   (c) for all links a, identify congestion durations:
      i. $k^{\text{e}} = K$
      ii. for $k = K - 1 \ldots 0$:
         A. if $\tau_a(k) = \tau_a^{\text{free}}$ then $k^{\text{e}} = k$
         B. $C_a(k) = \max\{0, (k^{\text{e}} - k) \cdot T - \tau_a^{\text{free}}\}$

This expression is straightforward to evaluated in a microsimulation context, where $dn = 1$ corresponds to the mass of a single agent and the difference in accumulated flows is easily evaluated by counting the agents leaving the considered link between t_0 and $t_{\text{e}}(t_0) - t_{\text{free}}$. A further simplification is obtained by replacing the accumulated flows in (5) by their linear approximations, which results for $dn = 1$ in

$$C(t_0) \approx t^{\text{e}}(t_0) - \tau^{\text{free}} - t_0. \quad (6)$$

An application of this result to a system optimal route assignment requires to calculate $C_a(t_0)$ for every link a and entry time t_0 in the network, and to add this term to the time-dependent link travel time that is evaluated in the route replanning of every agent. Alg. 2 outlines the arguably most straightforward implementation of this approach in a time-discrete multi-agent simulation.

4 Experimental Results

This section presents the result of a simulation-based comparison of the two presented routing approaches. The simulation setup is based on a real-world evacuation scenario for the Indonesian city of Padang. Padang faces high risk of being inundated by a tsunami wave. The city has approximately 1,000,000 inhabitants, with more than 300,000 people living in the highly endangered area with an elevation of less then 10 m above see level. An overview map of the city is shown in fig. 1 (left). The area more than 10 m above sea level is assumed to be safe (in dark color). A detailed description of the evacuation scenario can be found in [12].

Two simulations are conducted: *Run 1* implements the Nash equilibrium approach described in sec. 3.1. *Run 2* implements the system optimal approach described in sec. 3.2. Both simulations run on a network with 6,289 nodes and 16,978 unidirectional links. The synthetic population consists of 321,281 agents. This is the number of people living less than 10 m above see level. Both simulations are run for 200 iterations on a 3 GHz CPU running JAVA 1.5 on Linux. For *run 1* the overall runtime is 9:31 hours and for *run 2* it is 17:00 hours.

Fig. 2 (left) compares the learning progress of both approaches. In *run 1*, the average evacuation time per agent converges to 1718 seconds, and in *run 2* it

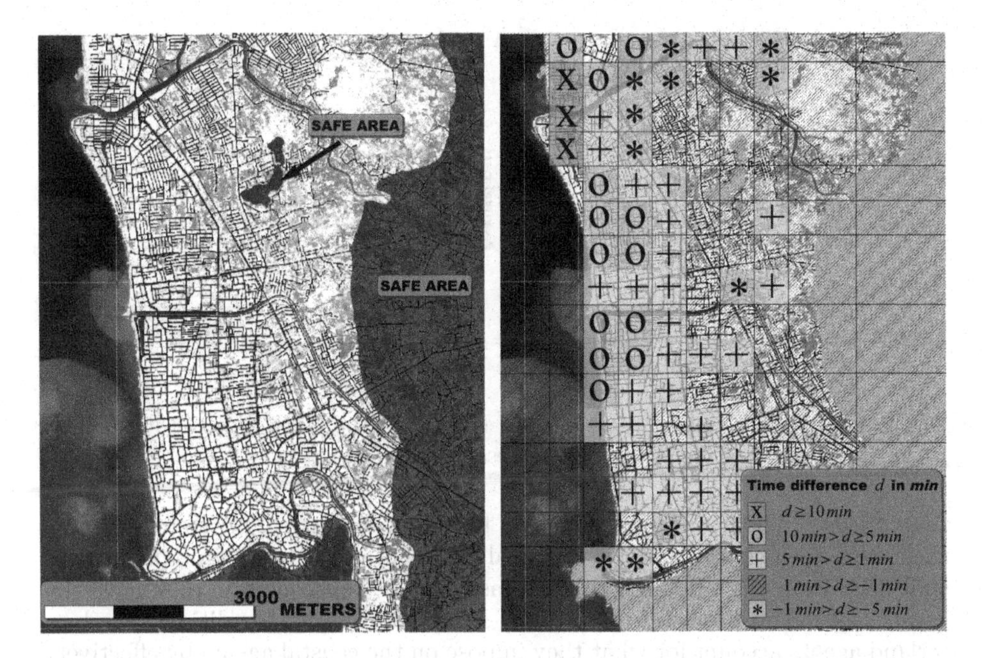

Fig. 1. Left: Overview map of downtown Padang. The safe area with an elevation of more than 10 m is in dark color, all other area is defined as unsafe. Right: Differences in evacuation time between Nash equilibrium approach and system optimal approach. In green parcels (X,O,+), the system optimal approach evacuates faster than the Nash approach, whereas red parcels (*) indicate the opposite.

converges to 1612 seconds[1]. This means that each agent gains on average 106 seconds in the system optimal approach. In both cases, the average evacuation time drops very fast in the first iterations, but from iteration 10 on it increases again. This effect is caused by the fact that in the first iterations not all agents manage to escape the tsunami, and agents that are caught in the flood wave are not considered in the calculation of the evacuation time. Since in the early iterations many agents starting in the coastal area with relatively long evacuation routes do not manage to escape, the average evacuation time is lower than during mid-iterations, where these agents have learned better evacuation routes.

Fig. 2 (right) compares the evacuation curves of *run 1* and *run 2* after 200 iterations of learning. The evacuation curve of *run 2* is steeper than the evacuation curve of *run 1*, which implies a higher outflow rate. The overall evacuation time of *run 2* is about 66 min, which is 3 min faster than in *run 1*. However, not all agents benefit from the system optimal approach. Fig. 1 (right) shows that mainly agents in the hinterland of Padang lose time in the system optimal approach, whereas many agents in the costal area of the city benefit by more

[1] The smoothing of the learning curves after iteration 150 results from a deactivation of the router, such that in iterations 150–200 the agents only select from previously generated routes, which stabilizes the simulation.

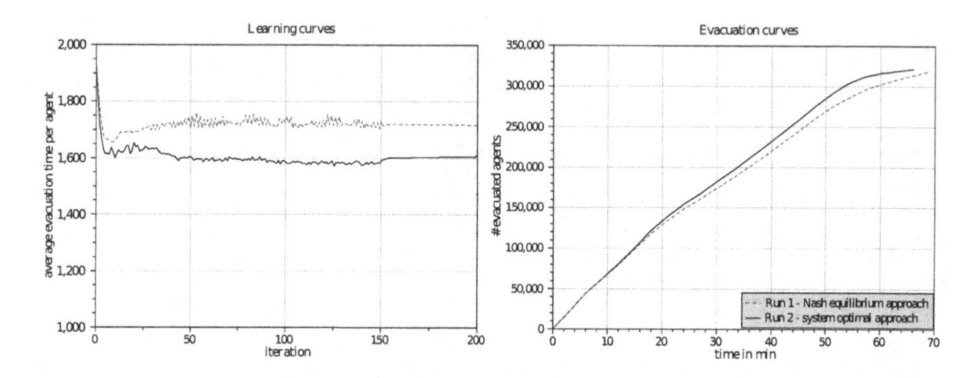

Fig. 2. Left: Average evacuation time per agent over the learning iteration number. Right: Comparison of the evacuation curves of *run 1* and *run 2*.

than 10 min. This can be explained in the following way: The agents starting their trips in the hinterland are technically in front of the multi-link queues that spill back from the safe area to the coastal area. Consequently, they have the greatest effect on the total travel time. In system optimal conditions, the hinterland agents account for what they impose on the coastal agents by effectively giving way to them.

5 Conclusion and Outlook

This article demonstrates that multi-agent simulations can be used to identify efficient evacuation strategies. Our results show that mathematically motivated cooperative routing solutions can be obtained with an acceptable computational overhead even in a purely simulation-based system. The presented cooperative routing approach, which approximates a system optimal solution, generates a substantially higher evacuation throughput than an alternative non-cooperative routing strategy. Even though the results from the system optimal approach do not reflect real human decision making, they serve as benchmark solutions and help to identify routing recommendations (e.g., placement of evacuation signs). The presented experiments with more than 300,000 evacuees show the feasibility of our approach even for large evacuation scenarios. Our ongoing research focuses on more precise system optimal routing strategies.

Acknowledgments

This project was funded in part by the German Ministry for Education and Research (BMBF), under grants numbers 03G0666E ("last mile") and 03NAPI4 ("Advest").

References

1. Ben-Akiva, M., Lerman, S.R.: Discrete choice analysis. The MIT Press, Cambridge (1985)
2. Bierlaire, M., Antonini, G., Weber, M.: Behavioral dynamics for pedestrians. In: Axhausen, K. (ed.) Moving through nets: The physical and social dimensions of travel. Elsevier, Amsterdam (2003)
3. Chen, X., Zhan, F.: Agent-based modeling and simulation of urban evacuation: Relative effectiveness of simultaneous and staged evacuation strategies. Paper 04-0329, Transportation Research Board Annual Meeting, Washington, D.C (2004)
4. Ferber, J.: Multi-agent systems. An Introduction to distributed artificial intelligence. Addison-Wesley, Reading (1999)
5. Gawron, C.: An iterative algorithm to determine the dynamic user equilibrium in a traffic simulation model. International Journal of Modern Physics C 9(3), 393–407 (1998)
6. Helbing, D., Farkas, I., Molnar, P., et al.: Simulation of pedestrian crowds in normal and evacuation situations. In: Proceedings of the 1st international conference on Pedestrian and Evacation Dynamics, 2001, Duisburg. Springer, Heidelberg (2002)
7. Hobeika, A., Kim, C.: Comparison of traffic assignments in evacuation modeling. IEEE Transactions on Engineering Management 45(2), 192–198 (1998)
8. Jha, M., Moore, K., Pashaie, B.: Emergency evacuation planning with microscopic traffic simulation. Paper 04-2414, Transportation Research Board Annual Meeting, Washington, D.C (2004)
9. Klüpfel, H., Meyer-König, T., Keßel, A., et al.: Simulating evacuation processes and comparison to empirical results. In: Fukui, M., et al. (eds.) Traffic and granular flow 2001, pp. 449–454. Springer, Heidelberg (2003)
10. Kwon, E., Pitt, S.: Evaluation of emergency evacuation strategies for downtown event traffic using a dynamic network model. Paper 05-2164, Transportation Research Board Annual Meeting, Washington, D.C (2005)
11. Lämmel, G., Grether, D., Nagel, K.: The representation and implementation of time-dependent inundation in large-scale microscopic evacuation simulations. Transportation Research Part C: Emerging Technologies. Corrected Proof: (in Press, 2009)
12. Lämmel, G., Rieser, M., Nagel, K., et al.: Emergency preparedness in the case of a tsunami – evacuation analysis and traffic optimization for the Indonesian city of Padang. In: Proceedings of the 4th international conference on Pedestrian and Evacuation Dynamics, 2008, Wuppertal. Springer, Heidelberg (2009)
13. Lu, Q., George, B., Shekhar, S.: Capacity constrained routing algorithms for evacuation planning: A summary of results. In: Bauzer Medeiros, C., Egenhofer, M.J., Bertino, E. (eds.) SSTD 2005. LNCS, vol. 3633, pp. 291–307. Springer, Heidelberg (2005)
14. Nash, J.: Non-cooperative games. Annals of Mathematics 54(2), 286–295 (1951)
15. Peeta, S., Mahmassani, H.: System optimal and user equilibrium time-dependent traffic assignment in congested networks. Annals of Operations Research 60, 81–113 (1995)
16. Peeta, S., Ziliaskopoulos, A.: Foundations of Dynamic Traffic Assignment: The Past, the Present and the Future. Networks and Spatial Economics 1(3), 233–265 (2001)
17. Schneider, V., Könnecke, R.: Simulating evacuation processes with ASERI. In: Proceedings of the 1st international conference on Pedestrian and Evacation Dynamics, 2001, Duisburg. Springer, Heidelberg (2002)
18. Sheffi, Y.: Urban Transportation Networks: Equilibrium Analysis with Mathematical Programming Methods. Prentice-Hall, Englewood Cliffs (1985)

Formalizing Joint Attention in Cooperative Interaction with a Virtual Human

Nadine Pfeiffer-Leßmann and Ipke Wachsmuth

Artificial Intelligence Group, Faculty of Technology
Bielefeld University
{nlessman,ipke}@techfak.uni-bielefeld.de

Abstract. Crucial for action coordination of cooperating agents, joint attention concerns the alignment of attention to a target as a consequence of attending to each other's attentional states. We describe a formal model which specifies the conditions and cognitive processes leading to the establishment of joint attention. This model provides a theoretical framework for cooperative interaction with a virtual human and is specified in an extended belief-desire-intention modal logic.

Keywords: cooperative agents, attention, alignment, BDI, modal logic.

1 Introduction

A foundational skill in human social interaction, joint attention is receiving increased interest in human-agent interaction. Attention has been characterized as an increased awareness [1] and intentionally directed perception [2] and is judged to be crucial for goal-directed behavior. Joint attention can be defined as simultaneously allocating attention to a target as a consequence of attending to each other's attentional states [3]. In contrast to joint perception (the state in which interactants are just perceiving the same object without further constraints concerning their mental states), the intentional aspect of joint attention has been stressed, in that interlocutors have to deliberatively focus on the same target while being mutually aware of sharing their focus of attention [2] [4].

The computational modeling of joint attention mechanisms or prerequisites thereof, such as perceptional attention focus, convincing gaze behavior, gaze following skills, has been addressed in cognitive robotics, e.g. [3] [5], and research on virtual humans and embodied conversational agents, e.g. [6] [7]. However, aspects of intentionality and explicit representation of the other's mental state are not accounted for in these approaches altogether.

In this paper, we address the cognitive challenges of joint attention in action coordination of cooperating agents [8]. According to Pickering and Garrod [9] successful communication is based on joint processes, called alignment, which realize action coordination between interlocutors without an explicit exchange of information states. In previous work we have argued [10] that one central condition of such alignment processes consists of joint attention and that activation of a dynamic working memory and a partner model are crucial constituents.

B. Mertsching, M. Hund, and Z. Aziz (Eds.): KI 2009, LNAI 5803, pp. 540–547, 2009.

We investigate joint attention in a cooperative interaction scenario with the virtual human Max, where the human interlocutor meets the agent face-to-face in 3D virtual reality. The human's body and gaze are picked up by infrared cameras and an eye-tracker [11]; e.g., Max can follow the human's gaze. The agent's mental state is modeled in the BDI (Belief-Desire-Intention) paradigm. In order to establish joint attention, the interlocutors need to be aware of each other's current epistemic activities. The human interlocutor's focus of attention is inferred from her overt behaviors, and focused objects are activated as salient in the agent's dynamic working memory; for detail cf. [10].

In this paper we describe a formal model which specifies the conditions and cognitive processes leading to the establishment of joint attention. This model provides a theoretical framework for a cooperative interaction scenario with the virtual human Max and the CASEC cognitive architecture introduced in [10]. In Section 2, we firstly introduce the use of activation values in modal logic and derive a definition of attention in Section 3. In Section 4, a formal definition of joint attention with regard to the required mental states is presented. In Section 5, we formally examine the action chain and skills involved bringing about the mental states requisite for joint attention. Section 6 presents a conclusion.

2 Formal Specification

To establish joint attention an agent must employ coordination mechanisms of understanding and directing the intentions underlying the interlocutor's attentional behavior, cf. [10]: The agent needs to (r1) track the attentional behavior of the other by gaze monitoring and (r2) derive candidate objects the interlocutor may be focusing on. Further, the agent has to (r3) infer whether attentional direction cues of the interlocutor are uttered intentionally. The agent has to (r4) react instantly, as simultaneity is crucial in joint attention and in response should (r5) use an adequate overt behavior which can be observed by its interlocutor.

Important in our approach is a dynamic working memory, which is inspired by Oberauer [12] who conceptualizes working memory in three successive levels characterized by increased accessibility for cognitive processes: (1) The activated part of long-term memory pre-selecting information over brief periods of time; (2) the region of direct access keeping a limited number of representational "chunks" available for ongoing cognitive processes; (3) the focus of attention holding the particular chunk selected for the immediate cognitive operation to be applied.

2.1 Beliefs

Our CASEC architecture (Cognitive Architecture for a Situated Embodied Cooperator) [10] adopts the BDI paradigm of rational agents [13] applying modal logic as a specification language, but additionally integrates a dynamic working memory. The formalism used to specify goals and beliefs builds on the possible worlds approach. We use a (doxastic) modal logic KD45 for modeling beliefs. In accordance with [13], we use the three modal connectives BEL, GOAL, and INTEND as atomic modalities.

Definition 1. *Any first-order formula is a state formula. If* φ_1 *and* φ_2 *are state formulae then also* $\neg\varphi_1$ *and* $\varphi_1 \vee \varphi_2$*. If* φ *is a formula then* $BEL(\varphi)$, $GOAL(\varphi)$, *and* $INTEND(\varphi)$ *are state formulae [13]. If i is an agent, then* $(BEL_i\, \varphi)$ *is an abbreviation denoting that agent i believes formula* φ *[14].*

Hereafter "formula" is to mean "state formula". To account for the dynamics of agent i's beliefs, we extend the formalism to include activation values (for further motivation cf. Section 3).

Definition 2. *If* $(BEL_i\, \varphi)$ *is a formula, then also* $(BEL_i\, \varphi\, a), a \in \Re^+$ *is a formula.* $Acti(BEL_i\, \varphi\, a) = a$ *returns the formula's current activation value a.*

Also terms are extended to contain an activation value:

Definition 3. *For a given formula* φ *with n terms, let* $t\_set(\varphi)$ *denote the set of terms of* φ, $t\_set(\varphi) := \{e_i | e_i \text{ term of } \varphi, i = 1, ..., n\}$*. Each term* e_i *with term value* $||e_i||$ *is augmented by an activation value a. Therefore we define* $\widehat{e}$ *to consist of:* $\widehat{e} := (||e||, a), a \in \Re^+$*. The function* $Acti(\widehat{e}) = a$ *returns the term's current activation value.*

Activiation values influence the beliefs' accessibility for mental operations. They are calculated by an ACT-R-like function for modelling recency effects and decay. Additionally, automatic activation impulses of different origins with own decay rates are included to model the overall saliency of a belief.

The activation value of a formula φ consists of the average of the contained terms' activations, $\#e_i$ denoting the number of terms (i=1,...,n):

$$Acti(\varphi) = \sum \frac{Acti(\widehat{e_i})}{\#e_i}, e_i \in t\_set(\varphi) \tag{1}$$

The set of current beliefs is defined as follows:

Definition 4. *Let* $Beliefs_i$ *denote the entire set of agent i's beliefs. Then we define* ${}^{cur}Bels_i := \{b_k | b_k \in Beliefs_i \wedge Acti(b_k) > \theta BEL_{acti}, k = 1, ..., n\}$*.*

θBEL_{acti} represents a threshold which is dynamically tuned so that only a limited number of items reside in the set of ${}^{cur}Bels_i$ modeling the region of direct access of Oberauer's working memory model (see Section 1). Figure 1 illustrates the extension of the classical set of beliefs to a dynamic model including activation values. Activation values can be seen as adding an additional dimension which allows for filtering mechanisms. Thus we model "increased awareness" by use of activation values for aligning a candidate set of mental operations to the current context as well as to the interaction partner.

In addition to the modal connectives introduced above, we follow [15] in adding HAPPENS and DONE to the atomic modalities. If α is an action then $(HAPPENS\, \alpha)$ states that action α will happen next and $(DONE\, \alpha)$ means that action α has happened. These basic temporal operators are augmented by the operator ";" responsible for describing event sequences e.g. $(\alpha; \beta)$ denotes that first action α and then action β is executed. Additionally, <> denotes the modal operator *possibly* and [ ] the modal operator *always* [14].

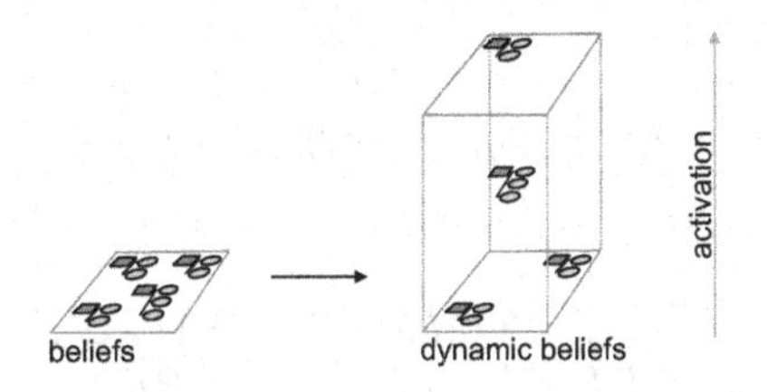

Fig. 1. Extending beliefs to dynamic beliefs with activation values

2.2 Goals - Intentions - Plans

Like [16] we see intentions as not reducible to beliefs and goals but as primitive modal connectives. However, they can be decomposed as follows (the modal operators PLAN, COMMIT are not formally introduced here).

Definition 5. *An intention is decomposed into the respective goal, the adopted plan and the commitment to use this plan as a means to achieve the goal:*

$$(INTEND_i\,\varphi) ::= (GOAL_i\,\varphi) \wedge (PLAN_i\,\varphi) \\ \wedge (COMMIT_i((GOAL_i\,\varphi), (PLAN_i\,\varphi)))$$

Whereas commitment is not directly relevant for the focus of attention, the parameters of the goal and the plan formulae directly apply to it. To cover the object related aspects of the formulae the function *t_set* (see Def. 3) is applied.

Definition 6. *The termsets of the modal connectives dissolve to the termset of the respective formula involved:* $t\_set(GOAL_i\,\varphi) := t\_set(\varphi)$, $t\_set(PLAN_i\,\varphi) := t\_set(\varphi)$. *The termset of an intention derives from the termset of the current goal:* $t\_set(INTEND_i\,\varphi) := t\_set(CurrentGoal_i(INTEND_i\,\varphi))$.

The *CurrentGoal* is the highest activated goal of the set of goals associated with the current intention. This set of goals consists of the intention's goal specification and the subgoals invoked in executing the adopted plan.

3 Defining the Focus of Attention

Like beliefs, also intentions and plans are qualified by activation values. We use activation values as a measure for saliency, i.e. an object with a higher activation value is more salient than one with a lower one. Whenever an object gets in the agent's gaze focus or is subject to internal processing, activation values are increased. That is, the set of ${}^{cur}Bels_i$ models *the region of direct access* proposed by Oberauer [12]. Depending on the processing step a new derived belief, a chosen intention, or an executed action of a plan corresponds to the focus of attention. We define the current belief and intention by use of activation values. The current plan step corresponds to the action of the currently adopted plan, an acyclic graph of nested goals covering the actions to be executed next:

Definition 7

$$^{cur}BEL_i := \{b_x | b_x \in {}^{cur}Bels_i \wedge \forall b \in {}^{cur}Bels_i \wedge b \neq b_x : Acti(b_x) > Acti(b)\}$$
$$^{cur}INT_i := \{n_x | n_x \in Ints_i \wedge \forall n \in Ints_i \wedge n \neq n_x : Acti(n_x) > Acti(n)\}$$

$$^{cur}PLAN\_STEP_i := \{action_x | (COMMIT_i(GOAL_i\ \varphi), {}^{cur}PLAN_i\ \varphi) \wedge action_x \in Acy\_graph({}^{cur}PLAN_i\ \varphi)) \wedge (HAPPENS_i\ action_x)\}$$

As these processes of perception and cognition run concurrently, we conjoin all three aspects in our concept of focus of attention.

Definition 8

$$ATT_i := \{t\_set({}^{cur}BEL_i) \cup t\_set({}^{cur}INT_i) \cup t\_set({}^{cur}PLAN\_STEP_i)\}$$

The focus of attention is part of dynamic working memory and is modulated by the changing beliefs and intentional states of the agent. Figure 2 illustrates the classical BDI model extended by the incorporation of activation values.

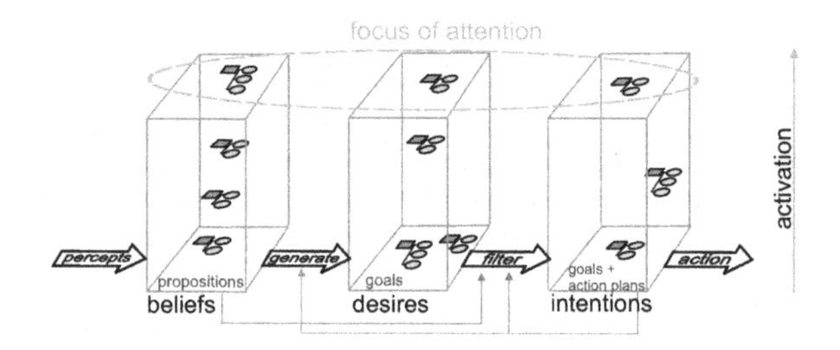

Fig. 2. BDI and Focus of Attention

4 A Definition of Joint Attention

In accordance with [2] we conceive of joint attention as an intentional process. Meeting the requirements of Sec. 2, we describe the mental state required for an agent i to believe in joint attention while focusing conjointly with its interlocutor j on a certain target ϑ (see Figure 3 for illustration and explanation next page).

Definition 9. $(BEL_i(JOINT\_ATT\ i\ j\ \vartheta))$ *iff*

1. *(being aware of other)*$BEL_i(ATT_j\ \vartheta) \wedge BEL_i(INTEND_j(ATT_j\ \vartheta))$
2. *(ascribing goal)* $BEL_i(GOAL_j(ATT_i\ \vartheta \wedge ATT_j\ \vartheta))$
3. *(adopting goal)* $GOAL_i(ATT_i\ \vartheta \wedge ATT_j\ \vartheta)$
4. *(feedback)* $BEL_i(BEL_j(ATT_i\ j))$
5. *(focus state)* $HAPPENS(< T(\vartheta) >_i \wedge < P(< T(\vartheta) >_j) >_i)$

(1) (being aware of other). By representing the explicit belief about the interlocutor's focus of attention $BEL_i(ATT_j\ \vartheta)$ the agent meets requirement (r1).

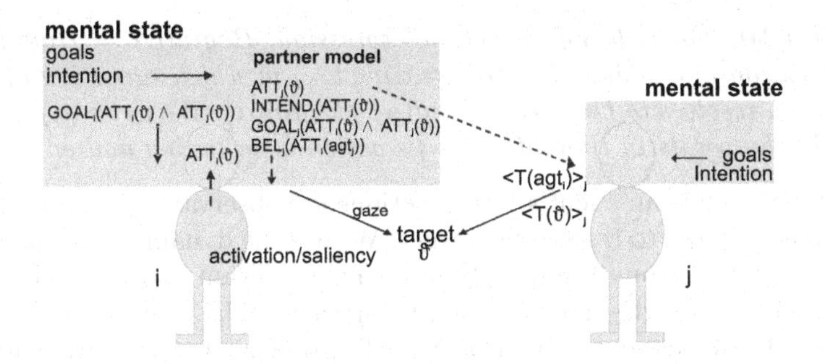

Fig. 3. Joint attention from agent i's perspective

To meet (r3) the agent additionally needs to infer whether the interlocutor intentionally draws its focus of attention on the object, $BEL_i(INTEND_j(ATT_j\,\vartheta))$.
(2) (ascribing goal). Agent i must believe that agent j has the goal that both agents draw their attention focus to the target $BEL_i(GOAL_j(ATT_i\vartheta \wedge ATT_j\vartheta))$. This belief can be evoked by an *initiate-act* of agent j e.g. by gaze-alternation.
(3) (adopting goal). The agent then needs to adopt the interlocutor's goal $GOAL_i(ATT_i\,\vartheta\ \wedge ATT_j\,\vartheta)$. To meet requirements (r4) and (r5), the agent as a recipient needs to employ an observable *respond-act.*
(4) (feedback). But for mutual belief, an additional *respond-act* is required from the initiator j so that agent i comes to believe $BEL_i(BEL_j(ATT_i\,j))$.
(5) (focus state). When agent i draws its focus of attention on the target ($< T(\vartheta) >_i$) while perceiving that its interlocutor also focuses on the target ($< P(< T(\vartheta) >_j) >_i$), then from agent i's perspective a joint attention state has been established. For definition of T (*test-if*) and P (*perceive-that*) see Sec. 5.

5 Grounding Modal Connectives in a Logic of Action

After defining the mental state required for joint attention, we need to specify the epistemic actions that lead to the respective beliefs and goals. To this end, a logic of action is required. Like [14] we adopt standard *propositional dynamic logic.* In this logic, epistemic actions of perceptual kind are applicable to all formulae (propositions and actions) but do not allow direct perception of mental states. However an agent can perceive overt actions of its interlocutor as well as propositions of objects. We adopt the definition of two epistemic actions of [14]:

- ***Perceive-that***: Action of perceiving some ϑ in the external world.
  $< P(\vartheta) >_j \varphi$ (always φ is true after agent j has perceived ϑ)
- ***Test-if***: (precursor of *Perceive-that*) Test-if actions observable and testable from other agents as they are expected to have an observable counterpart.

By default we assume that, whenever an agent perceives something, it believes what it has perceived: $[P(\vartheta)]_i \rightarrow (BEL_i\vartheta)$. As time constraints and coordination are crucial in joint attention, a representation of time is needed.

Definition 10. *For α being an action expression, $Begin(\alpha) :=$ time t_{begin} at which execution of α starts, $End(\alpha) :=$ time t_{end} at which execution of α ends. The duration resolves to $Dur(\alpha) := End(\alpha) - Begin(\alpha)$. We write $(\alpha)_{j[t_{begin}, t_{end}]}$ to describe the points in time of agent j's action α beginning and ending.*

Test-action. While an agent's *test-if* actions are observable [14], additional information is required to resolve the target. We use the dynamic working memory as a source of background information marking relevant objects and a partner model to account for the interlocutor's perspective. The candidate set of target objects are the objects in the interlocutor's line of gaze. Incorporating activation values in the reference resolution allows a fast and easy adjustment of the candidate set (meeting requirement (r2)). If the agent perceives the interlocutor's behavior as a *test-action* and is able to resolve a candidate object, the agent infers that the interlocutor's focus of attention must reside on that object.

$$< P(< T(\varphi) >_j) >_i \rightarrow (BEL_i(ATT_j\ \varphi)) \tag{2}$$

Beliefs about the interlocutor's focus of attention are updated dynamically, leading to new beliefs or increasing a belief's activation respectively. If the interlocutor focuses several times on an object (or for a long duration) the agent interprets this as the attention focus being *intentionally* drawn upon the target (cp. [10]):

$$< P(< T(\varphi) >_j) >_i ; < P(< T(\varphi) >_j) >_i \rightarrow (BEL_i(INTEND_j(ATT_j\ \varphi))) \tag{3}$$

Initiate-actions. One way to perform an *initiate-act* consists of gaze alternation. An object has to be the focus of attention for several times with additional short glances addressing the interlocutor inbetween (triadic relation).

$$\begin{aligned} &< P(< T(\varphi) >_j) >_i \ ; < P(< T(i) >_j) >_i ; < P(< T(\varphi) >_j) >_i ; \\ &< P(< T(i) >_j) >_i \rightarrow (BEL_i(GOAL_j(ATT_i\ \varphi)) \wedge (ATT_j\ \varphi)) \end{aligned} \tag{4}$$

Respond-actions. Respond-actions play a crucial role to backup the actions the agents have sought to perform. They can consist of *smiling at*, *focussing on*, and *nodding to* the interlocutor. The *respond-actions* can be applied to establish mutual understanding between the interlocutors. E.g. after agent i performed a respond-act, it checks whether agent j has noticed its response:

$$\begin{aligned} &(DONE(< T(j) >_i)_{[t_{end}]} \wedge (HAPPENS < P(< T(i) >_j) >_i)_{[t_{begin}]} \wedge \\ &(Dur(< T(j) >_i \geq 2s)) \wedge ((t_{begin} - t_{end}) \leq 5s) \rightarrow BEL_i(BEL_j(ATT_i\ j)) \end{aligned} \tag{5}$$

(Heuristics: Empirical research, not quoted here, has shown that the recipient's response to an agent initiating joint attention needs to take place in a 5s time frame, with a signal duration of more than 2s.)

6 Conclusion

We presented work on equipping a cooperative agent with capabilities of joint attention. To this end, a formal definition of joint attention has been introduced.

The required initiate- and respond-acts have been specified and grounded in a logic of action. The formalizations provide a precise means as to which requirements need to be met and which inferences need to be drawn to establish joint attention by aligning the mental states of cooperating agents. Implemented in the CASEC cognitive architecture [10] for our virtual human Max, they form the basis for the study of joint attention in a cooperative interaction scenario.

Acknowledgments. This research is supported by the Deutsche Forschungsgemeinschaft (DFG) in the Collaborative Research Center SFB 673.

References

1. Brinck, I.: The objects of attention. In: Proc. of ESPP 2003, Torino, pp. 1–4 (2003)
2. Tomasello, M., Carpenter, M., Call, J., Behne, T., Moll, H.: Understanding and sharing intentions: The origins of cultural cognition. Behavioral and Brain Sciences 28, 675–691 (2005)
3. Deak, G.O., Fasel, I., Movellan, J.: The emergence of shared attention: Using robots to test developmental theories. In: Proc. of the First Intl. Workshop on Epigenetic Robotics, Lund University Cognitive Studies, vol. 85, pp. 95–104 (2001)
4. Hobson, R.P.: What puts the jointness into joint attention? In: Eilan, N., Hoerl, C., McCormack, T., Roessler, J. (eds.) Joint attention: communication and other minds, pp. 185–204. Oxford University Press, Oxford (2005)
5. Doniec, M., Sun, G., Scassellati, B.: Active learning of joint attention. In: IEEE/RSJ International Conference on Humanoid Robotics (2006)
6. Kim, Y., Hill, R.W., Traum, D.R.: Controlling the focus of perceptual attention in embodied conversational agents. In: Proceedings AAMAS, pp. 1097–1098 (2005)
7. Gu, E., Badler, N.I.: Visual attention and eye gaze during multiparty conversations with distractions. In: Gratch, J., Young, M., Aylett, R.S., Ballin, D., Olivier, P. (eds.) IVA 2006. LNCS (LNAI), vol. 4133, pp. 193–204. Springer, Heidelberg (2006)
8. Kaplan, F., Hafner, V.: The challenges of joint attention. Interaction Studies 7(2), 135–169 (2006)
9. Pickering, M.J., Garrod, S.: Alignment as the basis for successful communication. Research on Language and Computation 4(2), 203–228 (2006)
10. Pfeiffer-Lessmann, N., Wachsmuth, I.: Toward alignment with a virtual human - achieving joint attention. In: Dengel, A.R., Berns, K., Breuel, T.M., Bomarius, F., Roth-Berghofer, T.R. (eds.) KI 2008. LNCS (LNAI), vol. 5243, pp. 292–299. Springer, Heidelberg (2008)
11. Pfeiffer, T.: Towards gaze interaction in immersive virtual reality. In: Virtuelle und Erweiterte Realität - Fünfter VR/AR Workshop, pp. 81–92. Shaker Verlag, Aachen (2008)
12. Oberauer, K.: Access to information in working memory: Exploring the focus of attention. J. of Exp. Psych.: Learning, Memory, and Cognition 28, 411–421 (2002)
13. Rao, A., Georgeff, M.: Modeling rational behavior within a BDI-architecture. In: Proc. Intl. Conf. on Principles of Knowledge Repr. and Planning, pp. 473–484 (1991)
14. Lorini, E., Tummolini, L., Herzig, A.: Establishing mutual beliefs by joint attention: towards a formal model of public events. In: Proc. of CogSc 2005 (2005)
15. Cohen, P., Levesque, H.: Intention is choice with commitment. Artificial Intelligence 42, 213–261 (1990)
16. Bratman, M.: Intention, Plans, and Practical Reason. Havard Univ. Press, Cambridge (1987)

Towards Determining Cooperation Based on Multiple Criteria

Markus Eberling

Department of Computer Science
University of Paderborn, Germany
markus.eberling@upb.de

Abstract. Selfish agents in a multiagent system often behave suboptimal because they only intend to maximize their own profit. Therefore a dilemma occurs between the local optimum and the global optimum (i.e. optimizing the social welfare). We deal with a decision process based on multiple criteria to determine with whom to cooperate. We propose an imitation-based algorithm working locally on the agents and producing high levels of global cooperation. Without global knowledge the social welfare is maximized with the help of local decisions. These decisions are influenced by different meme values which are transmitted through the population.

1 Introduction

Cooperation can be found in groups of humans or companies in a firm network. In both scenarios this cooperation leads to higher benefit for the whole group and to higher benefit for the individuals. Mostly, the group members have a common goal but different motivations to join them [9] or to stay in them [2]. Companies build networks to achieve their goals [8] and moreover good supply chains are helpful to produce qualitative products [11]. Reciprocal behavior is one of the characteristics of such networks [12]. Another aspect is altruism which is a kind of help without being payed for [1], although the help could produce costs [7,13]. The decision to cooperate or not is often based on different criteria like kin selection or subjective criteria such as social cues.

Cooperation is also needed in multiagent systems where the agents have a common goal. We want to model the cooperation determination with the help of memes each influencing some criterium. If all criteria are fulfilled, the agent will cooperate with an agent asking for help. We will see that our local algorithm leads to high cooperation rates with the help of imitation-based learning.

We model a multiagent system with agents having local knowledge and being assigned jobs to fulfill. These jobs require cooperation in most cases because the agents only have a limited set of skills which only allows them to complete a small fraction of all possible jobs. In our model cooperation will produce costs for agents who help others. But completing a job leads to a reward for the agent to which the job was assigned. Therefore we have a dilemma that the local best strategy leads not to the global optimum of optimizing the social welfare.

B. Mertsching, M. Hund, and Z. Aziz (Eds.): KI 2009, LNAI 5803, pp. 548–555, 2009.

Richard Dawkins firstly introduced the term *meme* in his book "The Selfish Gene" [3], which deals with cultural evolution. Since then many articles concerning learning with different kinds of memes and similar approaches have been published. A meme is a particular thought or idea which transfers itself from individual to individual. During this process a meme could be changed or mutated. Therefore, Dawkins used the term meme as an analogy to gene.

In [10] there is an approach of combat learning through imitation. Simple rules are broadcasted to other agents in the population. Based on a utility value new rules replace old rules in a rule set. In addition these rules are slightly changed through mutation. This imitation-based learning approach outperforms a previously used evolutionary algorithm.

In [5] and [4] David Hales experiments with agents equipped with one skill out of a specific skill set. Agents are given resources which they could harvest only, if the required skill of the resource matches their own skill. Otherwise the agent could pass the resource to another agent. The agents also have a so called tag $\tau \in [0, 1]$ and a threshold $0 \leq T \leq 1$. An agent D could only give a resource to another agent R if $|\tau_D - \tau_R| \leq T_D$ holds. Harvesting resources is rewarded with a payoff of 1 and searching for another agent is rewarded with a negative payoff where the height depends on the searching method. Hales' approach uses an evolutionary algorithm with tournament selection and slightly mutation. The results show high donation rates for good searching methods.

In [6] Hales presents a protocol for a decentralized peer-to-peer system called SLAC ("Selfish Link and Behavior Adaptation to produce Cooperation"). There, agents have simple skills and are awarded jobs. Each job requires a single skill, only. If the agent does not provide the required skill, it asks an agent from its neighborhood to complete the job. A boolean flag indicates whether or not an agent behaves altruistically. Altruists have to cooperate, if another agent asks for it. Completed jobs are rewarded with benefit of 1 and cooperation produces costs of 0.25. Agents are allowed to exchange jobs within their neighborhoods, which are modeled as a subset of the population. During the simulation agents are pairwise compared and the agent with lower utility value in a simulation round is changed. This means, that the neighborhood is totally destroyed and the agent creates a link to the better performing agent and copies its neighborhood and strategy (altruist or not). With some probabilities the agent also mutates its skill, the altruism flag and the neighborhood after this step. These drastic changes in the network structure lead to good results concerning the job completing rate.

The next section provides a formal description of our basic model while Sec. 3 will describe the considered scenario. Section 4 shows some experiments and discusses their results. The last section will draw a conclusion and ends with the presentation of an outlook to future work.

2 The Model

In this section we want to describe our basic model. To allow different granularities of jobs we model a job $j \in \mathcal{J}$ as a finite set of tasks $j = \{t_1, t_2, \ldots, t_n\}$

with $t_i \in \mathcal{T}$, $t_{min} \leq n \leq t_{max}$. $\mathcal{J}$ denotes the set of jobs and $\mathcal{T}$ a finite set of tasks in the system. A task $t = (s_t, p_t)$ is a pair of a required skill $s_t \in \mathcal{S}$ and a specific payoff $p_t \in \mathbb{R}_0^+$, where $\mathcal{S}$ is a finite set of possible skills. A task could only be handled by an agent a, if a offers the required skill in its skill set $\mathcal{S}_a$.

The agents are assigned to different jobs, which should be fulfilled. To finish a job, each task has to be completed. Therefore an agent has to search for cooperation partners in its neighborhood, if it does not incorporate all required skills. The neighborhood $\mathcal{N}_a$ of agent a is a subset of the population which is defined as a symmetric relation. All agents share a finite set of memes $\mathcal{M}$, with $|\mathcal{M}| = m$. A meme represents an opinion or idea like "I like wearing a hat". Each agent a has a m-dimensional vector $\mathcal{V}_a \in [0,1]^m$ which assigns values to the memes. The values could be interpreted as confidences on the memes' meanings. $\mathcal{V}_a^i = 1$ means that agent a agrees with meme i and $\mathcal{V}_a^i = 0$ means that the agent disagrees. To model different degrees of willingness to cooperate an agent a owns a threshold vector $T_a \in [0,1]^m$, which is the basis for decision making whether to cooperate or not. An agent a will only cooperate with an agent b if the inequality

$$|\mathcal{V}_a^i - \mathcal{V}_b^i| \leq T_a^i \tag{1}$$

holds for every meme $i \in \{1, 2, \ldots, m\}$. $\mathcal{V}^i$ is the i-th component of the meme value vector and T^i is the i-th component of the thresholds vector. It follows that all m inequalities constitute the criteria for cooperation. While the meme values are observable markers, the thresholds are not visible to others. If the inequality is met for every meme, we say that a is culturally related to b, denoted by $cr(a,b) = true$, since their meme values are very similar. Note that the cr-Relation is not symmetric and only computed from the viewpoint of a specific agent. Cooperation produces payoff dependent costs of $c_f \cdot p_t$ for a task t with a cost factor $c_f \in \mathbb{R}_0^-$. This models that the cooperating agent gives its computational power and resources to the other agent without being responsible for the job fulfillment. Since we currently do not consider load balancing and capacity constraints, an agent could handle an arbitrary number of tasks in a single step.

3 Scenario Description

In our scenario in each step $10 \cdot N$ jobs are generated and distributed to randomly chosen agents, N denoting the population size. If an agent is not able to fulfill all tasks, it searches for cooperation partners. After all jobs were generated and distributed to the agents an adaptation phase starts which consists of two parts. Each agent a calculates the elite set $\mathcal{E}_a$ of the ε best agents in its neighborhood depending on their utility values. If the agent is not in this set, it is said to be unsatisfied and adapts itself by moving its meme value vector into the direction of the best agent's vector to be more like it in the first part of this phase. Let a^* be the best agent of a's neighborhood. Then the adjusted meme value vector of a is calculated via

$$\mathcal{V}_a^{new} \leftarrow \mathcal{V}_a^{old} + \eta \cdot (\mathcal{V}_{a^*}^{old} - \mathcal{V}_a^{old}) \tag{2}$$

Algorithm 1. Simulation

1: Initialize N agents and their neighborhoods randomly
2: **loop**
3: Produce $10N$ jobs and allocate them to randomly chosen agents
4: **for all** agents $a \in \mathcal{A}$ **do**
5: $\mathcal{E}_a \leftarrow \varepsilon$ best agents of $\mathcal{N}_a \cup \{a\}$
6: **if** $a \notin \mathcal{E}_a$ **then**
7: $a^* \leftarrow$ best agent of $\mathcal{N}_a$
8: $\mathcal{V}_a \leftarrow \mathcal{V}_a + \eta \cdot (\mathcal{V}_{a^*} - \mathcal{V}_a)$ /* *adaptation* */
9: with probability $\mathcal{P}_\mathcal{N}$: replace r unrelated neighbors by r randomly chosen agents /* *social networking* */
10: **end if**
11: **end for**
12: **end loop**

where η is an important, exogenous parameter of the simulation, which specifies the adaptation strength. The idea behind this step is, that the better agent is believed to be more successful due to better meme values. Therefore the agent wants to change its values to be more like a^*. As an result the agent a is more likely to be culturally related to a^* than before this step.

After this part *social networking* will follow with a probability of $\mathcal{P}_\mathcal{N}$ as the second part of the adaptation phase. The agent chooses r agents out of its neighborhood which are not culturally related and replaces them by r randomly chosen ones from the population. If there are less than r agents, which are not culturally related, all of them are replaced. By replacing them the situation could not worsen but only improve for the agent. This is due to the fact that all agents, which are not culturally related to this agent are replaced by randomly chosen ones, with a non-zero probability of selecting culturally related agents. Note that only culturally related agents will help this agent, as stated above. The described approach is summarized in Algorithm 1.

4 Experiments and Results

In this section we describe some experiments and their results. We simulated 1000 agents over 200 generations with an initial neighborhood size of 20. Because of the social networking phase the neighborhoods could grow but we fixed them to a maximum size of 30. The growth of neighborhoods is due to the fact a chosen agent is not allowed to reject the new interaction link. Each agent was equipped with only one skill (i.e. $|\mathcal{S}_a| = 1$) and the whole skill set $\mathcal{S}$ contained five elements. We set the number of memes to $m = 2$. To ease the analysis of the results we fixed the number of tasks per job such that every job contains exactly three tasks ($t_{min} = t_{max} = 3$). The payoff for all tasks $t \in \mathcal{T}$ was set to $p_t = 1$. The probability for executing the social networking step was set to 0.01. These parameters were derived from previously hand-made experiments. Given the jobs containing three tasks and agents only have a single skill out of a set of five skills, it is easy to see that the probability of completing a job without cooperation is only 0.008. This means that high percentages of completed jobs necessarily imply high cooperation.

In this paper we want to examine the influence of the cost factor $c_f \in \{0.0, \underline{-0.25}, -0.5, -0.75, -1.0\}$ for the cooperation, the adaptation factor $\eta \in \{0.0, 0.1, 0.3, \underline{0.5}, 0.7, 0.9, 1.0\}$, the elite set size $\varepsilon \in \{2, \underline{4}, 6, 8\}$ and the number of replaceable agents $r \in \{\underline{1}, 2, 3, 4\}$. Based on a base set of parameter values we compare the algorithm's behavior for these varying parameters. The base setup is the set of underlined values. We expect that higher cooperation costs lead to lower levels of cooperation and thus to lower percentages of completed jobs. Since the jobs contain exactly three tasks, each with a payoff of 1, completed jobs are rewarded with 3 units of utility. $\eta = 0.0$ means that the meme value vector is not changed at all and $\eta = 1.0$ would mean copying the meme value vector of the best agent of the neighborhood. Both are extreme values for the adaptation step. The third examined parameter is the size of the elite set to calculate whether the agent is satisfied with the situation or not. The greater the elite set of the agent the lower the probability that the agent is not satisfied and intends to adapt. The probability of being unsatisfied is maximal for $\varepsilon = 2$ since only one other agent is allowed to have a higher utility value.

Figure 1 shows the results of all four simulation sets. The results are mean values of 30 individual runs all with different random seeds. The left hand side shows the percentage of completed jobs and the right hand side the percentage of jobs that could not be fulfilled because of missing culturally related agents. Note that in those cases the required skill was available.

As we can see in Fig. 1a there are two contrary developments. While the number of completed jobs increases for $c_f = 0$ and $c_f = -0.25$, this rate decreases for all $c_f < -0.25$. If cooperation is for free than there is high motivation for cooperation. We observe quite high values ($\sim 98\%$) of completed jobs. Low costs of about -0.25 lead to 95%. Note that $c_f = 0.0$ is not compatible with our environmetal constraints, but we wanted to show that cost-free cooperation is not that better in our approach. With increasing cooperation costs the motivation for cooperation decreases. Agents who did not help anybody but received a lot of help have very high utility values and become very attractive for adaptation to them. Therefore, the meme value vectors of agents with lower utility values are moved into this direction which leads to the observed development. Figure 1b shows the percentages of jobs that could not be completed because there were no culturally related agents in the neighborhood that offer the required skill. As we see this percentage converges to zero for $c_f = 0$ and $c_f = -0.25$ but always remains greater or equal 10% for all other settings. From this it follows that for $c_f < -0.25$ the agents neighborhoods are not able to become highly related.

A main idea of our model is the adaptation step described in Sec. 3. The strength of this adaptation depends on the parameter η. The results for different values of this parameter are shown in Fig. 1c. For better readability we omit the results for $\eta = 0.1$ and $\eta = 0.9$, because they show same behavior like for $\eta = 0.0$ and $\eta = 1.0$. The best results are achieved for $\eta = 0.5$. In all other experiments

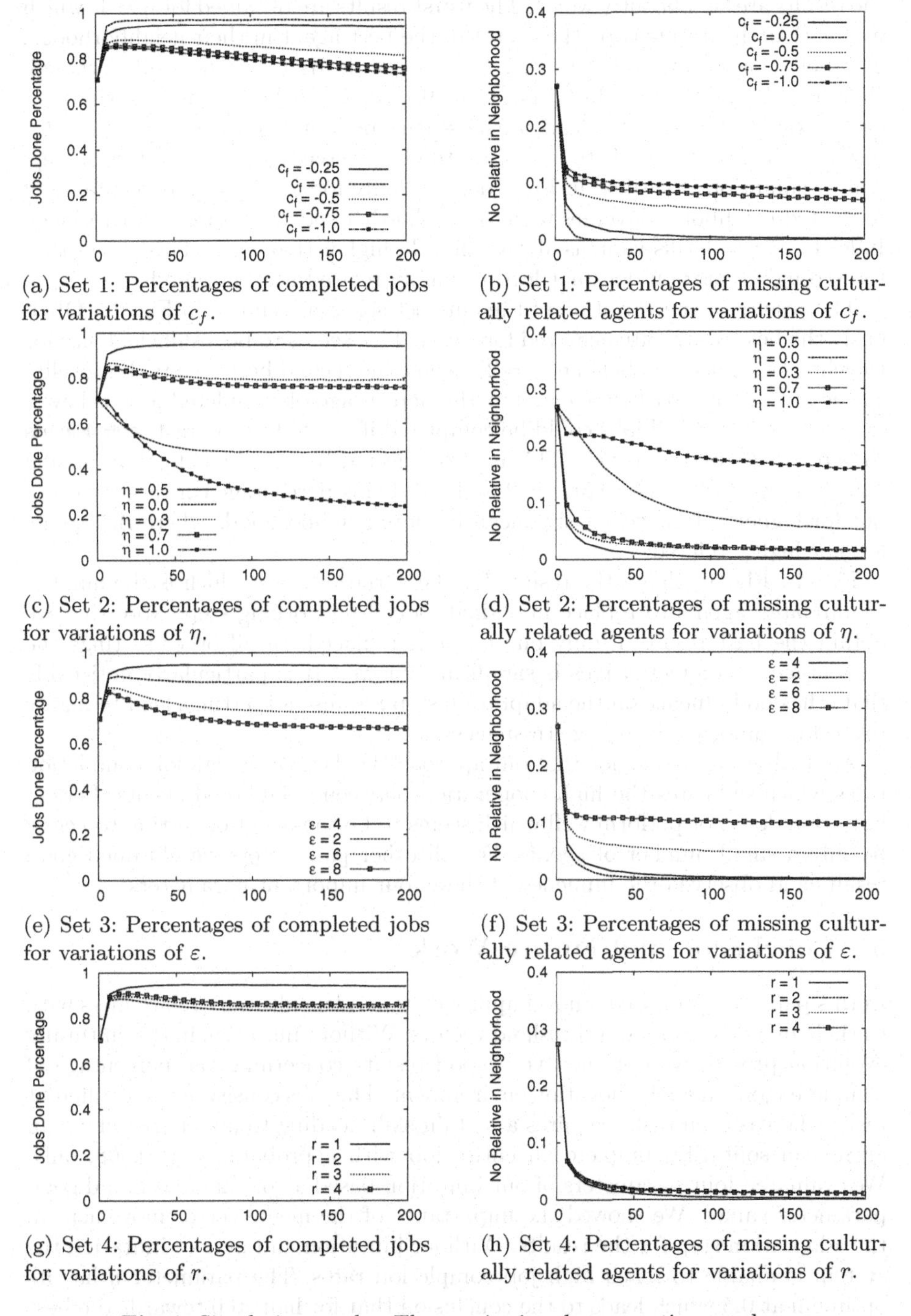

(a) Set 1: Percentages of completed jobs for variations of c_f.

(b) Set 1: Percentages of missing culturally related agents for variations of c_f.

(c) Set 2: Percentages of completed jobs for variations of η.

(d) Set 2: Percentages of missing culturally related agents for variations of η.

(e) Set 3: Percentages of completed jobs for variations of ε.

(f) Set 3: Percentages of missing culturally related agents for variations of ε.

(g) Set 4: Percentages of completed jobs for variations of r.

(h) Set 4: Percentages of missing culturally related agents for variations of r.

Fig. 1. Simulation results. The x-axis give the generation number.

the results are significantly worse. The worst results are obtained for $\eta = 1$, which means that the agents copy the vector of the best agent in their neighborhoods. The behavior of not adapting (i.e. $\eta = 0$) also leads to worse results, though they are higher than perfect imitation. 0.5 seems to be optimal as the next closest values, 0.3 and 0.7, lead to the second best and third best results. This is why we belief that half step sizes are the best choice for this adaptation step. This assumption is verified in Fig. 1d where only for $\eta = 0.5$ the percentages of unrelated neighbors converges to zero. In all other sets this state is not reached. Indeed, for $\eta = 1$ this value is always slightly higher than 20%. This shows that too strong adaptation does not lead to highly related neighborhoods.

The results of the variation of the elite set size ε are shown in Fig. 1e. Obviously the best results are achieved for $\varepsilon = 4$. This seems to be a sufficient size for the elite set. If it is smaller (i.e. $\varepsilon = 2$) we get the second best results. Large elite sets lead to more satisfied agents but the percentage of completed jobs is lower. Only about 70% of all jobs could be completed if $\varepsilon = 8$. In constrast, $\varepsilon = 4$ leads to percentages of about 95%. This is also reflected by the percentage of missing culturally related agents. For $\varepsilon = 2$ and $\varepsilon = 4$ this percentage converges to zero but for higher values we get significantly more neighborhoods where the agents are not highly related.

Finally, Fig. 1g shows the results for the variations of r which is the number of unrelated agents to replace in a single social networking step. Note that we obtain the best results if only one agent is replaced. In all other settings the percentage of completed jobs is significantly lower. It is particularly noticeable that r has no influence on the adaptation step because all settings lead to highly related neighborhoods as Fig. 1h suggests.

All in all we can conclude that our approach leads to very high job completion rates which indicates the high cooperation between neighbored agents. Perfect imitation does not perform well which seems to be the result of getting too close to only a small number of agents. For all other parameters we obtained good results and observed the influence of these four important parameters.

5 Conclusion and Future Work

In this paper we proposed a novel approach to model culturally relatedness with the help of meme value and threshold vectors. Without mutation in the culturally evolution process we obtained very good results concerning the percentage of completed jobs in a job allocation environment. The jobs consist of three different tasks whereas each tasks requires a certain skill, leading to a scenario where an agent can solitarily complete an entire job with a probability of 0.008, only. We evaluated four parameters of our imitation-based algorithm and found good parameter values. We showed the importance of the meme vector movement in the adaptation step for the results. Without imitation and perfect imitation the system is unable to reach high job completion rates. The parameter η has its optimum at 0.5 which leads to the conclusion that for half step towards the best agent the adaptation works best in our approach. For all other parameters the algorithm was proven to be very robust.

In future work we will examine the influence of the other parameters to get an idea of the parameter space and its optima. We also intend to formally analyze our system to get better insights into the systems behavior. After this we want to evaluate other adaptation steps, for example only adapting to some of the meme vector values. Another point of investigation is the application of the approach to more complex domains like the RoboCup Domain or other comparable domains. The next steps after this will also effect the model itself. Currently, the model leaks support for agent capacities, load balancing constraints and boundaries for parallel processing of tasks. We want to examine different scenarios, where the job requests are not uniformly and the agents are only allowed to handle a fixed number of tasks in parallel at once. These different scenarios will help to develop a better understanding for cooperation determination based on multiple criteria like the process is known from human societies.

References

1. Berkowitz, L., Macaulay, J.: Altruism and Helping Behavior: Social Psychological Studies of Some Antecedents and Consequences. Academic Press, New York (1970)
2. Buchanan, D., Huczynski, A.: Organisational Behaviour: An Introductory Text. Prentice-Hall, Pearson Education (1997)
3. Dawkins, R.: The Selfish Gene. Oxford University Press, Oxford (1976)
4. Hales, D.: Evolving specialisation, altruism, and group-level optimisation using tags. In: Sichman, J.S., Bousquet, F., Davidsson, P. (eds.) MABS 2002. LNCS (LNAI), vol. 2581, pp. 26–35. Springer, Heidelberg (2003)
5. Hales, D.: The evolution of specialization in groups. In: Lindemann, G., Moldt, D., Paolucci, M. (eds.) RASTA 2002. LNCS (LNAI), vol. 2934, pp. 228–239. Springer, Heidelberg (2004)
6. Hales, D.: Emergent group level selection in a peer-to-peer network. Complexus 3(1-3), 108–118 (2006)
7. Krebs, D.: Psychological approaches to altruism: An evaluation. Ethics 92 (1982)
8. Peitz, U.: Struktur und Entwicklung von Beziehungen in Unternehmensnetzwerken. PhD thesis, European Business School Oestrich-Winkel (2002)
9. Pennington, D.C.: The Social Psychology of Behavior in Small Groups. Psychology Group Ltd. (2002)
10. Priesterjahn, S., Eberling, M.: Imitation learning in uncertain environments. In: Rudolph, G., Jansen, T., Lucas, S., Poloni, C., Beume, N. (eds.) PPSN 2008. LNCS, vol. 5199, pp. 950–960. Springer, Heidelberg (2008)
11. Schmidt, D.: Unternehmenskooperationen in Deutschland. Deutscher Universitätsverlag, München (1997)
12. Sydow, J.: Strategische Netzwerke: Evolution und Organisation. Gabler. Habilitationsschrift Freie Universität Berlin (1992)
13. Wispé, L.: Altruism, sympathy, and helping: Psychological and sociological principles. Academic Press, New York (1978)

The SEASALT Architecture and Its Realization within the docQuery Project

Meike Reichle, Kerstin Bach, and Klaus-Dieter Althoff

Intelligent Information Systems Lab
University of Hildesheim
Marienburger Platz 22, 31141 Hildesheim, Germany
{reichle,bach,althoff}@iis.uni-hildesheim.de

Abstract. SEASALT (Sharing Experience using an Agent-based System Architecture LayouT) presents an instantiation of the Collaborating Multi-Expert Systems (CoMES) approach [1]. It offers an application-independent architecture that features knowledge acquisition from a web-community, knowledge modularization, and agent-based knowledge maintenance. The paper introduces an application domain which applies SEASALT and describes each part of the novel architecture for extracting, analyzing, sharing and providing community experiences in an individualized way.

1 Introduction

The development and expansion of Web 2.0 applications in the last years has resulted in the fact that formalized and structured documents have been largely replaced by individually structured and designed documents and experiences. Instead of using ready-made forms or templates to express their opinions, Web 2.0 participants present their experiences and ideas individually - for example via blog or forum posts, on mailing lists or in wikis. In order to keep up with the development towards more sophisticated social software applications, the techniques and approaches for intelligent information systems have to develop further as well. Traditional approaches like strictly structured monolithic data bases or highly specialized Text Mining approaches cannot deal sufficiently with the wealth of experiences provided in todays Word Wide Web.

In this paper we present a novel architecture for extracting, analyzing, sharing and providing community experiences. Our architecture is geared to real world scenarios where certain people are experts in special domains and the knowledge of more than one expert as well as the composition of a combined solution are required in order to solve a complex problem.

The core methodology for the realization of SEASALT (Sharing Experience using an Agent-based System Architecture LayouT), is Case-Based Reasoning (CBR) [2]. CBR has already been successfully applied in many industrial and academical applications [3,4]. Moreover, CBR is a technology for reusing experiences [5] and the technologies used within a CBR system can be customized according to a given domain.

B. Mertsching, M. Hund, and Z. Aziz (Eds.): KI 2009, LNAI 5803, pp. 556–563, 2009.

2 docQuery: An Application Based on SEASALT

Travel medicine is the prevention, management and research of health problems associated with travel play a major role alongside with individual aspects concerning the health status of the traveler and the desired destination. Therefore, information about the a traveler's home region as well as the destination region, the activities planned and additional conditions have to be considered when giving medical advice. Travel medicine starts when a person moves from one place to another by any kind of transportation and ends after returning home healthy. In case a traveler gets sick after a journey a travel medicine consultation might also be required.

Nowadays it is easier than ever to travel to different places, experience new cultures and get to know new people. In preparation for a healthy journey it is important to get a high quality and reliable answer on travel medicine issues. Both laymen and experts should get information they need and, in particular, they understand. For that reason we would like to introduce docQuery - a medical information system for travelers. Whether somebody travels frequently or occasionally, on business or for leisure, individually or with the whole family, docQuery should be able to provide individualized knowledge. The docQuery project focuses on high quality information that can be understood by everybody and maintained by a number of travel medicine experts, supported by intelligent methods executed by agents. Furthermore, the various and heterogeneous fields require independently organized knowledge sources. In comparison to traditional approaches that mostly rely on one monolithic knowledge resource, the docQuery system will adapt to the organization of knowledge given by the expert. An analysis of the expert's tasks shows that the information gathered from different channels (mailing lists, web forums, literature) has to be organized, analyzed, and synthesized before it can be provided. docQuery concentrates on a web community in which experts exchange and provide qualified information. docQuery can be used by inserting the key data on a travelers journey (like travel period, destination, age(s) of traveler(s), activities, etc.) and the system will prepare an individual composed information leaflet right away. The traveler can take the information leaflet to a general practitioner to discuss the planned journey. The leaflet will contain all the information needed to be prepared and provide detailed information if they are required. In the event that docQuery cannot answer the travelers question, the request will be sent to the expert community who will answer it.

3 The SEASALT Architecture

The SEASALT Architecture provides an application-independent architecture that features knowledge acquisition from a web-community, knowledge modularization, and agent-based knowledge maintenance. It consists of several components which will be presented in the following sections, ordered by their role within general information management, and exemplified using the docQuery project.

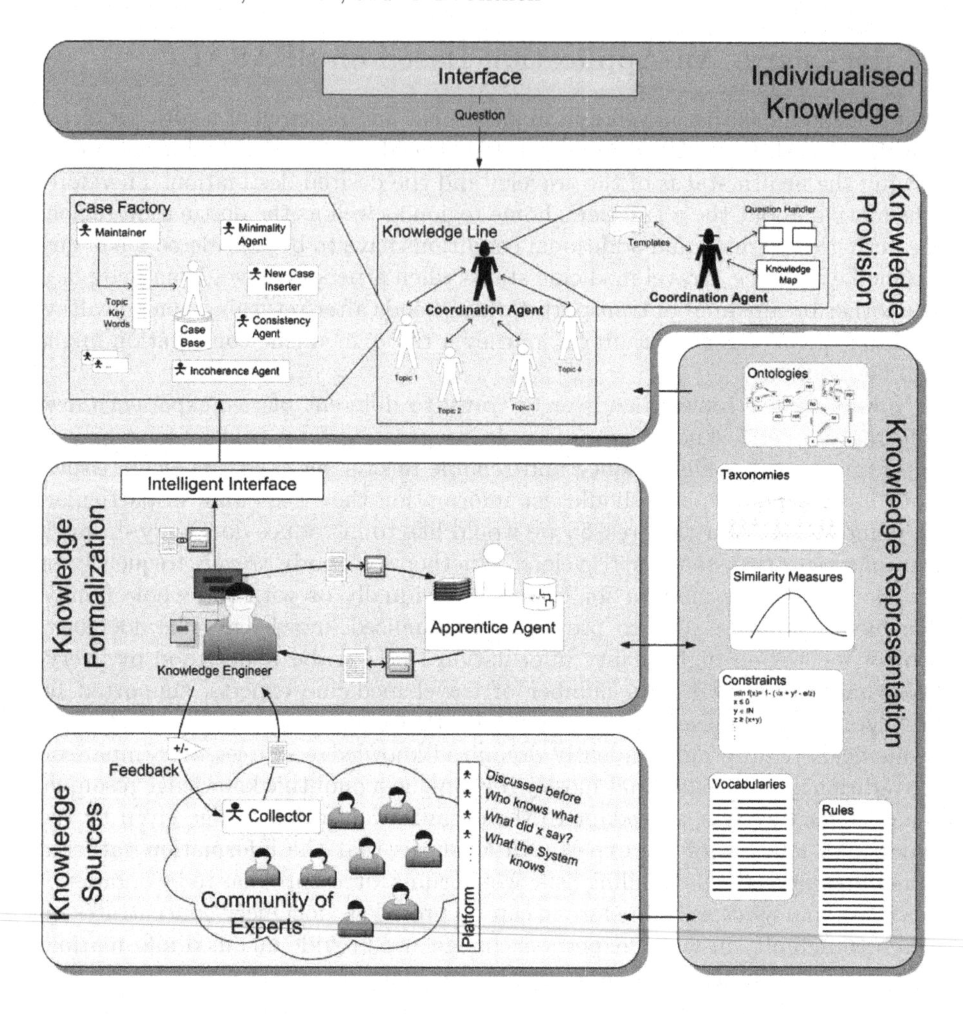

Fig. 1. The SEASALT architecture, the individual components are grouped into layers according to their function in knowledge management

3.1 Knowledge Sources

An interdisciplinary application domain such as travel medicine needs to draw information from numerous knowledge sources in order to keep up to date. Beyond traditional knowledge sources such as data bases and static web pages the main focus of SEASALT are Web 2.0 platforms. The SEASALT architecture is especially suited for the acquisition, handling and provision of experiential knowledge as it is provided by communities of practice and represented within Web 2.0 platforms [5]. Within our implementation of SEASALT we used a web forum software that was enhanced with agents for several different purposes. We chose a forum since it's a broadly established WWW communication medium and provides a low entry barrier even to only occasional WWW users. Additionally its contents can be easily accessed using the underlying data base. The forum itself serves as a communication and collaboration platform to the travel medicine community, which consists of professionals such as scientists and physicians who specialize in travel medicine and local experts from the health sector and private persons such as frequent travelers and globetrotters. The community uses the platform for sharing experiences, asking questions and doing general networking. The forum is enhanced with agents that offer content-based services such as the identification of experts, similar discussion topics, etc. and communicate by posting relevant links directly into the respective threads such as in [6].

The community platform is monitored by a second type of agents, the so called Collector Agents. These agents are individually assigned to a specific Topic Agent (see 3.2), their task is to collect all contributions that are relevant with regard to their assigned Topic Agent's topic. The Collector Agents pass these contributions on to the Knowledge Engineer and can in return receive feedback on the delivered contribution's relevance. Our Collector Agents currently use the information extraction tool TextMarker [7] to judge the relevance of a contribution. The Knowledge Engineer reviews each Collector Agent's collected contributions and implements his or her feedback by directly adjusting the agents' rule base.

The SEASALT architecture is also able to include external knowledge sources by equipping individual Collector Agents with data base or web service protocols or HTML crawling capabilities. This allows the inclusion of additional knowledge sources such as the web pages of the Department of Foreign Affairs or the WHO.

3.2 Knowledge Formalization

In order for the collected knowledge to be easily usable within the Knowledge Line (see 3.3) the collected contributions have to be formalized from their textual representation into a more modular, structured representation. This task is mainly carried out by the Knowledge Engineer. In the docQuery project the role of the Knowledge Engineer is carried out by several human experts, who carry out the Knowledge Engineer's tasks together. The Knowledge Engineer is the link between the community and the Topic Agents. He or she receives posts from the Collectors that are relevant with regard to one of the fields, represented by the Topic Agents, and formalizes them for insertion in the Topic Agents' knowledge bases using the Intelligent Interface. In the future the Knowledge Engineer will be additionally supported by the Apprentice Agent. The Intelligent Interface serves as the Knowledge Engineer's case authoring work bench for formalizing textual knowledge into structured CBR cases. It has been developed analogous to [8] and offers a graphical user interface that presents options for searching, browsing and editing cases and a controlled vocabulary.

The Apprentice Agent is meant to support the Knowledge Engineer in formalizing relevant posts for insertion in the Topic Agents' knowledge bases. It is trained by the Knowledge Engineer with community posts and their formalizations. The apprentice agent is currently being developed using GATE [9] and RapidMiner [10]. We use a combined classification/extraction approach that first classifies the contributions with regard to the knowledge available within the individual contributions using term-doc-matrix representations of the contributions and RapidMiner, and then attempts to extract the included entities and their exact relations using GATE. Considering docQuery's sensitive medical application domain we only use the Apprentice Agent for preprocessing. All its formalizations will have to be reviewed by the Knowledge Engineer, but we still expect a significantly reduced workload for the Knowledge engineer(s).

3.3 Knowledge Provision

SEASALT's knowledge provision is realized using the Knowledge Line approach [11]. The Knowledge Line's basic idea is a modularization of knowledge analogous to the modularization of software in the Product Line approach within software engineering [12]. Within the SEASALT architecture this knowledge modularization happens with regard to individual topics that are represented within the respective knowledge domain. Within the docQuery application domain travel medicine we identified the following topics: geography, diseases, pharmaceutics, constraints caused by chronic illnesses, vacation activities, local health facilities, and local safety precautions.

These topics are represented by Topic Agents. According to the SEASALT architecture the Topic Agents can be any kind of information system or service including CBR systems, databases, web services etc. Within docQuery we used the empolis Information Access Suite e:IAS [13], an industrial-strength CBR system, for realizing the individual agents [14]. We additionally extended the Topic Agents' CBR systems with Case Factories, which take care of the individual agents' case maintenance. The Case Factory approach is presented in more detail in [15]. Within SEASALT the Case Factory is used as a knowledge maintenance mechanism, comprising a number of agents that each carry out a simple maintenance task on an individual Topic agent's case base such as adding new cases, preserving consistency, or generalizing redundant cases [16].

The Topic Agents are orchestrated by a central Coordination Agent. The Coordination Agent receives a semi-structured natural language query from the user, analyses it using a rule-based question handler and subsequently queries the respective Topic Agents using incremental reasoning, that is using one agent's output as the next agent's input. In doing so the Coordination Agent's course of queries resembles the approach of a human amateur trying to answer a complex travel medical question. Confronted, for instance, with the question *"Which safety precaution should I take if I want to go diving in Alor for two weeks around Easter?"* and being no expert on the field an average person would first consult someone or something in order to find out that Alor is an Indonesian Island. Reading up on Indonesia the person would then find out that the rain season in Indonesia ends around Easter and that there is a hightened risk of Malaria during that time. The person would then look up information on Malaria and find out that the risk of contracting Malaria can be significantly reduced using prophylactic drugs. Knowing this he/she would the go on and acquire information on Malaria preventions and so on and so forth. This approach is mimicked by the Coordination Agent's approach. The world knowledge needed in order to carry out this incremental reasoning process is represented within the so called Knowledge Map, which provides formal representations of all Topic Agents and possible output/input connections encoded in a graph-like structure. The Coordination Agent's implementation is described in detail in [11], its theoretic foundations are described in [17]. Finally the Coordination Agent uses the query results and prefabricated templates to compose an information leaflet to be given to the user.

3.4 Knowledge Representation

Since the miscellaneous agents operating on the community platform (see Section 3.1), the Knowledge Engineer's tools (see Section 3.2) and the CBR systems of the individual Topic Agents (see Section 3.3) deal with the same knowledge domain(s), it makes sense to join their underlying knowledge models. This does not only greatly facilitate knowledge model maintenance but also allows for an easier interoperability between the individual components. SEASALT's knowledge representation includes rules, vocabulary, ontologies, and taxonomies, some of which were handmade for the purpose of the docQuery project, some are external, such as for instance WordNet (`http://wordnet.princeton.edu/`) or ICD10 (`http://www.who.int/classifications/icd/en/`).

3.5 Individualised Knowledge

The user interacts with docQuery via a web-based interface. The web based interface offers a semi-structured input in the form of different text fields used for entering information on the destination, the traveler, the time of travel and so on. The docQuery system provides individualized knowledge to its users by generating information leaflets as PDFs that only include information which is relevant to the respective user and its journey and can be used by the traveler to consult a physician for final advice and prescriptions.

4 Evaluation

A comparative evaluation of the SEASALT architecture in general is difficult, since the tasks within the docQuery project were executed manually until we started to introduce the system. Also we think that a purely local evaluation with regard to performance and runtime would be of little value to fellow researchers. Because of this we chose to do a practical evaluation within our first application domain travel medicine. The domain required a modularization of knowledge sources because the practitioners do not only use medical knowledge, but also need for instance regional and political information. This requirement is met by the concept of the Knowledge Line, because the Topic Agents represent an expert and their collaboration the composition of information leaflets. Especially the regional and political information have to be up-to-date and therefore we are able to extract knowledge about such topics from web communities and provide them within docQuery. The developed knowledge acquisition process optimizes the time until this information is available. Our application partner's current best practice is the manual assembling of information leaflets, mostly copy-pasting recurrent texts (like general information and warnings) from prepared templates and external sources. The application partner has been compiling these information leaflets for several years and has in the meantime optimized the process as far as possible. Using this approach a trained medical practitioner needs about an hour to create a complete leaflet. First tests have shown that the docQuery system offers a significant time saving and takes a lot of repetitive tasks from the medical practitioner. Even when counterchecking every generated leaflet and, if necessary, adding corrections or additional information the process of composition of information leaflets is significantly accelerated using docQuery.

5 Summary and Outlook

In this paper we presented the SEASALT architecture and described and exemplified its individual components using SEASALT's first instantiation, the docQuery project. The SEASALT architecture offers several features, namely knowledge acquisition from web 2.0 communities, modularized knowledge storage and processing and agent-based knowledge maintenance. SEASALT's first application within the docQuery project yielded very satisfactory results, however in order to further develop the architecture we are planning to improve it in several areas. One of these are the Collector Agents working on the community platform, which we plan to advance from a rule-based approach to a classification method that is able to learn from feedback, such as for instance CBR, so more workload is taken off the Knowledge Engineer. Also to this end more work will go into the Apprentice Agent, which is currently being developed. Another area of research that we currently look into are trust and provenance of information. SEASALT incorporates information from a large number of sources and we are currently looking into methods for making the source of the individual pieces of information more transparent to docQuery's users and thus improve the system's acceptance and trustworthiness. Finally we are planning to also apply the architecture in other application scenarios in order to further develop it and also ensure its general applicability in different application scenarios.

References

1. Althoff, K.D., Bach, K., Deutsch, J.O., Hanft, A., Mänz, J., Müller, T., Newo, R., Reichle, M., Schaaf, M., Weis, K.H.: Collaborative multi-expert-systems – realizing knowlegde-product-lines with case factories and distributed learning systems. In: Baumeister, J., Seipel, D. (eds.) KESE @ KI 2007, Osnabrück (September 2007)
2. Aamodt, A., Plaza, E.: Case-based reasoning: Foundational issues, methodological variations, and system approaches. AI Communications 1(7) (March 1994)

3. Bergmann, R., Althoff, K.D., Breen, S., Göker, M.H., Manago, M., Traphöner, R., Wess, S.: Selected Applications of the Structural Case-Based Reasoning Approach. In: Bergmann, R., Althoff, K.-D., Breen, S., Göker, M.H., Manago, M., Traphöner, R., Wess, S. (eds.) Developing Industrial Case-Based Reasoning Applications, 2nd edn. LNCS (LNAI), vol. 1612, pp. 35–70. Springer, Heidelberg (2003)
4. Bergmann, R., Althoff, K.D., Minor, M., Reichle, M., Bach, K.: Case-based reasoning - introduction and recent developments. Künstliche Intelligenz: Special Issue on Case-Based Reasoning 23(1), 5–11 (2009)
5. Plaza, E.: Semantics and experience in the future web. In: Althoff, K.-D., Bergmann, R., Minor, M., Hanft, A. (eds.) ECCBR 2008. LNCS (LNAI), vol. 5239, pp. 44–58. Springer, Heidelberg (2008)
6. Feng, D., Shaw, E., Kim, J., Hovy, E.: An intelligent discussion-bot for answering student queries in threaded discussions. In: IUI 2006: Proc. of the 11th Intl. Conference on Intelligent user interfaces, pp. 171–177. ACM Press, New York (2006)
7. Klügl, P., Atzmüller, M., Puppe, F.: Test-driven development of complex information extraction systems using textmarker. In: Nalepa, G.J., Baumeister, J. (eds.) KESE. CEUR Workshop Proceedings, CEUR-WS.org, vol. 425 (2008)
8. Bach, K.: Domänenmodellierung im textuellen fallbasierten schließen. Master's thesis, Institute of Computer Science, University of Hildesheim (2007)
9. Cunningham, H., Maynard, D., Bontcheva, K., Tablan, V.: GATE: A framework and graphical development environment for robust NLP tools and applications. In: Proc. of the 40th Meeting of the Association for Computational Linguistics (2002)
10. Mierswa, I., Wurst, M., Klinkenberg, R., Scholz, M., Euler, T.: Yale: Rapid prototyping for complex data mining tasks. In: Ungar, L., Craven, M., Gunopulos, D., Eliassi-Rad, T. (eds.) KDD 2006: Proc. of the 12th ACM SIGKDD international conference on Knowledge discovery and data mining, August 2006, pp. 935–940. ACM, New York (2006)
11. Bach, K., Reichle, M., Reichle-Schmehl, A., Althoff, K.D.: Implementing a coordination agent for modularised case bases. In: Petridis, M., Wiratunga, N. (eds.) Proc. of 13th UKCBR @ AI 2008, December 2008, pp. 1–12 (2008)
12. van der Linden, F., Schmid, K., Rommes, E.: Software Product Lines in Action - The Best Industrial Practice in Product Line Engineering. Springer, Heidelberg (2007)
13. empolis GmbH: Technical white paper e:information access suite. Technical report, empolis GmbH (September 2005)
14. Althoff, K.D., Reichle, M., Bach, K.: Realizing modularized knowledge models for heterogeneous application domains. In: Perner, P. (ed.) ICDM 2008. LNCS (LNAI), vol. 5077, pp. 114–128. Springer, Heidelberg (2008)
15. Althoff, K.D., Hanft, A., Schaaf, M.: Case factory – maintaining experience to learn. In: Roth-Berghofer, T.R., Göker, M.H., Güvenir, H.A. (eds.) ECCBR 2006. LNCS (LNAI), vol. 4106, pp. 429–442. Springer, Heidelberg (2006)
16. Bach, K.: docquery - a medical information system for travellers. Internal project report (September 2007)
17. Reichle, M., Bach, K., Reichle-Schmehl, A., Althoff, K.D.: Management of distributed knowledge sources for complex application domains. In: Proc. 5th Conference on Professional Knowledge Manegement - Experiences and Visions, WM 2009 (2009)

Case Retrieval in Ontology-Based CBR Systems

Amjad Abou Assali, Dominique Lenne, and Bruno Debray

University of Technology of Compiègne, CNRS
HEUDIASYC
{aabouass,dominique.lenne}@utc.fr
INERIS
bruno.debray@ineris.fr

Abstract. This paper presents our knowledge-intensive Case-Based Reasoning platform for diagnosis, COBRA. It integrates domain knowledge along with cases in an ontological structure. COBRA allows users to describe cases using any concept or instance of a domain ontology, which leads to a heterogeneous case base. Cases heterogeneity complicates their retrieval since correspondences must be identified between query and case attributes. We present in this paper our system architecture and the case retrieval phase. Then, we introduce the notions of similarity regions and attributes' roles used to overcome cases heterogeneity problems.

Keywords: Case-based reasoning, Ontology, Heterogeneous case base, Similarity measures, Similarity regions.

1 Introduction

In industrial sites concerned by the SEVESO directive, risks exist due to the presence of hazardous substances and processes. To reduce such risks, safety barriers are set up depending on the type of process and on the hazardous situation to control as well as on the local environmental conditions. However, barriers may not work properly, and thus hazards may arise. In such a case, industrial experts intervene to diagnose the barriers failure basing mostly on past failure experiences occurred in similar situations. The hypothesis of experts here is that "if a barrier did not work correctly in a past similar situation, it is strongly probable that it does not work, in the current situation, *for similar reasons*". To simulate the experts' activity, we use a Case-Based Reasoning (CBR) approach [1]. CBR is a problem-solving approach used to solve a new problem (target case) by remembering a previous similar situation (source case) and by reusing information and knowledge of that situation [2]. In the diagnosis activity, when no obvious cause of failure is found, experts may realize that more information is needed to find the right cause of failure. This led us to apply a conversational CBR approach where cases are enriched as experts advance in the diagnosis.

The system was first applied to the diagnosis of the failure of gas sensors installed in industrial plants. Such sensors are intended to detect certain gases

B. Mertsching, M. Hund, and Z. Aziz (Eds.): KI 2009, LNAI 5803, pp. 564–571, 2009.

so that if there is a leak somewhere in the site, a safety action can be undertaken. In this context, a case represents a diagnosis of the failure of a gas sensor in a given industrial environment.

In this paper, we introduce our platform COBRA (Conversational Ontology-based CBR for Risk Analysis), an ontology-based CBR platform. It allows to capitalize and reuse past failure experiences based on ontological models describing the domain and case structures. One of the interesting features of COBRA is that it allows a dynamic representation of cases; *i.e.,* users can define their cases' attributes at run time, which leads to a heterogeneous case base. However, this heterogeneity complicates the case retrieval phase. We describe in this paper the case retrieval phase as well as the similarity measures used. In addition, we present our approach to overcome the cases heterogeneity problems.

2 Related Work

The integration of general domain knowledge into knowledge-intensive CBR systems was an important aspect in several projects. In the CREEK architecture [3], we find a strong coupling between case-based and generalization-based knowledge. Thus, cases are submerged within a general domain model represented as a densely linked semantic network. Fuchs & Mille proposed a CBR modeling at the knowledge level [4]. They distinguished four knowledge models: 1) the conceptual model containing the concepts used to describe the domain ontology regardless of the reasoning process; 2) the case model that separates the case into three parts: problem, solution, and reasoning trace; 3) the models describing the reasoning tasks; 4) and the reasoning support models. Dìaz-Agudo & Gonzàlez-Calero [5] proposed a domain-independent architecture that helps in the integration of ontologies into CBR applications. Their approach is to build integrated systems that combine case specific knowledge with models of general domain knowledge. So, they presented CBROnto [6], a task/method ontology that provides the vocabulary for describing the elements involved in the CBR processes, and that allows to integrate different domain ontologies. CBROnto was reused later by jCOLIBRI, a powerful object-oriented framework for building CBR systems [7]. It splits the case base management into two concerns: persistence and in-memory organization, which allows different storage mediums of cases (text/XML files, ontology, *etc.*) accessed via specific connectors. However, jCOLIBRI does not allow the treatment of dynamic and heterogeneous case bases. In this work, we decided to keep the interesting aspects of jCOLIBRI, and to add our own layer that allows to work with heterogeneous case bases.

3 COBRA Architecture

Several CBR architectures have been presented in the literature [2,8]. Inspired by these architectures, COBRA architecture is composed of two main layers [9]:

- *Processes* layer: it contains the off-line process, *case authoring*, along with the reasoning processes: ELABORATE, RETRIEVE, DIAGNOSE, ENRICH, REVISE, and RETAIN. In the "diagnose" phase, the system tries to identify failure causes from the retrieved similar cases. If no cause is found, or even if the diagnosis proposed by the system was not validated by the user, the system asks the user to "enrich" his case description, and thus new similar cases can be retrieved. This cycle is repeated until a relevant diagnosis is proposed by the system, or no solution is found.

- *Knowledge containers* layer: it contains three main containers: vocabulary, case base, and similarity measures. In knowledge-intensive CBR systems, ontologies play an important role in representing these containers. They can be used as the vocabulary to describe cases and/or queries, as a knowledge structure where the cases are located, and as the knowledge source to achieve semantic reasoning methods for similarity assessment [7]. The vocabulary and the case base rely on two knowledge models, the domain and case models respectively.

3.1 Domain Model

This model represents the knowledge about safety barriers, our current domain of application, in an ontological structure. Following the classification proposed by Oberle [10], we have developed two types of ontologies : a core ontology that contains generic concepts about industrial safety such as EQUIPMENT, DANGEROUS PHENOMENON, SAFETY BARRIER, EFFECT, *etc.* A domain ontology describing the domain of safety barriers, in particular gas sensors. Its concepts are specializations of other concepts of the core ontology, and contains concepts such as: GAS SENSOR, SAFETY FUNCTION, ENVIRONMENTAL CONDITION, *etc.* Ontologies are represented in OWL Lite and have been developed by several experts of the INERIS, the French national expertise institute on industrial safety, with the help of an ontology expert [11].

3.2 Case Model

A case in our system represents a diagnosis experience, and thus consists of three main parts: a *description* part describing the context of the experience, a *failure mode* part describing the type of failure, and a *cause* part describing the different possible causes of this failure. A case in general is described by a pair (*problem*, *solution*). Accordingly, the *description* and *failure mode* parts of our model correspond to the *problem* part, and the *cause* part corresponds to the *solution* part. Inspired by the approach of jCOLIBRI, in order to enhance the communication between the case base and the domain model, the case model is represented within an ontology that integrates the domain model. This ontology contains the main following concepts (Figure 1): 1) CBR-CASE that subsumes the various case types that may exist in the system; 2) CBR-DESCRIPTION that subsumes the case main parts, *failure mode* and *cause*; and 3) CBR-INDEX allowing to integrate the domain model concepts used to describe cases.

Cases are thus represented by instances of the ontology and can have two types of attributes: *Simple attributes* corresponding to data-type properties of

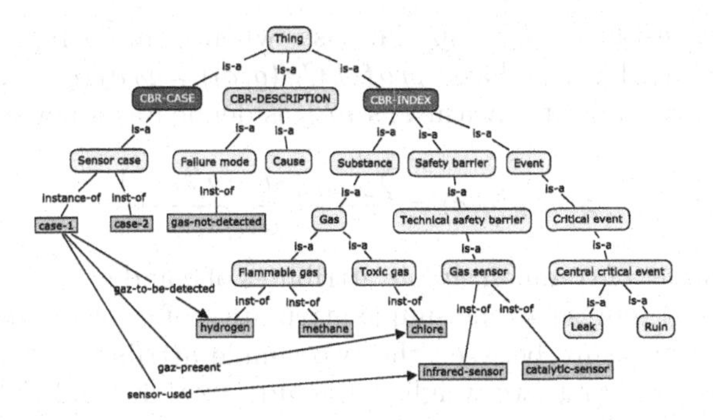

Fig. 1. Case model

the ontology, and *Complex attributes* corresponding to instances of the ontology that have, in turn, their own simple properties. In the case authoring phase, we realized that cases do not share always the same attributes. Thanks to the developed platform, experts were allowed to use any concept from the domain model to describe their cases [9]. This led to a heterogeneous case base, which complicated the case retrieval phase.

4 Case Retrieval

In this phase, similarity measures are employed to retrieve similar cases for a user query. Generally, with object-oriented case structures, similarity measures follow the "local-global principle" [12,13]. This principle is followed in our work since cases are represented by instances of the ontology. The similarity computation of two ontology concepts can be divided into two components [12,7]: *a concept-based similarity* that depends on the location of concepts in the ontology, and *a slot-based similarity* that depends on the fillers of the common attributes between the compared objects.

4.1 Weighted Similarity Measures

Sometimes, query attributes may not have the same importance (weight) in the similarity computation. In this work, weights can be assigned to simple attributes (IGNORE or EXACT), and/or to complex attributes (in the interval [0, 1]).

Let $Q = \{q_i : 1 \leq i \leq n, n \in \mathbb{N}^*\}$ be the user query, where q_i is a simple or a complex attribute, and let $\Omega = \{C_j : 1 \leq j \leq k, k \in \mathbb{N}^*\}$ be the case base, where $C_j = \{c_{jl} : 1 \leq l \leq m_j, m_j \in \mathbb{N}^*\}$. The concept-based similarity measure, sim_{cpt}, is defined as follows: For each complex attribute, $q \in Q$ and $c \in C$,

$$sim_{cpt}(q, c) = w_q * \frac{2 * prof(LCS(q, c))}{prof(q) + prof(c)} \tag{1}$$

where w_q is the weight assigned to q, $prof$ is the depth of a concept (or an instance) in the ontology hierarchy, and LCS is the Least Common Subsumer concept of two instances. In a special case, when q and c represent the same instance in the ontology, we have: $prof(LCS(q,c)) = prof(q)$.

The slot-based similarity measure, sim_{slt} is defined as follows:

$$sim_{slt}(q,c) = \frac{\sum_{s \in CS} sim(q.s, c.s)}{|CS|} \tag{2}$$

where CS is the set of common simple attributes of q and c, $|CS|$ is the set cardinality, $q.s$ (or $c.s$) represents the simple attribute s of q (or c), and $sim(q.s, c.s)$ is the similarity measure between the two simple attributes. For the moment, we consider only the first two weights (IGNORE, EXACT), and thus $sim(q.s, c.s)$ can be defined as:

$$sim(q.s, c.s) = \begin{cases} 1 & \text{if} \quad (w_{q.s} = exact) \wedge (v_{q.s} = v_{c.s}) \\ 0 & \text{otherwise} \end{cases}$$

where $w_{q.s}$ is the weight of the simple attribute $q.s$, and $v_{q.s}$ is its value.

Now, it is time to define the global similarity measure of the two complex attributes, q and c, which is given by the following formula [14]:

$$sim(q,c) = (1-\alpha) * sim_{cpt}(q,c) + \alpha * sim_{slt}(q,c) \tag{3}$$

where α is an experience parameter (at present, $\alpha = 0.4$).

To compute the similarity between a case and a given query, each complex attribute of the query is compared to its corresponding attribute in the case. In *homogeneous* case bases, all cases share the same predefined structure, and thus, corresponding complex attributes are already identified. On the other side, *heterogeneous* case bases contain cases with different structures. Therefore, before computing the similarity, we need to determine the corresponding complex attributes.

For each complex attribute $q' \in Q$, let $c' \in C_j$ be the corresponding complex attribute in the case $C_j \in \Omega$. To determine c', we considered first that it is the attribute with which q' has the maximum similarity in the case C_j:

$$sim(q', c') = \max_{1 \le l \le m_j} (sim(q', c_{jl})) \tag{4}$$

However, we realized that this definition of corresponding attribute is not sufficient, and that more conditions must be satisfied. The problem is that c' may have a maximum similarity with q' while, in reality, q' may have no corresponding attribute in C_j. The first improvement that we propose is to compare the obtained similarity $sim(q', c')$ with the maximum similarity obtained for q' within the whole case base. This leads to the following condition:

$$\frac{sim(q', c')}{\max_{1 \le j \le k, 1 \le l \le m_j} (sim(q', c_{jl}))} \ge \beta \tag{5}$$

where β is a specific threshold calculated after experimentations ($\beta = 0.6$).

This condition led to better results, but it was not however sufficient in special cases. Thus, we propose to consider the following interesting notions:

4.2 Similarity Regions

We consider that the similarity measures are not always applicable to each pair of concepts (or instances) of the ontology. For example, a gas instance must not be compared at all with an equipment instance. To prevent such comparisons, we propose to define the notion of *similarity regions*. A similarity region is a sub-branch of the ontology hierarchy where concepts and instances are comparable with each other (Figure 2). The definition of such regions is manual and depends on the target application. To compute the similarity between a query attribute and a case attribute, it must be verified first if these two attributes belong to the same similarity region.

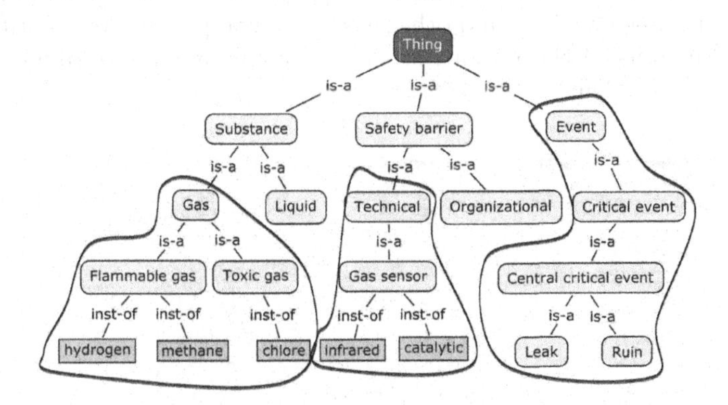

Fig. 2. Example of similarity regions

4.3 Roles of Attributes

For a given query attribute, several corresponding attributes may be found sometimes. Let's take, for example, the following excerpt of a case description: "At an industrial site, an infrared sensor was used to **detect** the *methane*. Other gases were **present** at the site such as the *hydrogen*. The sensor did not work as expected, and there was an explosion consequently". Now, let's suppose a query looking for cases where a gas sensor was used to **detect** the *hydrogen*. Following the approach presented till now, we find that the hydrogen of the case is the attribute corresponding to the hydrogen of the query. However, it is not the aimed result since we look for the hydrogen when it is the gas to be detected, which means that its actual corresponding attribute must be the methane.

To solve this ambiguity, we propose to describe the *role* of case attributes that may give rise to ambiguous situations. For example, the hydrogen of the case is a ***gas present*** at site, and the methane is the ***gas to be detected***. Thus, attributes that have same roles are identified first. Then, for the other attributes, we follow the proposed approach (Formulas (4), (5) along with similarity regions).

5 Results and Conclusion

This work led to the development of COBRA, a domain-independent CBR platform, as a JAVA application (Figure 3). It is based on jCOLIBRI, but it extends it with a layer that allows the treatment of dynamic and heterogeneous case bases. Thanks to the platform architecture, developing a new CBR system is made by supplying its domain ontology and reconfiguring some XML files. COBRA is used currently for two parallel objectives: to capitalize knowledge about gas sensor failures, and to provide support to experts and safety engineers to diagnose failure causes of gas sensors in industrial conditions. Experts can use any concept or instance of the domain model to describe their cases, which leads to a heterogeneous case base. Cases heterogeneity implicated that corresponding complex attributes, between a query and other cases, are complicated to identify. In this paper, we introduced the notions of similarity regions and attributes' roles to overcome this problem.

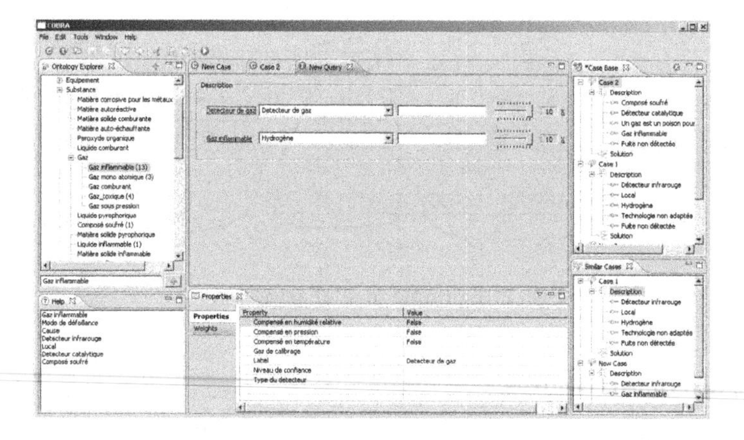

Fig. 3. COBRA platform

To evaluate the proposed approach, cases have been added to the case base. They correspond to real situations observed in industry or to tests realized in the INERIS during campaigns for qualification of sensors. These cases allowed us to do a preliminary validation of the case retrieval phase, which gave very satisfying results. The validation has been done to show the quality of results when adopting each of the improvements proposed; *i.e.,* Formula 5, similarity regions, and attributes' roles. We have also developed the "diagnose" phase of the CBR processes (out of the paper's scope). Then, more complex cases have to be added to the case base to allow validating our system. This validation will be done by experts at two levels:

- the first level concerns the CBR architecture: to what end the case structure and the reasoning processes are close to the real experts' activity? What other concepts should be added to the ontology? *etc.*

– the second level concerns the quality of diagnosis returned by the system for some given queries.

Acknowledgments. The present work has received the financial support of the French Ministry of Ecology and sustainable development. The authors also thank Mr. Sébastien Bouchet (INERIS) for his active participation to the development of the gas sensor ontology and the case base.

References

1. Riesbeck, C., Schank, R.: Inside Case-Based Reasoning. Lawrence Erlbaum Associates, Mahwah (1989)
2. Aamodt, A., Plaza, E.: Case-Based Reasoning: Foundational Issues, Methodological Variations, and System Approaches. AI Communications 7(1), 39–59 (1994)
3. Aamodt, A.: Explanation-Driven Case-Based Reasoning. LNCS, pp. 274–274 (1994)
4. Fuchs, B., Mille, A.: Une modélisation au niveau connaissance du raisonnement à partir de cas. In: L'Harmattan (ed.) Ingénierie des connaissances (2005)
5. Dìaz-Agudo, B., Gonzàlez-Calero, P.: An architecture for knowledge intensive CBR systems. In: Blanzieri, E., Portinale, L. (eds.) EWCBR 2000. LNCS (LNAI), vol. 1898, pp. 37–48. Springer, Heidelberg (2000)
6. Dìaz-Agudo, B., Gonzàlez-Calero, P.: CBROnto: a task/method ontology for CBR. In: Procs. of the 15th International FLAIRS, vol. 2, pp. 101–106 (2002)
7. Recio-Garcìa, J., Dìaz-Agudo, B., Gonzàlez-Calero, P., Sanchez, A.: Ontology based CBR with jCOLIBRI. Applications and Innovations in Intelligent Systems 14, 149–162 (2006)
8. Lamontagne, L., Lapalme, G.: Raisonnement à base de cas textuels: Etat de l'art et perspectives. Revue d'intelligence artificielle 16(3), 339–366 (2002)
9. Abou Assali, A., Lenne, D., Debray, B., Bouchet, S.: COBRA: Une plateforme de RàPC basée sur des ontologies. In: Actes des 20es Journées Francophones d'Ingénierie des Connaissances (IC 2009), Hammamet, Tunisie, May 2009, pp. 277–288 (2009)
10. Oberle, D.: Ontologies. In: Semantic Management of Middleware. Semantic Web and Beyond, vol. 1, pp. 33–53. Springer, Heidelberg (2006)
11. Abou Assali, A., Lenne, D., Debray, B.: Ontology development for industrial risk analysis. In: IEEE International Conference on Information & Communication Technologies: from Theory to Applications (ICTTA 2008), Damascus, Syria (April 2008)
12. Bergmann, R., Stahl, A.: Similarity measures for object-oriented case representations. In: Smyth, B., Cunningham, P. (eds.) EWCBR 1998. LNCS (LNAI), vol. 1488, p. 25. Springer, Heidelberg (1998)
13. Richter, M.: Similarity. In: Case-Based Reasoning on Images and Signals. Studies in Computational Intelligence, vol. 73, pp. 25–90. Springer, Berlin (2008)
14. Zhang, K., Tang, J., Hong, M., Li, J., Wei, W.: Weighted Ontology-Based Search Exploiting Semantic Similarity. In: Zhou, X., Li, J., Shen, H.T., Kitsuregawa, M., Zhang, Y. (eds.) APWeb 2006. LNCS, vol. 3841, pp. 498–510. Springer, Heidelberg (2006)

Behaviour Monitoring and Interpretation
A Computational Approach to Ethology

Björn Gottfried

TZI, University of Bremen

Abstract. Behaviour Monitoring and Interpretation is a new field that developed gradually and inconspicuously over the last decades. With technological advances made at the sensory level and the introduction of ubiquitous computing technologies in the nineties this field has been pushed to a new level. A common methodology of many different research projects and applications can be identified. This methodology is outlined as a framework in this paper and supported by recent work. As a result it shows how BMI automates ethology. Moreover, by bringing in sophisticated AI techniques, it shows how BMI replaces a simple behaviour-interpretation mapping through computational levels between observed behaviours and their interpretations. This is similar to the way of how functionalism and cognitive sciences enabled new explanation models and provided an alternative approach to behaviourism in the early days of AI. First research results can be finally given which back up the usefulness of BMI.

1 Introduction

This paper gives an introduction to the field of Behaviour Monitoring and Interpretation, BMI for short. BMI picks up problems investigated by ethologists: On the one hand, ethology is a branch of knowledge dealing with human character, with its formation and evolution. On the other hand, ethology is the scientific and objective study of animal behaviour especially under natural conditions [22]. BMI opens up a new trend in ethology, namely that ethological problems are tackled by means of a computational methodology. This opens up new chances and opportunities.

It is the ultimate goal of BMI to make intentions, desires, and goals explicit; more generally, every kind of information which is implicit in observable phenomena, in particular behaviours of man and beast. While this has been the realm of ethologists before computers entered the scene, technological advances approach problems in this field with electronic devices which get ever more accurate and precise, and even allow the acquired data to be evaluated with a great body of methods developed in computer science, especially Artificial Intelligence. The annual BMI workshop is a forum for discussing advances in this field [9,10].

Section 2 gives an overview of a number of applications which pertain to the BMI field, showing what BMI is about. At the same time, a framework is outlined according to which BMI applications usually adhere. On this basis, in Section 3, we are able to argue why it makes sense to introduce the notion of BMI. Conclusions drawn from these considerations are eventually provided in Section 4.

B. Mertsching, M. Hund, and Z. Aziz (Eds.): KI 2009, LNAI 5803, pp. 572–580, 2009.

2 The BMI Framework

A number of BMI applications are provided as an introduction to this field. A common framework and typical AI methods employed in this area are identified.

2.1 Example Scenarios

Two initial examples illustrate the objective of BMI. Behaviour patterns of mice living in a semi-natural environment are investigated by [18]. They compare motion behaviours of mice who have a genetic predisposition to develop Alzheimer's disease with their wild-type conspecifics. Aim of this project is the systematic support of behavioural observations by humans. For this purpose an RFID system is installed in the cage where the mice live in order to capture their positions. The cage comprises a number of different levels; at the highest level a video monitoring system is installed capturing more precisely the motion behaviours. The left part of Fig. 1 shows how this research project partitions into different abstraction layers with the bottom layer representing the behaviours of interest. By contrast, the top layer shows the interpretation of the observed behaviours. All intermediate layers mediate between observation and interpretation.

Another example, which at first sight looks somewhat different, fits into this very same schema. In [24] motion patterns of pedestrians going shopping are investigated. First, people are observed without knowing anything about it; for this purpose a hand-tracking tablet computer with a digital map is used in order to record the paths of the pedestrians in this map. Second, the people who have been observed get interviewed in order to let them tell a little bit about themselves, their intentions and social background. Finally, they get further tracked, this time equipped with a Bluetooth Smartphone or a GPS logger for indoor and outdoor tracking, respectively. The acquired data is analysed by clustering the obtained trajectories and by using speed histograms. As a result there are

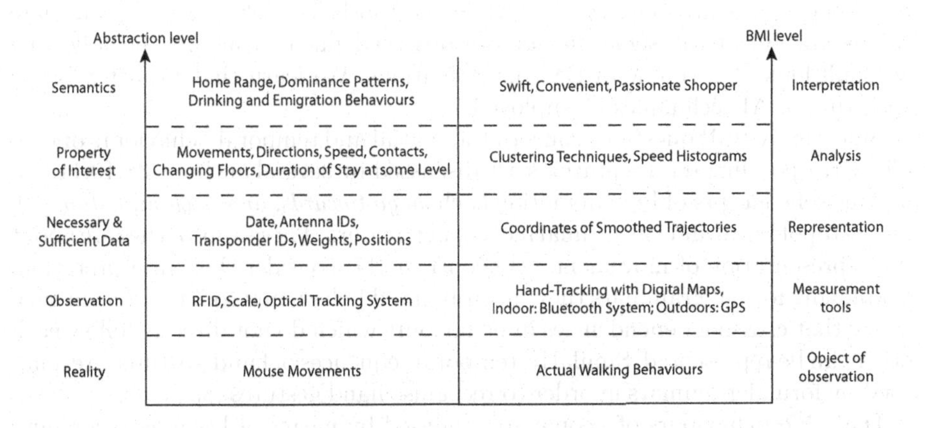

Fig. 1. Abstraction layers of two BMI scenarios

basically three types of shoppers identified, namely swift, convenient, and passionate shoppers. The right part of Fig. 1 summarises this study.

Both studies show the usual sequence of abstraction layers which can be found in each typical BMI application. The bottom layer represents the reality, and thus, the object of interest, in our examples, movements of mice and pedestrians. The second layer is about the observation of the object of interest, here by means of such techniques as RFID, Bluetooth, or GPS. The acquired data is frequently preprocessed (e.g. smoothing, detection of outlier) and data representations are chosen for storing the data in an appropriate structure, at the third layer; here, for trajectory positions and other positional information, such as RFID antennas as path landmarks. The fourth layer is about analysing the data, for example in order to determine the duration of staying at some level in the cage or to cluster similar trajectories of different pedestrians. Eventually, the top layer represents the result one is looking for, that is the home range of a mouse or its drinking behaviour or to tell apart swift shoppers from those who enjoy going shopping.

The conventional approach in ethology looks for a (direct) mapping between the bottom layer and the top layer. BMI is distinguished by automating this mapping. Additionally, as opposed to the research direction of behaviourism, BMI enables an arbitrary complex functionality mediating between observation and interpretation. For this purpose a multitude of AI techniques is employed, as we shall learn in the following.

2.2 The Employment of AI Techniques in BMI

From the point of view of AI the upper layers are of most interest. While the bottom layer just represents the object of interest, the second layer from below is about measurement tools. Smart Floors show that even this sensory level gets ever more intelligent [34]. The three top layers are basically about knowledge representation: while the third layer focuses on representational issues, the fourth layer is mainly about reasoning, and the top layer again about representations. The distinction between the third and fifth layers is that the former starts with representing the raw sensory data while the top layer seeks for an appropriate representation of the semantics of the observed phenomena. In the following we shall look at a couple of BMI investigations. We learn that in fact a broad spectrum of AI techniques is employed.

Since many BMI questions concern the spatial and temporal behaviours of people, corresponding representations are discussed by several authors [31,19,11,39]: [31] model concepts of human motion such as *go towards*, *approach* and *stop*, [19] model topological and [39] ordinal motion behaviours. These approaches are about the representation of movements. By contrast [17] consider their interpretation. Employing formal grammars they map observed behaviours to intentions in such a way that crossing dependencies of behaviours which do not directly follow each other can be represented. Similarly, temporal sequences of hand postures are analysed by formal grammars in order to recognise hand gestures [8].

Daily life behaviours of people are analysed by means of learning techniques to discover typical actions [5]; they look for repeating patterns of dependencies among sensor events. Learning techniques are also deployed by [12] for detecting high-level activities, such as morning care activitis like brushing one's teeth; they employ an unsupervised k-means clustering approach in combination with Hidden Markov Models for mapping cluster membership onto more abstract activities. As opposed to the previously discussed movement representations, those learning systems do also look at sensor events which are triggered when, for instance, a cabinet door is opened. In other words, BMI does not solely take into account the behaviours of people as they are directly observable, but also behaviours which are indirectly observable by means of changes in the environment.

A sophisticated behaviour recognition system based on video technology is investigated by [36]. Applied to dynamic indoor scenes and static building scenes, they implement a number of submodules: objects must be recognised, classified and tracked, qualitative spatial and temporal properties must be determined, behaviours of individual objects must be identified, and composite behaviours must be determined to obtain an interpretation of the scene as a whole. They describe how these tasks can be distributed over three processing stages (low-level analysis, middle layer mediation and high-level interpretation) to obtain flexible and efficient bottom-up and top-down processing in behaviour interpretation. [36] further point out that behaviour recognition appears to be a restricted topic with a focus on several different behaviour recognition tasks. They mention the following typical applications: vandalism in subway stations [40], thefts at a telephone booth [14], the filling up at a gas station [27], the identification of activities at the airport [37] or the placing of dishes on a table [15].

Particular complex monitoring systems are developed in the field of distributed smart cameras. In [1] the authors propose a vision-based framework to provide quantitative information of the user's posture. While quantitative knowledge from the vision network can either complement or provide specific qualitative distinctions for AI-based problems, these qualitative representations can offer clues to direct the vision network to adjust its processing operation according to the interpretation state. In this way they show by example that the proposed BMI framework is not a one-way street. One of their application examples is fall detection.

A well appreciated area for analysing typical BMI problems is soccer, the RoboCup community being indeed largely faced with analysing behaviours. In their paper [41] present a qualitative, formal, abductive approach, based on a uniform representation of soccer tactics that allows to recognise and explain the tactical and strategic behaviour of opponent teams based on past observations. A framework for argumentation and decision support in dynamic environments is investigated by [33]. This framework defines arguments which refer to conceptual descriptions of the given state of affairs. Based on their meaning and based on preferences that adopt specific viewpoints, it is possible to determine consistent positions depending on these viewpoints, allowing the interpretation of spatiotemporal behaviours.

While AI techniques frequently aim at automating the whole process, there are approaches which more rely on the skills of human beings. One such field is the area of *visual analytics* that combines automatic techniques to data analysis with humans skills, mainly vision and reasoning, and in this way relates to diagrammatic reasoning. The problem which is addressed in the context of visual analytics concerns mainly massive collections of data (as they are obtained throughout BMI applications) and the problem that it is hardly possible to do without human abilities to identify interesting patterns in the data. [3] consider the case of large amounts of movement data. In their scenario vehicles are equipped with GPS and they analyse typical paths and look for points of interest which can be derived purely from the observations.

While most approaches mentioned so far concentrate on the monitoring of the behaviours of individuals, we should also mention group behaviours, as of interest in the case of disaster management. While [44] provide a general classification scheme for behaviours of collectives, [13] investigate techniques to space-use analysis on a university campus. Regarding AI, interactions within groups, and in particular their simulation, is the realm of Multiagent Systems [38].

3 The Role of BMI in Related Areas

There are a lot of directions in AI and a multitude of areas where AI techniques play an essential role. The question arises whether it is necessary to talk about yet another area, as we do with BMI. In the following we will provide a couple of arguments for the integrating role of BMI.

First and foremost ubiquitous computing technologies have been devised in the last decade [42]. There is a fast development in this area which is mainly due to advances made with accurate and cheap sensor technologies. That is, with these technologies a fundamental basis for BMI applications has been established. Additionally the notion of ubiquitous computing closely relates to pervasive computing. The latter followed the former, is mainly coined by industry, and puts more emphasis on networks while ubiquitous computing is more human-centred. Both directions provide essential means for the BMI field.

Applications, as we have seen in the previous section, can be found in very different areas. One such area is the field of Ambient Intelligence (AmI) [6]:

> *The basic idea behind AmI is that by enriching an environment with technology (sensors, processors, actuators, information terminals, and other devices interconnected through a network), a system can be built such that based on the real-time information gathered and the historical data accumulated, decisions can be taken to benefit the users of that environment* [4].

BMI should be regarded as a subfield of AmI that deals with the *(real-time) information gathering* part and the evaluation of the gathered information. While AmI is a growing field that gets ever more complex it becomes useful to identify subfields, like BMI, that avoid losing track of things.

But there are other areas than AmI which can be found in different disciplines. For example, there is a broad community of geographers who are interested in investigating spatiotemporal phenomena of moving objects, e.g. in the context of wayfinding tasks [30,43]; and from the computational point of view [28,25,26]. Taking their work, it shows that there is a clear intersection with the AI community that investigates qualitative spatiotemporal phenomena [7]. BMI covers this area of intersection by focusing on the methodologies for investigating spatiotemporal behaviours.

Having motivated the notion of computational ethology, we should eventually mention behavioural science. This community not only employs ever new technologies as observational tools and for data analysis, but even recognises that research in this field often lacks a more formal approach [35]. BMI as the mediating field should aid in bringing the AI methodology for data analysis to the behavioural sciences, while the latter would stimulate further research at the formal but also sensory level. A couple of investigations can already be found that deal with genuinely ethological issues concerning both animal behaviours [16,18,20] and behaviours of humans [21,29,30,32,43].

One important issue, we must not oversee, is that in almost all discussed areas privacy concerns are to be considered. Ethical issues arise and have to be carefully taken into account. For instance, [2] argues for an approach which emphasises communication in the design and implementation of monitoring systems, allowing to find an acceptable balance between potential abuses and benefits.

4 Conclusions

BMI is the computational advancement of ethology that automates the gathering of data, data analysis and even data interpretation. Some of the used methods, including knowledge representation and interpretation techniques, show how in particular AI aids in this automation process. In detail, BMI

- supports precise and comprehensive descriptions of behavioural phenomena,
- automates the whole evaluation of behavioural phenomena,
- can take into account background knowledge as well as complex relationships,
- mediates between observation and prediction, and
- brings together behavioural sciences with new application fields.

The latter includes, for instance, Ambient Assisted Living, Ambient Intelligence, Smart Environments and everything that helps to better deal with everyday life problems, in particular of the elderly and challenged people. Some specific examples demonstrate the added value of BMI in comparison to non-automatic observational tools and the shallow mapping of behaviours to interpretations:

- long lasting systematic observations of behaviours [3,18],
- taking into account temporally disjoint behaviour relations [17],
- measuring behaviours indirectly [12],
- learning typical patterns of behaviours [5], and
- the bottom-up $\leftrightarrow$ top-down interplay within the analysis process [1,36].

Such ingenious means for analysing behaviours show the first important steps towards a computational theory to ethology. Future efforts are necessary in order to enable a more common view on the different BMI layers and to allow arbitrary complex behavioural events to be considered.

References

1. Aghajan, H., Wu, C.: From Distributed Vision Networks to Human Behavior Interpretation. In: BMI 2007, vol. 296, pp. 129–143. CEURS (2007)
2. Alder, G.S.: Ethical Issues in Electronic Performance Monitoring. Journal of Business Ethics 17, 729–743 (1998)
3. Andrienko, N., Andrienko, G.: Extracting patterns of individual movement behaviour from a massive collection of tracked positions. In: BMI 2007, vol. 296, pp. 1–16. CEURS (2007)
4. Augusto, J.C., Aghajan, H.: Editorial: Inaugural issue. Journal of Ambient Intelligence and Smart Environments 1(1), 1–4 (2009)
5. Aztiria, A., Izaguirre, A., Basagoiti, R., Augusto, J.C.: Autonomous Learning of User's Preferences improved through User Feedback. In: BMI 2008, vol. 396, pp. 87–101. CEURS (2008)
6. Boronowsky, M., Herzog, O., Lawo, M.: Wearable computing: Information and communication technology supporting mobile workers. it - Information Technology 50(1), 30–39 (2008)
7. Cohn, A.G., Hazarika, S.M.: Qualitative spatial representation and reasoning: An overview. Fundamenta Informaticae 43, 2–32 (2001)
8. Goshorn, R., Goshorn, D., Kölsch, M.: The Enhancement of Low-Level Classifications for AAL. In: BMI 2008, vol. 396, pp. 87–101. CEURS (2008)
9. Gottfried, B. (ed.): Behaviour Monitoring and Interpretation, vol. 296. CEURS Proceedings (2007)
10. Gottfried, B., Aghajan, H. (eds.): Behaviour Monitoring and Interpretation, vol. 396. CEURS Proceedings (2008)
11. Hallot, P., Billen, R.: Spatio-temporal configurations of dynamics points in a 1D space. In: BMI 2007, vol. 296, pp. 77–90. CEURS (2007)
12. Hein, A., Kirste, T.: Towards recognizing abstract activities: An unsupervised approach. In: BMI 2008, vol. 396, pp. 102–114. CEURS (2008)
13. Heitor, T., Tomé, A., Dimas, P., Silva, J.P.: Measurability, Representation and Interpretation of Spatial Usage in Knowledge-Sharing Environments. In: BMI 2007, vol. 296, pp. 43–61. CEURS (2007)
14. Hongeng, S., Bremond, F., Nevatia, R.: Representation and Optimal Recognition of Human Activities. In: CVPR. IEEE Press, Los Alamitos (2000)
15. Hotz, L., Neumann, B.: Scene Interpretation as a Configuration Task. Künstliche Intelligenz 19(3), 59–65 (2005)
16. Kaufmann, R., Bollhalder, H., Gysi, M.: Infrared positioning systems to identify the location of animals. In: ECPA-ECPLF, p. 721. ATB Agrartechnik Bornim / Wageningen Academic Publishers (2003)
17. Kiefer, P., Schlieder, C.: Exploring context-sensitivity in spatial intention recognition. In: BMI 2007, vol. 296, pp. 102–116. CEURS (2007)

18. Kritzler, M., Lewejohann, L., Krüger, A.: Analysing Movement and Behavioural Patterns of Laboratory Mice in a Semi Natural Environment Based on Data collected via RFID. In: BMI 2007, vol. 296, pp. 17–28. CEURS (2007)
19. Kurata, Y., Egenhofer, M.: The 9^+-Intersection for Top. Rel. between a Directed Line Segment and a Region. In: BMI 2007, vol. 296, pp. 62–76. CEURS (2007)
20. Laube, P., Imfeld, S., Weibel, R.: Discovering relative motion patterns in groups of moving point objects. IJGIS 19(6), 639–668 (2005)
21. Martino-Saltzman, D., Blasch, B., Morris, R., McNeal, L.: Travel behaviour of nursing home residents perceived as wanderers and nonwanderers. Gerontologist 11, 666–672 (1991)
22. Merriam-Webster Online Dictionary. ethology (2009), `http://www.merriam-webster.com/dictionary/ethology` (visited on April 7, 2009)
23. Miene, A., Visser, U., Herzog, O.: Recognition and prediction of motion situations based on a qualitative motion description. In: Polani, D., Browning, B., Bonarini, A., Yoshida, K. (eds.) RoboCup 2003. LNCS (LNAI), vol. 3020, pp. 77–88. Springer, Heidelberg (2004)
24. Millonig, A., Gartner, G.: Shadowing – Tracking – Interviewing: How to Explore Human ST-Behaviour. In: BMI 2008, vol. 396, pp. 42–56. CEURS (2008)
25. Muller, P.: A qualitative theory of motion based on spatio-temporal primitives. In: KR 1998, pp. 131–141. Morgan Kaufmann, San Francisco (1998)
26. Musto, A., Stein, K., Eisenkolb, A., Röfer, T., Brauer, W., Schill, K.: From motion observation to qualitative motion representation. In: Habel, C., Brauer, W., Freksa, C., Wender, K.F. (eds.) Spatial Cognition 2000. LNCS (LNAI), vol. 1849, pp. 115–126. Springer, Heidelberg (2000)
27. Nagel, H.H.: From image sequences towards conceptual descriptions. Image Vision Comput. 6(2), 59–74 (1988)
28. Peuquet, D.J.: It's about time: A conceptual framework for the representation of temporal dynamics in GIS. Annals of Assoc. of Am. Geogr. 84, 441–461 (1994)
29. Pollack, M.E.: Intelligent technology for an aging population: The use of ai to assist elders with cognitive impairment. AI Magazine 26(2), 9–24 (2005)
30. Rüetschi, U.-J., Timpf, S.: Modelling wayfinding in public transport. In: Freksa, C., Knauff, M., Krieg-Brückner, B., Nebel, B., Barkowsky, T. (eds.) Spatial Cognition IV. LNCS (LNAI), vol. 3343, pp. 24–41. Springer, Heidelberg (2005)
31. Shi, H., Kurata, Y.: Modeling Ontological Concepts of Motions with Projection-Based Spatial Models. In: BMI 2008, vol. 396, pp. 42–56. CEURS (2008)
32. Shoval, N., Isaacson, M.: Sequence Alignment as a Method for Human Activity Analysis in Space and Time. Annals of Assoc. of Am. Geogr. 97(2), 282–297 (2007)
33. Sprado, J., Gottfried, B.: Semantic argumentation in dynamic environments. In: ICEIS 2009. Springer, Heidelberg (2009) (to appear)
34. Steinhage, A., Lauterbach, C.: Monitoring Mov. Behav. by means of a Large Area Proximity Sensor Array. In: BMI 2008, vol. 396, pp. 15–27. CEURS (2008)
35. Taborsky, M.: The Use of Theory in Behavioural Research. Ethology 114, 1–6 (2008)
36. Terzic, K., Hotz, L., Neumann, B.: Division of Work During Behaviour Recognition. In: BMI 2007, vol. 296, pp. 144–159. CEURS (2007)
37. Thirde, D., Borg, M., Ferryman, J.M., Fusier, F., Valentin, V., Bremond, F., Thonnat, M.: A real-time scene understanding system for airport apron monitoring. In: ICVS 2006. IEEE Computer Society, Los Alamitos (2006)

38. Timm, I., Scholz, T., Krempels, K.-H., Herzog, O., Spaniol, O.: From Agents To Multiagent Systems. In: Kirn, S., et al. (eds.) Multiagent Engineering. Theory and Applications in Enterprises. International Handbooks on Information Systems, pp. 35–52. Springer, Heidelberg (2006)
39. Van De Weghe, N., Bogaert, P., Cohn, A.G., Delafontaine, M., De Temmerman, L., Neutens, T., De Maeyer, P., Witlox, F.: How to Handle Incomplete Knowledge Concerning Moving Objects. In: BMI 2007, vol. 296, pp. 91–101. CEURS (2007)
40. Vu, V.T., Bremond, F., Thonnat, M.: A novel algorithm for temporal scenario recognition. In: IJCAI (2003), pp. 1295–1302 (2003)
41. Wagner, T., Bogon, T., Elfers, C.: Incremental Generation of Abductive Explan. for Tactical Behavior. In: BMI 2007, vol. 296, pp. 117–128. CEURS (2007)
42. Weiser, M.: The Computer for the Twenty-First Century. Scientific American 265, 94–110 (1991)
43. Winter, S.: Route Adaptive Selection of Salient Features. In: Kuhn, W., Worboys, M.F., Timpf, S. (eds.) COSIT 2003. LNCS, vol. 2825, pp. 349–361. Springer, Heidelberg (2003)
44. Wood, Z., Galton, A.: Collectives and how they move: A tale of two classifications. In: BMI 2008, vol. 396, pp. 57–71. CEURS (2008)

Automatic Recognition and Interpretation of Pen- and Paper-Based Document Annotations

Marcus Liwicki, Markus Weber, and Andreas Dengel

Knowledge Management Department,
German Research Center for Artificial Intelligence DFKI GmbH
Trippstadter Strae 122, 67663 Kaiserslautern, Germany
{firstname.lastname}@dfki.de

Abstract. In this paper we present a system which recognizes handwritten annotations on printed text documents and interprets their semantic meaning. This system processes in three steps. In the first step, document analysis methods are applied to identify possible gestures and text regions. In the second step, the text and gestures are recognized using several state-of-the-art recognition methods. In the fourth step, the actual marked text is identified. Finally, the recognized information is sent to the Semantic Desktop, the personal Semantic Web on the Desktop computer, which supports users in their information management. In order to assess the performance of the system, we have performed an experimental study. We evaluated the different stages of the system and measured the overall performance.

1 Introduction

The *paperless office*, i.e., the philosophy of working with minimal amount of paper and converting everything into digital documents, was predicted more than thirty years ago. However, today we have not replaced all paper-based workflows with digital counterparts. In contrast, there is still an increasing amount of paper used by humans in their everyday work. Using paper is motivated by several issues. First, the user is not bound to a specific electronic device or to a specific medium. Second, paper is portable, allowing for making notes anytime and anywhere. Furthermore, among other issues, writing on paper is more natural to most persons making the pen their preferred writing instrument for tasks such as brainstorming, collaborative work or reviewing documents [1].

Especially in workflows like annotating documents, users often tend to print the document on paper, because using a pen makes it easier and faster for them to attach their thoughts to the document. However, the interpretation of the notes is often more complex compared to the situation of digital comments. If the annotations have a meaning for the marked text, the problem of mapping the paper to the digital counterpart arises. To overcome these drawbacks, several document analysis approaches have been proposed which automatically map the documents, process human-made notes, and transform them into digital format [2,3]. They use cameras, Wacom Graphics Tablets[1], ultrasonic positioning, RFID antennas, bar code readers, or Anoto's Digital Pen and

[1] http://www.wacom.com

B. Mertsching, M. Hund, and Z. Aziz (Eds.): KI 2009, LNAI 5803, pp. 581–588, 2009.

Paper technology[2]. The Anoto technology is particularly interesting because it is based on regular paper and the recording of the interactions is precise and reliable.

The Semantic-eInk system is motivated by the following scenario. A knowledge worker is flying to a meeting and still has to read and annotate relevant documents. As working with a laptop is not alway possible during a flight, she prefers to edit and annotate document using pen and paper. During the flight, the knowledge worker annotates the printed documents with her digital pen, which records the absolut position on the paper. She marks the important parts for her personal knowledge base. Furthermore she writes down some concepts which should be connected to this document and adds comments to be processed later. After the flight she synchronizes the pen data with her computer. The Semantic-eInk system analyses the handwriting data and the gestures and interprets the data with the knowledge of the text in between the annotations. After the analyzed information is added to the Semantic Desktop and the knowledge worker can access the information without manually entering the data.

In this paper we present the Semantic e-Ink system, which automatically processes handwritten annotations on printed documents and interprets the semantic information of these annotations. This information is expressed through formal semantics using the individual's vocabulary, and integrated into the personal knowledge base, the Semantic Desktop[4]. The integration makes this knowledge searchable, reusable, sharable and gives a context for its interpretation. Semantic eInk extends the Semantic Desktop with a new input modality, interactive paper. Thus it supports personal knowledge work on paper.

The motivation for integrating a document analysis system into the Semantic Desktop is two-fold. First, the Semantic Desktop is enhanced with a novel input modality, allowing to maintain the knowledge base with paper-based annotations. Second, knowledge from the Semantic Desktop could be used to improve the results of the document analysis and handwriting recognition part. We believe that integrating semantic information into the recognition is an important step for improving handwriting recognition systems.

An early prototype of Semantic eInk has been proposed in an extended abstract [5]. In this paper, the prototype has been extended to work with any kind of electronic documents and the document analysis tools have been improved, which is also one of the contributions of this paper. A further contribution is a first experimental evaluation on real data, which gives insights to the performance of the system. Finally, this paper suggests possible improvements based on the experimental results.

The rest of this paper is organized as follows. In the next section, an overview of the system is given. The methods used for analyzing the document and recognizing the text are summarized in Section 3. Then, Section 4 presents the experiments and their results. Finally, Section 5 draws some conclusions and gives an outlook to future work.

2 System Overview

With the Semantic eInk system, the user can make two types of annotations (see Fig. 1), marks (of text), and comments. To mark a text, the user can mark the desired text

[2] http://www.anoto.com

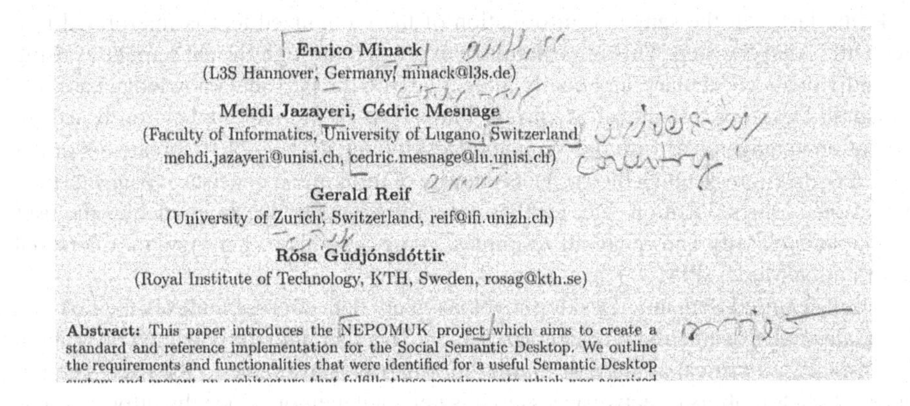

Enrico Minack
(L3S Hannover, Germany, minack@l3s.de)

Mehdi Jazayeri, Cédric Mesnage
(Faculty of Informatics, University of Lugano, Switzerland
mehdi.jazayeri@unisi.ch, cedric.mesnage@lu.unisi.ch)

Gerald Reif
(University of Zurich, Switzerland, reif@ifi.unizh.ch)

Rósa Gudjónsdóttir
(Royal Institute of Technology, KTH, Sweden, rosag@kth.se)

Abstract: This paper introduces the NEPOMUK project which aims to create a standard and reference implementation for the Social Semantic Desktop. We outline the requirements and functionalities that were identified for a useful Semantic Desktop

Fig. 1. Annotations on a document which are interpreted by Semantic eInk

passage with right angle strokes ("⌈" "⌉") (see the right-angle strokes around the first authors' name in Fig. 1 for an example), or by drawing a straight line (up or down) next to the text. Subsequently, he or she can write some text as an annotation to that marked text (e.g., "author" in Fig. 1). To make a comment, the user can write a text at any place in the document without marking a text beforehand.

The Semantic eInk system performs in four stages. First, the raw ink data is processed to determine which strokes are gestures and which strokes are handwritten annotations. Second, the individual gesture and text notes are recognized. Next, the marked text is identified. Finally, the recognized text and the marked text are semantically analyzed, i.e., the meaning of the annotations is interpreted and the personal knowledge base is updated. This section will describe the individual steps with more detail.

In order to retrieve written annotations from paper documents, our system builds on the interactive paper (iPaper) framework [6]. The iPaper framework enables the capture of user actions on paper and the activation of corresponding digital services or access to digital resources. The documents become interactive at print time by augmenting the paper with a special Anoto dot pattern. The Anoto pattern represents an absolute positioning system based on (x,y) coordinates that can be read by special digital pens such as the Magicomm G303[3]. After decoding the pattern, the pen transmits the (x,y) positions to a computer either through a wireless Bluetooth connection or by docking it on a docking station and storing the information in a file.

The next two steps, i.e., ink identification and recognition are described in the next section. For the identification of the marked text, the pen-based annotation has to be associated with a specific word, sentence, or section within the digital document. Therefore, precise mapping between the digital content and its printed version has to be performed. In this paper we apply a commercial OCR on the pdf-version of any printed document and store the direct layout mapping in a specific XML-file. This task is automated by an Windows printer driver. Because of the perfect quality of the documents, we have detected no OCR errors during our experiments described in Section 4.

[3] http://www.magicomm.co.uk

In the last step the semantic information of the recognized text is interpreted in a Semantic Analysis step. This information is expressed through formal semantics using the individuals vocabulary, and finally integrated into the personal knowledge base, the Semantic Desktop. In contrast to current limitations in file and application based information management, with the Semantic Desktop the user is able to create his or her own classification system which reflects the way of thinking: it consists of projects, people, events, topics, locations etc. Furthermore, the Semantic Desktop enables the user to annotate, classify and relate all resources, expressing his or her view in a Personal Information Model (PIMO) [7].

A full-featured Semantic Desktop involves many data sources, indexes the text and metadata of all documents and categorizes them. Together with the collected facts about the instances, a critical amount of digital information is available to the user. The Semantic Desktop offers effective access to this personal memory since the information is not only personalized, facilitating navigational search, but also machine readable. Thus it assists the user in his or her every-day activities and the extension and maintenance of the PIMO.

The existing resources and relations of the PIMO are matched with the recognized annotations. If a reasonable interpretation is found, the new relations and instances are added to the PIMO. Otherwise, the annotation is added as a comment to the document. Since the main focus of this paper lies on the document analysis and evaluation, no further description of the semantic analysis is presented here. For more details on this particular step, refer to [5].

3 Document Analysis and Handwriting Recognition Methods

The processing system for handwritten annotations build on recent advantages in handwriting recognition and gesture recognition. The domain of handwriting recognition has traditionally been divided into on-line and off-line recognition. In off-line recognition the text to be recognized is captured by a scanner and stored as an image, while in the on-line mode the handwriting is produced by means of an electronic pen or a mouse and acquired as a time-dependent signal. Good progress has been achieved in both on-line and off-line recognition [8,9,10].

The digital pen is able to record the sequence information of the strokes. Hence, in this paper the data is available in on-line format. The task of handwriting processing and recognition is considered to be easier for this format. Nowadays, commercial applications are available on the market, which are able to recognize complete handwritten text lines with an acceptable recognition accuracy.

Our system supports the use of the recognizer from Vision Objects [11][4] or the recognizer from Microsoft [12][5]. A short description of the concepts of both recognizers follows. After normalizing and pre-segmenting the handwriting into strokes and sub-strokes, both recognizers extract feature sets using a combination of on-line and

[4] http://www.visionobjects.com

[5] http://www.microsoft.com/windowsxp/tabletpc/default.mspx

off-line information. These feature sets are subsequently processed by a set of character classifiers, which use Neural Networks and other pattern recognition paradigms. The training set contains ink samples from thousands of people with a diverse range of writing styles.

For the recognition of gestures, we use the E-Rubine algorithm [13]. It is an extended version of the Rubine [14] algorithm, which automatically learns to recognize gestures by examples, instead of describing them by a rule-based program. The extended version introduces 11 additional single and multi-stroke features to the original Rubine feature set (13 features). The weighted sum of the features is then taken as a confidence value. For further details on the algorithm and features refer to [14] and [13], respectively.

The following gestures are supported by our system (compare to the allowed marks in Section 2): *up-right*, *down-right*, *left-down*, and *left-up* (for marking the left end of the text), *up-left* or *down-left*, *right-down*, and *right-up* (for marking the right side of the text); *up* and *down* (for side-marks).

As mentioned in Section 2, it has to be determined which strokes are gestures and which strokes are handwritten annotations. Therefore, we apply a set of heuristics. First, succeeding strokes which are close to one another are put into the same group in the annotation queue. This prevents the system from recognizing small strokes like "T"-bars as a gesture. Next, we sequentially apply the following procedure to each group of strokes g_i:

1. If the gesture recognizer is sure that g_i is a gesture, it is classified as one.
2. Otherwise, the g_i is handeled as text and the succeding groups are added to g_i if they are likely to continue the text line, i.e., if they are in the same y-area, close to g_i in x-direction, and the text recognizer's confidence does not drop.

This procedure incorporates *a priori knowledge* about the handwritten text as well as the recognizers confidences.

4 Experiments and Results

We have done a set of experiments to assess the performance of the overall Semantic eInk system and particular parts of it. For acquiring the data, we have asked four volunteers to make about 100 meaningful annotations on a set of printed documents, resulting in a total 400 annotations. The users were asked to mark the text with the gestures mentioned in Section 3 and subsequently making the handwritten annotation (mostly only one word per annotation). The data has been transcribed manually. Therefore the data set is not very large. However, it is already enough data to make a rough estimation of the performance and to find the bottlenecks of the system. Note that no fine-tuning of any parameters has been performed and no recognizer has been trained on writer-specific data. This ensures an objective evaluation.

For the recognition of the gestures, we used the iGesture framework [13]. The Microsoft handwriting recognition engine has been used for text recognition. The identification of the strokes has been performed as described in Section 3. A Web-Service was implemented to interpret the recognized annotations and send them to the Semantic Desktop.

First, the overall system performance has been evaluated. The recognition results of the document analysis parts appear in Table 1. An annotation is only treated as correctly recognized, if the first three stages described in Section 2 have been performed correctly, i.e., if the identification of the strokes, the recognition of the handwriting and gestures, and the identification of the marked text. From the 400 annotations, 76% have been recognized correctly. Note that due to the wrong recognition of some gestures, some annotations have been skipped. During the semantic analysis step all annotations have been interpreted. Those annotations which were recognized correctly were also interpreted in a reasonable way. The remaining annotations usually did not match any PIMO-resource and were added as a comment. It is an open issue in the domain of knowledge management to objectively say which annotations were interpreted correctly, because different users might have different opinions on that.

To further analyze the individual parts, we have measured the performance of the gesture and handwriting recognition. The results of the gesture recognition part are presented in Table 3. As can be seen, 91% of the gestures were correctly identified and no gesture was wrongly recognized. However, 4% of the gestures were rejected and treated as handwriting, because the E-Rubine algorithm had a low confidence. In some cases two succeeding gestures were merged into one gesture and treated as handwriting. An illustration of such a case can be seen in Fig. 2. The two encircled right-angle strokes are close to one another and recognized as 72. It is interesting to note that 2% of the gestures have not been recorded, i.e., the Anoto pen did not capture the strokes. This often happens when the user holds the pen incorrectly or when the paper is not completely flat. Usually, the pen gives a direct feedback by vibrating. However, sometimes the users ignored this feedback.

The recognition rates of the handwriting recognition part appear in in Table 2. About 84% of the words were recognized correctly. This recognition rate might be too small for practical usefulness. To analyze the potential of improvement, we have further investigated the n-best list of the recognizer. An n-best list contains the n top recognition alternates of the recognizer. Already 98% of the correct words were present in the 5-best list and more than 99% were in the 10-best list. Furthermore, most of the incorrect words had nothing to do with the annotated document or with the knowledge domain

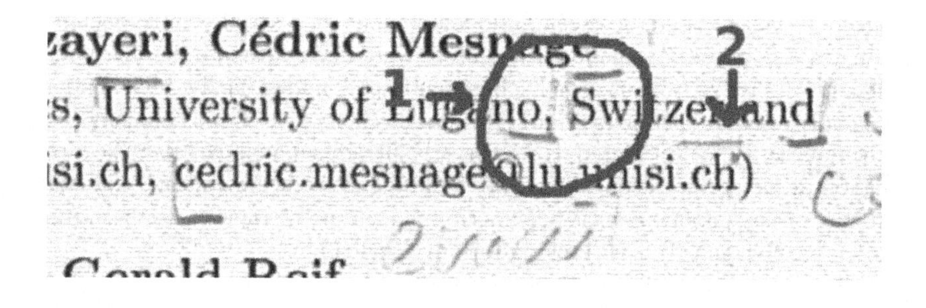

Fig. 2. Illustration of recognition errors: two gestures were connected (1), one gesture was wrongly recognized (2)

Table 1. Ovr. results

	Percentage
correct	76 %
incorrect	21 %
skipped	4 %

Table 2. Hwr. results

	Percentage
correct	84 %
incorrect	15 %
not recorded	1 %

Table 3. Gesture results

	Percentage
correct	91 %
incorrect	0 %
rejected	4 %
wrong merging	3 %
not recorded	2 %

around the document. Therefore we conclude that integrating the semantic information of the document could help to improve the recognition results.

5 Conclusions and Future Work

In this paper we presented the Semantic eInk system, which is able to automatically process handwritten annotations on printed documents. Furthermore it interprets the semantic information of these annotations and sends them to the Semantic Desktop, the personal SemanticWeb. To understand the meaning of the marked text, the paper is mapped to its electronic counterpart. Using our extension of the iPaper framework, the system is able to handle any electronic document format.

To assess the performance of the Semantic eInk system, we have performed recognition experiments on real annotation data. Finally, about three out of four annotations were recognized and interpreted correctly. A deeper analysis has shown that the gesture recognition is quite reliable. Most mistakes result from incorrect stroke identification. To overcome this problem, we plan to use a dynamic programming algorithm similar to the one developed for circuit recognition in [15]. The unconstrained handwriting recognition is the hardest part of the system. Initial tests with other recognizers did not achieve significantly higher results. A high potential lies in using the semantic information directly for this step, because more than 99% of the words appeared in the 10-best list and mostly the correct words also appeared in the annotated text or in the personal knowledge base. The use of this knowledge is a promising direction for future work.

The environment for data acquisition was quite restricted. Currently, we are collecting a large database with real annotations from more than a hundred writers, where only a few restrictions are made for the annotations. In Future Work we plan to further support other input devices, such as the iLiad[6] from iRex Technologies. This particular device integrates electronic ink and Wacom technology for viewing and annotating documents.

Acknowledgements

This work was supported by the Stiftung Rheinland-Pfalz für Innovation. We would like to thank Dr. Beat Signer for making the iPaper and iGesture packages available for us. Furthermore, we thank Nadir Weibel for answering Anoto-related questions.

[6] `http://www.irextechnologies.com` [Last access: 2009-04-23].

References

1. Sellen, A.J., Harper, R.H.R.: The Myth of the Paperless Office. MIT Press, Cambridge (2001)
2. Rodríguez, J., Sánchez, G., Lladós, J.: A pen-based interface for real-time document edition. In: International Conference on Document Analysis and Recognition, vol. 2, pp. 939–943 (2007)
3. Weibel, N., Ispas, A., Signer, B., Norrie, M.C.: PaperProof: a paper-digital proof-editing system. In: Conference on Human Factors in Computing Systems, pp. 2349–2354. ACM, New York (2008)
4. Sauermann, L.: The semantic desktop - a basis for personal knowledge management. In: Maurer, H., Calude, C., Salomaa, A., Tochtermann, K. (eds.) Proceedings of the I-KNOW 2005. 5th International Conference on Knowledge Management, pp. 294–301 (2005)
5. Liwicki, M., Schumacher, K., Dengel, A., Weibel, N., Signer, B., Norrie, M.C.: Pen and paper-based interaction with the semantic desktop. In: Handout of 8th Int. Workshop on Document Analysis Systems, IAPR, o.A., September 2008, pp. 2–5 (2008)
6. Norrie, M.C., Signer, B., Weibel, N.: General framework for the rapid development of interactive paper applications. In: Proc. of CoPADD 2006, 1st Int. Workshop on Collaborating over Paper and Digital Documents, pp. 9–12 (2006)
7. Sauermann, L., van Elst, L., Dengel, A.: Pimo - a framework for representing personal information models. In: Pellegrini, T., Schaffert, S. (eds.) Proceedings of I-Semantics 2007, JUCS, pp. 270–277 (2007)
8. Plamondon, R., Srihari, S.N.: On-line and off-line handwriting recognition: a comprehensive survey. IEEE Trans. Pattern Analysis and Machine Intelligence 22(1), 63–84 (2000)
9. Tappert, C.C., Suen, C.Y., Wakahara, T.: The state of the art in online handwriting recognition. IEEE Trans. Pattern Analysis and Machine Intelligence 12(8), 787–808 (1990)
10. Vinciarelli, A.: A survey on off-line cursive script recognition. Pattern Recognition 35(7), 1433–1446 (2002)
11. Knerr, S., Anisimov, V., Baret, O., Gorsky, N., Price, D., Simon, J.: The A2iA INTERCHEQUE system: Courtesy amount and legal amount recognition for French checks. Int. Journal of Pattern Recognition and Artificial Intelligence 11(4), 505–548 (1997)
12. Pittman, J.A.: Handwriting recognition: Tablet pc text input. Computer 40(9), 49–54 (2007)
13. Signer, B., Kurmann, U., Norrie, M.C.: iGesture: a general gesture recognition framework. In: Proc. of ICDAR 2007, 9th Int. Conference on Document Analysis and Recognition, pp. 954–958 (2007)
14. Rubine, D.: Specifying gestures by example. In: SIGGRAPH 1991: Proceedings of the 18th Annual Conference on Computer Graphics and Interactive Techniques, pp. 329–337 (1991)
15. Feng, G., Viard-Gaudin, C., Zhengxing, S.: electric circuit diagram understanding using 2d dynamic programming. In: 11th Int. Conf. on Frontiers in Handwriting Recognition (2008)

Self-emerging Action Gestalts for Task Segmentation

Michael Pardowitz, Jan Steffen, and Helge Ritter

Universität Bielefeld, AG Neuroinformatik
{mpardowi,jsteffen,helge}@techfak.uni-bielefeld.de

Abstract. Task segmentation from user demonstrations is an often neglected component of robot programming by demonstration (PbD) systems. This paper presents an approach to the segmentation problem motivated by psychological findings of gestalt theory. It assumes the existence of certain "action gestalts" that correspond to basic actions a human performs. Unlike other approaches, the set of elementary actions is not prespecified, but is learned in a self-organized way by the system.

1 Introduction

Programming a robot to achieve an individual task is a complex problem. One promising way to ease this is to equip cognitive robots with task learning abilities, that lets them learn a task from demonstrations of naive (non-expert) users. This paradigm is widely known as *Programming by Demonstration (PbD)* or *Imitation Learning*. Although several systems for Programming by Demonstration have been proposed , the problem of task segmentation has only received minor attention. The decomposition of a task demonstration into its elements was tackled only in problem specific ways, and there is no consensus on how to choose the basic actions yet.

[1] applies hand-crafted rules to detect state transitions from video sequences. Segments are characterised through stable contact points between the objects recognized in the scene. More formalized models use Hidden-Markov-Models (HMMs) to segment walking or grasping actions from motion-capture data [3]. A taxonomy of action primitives is presented in [4]. These primitives of action (mainly concerned with grasping) are learned in a supervised way which allows to classify each frame of a task demonstration and to construct task segments from those classifications. These segments have been transformed into petri-nets for execution on a humanoid robot [10]. A similar way is proposed in [2] where a user demonstration is segmented based on the most likely primitives performed in each timestep. A new attempt to use gestalt-based neural methods for task segmentation was recently presented in [6]. It was inspired by the work on visual segmentation cited above, and transferred the concept of visual gestalts to the human action domain. Our work presented here is based on the same model, namely the Competitive Layer Model (CLM), but will not rely on the supervised learning approach that was used to determine the shape of the applied action gestalts.

B. Mertsching, M. Hund, and Z. Aziz (Eds.): KI 2009, LNAI 5803, pp. 589–596, 2009.

In this paper we propose a method for the unsupervised learning of task elements and a task decomposition method. It is based on an idea that is transferred from the visual segmentation domain to the task segmentation field. Namely, we extend the idea of perceptual grouping via gestalt rules from the domain of visual patterns into the domain of spatio-temporal processes arising from actions. We extend this line of thinking more deeply into the realm of task decomposition and to explore the power of Gestalt laws for characterizing *good action primitives* for task decomposition. Such an approach can connect the so far primarily perception-oriented Gestalt approach with more current ideas on the pivotal role of the action-perception loop for representing and decomposing interactions.

The following section will review the literature on action and image segmentation. After that, section 2 will introduce the computational model for gestalt based action segmentation used in the experiments in this paper. Section 4 describes a learning method to construct such models from human demonstrations in a supervised learning setup. Section 4 reports the hardware setup and preprocessing steps and the results obtained from our experiments. Finally, we conclude this paper with a discussion and an outlook on future work.

2 The Competitive Layer Model

The Competitive Layer Model (CLM) [9] consists of L neuron layers. Each layer $\alpha = 1 \ldots L$ acts as a "feature map" associating its positions t with feature value combinations m_t from the chosen feature space V. In our specific application, t represents the point of time where a feature vector m_t was recorded. Figure 1 shows a simple example of a CLM with $L = 3$ and $N = 4$: Four input feature vectors are extracted from an input trajectory. A single neuron represents that trajectory segment in each layer. The neurons are connected laterally and columnarly. Identical positions t in different layers share the same input line and receive a (usually scalar) input activity h_t. In the simplest case we assume that the (prespecified) feature maps are layer-independent (no α index) and are implemented in a discretized form by N linear threshold neurons with activities $x_{t,\alpha}$ located at the same set of discrete positions r in each layer α. Therefore, the system consists of L identical feature maps that can be activated in parallel.

The idea of the CLM is to use this coding redundancy to introduce a competitive dynamics between layers for the coding of features in its input pattern h_t. The outcome of this competition partitions the features comprising the input into disjoint groups called gestalts, with each group being characterized as one subset of features coded by activity in the same "winning layer".

The competitive dynamics is based on two different sets of connections between the neurons of the CLM. Two timesteps t and t' belong to the same segment if they show simultaneous activities $x_{t\alpha} > 0$ and $x_{t'\alpha} > 0$ in a certain layer α. In each layer, the neurons are fully connected with symmetric weights $f_{tt'} = f_{t't}$. The values of $f_{tt'}$ establish semantic coherence of features. Positive values indicate feature compatibility through excitatory connections while negative values express incompatibility through inhibitory connections.

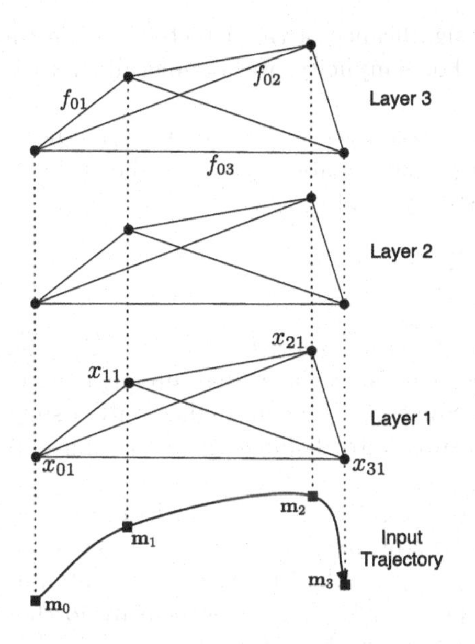

Fig. 1. The Competitive Layer Model Architecture for an input trajectory with $L = 3, N = 4$. The solid and dotted lines represent lateral and columnar interactions respectively. Lateral connections are labelled with their according weights $f_{tt'}$. The input features $m_0 \dots m_3$, the neurons of layer 1 and the lateral interactions for neuron 0 in layer 3 are labeled with the corresponding notation.

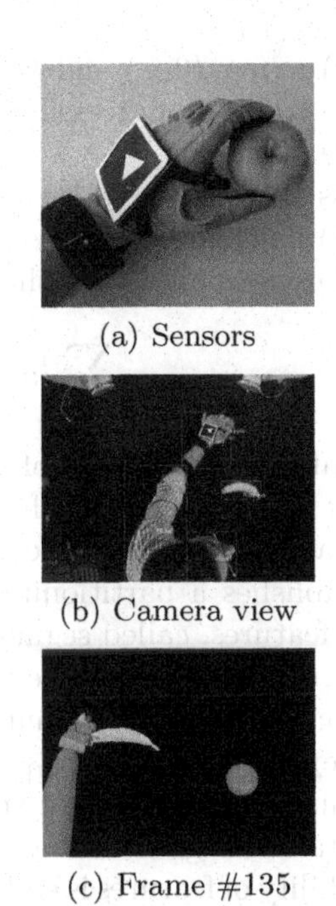

(a) Sensors

(b) Camera view

(c) Frame #135

Fig. 2. Sensors and Key Frames. (a) Sensing devices: A Cyberglove for hand posture tracking together with a AR-Toolkit Marker mounted on the back of the hand. (b) The marker and the object are tracked with an overhead camera. Marker and color blob tracking is performed. (c) Sample frame of recorded sequence in 3D-visualization.

The purpose of the layered arrangement and the columnar interactions in the CLM is to enforce a dynamical assignment of the input features to layers that respects the contextual information stored in the lateral interactions $f_{tt'}$. This assignment segments the input into partitions of matching features which links each feature t with its unique label $\alpha(r)$. A columnar Winner-Takes-All (WTA) circuit realizes this unique assignment using mutual symmetric inhibitory interactions with strength $J > 0$ between neural activations $x_{t\alpha}$ and $x_{t\beta}$. Due to the WTA coupling, only one neuron from a single layer can be active in every column, as soon as a stable equilibrium state of the CLM is reached.

Equation (1) combines the inputs with the lateral and columnar interactions into the CLM dynamics (see [7]):

$$\dot{x}_{t\alpha} = -x_{t\alpha} + \sigma \left(J(h_t - \sum_{\beta} x_{t\beta}) + \sum_{t'} f_{tt'} x_{t'\alpha} \right) . \qquad (1)$$

Here is $\sigma(x) = \max(0, x)$ and h_t is the significance of the detected feature t as obtained by some preprocessing steps. For simplicity, we assume all h_t to be equal to one in this paper.

A process that updates the neural activations according to the dynamics of equation (1) converges towards several possible stable states, as shown in [8]. These stable states all satisfy the consistency conditions

$$\sum_{t'} f_{tt'} x_{t'\beta} \leq \sum_{t'} f_{tt'} x_{t'\hat{\alpha}(t)}, \tag{2}$$

which indicate the assignment of a feature t to the layer $\hat{\alpha}(t)$ with the highest lateral support for that feature. This corresponds to the layer that already contains the most features t' compatible with t. Since every column t has only a single $\hat{\alpha}(t)$, $\hat{\alpha}$ establishes a partitioning of features into disjunctive sets of mutually compatible features, called segments.

Given a ground truth for the segmentation $\alpha^*(t)$, e.g. obtained from manual segmentation performed by a human instructor, on can calculate the segmentation accuracy $a_{\hat{\alpha}}$ as $a_{\hat{\alpha}} = \arg_{p \in P} \frac{1}{N} \sum_t \delta_{p(\hat{\alpha}),\alpha^*}$, with P being the set containing all permutation functions over the set $\{1, \ldots, L\}$. $a_{\hat{\alpha}}$ gives a measure for how good an obtained segmentation $\hat{\alpha}$ is, compared with $\alpha^*(t)$.

Compatibility of features is coded in the lateral inhibitory or excitatory weights $f_{tt'}$. They represent the knowledge about the shape of the action gestalts or segments that are to be found in the demonstration sequence. The next section describes, how good lateral interactions can be learned in a self-organized way.

3 Self-emergence of Action Gestalts

As the preceding section stated, the CLM converges towards a segmentation that is determined by the lateral interactions $f_{tt'}$. These encode the characteristics of the underlying action gestalts and can be learned in a supervised way, as [6] has shown. However, this approach requires training examples, which are manually labelled task demonstration sequences. Humans, when faced with a new task, clearly do not need such structured data. They can infer both the elementary actions (action gestalts) and their sequence from a single demonstration. This section proposes a method for the self-organization of action gestalts by tuning the lateral connection weights of a CLM as introduced in section 2.

All lateral connections $f_{tt'}$ are initialized with small random values drawn from a zero-mean Gaussian distribution with variance σ_{init}. After that starts a process of alternating CLM dynamics execution as stated in section 2 and weight update according to the following update rule:

$$\Delta f_{tt'} = \eta \cdot f\left(-\frac{d_s(t,t')^2}{\sigma_s}\right) \cdot f\left(-\frac{d_t(t,t')^2}{\sigma_t}\right) \cdot f\left(-\frac{d_f(t,t')^2}{\sigma_f}\right) \tag{3}$$

where $f(x) = 2\tanh(x) + 1$ represents a nonlinear activation function, η the learning rate, $\sigma_s, \sigma_t, \sigma_f$ scalar width constants that influence the sensitivity of

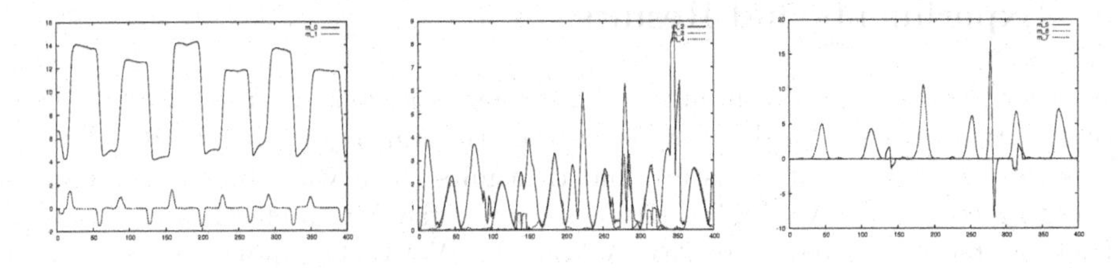

Fig. 3. Extracted features. (a) m_0, m_1 (b) $m_2 - m_4$ (c) $m_5 - m_7$.

the algorithm to differences in the segmentation, time and feature domain, and d_s, d_t, d_f various distances between two time-frames t and t'. These are described as follows:

- d_s measures the segmentation distance of two time-frames t, and t'. It is defined as

$$d_s(t,t') = \begin{cases} 0, & \text{if } \arg\max_\alpha x_{t\alpha} = \arg\max_\alpha x_{t'\alpha} \\ 1, & \text{otherwise,} \end{cases}$$

  that is, if two time-frames are segmented into the same CLM layer, they have a very low distance $d_s(t,t')$, while time-frames assigned to different segments have a higher distance. This has the effect, that time-frames that have been grouped together by the last iteration of the CLM dynamics receive higher values for $f_{tt'}$. On the long run, these time-frames tend to be grouped to the same group again in future CLM dynamics executions, leading to the emergence of stable gestalts.
- d_t measures the temporal distance and is defined as $d_t(t,t') = ||t - t'||$. This reflects the fact that action gestalts are coherent over time and have a certain maximal duration that can be influenced by tuning the parameter σ_t. Dropping the temporal distance term from equation (3) would lead to cluttered segments with "holes". Instead, the inclusion of this term is motivated by the fact that action gestalts should be completed in a single sweep.
- d_f measures the feature distance. It is defined as the mahalanobis distance $d_f(t,t') = \sqrt{(\mathbf{m}_t - \mathbf{m}_{t'})^T \mathbf{C}^{-1} (\mathbf{m}_t - \mathbf{m}_{t'})}$ between the feature vectors $\mathbf{m}_t$ and $\mathbf{m}_{t'}$ associated with the time-frames t and t', where $\mathbf{C}$ the covariance matrix of all feature vectors $\mathbf{m}_t$. The feature distance term d_f in equation (3) guarantees that the emerging gestalts are grounded in the physical sensor values or feature vectors, respectively.

Combining those three measures of similarity by multiplication leads to a iterative process of weight adaption according to equation (3) alternating with the neural CLM dynamics according to equation (1). The former produces a new set of lateral weights, which are used by the latter to compute a new segmentation of the input sequence, that in turn influence the next set of lateral weights.

4 Experiments and Results

In order to obtain data from human task demonstrations, we used the following setup: An Immersion Cyberglove [5] senses the finger joint angles in order to capture the human hand posture and transmits them via a bluetooth connection to a computer. An ARToolKit Marker was fixed with elastic straps on the back of the hand (see figure 2a). A Sony DFW-VL500 Firewire camera with a resolution of 640x480 pixels tracked the scene from an overview perspective (see figure 2b). This setting allowed us to locate a person's hand in 6D-space (from the ARToolKit Marker) together with its posture (from the Cyberglove sensor readings) and estimate the actions the user performs. Figure 2a shows this setup.

Equipped with the sensors described above, the user faced an environment which contained various objects (i.e. apples, bananas). To obtain information on the position and movements, we used methods for tracking colored blobs to record object movements in the same overhead camera images that were used for the ARToolKit tracking. The results are shown in figure 2(b).

Several sequences with lengths N varying between 350 and 400 have been recorded with this sensory setup. A sample from one sequence used for the experiments displayed in figure 2 (c) as a 3D reconstruction of the obtained data. The hand position and posture together with the object positions yielded enough information to extract the following features (see also figure 3):

1. The sum of the joint angles of the human hand: $m_0(t) = \sum_i \theta_i(t)$. This gives a measure for the degree of opening/closing of the hand.
2. The derivative of m_0: $m_1(t) = \dot{m}_0(t)$. This gives large values at times when the user opens or closes a grip.
3. The velocities of the hand and the objects $m_2(t) = |\mathbf{v}_{hand}(t)|, m_3 = |\mathbf{v}_{obj1}(t)|, m_4 = |\mathbf{v}_{obj2}(t)|$. The velocities tend to remain positive during connected segments and disappear only at points where the goal context of the user changes.
4. The co-occurrence of parallel movements is calculated using the scalar product of the movements of the hand with an object or between the two objects respectively: $m_5(t) = \mathbf{v}_{hand}(t)^T \cdot \mathbf{v}_{obj1}(t)$, $m_6(t) = \mathbf{v}_{hand}(t)^T \cdot \mathbf{v}_{obj2}(t)$, $m_7(t) = \mathbf{v}_{obj1}(t)^T \cdot \mathbf{v}_{obj2}(t)$, These features give large values for segments where an object moves in the same direction as the hand, or the objects move parallel to each other.

These features are computed for each frame of a demonstration sequence. Stacking all features of a time-frame t into a single feature vector $\mathbf{m}_t = [m_0(t), \ldots, m_7(t)]^T$ yields the inputs to the CLM depicted in figure 1 and the gestalt emergence algorithm described in section 3.

The gestalt emergence algorithm (see section 3) was run with fixed values of $0.5, 500, 50$ and 0.02 for the constants $\sigma_s, \sigma_t, \sigma_f$ and η, respectively. It was run for 100 iterations. In each of these iterations there were $1 \cdot 10^7$ neural update steps according to equation (1) during the CLM dynamics execution phase. After

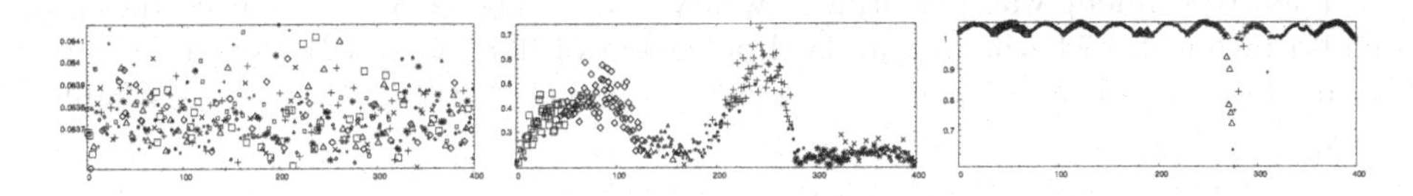

Fig. 4. Activations $x_{t\alpha}$ of the CLM neurons after dynamics execution phase. The x-axis represents the feature index t, while the activation of the winning neuron $x_{t\alpha}$ is depicted on the y-axis (note the different scale!). The winning layer is coded in the symbol: each layer is represented by an unique combination of symbol (square, diamond, triangle, cross, plus, star) and size (small, large).

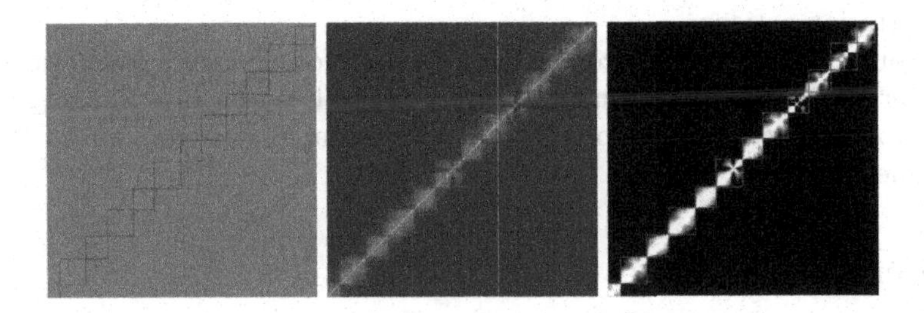

Fig. 5. Emergence of action gestalts in lateral weights. Each pixel at position (t, t') represents the value of the lateral connection weight $f_{tt'}$, that is the x-axis corresponds to values of t, while the values for t' can be found on the y-axis. Black pixels correspond to a weight value for $f_{tt'}$ of -1, white pixesl to weights of 1. The squares surrounded by red lines indicate areas which would be completely white for perfect action gestalts that can be obtained from manual segmentation, that is, in each of those areas the equation $\alpha^*(t) = \alpha^*(t')$ holds.

that, the lateral weights of the CLM were re-estimated according to equation (3), with the number of time-frames N being 400 and $L = 12$ layers, and finally, the next iteration started.

The resulting neural activations after the CLM dynamics phase are shown in figure 4 for selected iterations. One can see, that the neural network finishes with very low and random activations in the initial iterations which correspond to unstable and highly cluttered segments. In iteration 30 one can see first segments emerging, and finally, the process converges towards sharply delimited segments. These correspond to stable action gestalts.

The development of the lateral connection weights is shown in figure 5. Initially, the weights are initialized randomly to values close to zero, which corresponds to a grey color. Iterating the weight update rule of equation (3), the lateral weights become more distinctive. The small white areas around the diagonal after 100 iterations (figure 5c) indicate the areas with a high probability of being grouped into the same segment.

This experiment was performed twenty times. The segmentation accuracies had a mean of 0.74 with a standard deviation of 0.07. Given the unsupervised setup and the sensory inaccuracies of the experiments[1], this is a quite remarkable rate.

5 Conclusion

This paper showed a method the autonomous emergence of action symbols (called gestalts) from time-series observations. We presented an approach that was based on a dynamic recurrent neural network and an iterative weight update rule and validated it on a task demonstration sequence featuring different object transportation actions.

We found good convergence behavior in ptractice, though a theoretical proof of convergence is yet to be found. Future work will concentrate the autonomous estimation of feature extractors m_i and speed-related shortcomings of the proposed method.

References

1. Arsenio, A.M.: Learning task sequences from scratch: applications to the control of tools and toys by a humanoid robot. In: CCA, vol. 1, pp. 400–405 (2004)
2. Bentivegna, D.C.: Learning from Observation Using Primitives. PhD thesis, Georgia Institute of Technology (July 2004)
3. Calinon, S., Guenter, F., Billard, A.: On learning the statistical representation of a task and generalizing it to various contexts. In: ICRA (2006)
4. Ehrenmann, M., Zöllner, R., Rogalla, O., Vacek, S., Dillmann, R.: Observation in programming by demonstration: Training and exection environment. In: HUMANOIDS (2003)
5. Immersion. CyberGlove II Wireless Glove (2009)
6. Pardowitz, M., Haschke, R., Steil, J., Ritter, H.: Gestalt-Based Action Segmentation for Robot Task Learning. In: HUMANOIDS (2008)
7. Weng, S., Wersing, H., Steil, J., Ritter, H.: Learning lateral interactions for feature binding and sensory segmentation from prototypic basis interactions. IEEE TNN 17(4), 843–863 (2006)
8. Wersing, H., Beyn, W.-J., Ritter, H.: Dynamical stability conditions for recurrent neural networks with unsaturating piecewise linear transfer functions. Neural Comput. 13(8), 1811–1825 (2001)
9. Wersing, H., Steil, J.J., Ritter, H.J.: A competitive-layer model for feature binding and sensory segmentation. Neural Computation 13(2), 357–387 (2001)
10. Zöllner, R., Asfour, T., Dillmann, R.: Programming by demonstration: Dual-arm manipulation tasks for humanoid robots. In: IROS (2004)

[1] Visual inspection yielded four events where the vision component lost either the marker or an color blob at time-frames 145, 280, 315 and 350, longing for about 10 erroneuos frames each, which corresponds to 10% inaccurate frames.

Prediction and Classification of Motion Trajectories Using Spatio-Temporal NMF

Julian P. Eggert[2], Sven Hellbach[1], Alexander Kolarow[1], Edgar Körner[2], and Horst-Michael Gross[1]

[1] Ilmenau University of Technology, Neuroinformatics and Cognitive Robotics Labs, POB 10 05 65, 98684 Ilmenau, Germany
sven.hellbach@tu-ilmenau.de
[2] Honda Research Institute Europe GmbH, Carl-Legien-Strasse 30, 63073 Offenbach/Main, Germany
julian.eggert@honda-ri.de

Abstract. This paper's intention is to present a new approach for decomposing motion trajectories. The proposed algorithm is based on non-negative matrix factorization, which is applied to a grid like representation of the trajectories. From a set of training samples a number of basis primitives is generated. These basis primitives are applied to reconstruct an observed trajectory. The reconstruction information can be used afterwards for classification. An extension of the reconstruction approach furthermore enables to predict the observed movement into the future. The proposed algorithm goes beyond the standard methods for tracking, since it does not use an explicit motion model but is able to adapt to the observed situation. In experiments we used real movement data to evaluate several aspects of the proposed approach.

Keywords: Non-negative Matrix Factorization, Prediction, Movement Data, Robot, Motion Trajectories.

1 Introduction

The understanding and interpretation of movement trajectories is a crucial component in dynamic visual scenes with multiple moving items. Nevertheless, this problem has been approached very sparsely by the research community. Most approaches for describing motion patterns, like [1], rely on a kinematic model for the observed human motion. This causes the drawback that the approaches are difficult to adapt to other objects. Here, we aim at a generic, model-independent framework for decomposition, classification and prediction.

Consider the simple task for a robot of grasping an object which is handed over by the human interaction partner. First of all, the grasping task has to be recognized by the robot. We assume that this information can be gained from the motion information. Furthermore, to avoid a purely reactive behavior, which might lead to 'mechanical' movements of the robots, it is necessary to predict the further movement of the human's hand.

B. Mertsching, M. Hund, and Z. Aziz (Eds.): KI 2009, LNAI 5803, pp. 597–606, 2009.

In [2] an interesting concept for a decomposition task is presented. Like playing a piano a basis alphabet – the different notes – are superimposed to reconstruct the observation (the piece of music). The much less dimensional description of when each basis primitive is used, can be exploited for further processing. While the so-called piano model relies on a set of given basis primitives, our approach is able to learn these primitives from the training data.

Beside the standard source separation approaches, like PCA and ICA, another promising algorithm exists. It is called non-negative matrix factorization (NMF) [3]. Because of its non-negative character it is well suited for e. g. audio source separation. The system of basis vectors which is generated by the NMF is not orthogonal. This is very useful for motion trajectories, since one basis primitive is allowed to share a common part of its trajectory with other primitives and to specialize later.

The next section introduces the standard non-negative matrix factorization approach and two extensions that can be found in the literature. In section 3 the new approach for decomposing motion trajectories is presented. The experiments with their conditions and results are presented in section 4, while the paper concludes in section 5.

2 Non-negative Matrix Factorization

Like other approaches, e. g. PCA and ICA, non-negative matrix factorization (NMF) [3] is meant to solve the source separation problem. Hence, a set of training data is decomposed into basis primitives: $\mathbf{V} \approx \mathbf{W} \cdot \mathbf{H}$ Each training data sample is represented as a column vector $\mathbf{V}_i$ within the matrix $\mathbf{V}$. Each column of the matrix $\mathbf{W}$ stands for one of the basis primitives. In matrix $\mathbf{H}$ the element H_i^j determines how the basis primitive $\mathbf{W}_j$ is activated to reconstruct training sample $\mathbf{V}_i$. The training data $\mathbf{V}$ can only be approximated by the product of $\mathbf{W}$ and $\mathbf{H}$. This product will be referred to as reconstruction $\mathbf{R} = \mathbf{W} \cdot \mathbf{H}$ later.

Unlike PCA or ICA, NMF aims to a decomposition, which only consists of non-negative elements. This means that the basis primitives can only be accumulated. No primitive exists which is able to erase a 'wrong' superposition of other primitives. This leads to a more specific set of basis primitives, which is an advantage for certain applications, e. g. face recognition [4].

For generating the decomposition, optimization-based methods are used. Hence, an energy function E has to be defined:

$$E(\mathbf{W}, \mathbf{H}) = \frac{1}{2} \|\mathbf{V} - \mathbf{W} \cdot \mathbf{H}\|^2 \tag{1}$$

By minimizing the energy equation, it is now possible to achieve a reconstruction using the matrices $\mathbf{W}$ and $\mathbf{H}$. This reconstruction is aimed to be as close as possible to the training data $\mathbf{V}$. No further constraints are given in the standard formulation of the NMF. As it can be seen in equation 1, the energy function depends on the two unknown matrices $\mathbf{W}$ and $\mathbf{H}$.

Since both matrices usually have a large number of elements, the optimization problem seems to be an extensive task. Fortunately, each training sample can be regarded independent from the others. Furthermore, both matrices are adapted in an alternating fashion. This helps to reduce the number of dimensions for the optimization process and allows a training regime with fewer examples. The algorithm is formulated in the following description in vectorwise notation:

1. Calculate the reconstruction
$$\mathbf{R}_i = \sum_j H_i^j \mathbf{W}_j \tag{2}$$
2. Update the activities
$$H_i^j \leftarrow H_i^j \odot \left(\mathbf{V}_i^T \mathbf{W}_j \oslash \mathbf{R}_i^T \mathbf{W}_j\right) \tag{3}$$
3. Calculate the reconstruction with the new activities
$$\mathbf{R}_i = \sum_j H_i^j \mathbf{W}_j \tag{4}$$
4. Update the basis vectors
$$\mathbf{W}_j \leftarrow \mathbf{W}_j \odot \left(\sum_i H_i^j \mathbf{V}_i \oslash \sum_i H_i^j \mathbf{R}_i\right) \tag{5}$$

Where the operations $\odot$ and $\oslash$ denotes a component-wise multiplication and division. For each of the training samples steps 1 to 4 are processed. The training samples are iterated until a defined convergence criterion is reached. For the criterion the energy function can be used in a usual fashion.

2.1 Sparse Coding

As it could be seen in equation 1 the energy function is formulated in a very simple way. This results in a decomposition, which is quite arbitrary with no further characteristics. This can lead, for example, to redundant information. Especially, if the number of basis primitives is chosen higher than needed to decompose the given training data. To compensate this drawback, it is useful to apply a constraint which demands a sparse activation matrix, like it was introduced in [5]. This avoids the fact, that several basis primitives are activated at the same time, and hence are being superimposed.

$$E(\mathbf{W}, \mathbf{H}) = \frac{1}{2} \left\|\mathbf{V} - \mathbf{W} \cdot \mathbf{H}\right\|^2 + \lambda \sum_{i,j} H_i^j \tag{6}$$

The influence of the sparsity constraint can be controlled using the parameter λ. In this paper, we only discuss a special case for the sparsity term. A more detailed discussion can be found in [5]. The algorithmic description is similar to the one of the standard NMF. The only thing that has to be considered is that the basis primitives need to be normalized. The term $\lambda \sum_{i,j} H_i^j$ together with normalization and non negativity ensures sparsety in the sence of peak distribution around low values.

2.2 Transformation Invariance

Beside the sparsity constraint another extension to NMF has been published in [6]. The concept of transformation invariance allows to move, rotate, and scale the basis primitives for reconstructing the input. In this way, we do not have to

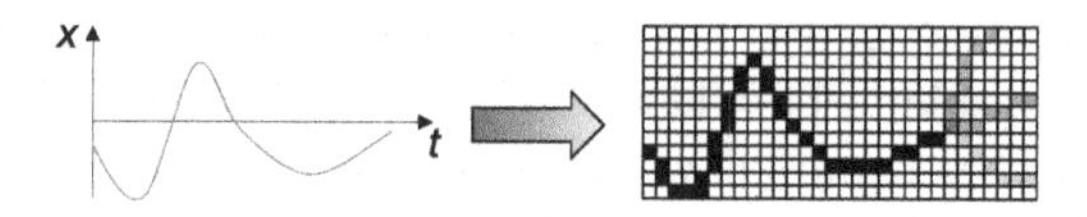

Fig. 1. Motion Trajectories are transferred into a grid representation. A grid cell is set to 1 if it is in the path of the trajectory and set to zero otherwise. During the prediction phase multiple hypotheses can be gained by superimposing several basis primitives. This is indicated with the gray trajectories on the right side of the grid.

handle each possible transformation using separate basis vectors. This is achieved by adding a transformation $\mathbf{T}$ to the decomposition formulation: $\mathbf{V} \approx \mathbf{T} \cdot \mathbf{W} \cdot \mathbf{H}$.

However the activation matrix $\mathbf{H}$ has to be adapted in a way that each possible transformation carries its own activation. Hence, the matrix $\mathbf{H}$ becomes an activation tensor $\mathbf{H}^{\mathbf{m}}$, while $\mathbf{m}$ is a vector describing the transformation parameters (rotation, scaling and translation).

$$\mathbf{V}_{\mathbf{i}} \approx \sum_{j} \sum_{\mathbf{m}} H_i^{j,\mathbf{m}} \cdot \mathbf{T}^{\mathbf{m}} \cdot \mathbf{W}_{\mathbf{j}} \tag{7}$$

For each allowed transformation the corresponding activity has to be trained individually.

3 Decomposing Motion Trajectories

For being able to decompose and to predict the trajectories of the surrounding dynamic objects, it is necessary to identify them and to follow their movements. For simplification, a tracker is assumed, which is able to provide such trajectories in real-time. A possible tracker to be used is presented in [7]. The given trajectory of the motion is now interpreted as a time series $\mathcal{T}$ with values $\boldsymbol{s}_i = (x_i, y_i, z_i)$ for time steps $i = 0, 1, \ldots, n-1$: $\mathcal{T} = (\mathbf{s}_0, \mathbf{s}_1, \ldots, \mathbf{s}_{n-1})$.

It is now possible to present the vector $\mathcal{T}$ directly to the NMF approach. But this could result in an unwanted behavior. Imagine two basis primitives, one representing a left turn and another representing a right turn. A superposition of those basis primitives would express a straight movement. However, we would need to express either a left or a right turn.

The goal is to have a set of basis primitives, which can be concatenated one after the other. Furthermore, it is necessary for a prediction task to be able to formulate multiple hypotheses. For achieving these goals, the x-t-trajectory is transferred into a grid representation, as it is shown in figure 1. Then, each grid cell (x_i, t_j) represents a certain state (spatial coordinate) x_i at a certain time t_j. Since most of the state-of-the-art navigation techniques rely on grid maps [8], the prediction can be integrated easily. This 2D-grid is now presented as image-like input to the NMF algorithm using the sparsity constraint as well as transformation invariance (See section 2.1 and 2.2 respectively). Using the grid representation of the trajectory also supports the non-negative character of the basis components and their activities.

It has to be mentioned, that the transformation to the grid representation is done for each of the dimensions individually. Hence, the spatio-temporal NMF

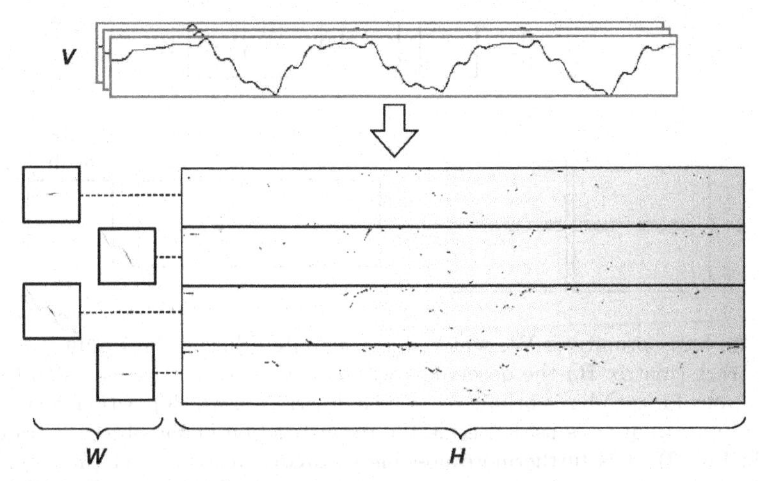

Fig. 2. Training with Spatio-Temporal NMF. Given is a set of training samples in matrix **V**. The described algorithm computes the weights **W** and the corresponding activities **H**. Only the weights are used as basis primitives for further processing.

has to be processed on each of these grids. Regarding each of the dimensions separately is often used to reduce the complexity of the analysis of trajectories (compare [9]). Theoretically, the algorithm could also handle multi-dimensional grid representation.

While applying an algorithm for basis decomposition to motion trajectories it seems to be clear that the motion primitives can undergo certain transformations to be combined to the whole trajectory. For example, the same basis primitive standing for a straight move can be concatenated with another one standing for a left turn. Hence, the turning left primitive has to be moved to the end of the straight line, and transformation invariance is needed while decomposing motion data. For our purposes, we concentrate on translation. This makes it possible to simplify the calculations and to achieve real time performance. Using only the grid like approach, adjacent grid cells are just different dimensions in the grid vector just as completely remote cells are. Adding translation invariance guarantees that adjacent cells have a neighbouring character.

It is known for each basis primitive,how they have to be combined to reconstruct the input data. Hence, one of the algorithm's outputs is a description of the sequence of the basis primitives which can be used as input for a classifier.

The sparse coding constraint helps to avoid trivial solutions. Since the input can be compared with a binary image, one possible solution would be a basis component with only a single grid cell filled. The trajectory is then simply copied into the activities.

3.1 Learning Phase

The goal of the learning phase is to gain a set of basis primitives which allow to decompose an observed and yet unknown trajectory (see Fig. 2). As it is

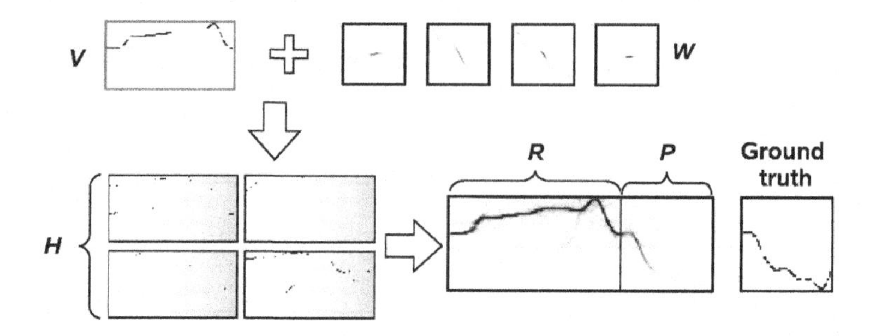

Fig. 3. The basis primitives **W**, which were computed during the training, are used to reconstruct (matrix **R**) the observed trajectory **V**. This results in a set of sparse activities – one for each basis primitive – which describe on which position in space and time a certain primitive is used. Beside the reconstruction of the observed trajectory (shown in Fig. 3), it is furthermore possible to predict a number of time steps into the future. Hence, the matrix **R** is extended by the prediction horizon **P**. Right to the prediction **P** the ground truth is plotted to get an idea of the prediction quality.

discussed in section 3, the training samples are transferred into a grid representation. These grid representations are taken as input for the NMF approach and are therefore represented in matrix **V**. On this matrix **V** the standard NMF approach, extended by the sparsity constraint and by translation invariance, is applied. The algorithm is summarized in Fig. 4.

Beside the computed basis primitives, the NMF algorithm also provides the information of how each of the training samples can be decomposed by the basis primitives. To be able classify the trajectory later, this information is used for training the classifier. Since trajectories and hence the calculated activities are temporal sequences, a recurrent neural network seems to be an adequate classifier. The network is trained by presenting a discrete time step of the activities (i.e. a single column in the matrix).

3.2 Application Phase

As it is indicated in Fig. 3, from the learning phase a set of motion primitives is extracted. During the application phase, we assume that the motion of a dynamic object (e. g. a person) is tracked continuously. For getting the input for the NMF algorithm, a sliding window approach is taken. A certain frame in time is transferred into the already discussed grid like representation. For this grid the activation of the basis primitives is determined. The algorithm is identical to the one sketched in Fig. 4 beside that steps 4 and 5 can be skipped.

The resulting activities are now used as input for the classifier, which was adapted to the activities in the learning phase. In this way, a classification result (e. g. the interest for interaction or the performed action) can be assigned to the regarded time window.

The standard approach to NMF implies that each new observation at the next time step demands a new random initialization for the optimization problem.

1. Normalize the basis vectors according to

$$\overline{\mathbf{W}}_j = \mathbf{W}_j \cdot \|\mathbf{W}_j\|^{-1} \quad (8)$$

2. Calculate the reconstruction

$$\mathbf{R}_i = \sum_j \sum_{\mathbf{m}} H_i^{j,\mathbf{m}} \mathbf{T}^{\mathbf{m}} \overline{\mathbf{W}}_j \quad (9)$$

3. Update the activities

$$H_i^{j,\mathbf{m}} \leftarrow H_i^{j,\mathbf{m}} \odot \left(\mathbf{V}_i^T \mathbf{T}^{\mathbf{m}} \overline{\mathbf{W}}_j \oslash \mathbf{R}_i^T \mathbf{T}^{\mathbf{m}} \overline{\mathbf{W}}_j \right) \quad (10)$$

4. Calculate the reconstruction with the new activities

$$\mathbf{R}_i = \sum_j \sum_{\mathbf{m}} H_i^{j,\mathbf{m}} \mathbf{T}^{\mathbf{m}} \overline{\mathbf{W}}_j \quad (11)$$

5. Update the basis vectors

$$\mathbf{W}_j \leftarrow \mathbf{W}_j \odot \frac{\sum_i \sum_{\mathbf{m}} H_i^{j,\mathbf{m}} \mathbf{V}_i^T \mathbf{T}^{\mathbf{m}} + \overline{\mathbf{W}}_j \overline{\mathbf{W}}_j^T \sum_i \sum_{\mathbf{m}} H_i^{j,\mathbf{m}} \mathbf{R}_i^T \mathbf{T}^{\mathbf{m}}}{\sum_i \sum_{\mathbf{m}} H_i^{j,\mathbf{m}} \mathbf{R}_i^T \mathbf{T}^{\mathbf{m}} + \overline{\mathbf{W}}_j \overline{\mathbf{W}}_j^T \sum_i \sum_{\mathbf{m}} H_i^{j,\mathbf{m}} \mathbf{V}_i^T \mathbf{T}^{\mathbf{m}}} \quad (12)$$

Fig. 4. Algorithmic description of the Spatio-temporal NMF

Since an increasing column number in the grid representation stands for an increase in time, the trajectory is shifted to the left while moving further in time. For identical initialization, the same shift is then reflected in the activities after the next convergence. To reduce the number of iterations until convergence, the shifted activities from the previous time step are used as initialization for the current one.

To fulfill the main goal discussed in this paper – the prediction of the observed trajectory into the future – the proposed algorithm had to be extended. Since the algorithm contains the transformation invariance constraint, the computed basis primitives can be translated to an arbitrary position on the grid. This means that they can also be moved in a way that they exceed the borders of the grid. Up to now, the size of reconstruction was chosen to be the same size as the input grid. Hence, using the standard approach means that the overlapping information has to be clipped. To be able to solve the prediction task, we simply extend the reconstruction grid to the right – or into the future (see Fig. 3). So, the previously clipped information is available for prediction.

As discussed, a classifier is trained during the learning phase using the activities of the training samples. In the application phase the activities for the current observation is calculated. The last activity column is used as input for the classifier. So a classification result is gained for each time step on-line.

4 Evaluation

Taking a closer look at the example scenario from introductory section 1 reveals that a robust identification and tracking of the single body parts is needed. To be

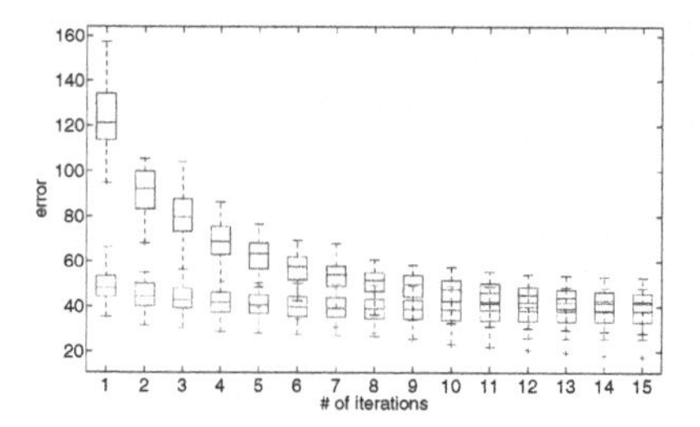

Fig. 5. Box whiskers plot showing the convergence characteristics of the energy function (see eqn. 1) for 15 iteration steps. For the upper (blue) plot the activities are initialized randomly after each shift of the input data. For the lower (red) curve the activities from the previous computations are shifted and used as initialization.

comparable and to avoid errors from the tracking system influencing the test results, movement data from the Perception Action Cognition Lab at the University of Glasgow [10] is used. The data contains trajectories from 30 persons is recorded performing different actions in different moods. The movement data has a resolution of 60 time steps per second, so that an average prediction of about 50 steps means a prediction of 0.83 seconds into the future. Since most trackers work with a lower resolution, a prediction further into the future is still possible.

In the next subsections, two aspects of the proposed algorithm are investigated in detail. First, it is shown that activity shifting brings a great benefit towards real time performance. Afterwards the focus is set to the quality of the prediction part.

For the experiments, the size of the basis primitives was chosen to be 50×50 grid cells The input grid size during the learning phase was set to 500×50 for each of the trajectories and to 100×50 during application phase.

4.1 Activity Shifting

In section 3.2 it has been mentioned that the information from the previous time step can be used as initialization for the current one. Figure 5 shows the energy function, which is defined in equation 1, for both possibilities of initialization. It is plotted only for a low number of iteration steps (up to 15), since already there the effect can be observed. For the upper (blue) plot a random initialization of the activities was used. For the lower (red) curve the activities from the previous computations are shifted and used as initialization. It can clearly be seen that the convergence is faster by a number of about 10 steps in average.

4.2 Prediction

For evaluating the quality of the prediction, the prediction is compared with the grid representation of the actual trajectory **G**. For each occupied grid cell the value of the columnwise normalized prediction is added. The sum is divided by the length of the trajectory:

$$S_{GT} = |T|^{-1} \sum_{t \in T} \mathbf{P}_t^T \cdot \mathbf{G}_t \cdot \left(\sum_i \mathbf{G}_t^i\right)^{-1} \quad (13)$$

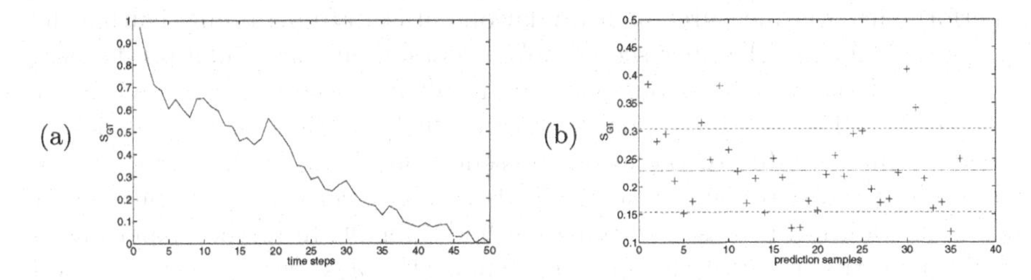

Fig. 6. (a) The mean correlation S_{GT} (see eqn. 13) between the ground truth trajectory and the prediction is plotted for each time step of the prediction horizon. A fit value of 1.0 stands for a perfect prediction over the whole prediction horizon. As it is expected the accuracy of the prediction decreases for a longer prediction period. (b) The plot shows the prediction accuracy for predictions along a sample trajectory. The 36 predictions were performed at each tenth time step of the chosen trajectory. A fit value of 1.0 stands for a perfect prediction over the whole prediction horizon. The constant and dotted lines (red) indicate mean and variance respectively.

The normalization of the prediction is done separately for each time slice (column in the grid).

The basis primitives can at most be shifted by their width out of the reconstruction grid **R**. So the maximal size of the prediction horizon equals the width of the basis primitives. Practically this maximum can not be reached, because the basis primitives need a reliable basis in the part where the input is known. Nevertheless, we have chosen to use the theoretical maximum as basis for the evaluation.

The results are depicted in Fig 6. The first plot (Fig. 6(a)) shows the expected decrease of the average prediction quality over the prediction horizon. Nevertheless, the decrease is smooth and no sudden collapses can be observed. For Fig. 6(b) an example trajectory has been selected for the reasons of clearness. The plot is intended to show how the algorithm behaves in practical applications. The predictions were performed at each tenth time step of the chosen trajectory. A fit value of 1.0 stands for a perfect prediction over the maximum prediction horizon, with only a single hypothesis for the prediction. The value decreases significantly with multiple hypotheses being present.

4.3 Classification Task

The goal of the classifier is to classify the current action performed by the proband (throw, walk, knock and lift). The input size is reduced by only using only a single limb (the right wrist). Using half of the data set the basic primitives were trained (twelve for each spatial dimension). As classifier we use a Layer-Recurrent Neural Network in MATLAB, which is a multi-layer version of the well known Elman network. The network was initialised with an input-size of 1800 neurons, 20 recurrent layers, sigmoid transfer function, and an output-size of 4 neurons (binary coded output, one neuron for each class). The classifier was trained using standard back-propagation with MSE. In order to evaluate the classifier during training, the error on a test dataset is computed. Additionally the performance was measured on a validation dataset after training. During the first experiments the classifier scored 40% accuracy on the validation dataset. The poor performance of the classifier may result from the still very high input dimensionality. The classifier is not able to profit from the sparse activation of the basic primitives due to the grid representation. Additionally the grid representation of the sparse input vector leads to many zeros in the input vector which makes a gradient based learning method difficult. In further experiments we will exploit the sparse input representation for the classifier using the position of the maximum activation. The compacter encoding of the input should result into better performance of the classifier.

5 Conclusion and Outlook

This paper presented a new approach for decomposing motion trajectories using non-negative matrix factorization. To solve this problem, sparsity constraint and transformation invariance have been combined. The trajectories were then decomposed using a grid-based representation. It could be demonstrated that the concept of activity shifting clearly decreases the number of iterations needed until convergence. Furthermore it was shown that the proposed algorithm is able to predict the motion into the future. The prediction occurs by a superposition of possible trajectory alternatives, yielding a quasi-probabilistic description. At this point, the information about the sparse activation of the basis primitives was used only for reconstruction purposes, even though it contains significant information about the global motion. The classification task is still insufficent and should be further improved as discussed. Furthermore, different kinds of networks need to be investigated. Nevertheless it could be shown exemplarily, how the suggested motion trajectory representation can be used to solve a classification task.

References

1. Hoffman, H., Schaal, S.: A computational model of human trajectory planning based on convergent flow fields. In: Meeting of the Soc. of Neuroscience (2007)
2. Cemgil, A., Kappen, B., Barber, D.: A generative model for music transcription. IEEE Transactions on Speech and Audio Processing 14, 679–694 (2006)
3. Lee, D.D., Seung, H.S.: Algorithms for non-negative matrix factorization. Advances in Neural Information Processing 13, 556–562 (2001)
4. Rajapakse, M., Wyse, L.: NMF vs ICA for face recognition. In: Guo, M. (ed.) ISPA 2003. LNCS, vol. 2745, pp. 605–610. Springer, Heidelberg (2003)
5. Eggert, J., Körner, E.: Sparse Coding and NMF. In: IJCNN, pp. 2529–2533 (2004)
6. Eggert, J., Wersing, H., Körner, E.: Transformation-invariant representation and NMF. In: IJCNN, pp. 2535–2539 (2004)
7. Otero, N., Knoop, S., Nehaniv, C., Syrdal, D., Dautenhahn, K., Dillmann, R.: Distribution and Recognition of Gestures in Human-Robot Interaction. In: ROMAN 2006, September 2006, pp. 103–110 (2006)
8. Elfes, A.: Using Occupancy Grids for Mobile Robot Perception and Navigation. Computer 12(6), 46–57 (1989)
9. Naftel, A., Khalid, S.: Classifying spatiotemporal object trajectories using unsupervised learning in the coefficient feature space. MM Syst. 12(3), 227–238 (2006)
10. http://paco.psy.gla.ac.uk/data_ptd.php

A Manifold Representation as Common Basis for Action Production and Recognition

Jan Steffen, Michael Pardowitz, and Helge Ritter

Neuroinformatics Group, Faculty of Technology, Bielefeld University, Germany
{jsteffen,mpardowi,helge}@techfak.uni-bielefeld.de

Abstract. In this paper, we first review our previous work in the domain of dextrous manipulation, where we introduced *Manipulation Manifolds* – a highly structured manifold representation of hand postures which lends itself to simple and robust manipulation control schemes.

Coming from this scenario, we then present our idea of how this generative system can be naturally extended to the recognition and segmentation of the represented movements providing the core representation for a combined system for action production and recognition.

1 Introduction

In the field of humanoid robotics, two of the key challenges are the production of naturally looking movements on the one hand and the recognition of observed movements or their segmentation into several meaningful subparts on the other hand. Whereas these two problems are complementary and usually addressed independently from each other, we believe that they are indeed closely related and that is beneficial to base their handling on one and the same core representation of the underlying – observed or produced – movements. Whereas such common basis for action and perception could not be established yet in the field of robotics, it is widely known from neurophysiology where research on monkey brains reports from mirror neurons in the premotor cortex which not only show activity during the monkey's own excitations but as well during observations of the same actions performed by another monkey (e.g. [7]).

With our previous work, we approached this problem from the production side using motion capture data from human demonstration. In this domain, many recent approaches focus on the Gaussian Processes Latent Variable Model (GPLVM, [4]) and variants. For example, Bitzer et al. [1] propose a methodology for learning and synthesising classes of movements using the GPLVM and identify robust latent space control policies which allow for generating novel movements. In another approach, Urtasun et al. [12] extend the GPLVM in order to learn interpretable latent directions and transitions between motion styles.

Whereas such approaches yield very promising results for reproducing and synthesising motion capture data, it is not clear how they can be extended for motion recognition. For our work, we were thus looking for a method that enables us to directly reinforce a clear and predefined structure of the latent variables

B. Mertsching, M. Hund, and Z. Aziz (Eds.): KI 2009, LNAI 5803, pp. 607–614, 2009.

Fig. 1. Example of a hand posture sequence corresponding to a training manipulation

which then lends itself to simple and robust control schemes. The knowledge about this clear structure – besides of using it for motion *generation* – then can also be exploited for *recognition* afterwards. In the context of manifold generation, we presented modifications to a recent approach to non-linear manifold learning, namely *Unsupervised Kernel Regression* (UKR, [5]), which either allow for directly incorporating prior knowledge in a constructive manner [8] or in an automatic learning scenario [9]. As shown in [8,9], the resulting *Manipulation Manifolds* then provide the desired highly structured latent spaces and can be used as basis for reproduction and synthesis of the represented movement class.

In this paper, we present our idea of how this generative system can be naturally extended to the recognition of the represented movements.

The paper is organised as follows: After a description of the training data (Sec. 2), we briefly review UKR (Sec.3) and the two methods to generate the *Manipulation Manifolds* (Sec. 4 and Sec. 5). Section 6 then describes the manifold characteristics which are exploited in Section 7 for the motion production. Section 8 finally presents our idea of a recognition system based on this representation followed by a conclusion in Section 9.

2 Manipulation Data

The training data consist of sequences of hand postures (each a 24D joint angle vectors) recorded with a data glove during cap turning movements for different cap radii ($r = 1.5cm$, $2.0cm$, $2.5cm$, $3.0cm$ and $3.5cm$) in a physics-based computer simulation (e.g. Fig. 1). For each radius, we produced five to nine sequences of about 30 to 45 hand postures each – in total 1204 for all sequences.

3 Unsupervised Kernel Regression

UKR is a recent approach to learning non-linear continuous manifolds, that is, finding a lower dimensional (latent) representation $\mathbf{X} = (\mathbf{x}_1, \ldots, \mathbf{x}_N) \in \mathbb{R}^{q \times N}$ of a set of observed data $\mathbf{Y} = (\mathbf{y}_1, \ldots, \mathbf{y}_N) \in \mathbb{R}^{d \times N}$ and a corresponding functional relationship $\mathbf{y} = \mathbf{f}(\mathbf{x})$. UKR has been introduced as the unsupervised counterpart of the Nadaraya-Watson kernel regression estimator by Meinecke et al. in [5]. Further development has lead to the inclusion of general loss functions, a landmark variant, and the generalisation to local polynomial regression [3]. In its basic form, UKR uses the Nadaraya-Watson estimator [6,13] as smooth mapping from latent to observed data space ($K_{\mathbf{H}}$: Kernel with bandwidth $\mathbf{H}$):

$$\mathbf{f}(\mathbf{x}) = \sum_{i=1}^{N} \mathbf{y}_i \frac{K_{\mathbf{H}}(\mathbf{x} - \mathbf{x}_i)}{\sum_j K_{\mathbf{H}}(\mathbf{x} - \mathbf{x}_j)} \tag{1}$$

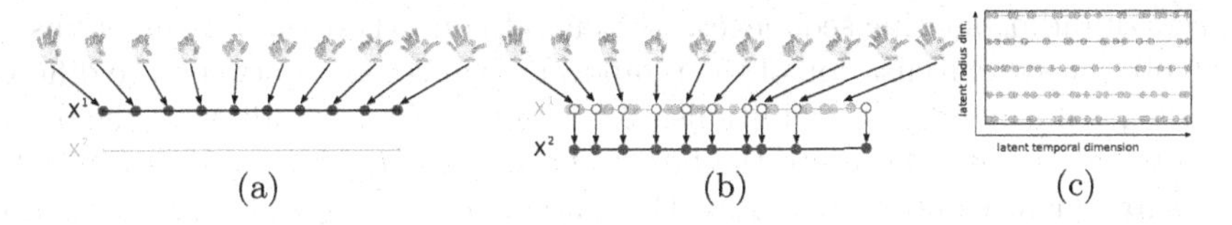

Fig. 2. Schematic description of different steps in the manifold construction process

$\mathbf{X} = \{\mathbf{x}_i\}, i = 1..N$ now plays the role of input data to the regression function (1) and is treated as set of *latent parameters* corresponding to $\mathbf{Y}$. As the scaling and positioning of the $\mathbf{x}_i$'s are free, the formerly crucial bandwidths $\mathbf{H}$ become irrelevant and can be set to 1.

UKR training, that is finding optimal latent variables $\mathbf{X}$, involves gradient-based minimisation of the reconstruction error:

$$R(\mathbf{X}) = \frac{1}{N}\sum_i \parallel \mathbf{y}_i - \mathbf{f}(\mathbf{x_i};\mathbf{X}) \parallel^2 = \frac{1}{N} \parallel \mathbf{Y} - \mathbf{YB}(\mathbf{X}) \parallel_F^2 . \tag{2}$$

Here, $\mathbf{B}(\mathbf{X})$ with $(\mathbf{B}(\mathbf{X}))_{\mathbf{ij}} = \frac{K(\mathbf{x}_i - \mathbf{x}_j)}{\sum_k K(\mathbf{x}_k - \mathbf{x}_j)}$ is an $N \times N$ *basis function matrix*.

To avoid poor local minima, i.e. PCA [2] or Isomap [11] can be used for initialisation. These eigenvector-based methods are quite powerful in uncovering low-dimensional structures by themselves. Contrary to UKR, however, PCA is restricted to linear structures and Isomap provides no continuous mapping.

To avoid a trivial solution by moving the $\mathbf{x}_i$ infinitively apart from each other ($\mathbf{B}(\mathbf{X})$ becoming the identity matrix), several regularisation methods are possible [3]. Most notably, leave-one-out cross-validation (LOO-CV: reconstructing each $\mathbf{y}_i$ without using itself) is efficiently realised by zeroing the diagonal of $\mathbf{B}(\mathbf{X})$ before normalising its column sums to 1. The inverse mapping $\mathbf{x} = \mathbf{f}^{-1}(\mathbf{y};\mathbf{X})$ can be approximated by $\mathbf{x}^\star = \mathbf{g}(\mathbf{y};\mathbf{X}) = \arg\min_{\mathbf{x}} \|\mathbf{y} - \mathbf{f}(\mathbf{x};\mathbf{X})\|^2$.

4 Manifold Construction

A simple but effective approach to generating a *Manipulation Manifold* is to construct the final manifold out of several sub-manifolds each realising a manipulation movement for one motion parameter (cp. [8]). In the example of turning a bottle cap, we incorporate the progress in time of the movement and the radius of the cap. The construction of the final manifold is performed iteratively starting with a sequence associated with the smallest radius. The latent parameters of the first 1D-UKR manifold are equidistantly distributed in a predefined interval according to the chronological order of the hand postures (Fig.2(a)). The second sequence of the same radius then is projected pointwise into the latent space of the previous 1D-manifold. By dint of this projection, we approximate a synchronisation of the temporal advance of the two movements. In the next step, we combine those data to a new UKR manifold by extending the sets of observed data and latent parameters by the new sequence data (cp. Sec.3: $\mathbf{Y}$ and $\mathbf{X}$).

Repeating this step for all sequences of one radius yields a radius-specific 1D manifold representing a generalised movement. Thus, by applying this method to all sets of radius-specific sequences, we generate one 1D manifold per radius. To promote the synchronisation of the temporal advances also between the different radius-specific manifolds, we initialise the manifolds of new radii with the projection of the observed data into the latent space of the previous radius (cp. 2(b)). The subsequent combination of all 1D-manifolds to one 2D-manifold representing the complete movements for all training radii then is realised by expanding each 1D latent parameter by a second dimension denoting the appropriate radius corresponding to the associated training sequence (2(c)).

Using the radius information automatically results in the correct ordering of the latent parameters in the new dimension. Whereas this last step always requires meta knowledge about the training data, at the same time, it provides a simple and effective way of incorporating prior knowledge into the manifold generation procedure. Another benefit of directly using observed or predefined meta data (like the cap size) as values of latent parameter dimensions is that new data recorded after the initial training can directly be added to the manifold in the same way and then serve to locally refine the manifold structure.

5 Manifold Learning

In several cases, it is desirable to automatically learn the *Manipulation Manifolds* from training sequences instead of construct them in the described way. We thus presented extensions to original UKR training to provide implicit mechanisms for incorporating given knowledge about training data structures [10].

To take the periodic nature of the data sequences into account, we there further extended original UKR by allowing for different univariate kernels K_l (with dimension-specific parameters Θ_l) for different latent dimensions l (cp. Eq.2):

$$(\mathbf{B}(\mathbf{X}))_{\mathrm{ij}} = \frac{\prod_{l=1}^{q} K_l(x_{i,l} - x_{j,l}; \mathbf{\Theta}_l)}{\sum_k^N \prod_{l=1}^{q} K_l(x_{k,l} - x_{j,l}; \mathbf{\Theta}_l)}. \tag{3}$$

As kernel for the non-periodic dimensions, we use a standard Gaussian kernel with (inverse) bandwidth parameter Θ: $K_g(x_i - x_j; \Theta) = \exp\left[-\frac{1}{2}\Theta^2(x_i - x_j)^2\right]$, and for the periodic dimensions, we proposed a sin^2 kernel, periodic in $[0; \pi]$, again with parameter Θ: $K_{\circlearrowleft}(x_i - x_j; \Theta) = \exp\left[-\frac{1}{2}\Theta^2 \sin^2(x_i - x_j)\right]$.

The two key features provided by the construction described in the last section are that (a) the chronological order of the training sequences is reflected in the corresponding latent variables and (b) the latent representations of the training sequences have constant values in the latent radius dimension (assuming that the underlying movement parameters do not change within the sequences.)

In the automatic learning case, we approximate these two features by means of penalty terms to the standard loss function (Eq.2):

(a) In the periodic case using *closed* sequences of training data ($\mathbf{x}_0^\sigma = \mathbf{x}_{N_\sigma}^\sigma$), we can express this as regularisation of the sum of successor distances:

$$E_{cseq}(\mathbf{X}) = \sum_{\sigma=1}^{N_S} \sum_{i=1}^{N_\sigma} \sin^2(x_{i,d_t}^\sigma - x_{(i-1),d_t}^\sigma). \tag{4}$$

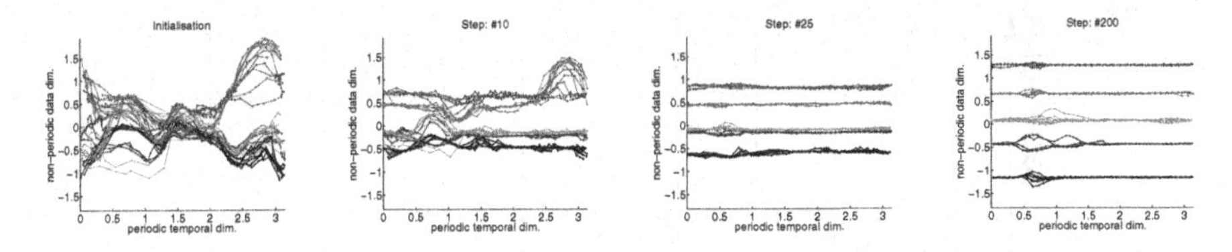

Fig. 3. Development of UKR latent variables after 0, 10, 25, and 200 steps. Connected points represent training sequences; different colours correspond to different cap radii

for sequences $\mathcal{S}_\sigma = (\mathbf{y}_1^\sigma, \mathbf{y}_2^\sigma, \ldots, \mathbf{y}_{N_\sigma}^\sigma)$, $\sigma = 1..N_S$ with corresponding latent parameters $(\mathbf{x}_1^\sigma, \mathbf{x}_2^\sigma, \ldots, \mathbf{x}_{N_\sigma}^\sigma)$. d_t denotes the latent time dimension.

(b) is realised by penalising high variances in the parameter dimensions $k \neq d_t$:

$$E_{pvar}(\mathbf{X}) = \sum_{\sigma=1}^{N_S} \sum_{k \neq d_t} \frac{1}{N_\sigma} \sum_{i=1}^{N_\sigma} \left(x_{i,k}^\sigma - \langle x_{\cdot,k}^\sigma \rangle\right)^2 \tag{5}$$

The resulting overall loss function then can be denoted as:

$$E(\mathbf{X}) = R(\mathbf{X}) + \lambda_{cseq} \cdot E_{cseq}(\mathbf{X}) + \lambda_{pvar} \cdot E_{pvar}(\mathbf{X}). \tag{6}$$

Fig.3 visualises an exemplary development of the latent variables using this method and the training data described in Sec.2. For further details see [10,9].

6 Characteristics of the Manipulation Manifold

Figure 4 visualises an exemplary *Manipulation Manifold*. As result of the learning (or construction), the horizontal dimension corresponds to the temporal aspect of the movement and the vertical dimension describes the cap size as motion parameter (please consider the video referenced in Fig. 4). Like this, it forms a representation of the movement of turning a bottle cap for different cap sizes that fulfils our goal of fitting to the desired simple control strategy: the represented movement can be produced by projecting a linear trajectory that follows the time dimension in latent space into hand posture space.

This characteristics is realised by distorting the natural topology of the latent space such that those parts of the movement which are independent of the cap radius – and thus very similar for all radii (i.e. Fig. 4, the backward movement of the hand: columns 1-3) – are pushed away from each other to span the same latent radius range as the rest of the sequence. In contrast to this, with purely unsupervised learning, the similar parts would collapse to thin regions in latent space. This however would make the targeted control scheme impossible.

Indeed, whereas the distortion is beneficial for the production of motions, it poses some problems for the inverse direction, i.e. projecting hand posture sequences into latent space. In that case, whereas the temporal information is robust, the projection is strongly non-robust in the parameter (radius) dimension in the described radius-independent parts as the corresponding hand postures are fairly similar for the whole range of latent radii for a specific point in time and the result of $\mathbf{g}(\mathbf{y})$ can heavily vary for small changes of $\mathbf{y}$.

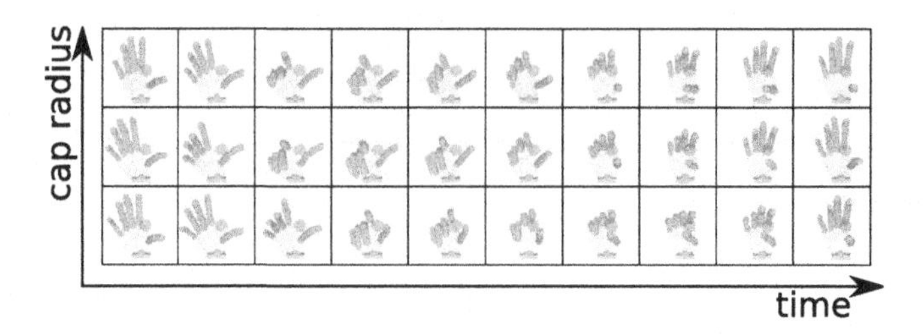

Fig. 4. Visualisation of the training result in the hand posture space. The depicted postures correspond to the reprojections $\mathbf{f}(\mathbf{x};\mathbf{X})$ of regularly sampled positions $\mathbf{x}$ in the trained latent space. Please consider also the video available under `http://www.techfak.uni-bielefeld.de/~jsteffen/mov/ki2009/upkrturn/`

7 Motion Production and Synthesis

The clear structure of the latent space enables the use of a very simple controller in order to synthesise the represented movements: The algorithm starts in an initial hand posture corresponding to a fixed latent position on the 'maximum radius' border of the latent space in a temporal position where the fingers have contact with the cap. The motion controller then is subdivided into two different phases of orthogonal, straight navigations through the latent space: (a) Grasping the cap is realised by a straight navigation in direction of decreasing radii following the radius dimension until thumb, fore finger and middle finger have contact. (b) The manipulation – during which the adapted radius is fixed – is performed by navigating through the latent space following the temporal dimension. Please consider also the corresponding video[1] and [9] for further details.

8 Towards Motion Recognition and Segmentation

The recognition approach takes the inverse direction to the motion production described in the last section: instead of projecting latent trajectories into hand posture space in order to determine a sequence of intermediate target hand postures, we now observe such sequences and use it as input. By projecting them into latent space ($\mathbf{g}(\cdot)$) and back to hand posture space ($\mathbf{f}(\mathbf{g}(\cdot))$, cp. Sec. 3), we obtain means to define features which express the degree of compatibility of the observed postures and the manifold:

a) The compatibility of single observations with the manifold can be expressed with the self-reconstruction error of the observations yielding a measure for the similarity of observed and best-matching represented posture in the manifold:

$$C_{rec}(\mathbf{y}^\star;\mathbf{Y}) = -1 + 2 \cdot \exp\left(-\Delta^T \Delta\right) \in [-1;1] \tag{7}$$

where $\Delta(\mathbf{y}^\star) = \mathbf{y}^\star - \mathbf{f}(\mathbf{g}(\mathbf{y}^\star))$ is the self-reconstruction error of observation $\mathbf{y}^\star$.

b) The *temporal* compatibility of a single observation with its preceding observations (*history*) can be expressed by the relative positions of the representations

[1] http://www.techfak.uni-bielefeld.de/∼jsteffen/mov/ki2009/upkrmanip/

of the current and preceding observations in latent space: if the single observations of the input sequence are compatible with the manifold (in the sense of (a)), then, the compatibility of the chronological order of their latent representations with the manifold can be expressed as the sum over the distances between projections of successive data. As a measure for the compatibility of the observation $\mathbf{y}_{t-h}$ in the history of $\mathbf{y}_t$, we thus define:

$$c_{hist}(h,t) = \frac{1}{2}\cos(\delta_{h,t}) + \frac{1}{2}C_{rec}(\mathbf{y}_{t-h};\mathbf{Y}) \tag{8}$$

where $\delta_{h,t} = mod_{\pi}(\mathbf{g}(\mathbf{y_{t-h-1}}) - \mathbf{g}(\mathbf{y_{t-h}}))$ is the *directed* temporal difference of the latent space projections $\mathbf{g}(\cdot)$ of the historic observation $\mathbf{y}_{t-h}$ and its predecessor $\mathbf{y}_{t-h-1}$ (taking the period π of the dimension into account). C_{rec} again is the self-reconstruction error described in (a).

For the compatibility of the whole history of $\mathbf{y}_t$ of length H, we define:

$$C_{hist}(H,t) = \frac{\sum_{h=1}^{H} \gamma^h c_{hist}(h,t)}{\sum_{h=1}^{H} \gamma^h} \tag{9}$$

where $\gamma \in [0;1]$ is the discount factor for historic observations. As C_{rec}, C_{hist} can take values in $[-1;+1]$ whereas -1 corresponds to maximally incompatible and $+1$ to maximally compatible with the underlying UKR manifold.

The combination of (a) and (b) with $\lambda \in [0;1]$ to an overall compatibility measures yields:

$$C = \lambda C_{rec} + (1-\lambda)C_{hist} \in [-1;+1] \tag{10}$$

Like this, C gives a measure for the compatibility of the observation together with its history with the underlying manifold. In other words, C realises a measure to quantify the appropriateness of the candidate manifold to reproduce the observation and the history. By means of λ, the importance of single observations versus observation history can be specified according to the requirements of the task. The classification of the observation to one of several candidate classes then can be realised as a winner-takes-all mechanism that works on the results of all UKR manifolds.

The special strength of this approach is that this compatibility is a pointwise measure (incorporating few historic hand postures) and is thus independent of a fixed data window. In addition, this enables the method to work on inhomogeneous data sequences which consist of more than one movement and hence enables the use for a segmentation of such sequences into several candidate motions or even only motion parts.

9 Conclusion

In the field of humanoid robotics, two of the key challenges are the production of naturally looking movements on the one hand and the recognition of observed movements or their segmentation into several meaningful subparts on the other hand. While these problems are basically complementary, they are closely connected to each other. In this paper, we presented our idea to base both task on one and the same core representation – namely the *Manipulation Manifolds* consisting of Structured UKR manifolds – in order to benefit from such common representation and express the inherent connection of the tasks also on the level of their representations.

After a revision of the basic method and the generation and use of the manifolds for the action production part of the system presented in previous work, we then presented our basic plan to perform recognition and segmentation tasks on the basis of the same representation. For this part, only initial evaluation has been done. Indeed, the results are very promising and we are convinced that an elaborate evaluation will help us to further refine our approach.

Acknowledgement. This work has been carried out with support from the German Collaborative Research Centre "SFB 673 - Alignment in Communication" granted by the DFG and from the German Cluster of Excellence 277 "Cognitive Interaction Technology" (CITEC).

References

1. Bitzer, S., Havoutis, I., Vijayakumar, S.: Synthesising Novel Movements through Latent Space Modulation of Scalable Control Policies. In: Asada, M., Hallam, J.C.T., Meyer, J.-A., Tani, J. (eds.) SAB 2008. LNCS (LNAI), vol. 5040, pp. 199–209. Springer, Heidelberg (2008)
2. Jolliffe, I.T.: Principal Component Analysis, 2nd edn. Springer, New York (2002)
3. Klanke, S.: Learning Manifolds with the Parametrized Self-Organizing Map and Unsupervised Kernel Regression. PhD thesis, Bielefeld University (2007)
4. Lawrence, N.: Probabilistic Non-linear Principal Component Analysis with Gaussian Process Latent Variable Models. Machine Learning Research 6 (2005)
5. Meinicke, P., Klanke, S., Memisevic, R., Ritter, H.: Principal Surfaces from Unsupervised Kernel Regression. IEEE Trans. PAMI 27(9) (2005)
6. Nadaraya, E.A.: On Estimating Regression. Theory of Probability and Its Appl. (9) (1964)
7. Rizzolatti, G., Fabbri-Destro, M., Cattaneo, L.: Mirror neurons and their clinical relevance. Nat. Clin. Pract. Neuro. 5(1), 24–34 (2009)
8. Steffen, J., Haschke, R., Ritter, H.: Towards Dextrous Manipulation Using Manifolds. In: Proc. Int. Conf. on Intelligent Robots and Systems, IROS (2008)
9. Steffen, J., Klanke, S., Vijayakumar, S., Ritter, H.: Realising Dextrous Manipulation with Structured Manifolds using Unsupervised Kernel Regression with Structural Hints. In: ICRA Workshop: Approaches to Sensorimotor Learning on Humanoid Robots (May 2009) (to appear)
10. Steffen, J., Klanke, S., Vijayakumar, S., Ritter, H.: Towards Semi-supervised Manifold Learning: UKR with Structural Hints. In: Proc. WSOM (June 2009) (ta)
11. Tenenbaum, J.B., de Silva, V., Langford, J.C.: A global geometric framework for nonlinear dimensionality reduction. Science 290(5500), 2319–2323 (2000)
12. Urtasun, R., Fleet, D., Geiger, A., Popovic, J., Darrell, T., Lawrence, N.: Topologically-Constrained Latent Variable Models. In: Proc. ICML (2008)
13. Watson, G.S.: Smooth Regression Analysis. Sankhya, Ser. A 26 (1964)

An Argumentation-Based Approach to Handling Inconsistencies in DL-Lite*

Xiaowang Zhang[1,2] and Zuoquan Lin[1]

[1] School of Mathematical Sciences, Peking University, Beijing 100871, China
[2] School of Mathematical Sciences, Anhui University, Hefei 230039, China
{zxw,lzq}@is.pku.edu.cn

Abstract. As a tractable description logic, DL-Lite provides a good compromise between expressive power and computational complexity of inference. It is therefore important to study ways of handling inconsistencies in tractable DL-Lite based ontologies, as classical description logics break down in the presence of inconsistent knowledge bases. In this paper, we present an argumentation-based approach to dealing with inconsistent DL-Lite based ontologies. Furthermore, we mainly develop a graph-based algorithm to implement paraconsistent reasoning in DL-Lite.

1 Introduction

Description logic Lite (DL-Lite) presented by D.Calvanese et al in [1] is a fragment of expressive description logics (DLs) which are the logical foundation of current ontology languages such as *Ontology Web Language* (OWL) (see [2]) with assertions and inverse roles. DL-Lite has ability enough to capture the main notions of conceptual modeling formalism used in databases and software engineering such as *Entity-Relationship Model* (ERM[1]) and *Unified Modeling Language* (UML[2]) class diagrams(see [1]). It is usual that DL-ontologies contain classically inconsistent information because many reasons possibly cause the occurrence of inconsistencies such as modeling errors, migration from other formalisms, merging ontologies and ontologies evolution (see [3]). However, DL-Lite fails to tolerate inconsistencies. Therefore, handling inconsistencies in DL-Lite becomes an important issue in the field of artificial intelligence in recent years.

Roughly, there are two fundamentally different approaches to handling inconsistencies in DLs. The first is based on the assumption that inconsistencies indicate erroneous data. Based on this assumption, inconsistencies should be repaired in order to obtain a consistent DL-knowledge base or (ontology) [4,5]. Unfortunately, we may lose useful information so that we might not obtain more trustworthy conclusions from those inconsistent information. Another approach, called *paraconsistent approach*, is not to simply avoid the inconsistencies but to apply a non-standard reasoning method to obtain

* This work is supported by the major program of National Natural Science Foundation of China.

[1] http://www.csc.lsu.edu/~chen/

[2] http://www.omg.org/technology/documents/formal/uml.htm

B. Mertsching, M. Hund, and Z. Aziz (Eds.): KI 2009, LNAI 5803, pp. 615–622, 2009.

meaningful answers [6,7]. A more prominent one of them is based on the use of additional truth values standing for *underdefined* (i.e. neither true nor false) and *overdefined* (or *contradictory*). However, the capability of those paraconsistent reasoning are far weaker. For instance, there exists some inference rules which might not be valid non-standard reasoning. Moreover, though there are some different characteristics among those existing approaches, the common problem is that inconsistencies are not analyzed in detail further but either isolated or ignored in reasoning. In philosophy, there is a classical theory called *argumentation theory (or argumentation)*, which embraces the arts and sciences of civil debate, dialogue, conversation and persuasion. P.Dung researches the fundamental mechanism, humans use in argumentation, and explores ways in computer science (see [8]). M.Elvang and A.Hunter [9] have presented an argumentative framework which is different from P.Dung's argumentative framework [8] to resolve inconsistencies in propositional logic. In recent years, there are some work about introducing argumentation theory into ontologies. For instance, s C.Tempich et al [10] apply argumentation theory to deal with conflicts occurring in *Distributed Ontologies* and S.A.Gomezte al [11] employ argumentation theory to handle inconsistencies in *Defeasible Logic Programs* (DeLP)[3]. In addition, D.Grossi [12] applies well-investigated *modal logics*[4] to provide formal foundations to specific fragments of argumentation theory.

In this paper, based on X.Zhang et al [13], we introduce M.Elvang and A.Hunter's *argumentation theory* for DL-Lite to handle inconsistencies. The main innovations and contributions of this paper can be summarized as follows:

- defining semantically the notion called *quasi-negation* to characterize the semantic reverse of axioms in DLs; and introducing the argumentation theory for DL-Lite by defining some basic notions, namely, *argument, undercut, maximal conservative undercut* and *canonical undercut*;
- presenting an argumentative framework, which is composed of canonical undercuts, to demonstrate the structure of arguments in ontologies; and providing an argumentative semantics based on *binary argumentation* for DL-Lite and discussing two basic inference problems, namely, satisfiability of concepts and inference problems based on our argumentative semantics in DL-Lite;
- developing a graph-based algorithm to search arguments in DL-Lite.

This paper is structured as follows. Section 2 reviews briefly DL-Lite. Section 3 introduces argumentative semantics for DL-Lite. Section 4 develops a graph-based approach to searching arguments in DL-Lite ontologies. Section 5 concludes this paper and discusses our future work. Due to the space limitation, all proofs and applied examples are omitted but they are available in a TR[5].

2 Preliminaries

In this section, we briefly introduce DL-Lite$_{core}$ and querying problems in DL-Lite ontologies because DL-Lite$_{core}$ is the core language in the family of DL-Lite (see[1]).

[3] http://cs.uns.edu.ar/~ajg/

[4] http://plato.stanford.edu/entries/logic-modal/

[5] http://www.is.pku.edu.cn/~zxw/publication/TRADLLite.pdf

In DL-Lite$_{core}$, basic concepts are defined as follows:

$$B := A \mid \exists R \qquad R := P \mid P^- \qquad C := B \mid \neg B \qquad E := R \mid \neg R$$

where A is a concept name; P is an atomic role; and P^- is the inverse role of the atomic role P respectively.

A DL-Lite$_{core}$ *ontology* $\mathcal{O}$ contains two parts , namely *ABox* and *TBox*, where ABox is used to represent extensional information and TBox is used to represent intensional knowledge. DL-Lite$_{core}$ TBox axioms are of the form $B \sqsubseteq C$ and DL-Lite$_{core}$ ABox axioms are of the form $C(a)$, $R(a,b)$ where a, b are constants.

An interpretation $\mathcal{I} = (\Delta^{\mathcal{I}}, \cdot^{\mathcal{I}})$ consists of a first structure over $\Delta^{\mathcal{I}}$ with an interpretation function $\cdot^{\mathcal{I}}$ mapping a concept to a subset of $\Delta^{\mathcal{I}}$, a role name to pair of a subset of $\Delta^{\mathcal{I}}$ and an individual to a member of $\Delta^{\mathcal{I}}$ and satisfying the following equations:
(1) *negation*: $(\neg B)^{\mathcal{I}} = \Delta^{\mathcal{I}} \setminus B^{\mathcal{I}}$;
(2) *inverse role*: $(P^-)^{\mathcal{I}} = \{(b^{\mathcal{I}}, a^{\mathcal{I}}) | (a^{\mathcal{I}}, b^{\mathcal{I}}) \in P^{\mathcal{I}}\}$;
(3) *exists restriction*: $(\exists P)^{\mathcal{I}} = \{x \mid \exists y, (x,y) \in P^{\mathcal{I}}\}$.

An interpretation $\mathcal{I}$ is a model of an/a inclusion axiom (concept axiom, role axiom) $B \sqsubseteq C$ ($C(a)$, $R(a,b)$) if $B^{\mathcal{I}} \subseteq C^{\mathcal{I}}$ ($a^{\mathcal{I}} \in C^{C}$, $(a^{\mathcal{I}}, b^{\mathcal{I}}) \in R^{\mathcal{I}}$) , denoted by $\mathcal{I} \models B \sqsubseteq C$ ($\mathcal{I} \models C(a)$, $\mathcal{I} \models R(a,b)$) respectively. An interpretation is a model of TBox $\mathcal{T}$ (ABox $\mathcal{A}$) if the interpretation is a model of all axioms in $\mathcal{T}$ ($\mathcal{A}$), denoted by $\mathcal{I} \models \mathcal{T}$ ($\mathcal{I} \models \mathcal{A}$) respectively. An ABox is *consistent* if it has at least one model, *inconsistent* otherwise. An ABox is *coherent* w.r.t. a TBox if any model of the TBox is a model of the ABox, *incoherent* otherwise. It easily shows that if an ABox is coherent w.r.t. a TBox then there exists a $B_1 \sqsubseteq \neg B_2$ in the TBox and an individual a such that the axioms $B_1(a)$ and $B_2(a)$ belong to the ABox. An ontology is *inconsistent* iff either its ABox is inconsistent or its ABox is incoherent w.r.t. its TBox. An ontology $\mathcal{O}$ *entails* an axiom α iff any model of $\mathcal{O}$ is a model of α, denoted by $\mathcal{O} \models \alpha$.

In DL-Lite, the *unique name assumption* (UNA) means that different names always refer to different entities, that is to say, $a \neq b$ iff $a^{\mathcal{I}} \neq b^{\mathcal{I}}$ for any interpretation $\mathcal{I}$. In this paper, we base on UNA.

In DL-Lite, there are two basic entailment problems as follows: (1) *instance checking*: an individual a in an ABox $\mathcal{A}$ is an instance of a concept C w.r.t. a TBox $\mathcal{T}$ if $a^{\mathcal{I}} \in C^{\mathcal{I}}$ for all models $\mathcal{I}$ of $\mathcal{A}$ and $\mathcal{T}$ (denoted with $\mathcal{A} \models_{\mathcal{T}} C(a)$); and (2) *subsumption*: a concept B is subsumed by a concept C w.r.t. a TBox $\mathcal{T}$ if $B^{\mathcal{I}} \subseteq C^{\mathcal{I}}$ for every model $\mathcal{I}$ of $\mathcal{T}$. Based on practical quality of DL-Lite, there are five subsumption problems which are also important in ER and UML, namely, *ISA* ($A_1 \sqsubseteq A_2$), *class disjointness* ($A_1 \sqsubseteq \neg A_2$), *role-typing* ($\exists P \sqsubseteq A_1$ or $\exists P^- \sqsubseteq A_2$), *participation constraints* ($A \sqsubseteq \exists P$ or $A \sqsubseteq \exists P^-$), *non-participation constraints* ($A \sqsubseteq \neg\exists P$ or $A \sqsubseteq \neg\exists P^-$).

A *conjunctive query* $\mathbf{q}(\mathbf{x}) \leftarrow \exists \mathbf{y}.conj(\mathbf{x},\mathbf{y})$, where $conj(\mathbf{x},\mathbf{y})$ is conjunction of atoms whose predicates are atomic concepts and roles of an ontology $\mathcal{O}$, and whose variables are among $\mathbf{x}$ and $\mathbf{y}$, is interpreted in $\mathcal{I}$ as the set $\mathbf{q}^{\mathcal{I}}$ of the tuples $\mathbf{c} \in \Delta^{\mathcal{I}} \times \cdots \times \Delta^{\mathcal{I}}$ such that, when we substitute the variables $\mathbf{x}$ with the constants $\mathbf{c}$, the formula $\exists \mathbf{y}.conj(\mathbf{x},\mathbf{y})$ evaluates to true in $\mathcal{I}$. $\mathbf{t}$ is a certain answer to $\mathbf{q}$ in $\mathcal{O}$, if $\mathcal{O} \models \mathbf{q}(\mathbf{t})$ where $\mathbf{q}(\mathbf{t})$ is the sentence obtained from the body of $\mathbf{q}$ by replacing its distinguishes variables by contains in $\mathbf{t}$. The set of certain answer to $\mathbf{q}$ in $\mathcal{O}$ is denoted by $ans(\mathbf{q}, \mathcal{O})$.

3 Argumentation for DL-Lite

In the section, based on X.Zhang et al [13], we introduce argumentation theory for DL-Lite. Firstly, we introduce some basic definitions of argumentation for DL-Lite.

Let α and β be axioms and $\mathcal{O}$ be an ontology in DL-Lite. In DLs, there are two logical connectives "*conjunction*" ($\wedge$) and "*disjunction*" ($\vee$) defined as follows: (a) $\mathcal{O} \models \alpha \wedge \beta$ iff $\mathcal{O} \models \alpha$ and $\mathcal{O} \models \beta$; and (b)$\mathcal{O} \models \alpha \vee \beta$ iff $\mathcal{O} \models \alpha$ or $\mathcal{O} \models \beta$. If α, β be axioms in DL-Lite, then $\alpha \wedge \beta$ (or $\alpha \vee \beta$) is called *conjunction of axioms* (or *disjunction of axiom*).

Definition 1. *Given a DL-Lite ontology $\mathcal{O}$, concepts B, C, an individual a and two axioms α, β, a* **quasi-negation**, *written by $\sim$ on axioms or disjunction (conjunction) of axioms is defined as follows:*

(1) $\mathcal{O} \models\sim C(a)$ *iff* $\mathcal{O} \models \neg C(a)$; (2) $\mathcal{O} \models\sim R(a,b)$ *iff* $\mathcal{O} \models \neg R(a,b)$;
(3) $\mathcal{O} \models\sim (B \sqsubseteq C)$ *iff* $\mathcal{O} \models B \sqcap \neg C(\iota)$ *for some individual* ι *in* $\mathcal{O}$;
(4) $\mathcal{O} \models\sim (\alpha \wedge \beta)$ *iff* $\mathcal{O} \models\sim \alpha \vee \sim \beta$; (5) $\mathcal{O} \models\sim (\alpha \vee \beta)$ *iff* $\mathcal{O} \models\sim \alpha \wedge \sim \beta$.

Axioms, conjunction of axioms, disjunction of axioms and their quasi-negation are called *extended axioms*. In this paper, extended axioms are not members of DL-Lite ontologies but are taken as the consequence of arguments (defined later).

3.1 Arguments in DL-Lite

Definition 2. *Let Φ be a set of axioms and α be an extended axiom in DL-Lite. An* argument *is a pair $\langle\Phi, \alpha\rangle$, denoted by $\mathbf{A} = \langle\Phi, \alpha\rangle$, such that:* (1) *$\Phi$ is consistent;* (2) *$\Phi \models \alpha$; and* (3) *no $\Phi' \subset \Phi$ satisfies $\Phi' \models \alpha$. We say that $\langle\Phi, \alpha\rangle$ is an argument for α. We call α the* consequence *of the argument and Φ the* support *of the argument, denoted by $Support(\langle\Phi, \alpha\rangle) = \Phi$ and $Consequence(\langle\Phi, \alpha\rangle) = \alpha$.*

An argument $\langle\Phi, \alpha\rangle$ is *more conservative* than an argument $\langle\Psi, \beta\rangle$ iff $\Phi \subseteq \Psi$ and $\beta \models \alpha$, written by $\langle\Psi, \beta\rangle \preceq_c \langle\Phi, \alpha\rangle$. If $\alpha \not\models \beta$ then $\langle\Phi, \alpha\rangle$ is called *strictly more conservative* than an argument $\langle\Psi, \beta\rangle$, written by $\langle\Psi, \beta\rangle \prec_c \langle\Phi, \alpha\rangle$. In short, we define a pre-order relationship on arguments using conservative relationship. An argument $\langle\Psi, \beta\rangle$ is a *defeater* of an argument $\langle\Phi, \alpha\rangle$ such that $\beta \models\sim (\phi_1 \wedge \ldots \wedge \phi_n)$ for some $\{\phi_1, \ldots, \phi_n\} \subseteq \Phi$. An *undercut* for an argument $\langle\Phi, \alpha\rangle$ is an argument $\langle\Psi, \sim (\phi_1 \wedge \ldots \wedge \phi_n)\rangle$ where $\{\phi_1, \ldots, \phi_n\} \subseteq \Phi$. Two arguments $\langle\Phi, \alpha\rangle$ and $\langle\Psi, \beta\rangle$ are *equivalent* iff Φ is logically equivalent to Ψ and α is logically equivalent to β. Clearly, if two arguments are equivalent then either each is more conservative than the other or neither is. $\mathbf{A}'$ is a *maximally conservative defeater* of $\mathbf{A}$ iff $\mathbf{A}'$ is a defeater of $\mathbf{A}$ for any defeater $\mathbf{A}''$ for $\mathbf{A}$ such that $\mathbf{A}'' \prec_c \mathbf{A}'$. A *maximally conservative undercut* of $\langle\Phi, \alpha\rangle$ is defined analogously. An argument $\mathbf{A}'$ is a *canonical undercut* of $\mathbf{A}$ iff $\mathbf{A}'$ is a maximally conservative undercut of $\mathbf{A}$.

The following theorem availably provides a norm form of undercuts via *quasi-negation*.

Theorem 1. *$\langle\Psi, \sim (\phi_1 \wedge \cdots \wedge \phi_n)\rangle$ is a canonical undercut for an argument $\langle\Phi, \alpha\rangle$ iff it is an undercut for $\langle\Phi, \alpha\rangle$ and $\{\phi_1, \ldots, \phi_n\}$ is the enumeration of Φ.*

3.2 Argumentative Semantics

Given α is an axiom, an *argument tree* for α is a tree where the nodes are arguments such that (a) the root is an argument for α; (b) for no node $\langle \Phi, \beta \rangle$ with ancestor nodes $\langle \Phi_1, \beta_1 \rangle \dots \langle \Phi_n, \beta_n \rangle$ is Φ a subset of $\Phi_1 \cup \dots \cup \Phi_n$; and (c) the children nodes of a node **A** consist of all canonical undercuts for **A** that obeys item (b). An *argumentative framework* of an axiom α is a pair of sets $\langle \mathcal{P}, \mathcal{C} \rangle$ where $\mathcal{P}$ is the set of argument trees of α and $\mathcal{C}$ is the set of argument trees for $\sim \alpha$.

If $\mathbf{A}_1, \mathbf{A}_2$ and $\mathbf{A}_3$ are three arguments such that $\mathbf{A}_1$ is undercut by $\mathbf{A}_2$ and $\mathbf{A}_2$ is undercut by $\mathbf{A}_3$ then $\mathbf{A}_3$ is called a *defence* for $\mathbf{A}_1$. We define the "*defend*" relation as the transitive closure of "*being a defence*". An argument tree is said to be *successful* iff every leaf defends the root node. The *categorizer* is a function, denoted by $\mathbf{c}$, from the set of argument trees to $\{0, 1\}$ such that $\mathbf{c}(T) = 1$ iff T is successful. The *categorization* of a set of trees is the collections of their values by categorizer function on them. The *accumulator* of a query α is a function, denoted by $\mathbf{a}$, from categorizations to the set $\{(1,1), (1,0), (0,1), (0,0)\}$. Let $\langle X, Y \rangle$ be a categorization of argumentation framework of an axiom α, then $\mathbf{a}(\langle X, Y \rangle) = (w(X), w(Y))$ where $w(Z) = 1$ iff $1 \in Z$. A *valuation function* $\mathbf{v}$ mapping a set of axioms to a set $\{B, t, f, N\}$ (which is based on *binary argumentation theory* (see [9])), where $\langle X, Y \rangle$ is a categorization of argumentation framework of α in $\mathcal{O}$, is defined as follows:

(1) $\mathbf{v}(\alpha) = t$ iff $\mathbf{a}(\langle X, Y \rangle) = (1, 0)$; (2) $\mathbf{v}(\alpha) = B$ iff $\mathbf{a}(\langle X, Y \rangle) = (1, 1)$;
(3) $\mathbf{v}(\alpha) = f$ iff $\mathbf{a}(\langle X, Y \rangle) = (0, 1)$; (4) $\mathbf{v}(\alpha) = N$ iff $\mathbf{a}(\langle X, Y \rangle) = (0, 0)$.

Definition 3. *Let $\mathcal{O}$ be a DL-Lite ontology. A set of axioms $\mathcal{M}$ is an* argumentative model *of $\mathcal{O}$ if for every axiom α of $\mathcal{M}$ there exists a successful argument tree of α in $\mathcal{O}$. If $\alpha \in \mathcal{M}$, then we call $\mathcal{O}$* **argumentatively entails** *α, denoted by $\mathcal{O} \models_a \alpha$.*

Theorem 2. *Given an ontology $\mathcal{O}$ and an axiom α, the following properties are equivalent each other.*
(1) $\mathcal{O} \models_a \alpha$. (2) $\mathbf{v}(\alpha) \in \{\top, t\}$. (3) $\mathbf{c}(T) = 1$ *for an argument tree T of α.*

Theorem 3. *Given a consistent ontology $\mathcal{O}$ and an axiom α, then $\mathcal{O} \models_a \alpha$ iff $\mathcal{O} \models \alpha$.*

However, an ontology $\mathcal{O}$ is inconsistent, then "$\models_a$" is weaker than "$\models$". Clearly, $\mathcal{O} \models_a \sim \alpha$ doesn't possibly hold if $\mathcal{O} \not\models_a \alpha$ for any axiom α. $\mathcal{O} \models_a \beta$ is not inferred from $\mathcal{O} \models_a \alpha$ and $\mathcal{O} \models_a \sim \alpha \vee \beta$. Since $\{\alpha, \sim \alpha\} \not\models_a \beta$ for any query β, we conclude that "$\models_a$" is non-trivial.

4 Algorithm for Searching Arguments in DL-Lite

In this section, we present a graph-based approach to search arguments in DL-Lite ontologies by extending V.Efstathiou and A.Hunter's graph technique in [14].

A *disjunct function* mapping from a set of formulae to a set of sets of clauses in DL-Lite$_{core}$, is defined as follows:

(1) $Disjunct(B(a)) = \{B(a)\}$;
(2) $Disjunct(P(a, b)) = \{\exists P(a), \exists P^{-}(b)\}$;
(3) $Disjunct(B_1 \sqsubseteq B_2) = \{\neg B_1, B_2\}$.

Let $\mathcal{O}$ be a DL-Lite ontology, we write $Disjunct(\mathcal{O}) = \{Disjunct(\alpha)|\alpha \in \mathcal{O}\}$. Given an ontology $\mathcal{O}$, axioms ϕ, ψ, a concept C and an individual a, then the attack set between ϕ and ψ is defined as follows: $Attackset(\phi, \psi) = \{C(a) \mid C(a) \in Disjunct(\phi)$ and (either $\neg C(a) \in Disjunct(\psi)$ or $\neg C \in Disjunct(\psi))\} \cup \{C \mid C \in Disjunct(\phi)$ and $\neg C \in Disjunct(\psi)\}$. $Attack(\phi, \psi) = \beta$ if $Attackset\ (\phi, \psi) = \{\beta\}$ for some β; otherwise, $Attack(\phi, \psi) = null$. We conclude that the *attack set* of a pair ϕ, ψ is the collection of all members which cause the occurrence of inconsistencies between ϕ and ψ.

In the following, we introduce an *attack graph* based on attack set and the relation of attack for ontologies $\mathcal{O}$ in DL-Lite$_{core}$. Let $\mathcal{O}$ be an ontology and a be an individual.

- The graph for $\mathcal{O}$, denoted by $Graph(\ \mathcal{O}) = (V, E)$, where $V = \mathcal{O}$ and $E = \{(\phi, \psi) \mid$ there is a $\beta \in Disjunct(\phi)$ such that $Attack(\phi, \psi) = \beta\}$.
- A sequence of nodes is called a *path* in $Graph(\mathcal{O})$ if from each of its nodes there is an edge to the next node in the sequence.
- $Graph(\mathcal{O})$ is *connected* iff for any two nodes n_1 and n_2 in $Graph(\mathcal{O})$, there exists a path from n_1 to n_2 or a path from n_2 to n_1 in $Graph(\mathcal{O})$.
- An *attack graph* w.r.t. an individual a for $\mathcal{O}$, denoted by $AttackGraph(\mathcal{O}, a) = (V, E)$, is the union of all connected subgraphs of $Graph(\mathcal{O})$ which are obtained by deleting all nodes with the form of $B(b)$ where $a \neq b$ based on UNA and B is some concept from $Grpah(\mathcal{O})$ and all edges connecting it.
- An *attack graph* for $\mathcal{O}$, denoted by $AttackGraph(\mathcal{O})$, is a graph $(\mathcal{V}, \mathcal{E}, \mathcal{W})$. It is the union of all attack graphs w.r.t. individuals by adding signed direction edges to connect two attack graph w.r.t. individuals. That is to say, $\mathcal{V} = \{node \mid node \in AttackGraph(\mathcal{O}, a)$ and $a \in U_{\mathcal{A}}\}$, where $U_{\mathcal{A}}$ is a set of individuals that occur in ABox $\mathcal{A}$ of $\mathcal{O}$, and $\mathcal{E} = \{edge \mid edge \in AttackGraph(\mathcal{O}, a)$ and $a \in U_{\mathcal{A}}\}$. $\mathcal{W} = \{\langle \exists P(a), \exists P^-(b) \rangle \mid P(a,b) \in \mathcal{O}$ for some role $P\}$. Furthermore, $AttackGraph(\mathcal{O}, a)$ is a *component* of $AttackGraph(\mathcal{O})$.
- A node $Node_\phi$ is *close* if for each $\beta \in Disjunct(\phi)$ there is a $Node_\psi \in V$ such that $Attack(\phi, \psi) = \beta$. $AttackGraph\ (\mathcal{O}, a) = (V, E)$ is *close* if each node in $AttackGraph\ (\mathcal{O}, a)$ is closed. Algorithm 1 is applied to decide whether a node is closed or not in the graph of $AttackGraph(\mathcal{O}, a)$.
- A node is a *leaf* if the node has only one edge connected with another node.
- A *close graph* w.r.t. an individual a for $\mathcal{O}$, denoted $Close(\mathcal{O}, a)$, is the largest subgraph of $AttackGraph(\mathcal{O}, a)$ which is close. A *close graph* for $\mathcal{O}$, denoted $Close(\mathcal{O})$, is the largest subgraph of $AttackGraph(\mathcal{O})$ which is close.
- A graph is a *focal graph* of ϕ in $\mathcal{O}$, denoted $Focal(\mathcal{O}, \phi)$, if there is a component X in $Close(\mathcal{O})$ containing the node ϕ, then $Focal(\mathcal{O}, \phi) = X$, otherwise $Focal(\mathcal{O}, \phi)$ is an empty graph. In short, a focal graph is a subgraph of close graph for a KB which is specified by a formula and corresponds to the part of the close graph that contains the formula. The focal graph is introduce to find the scope of arguments for queries in an ontology.

For a graph $G = (V, E)$, let the function $Nodes(G)$ return the set of formulae in corresponding to the nodes of the graph, i.e., $Nodes(G) = V$. Then, let $Zone(\mathcal{O}, \alpha) = \{\phi \in \mathcal{O} \mid \phi \in Nodes(Focal(\mathcal{O} \cup \{\neg\alpha\}, \alpha)\}$. For a given query $C(a)$, $Zone(\mathcal{O}, C(a))$ is obtained by using Algorithm 2.

Algorithm 1. $isClose(\mathcal{O}, a, \phi)$ **return** a boolean value "*true*" or "*false*"

```
Input: an ontology O, an individual a and a formula φ
Output: "true" or "false"
if φ is the form of B(a) then
   if there exists ψ in AttackGraph(O, a) such that Attack(φ, ψ) = B(a) then
      return true ("true" means each literal of φ has an edge with other nodes.)
   end if
else
   (Here φ has the form of B ⊑ C)
   if there exist ψ1 and ψ2 where ψ1 ≠ ψ2 in AttackGraph(O, a) such that
   Attack(φ, ψ1) = ¬B and Attack(φ, ψ2) = C then
      return true
   end if
end if
return false ("false"means each literal of φ has no edge with other nodes.)
```

Algorithm 2. $GetZone(\mathcal{O}, C(a))$ **return** a set S of nodes

```
Input: an ontology O and a query C(a)
Output: a set of nodes of Focal(O, C(a))
G ← AttackGraph(O ∪ {¬C(a)}, a)
S ← Nodes(G)
if C(a) ∉ O or ¬isConnected(G, Node_C(a)) then
   return ∅
else
   for all nodes Node which is not ¬C(a) in S do
      if isClose(O, a, Node) = false then
         S ← S\Node
      end if
   end for
end if
return S
```

Given an ontology $\mathcal{O}$ and a conjunctive query $\alpha = \psi_1 \wedge \cdots \wedge \psi_m$, Let

$$SupportBase(\mathcal{O}, \alpha) = \bigcup_{i=1}^{m} Zone(\mathcal{O}, \psi_i).$$

We conclude that the notion of $SupportBase$ is an extension of the notion of $Zone$ and $SupportBase$ is $Zone$ when $m = 1$.

Theorem 4. *Let $\mathcal{O}$ be an ontology and $\alpha = \psi_1 \wedge \cdots \wedge \psi_m$ be a query. If $\langle \Phi, \alpha \rangle$ is an argument for α in $\mathcal{O}$ then $\Phi \subseteq SupportBase(\mathcal{O}, \alpha)$. Specially, if $m = 1$ then $\Phi \subseteq Zone(\mathcal{O}, \alpha)$.*

A subgraph G of $Closed(\mathcal{O})$ is minimal closed graph if any subgraph of G is not closed. We give an approach to finding minimal closed graph in focal graph to searching arguments for queries by the following theorem.

Theorem 5. *Let $\mathcal{O}$ be an ontology and α be a query. Then $\langle \Phi, \alpha \rangle$ is an argument for α in $\mathcal{O}$ iff $AttackGraph(\Phi \cup \{\neg\alpha\})$ is a minimal closed subgraph of $Focal(\mathcal{O}, \alpha)$.*

5 Conclusions

In this paper, we introduced the argumentation theory presented in [13] for DL-Lite to handle inconsistencies. We presented an argumentative semantics based on an argumentation framework for DL-Lite. Our argumentative semantics holds a certain degree of justifiability by analyzing all arguments which cause inconsistencies. We mainly developed a graph-based approach by extending V.Efstathiou and A.Hunter's graph technique for searching arguments in propositional logic [14] to search arguments in DL-Lite ontologies feasibly. The most difficult problem of our approach is which techniques we apply to search argument by focusing a limited scope from an infinite ontology. However, our approach is not efficient enough in the worst-case. Finding an efficient approach to searching arguments in expressive DLs will be our future work.

References

1. Calvanese, D., Giacomo, G.D., Lenzerini, M., Rosati, R., Vetere, G.: DL-Lite: Practical reasoning for rich DLs. In: Proc. of DL 2004, Canada. CEUR 104, vol. 104, CEUR-WS.org (2004)
2. McGuinness, D.L., van Harmelen, F.: OWL Web Ontology Language Overview (February 2004), `http://www.w3.org/TR/owl-features/`
3. Besnard, P., del Cerro, L.F., Gabbay, D.M., Hunter, A.: Logical handling of inconsistent and default information. In: Uncertainty Management in Information Systems, pp. 325–342 (1996)
4. Huang, Z., van Harmelen, F., ten Teije, A.: Reasoning with inconsistent ontologies. In: Proc. of IJCAI 2005, UK, pp. 454–459. Professional Book Center (2005)
5. Eiter, T., Ianni, G., Schindlauer, R., Tompits, H., Wang, K.: Forgetting in managing rules and ontologies. In: Proc. of WI 2006, China, pp. 411–419. IEEE Computer Society, Los Alamitos (2006)
6. Ma, Y., Hitzler, P., Lin, Z.: Paraconsistent reasoning for expressive and tractable description logics. In: Proc. of DL 2008, Germany. CEUR 353, CEUR-WS.org (2008)
7. Zhang, X., Xiao, G., Lin, Z.: A tableau algorithm for handling inconsistency in OWL. In: Proc. of ESWC 2009, Greece. LNCS, vol. 5554, pp. 399–413. Springer, Heidelberg (2009)
8. Dung, P.M.: On the acceptability of arguments and its fundamental role in nonmonotonic reasoning, logic programming and n-person games. Artif. Intell. 77(2), 321–358 (1995)
9. Elvang-Gøransson, M., Hunter, A.: Argumentative logics: Reasoning with classically inconsistent information. Data Knowl. Eng. 16(2), 125–145 (1995)
10. Tempich, C., Pinto, H.S., Sure, Y., Staab, S.: An argumentation ontology for distributed, loosely-controlled and evolving engineering processes of ontologies. In: Gómez-Pérez, A., Euzenat, J. (eds.) ESWC 2005. LNCS, vol. 3532, pp. 241–256. Springer, Heidelberg (2005)
11. Gomez, S.A., Chesnevar, C.I., Simari, G.R.: An argumentative approach to reasoning with inconsistent ontologies. In: Proc. of KROW 2008, Australia, pp. 11–20. CRPIT 90, ACS (2008)
12. Grossi, D.: Doing Argumentation Theory in Modal Logic. Technical Report of ILLC. PP-2009-24 (2009), `http://www.illc.uva.nl/Publications/`
13. Zhang, X., Zhang, Z., Lin, Z.: An argumentative semantics for paraconsistent reasoning in description logic ALC. In: Proc. of DL 2009, UK. CEUR 477, CEUR-WS.org (2009)
14. Efstathiou, V., Hunter, A.: Focused search for arguments from propositional knowledge. In: Proc. of COMMA 2008, France. FAIA 172, pp. 159–170. IOS Press, Amsterdam (2008)

Thresholding for Segmentation and Extraction of Extensive Objects on Digital Images

Vladimir Volkov

State University of Telecommunications
Saint-Petersburg, Russia
vl_volk@mail.ru

Abstract. The threshold setting problem is investigated for segmentation and extraction of extensive objects on digital images. Image processing structure is considered which includes thresholding for binarization. A new method for dynamic threshold setting and control is proposed which is based on the analysis of isolated fragments to be extracted in the making of segmentation. Extraction of extensive objects is obtained by the use of sequential erosion of isolated fragments on images. Number of points deleted is used for dynamic thresholding. The method proposed has optimality properties.

Keywords: Segmentation, extraction, thresholding, erosion.

1 Introduction

Segmentation is a general problem of image partitioning to separate objects, parts or segments [1,2,3,4,5,6,7]. In many cases useful objects are extensive and segmentation is based on the property of homogeneity. This idea is developed in this paper and new method is proposed for extraction extensive objects with prescribed sizes.

Thresholding is a necessary operation which gives binary image for segmentation and classification. Automatic setting of threshold is the main problem for qualitative segmentation. Well known global and local threshold evaluations was categorized [2] into six groups: histogram shape-based [3] , spatial [4] , local properties of the point, clustering-based, entropy-based and object attribute-based methods. Almost all quality functions for thresholding are based on the original image properties, only the last two methods use a binarized image for cross-entropy or similarity measures calculations. There are no methods which analyze results of segmentation and use them for choosing a threshold.

The aim of this paper is to investigate the influence of a threshold level to the results of segmentation, and to propose a method for threshold setting and control. The method proposed is new, it uses the results of segmentation for threshold evaluation. It is applicable both to global thresholding and to evaluation of local threshold in the sliding window.

B. Mertsching, M. Hund, and Z. Aziz (Eds.): KI 2009, LNAI 5803, pp. 623–630, 2009.

2 Image Processing Structure

A typical processing structure for image segmentation and object extraction may be divided into three stages: pre-filtering, threshold processing (binarization) and processing with binary images. Registered digital images are represented by digital values on a rectangular grid of pixels.

Pre-filtering includes linear filtering and non-linear transforms. It usually uses local sliding windows which are determined from the size of local stationarity of the initial image. It allows eliminating slow nonstationarities and nonhomogenuities, then it becomes possible to use the only global threshold for binarization. An equivalent procedure may be realized by the use of local thresholding without pre-filtering. The following processing for image segmentation includes logical and morphological operations, such as dilation and erosion [1].

This division of overall image processing is very useful though sometimes logical operations may be included in pre-filtering and may forego threshold processing [5] .

3 Problem Statement and Method of Solving

Suppose there is a classification problem, so that objects have to be extracted belong to certain classes. But these classes have no precise description and this is the object of investigation. One of the main features for object segmentation is its extension. Useful objects are usually extensive, after binarization they consist of connected points, and look like lines or regions.

Solving of segmentation problem includes thresholding and selection of objects with different extensions. Thresholding gives the binary image and the following segmentation is realized with a special method of erosion. Dynamical threshold setting is the problem which should be solved with taking into account results of segmentation.

For this purpose a hierarchy of isolated fragments is proposed which relates to extension and orientation. Segmentation comes to a special method of sequential erosion of isolated fragments with increasingly rising sizes.

It is evident that quality of segmentation depends on the threshold. Low threshold levels give much noise, and the following processing becomes inefficient, too high levels result in destroying of useful objects which can be split up to small fragments. The best threshold should be set after analysis of segmented fragments with the use of some quality indicator for extraction and segmentation.

A degree of segmentation is characterized by the effectiveness of erosion which is defined as a relative number of points deleted at each step. This number is used as the attribute for dynamic threshold setting.

4 Pre-filter Processing

Pre-filter F is usually used for smoothing and differentiation to avoid nonhomogeneity in images. It has the local sliding window (mask) $S_{i,j}$, which size corresponds to a homogeneity region of the initial image and (i, j) are coordinates of the centre of the sliding window.

There are two basic techniques for differentiation in local windows: gradient and Laplacian [1,6,7]. Two pre-filtering algorithms were used here. The first $y_{i,j}^A$ is standard linear Laplacian and uses a sample average in 9x9 sliding window. The second $y_{i,j}^M$ is non-linear and uses a sample median in the same window instead of the sample average:

$$y_{i,j}^A = x_{i,j} - A_{i,j} \text{ and } y_{i,j}^M = x_{i,j} - M_{i,j}, \tag{1}$$

$$\text{where } A_{i,j} = (1/N)\sum_k x_k,\ M_{i,j} = med(x_k),\ k \in S_{i,j} \tag{2}$$

Here k designates a couple of pixel coordinates around the point (i, j) , and the mask $S_{i,j}$ does not contain the central element (i, j) .

Fig. 1 shows the original infrared image of ships on the sea after registration and result of pre-filtering with sample average in 9x9 sliding window. Images after pre-filtering may be viewed after the corresponding shift-scale transformation.

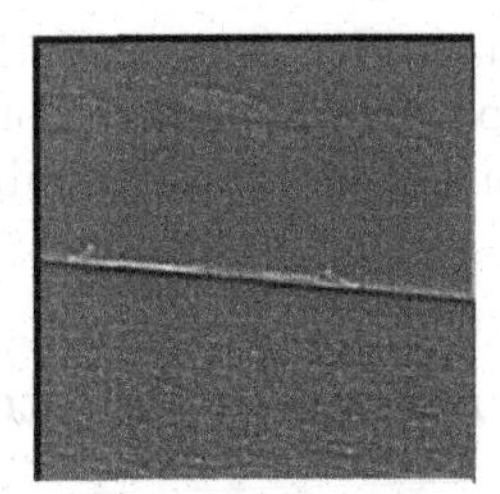

Fig. 1. The original infrared image of two ships on sea and pre-filtered image with the use of sample average in 9x9 window

5 Threshold Processing

This stage of thresholding is very important. An incorrect threshold level may result in irreversible losses of information. Suppose we are interested in light objects, and processing results in high level for pixels if a threshold level is exceeded, and a zero level otherwise.

Fig. 2 illustrates changes of binary images upon threshold variations from a low level to a high level. As it may be easily observed, a low threshold is not useful because of segmentation problems: useful and noisy objects do not differ in extension. As far as a threshold level rises, differences between extensions of objects start to appear, but at high levels we can see destroying of useful extended objects.

The best threshold level gives maximal differences in extensions between useful and noisy objects.

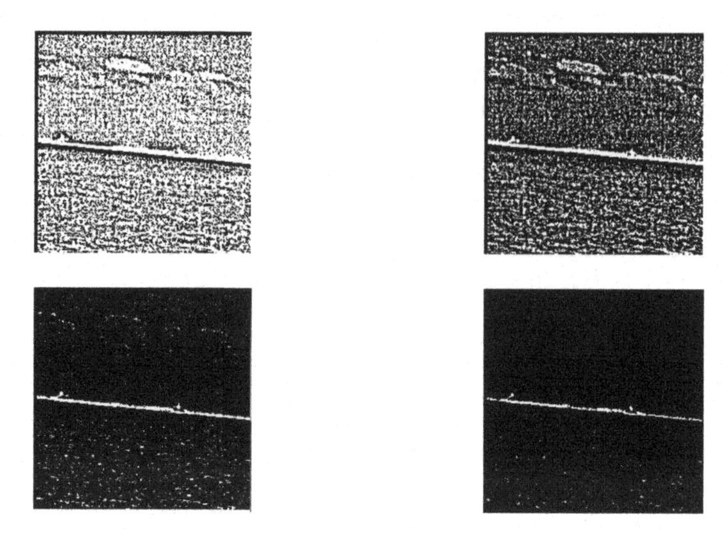

Fig. 2. Results of threshold processing for four threshold levels from a low level (left top) to a high level (right bottom)

The main idea is to set a threshold level according to segmentation results. For this purpose a hierarchy of isolated fragments is proposed, and effectiveness of erosion is inserted as an indicator of segmentation degree. It may be used for threshold setting and control.

6 Hierarchy of Isolated Fragments to Be Extracted

Our aim is to find out attributes of image which characterize extension properties of objects and allow us to control a threshold level. Suppose we have binary image after thresholding. The concept of the characterizing mask for isolated fragments is introduced which is used for analysis of segmentation results.

It relates to definition of continuity and adjacency of pixels, in this paper usual definitions are used [1].

Fragments are isolated if they have no mutual pixels. Suppose an isolated fragment consists of several adjacent pixels on the binary image. Extension properties of isolated fragments may be characterized by sizes of minimal rectangular mask which entirely covers this fragment. Objects may include several isolated fragments. Extensive objects usually contain extensive fragments.

Isolated points have the characterizing mask 1x1, isolated pairs may have characterizing masks 1x2, 2x1 and 2x2 corresponding to their orientations. An extensive object with horizontal orientation which has more sizes but entirely covers by mask 2x3, has the characterizing mask. A similar object with vertical orientation covers by characterizing mask 3x2, and so on.

The simplest hierarchy which will be considered uses only square characterizing masks, so that we have 1x1, 2x2, 3x3 and so on characterizing masks of extensive objects. This hierarchy does not take into account orientations of fragments.

7 Segmentation of Isolated Fragments on Images and Obtaining the Indicator for Threshold Setting and Control

The method of segmentation proposed includes sequence of erosions of isolated fragments with rising extensions. Deleted fragments have their orientations and this information may be useful for further analysis. For simplicity only the simplest hierarchy of isolated fragments was used here.

Results of isolated fragments erosion with characterizing masks up to 9x9 are presented in Fig. 3 for four threshold levels considered.

Number of points deleted at each stage of erosion is the useful indicator for threshold setting and control. At the first stage only isolated points are extracted and deleted from the binary image. At the second stage it is done with isolated objects of the characterizing mask 2x2, at the third stage - with the mask 3x3, and so on.

Suppose several stages of erosion have been completed and we have numbers of deleted points at each stage. As the total number of deleted points varies with respect to a threshold level it needs to normalize points deleted at each stage to this total number. It may be considered as the effectiveness of erosion.

Results for the image considered are collected in Fig. 4 below. They show effectiveness of erosion calculated as a function of threshold value for different sizes of isolated objects deleted at each step of erosion.

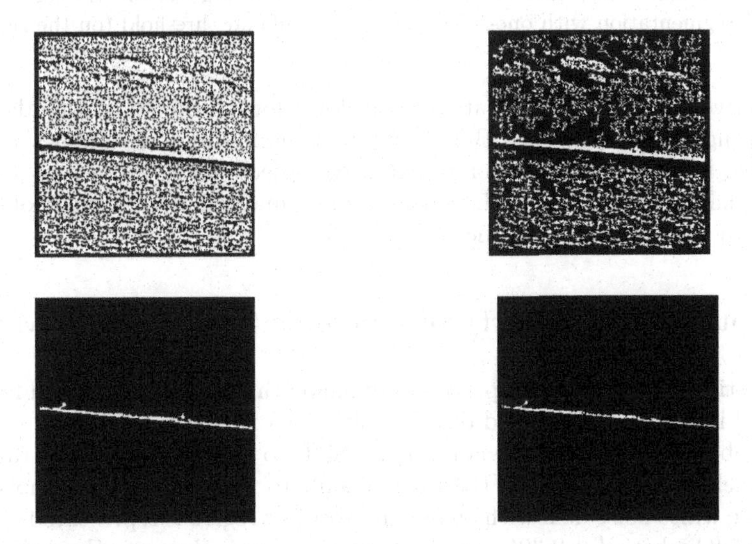

Fig. 3. Results of extensive objects extraction by the use of erosion of isolated fragments with masks up to 9x9 for four threshold levels (left top picture corresponds to a low threshold level, right bottom picture corresponds to a high level)

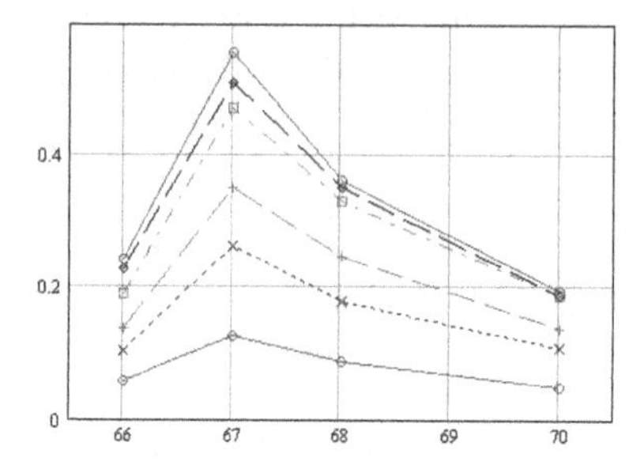

Fig. 4. Effectiveness of erosion with respect to the threshold value. Curves locate from the bottom to the top with increasing characterizing mask sizes: 1x1, 2x2, 3x3, 5x5, 7x7, 9x9.

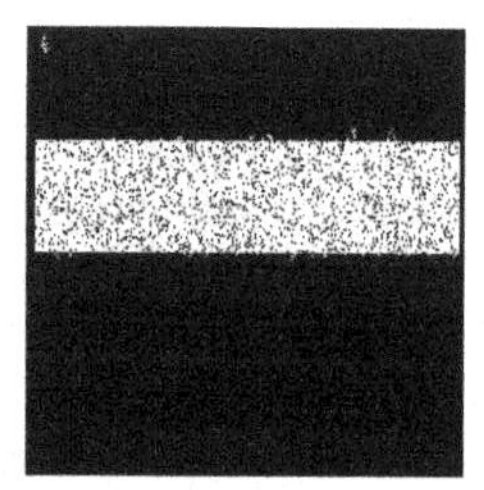

Fig. 5. Gaussian white noise field with extensive strip regions (on the left) and the result of segmentation with one-sided maximal likelihood threshold (on the right)

Effectiveness of erosion is rather small for a low threshold and tends to rise with rising threshold levels. But it has maximum which determines the best threshold level for segmentation of extensive objects those extensions are more than a characterizing size of the corresponding mask. The best threshold corresponds to the left picture in the bottom row.

8 Optimality Property of Dynamic Thresholding Method

To investigate optimality properties of dynamic thresholding Gaussian field was modeled in which an extended object looks like a horizontal strip.

The object differs from the noise by its shift, so that signal-to-noise ratio d in each pixel may be defined as the ratio of shift to mean deviation of noise. The Gaussian white noise field with an extensive object region is represented by Fig. 5 (on the left) where $d = 2.326$, result of thresholding is shown to the right of Fig. 5 for maximal likelihood threshold which is equal to $d/2 = 1.116$. This threshold value minimizes the total error probability for detection of positive shift in each pixel. Fig. 6 shows that the effectiveness of erosion has its maximum near this value. Fore pure white noise $d = 0$, and also from Fig. 7 we can see that the effectiveness of erosion has no maxima.

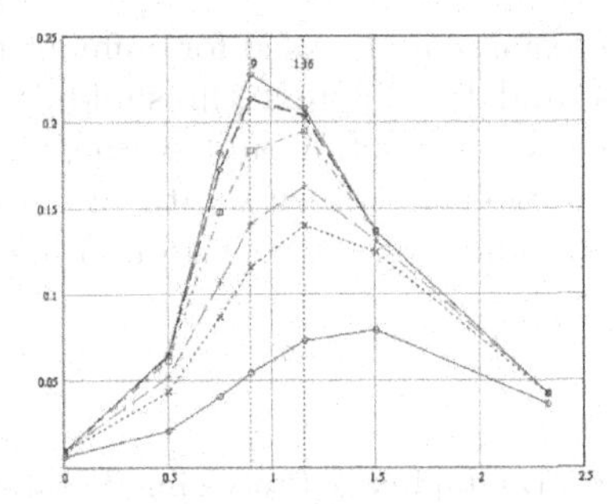

Fig. 6. Effectiveness of erosion for presence of extensive region with respect to the threshold value. Curves locate from the bottom to the top with increasing characterizing mask sizes: 1x1, 2x2, 3x3, 5x5, 7x7, 9x9.

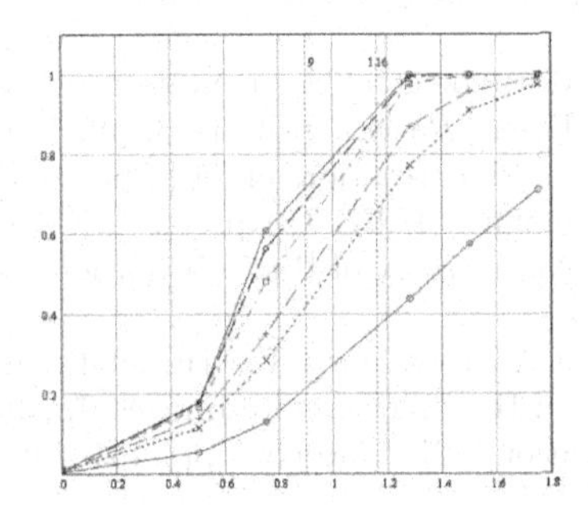

Fig. 7. Effectiveness of erosion for pure noise with respect to the threshold value. Curves locate from the bottom to the top with increasing characterizing mask sizes: 1x1, 2x2, 3x3, 5x5, 7x7, 9x9.

9 Conclusions

The general method of segmentation is described which is based on erosion of isolated objects on the binary image. It allows extracting extensive objects with different extension and orientation.

Threshold processing results in binary image, and the following extraction of extensive objects includes erosion of small isolated fragments. The hierarchy of isolated fragments is proposed for analysis of extent of erosion.

Analysis of deleted fragments is useful for obtaining indicators for threshold setting and control. It may be used for comparing binarized images. The effectiveness of erosion is introduced as the relative number of points deleted at each stage of erosion. The best threshold level should give maximal effectiveness of erosion for a given size of characterizing mask.

Setting of the threshold has optimality property which was checked by modeling Gaussian field with extensive object region. Thresholds evaluated are settled near the value of optimal maximal likelihood threshold for detection of shift on Gaussian field.

The effectiveness of erosion may be used for comparative analyses of different pre-filtering algorithms, and for different thresholding methods. This method of dynamic thresholding for extraction and segmentation is oriented to final result of segmentation of extensive objects with prescribed sizes. With evident modifications it may be applied for segmentation of 3D images.

References

1. Gonzalez, R., Woods, R.: Digital Image Processing. Prentice Hall, New Jersey (2005)
2. Sezgin, M., Sankur, B.: Survey over image thresholding techniques and quantitative performance evaluation. Journal of Electronic Imaging 13(1), 146–165 (2004)
3. Otsu, N.: A threshold selection method from gray-level histograms. IEEE Trans. SMC 9, 62–66 (1979)
4. Weszka, J., Rosenfeld, A.: Threshold Evaluation Techniques. EEE Trans. SMC-8, 622–629 (1978)
5. Akcay, H., Aksoy, S.: Morphological Segmentation of Urban Structures. In: Urban Remote Sensing Joint Event, Paris, April 11-13, pp. 1–6 (2007)
6. Volkov, V.: Detection and extraction of objects on opto-electronic images. In: Proceedings of St Petersburg IEEE Chapters. IEEE Russian Northwest Section Int. Conf. "The 110th Anniversary of Radio Invention" V.II St. Petersburg, Russia, pp. 128–133 (2005)
7. Volkov, V., Makarenko, A., Rogachev, V., Turnetsky, L.: Characteristics Analysis and Digital Processing of Heating Discrete Images. In: 10 Int. Conf. "Digital Signal Processing and Applications" X-2 Moscow, pp. 485–488 (2008)

Agent-Based Pedestrian Simulation of Train Evacuation Integrating Environmental Data

Franziska Klügl[1], Georg Klubertanz[2], and Guido Rindsfüser[2]

[1] Modeling and Simulation Research Center, Örebro University
Örebro, Sweden
franziska.klugl@oru.se
[2] Emch& Berger, AG Bern
Bern, Switzerland
{georg.klubertanz,guido.rindsfueser}@emchberger.ch

Abstract. Simulating evacuation processes forms an established way of layout evaluation or testing routing strategies or sign location. A variety of simulation projects using different microscopic modeling and simulation paradigms has been performed. In this contribution, we are presenting a particular simulation project that evaluates different emergency system layout for a planned train tunnel. The particular interesting aspect of this project is the integration of realistic dynamic environmental data about temperature and smoke propagation and its effect on the agents equipped with high-level abilities.

1 Introduction

Simulating evacuation processes on the level of pedestrians forms a very effective mean of testing the layout of a newly built or planned environments, as well as for evaluating routing or information strategies. Thus, during the last years many – also commercial – simulators for microscopic evacuation simulation became available. One can even observe that simulations are applied for almost all new major public building – sport stadium or railway stations.

Although the microscopic way of simulation allows to capture interactions between pedestrians – usually for collision avoidance – as well as increased heterogeneity in movement speeds, previously performed studies are not fully realistic as environmental conditions are hardly considered beyond reduced perception radius. In this contribution we are describing a particular evacuation project in a train tunnel. A realistic fire and smoke model was integrated to an agent-based simulation. As a consequence, the pedestrian agent model was extended for flexibly reacting to perceptions, communications (and beliefs) about exit directions, heat or smoke concentrations. Thus, an agent-based evacuation simulation was created that demonstrates the usefulness of the intelligent agent concept in evacuation simulation.

The remainder of this contribution is structured as follows: We first give a short overview over the current state of art in pedestrian simulation, especially for evacuation scenarios, followed by a sketch of our environmental and agent

B. Mertsching, M. Hund, and Z. Aziz (Eds.): KI 2009, LNAI 5803, pp. 631–638, 2009.

model. In section 4 we illustrate the dynamics of the model. As a conclusion, we give a summary as well as we discuss issues like validation and model reuse.

2 State of the Art in Microscopic Pedestrian and Evacuation Simulation

Agent-based pedestrian models have received a lot of attention during the last decade. Pedestrian dynamics are different from vehicle-based traffic due to a variety of properties [1]: pedestrians can immediately stop their movement, move in full speed from one moment to the other, possess no restrictions in turning behavior and are not forced to follow lanes or move in certain restricted areas like roads. This results in more degrees of freedom and potentially more complex simulated behavior compared to vehicle-based traffic.

Basically three types of microscopic pedestrian simulation models have been proposed:

1. Force-based models, like the *social force model* [2] are based on the assumption that the movement of a pedestrian can be modeled similar to particle moving in attractive and repelling force fields emitted by the destination locations or static and dynamic obstacles. A recent survey can be found in [3].
2. Cellular automata based approaches (such as [4], [5]) rely on discrete spatial representations. The state of a cell represents the existence of a pedestrian on that cell. Additionally it captures static and dynamic potential fields away from obstacles and towards the exit for guiding the movement of the pedestrian. A recent review of cellular automata-based approaches to microscopic pedestrian simulation can be found in [6] with a special focus on evacuation scenario. Cellular automata-based microscopic pedestrian simulations was successfully applied for evaluating evacuation times and standards for newly build and planned (public) building or for cruise ships.
3. Agent-based pedestrian simulations (such as [7], [8], [9]) contain more details enabling richer environmental structures with multiple destinations and agent activities. Bandini et al. [10] combine a cellular automata with agents moving upon it.

   More complex agent architectures have been proposed by Dijkstra et al. [11] and Timpf [12]. The first suggests an agent architecture that is able to operate in rich environment with a representation of realistic properties of shape and dynamics; the latter suggests an agent model based on spatial cognition research integrating realistic wayfinding and navigation in space. The model described in the following is simpler than these approaches, but exhibits flexible behavior not induced by potential fields, but by explicit reasoning about perceptions.

   Evaluation of evacuation strategies is also a major application area of agent-based pedestrian simulation. Lämmel et al. [13] present a large scale evacuation study of Zürich in case of a dam break of a nearby lake. This scenario is based on a meso-level movement of pedestrians derived from queueing simulation on links. The routing of agents is here based on shortest path algorithms without adaptations during movement. Nakajima et al. [14] is also based on a meso-level of movement on links with a flexible re-selection of the next link based on perceived damage or proximity of a shelter. In contrast to the latter approaches, the agents described here trigger re-decision of the next exit during movement not just when the end of a link is reached, but may turn around; additionally there is interaction between agents, not potential instructions from outside.

3 The Evacuation Model

A train evacuation simulation was developed as a collaboration between the agent-based simulation group at the University of Würzburg (meanwhile located at Örebro University) and the division Mobility and Transport of the Emch& Berger, AG, Bern, by order of a Italian-French Consortium responsible for planing the new railway track between Turin and Lyon. The basic objective was to find out which of a set of given emergency layouts is best in terms of evacuation times versus construction costs. From research side, the project was interesting because realistic environmental conditions and reactions of simulated pedestrians had to be integrated into an existing pedestrian model [9]).

3.1 Environmental Model

The environmental model assuming continuous space was based on realistic geometric extensions of the interior of the trains and different variants of the emergency exit system outside the train. Two coupled, with 788 simulated travelers fully occupied TGV-trains were modeled with an assumed fire at the motor of the second train. The eight scenarios were different in terms of distance between emergency tunnels (400 to 200m), breadth of the evacuation platform in the train tunnel (1,2m or 1,6m), breadth of the exit doors to the emergency tunnel (1,4m or 2,4m) and breadth of the emergency tunnel itself (2,4m or 4m). At time point zero of the simulation the trains were stopping at a given position in the tunnel due to a defect and fire in the motor of the second TGV. A particular challenge was imposed by the interior of the TGV trains with their seats in different orientations, narrow legroom between seats, the geometry of the bistro wagon and the overall narrow conditions for the two wheelchair that had to be considered. The screenshot in figure 1 (a) illustrates this showing a part of the interior after 16 simulated seconds. The continuous space for movement was augmented with a topological graph structure (figure 1 (b)) that was used by the agents for planing their route.

On the emergency platform on that the simulated pedestrians have to walk in the tunnel to reach the emergency exits, detailed environmental conditions are integrated. Data for dynamic temperature and CO concentrations – the latter as a representation of smoke also influencing the local perception range of the simulated pedestrians – were provided as a table with a spatial discretization of 30m and a temporal discretization of 15 seconds – one second is the basic time

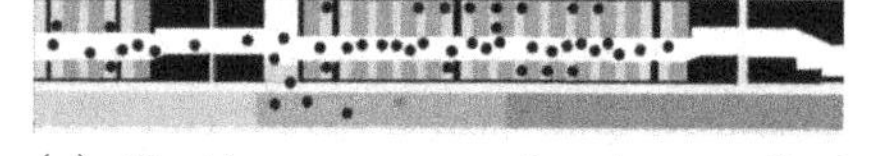

(a) Continuous map for locomotion and local perception

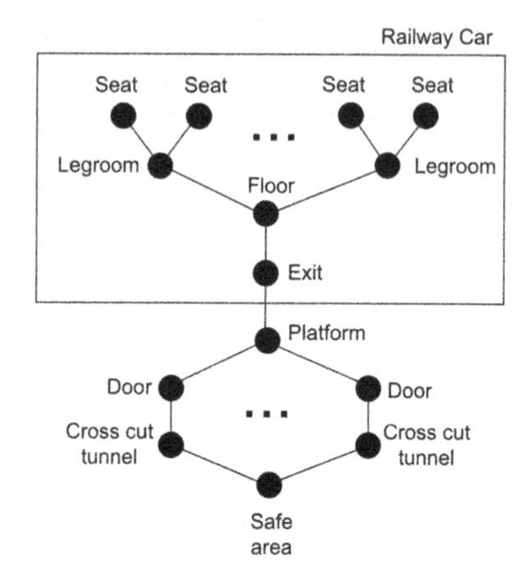

(b) Topological graph for individual path planning

Fig. 1. Spatial representation used in environmental model

resolution of the model. The fire model was simulated separately from the agent-based simulation. Cross sections at different heights were tested. The continuous space of the emergency platform was discretized into plates of the appropriate size. Each of this plates was associated with data series for temperature and CO concentration. For the perception of local temperature and smoke, the plate with the largest intersection with the agents was selected and the current value given to the agent. Signs directing towards the next exit are positioned with 25m distance.

3.2 Agent Decision Making and Behavior

The agent architecture used in this evacuation simulation is a two-layer architecture consisting of a layer controlling the collision-free locomotion and a "cognitive" part containing processes for high-level path planning. Figure 2 sketches this overall architecture of a pedestrian agent.

Agents are assigned to one of three different age classes that determines the individually desired speed. Data about age classes was given by the awarding authority, data about speed was taken from literature; 10% noise was applied for more realism. The default perception range was set to 7m due to the darkness in the tunnel. A simulated pedestrian is represented by a circle with 40cm diameter.

In the following, the high-level reasoning will be shortly described. The locomotion level contains a rule-based collision avoidance based on calculating individually possible avoidance directions and step sizes. This locomotion model was already used in a railway station simulation [9].

After perceiving an evacuation signal, every pedestrian agent stands up and heads towards the nearest train exit. It was assumed that there is no individual

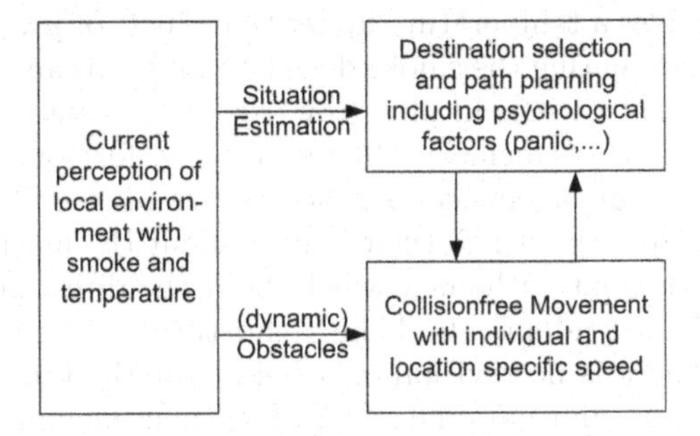

Fig. 2. Overall architecture of a pedestrian agent during the TGV evacuation simulation

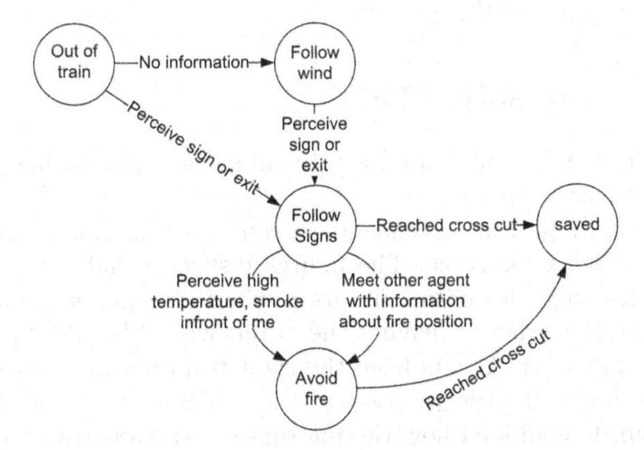

Fig. 3. States of orientation respectively information of a pedestrian agent during the TGV evacuation simulation

delay in reaction times as we did not had reliable data for delaying the reaction. Every agent knows the next exit, but may change to another exist in its back if there is congestion in front and it has enough space for turning and moving towards the other exit. An agent beliefs that there is congestion when it does not move for a given number of seconds.

When leaving the train, an agent had to decide about their movement direction. In contrast to other microscopic evacuation simulations the destination for movement was not a priori given to the agents. Without any information, the agents follow the (given) direction of the ventilation air flow in the tunnel and start moving. While moving, agents continuously scan their local environment: When noticing an exit sign indicating the next emergence crosscut or perceive an exit door, it becomes "informed" about the reasonable direction to the next safe zone. While movement the environmental conditions change according to the given data table. These environmental conditions influence the agent behavior and destination choice while moving on the emergency platform:

If an agent perceives a temperature larger than 70°C or a CO density of more than 3000*ppm* – representing the smoke density –, it turns and flees independent of the distance to the exit it is heading towards. The agent adds a belief about the position of the fire to its memory. On a narrow platform with high densities of pedestrian agents turning agents cause a breakdown in overall crowd movement. In this case agents are "shouting" their beliefs about the fire position – in form of a multicast. If an agent without a belief about the direction towards the fire receives the warning, it adopts the beliefs and turns as well if it was heading towards the fire. There is no revision of belief about the fire position. Figure 3 illustrates the different informational states of an agent moving on the emergency platform.

During their movement on the emergency platform, agents may be harmed by the heat or smoke. Movement speed is reduced and in the worst case the agents are transformed to immobile obstacles.

4 Example Simulation Run

For illustration of the simulated crowd dynamics we are showing a sequence of states from one particular run.

The collection of partial screenshots in figure 4 illustrates that the overall dynamics are at least plausible. The figures just show half of the trains. The simulation starts with the initial situation where all agents are sitting in full train. One after the other is leaving the train, while the agents are queueing up within the trains (a). Agents from the right train heading towards the cross cut in the middle notice the position of the fire first, turn and communicate the fire position. It is hidden how the information spreads with turning agents, all agents on the right are fleeing the fire (b). Critical situation in the middle: some agents almost reached the emergency tunnel, others are already fleeing the fire (c); Agents from both trains are heading away from the fire, towards exits outside of the figure (d). After about 10 minutes of simulated time all simulated pedestrians have left the platform (not shown in figure 4).

5 Evaluation

Validation of such simulation runs is difficult; No data from evacuation experiments with real humans were available. Thus, only face validation by experts was done. In addition to frequent plausibilization by expert reviews, we successfully simulated relevant test scenarios of the RiMEA (*www.rimea.de*) guideline for evacuations simulations. The RIMEA consortium defined a set of test situations that a evacuation simulation should pass. These address mostly low level aspects of locomotion behavior, but also simple adaptive exit choice.

For the particular project all parameters such as the threshold for recognizing the fire position or for affecting the simulated pedestrians health were taken from a reference model of the tunel, literature or given by experts. A sensitivity analysis is still missing.

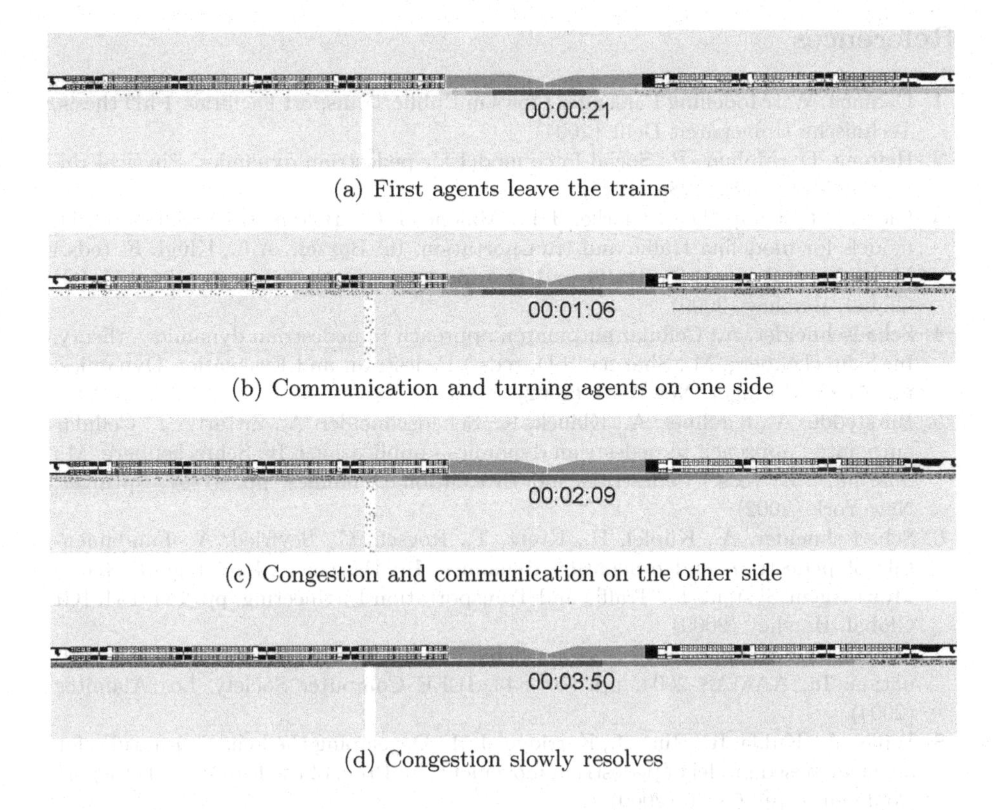

(a) First agents leave the trains

(b) Communication and turning agents on one side

(c) Congestion and communication on the other side

(d) Congestion slowly resolves

Fig. 4. The basic simulation experiment: situation at different stages; grey agents have no destination, green agents follow signs, red (darker) agents flee from he believed position of fire.

6 Conclusion

In this contribution we presented an agent-based evacuation model applied to a particular scenario where trains are stopped in a tunnel due to a fire. The simulation objective was to test emergency layout under realistic conditions. It turned out that an agent-based approach with agents capable of communication and flexible decision making is necessary for tackling this simulation objective. Thus, this contribution describes a successful AI application that would not be possible with simpler crowd simulation approaches.

An important issue is the reuse of the model in other evacuation scenarios. On one hand, reuse of simulation models of this size and complexity is a question of modeling efficiency and cost, on the other hand testing the model in scenarios with other geometry or fire models makes implicit assumptions explicit. We are currently working on a generic pedestrian model that is based on this project, but also allows more complex outlines.

References

1. Daamen, W.: Modelling Passenger Flows in Public Transport Facilities. PhD thesis, Technische Universiteit Delft (2004)
2. Helbing, D., Molnar, P.: Social force model for pedestrian dynamics. Physical Review E 51(5), 4282–4286 (1995)
3. Oleson, R., Kaup, D.J., Clarke, T.L., Malone, L.C., Boloni, L.: Social potential models for modeling traffic and transportation. In: Bazzan, A.L., Klügl, F. (eds.) Multi-Agent Systems for Traffic and Transportation Engineering, pp. 155–175. IGI Global, Hershey (2009)
4. Schadschneider, A.: Cellular automaton approach to pedestrian dynamics - theory. In: Schreckenberg, M., Sharma, S.D. (eds.) Pedestrian and Evacuation Dynamics, pp. 75–85. Springer, New York (2002)
5. Burstedde, A., Kirchner, A., Klauck, K., Schadschneider, A., Zittartz, J.: Cellular automaton approach to pedestrian dynamics - applications. In: Schreckenberg, M., Sharma, S.D. (eds.) Pedestrian and Evacuation Dynamics, pp. 87–97. Springer, New York (2002)
6. Schadschneider, A., Küpfel, H., Kretz, T., Rogsch, C., Seyfried, A.: Fundamentals of pedestrian and evacuation dynamics. In: Bazzan, A.L., Klügl, F. (eds.) Multi-Agent Systems for Traffic and Transportation Engineering, pp. 124–154. IGI Global, Hershey (2009)
7. Osaragi, T.: Modeling of pedestrian behavior and its applications to spatial evaluation. In: AAMAS 2004, pp. 836–843. IEEE Computer Society, Los Alamitos (2004)
8. Willis, A., Kukla, R., Hine, J., Kerridge, J.M.: Developing the behavioural rules for an agent-based model of pedestrian movement. In: Proc. of the European Transport Conference, pp. 69–80 (2000)
9. Klügl, F., Rindsfüser, G.: Large scale agent-based pedestrian simulation. In: Petta, P., Müller, J.P., Klusch, M., Georgeff, M. (eds.) MATES 2007. LNCS (LNAI), vol. 4687, pp. 145–156. Springer, Heidelberg (2007)
10. Bandini, S., Federici, M.L., Vizzari, G.: Situated cellular agents approach to crowd modeling and simulation. Cybernetics and Systems 38(7), 729–753 (2007)
11. Dijkstra, J., Jessurun, J., de Vries, B., Timmermans, H.: Agent architecture for simulating pedestrians in the built environment. In: Bazzan, A.L.C., Chaib-Draa, B., Klügl, F., Ossowski, S. (eds.) 4th Agents in Traffic and Transportation at AAMAS 2006, Hakodate Japan, pp. 8–16 (2006)
12. Timpf, S.: Towards simulating cognitive agents in public transport systems. In: Bazzan, A.L., Klügl, F. (eds.) Multi-Agent Systems for Traffic and Transportation Engineering, pp. 176–190. IGI Global, Hershey (2009)
13. Lämmel, G., Rieser, M., Nagel, K.: Bottlenecks and congestion in evacuation scenarios: A microscopic evacuation simulation for large-scale disasters. In: Bazzan, A.L.C., Klügl, F., Ossowski, S. (eds.) Proc. of 5th Workshop on Agents in Traffic and Transportation, at AAMAS 2008, Estoril, PT, pp. 54–61 (2008)
14. Nakajima, Y., Shiina, H., Yamane, S., Ishida, T., Yamaki, H.: Disaster evacuation guide: Using a massively multiagent server and gps mobile phones. In: 2007 International Symposium on Applications and the Internet (SAINT 2007), Hiroshima, Japan, January 15-19 (2007)

An Intelligent Fuzzy Agent for Spatial Reasoning in GIS

Rouzbeh Shad, Mohammad Saadi Mesgari, Hamid Ebadi, Abbas Alimohammadi, Aliakbar Abkar, and Alireza Vafaeenezhad

Faculty of Geodesy and Geomatics Eng. K.N.Toosi University of Technology
No 1346, Mirdamad cross, Valiasr st., Tehran, Iran
Rouzbeh_Shad@yahoo.com, mesgari@kntu.ac.ir,
Hamid_Ebadi@kntu.ac.ir, Alimoh_abb@yahoo.com,
Abkar@kntu.ac.ir, Vafaee78@yahoo.com

Abstract. In this paper, an intelligent fuzzy agent can identify the values of risks and the environmental damages of the smoke plumes. When smoke plumes move: data extractor extracts the fuzzy areas form NOAA satellite images, spatial decision support system updates information from data base, and topological simulator computes the strength and type of topological relationships and sends the extracted information to a designed knowledge based system. A fuzzy inference subagent infers the information provided by data extractor subagent, topological simulator subagent and knowledge base, and sends the results back to the spatial decision support subsystem. The risk amounts for pixel elements of the forest area are computed and dangerous sites are specified based on the spatial decision support system in GIS environment. Then, a genetic learning agent tries to generate and tune the spatial knowledge bases for the next risk calculation. By the experimental results, the designed system provides flexibility, efficiency and robustness for air pollution monitoring.

Keywords: Spatial reasoning, Agent, Fuzzy, GIS.

1 Introduction

Agents are one of the most popular objects for searching solutions to realistic computational problems characterized by incomplete information and autonomy in dynamic and distributed spaces. Different definitions for an agent presented by researchers; for example Ferber said "An agent is a program or spatial package that is capable for acting in an environment and can communicate with other agents" [1]. Generally, we can say an intelligent agent should behave in a SMART (Specific, Measurable, Attainable, Realistic, Time bound) manner and generate different judgment results by using various parameter settings and training sets. These agents have several characteristics such as Autonomy, Mobility, Social ability, Reactivity and Proactiveness [2; 3] and typically deals with dynamic, incomplete and uncertain domains of decision making problems where mathematical methods can not perform well, because of ill-structured forms. Ill-structured forms are complex and dynamic decision making problems which included incomplete or indefinite goals, objectives, criteria and alternatives. Wide ranges of these problems are appeared in the spatial applications such as environmental modeling, ecological management, land use planning and etc [4; 5; 6].

B. Mertsching, M. Hund, and Z. Aziz (Eds.): KI 2009, LNAI 5803, pp. 639–647, 2009.

[7; 8; 9] seem to be first spatial systems, employing agent technology for spatial applications. They used agents for fast creation of robust, scalable and seamless access to nomadic services. In [10], an automated generalized agent-based system has been developed for a digital personal mobile tourist guide using GIS, databases, natural language processing, intelligent user interfaces and knowledge representation parts. Also, [11] has implemented an agent-based architecture providing adaptive services using intelligent agents that learns the key characteristics quite quickly, including spatio-temporal variations for wireless networks. In another research area, [12] presents the IMA architecture which is aimed at replacing the monolithic approach to spatial systems with a dynamic, lean, and customizable system supporting spatially-oriented applications.

But, the mentioned systems proposed by authors in above, have the lacks of flexible uncertain spatial behavior learning and reasoning mechanism. Our intelligent agent based system is more powerful, because of uncertain spatial reasoning and learning capabilities. This system makes use of a spatial knowledge base and uncertain algorithms to carry out goals defined by human developers or runtime users in a GIS. Geographic information system (GIS) provides the prospect of monitoring dynamic variations in the phenomena with indefinite boundaries by different criteria and factors. Also, our system presents a new way for analyzing topological relations between uncertain spatial phenomena in an online environmental alarm system. The necessity of this subject is stated by the fact that environmental pollution, as a result of oil wells, adversely affects forest areas.

To satisfy the needs of quick response, this study presents an intelligent fuzzy agent system using real-time information to enhance production decisions. For this purpose the designed intelligent spatial fuzzy agent system (ISFA) contains: a data extractor subagent (DEA), fuzzy topological simulator subagent (FTSA), fuzzy inference subagent (FIA) and a genetic learning subagent (GLA), to carry out the risk values. Moreover, the spatial decision support subsystem (SDSS) will process the information and order the risk areas for final decisions in overall snapshots. In addition, spatial data base (SDB) and knowledge base (KB) will store the extracted and predefined information and rules for mentioned application.

2 Intelligent Spatial Fuzzy Agent Architecture

Our system contains a sophisticate user interface, a spatial data base, different active agents, a knowledge base, a spatial decision support and a GIS. Designing this system would not be possible if integration of each component required extensive programming. Then, we decided to build an intelligent agent in object oriented programming language. In this kind of programming, an object can respond to the request of any other objects that knows how to address it. Figure 1 shows the general architecture of this system.

In Figure 1, While satellite images (NOAA-AVHRR) and different spatial data (Iran political boundary, Forest areas, DTM, Soil types, Synoptic stations and etc) are stored in the SDB, the DEA starts extracting fuzzy boundaries of smoke plumes and target forest area based on different sample points and spatial images, then, immediately sends them to FTSA for simulating the topological relations based on fuzzy topological matrixes. The derived linguistic terms extracted by FTSA are sent to the KB and FIA to save as fuzzy rules and infer by the other experts predefined rules for the application.

This mean, the parameters of the fuzzy variables that represent the behavior of smoke plumes and forest area will be stored in the KB for FIA. FIA retrieves the KB to get the fuzzy risk values of all the forest area for computing the risk and possibility and sends the computing results to SDSS. Also, SDB sends the required information including Soil types and Vegetation Species to the SDSS for ranking and ordering alternatives. Finally, SDSS will refer to the FIA and SDB to get the dangerous risk area and announce the GIS using sound. Furthermore, the actual decisions of the risk area will be stored into KB for genetic learning behavior. To solve the contradictory problem, the GLA will judge the consistency of the training data set retrieved from the KB first. If there is a case with different output for the same inputs then GLA will discard the contradictory information.

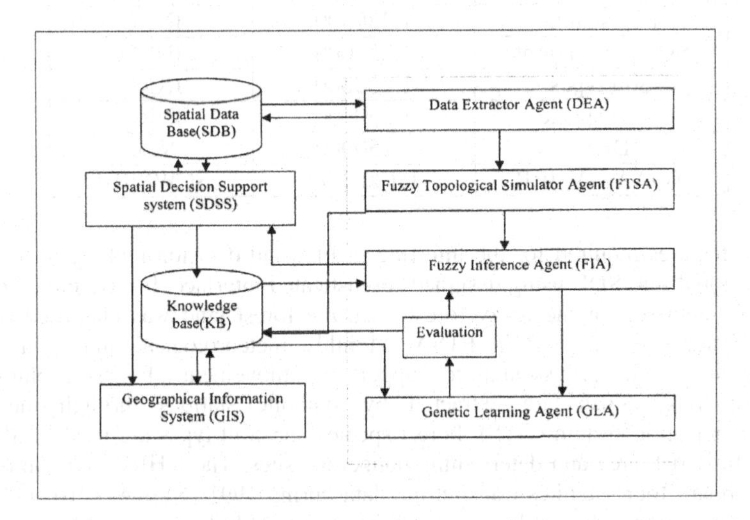

Fig. 1. The ISFA architecture

3 Case Study and Related Data

Our case study is located at 32° N, 53° E in the Middle East, between Kuwait and Iran territory and support the assumption that smoke plumes from Kuwait reached the territory of Iran during 1991 Persian Gulf War. It has been reported that nearly 700 oil wells were set on fire starting on 19 February 1991 for which the last fir extinguished on 2nd November 1991. During the peak period of the fires, the wells were emitting about 5000 tones of smoke per day [13]. Oil pollution movement via south west of Iran could be confirmed using NOAA-AVHRR midday images. The polluted inland areas can be outline as natural resources such as forest lands. This pollutant impacts on the: biological, physical and chemical characteristics of soil, the amount of acid rain falls and increasing heavy metals through the forest lands. The main objective of our study is to track, estimate and evaluate the risk values in terms of natural forest cover due to the mentioned atmospheric pollution quickly. This purpose would be possible if an integrated intelligent system designed using a variety of tools for making decision available. Thus, using GIS and Remote sensing data are essential for the identification of dangerous sites prior to undertaking further analyses or field investigations. GIS spatio-temporal data sets of Forest area, Political boundary, Synoptic stations (for Wind direction and climate), Soil types, Forest species, Digital elevation models (DEMs) and satellite NOAA-AVHRR daily data were collected for the southwest of Iran from different sources, such as Iranian Natural Resources Organization (INRO), Iranian National Centre for Oceanography (INCO), National Cartographic Center (NCC), and Soil Conservation and Watershed Management Research Center (SCWMRR) during 1991. They were summarized as shown in Table 1.

Table 1. Spatial data used in the study

Data	Scale	Data Source
Forest area	1:250000	INRO
Synoptic stations	1:250000	INCO
Soil types	1:250000	INRO
Forest species	1:250000	INRO
DEM	1:250000	NCC
NOAA-AVHRR	1:1000000	SCWMRR

These features required for the inferring in FIA and decision making in SDSS and they are stored into SDB using designed sophisticated interface. For example, to obtain the spatial data sets of the fuzzy forest area, the forest land was classified by DEA around features to be used in FTSA. Monthly meteorological and oceanography information, which are essential to support the movement of smoke plumes and estimating risk values, are extracted by synoptic stations through the global telecommunication system (GTS). Forest species and Soil types are used in SDSS for ranking final risk area and determining dangerous sites. The AHRR (1.1 km at nadir) sensors aboard four satellites transmitting data during 1991 (NOAA-9, 10, 11 and 12) were capable of providing at least four images every 24th. In this way, they provided a better opportunity to follow the dynamics of smoke plumes than satellite with longer intervals. The available spectral channels have simply demonstrated in Table 2.

Table 2. Characteristics of NOAA Images

Spectral Bands	NOAA-10	NOAA-9,11,12
Band-1	0.58-0.68(visible)	0.58-0.68(visible)
Band-2	0.725-1.10(near-infrared)	0.725-1.10(near-infrared)
Band-3	3.55-3.93(thermal-infrared)	3.55-3.93(thermal-infrared)
Band-4	10.50-11.50(thermal-infrared)	10.3-11.3(thermal-infrared)
Band-5	10.50-11.50(thermal-infrared)	11.50-12.50(thermal-infrared)

For the purpose of detection and estimation of smokes and reducing the time of processing, we used channel 2 midday images with minimum shadow and shading. In addition, persistency and accumulated smoke density which are two important parameters for risk estimation derived by DEA using NOAA series images and saved in SDB for further analysis.

4 Implementation Results

The intelligent spatial fuzzy agent (ISFA) user interface contains four subagents including data extractor subagent (DEA), fuzzy topological simulator subagent (FTSA), fuzzy inference subagent (FIA) and genetic learning subagent (GLA) to assist the SDSS including spatial decision subagent (SDA). Each subagent in the proposed system possesses both the extracted and the predefined knowledge to perform a particular step in decision making process. The designed ISFA user interface, consisted of control codes written in Mathlab and a large set of VB.NET modules, allows users to choose the required functions to be used in the system.The extracted knowledge for a task is captured in procedural codes written in VB.NET. Often these are procedures for executing a piece of software developed by a simulation subsystem, a geographical information subsystem, a knowledge based subsystem, or spatial decision support subsystem. In some cases, some of designed subagents know how to operate class of models (for example, spatial data extractor and fuzzy topological simulator have this capability). DEA and FTSA gather the spatial information and fuzzy topological relations from spatial data and satellite images and send them to FIA and SDSS. Also, DEA receives the computed results of SDSS and sends them to FIA. Data extractor subagent can extract the required data for FTSA, FIA and SDSS based on various smoke plumes and environmental treatments. The DEA responds to the stored and portal received data by the interface and helps users set up their preference and required parameters in different snapshots. While a snapshot is started, DEA connected to the SDB and uses preference module to identify the fuzzy spatial regions of forest area and smoke plumes. These computed regions are sent to FSTA for deriving topological linguistic terms between regions which are used in FIA for inferring. Also, DEA determines fuzzy values of persistency, wind direction, amount of smoke, distance, and inclusion index in each snapshot for every pixel of the classified regions and updates knowledge base by derived data. Moreover, it can connect to user preference module for determining fuzzy weights of Soil types and Forest species layers semantically for ranking risk areas in SDSS. Users can recall and revise this treatment knowledge base later and reuse it for other risk units. FTSA is responsible for constructing Fuzzy topological relations between fuzzy smoke plumes and forest areas by simulating the Fuzzy 9-intersection matrixes [14]. Therefore, the relations can be described using the terms of 'Disjoint', 'Touch', 'Equal', 'Contain', 'Cover' and 'Overlap'. The FTSA is responsive to change in the values of these variables with respect to the conditions in the real world and try to update knowledge base using simulation results. Moreover, this subagent can compute the inclusion index parameter to add more quantitative information to the topological descriptions [15]. The overall architecture of Fuzzy Inference Agent (FIA) in the ISFA is consisted of three layers. There are three kinds of nodes in this model: input linguistic term nodes, rule nodes, and output linguistic term nodes. In layer1, a fuzzy linguistic term node represents a fuzzy variable and the mapping degree of it. The nodes in the first layer just directly transmit input values to next layer to constitute a condition specified in some rules.

In layer2, a rule node represents a rule and decides the final firing strength of that rule during inferring. In our model, 22 rules are defined by domain expert's knowledge previously (Figure 2). Hence, the rule nodes perform the fuzzy AND operation.

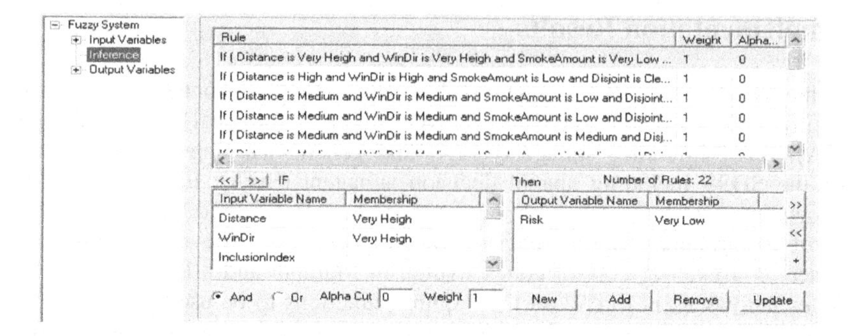

Fig. 2. Designed FIA rules

In layer3, an output linguistic term node shows a fuzzy output variables resulted by inferring. In this layer the output fuzzy variable for FIA is denoted the risk values for each smoke plume at the forest area. The output term node performs the fuzzy OR operation to integrate the fired rules which have the same consequence and then uses the Centroid operators for defuzzyfication.

After each snapshot, the solution result is stored in the KB and SDB. GLA retrieves the recorded data to encode each solution and evaluate them for creating next new solutions. In here, we used the approach proposed by [16] to learn the solutions and rules for every snapshot. In KB and SDB, fuzzy variables and linguistic modifiers are coded, then, each chromosome is composed of these two parts for each smoke plume i. After that, dynamic restrictions which are used to preserve meaningful fuzzy sets and improve learning efficiency are applied. The initial population for the gene pool is composed of four groups considering the randomized and original fuzzy variables and linguistic modifiers chromosomes (two-by-two). The chromosome in the current population is evaluated by the fitness function. If the evaluation does not satisfy the fitness function, then elitism is used in GLA. In the beginning of selection, the best chromosomes in the current population are selected to the new population without crossover and mutation. The remaining chromosomes in the new population are selected by the Roulette Wheel Selection mechanism [17]. After the elitism selection, the one-point crossover method is adopted and a crossover point is randomly set. The portions of the two chromosomes beyond this cut-off point to the right are to be exchanged to form the offspring. An operation rate with a typical value between 0.7 and 1.0 is normally used as the probability of crossover. The mutation process is applied to each offspring individually after the crossover exercise. It alters each gene randomly with a typical probability value of less than 0.1. The probability parameters of crossover and mutation are critically dependent upon the nature of the objective function. An objective function is a measuring mechanism that is used to evaluate the status of chromosome. The objective (fitness) function used here is to minimize the mean square error (MSE) as follows:

$$\text{MSE}=\frac{1}{2n}\sum_{t=1}^{n}(O_{in}-O_{ind})^2 \tag{1}$$

Where n denotes the number of the training data for smoke plume i, O_{in} denotes the output of FIA for the nth training data, and O_{ind} denotes the desired output of the nth training data for smoke plume i. It is necessary to note that computational value in each snapshot is introduced as input fuzzy variable in next snapshot. Thus, it can be said that computational results in various periods will be dependent on each other through entering the results obtained from inference to the knowledge base. MSEs indicate that performance of GL is better than Cordon, with GL (0.9, 0.05) having better performance than others.

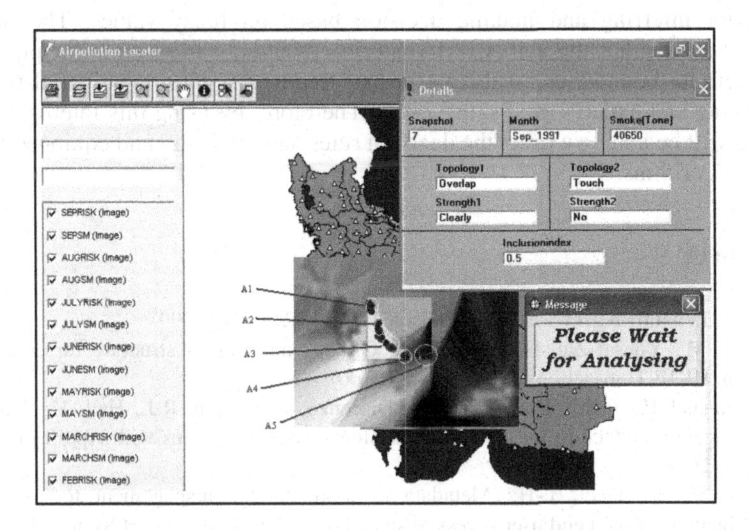

Fig. 3. Final risk assessment unites for the forest area during Persian Gulf War

Spatial decision making agent uses user preference module and fuzzy rule sets to estimate and judge how well a final risk unit satisfies expert's goals. Final risk unites are achieved by accumulating different snapshots and learning results during a period of time (in here 1991). For example, daily snapshots of smoke plumes give a risk area on the forest region where accumulated monthly and monthly results accumulated yearly. Four fuzzy categories: fails, nearly passes, barely passes, and passes indicate how well a goal is met in every set of snapshots. Spatial decision analysis subagent can perform goal satisfaction analysis on any sets of snapshots representing the risk areas at every pixels of forest in time. SDB and KB provide fuzzy risk units, soil types and forest species for the used application. DEA, user preference module and FIA provide the required information for fuzzy ranking of alternatives.

Final assessments are presented as fuzzy set values or intervals, then, probabilistic method which is more attractive and considers minimum sets of preliminarily assumptions [18] is applied for final fuzzy ranking. Figure 3 demonstrates the final risk assessment unites during 1991 for 5 classified risk areas (A1, A2, A3, A4, A5) during 1991.

5 Conclusion and Remarks

The designed intelligent fuzzy agent system includes online uncertain analysis of air pollution and its impacts on the environmental phenomena, then, is tested for the southwest of Iran during Persian Gulf (1991). This system contains five subagents including DEA, FTSA, FIA, GLA, and SDA to perform the intelligent air pollution support task. Moreover, a spatial user interface to evaluate the spatial results of proposed system is also constructed. Case studies using the ISFA decision process on the ranging of south west of Iran have been initiated and provide agents by flexible manner for inferring and making decision based on fuzzy values. The proposed intelligent system has the ability of monitoring dynamic variations in the phenomena with indefinite boundaries, analyzing spatial treatments, uncertain spatial reasoning and learning and expert's goal satisfaction. Therefore, by using this intelligent agent, the users will be able to extract the decision rules automatically and equipped them by learning algorithms.

References

1. Ferber, J.: Multi-Agent Systems. Addison-Wesley, Reading (1999)
2. Sarker, M., Yousaf-Zai, H., Yousaf-Zai, Q.F.: A multi-agent structure for collaborative design. HKIE Transactions 13(3), 44–48 (2006)
3. Maturana, F.P., Tichy, P., Slechta, P., Discenzo, F., Staron, R.J., Hall, K.: Distributed multi-agent architecture for automation systems. Expert Systems with Applications 26(1), 49–56 (2004)
4. Lawrence, A., Traci, J.-H.: Metadata as a knowledge management tool: supporting intelligent agent and end user access to spatial data. Decision Support System 32, 247–264 (2002)
5. Nute, D., Potter, W.D., Maier, F., Wang, J., Twery, M., Rauscher, H.M., Knopp, P.S., Thomasma, D.M., Uchiyama, H., Glende, A.: NED-2: an agent-based decision support system for forest ecosystem management. Environmental Modeling & Software 19, 831–843 (2004)
6. Qiu, F., Li, B., Chastain, B., Alfarhan, M.: A GIS based spatially explicit model of dispersal agent behavior. Forest Ecology and Management 254, 524–537 (2008)
7. Laukkanen, M., Helin, H., Laamanen, H.: Tourists on the move. In: Klusch, M., Ossowski, S., Shehory, O. (eds.) CIA 2002. LNCS (LNAI), vol. 2446, pp. 36–50. Springer, Heidelberg (2002)
8. Schmidt-Belz, B., Poslad, S., Nick, A., Zipf, A.: Personalized and location-based mobile tourism services. In: Workshop on Mobile Tourism Support Systems, in conjunction with the Fourth International Symposium on Human Computer Interaction with Mobile Devices, Pisa, Italy, pp. 18–20 (2002)
9. Zipf, A.: User-adaptive maps for location-based services (LBS) for tourism. In: Proceedings of the 9th International Conference for Information and Communication Technologies in Tourism, Austria, Innsbruck, pp. 183–197 (2002)
10. Lamy, S., Ruas, A., Demazeau, Y., Baeijs, C., Jackson, M., Mackaness, W., Weibel, R.: Agent Project: Automated Generalisation New Technology. In: Proceedings of the 5th EC-GIS Workshop, Stresa, Italy (1999)

11. Misikangas, P., Makela, M., Raatikainen, K.: Predicting quality-of-service for nomadic applications using intelligent agents. Agent Technology for Communication Infrastructures 15, 197–208 (2001)
12. Gervais, E., Liu, H., Nussbaum, D., Roh, Y.S., Sack, J.R., Yi, J.: Intelligent map agents-An ubiquitous personalized GIS. Photogrammetry & Remote Sensing 62, 347–365 (2007)
13. Jalali, N., Nooozi, A., Abkar, A.: Tracking of Oil Spills and Smoke Plumes of Kuwait's Oil Well Fires to the Coast and Territory of I.R. of Iran as Result of the 1991 Persian Golf War. In: International Institute for Aerospace Survey and Earth Sciences, Netherland, ITC, pp. 1–20 (1998)
14. Egenhofer, M.J., Franzosa, R.D.: Point-set topological spatial relations. International Journal of Geographical Information Systems 5(2), 161–174 (1991)
15. Bouchon-Meunier, B., Rifqi, M., Bothorel, S.: Towards general measures of comparison of objects. Fuzzy Sets and Systems 84, 143–153 (1997)
16. Cordon, O., Herrera, F., Villar, P.: Generating the knowledge base of a fuzzy rule-based system by the genetic learning of the data base. IEEE Transactions on Fuzzy Systems 9(4), 667–674 (2001)
17. Man, K.F., Tang, K.S., Kwong, S.: Genetic Algorithm: Concepts and design. Springer, London (1999)
18. Sewastianow, P., Rog, P.: Two-objective method for crisp and fuzzy interval comparison in optimization. Computers and Operations Research 33, 15–31 (2006)

Learning Parametrised RoboCup Rescue Agent Behaviour Using an Evolutionary Algorithm

Michael Kruse, Michael Baumann, Tobias Knieper, Christoph Seipel, Lial Khaluf, Nico Lehmann, Alex Lermontow, Christian Messinger, Simon Richter, Thomas Schmidt, and Daniel Swars

Students of Computer Science, University of Paderborn
{meinert,mbaumann,tknieper,seipel,klial,nlehmann,alesha,michri, richter,sup,swars}@mail.upb.de

Abstract. Although various methods have already been utilised in the RoboCup Rescue simulation project, we investigated a new approach and implemented self-organising agents without any central instance. Coordinated behaviour is achieved by using a task allocation system. The task allocation system supports an adjustable evaluation function, which gives the agents options on their behaviour. Weights for each evaluation function were evolved using an evolutionary algorithm. We additionally investigated different settings for the learning algorithm. We gained extraordinary high scores on deterministic simulation runs with reasonable acting agents.

1 Introduction

The RoboCup Rescue project [1] is an international competition aiming to improve the technology and strategy of robots in disaster areas. It was founded in 1999 as a reaction to the great earthquake in Kobe City, Japan. Construction of real robots to be used in such disaster areas is covered as well as their behaviour in a computer-simulated environment.

The RoboCup Rescue Simulation part consists of a robotics, a simulation and an infrastructure league. The robot league deals with the development and construction of autonomous robots, which are supposed to operate in a real-life disaster scenario. Contrarily, the simulation league investigates various challenges to design multi-agent systems.

The challenge in the simulation league is the coordination of heterogeneous agents with limited global knowledge and communication capabilities. Heterogeneous means that some actions cannot be done by all agent types. The agents just knows the environment around themselves and communication happens with limited bandwidth and relatively high latency. But the common goal of all agents is to rescue as many people and to extinguish as many fires as possible.

In our work, we focus on the development of an architecture to facilitate coordinated behaviour of heterogeneous agents. The architecture is defined by two aspects: infrastructure and controlling algorithms. The latter is responsible for the way the agents perform their actions. Thereby the emphasis was made on

B. Mertsching, M. Hund, and Z. Aziz (Eds.): KI 2009, LNAI 5803, pp. 648–655, 2009.

the application of artificial intelligence, but without using any central instance. Each agent maintains its own knowledge base and makes its decisions locally. In order to achieve this, we designed an architecture, which provides support for different functionalities like communication, information distribution and, focused in this paper, task allocation.

In the described approach we model the behaviour of agents by a task allocation system, which comprises a set of predefined behaviour patterns. A task allows to define a simple behaviour as well as a more complex one. For example, a simple task might be to move to a specific location whereas a more sophisticated task could define the behaviour of rescuing a civilian. In the environment of RoboCup Rescue, it is difficult to decide which task an agent should perform in a specific situation. To overcome this problem, we allow a learning algorithm to adjust the importance of a task for each particular state of the environment. A state is defined by the local knowledge base of the agent. Moreover, as a task can define a complex behaviour, it is not possible to determine in which way it should be performed. For instance, it might make more sense to rescue a nearby civilian with a small amount of health instead of rescuing a healthier civilian which is further away. The problem is solved by introducing task parameters, which can be learned by an evolutionary algorithm. The learning algorithm adjusts not only the order in which the tasks are performed, but also the task behaviour in particular.

This paper is a summary of our work, which is described in more detail in our elaboration [2]. The next section of this paper gives an overview on other work that has been done in connection with the RoboCup Rescue simulation. Sect. 3 then describes our approach in detail, in particular the design of our agents and the tasks, how they behave and what aspect of their behaviour is learned. The evolutionary algorithm we use as our learning method is described in Sect. 4, and its results are discussed in Sect. 5. In Sect. 6 we conclude our approach and suggest topics for further research in continuation of our approach.

2 Related Work

The RoboCup Rescue simulation project promotes research on distributed strategies for multi-agent systems in simulated disaster scenarios. Complex cooperative multi-agent systems such as RoboCup Rescue can be investigated using different approaches. Some authors concentrate on the general problems, such as hardware components or communication protocols, for example in [3].

The closest previous work to our approach we found is [4]. Both approaches use an evolutionary algorithm to learn weights, which control the agents' behaviour. The difference is that the authors of [4] use weights of a neural network and conclude that a hand-coded implementation is still better. In contrast, we use these weights to choose between several competing hand-coded implementations. Thus, we combine the best out of both strategies.

One of the most frequently discussed problems is whether a centralised or decentralised coordination strategy is better. In [5] the authors compare both coordination strategies and show that, in the case of a simple task, it is better to use a centralised approach. In order to increase efficiency, the authors suggest implementing standardisation rules like the partitioning of the map. In our work we also resort to the partitioning of the whole map into areas.

The authors of [6] present an approach based on a dynamic task allocation problem and use it to implement a centralised solution. In this context, a task is for instance a fire in a burning building that has to be extinguished. The centre agents are used as auctioneers and platoon agents make bids to be assigned to jobs.

3 Agents

There are four types of agents in the RoboCup Rescue simulation: first, there exist *centre agents* that are only used as communication relays in our approach. Second, *fire brigade agents* can extinguish fires, third, *police force agents* are able to clear blocked roads and finally, *ambulance team agents* can dig out buried civilians and carry them to a refuge.

We implemented the latter three agent types with a task based behaviour. Its main idea is that every agent has to select the best task per timestep. A task can be a simple action like "go to the refuge to refill your water" or a more sophisticated plan like "rescue civilian #42". Each task includes an evaluation function that defines how valuable it is to execute the task in the current situation. Any information of the simulation state, like positions of agents or distances to targets, can be used as a parameter to influence the evaluation value.

Second to the dynamic information drawn from the current state of the simulation, we use weights to control the influence of a particular parameter. Those weights are learned by an offline learning algorithm which is discussed in detail in Sect. 4.

For instance, the weight *ExtinguishBaseEvaluation* describes the general importance of the task that is responsible for extinguishing a fire. Additional weights called *ExtinguishImportanceOfDistance* and also *ExtinguishImportance-OfFieryness* define how much the parameters "distance from current position to the fire to extinguish" and the fire's fieriness respectively, influence the base evaluation value. That is, the closer the distance to the fire and the higher the weight *ExtinguishImportanceOfDistance* is, the higher the final evaluation value for this task.

Some tasks are most useful when they are executed by one agent only. For instance, the task called *ExploreArea* should be performed by one single agent per (map-)area at the same time because two agents at the same position do not explore more than a single one. Therefore, the agents need to know what the other agents do. We call the aim of a task a job. Multiple tasks can contribute to the same job. The information which agent works on which job is distributed to all agents via the communication capabilities offered by the simulation environment. This information is used during the evaluation of tasks such that the ideal combination of agents work on a specific job.

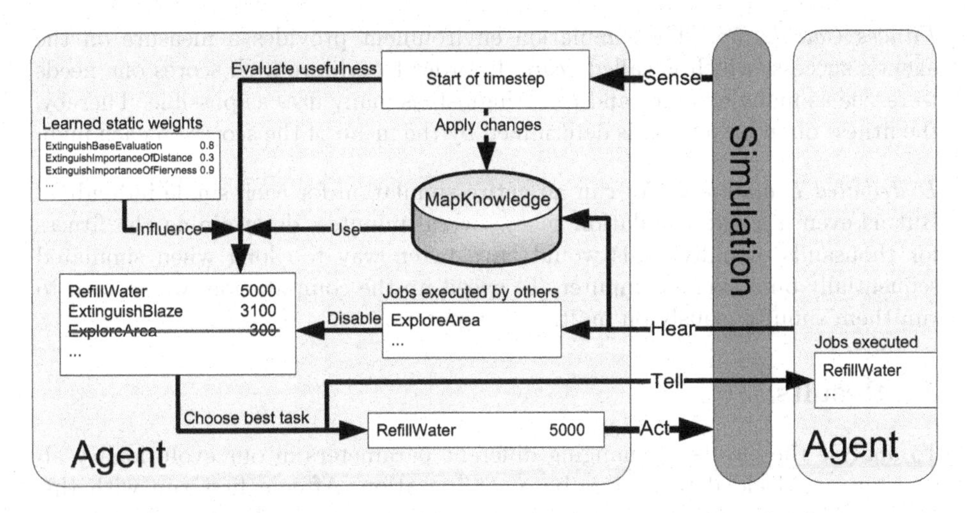

Fig. 1. At the beginning of a timestep, an agent receives the changes on the map within its vision range. The combined knowledge about the environment is used to evaluate the usefulness of the agent's tasks. Tasks for jobs that are already executed by another agent are not considered this round. The task with the highest evaluation value is chosen and decides which actions the agent performs this timestep. Additionally, the other agents are informed about the chosen task and the changes on the map.

The outcome of this is an implicit cooperation between the agents. For example, the agents executing *ExploreArea* will distribute to different areas of the map. To illustrate the design, Fig. 1 describes the steps which every agent performs in each timestep.

4 Learning Algorithm

We developed an evolutionary algorithm to evolve the behaviour of our agents which in particular depends on the order of task execution. Therefore, the learning algorithm adjusts the weights of the parameters of the evaluation functions to continuously improve the agent's performance. One set of weights is called an individual. An individual contains the weights for all agents, i.e. all agents in the simulation use the same set of weights. Our approach computes an individual that leads the multi-agent system to behave efficiently. For this, we employ a generic evolutionary algorithm as presented in [7] with the settings described in the following.

Individual Structure. The evolutionary algorithm is applied to a pool of individuals, where each individual consists of three agent type specific chromosomes and one common chromosome. The common chromosome contains the weights of the parameters, which are used by all agent types.

Fitness Calculation. The simulation environment provides a measure on the agent's success, which is called *score*. In order to achieve high scores one needs to rescue as many civilians and to extinguish as many fires as possible. Thereby, the fitness of an individual is determined by the mean of the scores on each map.

Distributed Evaluation. We run an entire simulation for each single individual. But as even a single simulation takes several minutes, determining the fitness for thousands of individuals would have taken way too long when simulated sequentially on a single computer. To speed up the computation, we decided to run them simultaneously on multiple computers.

5 Results

To analyse the effects of changing different parameters in our evolutionary algorithm, we decided to fix one base configuration. After a first run with this base configuration we changed different parameters, but only one parameter at a time. In this way we were able to monitor the effects of this single change. For instance, we decided to use different population sizes, different crossover methods, various mutation configurations and others. In the following we introduce the base configuration in detail and afterwards the mentioned experiments are presented. All results were determined by single deterministic simulation runs on the map VC. We decided to use a deterministic simulation environment to get reliable scores and to reduce the computational effort.

After the initial experiments, we built an algorithm that incorporates various changes of the base configuration by combining promising changes to get a high potential evolutionary algorithm.

5.1 Base Configuration

The base configuration contains the following methods and parameters. The first experiments were executed to determine the population size whereas 30 turned out to be most promising. We chose a roulette wheel selection to build the parent pool and an one-point crossover per chromosome. We investigated several mutation configurations and chose a mutation rate of 2%. When a variable is mutated, it is set with 10% probability (the *random rate*) to a uniformly distributed random variable. With a probability of 90%, a normally distributed random variable with mean 0 and standard deviation 0.02 is added to the old value. If the new value exceeds the codomain of our weights [0,1], the corresponding boundary is chosen. Finally, we decided to use an approach with elitism, that is, the best individual is carried to the next generation without alteration.

5.2 Experiments

Population Size. We chose a population size of 30 after performing experiments with 20, 30 and 40 individuals. This is a good compromise between computation time and the performance of the evolutionary algorithm.

Mutation. Experiments with a mutation rate of 4% did not lead to better results. We also tried to disable the random mutation, but the best individual did not reach a promising score. Furthermore, we did not eliminate the randomness completely, but we analysed the effect of decreasing its influence. The starting mutation rate and random rate were set to 6% and 50%, respectively. In generation t, the mutation rate is $mr_{start} \cdot 0.99^t$ and the random rate is $mrr_{start} \cdot 0.985^t$. Our experiments showed that a decreasing mutation rate with constant random mutation is the best choice.

Fitness Adjustment. To control the selection pressure in the roulette wheel selection, we implemented a changed fitness calculation that sets the fitness of the individuals to the difference between the individual's score and the score of the worst individual. Formally the fitness of the i-th individual is computed as $fitness(i) = score(i) - min\{score(1), \ldots, score(n)\}$. This ensures that weak individuals are chosen with a lower probability whereas the weakest one is never selected.

Parent Pool Restriction. We tested the influence of restrictions in the parent selection: in the base configuration every individual has the chance to become a parent while in this setting only the best 50% of the generation is allowed to enter the parent pool. It turned out that the average score in this experiment is significantly better and more stable than the base configuration.

Selection. Moreover, we executed experiments with different selection methods. In one setting we removed the mechanism to save the best individual from the last generation, in another one the parent pool consisted of individuals selected by a tournament selection. The tournament selection turned out to be more stable than the roulette wheel selection, while the approach without elitism did not converge.

5.3 Combined Configurations

After examining the preliminary results, we evaluated how multiple parameter changes at the same time affect the performance of the evolutionary algorithm. Therefore, we chose the most promising parameter changes and merged them into two new configurations. Both inherit the adjusted fitness and the decreasing mutation rate with constant random mutation. One algorithm additionally uses the restricted parent pool. We dismissed the tournament selection although it looked promising in favour of the adjusted fitness calculation which requires the roulette wheel selection.

We consider both algorithms superior to the base configuration, while the configuration with the parent pool restriction outperformed all others as shown in Fig. 2a.

Furthermore, we tested how our best configuration with adjusted fitness, restricted parent pool and decreasing mutation rate performs, when the simulation runs on the maps Foligno, Kobe and VC simultaneously. Each map requires a

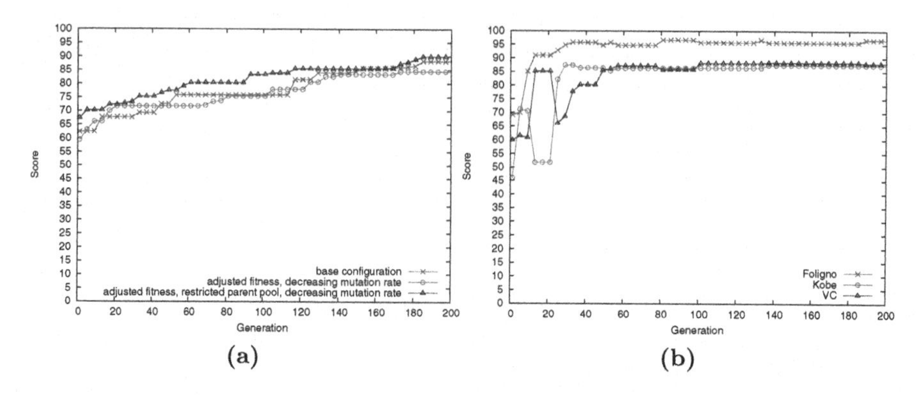

Fig. 2. Figure 2a shows a comparison of the best individuals in the combined configurations on the map VC. Figure 2b shows the best individuals in a combined configuration on various maps shown separately for the maps Foligno, Kobe and VC.

special behaviour and interaction of the agents. As we anticipated, the convergence is similar to the previous experiment. In Fig. 2b one can observe that an increase of the score on one map results in a decrease on another one. For instance, around generation 20, there is a change that led to a better score on VC, but to a major drawback on Kobe. Although, the algorithm successfully resolved this problem and converged to high scores on all maps.

The individual which reaches the highest scores learned that civilians around burning buildings should be rescued first. Another behaviour which shows up is that smaller fires are extinguished with multiple fire brigades before larger ones. The fire brigades also try to extinguish just one fire at once instead of dividing the water onto several buildings.

The scores on these maps are impressive compared to those reached in the official RoboCup Rescue competitions. One possible reason is the use of a different version of the simulation environment (we used 0.49.9). On the other hand, the final score highly depends on the indeterminism in the simulation (which seems to be unintended because by default it sets a random seed, hence our deterministic modification). The score of the winning individual falls rapidly if ran on a different map or just with a different random seed. The indeterministic simulation is too chaotic for any meaningful comparison between different runs.

6 Conclusion and Future Work

In our work we implemented a task allocation system to control the agents' behaviour. In combination with an evolutionary algorithm, we reached very high scores on multiple maps for a deterministic simulation.

To further improve the learned behaviour, one could try to enhance the reactiveness of the agents by adding some sort of online learning mechanism. This would allow the agents to react on unexpected situations that were not covered during learning.

Furthermore, the evaluation of the tasks could use abstracted map parameters to reduce the dependence to the actual map. Examples are the number of buried or endangered civilians, the quantity and types of neighbouring agents and the ratio of the explored map area. With this additional information, one could also realise an event-triggered approach to switch between different tactics when these parameters change.

It might be beneficial to implement more tasks for the same jobs that compete with each other. For example, several tasks for extinguishing fires with different tactics from which the evolutionary algorithm can choose one for each situation. Moreover, techniques like dynamic programming or neural networks could be able to evolve unforeseen task that could be added to the system.

The results imply that more runs on random maps and different random seeds are necessary to make the individuals less dependent on few situations only, but with formerly unknown maps, too.

References

1. RoboCup Rescue Project, `http://www.robocuprescue.org`
2. PaderRescue: Project-group for Agent-based Disaster Management and Emergent Realtime Rescue for Emergency SCenarios in Uncertain Environments, University of Paderborn, `http://wwwcs.uni-paderborn.de/cs/ag-klbue/de/research/PaderRescue/pgrescue.html`
3. Stef, B.M.P., Maurits, L.F., Visser, A.: The High-Level Communication Model for Multi-agent Coordination in the RoboCupRescue Simulator. In: Polani, D., Browning, B., Bonarini, A., Yoshida, K. (eds.) RoboCup 2003. LNCS (LNAI), vol. 3020, pp. 503–509. Springer, Heidelberg (2004)
4. Ohta, M.: An Implementation of Rescue Agents with Genetic Algorithm. In: 2nd International Workshop on Synthetic Simulation and Robotics to Mitigate Earthquake Disaster, Lisbon (2004), `http://www.rescuesystem.org/robocuprescue/SRMED2004/MOhta.pdf`
5. Paquet, S., Bernier, N., Chaib-draa, B.: Comparison of Different Coordination Strategies for the RoboCupRescue Simulation. In: Orchard, B., Yang, C., Ali, M. (eds.) IEA/AIE 2004. LNCS (LNAI), vol. 3029, pp. 987–996. Springer, Heidelberg (2004)
6. Nair, R., Ito, T., Tambe, M., Marsella, S.: Task Allocation in the RoboCup Rescue Simulation Domain: A Short Note. In: Birk, A., Coradeschi, S., Tadokoro, S. (eds.) RoboCup 2001. LNCS (LNAI), vol. 2377, pp. 751–754. Springer, Heidelberg (2002)
7. Eiben, A., Smith, J.: Introduction to Evolutionary Computing. Springer, Heidelberg (2003)

Heuristics for Resolution in Propositional Logic

Manfred Kerber

Computer Science, University of Birmingham
Birmingham B15 2TT, England
http://www.cs.bham.ac.uk/~mmk

Abstract. One of the reasons for the efficiency of automated theorem systems is the usage of good heuristics. There are different semantic heuristics such as set of support which make use of additional knowledge about the problem at hand. Other widely employed heuristics work well without making any additional assumptions. A heuristic which seems to be generally useful is to "keep things simple" such as prefer small clause sets over big ones. For the simple case of propositional logic with three variables, we will look at this heuristic and compare it to a heuristic which takes the structure of the clause set into consideration. In the study we will take into account the class of all possible problems.

1 Introduction

It is typically difficult to study the value of a heuristic on the class of all problems. Rather a heuristic is studied on a set of challenge problems and if it is beneficial on those it is assumed to be valuable in a wide class of related problems. This is a valid approach since it offers some insights into the quality of a heuristic. However it does not say much on the general benefit when applying a heuristic to the class of all problems in a particular logic.

In this work we will investigate a few heuristics for a very simple logic, namely propositional logic with three variables ($PROP_3$). There is justified hope that this can be extended to propositional logic with four variables in a complete study. However, a complete study of logics with more than four variables looks beyond reach at the moment, since the set of possible clause sets grows hyper-exponentially, actually with n variables, there are 2^{3^n} different clause sets (without tautologies). This means even for $n = 4$ it is impossible to consider all 2417851639229258349412352 different clause sets. For $n = 3$ there are still 134217728 different clause sets. As reported in [3], by applying symmetry reductions (based on permutations and flips of truth values) and traditional theorem proving reductions (subsumption and purity) the number of cases for $n = 3$ is reduced to just 411 cases. Unfortunately the reduced class grows rapidly as well and for $n = 4$ the number of cases in the reduced class is already in the millions. For any $n > 4$ even the reduced class looks beyond practical computability.

Still, the reduction allows to study the whole class of problems for $n = 3$ and see whether some special heuristics are beneficial on the whole class altogether. As we will see, an instance of the heuristic to "Keep It Short and Simple"

B. Mertsching, M. Hund, and Z. Aziz (Eds.): KI 2009, LNAI 5803, pp. 656–663, 2009.

(KISS) is overall beneficial. The concrete instance is that among all possible steps always one is chosen so that the resulting clause set (after reduction) has minimal cardinality. This shows that in this propositional logic, there is actually a "free lunch." Or to put it differently, this heuristic is beneficial over the whole class and not only over a subclass. Although the heuristic KISS is beneficial it is not optimal in the sense that it prescribes always best possible choices. From this follows the question whether it is possible to improve on KISS. In order to do so we will have to clarify what we mean by a heuristic. If we put arbitrary resources into a heuristic we can first build a complete analysis and then just follow a best route. This, however, should not count as a heuristic, because it is computationally (with respect to time and space) prohibitively expensive.

In the next section we will introduce the general framework. Then the results of [3] will be summarized. We will build a new heuristic and evaluate it as well as KISS on the whole class of $PROP_3$.

2 Resolution, (Un)Satisfiability, Heuristics

The goal of this work is to find properties of heuristics in theorem proving, and in particular in resolution theorem proving. Ultimately we would like to find heuristics for resolution in first-order logic and prove properties for them which go beyond proofs of the kind that a restriction strategy is complete and does well on a set of challenge problems. For this reason we will use the resolution calculus in order to establish whether a clause set is satisfiable or not.

We assume that the problem is given in form of a set of clauses (which have to be made simultaneously true). A clause is a disjunction of literals and a literal is either a propositional logic variable or its negation. Since tautologies do not contribute to the problem and can be easily detected, we assume that all our clauses are free of tautologies. In the study we apply the propositional logic resolution rule. The question of finding a short or a long proof is the question of selecting two good clauses for applying the rule. That is, a heuristic can be viewed as a procedure which restricts the choice of possible resolution steps. This type of heuristic is also called a restriction strategy in theorem proving. Typically there is more than one resolution possibility before applying a restriction strategy and fewer but still more than one after applying it. Among all those remaining we assume that a theorem prover makes a random choice. Initially and always after the application of the resolution rule the procedure applies eagerly the reductions of subsumption and purity. (Without loss of generality we call all the propositional logic variables X_i.)

Subsumption means: if there are clauses in the clause set so that one subsumes the other then the subsumed clause can be and is deleted from the clause set. For instance, $X_1 \vee X_2 \vee \neg X_3$ is subsumed by $X_1 \vee \neg X_3$, since the information in $X_1 \vee \neg X_3$ entails $X_1 \vee X_2 \vee \neg X_3$. Hence $X_1 \vee X_2 \vee \neg X_3$ can be deleted from a corresponding clause set in the presence of $X_1 \vee \neg X_3$.

Purity means that a particular propositional logical variable occurs in the whole clause set either exclusively positively (e.g., X_1) or exclusively negatively (e.g., $\neg X_1$). In the first case all clauses containing X_1 can be satisfied by assigning `true` to X_1, in the second case by assigning `false` to it. Since this has no impact on the rest of the problem formulation, the corresponding clauses can be deleted.

Furthermore we reduce the class of problems by forming equivalence classes of structurally equivalent clause sets with respect to flipping truth values (e.g., $\{X_1 \vee X_2 \vee X_3, \neg X_1, \neg X_2 \vee X_3\}$ is structurally equivalent to $\{\neg X_1 \vee X_2 \vee X_3, X_1, \neg X_2 \vee X_3\}$) or permuting the names of the X_i (e.g., $\{X_1 \vee X_2 \vee X_3, \neg X_1, \neg X_2 \vee X_3\}$ is structurally equivalent to $\{X_1 \vee X_2 \vee X_3, \neg X_2, \neg X_1 \vee X_3\}$). For details, see [3].

If in the process of deleting clauses from the clause set all clauses are deleted then the remaining clause set (the clause set consisting of no clauses at all) is trivially satisfiable. This means that the original clause set was satisfiable. If, however, by applying the resolution rule, the empty clause is derived then the corresponding clause set is unsatisfiable and so is the original clause set.

When we want to compare heuristics with respect to the full class we have to make a choice what we mean by full class and how the comparison should look like. We assume that the simplifications which are usual in theorem proving, subsumption and purity deletion, are applied. In the class of the remaining clause sets, every clause set goes with a multiplicity, that is, a number which states the size of the equivalence class. For instance, the empty clause set and the clause set consisting only of the empty clause stand only for themselves each. The (satisfiable) clause set $\{\neg X_2 \vee \neg X_3, X_2 \vee X_3, \neg X_1 \vee \neg X_3, \neg X_1 \vee \neg X_2 \vee X_3, X_1 \vee \neg X_2 \vee X_3, X_1 \vee X_2 \vee \neg X_3\}$ on the other hand stands for in total 48 different structurally equivalent clause sets which are all subsumption and purity free, and which can be obtained from it by permutations and flipping truth values (see [3]). When we take an average of the benefit (or damage) of a heuristic then we will take the weighted average with respect to the cardinality of the equivalence class.

3 Lengths of Proofs

In order to compute the lengths of proofs we first generate the 411 representatives for all clause sets with three variables (as described in [3]). Next these are sorted with the two trivial clause sets, the empty set (represented by `NIL`) and the clause set consisting of the empty clause (represented by `("###")`), coming first and second. All other clause sets are ordered with respect to one-step application of the resolution rule. A clause set always comes after all clause sets which it results in by applying resolution once. This way we get a DAG (Directed Acyclic Graph) with two connectivity components, the satisfiable and the unsatisfiable clause sets. Note that we represent clauses in form of strings of length 3 over the alphabet `{0,1,#}` in the following form, a string `"01#"` stands for $\neg X_1 \vee X_2$ and `"111"` for $X_1 \vee X_2 \vee X_3$ and so on. That is, in such a string the ith position being `0` means that X_i occurs negatively, being `1` that it occurs positively, and `#` that it does not occur at all. Hence the empty clause is represented as `"###"`.

In Table 1, some of the 411 representatives for all clause sets (without tautologies) are displayed. The first column contains the running number, the second the multiplicity (that is, the cardinality of the corresponding equivalence class),

Table 1. Some of the 411 structurally different clause sets

```
Nr Mul Clause set                                        Resolvents STEP-Nr Sat
( 0   1 NIL                                                     (0) 0 0   0 Y)
( 1   1 ("###")                                                 (1) 0 0   0 N)
( 2   4 ("000" "111")                                           (0) 1 1   1 Y)
( 3   6 ("11#" "00#")                                           (0) 1 1   1 Y)
( 4   3 ("##0" "##1")                                           (1) 1 1   1 N)
( 5  24 ("000" "011" "100")                                     (0) 1 1   1 Y)
( 6   8 ("000" "011" "101")                                     (0) 1 1   1 Y)
(10  12 ("10#" "01#" "00#")                                     (0) 1 1   1 Y)
(11  24 ("##0" "00#" "111")                                     (3) 2 2   2 Y)
(12  12 ("#11" "#0#" "##0")                                     (4) 2 2   2 N)
(13   8 ("000" "001" "010" "100")                               (0) 1 1   1 Y)
(14  24 ("000" "001" "010" "101")                             (7 0) 1 2 3/2 Y)
(24   3 ("##0" "##1" "#0#" "#1#")                               (1) 1 1   1 N)
(25  24 ("#00" "#01" "00#" "10#" "010")                         (0) 1 1   1 Y)
(26  48 ("#00" "#01" "0#0" "00#" "110")                        (0 10) 1 2 3/2 Y)
(27  48 ("#00" "#01" "0#0" "00#" "111")                        (11 3) 2 3 5/2 Y)
(28  24 ("##0" "#01" "0#1" "01#" "10#")                         (0) 1 1   1 Y)
(29  12 ("##0" "##1" "00#" "01#" "10#")                       (1 4) 1 2 3/2 N)
(30  12 ("##0" "##1" "#0#" "0##" "11#")                      (1 24) 1 2 3/2 N)
(410 1 ("000" "001" "010" "011" "100" "101"
                                  "110" "111") (409) 7 18
                     4995158050284531262459/462508951339008000000 N)
```

the third a representing clause set, the fourth the running numbers of all possible clause sets resulting from one-step resolution (representing a DAG), the fifth the minimal number of steps to arrive either at the empty clause set or the empty clause, the sixth the corresponding maximal number, the seventh the average number, and finally the eighths the information of whether the clause set is satisfiable or unsatisfiable. With our convention from above, we read in clause set 10, for instance, ("10#" "01#" "00#") as $\{X_1 \vee \neg X_2, \neg X_1 \vee X_2, \neg X_1 \vee \neg X_2\}$ and in clause set 11, ("##0" "00#" "111") as $\{\neg X_3, \neg X_1 \vee \neg X_2, X_1 \vee X_2 \vee X_3\}$.

The minimal number of resolution steps which is necessary to decide whether a clause set is satisfiable or not is stored in the fifth column. It is computed recursively by taking the minimal number of steps for the first two clause sets as zero and for each other as the minimum of all minimal numbers of steps necessary for all possible one-step results plus one. For instance, for the value of clause set 27 we look up the minimal numbers of clause sets 11 and 3 as 2 and 1 step(s), respectively, the minimum is 1 step, hence the value for 27 is 2. Likewise the maximal number (in the sixth column) is computed as the maximum of all maximal numbers of steps necessary for all one-step results plus one, that is, for 27 it is $\max(2, 1) + 1 = 3$. For the average number of steps the average of the numbers of steps of the results plus one is taken. For instance, the average number for clause set 30 is computed as the average of clause set 1 with value 0 and clause set 24 with value 1, that is average 1/2, hence the value is 3/2. If we take the weighted average of the corresponding numbers we get for $n = 3$:

best	random	worst	fewest	most
$\frac{7997}{3174} \approx 2.52$	$\frac{2250875353942862331559161 83}{5284812281580041011200000 0}$			

That is, if we have an oracle which tells us the best steps to take then we need on all problems on average approximately 2.52 steps (in the class of 12696 possible clause sets, which are free from subsumed or pure clause sets). Making random choices we need approximately 4.26 steps, and if the oracle always chooses the worst steps it takes on average approximately 7.58 steps. The most difficult clause set requires at best 7 steps and at worst 18.

4 Heuristics

In this section we will look at three different heuristics, KISS, anti-KISS, and INNER (a new heuristic built on the inner product). We will first describe the three heuristics and in the next section describe how resolution behaves using them. (Other heuristics are currently under investigation.)

KISS (Keep It Short and Simple) is a heuristic used from the early days of theorem proving. In the concrete instance we use it, it means, if there are more than one resolvent possible from a given clause set, then choose randomly one which results – after purity and subsumption reduction – in a *shortest* clause set (that is, a clause set with lowest cardinality).

Anti-KISS does the opposite, that is, it will always choose randomly one from the longest clause sets (biggest cardinality).

For INNER we define an inner product of clauses as follows. For any two clauses $C =$ "$c_1 c_2 \cdots c_n$" and $C' =$ "$c'_1 c'_2 \cdots c'_n$" (and for a clause set S) the inner product is defined as

$$C \cdot C' = \sum_{i=1}^{n} c_i \cdot c'_i \text{ and } \hat{S} = \frac{\sum_{C,C' \in S} (C \cdot C')^2}{\sqrt{|S|}}$$

For the empty clause set we define the $\hat{\emptyset} = 0$. The product of two characters is defined by four real-valued constants v_1, v_2, v_3, and v_4:

$$\begin{array}{lll} \texttt{\#} \cdot \texttt{\#} = v_1 & \texttt{0} \cdot \texttt{\#} = v_3 & \texttt{1} \cdot \texttt{\#} = v_3 \\ \texttt{\#} \cdot \texttt{0} = v_3 & \texttt{0} \cdot \texttt{0} = v_2 & \texttt{1} \cdot \texttt{0} = v_4 \\ \texttt{\#} \cdot \texttt{1} = v_3 & \texttt{0} \cdot \texttt{1} = v_4 & \texttt{1} \cdot \texttt{1} = v_2 \end{array}$$

In experiments we found (by local search) that $v_1 = 4.00$, $v_2 = -4.30$, $v_3 = 2.39$ and $v_4 = 2.74$ are good values which form a local optimum. Note that the inner product is defined so that it is firstly symmetric and secondly invariant with respect to flips of truth values ($\texttt{\#} \cdot \texttt{0} = \texttt{\#} \cdot \texttt{1}$, $\texttt{0} \cdot \texttt{1} = \texttt{1} \cdot \texttt{0}$, and $\texttt{0} \cdot \texttt{0} = \texttt{1} \cdot \texttt{1}$) and permutations (because of the commutativity of summation). This invariance is important, since we consider only representatives of clause sets rather than the full classes of all clause sets and this property guarantees that the product is well-defined (that is, independent of the particular representative we choose).

The INNER heuristic restricts the possible resolvents to choose from the ones for which the following weighted sum of the resulting clause set S is minimal:

$$\mathit{INNER}(S) = \alpha \cdot \hat{S} + \beta \cdot W(S)$$

$W(S)$ is defined similarly to the number of clauses in a clause set. However, a clause with i # symbols counts for 2^i clauses. Experimentally again $\alpha = 1.1$ and $\beta = -1.0$ have turned out to be good values, which are locally optimal.

Why are the inner product and the #-weighted cardinality of a clause set relevant? The length of a clause set is relevant in that 'the more different clauses are in a clause set the more likely it is that the clause set is unsatisfiable'. A high number of hashes means that the clause is in the directed acyclic graph of all clauses close to the bottom most elements (empty clause set and empty clause). The inner product between different characters is relevant as to the superficial structure of clauses.

Note that all three heuristics, KISS, anti-KISS, and INNER can be computed in polynomial time (obviously KISS and anti-KISS are much easier to compute than INNER). For each it is necessary to first compute all possible resolvents of a clause set (which is feasible for this study but will typically be avoided in actual theorem proving). For KISS and anti-KISS, just the length of each of those is needed. This can be computed in constant time (times the number of resolvents). For INNER we need to compute sums which depend on the cardinality of the clause sets (quadratic) and the length of the strings (linear). As remarked at the end of the introduction, a heuristic needs to have a reasonable complexity in order to be useful. While in the current study we have computed complete information for $PROP_3$, we do not take this as heuristic information, but as information for the study of the heuristics. Obviously, the hope is that the heuristics found here will turn out useful without the need for a complete analysis. In the next section we will compute the quality of the three heuristics in $PROP_3$ with respect to three different measures.

5 Findings

In this section we will study three different measures of the heuristics (KISS, anti-KISS, and INNER) and compare them to the best possible, average, and worst possible performance. For INNER, we use in this the values $v_1 = 4.00$, $v_2 = -4.30$, $v_3 = 2.39$ and $v_4 = 2.74$ as well as $\alpha = 1.1$ and $\beta = -1.0$. These turn out to be optimal in the sense that the values reported below cannot be improved by small local changes to any individual value. However, it may still be the case that these values form only a local optimum and that there are better values, which are significantly different from the chosen ones.

We study in the following three criteria for performance and give results for all clause sets as well as satisfiable ones and unsatisfiable clause sets separately.

- What is the average performance, that is, how many resolution steps does it take to arrive either at the empty clause set or the empty clause on average, where the average is taken as the weighted average with the multiplicities specified in the clause sets?
- In how many cases of the 411 structurally different cases (or in all 12696 cases) does the heuristic not result in an optimal derivation?

– What is the maximal difference between the optimal performance and the actual performance following the heuristic?

If we compare first the average performance computed as the weighted average of steps necessary, we get (rounded values computed as rational numbers):

Heuristic	worst	anti-KISS	average	KISS	INNER	best
all	7.58	7.46	4.26	2.70	2.62	2.52
SAT	5.22	5.07	3.11	1.81	1.78	1.74
UNSAT	10.42	10.35	5.64	3.76	3.62	3.46

This means that KISS is quite good with a performance of 93%. INNER improves on this to 96%, however, at the price that it is not easily understandable, and that the parameters have to be selected. anti-KISS is almost as bad as it gets.

If we compare in how many cases the heuristic does not give the best possible result we get:

Heuristic	worst	anti-KISS	average	KISS	INNER	best
all	362 (11867)	362 (11867)	362 (11867)	96 (2813)	42 (1216)	0 (0)
SAT	162 (6152)	162 (6152)	162 (6152)	23 (740)	9 (288)	0 (0)
UNSAT	200 (5715)	200 (5715)	200 (5715)	73 (2073)	33 (928)	0 (0)

The two rows are out of the 411 different equivalence classes and in brackets the total number of 12696 cases altogether, respectively. That is, with this measure again KISS is a very useful heuristic, but can be significantly improved on by INNER.

Finally we look at the maximal difference between the optimal number of steps and the actual number of steps needed with the heuristic. We get:

Heuristic	worst	anti-KISS	average	KISS	INNER	best
all	11	11	4.40	1.50	1	0
SAT	7	7	3.61	1.35	1	0
UNSAT	11	11	4.40	1.50	1	0

In all these cases we find the order: worst $\leq$ anti-KISS $\leq$ average $<$ KISS $<$ INNER $<$ best. According to the first criterion we find that the worst choice is only slightly worse than anti-KISS, which is much worse than random selection. Random selection in turn is much worse than KISS. KISS is slightly worse than INNER, which in turn is almost optimal.

6 Conclusion

This work was partly inspired by work on symmetry in constraint satisfaction problems, in particular by the work of Frisch et al. [2], where permutations and changes of polarity of boolean variables play a major role. Likewise the studies in the complexity of classifier systems [4] led to a better understanding of representations of clause sets. The work by Pearl [5] is a numerical study of heuristics in general, while Ben-Sasson and Wigderson [1] investigate properties of short proofs in resolution. In traditional theorem proving typically the completeness of a restriction or selection strategy is proved (since, e.g. [6]), however, the usefulness is typically established by evaluating them against test cases (as found in the TPTP [7], for instance).

In the current work we could establish that heuristics can be useful (and in consequence also harmful) over the class of *all* problems in PROP_3. A very simple heuristic (KISS) is very effective. Although it is very good and close to an optimal behaviour, it is not optimal and the author could not find an efficient heuristic which is. However, it was possible to improve on it by the INNER heuristic by more than 40% with respect to the average number of steps needed. Obviously the work reported here can be extended in various ways. A real test for the usefulness of KISS and INNER will be to apply them systematically in PROP_4 and sample them in PROP_n for $n > 4$. This is work in progress. Unfortunately even the reduced class of structurally different clause sets has a cardinality which goes in the millions, this involves time and space issues. The time issues are mainly solved (using hashtables), the space issue is under consideration. The thesis which will be tested is that the heuristics KISS and INNER will be useful in PROP_4 as well (and for more propositional logic variables), but that the usefulness will be reduced. Possibly the parameters for INNER will have to be changed. Also other heuristics are under investigation.

In PROP_3, the most difficult clause set (the biggest number of resolutions steps necessary) is `("000" "001" "010" "011" "100" "101" "110" "111")`. It requires in the worst case 18 steps, in the best case 7. With INNER we need 7 steps, 7.2 with KISS, 18 with anti-KISS. This problem can be easily generalized for higher n in PROP_n. To study the behaviour of KISS for solving this problem in bigger n is also left for future work.

References

1. Ben-Sasson, E., Wigderson, A.: Short proofs are narrow–resolution made simple. Journal of the ACM 48(2), 149–169 (2001)
2. Frisch, A.M., Jefferson, C., Hernandez, B.M., Miguel, I.: The rules of constraint modelling. In: Proc. of the 19th IJCAI, pp. 109–116 (2005)
3. Kerber, M.: Normalization issues in mathematical representations. In: Autexier, S., Campbell, J., Rubio, J., Sorge, V., Suzuki, M., Wiedijk, F. (eds.) AISC 2008, Calculemus 2008, and MKM 2008. LNCS (LNAI), vol. 5144, pp. 494–503. Springer, Heidelberg (2008)
4. Kovacs, T., Kerber, M.: A study of structural and parametric learning in XCS. Evolutionary Computation Journal 14(1), 1–19 (2006)
5. Pearl, J.: Heuristics – Intelligent Search Strategies for Computer Problem Solving. Addison-Wesley, Reading (1984)
6. Chang, C.L., Lee, R.C.T.: Symbolic Logic and Mechanical Theorem Proving. Academic Press, New York (1973)
7. Sutcliffe, G.: The TPTP Problem Library and Associated Infrastructure. The FOF and CNF Parts, v3.5.0. Journal of Automated Reasoning (to appear, 2009)

Context-Aware Service Discovery Using Case-Based Reasoning Methods

Markus Weber[1], Thomas Roth-Berghofer[1,2], Volker Hudlet[2], Heiko Maus[1], and Andreas Dengel[1,2]

[1] Knowledge Management Department,
German Research Center for Artificial Intelligence DFKI GmbH
Trippstadter Straße 122, 67663 Kaiserslautern, Germany
[2] Knowledge-Based Systems Group, Department of Computer Science,
University of Kaiserslautern, P.O. Box 3049, 67653 Kaiserslautern
{firstname.lastname}@dfki.de, v_hudlet@cs.uni-kl.de

Abstract. This paper presents an architecture for accessing distributed services with embedded systems using message oriented middleware. For the service discovery a recommendation system using case-based reasoning methods is utilized. The main idea is to take the context of each user into consideration in order to suggest appropriate services. We define our context and discuss how its attributes are compared.

The presented prototype was implemented for Ricoh & Sun Developer Challenge. Thus the client software was restricted to Ricoh's Multi Functional Product as an embedded system. The similarity functions were designed and tested using myCBR, and the service recommender application is based on the jCOLIBRI CBR framework.

Keywords: Case-Based Reasoning, context, service discovery, myCBR, jCOLIBRI.

1 Introduction

The Ricoh & Sun Developer Challenge[1] is a programming contest where students invent and implement innovative applications for Ricoh's Multifunctional Product (MFP)[2]. The MFP is an office machine that incorporates the functionality of several devices. It combines printer, scanner, photocopier, fax and e-mail functions. It offers developers Ricoh's Embedded Software Architecture SDK (SDK/J)[1] for implementing and delivering customized Java-based solutions hosted on Ricoh MFPs. A large touch screen is the main interaction device.

In order to decouple business logic from the MFP, we decided to come up with a distributed architecture to access services. Obviously a lot of applications for MFPs focus on document handling. Choosing a centralized approach seems to be reasonable for implementing services. A service, implemented and running on a server machine, is not as limited as one on an embedded system. Furthermore the

[1] http://edu.ricoh-developer.com/contest/open/index.jsp
[2] http://www.ricoh.de/products/multifunction/mediumworkgroup/mpc2550.xhtml

B. Mertsching, M. Hund, and Z. Aziz (Eds.): KI 2009, LNAI 5803, pp. 664–671, 2009.

services are independent of the programming language offered on the embedded system, as the business logic is implemented on a server machine.

Such a service-oriented architecture will grow quite fast if a suitable software development kit is available for a community or other vendors, as can be seen by such current trends as the Apple Appstore[3] for the iPhone[4]. So service discoverability is an issue in such a service oriented architecture. There is a need for an intelligent way of recommending services, especially considering the end user at an MFP is confronted with limited time and interaction convenience.

Let us assume the following scenario: A business man is traveling to a conference. At the airport he might be interested in information related to *his location, the airport*, such as offers of the duty free shop. Arriving at the hotel he, in the role of *a hotel guest*, might be interested in information about the hotel services. *In the evening* he might be looking for a good restaurant and in the morning at the conference he, now in the role of a conference participant, may want to read news about the financial market.

In this scenario the context is a good indicator which service might be helpful. Case-based reasoning (CBR) is used to implement content-based recommender systems (see, for example, [2]). Context information is stored as a case. One of the advantages of CBR systems is that they are capable of learning. Service recommendation is improved by taking user feedback into consideration.

In this paper we discuss the design of our architecture and its implementation. We present a service recommender that takes the context of the user into account for suggesting services using case-based reasoning methods. Our prototype focuses on the MFP as a service consumer. In Section 2 we will discuss the context, and in Section 3 we present the suggested architecture. The service recommendation using a case-based reasoning approach will be described in Section 4. Finally, there will be an outlook on further work.

2 Context

In our work we decided to use the context as an indicator to suggest services that fit to current needs of the user. In mobile computing context-awareness is an important research topic, as modern human mobile computing becomes more important. Modern mobile devices are capable of accessing online services via GPRS, UMTS, WLAN, and so forth, hence the need for personalised and adaptive information services is rapidly increasing.

Schmidt et al. [3] investigated in their work how to specify context in mobile computing. They introduced a working model for context and discuss mechanisms to acquire context beyond location, and application of context-awareness in ultra-mobile computing. Therefore they investigated the utility of sensors for context-awareness and present two prototypical implementations. In [4], Kofod-Petersen describes an approach to facilitate the use of contextual information in order to improve the quality of service in a mobile ambient environment. So

[3] `http://www.apple.com/de/iphone/appstore/`

[4] `http://www.apple.com/de/iphone/`

context-awareness seems to be a promising approach for the MFP, as information about the MFP's environment is easy to measure. Hence we need to define the term context and how our context looks like.

A popular definition of context is given by Dey [5]: "Context is any information that can be used to characterize the situation of an entity. An entity is a person, place, or object that is considered relevant to the interaction between a user and an application, including the user and applications themselves." For our purposes we could take the definition quite literally. We looked at the context of an MFP and investigated which information could be automatically measured in order to characterize a situation and found that the attributes *user role, daytime, device,* and *type of location* were appropriate for defining our context.

Identification of the *user role* is not a trivial task. For the prototype we assumed that the user is registered at a service provider and some initial knowledge about his role exists, such as his job or his hometown. To improve the result of this simple reasoning method, the registry can ask the user for feedback and propose possible roles to choose from. By using simple assumptions like if someone uses an MFP in a city that is far away from his hometown, he might be there as a tourist, or on a business trip.

Measuring *daytime* is a trivial task, as the devices have an internal clock. For our purpose, we divided a day into five intervals: Morning (6:00–11:00am), Noon (11:00am–2:00pm), Afternoon (2:00–6:00pm), Evening (6:00–11:00pm), and Night (11:00pm–6:00am). During the day the needs of a user might change. For instance around noon or in the evening a restaurant guide might be more likely to be helpful as in the middle of the night.

Even though the focus of the developer challenge was on MFP devices, the services could also be used by mobile devices, as already addressed. Thus the *type of the device* is an important issue. Services producing larger documents are more interesting for e-readers or MFP devices as for smart phones, as reading larger texts on a smart phone can be unpleasant.

The *type of location* has to be set by the administrator of the device. Therefore a predefined set of possible types has to be available, such as public place, office or private household. According to the location several services could be excluded as the environment is not adequate, for instance a banking service on a public place. Another example could be a hotel, in which tourist services, restaurant and entertainment guides would be appropriate.

Next we take a closer look at the system's architecture and give an overview where context and CBR methods can improve quality of the system.

3 Architecture Overview

Our framework is a distributed architecture for accessing services with an MFP or a mobile device, respectively. The business logic is implemented in services running on server machines. The device just provides the user interface and communicates with the services. Figure 1 illustrates schematically the system and its components.

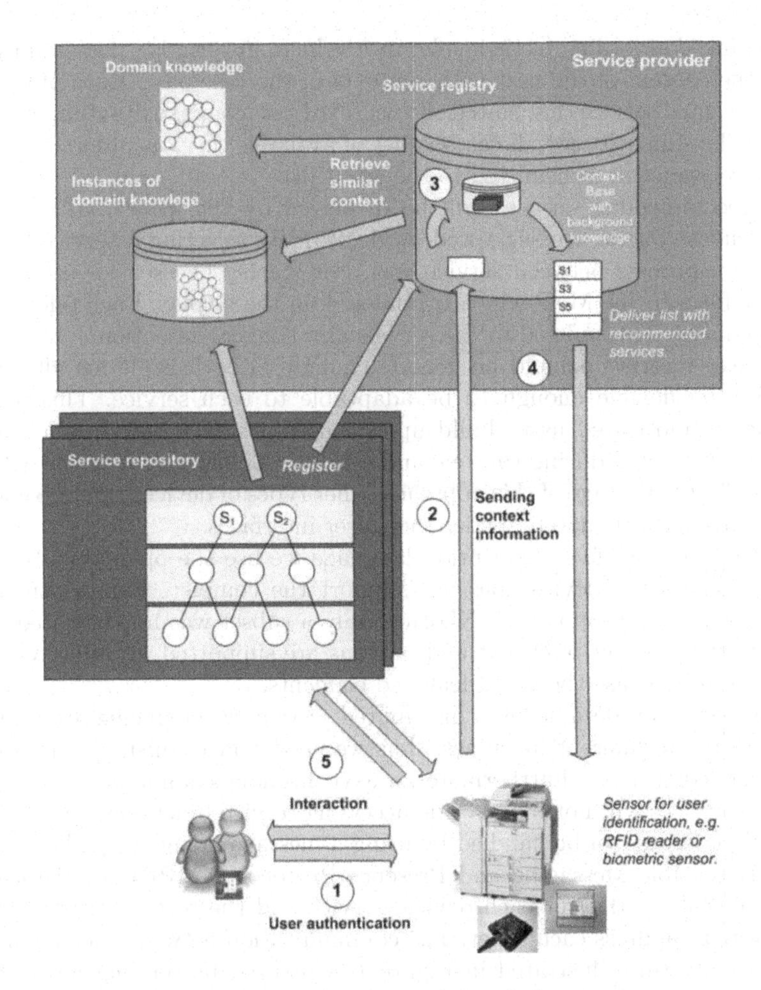

Fig. 1. Technical overview of the whole system

The *service provider* is the entity that offers services and the service registry. Users of the system have to be registered at the service provider as additional services can only be provided to registered users. Each user creates a profile with knowledge about his interests and some meta information about him, such as his job, his city, and so on.

In order to authenticate users (1), we suggest using RFID card systems (such solutions are widely available for MFPs) or biometric sensors on MFP devices, as no typing is needed for user password authentication. Context information, described in detail in Section 2, will be transmitted to the *service registry* (2).

The *service registry* is the main component in the architecture as it manages the services and their recommendation. Each service, either of the service provider itself or other providers, has to register at the service registry. The recommendation is realized by using a case-based reasoning system (3). Services will be ranked according to their relevance, which, in turn, is calculated by comparing the context of the user with contexts in the case base. Each of the cases has an assigned service that suited the context best in a past situation. After calculating the similarity for all cases, a list of available services ordered according to their relevance is transmitted to the MFP (4).

The recommended services will be displayed on the MFP and the user is able to choose one of the suggested service (5). By selecting a service the communication process between service and user starts. The service sends a user interface form to the MFP which is rendered on the display. Each selection will be immediately transmitted to the service and triggers an action.

The user interface should be generic enough to be feasible for all services, but must be flexible enough to be adaptable to each service. This dynamic requirement motivated us to build upon a user interface description language which is interpreted during runtime and rendered on the screen of the MFP. As an additional advantage of this approach other types of devices can also consume the services and adequately render their user interfaces.

For the user interface description language we use the open W3C standard XForms [6]. As the device does not support the complete bandwidth of user interface elements proposed by XForms, only a subset was implemented on the MFP. At the moment only text and buttons are supported. In later versions it is possible to increase the set of featured elements.

As our system is designed as a distributed system, we are facing several issues. Services run on different machines, thus we need a mechanism to discover the recommended services. Furthermore an asynchronous communication is better suited as devices will not block while accessing service functionality.

All these issues can be handled by using a message oriented middleware such as the Extensible Messaging and Presence Protocol (XMPP) [7]. We assign a unique identifier (JID) to each device, service and the service registry to provide a way to address each other. The communication between each component in the architecture illustrated in Figure 1 is realized by sending messages. As XMPP is an open protocol based on XML there are several libraries for various programming languages[5] available.

The device starts the communication process by sending the context to the service registry. This triggers the reasoning process and a list of recommended services and their respective JID is replied to the device. The user now selects one of the recommended services. By using the JID the device can address the service and directly start the communication process with the service, without knowing the actual host where it is running. As the routing is handled by the XMPP protocol even using a service running behind a firewall is possible.

After discussing the components from a top-level perspective, we will have a closer look at the system. As the service registry is the central component of the system it will be introduced in the following section.

[5] http://en.wikipedia.org/wiki/List_of_XMPP_library_software

4 Service Recommender

The main function of the service registry is the discovery of appropriate services. We interpret the context of the user as problem and the service as solution. By taking the current context into consideration the system calculates a relevance for the cases, stored in the case base. As the main idea of CBR is to learn from experience we see the service registry as an experienced advisor that knows about several contexts and can tell the user which services are available in order to help serving his needs.

Firstly each service has to register at the service registry and provide information, such as service name, a service identifier, a service description with focus on the user, the category of the respective service, and initial context. For the service identifier the Jabber ID is used.

The initial context is needed to solve the bootstrapping problem, as we need cases for our case base. So developers of the service have to think about a reasonable context, where the service is helpful. This context is a prototypical for this service and will be relevant to find the service in its initial state. For instance, a restaurant guide could be helpful at a public place at noon or in the evening. This common case will be added to the case base.

During the retrieval process and the assembly of the service list, we also have the opportunity to apply filters. For instance there could be certain limitations for the services, e.g., a service might not be allowed to be accessed in a public place, because confidential data is accessed. Thus services can be excluded by the application of a set of filters.

We also needed to design similarity functions that compare context information. In order to compare the user role we decided to use an asymmetric table similarity. For instance we need to find a measure how similar a student is to a professor or a business man to a tourist. The asymmetric character becomes obvious if we compare a student and a tourist. In our example, the student is more similar to a tourist, as they might have the same interests. If we compare a tourist to a student, the similarity might be even 0.0, as a tourist is not interested in services offered by a university. In our prototype the similarity values are defined by a domain expert.

For daytime we defined time intervals and gave them a symbolic value, such as MORNING, NOON and so on. Thus we are able to compare them with a symmetric table similarity where MORNING is more similar to NOON as to AFTERNOON, and vice versa.

The location itself is just matched exactly. A more sophisticated solution would be to compare cities by using an ontology. Cities can be clustered according to special characteristics and arranged in a taxonomy. But as the location is just important for services, if they provide location based information, we just implemented the simple approach. Currently the more important attribute is the type of the location as it describes the location in a more common sense. A hotel lounge in New York is quite similar to a hotel lounge in Rome. Thus this information is more important for service recommendation. We chose different symbols such as HOTEL-LOUNGE, KIOSK, OFFICE, and so forth and arranged them in a taxonomy using abstract terms to group them such as PUBLIC-PLACE.

In our prototype the device attribute is less important, as we only implemented a client for a specific MFP device. Nevertheless, the basic idea of the architecture is to take different devices into consideration. A service that might be appropriate for MFP devices, can be inappropriate for smart phones, because the generated document is too large to read it on a small display. In order to group similar types of devices a taxonomy is defined, with abstract terms like MOBILE-DEVICE.

The global similarity is a weighted average of local similarities. After a similarity value for the retrieved cases is calculated and ordered according their relevance, a set of services has to be transfered to the requesting device.

If none of the suggested service is appropriate for the user, he is able to browse a category based listing of the services. Even though the calculated relevance might be low the service might be appropriate in the users current context. Thus in the revise step of the CBR cycle [8] the user decides by selecting a service that it fits to his current context. As the selected service and the current context are transmitted to the service registry, the system can learn additional contexts where the service might be applicable in the retain step. Thus the relevance of the service will be higher if a similar context comes up.

For the implementation of the prototype we chose myCBR[6], as it supports rapid prototyping to design and test the similarity functions [9]. Furthermore these functions can be exported and used within the jCOLIBRI[7] CBR Framework [10] quite easily, which was used to implement the service recommender.

5 Conclusion and Outlook

In this paper we have shown an application framework which has been developed within the scope of the The Ricoh & Sun Developer Challenge programming contest. This framework enables the creation and deployment of context-aware services that can be consumed by devices, especially the MFP. A prototype has been implemented and made second place in the contest.

As the architecture and utilized technologies are not limited to the MFP, the architecture could be extended to access services with mobile devices. So instead of printing a document, it could also be shown on a mobile device, such as the e-reader iRex Iliad[8] or on a smart phone such as Apple's iPhone.

By offering a service development kit to other vendors or a community an application market platform can grow quite fast, ideally like the Apple App Store. Thus an intelligent way to recommend services could be helpful to find appropriate services. Crawling through categories is time-consuming and an appropriate service might even not be found. Therefore using CBR methods seems to be a promising approach in this scenario, as it is an intuitive way to use experience

[6] `http://mycbr-project.net/`

[7] `http://gaia.fdi.ucm.es/projects/jcolibri/`

[8] `http://www.irextechnologies.com/products/iliad`

with usage of this service. Furthermore the system is able to learn from user feedback, as mentioned in Section 4.

A further development which supports other (mobile) devices is possible, as we build upon a programming language independent communication stack and user interface, but this remains a future task. As we have shown the core feature, the service recommender takes advantages of case based reasoning methods and frameworks. By the use of these and also by the use of the current surrounding context the recommender can propose suitable services. In future work the recommender will be refined to offer even better service matches.

Another task will be evaluate the performance of the system. As there are just two prototypical services available, there was not enough data to evaluate the system yet. A larger number of services is required to compare the performance with conventional category browsing or other recommendation systems.

References

1. Ricoh: White paper: Embedded software architecture sdk (2004)
2. Bridge, D.: Product recommendation systems: A new direction. In: Workshop on CBR in Electronic Commerce at The International Conference on Case-Based Reasoning, ICCBR 2001, 79–86 (2001)
3. Schmidt, A., Beigl, M., Gellersen, H.W.: There is more to Context than Location. Computers & Graphics Journal 23, 893–901 (1999)
4. Kofod-Petersen, A., Mikalsen, M.: An Architecture Supporting implementation of Context-Aware Services. In: Floréen, P., Lindén, G., Niklander, T., Raatikainen, K. (eds.) Workshop on Context Awareness for Proactive Systems (CAPS 2005), Helsinki, Finland, June 2005, pp. 31–42. HIIT Publications (2005)
5. Dey, A.K.: Understanding and using context. Personal and Ubiquitous Computing 5, 4–7 (2001)
6. W3C: Xforms 1.0, 3 edn. (2007), `http://www.w3.org/TR/xforms` (Last accessed: 2009-02-09)
7. Saint-Andre, P.: Extensible Messaging and Presence Protocol (XMPP): Core. RFC 3920 (Proposed Standard) (October 2004)
8. Aamodt, A., Plaza, E.: Case-Based Reasoning: Foundational Issues, Methodological Variations, and System Approaches. AICom - Artificial Intelligence Communications 7(1), 39–59 (1994)
9. Stahl, A., Roth-Berghofer, T.: Rapid prototyping of cbr applications with the open source tool mycbr. In: Althoff, K.-D., Bergmann, R., Minor, M., Hanft, A. (eds.) ECCBR 2008. LNCS (LNAI), vol. 5239, pp. 615–629. Springer, Heidelberg (2008)
10. Recio-García, J.A., Díaz-Agudo, B., González-Calero, P.A.: Prototyping recommender systems in jcolibri. In: ACM Conference On Recommender Systems RecSys 2008, pp. 243–250 (2008)

Forward Chaining Algorithm for Solving the Shortest Path Problem in Arbitrary Deterministic Environment in Linear Time - Applied for the Tower of Hanoi Problem

Silvana Petruseva

Mathematics Department, Faculty of Civil Engineering,
St.Cyril and Methodius University, Partizanski Odredi 24, P.O. Box 560,
1000 Skopje, Macedonia
silvanap@unet.com.mk

Abstract. The paper presents an application of an algorithm which solves the shortest path problem in arbitrary deterministic environment in linear time, using OFF ROUTE acting method and emotional agent architecture. The complexity of the algorithm in general does not depend on the number of states n, but only on the length of the shortest path, in the worst case the complexity can be at most O (n). The algorithm is applied for the Tower of Hanoi problem.

1 Introduction

Emotion has been identified as one of the key elements of the intelligence and of the adaptive nature of human beings [8]. The emotional based agent learning tries to develop artificial mechanisms that can play the role which the emotions play in human life. These mechanisms are called "artificial emotions". In the last 20 years there have been several pointers toward the issue of feelings and emotions as needed features for developing an artificial intelligent creature [2], [4], [5], [6], [11].

We use the *consequence-driven systems* approach [4],[5],[6] to emotional learning. *Consequence-driven systems theory* is an attempt to understand the *agent personality*, it tries to find architecture that will ground the notions such as motivation, emotion, learning, disposition, anticipation and behaviour. It originated in 1981 [3], [4] in an early reinforcement learning research effort to solve the assignment of credit problem using a neural network. Resulting agent architecture was the Crossbar Adaptive Array architecture (CAA). The CAA approach introduced the concept of *state evaluation as feeling*, which it used as an internal reinforcement entity. The consequence-driven systems theory assumes that the agent should always be considered as a three–environment system: The agent expresses itself in its *behavioural environment* where it acts. It has its own *internal environment* where it synthesizes its behaviour, and it also has access to a *genetic environment* from where it receives its initial conditions for existing in its behavioural environment. The genetic environment contains the initial values W of the *crossbar learning* memory and properly reflects the desirable or undesirable situations in the behavioural environment. Having initial emotional values of the desirable situations, and using the emotional value backpropagation mechanism, the CAA will learn how to reach those situations. The main

B. Mertsching, M. Hund, and Z. Aziz (Eds.): KI 2009, LNAI 5803, pp. 672–679, 2009.

module of the CAA architecture is its memory module, a connectionist weight matrix W. Each memory cell w_{aj} represents emotional value toward performing action *a* in the situation *j*. It can be interpreted as tendency, disposition, motivation, expected reward, or by other terms, depending on the considered problem to which CAA is used as solution agent architecture. From the emotional values for performing actions in a situation *k*, CAA computes emotional value v_k of being in the situation *k*. Having v_k, CAA considers it as a consequence of the action *a* performed previously in a situation *j*. So, it learns that performing *a* in *j* has as a consequence emotional value v_k, and on that basis it learns the desirability of performing *a* in *j*. Using *crossbar computation* over the *crossbar elements* w_{aj}, CAA performs its *crossbar learning method*, which has 4 steps:

1) state *j*: choose an action in state j, $a_j=Afunc\{w_{*j}\}$. Let the environment return situation *k*;
2) state *k*: feel the emotion being in state *k*: $v_k=Vfunc\{w_{*k}\}$;
3) state *j*: learn the emotion toward *a* in *j*: $w_{aj}=Ufunc(v_k)$;
4) change state: *j=k*, go to 1;

where w_{*k} means that computation considers all the possible desirability values of the k- th column vector. The crossbar learning method was a forerunner [1] to the Q-learning method [13]. The crossbar learning method above is classified as ON ROUTE acting method. Another variant of the CAA method is the so called OFF-ROUTE acting method which differs from the ON-ROUTE only in the 4-th step, which is: *4) choose j; go to 1*; the steps 1, 2 and 3 can be applied at any state; the state can be chosen randomly or some next–state choosing policy can be used.

The concept of underground dungeon can be used for describing the two methods. Assume dungeon (graph) with *n* rooms underground. If the graph has only one entry and one exit to the earth surface above, the only way to find an existing path is to go underground and search (ON ROUTE). If each room has access from the surface, a path can be found underground by visiting each room directly (OFF-ROUTE).

2 OFF ROUTE Forward Chaining Algorithm and the Modified CAA Architecture – Applied for the Tower of Hanoi Problem

The algorithm described below can be applied for solving the shortest path problem in *arbitrary deterministic environment* with *n* states using OFF ROUTE acting method. An assumption is that there is a path from the initial state to the goal state, and the maximal number of actions which the agent can undertake in every state is the constant *m* ($m<n$).

The agent starts from the starting state and should learn to move along the shortest path to the goal state. From each state the agent can undertake one of possible *m* actions, which can lead to another state or to a certain obstacle. By $succ(a_i, s)$ we shall denote the next state which is achieved from *s*, or the obstacle which can appear after executing the action a_i in *s*.

The agent uses *direct learning* which implements breadth-first searching, based on real experience, combined with *planning* based on simulated experience. The word "planning" is used to refer to any computational process that takes a model as input and produces or improves a policy for interacting with the modelled environment. The model is used to *simulate* the environment and produce *simulated experience.* The difference between learning and planning is that whereas planning uses simulated experience generated by a model, learning methods use real experience generated by the environment [12].

The knowledge about the connections of the states in the environment is considered as a part of the model of the environment. The agent learns partial model of the environment through the real experience.

The agent architecture used is shown in Fig.1. It is the modified CAA architecture explained in sec.1. The memory module W is widen with the memory elements *ps[j]* and *pac[j]* (j=1,..,n). *ps[j]* is the element for memorizing the first predecessor state *s* from which the state *j* is achieved, and *pac[j]* is the element for memorizing the action chosen in that state *s,* which leads to *j*. These elements are used for learning the partial model of the environment in the phase of the *model learning.* This phase is being accomplished simultaneously with the phase of direct learning. When the agent finds the goal state, the simulated experience is being generated using these elements. These elements are used for finding the optimal path from the initial state to the goal state, which is being marked backward - from the goal state to the initial state.

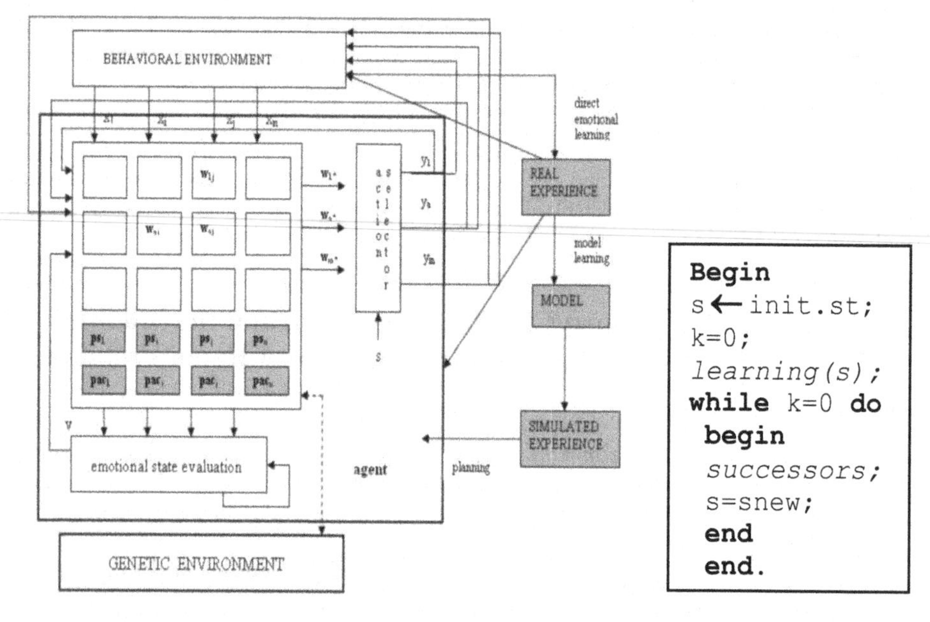

Fig. 1. Modified CAA architecture. Modifications referring to the original CAA are colored

Fig. 2. The forward chaining algorithm

The three learning functions explained in sec.1 are given in the following way:
1.The function Afunc for choosing an action in a state *j* is :

a) In exploring phase until the goal state is found, in every state the agent chooses and explores every action from that state one by one: y(j)= a (a=1,..m).

b) when the goal is found and the simulated experience is generated from the partial model, the agent uses this simulated experience in the phase of planning, moving through the optimal path, by choosing the actions in every state from the initial to the goal state with the function: $y(j)=\text{argmax}_a(w[a,j])$.

2. The function Vfunc for computing the emotional state of the agent in a state *k* is:

$$V(k)=\begin{cases}-100 & \text{if the state k is undesirable}\\ V(j)+1 & \text{otherwise (j is the first predecessor of k)}\end{cases}$$

3. The function Ufunc for updating the memory elements w[a,j] is given by:

$$w[a,j]=\begin{cases}V(k) & \text{if the action a in j leads to state k which is not passed before}\\ -100 & \text{if the action a in j leads to an obstacle or previously learned state}\end{cases}$$

Initially, all V(j)=0, w[a,j]=0 (j=1,..n, a=1,..m), except for the undesirable states *j* w[a,j]= -1, (a=1,..m), and for the goal state *g*, w[a,g]=1 (a=1,..m). The *forward chaining algorithm* is in Fig. 2 and the procedures for the algorithm are in Tab. 1.

After taking an action in *j*, if the successor state *k* is not an undesirable state, or previously passed state, the agent 1)goes in that state; 2) updates the value V(k) from V(k)=0 to V(k)=V(j)+1; 3) in *k* the agent memorizes (learns) the previous state *j* and the action which leads to *k* ; 4) updates the value w[a,j]=V[k] and returns to *j*. The value w[a,j]=V[k] means that by taking the action *a* in *j* the agent will learn that the shortest path from the initial state to the next state *k* is V[k]. After exploring all actions from the initial state *s*, the agent has found (learned) all consequence states *ss* with V(ss)=1 (the initial state has V(s) = 0). After that, the agent explores every state **ss** with V(ss)=1 one by one in the same way. So, after being in all these states, (with V(ss)=1 and exploring their actions), the agent finds all states *ss* with V(ss) = 2, and explores each of these states, and so on. When the agent finds the goal state, this exploring phase finishes and the simulated experience is being generated from the partial model which the agent has learned: Starting from the goal state every memory element w[i,j], for the action *i* and state *j* which were memorized in the phase of model learning, gets some high value (greater than *n,* for example 100). In this way the shortest path is marked from the goal state to the initial state. In the planning phase the agent moves through the optimal path using this simulated experience.

Wittek [14] explains that the human emotions are energy complexes which are composed of pictures, whose elements are thoughts, words, and acting from the past. They are consequences of our previous thinking, speaking and acting (previous experience) and precisely express our way of thinking and acting from the past.

In this approach toward emotional learning, the emotional value V(s) expresses the consequence of the agent's actions: taking an action from the previous state and being in the state *s*, the agent learns in *s* that the shortest path from the starting state is V(s). V(s) summarize the whole acting of the agent from the initial state to the state *s*. This algorithm originates from the polynomial algorithm [10] with speed $O(n^2)$,

Table 1. The procedures for the algorithm: *learning, successors, simulating-exp, planning*

```
procedure learning(s)
begin
 i=0;k=0;
  repeat
   i=i+1;
   take the action a_i in s:y[s]=a_i
   compute ss = succ(a_i,s);
   if ss is not undesirable state,
   obstacle^1 or previously learned
   state, then
   begin
   go in ss and learn the emotional
   value in ss: V(ss)=V(s)+1;
   learn the predecessor state s
   and the action a_i which leads to
   ss: Ps(ss)=s;    Pac(ss)=a_i;
   update the emotional value toward
   taking the action a_i in s:
     w[a_i,s]=V(ss);
   return to state s;
   end;
    if ss is a goal state then
     begin
      k=1;
      simulating-exp;
      planning;
      end
     else
     begin
     if ss is undesirable state
       then V[ss]= - 100;
       if the action leads to
       an obstacle,previously learned
       or undesirable state then
       w(a_i,s) = -100;
      end;
    until k=1
   end.
```

```
Procedure successors
 Begin
 ∀state s_1 for
   which
   V(s_1)=V(s)+1:
 repeat
   choose state s1
   snew = s_1;
   learning (snew);
 until k=1
 end.
```

```
Procedure
simulating-exp
begin
s'=goal state;
repeat
 subg= ps[s'];
 act=pac[s'];
 w[act,subg]=100 ;
 s'=subg;
until s'=start state
end.
```

```
procedure planning
begin
 s=starting state;
 repeat
  a=argmax w(a_i,s);
      a_i
  ss=succ(a,s);
  s=ss;
 until s=goal state
end.
```

which is obtained from exponential algorithm [9] when the dynamic programming is applied. The emotion function V(s) for the exponential algorithm measures the consequence of the agent action generally (only with 3 values: 1 (good), 0 (neutral), -1 (bad). For the polynomial algorithm V(s) measures the consequence of the agent

[1] Obstacle can be any kind of bar or a wall which the agent can encounter taking one action.

actions starting from the goal state, but very precisely (with exact value of the distance from the goal) whereby drastic improvement in the speed of convergence is obtained. Due to this observation, the idea was to propose algorithm whose emotion function will measure the consequences of the agent actions precisely starting even from the initial state - in this way this function implements very important property of human emotions. Thus, this linear-time algorithm was obtained.

Theorem 1: The extended *CAA* architecture with the *forward chaining algorithm* finds one shortest path from the initial state to the goal state in arbitrary deterministic environment in time which (without the operations for initiation) in general does not depend on *n,* the number of states, but only on the length of the shortest path, (it is task dependant), but in the worst case, depending on the position of the goal state, it can be at most O(n). The only assumption is that there is a path from the initial state to the goal state, and the maximal number of actions in each state is m $(m<n)$.

The proof of this theorem can be found in [11].

This algorithm is *uninformed* and its memory usage is at most O(n(m+k)) (m,k, constants, m<n) (the action and the emotion are being computed as neuron potentials and do not use memory space).

The algorithm can be applied to arbitrary *deterministic* environments, but variants of the algorithm should be explored for *non-deterministic* environments.

The algorithm finds *one* shortest path, and with some minimal corrections it will find *all* shortest paths.

This algorithm can be applied to any kind of problem which can be reduced to shortest path problem in arbitrary graph with n nodes, the only assumption is existing path from the initial to the goal node and maximal number of arcs from a node is m<n.

The most popular shortest path algorithm A* is *informed*. The time complexity of A* depends on the heuristic. In the worst case the number of nodes expanded is exponential in the length of the shortest path, end it is polynomial when the heuristic meets certain condition. The special case of A*, Dijkstra algorithm, runs in time $O(n^2)$. More problematic than its time complexity is A* memory usage. In the worst case it must remember an exponential number of nodes [15].

Now we shall present the application of the algorithm to the Tower of Hanoi problem.

The well-known problem called Tower of Hanoi (TH) is given in Fig.3a.

It has been pointed out that the graph of the problem has a fractal structure [7] and here we will use that graph as environment to which we will apply our algorithm. If k is number of disks, the adjacency matrices for *k*=1 and *k*=2 are shown in Fig. 3 (b and c). For each *k* they are square matrices of order 3^k, and their analytic form for k=1and k=2 is:

$$M_3 = \begin{bmatrix} 0 & 1 & 1 \\ 1 & 0 & 1 \\ 1 & 1 & 0 \end{bmatrix} \qquad M_{3^2} = \begin{bmatrix} M_3 & E_3(3,1) & E_3(2,1) \\ E_3(1,3) & M_3 & E_3(2,3) \\ E_3(1,2) & E_3(3,2) & M_3 \end{bmatrix}$$

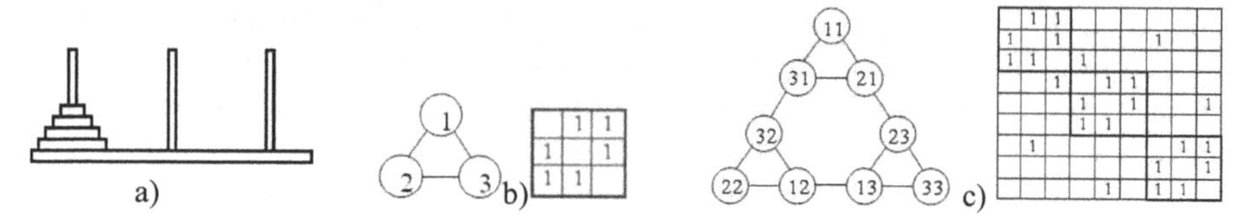

Fig. 3. a) Tower of Hanoi for 4 disks; graph and the corresponding adjacency matrix for the problem space for TH with: b)one disk; c) two disks (the empty squares in the matrices are 0)

For $k \geq 2$ every graph has 3 main triangles: upper – we enumerate it with 1, lower left (2), and lower right (3). If $T_{il}(k)$ and $T_{ir}(k)$ (i=1,2,3) are the ordinal numbers of the states in the lower left and the lower right top of the triangle *i* (i=1,2,3) of k-disk TH, then for $k \geq 2$ the adjacency matrix is:

$$M_{3^k} = \begin{bmatrix} M_{3^{k-1}} & E_{3^{k-1}}(T_{1r}(k-1)+3^{k-2}+3^{k-2},1) & E_{3^{k-1}}(T_{1l}(k-1)+3^{k-2},1) \\ E_{3^{k-1}}(1,T_{1r}(k-1)+3^{k-2}+3^{k-2}) & M_{3^{k-1}} & E_{3^{k-1}}(T_{1r}(k-1)+3^{k-2}+3^{k-2},T_{1l}(k-1)+3^{k-2}) \\ E_{3^{k-1}}(1,T_{1l}(k-1)+3^{k-2}) & E_{3^{k-1}}(T_{1r}(k-1)+3^{k-2}+3^{k-2},T_{1l}(k-1)+3^{k-2}) & M_{3^{k-1}} \end{bmatrix}$$

where $E_{3^{k-1}}(i,j)$ is a square block matrix of order 3^{k-1} whose elements are all zero, except the element $e(i,j)=1$.

Initially: $T_{1l}(1)= T_{1r}(1)=1$; $T_{2l}(1)= T_{2r}(1)=2$; $T_{3l}(1)=T_{3r}(1)=3$ (for k=1 the graph has only one triangle. The other $T_{il}(k)$ and $T_{ir}(k)$ ($k \geq 2$) are computed in the following way:

$T_{1l}(k)=T_{3r}(k-1)$; $T_{1r}(k)=T_{2l}(k-1)$; $T_{3r}(k)=T_{2r}(k)+3^{k-1}$; $T_{2l}(k)=T_{1l}(k)+3^{k-1}$; $T_{2r}(k)=T_{1r}(k)+3^{k-1}$.

Theorem 2: The modified CAA with *the forward chaining algorithm* using OFF ROUTE acting finds the shortest path in the graph of the problem space for *k*-disk Tower of Hanoi in linear time **O**(n)(*n*-number of states); but depending on *k* the time is exponential.

Proof: For k=2 disks:

$$\begin{aligned} F(n) &< a + 2a + a + bi = 4a + bi = (9-4-2+1)a + b(2^2-1) = \\ &= a(3^2-2^2-2^{2-1}+1)+b(2^2-1) = a(n-2^k-2^{k-1}+1)+b(2^k-1) \end{aligned} \quad (1)$$

where *a* is a constant - the maximal number of operations for exploring all actions in a state and going to another state. The number 4 (from the term after the first sign =) is the total number of passed states, states in which the agent has been and has explored their actions. *bi* (b-const.) is the number of operations for generating the simulated experience and for moving through the shortest path in the phase of planning, *i* is the length of the shortest path. The number 4 is obtained as the total number of states 3^2 minus the number of states in the last row of the triangle of the state space which has 2^2 states, minus the number of the states in the penultimate row which has 2^{2-1} states, and 1 is the predecessor of the goal state. 2^2-1 is the length of the shortest path.

Since for every *k* the graph is fractal, and the agent always explores number of states which is equal to: all states *n* minus number of the states from the last row (which for every *k* has 2^k states), minus number of states from the penultimate row (with 2^{k-1} states) plus 1 (the state -predecessor of the goal), the estimation for every k is the same:

$$F(n) < a(n-2^k-2^{k-1}+1)+b(2^k-1) \quad (2)$$

So, the complexity for every k is O(n) – linear function of n – the number of the states of the problem space, but it is exponential function of k – the number of disks.

3 Conclusion

The paper presents an application of a linear-time algorithm for solving a shortest path problem in arbitrary deterministic environment using OFF ROUTE acting method. The algorithm implements an approach toward emotion learning and is applied for solving the Tower of Hanoi problem. The modified CAA agent architecture is used as a problem solver. The problem is solved in time which is linear function of n - the number of states, but it is exponential function of k - the number of disks.

References

1. Barto, A.: Reinforcement learning. In: Omidvar, O., Elliot, D. (eds.) Neural systems for control, pp. 7–30. Academic Press, San Diego (1997)
2. Botelho, L.M., Coelho, H.: Adaptive agents: emotion learning. In: AAAI, pp. 19–24 (1998)
3. Bozinovski, S.: A Self-learning System using Secondary Reinforcement. ANW Report, November 25 (1981), COINS, University of Massachusetts at Amherst (1981)
4. Bozinovski, S.: A self learning system using secondary reinforcement. In: Trappl, E. (ed.) Cybernetics and Systems Research, pp. 397–402. N. Holland Publishing Company, Amsterdam (1982)
5. Bozinovski, S.: Crossbar Adaptive Array: The First Connectionist Network that Solved the Delayed Reinforcement Learning Problem. In: Dobnikar, A., Steele, N., Pearson, D., Alberts, R. (eds.) Artificial Neural Networks and Genetic Algorithms, pp. 320–325. Springer, Heidelberg (1999)
6. Bozinovski, S.: Anticipation Driven Artificial Personality: Building on Lewin and Loehlin. In: Butz, M.V., Sigaud, O., Gérard, P. (eds.) Anticipatory Behavior in Adaptive Learning Systems. LNCS (LNAI), vol. 2684, pp. 133–150. Springer, Heidelberg (2003)
7. Bozinovski, S.: The Artificial Intelligence (In Macedonian). Gocmar Press, Bitola (1994)
8. Damasio, A.: Descartes' Error: Emotion. Reason and the Human Brain, Grosset (1994)
9. Petruseva, S.: Comparison of the efficiency of two algorithms which solve the shortest path problem with an emotional agent. International Yugoslav Journal of Operations Research (YUJOR) 16(2), 211–226 (2006a)
10. Petruseva, S.: Consequence programming: Solving a shortest path problem in polynomial time using emotional learning. International Journal of Pure and Applied Mathematics (IJPAM) 29(4), 491–520 (2006b)
11. Petruseva, S.: Emotion learning: Solving a shortest path problem in an arbitrary deterministic environment in linear time with an emotional agent. International Journal of Applied Mathematics and Computer Science (AMCS) 18(3), 409–421 (2008)
12. Sutton, R., Barto, A.: Reinforcement Learning: An Introduction. MIT Press, A Bradford Book (1998)
13. Watkins, C.: Learning from Delayed Rewards. Ph.D. Thesis. King's College, Cambridge (1989)
14. Wittek, G.: Me, Me, Me, the spider in the web, The Law of correspondence, and the Law of projection. Verlag DAS WORT, Marktheidenfeld-Altfeld (1995)
15. Wikipedija, the free encyclopedia, `http://en.wikipedija.org/wiki/A-star_ algorithm`

Early Top-Down Influences in Control of Attention: Evidence from the Attentional Blink

Frederic Hilkenmeier, Jan Tünnermann, and Ingrid Scharlau

Paderborn University, Warburger Str. 100, 33098 Paderborn, Germany
frederic.hilkenmeier@uni-paderborn.de

Abstract. The relevance of top-down information in the deployment of attention has more and more been emphasized in cognitive psychology. We present recent findings about the dynamic of these processes and also demonstrate that task relevance can be adjusted rapidly by incoming bottom-up information. This adjustment substantially increases performance in a subsequent task. Implications for artificial visual models are discussed.

Keywords: visuo-spatial attention, top-down control, task relevance, artificial visual attention, attentional blink.

1 Introduction

According to widespread assumptions in cognitive psychology and modeling research, attention facilitates processing of attended compared to unattended information. This is achieved by selecting relevant or conspicuous information or disregarding information which is not relevant to the current task or visually unobtrusive. Basically, attended information is processed in more detail [1] or quicker [2] than unattended information. In contrast to unattended information it can be part of higher level object representations [3], or be perceived consciously [4]. Although the general finding - a boost of attended information - is undisputed, the control mechanisms are a matter of debate. How far, for instance, is attention controlled by visual saliency of a stimulus, and how much can be done by current intentions [5] [6]?

Here, we want to draw attention towards two current lines of thinking. First, attending to a stimulus may come by a cost that is overlooking information in the close temporal vicinity of an attended target (attentional blink) [7]. Second, the role of top-down task relevance in the control of attention cannot be stressed too much [8] [9], at least after the initial transient bottom-up saliency [10].

In an attentional blink task, observers watch simple dynamic scenes consisting of a fast stream of single items in the same spatial location, as illustrated in Figure 1. These items are either distractors (e.g. letters) or targets (e.g. digits). In this setup, observers typically have no difficulties reporting the first target digit ($T1$), but fail in identifying the second target ($T2$) if it is presented between 200 to 500 ms after $T1$. It is as if attention, analogous to the lid closure of an

B. Mertsching, M. Hund, and Z. Aziz (Eds.): KI 2009, LNAI 5803, pp. 680–686, 2009.

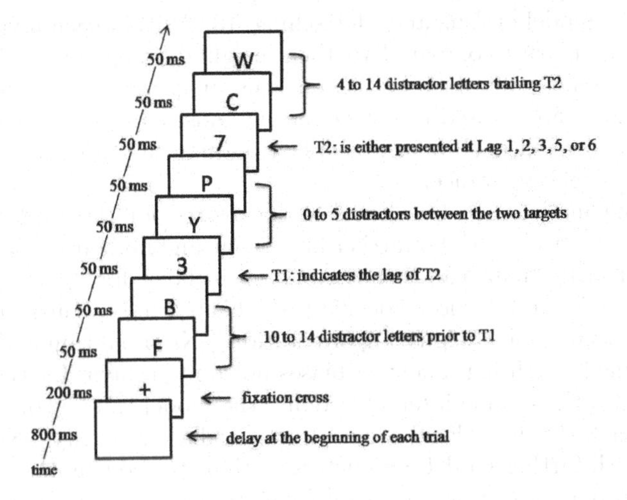

Fig. 1. Example of stimulus sequences. Here, $T2$ follows $T1$ at lag 3 (after 150ms).

eye blink, briefly switched off before new information can be processed [11]. The precise adaptive significance behind this attentional deficit is still unknown. Interestingly, attention seems to consist of two mechanisms: a boost (facilitation) after a target has been detected, and a bounce (inhibition) after a distractor has been detected. $T2$ suffers from the attentional inhibition, which was originally directed to the distractor preceding $T2$. The approach to define attention as two-fold is quite novel [9], but is nevertheless already impressive in explaining a number of empirical results from different spatial and temporal phenomena.

More important in the present context is that this attentional allocation depends upon the current task, not on bottom-up input like intensity or color. Of course, the cognitive system requires this bottom-up input to decide whether a stimulus matches the task set or not, but the deployment of attention is modulated by task relevance: stimuli that match the criterion get boosted, whereas stimuli that do not match the task set (or even match a distractor set; see [12]) get bounced. This importance of task relevance was impressively shown by Nieuwenstein [13]: A distractor preceding $T2$ that shares a feature with $T2$ can significantly reduce the attentional blink deficit. For example, a red distractor letter preceding a green digit $T2$ is an effective cue when the task is to look for red and green digits, but the same red cue is ineffective when the task is to look for only green digits.

These recent findings in cognitive psychology emphasize task relevance as key element for attentional deployment: depending on the current task set, the exact same stimuli can either be inhibited or facilitated. In contrast, classic computational models [14] [15] understand spatial attention as a stimulus-driven, bottom-up process of feature-integration [16]. The input scene is processed in maps of simple features. In the Itti and Koch model [14], these are color, intensity, and orientation, but others may be appropriate, such as symmetry, size, and eccentricity in the model of Aziz and Mertsching [15]. Within each map, stimuli with a certain singularity compared to their neighbors are more salient and therefore receive higher values. These single feature maps are combined into one saliency map which can be used to determine the most conspicuous location in the scene and guide attention. So far no explicit mechanisms integrate top-down influences, such as task relevance.

The mentioned models can be viewed as hardwired for the task "find the most conspicuous feature" with no further knowledge and influences. A weighted combination of feature maps can bias the system in favor for certain features, tweaking the model to fulfill more specific tasks like: "Find a conspicuous feature with a high color saliency ignoring orientation". Navalpakkam and Itti [17] suggest fine-grained weighting, allowing biases not only in favor for the feature in general, but also for specific intervals. Again, the model can be configured to perform narrower tasks like: "Find a dot of intermediate intensity." Setting up weights, along with further model parameters, such as the constants depicted in table 1 in [15], provides a priori task configurability. For many applications in computer vision the described models may be sufficient. However, from a biological point of view they do not fully depict the early and parallel influences of top-down knowledge.

Here, we want to demonstrate that a specific stimulus which has relevant information for a subsequent task can rapidly be processed and used for top-down allocation of attention within the task. This means that the top-down allocation can be adjusted dynamically within virtually no time, which is clearly at odds with a rather static view of saliency maps.

2 Experiment

In the present attentional blink experiment, we integrate information about the temporal position of target $T2$ into the identity of target $T1$. The targets are digits, with $T1$ being a valid cue informing about the target onset asynchrony (i.e. the specific temporal onset of $T2$). There are two conditions: In the baseline condition the participants are unaware of the fact that the cue informed about the target onset asynchrony. In the top-down condition, the instructions emphasize the relationship between $T1$ identity and temporal position of $T2$. A diminished attentional blink in the top-down condition would support our idea that the identity of a leading target stimulus is rapidly available for top-down usage in attention. This is found in other paradigms and with other dependent measures like temporal order judgments [18], event-related brain potentials, reaction times [6], or MEG data [19] as well.

2.1 Apparatus, Stimuli, and Procedure

Participants sat in a dimly lit room with the center of a 19" CRT monitor at eye level. Viewing distance was 50 cm, set by a chin rest. The resolution was 800 x 600 pixels at 100 Hz. The experimental program was written in MATLAB

7.5.0 and made use of the PsychToolbox [20]. Stimuli were displayed as black symbols on a medium grey background. The distractors were letters from the Roman alphabet; the targets were the digits 1-9. Stimuli subtended about 1 degree in visual angle. Each item was presented for 40 ms and separated from the next by an interstimulus interval of 10 ms, resulting in a presentation rate of 20 items/sec. Stimuli were presented at the center of the screen (Figure 1). The forty students that participated in this experiment and were paid for their participation were split into two groups. In both groups, the first target digit was a 100% valid cue about the lag $T2$ was presented in. This means when $T1$ was a "1", $T2$ immediately followed $T1$ (lag 1). When $T1$ was a "2", $T2$ was presented at lag 2, with one intervening distractor between the targets, and so on. The identity of the second target was selected randomly. The instruction for the 20 observers in the top-down group emphasized these facts, whereas the 20 observers in the baseline group were told that the two target digits were always random. The lags 1,2,3,5 and 6 were included in this experiment, that is, the maximal target onset asynchrony was 300 ms. Each lag was repeated 20 times. A fixation cross was presented at the center of the screen for 200 ms followed by a rapid serial stream consisting of 26 items. The first target was the 11th to 15th character in a stream. The second target was followed by 4 to 14 distractors. The targets within a trial were always different. After each trial, the observers were asked to identify the two targets in the order they appeared in by pressing the corresponding keys on the keypad. In case they had not recognized one or both targets, they were encouraged to guess. Observers were allowed to rest whenever they wanted.

3 Results and Discussion

Figure 2 shows the identification rate of $T2$ given $T1$ identified correctly ($T2|T1$), separately for the top-down and baseline conditions. The attentional blink is attenuated, but not abolished, when participants are instructed to use the identity information of the first target as a cue for the temporal position of the second target. This means that observers have fast access to the identity of the first target and were also able to use this information for a virtually instant redeployment of attention. As can be seen from Figure 2, the first significant difference between the top-down and the baseline condition is at lag 1, that is 50 ms after $T1$ onset. We therefore can conclude that the information about $T1$ is available to top-down usage within 50 ms.

3.1 Statistics

In statistical terms the attenuated attentional blink is shown by a significant main effect for the within-subjects factor of lag ($F[4, 190] = 21.95, p < 0.01$) and also a significant main-effect for the between-subjects factor condition (top-down vs. baseline; $F[1, 190] = 21.6, p < 0.01$) in a two-way analysis of variance. The

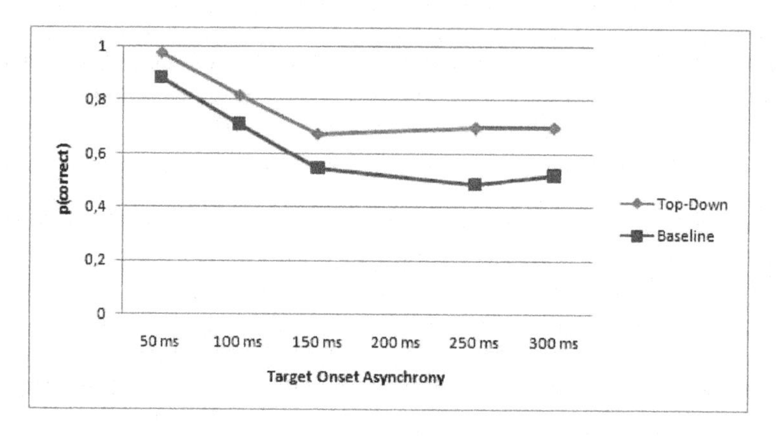

Fig. 2. Conditional accuracy of $T2|T1$ as a function of lag in steps of 50 ms. Order reversals are counted as correct.

interaction was not significant ($F[4,190] = 0.32, p = 0.86$). Additional independent one-tailed two-sample t-tests between the top-down and the baseline condition revealed a significantly better $T2|T1$ performance at all lags ($t[38] = 2.2$, $p = 0.02$; $t[38] = 1.9$, $p = 0.03$; $t[38] = 2.2$, $p = 0.02$; $t[38] = 1.7$, $p = 0.047$; $t[38] = 2.6$, $p < 0.01$ and $t[38] = 2.2$, $p = 0.02$ for lags 1-6, respectively).

4 General Discussion

To sum up: Intentions play a vital role in the control of attention. Top-down information kicks in very early in information processing. This is a central finding in actual current research and has been shown in other attentional paradigms like spatial cueing as well: A feature singleton only captures attention when its feature matches a currently active task set, but not when it is irrelevant for the task [5]. These findings clearly indicate that attention is modulated solely [6], or at least primarily [5] [18], via top-down information about the current task. But task demands not only influence which stimuli get preferred processing. In the present experiment we demonstrated that the task can be adjusted dynamically. This top-down adjustment is due to the characteristics of a stimulus that is just being processed, indicating that the brain is capable of rapid evaluation of a stimulus' behavioral relevance. These early attentional effects are not considered in the computational models discussed so far and are, at least in our view, difficult to implement.

Hamker's model [21] [22] takes a somewhat different approach: To account for rapid attentional effects, he proposes a computational attention model which utilizes feature feedback and spatially organized reentry to implement task relevance in a way that matches biological observations. Bottom-up information from feature maps meet a target template which is held in a functional block representing working memory. This initiates a dynamic recognition process and enhances features that match the target template and filters out information that is irrelevant when fed back to the original feature maps. This happens for feature dimensions, such as orientation and color. The activation in feature maps is also projected to a functional block, which simulates a brain area that programs eye movements in human vision. Here a competition between the active regions takes place, until the system stabilizes with high activation in one region and low everywhere else. This information reenters the feature map, now enhancing specific locations in addition to the feature specific enhancements mentioned before. This step-by-step description might lead to misinterpretation of the model as a sequential system, which it is not. All components act in parallel and continuously, implicitly emerging attention.

Hamker's intention to investigate in biological plausible top-down interaction, rather than developing a usable model for computer vision, results in a minimal model to run on simplified inputs. Because of that, direct application in computer vision is not possible. Nevertheless, Hamker and colleagues [23] [19] have successfully tested this model in experimental work.

In conclusion, attentive vision is an active, dynamic and constructive process, as demonstrated by a variety of findings presented throughout this article. If stimuli provide information that is relevant for the current task, this information can rapidly be used to adjust the task and substantially increase performance. In our view, this shows that investigating the nature of top-down influences is valuable for a biologically inspired model of artificial visual attention.

Acknowledgments. Supported by DFG grants Ne 366/7-2 to Ingrid Scharlau and Scha 1515/1-1 to Ingrid Scharlau and Ulrich Ansorge.

References

1. Broadbent, D.E.: Perception and Communication. Pergamon, London (1958)
2. Posner, M.I.: Orienting of attention. The Quarterly Journal of Experimental Psychology 32, 3–25 (1980)
3. Kahneman, D., Treisman, A., Gibbs, B.J.: The reviewing of object files: Object-specific integration of information. Cognitive Psychology 24, 175–219 (1992)
4. Simons, D.J., Chabris, C.F.: Gorillas in our midst: sustained inattentional blindness for dynamic events. Perception 28, 1059–1074 (1999)
5. Folk, C.L., Remington, R.W., Johnston, J.C.: Involuntary covert orienting is contingent on attentional control settings. Journal of Experimental Psychology: Human Perception and Performance 18, 1030–1044 (1992)
6. Ansorge, U., Kiss, M., Eimer, M.: Goals-driven attentional capture by invisible colours: Evidence from event-related potentials. Psychonomic Bulletin & Review (in press)
7. Chun, M.M., Potter, M.C.: A two-stage model for multiple target detection in rapid serial visual presentation. Journal of Experimental Psychology: Human Perception and Performance 21, 109–127 (1995)
8. Ansorge, U., Neumann, O.: Intentions determine the effect of invisible metacontrast-masked primes: Evidence for top-down contingencies in a peripheral cuing task. Journal of Experimental Psychology: Human Perception and Performance 31, 762–777 (2005)

9. Olivers, C.N., Meeter, M.: A boost and bounce theory of temporal attention. Psychological Review 115, 836–863 (2008)
10. Van Zoest, W., Donk, M.: Goal-driven modulation as a function of time in saccadic target selection. The Quarterly Journal of Experimental Psychology 61, 1553–1572 (2008)
11. Raymond, J.E., Shapiro, K.L., Arnell, K.M.: Temporary suppression of visual processing in an rsvp task: an attentional blink? Journal of Experimental Psychology: Human Perception and Performance 18, 849–860 (1992)
12. Olivers, C.N.L., Watson, D.G.: Input control processes in rapid serial visual presentation: Target selection and distractor inhibition. Journal of Experimental Psychology: Human Perception and Performance 32, 1083–1092 (2006)
13. Nieuwenstein, M.R.: Top-down controlled, delayed selection in the attentional blink. Journal of Experimental Psychology: Human Perception and Performance 32, 973–985 (2006)
14. Itti, L., Koch, C.: Computational modelling of visual attention. Nature Reviews Neuroscience 2, 194–203 (2001)
15. Aziz, M.Z., Mertsching, B.: Fast and robust generation of feature maps for region-based visual attention. IEEE Transactions on Image Processing 17, 633–644 (2008)
16. Treisman, A.M., Gelade, G.: A feature-integration theory of attention. Cognitive Psychology 12, 97–136 (1980)
17. Navalpakkam, V., Itti, L.: Top-down attention selection is fine-grained. Journal of Vision 6, 1180–1193 (2006)
18. Scharlau, I., Ansorge, U.: Direct parameter specification of an attention shift: evidence from perceptual latency priming. Vision Research 43, 1351–1363 (2003)
19. Bröckelmann, A.K., Junghöfer, M., Scharlau, I., Hamker, F.H.: Reentrant processing from attentional task sets: Converging support from magnetoencephalography and computational modeling (in prep.)
20. Brainard, D.H.: The psychophysics toolbox. Spatial Vision 10, 433–436 (1997)
21. Hamker, F.: A computational model of visual stability and change detection during eye movements in real-world scenes. Visual Cognition, 1161–1176 (2005)
22. Hamker, F.H.: Modeling feature-based attention as an active top-down inference process. Bio Systems 86, 91–99 (2006)
23. Hamker, F.H.: A dynamic model of how feature cues guide spatial attention. Vision Research 44, 501–521 (2004)

Probabilistic Models for the Verification of Human-Computer Interaction

Bernhard Beckert and Markus Wagner

Department of Computer Science, University of Koblenz, Germany
beckert@uni-koblenz.de, wagnermar@uni-koblenz.de

Abstract. In this paper, we present a method for the formalization of probabilistic models of human-computer interaction (HCI) including user behavior. These models can then be used for the analysis and verification of HCI systems with the support of model checking tools. This method allows to answer probabilistic questions like "what is the probability that the user will unintentionally send confidential information to unauthorized recipients." And it allows to compute average interaction costs and answer questions like "how much time does a user on average need to send an email?"

1 Introduction

Interaction between computer and user is a two-way communication process, where the user enters commands and the system responds to the input. This is referred to as interactive control [1]. Unfortunately, this control does not always work perfectly and errors in human-computer interaction (HCI) occur. A huge number of errors in HCI can be traced back to user interfaces that are insufficiently secured against these errors, irrespective of whether the error is caused (a) by the user – intentionally or unintentionally – interacting incorrectly with the computer, (b) the computer reacting incorrectly to the user's input, or (c) the user interpreting the computer's output incorrectly. The consideration of user errors and their overall impact on the system forms an important part of an analysis of a system's usability, safety, and security. As a result, the designers of systems often consider human frailty and try to reduce errors and usability problems. A popular approach is to stick to informal lists of design rules [1,2,3].

In our approach to the analysis of HCI, three model components are combined: (1) a model of the user's behavior, (2) a model of the system's behavior, and (3) a model of the user's assumptions about the system's state (based on the user's inputs and the systems reactions). The advantage of our method is that each model component can be probabilistic. This allows for the specification of errors in the user behavior, in the application, and in the way the user interprets the application's reactions. The goal of our technique is to prove properties of the interaction by means of probabilistic model checking – or more specifically, to formally prove that, considering all the possible ways of HCI in a given scenario, the given probabilistic requirements are fulfilled.

B. Mertsching, M. Hund, and Z. Aziz (Eds.): KI 2009, LNAI 5803, pp. 687–694, 2009.

The basis of our work is the (non-probabilistic) method of Beckert and Beuster [4] for the formalization, analysis, and verification of user interfaces. It is based on GOMS [5], which is a well-established user modeling method. The GOMS models are formalized and augmented with formal models of the application and the user's assumptions about the application's state.

2 Cost and Probability of HCI Sequences

Human-computer interactions can be split into sequences of actions. In our approach, each action of the user or (re-)action of the system is represented by a transition in the HCI model. To take the different cost of actions into consideration, the notion of *action cost* is introduced, which can be used to represent quantities like "amount of time", "amount of money," or "amount of resources." Action costs are attached to the corresponding transition in the model. Later, this is used to reason about the effects of minor errors, such as mistyping characters when writing an email, and major errors, such as sending confidential data to unauthorized recipients. Whether an interaction is an error or not depends on what the goal of the interaction is. Given a set of goal states, an action is an error if it increases the total cost required to reach the goal or makes it impossible to reach the goal. In the latter case, the error cannot be undone, in the former case, additional or more costly actions are required to undo the effect of the error, i.e., an *error cost* is introduced.

Based on the total interaction cost and on the assumption that there is a limit on the acceptable cost of the interaction, the sequences of actions that start in the initial state and end in a goal state can be classified into three categories:

1. optimal sequences with minimal total cost,
2. acceptable sequences with total cost that do not exceed the limit, and
3. unacceptable sequences with total costs that exceed the user's budget.

The analyst of an HCI scenario may be interested in computing the probabilities or the expected cost of reaching certain states. This can be, e.g., a single state, a set of goal states, or all the final states of the interaction. Similarly, the analyst may be interested in verifying that certain states, or set of states, are (not) reached with a certain probability. With our method, this all can be done.

Regarding the action's probabilistic values that are needed for the construction of the interaction model, the analyst applying our method may use estimates or experimental data. Also, databases containing probabilities of different kinds of human errors may be used (e.g. [6,7]). This data comes from many sources, such as nuclear power plants, simulator studies, and laboratory experiments. A method is provided for combining the data in order to produce estimations of the erroneous executions of tasks.

3 Basis: The Non-probabilistic Model

In this section, we present the non-probabilistic basis of our model. Following [4], we assume that a user interacts with a system based on what his or her assumptions of the application's configuration are, including assumptions about the internal state and relevant data. In terms of Linear Temporal Logic (LTL), always correctly interprets the system's state if:

$$\mathbf{G}((a_0 \leftrightarrow c_0) \wedge (a_1 \leftrightarrow c_1) \wedge \cdots \wedge (a_n \leftrightarrow c_n)) ,$$

where $a_0, \ldots, a_n$ are the critical properties of the application, and $c_0, \ldots, c_n$ are the user's assumptions about whether these properties hold or not.

If this formula does not hold, the user is error-prone. Then, the following scenarios are possible: (1) Parts of the user's assumptions are wrong, and (2) the user's assumptions are incomplete (over abstraction). For example, in the first scenario the user could assume that the software is in another state that allows other actions to be executed. In the second scenario, the user could be unaware of the necessity to take certain action to avoid high costs. But even if the above formula holds, the user can still execute erroneous actions. Even though he or she could have known better (having the right assumptions about the system state), missing knowledge about what the most cost-effective actions are may lead to errors.

In [4], the model of the interaction is the result of the combination of three model components:

1. a formal GOMS model describing the user behavior,
2. a component representing the user's assumptions of the software's state, and
3. a component representing the application itself.

Input Output Labeled Transition Systems (IOLTS) are used to model these components.

Definition 1. *An IOLTS is a tuple $L = (S, \Sigma, s_0, \rightarrow)$ where S is a set of* states, *$s_0 \in S$ is an* initial state, *$\rightarrow \subseteq S \times \Sigma \times S$ is a* transition relation, *and Σ is a set of* labels *with $\Sigma = \Sigma? \cup \Sigma! \cup \Sigma I$. We call $\Sigma?$ the* input alphabet, *$\Sigma!$ the* output alphabet, *and ΣI the* internal alphabet.

For example, whenever the user executes an action, the corresponding state transition is performed. The corresponding edge in the automaton is annotated with a label denoting that action.

The transition systems are combined by *mutual composition.* In mutual compositions of IOLTSs L_a and L_b, the output of L_a serves as input for L_b, and the output of L_b serves as input of L_a, which is illustrated in Figure 1.

Definition 2. *Let $L_a = (S_a, \Sigma_a, s_{0a}, \rightarrow_a)$ and $L_b = (S_b, \Sigma_b, s_{0b}, \rightarrow_b)$ be two IOLTSs. We assume the input and output alphabets of L_a and L_b to consist of internal and external subsets, where the internal input is denoted with $\Sigma?I$, the external input with $\Sigma?E$, the internal output with $\Sigma!I$, and the external output with $\Sigma!E$. And we demand that these subsets are chosen such that $\Sigma!I_a = \Sigma?I_b$ and $\Sigma!I_b = \Sigma?I_a$. Then, the* mutual composition *$(L_a||_m L_b) = (S, \Sigma, s_0, \rightarrow)$ of*

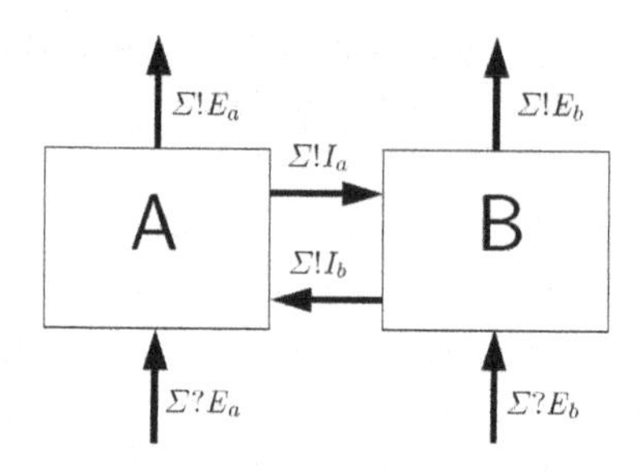

Fig. 1. Mutual composition of two IOLTSs

L_a and L_b is defined by:

$$
\begin{aligned}
S &= S_0 \times S_1 \\
\Sigma? &= \Sigma?E_a \cup \Sigma?E_b \\
\Sigma! &= \Sigma!E_a \cup \Sigma!E_b \\
\Sigma I &= \Sigma I_a \cup \Sigma I_b \cup \Sigma!I_a \cup \Sigma!I_b \\
s_0 &= (s_{0a}, s_{0b}) \\
\rightarrow &= \{(s_a, s_b), \sigma, (s'_a, s_b)) \mid s_a \xrightarrow{\sigma}_a s'_a \textit{ with } \sigma \in \Sigma?E_a \cup \Sigma!E_a \cup \Sigma I_a\} \cup \\
&\quad \{(s_a, s_b), \sigma, (s_a, s'_b)) \mid s_b \xrightarrow{\sigma}_b s'_b \textit{ with } \sigma \in \Sigma?E_b \cup \Sigma!E_b \cup \Sigma I_b\} \cup \\
&\quad \{(s_a, s_b), \sigma, (s'_a, s'_b)) \mid s_a \xrightarrow{\sigma}_a s'_a \textit{ and } s_b \xrightarrow{\sigma}_b s'_b \textit{ with} \\
&\qquad \sigma \in \Sigma!I_a \cup \Sigma!I_b = \Sigma?I_b \cup \Sigma?I_a\}
\end{aligned}
$$

The mutual composition of the three aforementioned model components provides the complete model of the interaction, making complete formal modeling possible (see Figure 2).

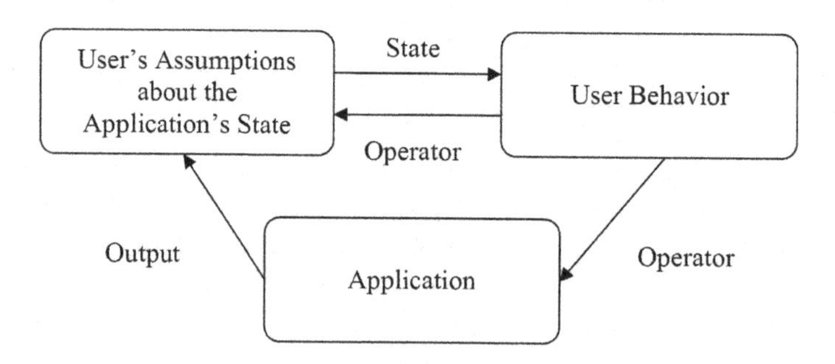

Fig. 2. Basic model of the interaction

4 The Probabilistic Extension

For our probabilistic extension of the models described in the previous section, we introduce several modifications. In order to incorporate probabilistic values as well as the idea of costs, variations of the IOLTS are introduced. Thus, the state transitions of the resulting models contain information about the actions' costs and the actions' probabilities.

A *Probabilistic Input Output Labeled Transition System* (PIOLTS) is an IOLTS, where the set Σ of labels consists of tuples (l, p) where l is the label denoting the action and p is the probability of that action $(0 \leq p \leq 1)$. Similarly, a *Valued Input Output Labeled Transition System* (VIOLTS) is an IOLTS, where Σ consists of tuples (l, v) where v is the cost assigned to that action $(v \geq 0)$. Finally, a *Probabilistic Valued Input Output Labeled Transition System* (PVIOLTS) is a combination of both variants, where Σ consists of tuples (l, p, v).

To model the user's behavior, we use a PVIOLTS. Thus, not only the probabilistic behavior is modeled, but also "personal" costs. The application's model is extended to a VIOLTS by numerical values representing the costs for the execution of each single step. Probabilities can be added to model non-deterministic application behavior. The user's assumptions are modeled using a PIOLTS, and costs can be added if needed. As sources for the statistical data, we suggest to use databases as mentioned in Section 2. However, this is not mandatory and the analyst can come up with his/her own values. Now, the mutual composition of these three components provides the extended model of the HCI, incorporating probabilistic user behavior as well as the interaction cost.

The probabilistic models resulting from our method can be used to prove quantitative aspects of the interaction. In general, probabilistic model checking is an automatic formal verification technique for the analysis of systems that exhibit stochastic behavior [8]. In our setting, the models can be represented as Discrete-Time Markov chains. Using Markov chains instead of Bayesian networks, which are another class of probabilistic graphical models, allows us to use cyclic structures. The properties of the HCI are expressed in Probabilistic Computation Tree Logic (PCTL) [9] and verified using the probabilistic model checker PRISM [10].

5 Example

In this section, we present an example that was implemented as proof of concept for our method. Due to the page limit, we only present the final model of the HCI, which is used to verify quantitative properties.

We model the interaction of a user with an email client. The user intends to write a confidential email and send it to *Alice*. However, with a certain probability, he/she performs an erroneous action choosing *Bob* instead, which results in high costs as confidential information is disclosed. Then, among others, the following erroneous interaction sequences are possible (with certain probabilities):

- The user accidentally selects *Bob* as the addressee. The application consequently reacts with setting *Bob* to be the recipient. Based on the output provided by the application, the user's assumptions about the applications's state, in particular the addressee, are not met and the user notices the error. The user corrects the error. Then, compared to the non-erroneous situation, the total cost of the interaction is increased by the amount it takes to select and set an addressee, and to notice the mistake.
- The user accidentally selects *Bob*, and the application sets *Bob* as recipient. In contrast to the previous sequence of actions, the user interprets the application's output incorrectly, assuming that *Alice* is selected. Thus the user fails to notice the first error, and as a second error sends the email to *Bob*. Here, the error cost is the cost of sending confidential data to the wrong recipient.
- The user accidentally selects *Bob*, the application sets *Bob*, and the user notices the error based on the application's output. However, the user reacts with a second error by sending the email instead of correcting the address. Again, the error cost is the cost of sending confidential data to the wrong recipient.

Once the model of the interaction is constructed using our method, properties of the interaction can be formulated in PCTL and then checked by PRISM. Figure 3 shows part of the PVIOLTS model of the interaction. For example,

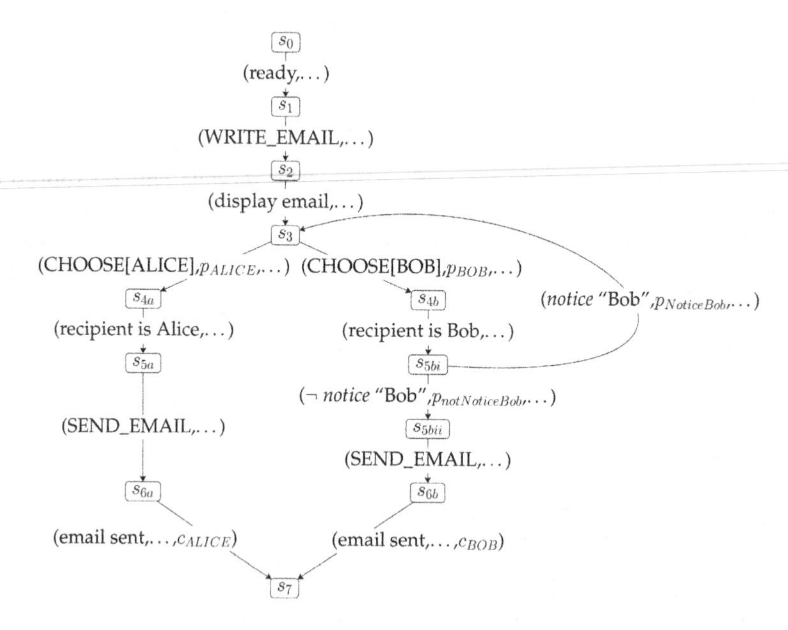

Fig. 3. Part of the PVIOLTS of the mailing example. Labels written in uppercase denote user's inputs, lowercase denote the software's outputs. The interpretation of the application's output by the user is included only for the case that *Bob* is chosen as the recipient.

the formula $P_{\geq 0.95}[\Diamond sentTo(Alice)]$ expresses that the email is sent to *Alice* in at least 95% of the interactions. To express that "with a probability of at most 0.02, *Bob* is selected as the addressee and the error is not corrected", the property $P_{\leq 0.02}[\Diamond(chosen(Bob) \wedge (chosen(Bob)\ U\ sentTo(Bob)))]$ can be used. With PRISM's support of cost analysis, the average total cost for sending an email can be computed by the query $R =? [F\ email\_sent]$. Due to the possibility of erroneous interaction, the average total cost here will be higher than in a scenario where errors do not result in additional cost.

A way to lower the probability of sending the email to *Bob* would be to introduce additional dialogue boxes that require the user to confirm the address selection. If the user is modeled to react "reasonably" to a confirmation request with a high probability (as opposed to, e.g., blindly confirming the selection), it can be proven that the probability to send the confidential email to *Bob* is indeed lowered.

6 Modeling Changing Probabilities

In the probabilistic HCI models described so far, the probabilities and costs are fixed for each action. In certain situations, however, it is useful to model changing probabilities and costs. This can be done by adding a set S of variables to the states of the transition system whose values can be changed by state transitions. Probabilities and costs can then be modeled as function of the values of these state variables, which greatly improves the expressiveness of our approach. For example, the probability $p(a)$ of an action a becomes a probability $p(a, S)$. The variables in S can be flags, counters, etc. For first experiments, this concept of changing probabilities has been implemented.

Some examples for the use of this extension in the context of the email scenario from the previous section are:

- different probabilities for an action can be used for different levels of complexity of a confirmation:

$$p(a, S) = \begin{cases} 0.7 & \text{if } S.confirmationType = simple \\ 0.9 & \text{if } S.confirmationType = complex \end{cases}$$

- a "learning" user can be modeled by increasing the probability for the correct action over time:

$$p(a, S) = \begin{cases} 0.6 & \text{and} & \{S.learned := true\} & \text{if } \neg S.learned \\ 0.8 & \text{and} & \{S.learned := true\} & \text{if } S.learned \end{cases}$$

- an increasingly bored and thus "inattentive" user can modeled by decreasing this probability over time:

$$p(a, S) = \frac{1}{S.inattentionLevel + 2} + 0.4 \quad \text{and} \quad \{S.inattentionLevel{+}{+}\}$$

yields the sequence $0.9, 0.73, 0.65, 0.6, \ldots$ if $S.inattentionLevel = 0$ initially.

Confirmations are modeled on the user side as well as the application side. While they decrease total cost of the application's actions as unintended actions are avoided, they increase total cost of the user's action as he has to interpret the confirmation request and react to it. Using our method, one can analyse the trade off between the cognitive workload that is imposed on the user by complex confirmations, and the average cost that is saved.

7 Conclusion

In this paper we have introduced a method for the formalization of probabilistic models of human-computer interaction (HCI) including user behavior. These models can then be used for the analysis and verification of HCI systems with the support of model checking tools. This allows to answer questions, such as "what is the probability that the user will send confidential information to unauthorized recipients?" and to verify the corresponding properties of the HCI model.

Our method can help to develop user interfaces by avoiding the trap of having to do human error analysis at the latest stages in the design process. Furthermore, it can help a designer to validate assumptions about human performance. By setting up several scenarios, the analyst is able to discover the impact of alternative designs on the expected cost of HCI. Thus, formal modeling and an examination of the expected costs can together contribute to the design of user interfaces, which have to be robust to the error-prone behavior of humans.

References

1. Avery, L.W., Sanquist, T.F., O'Mara, P.A., Shepard, A.P., Donohoo, D.T.: U.S. Army weapon systems human-computer interface style guide. Version 2 (2007)
2. Shneiderman, B.: Designing the User Interface, 3rd edn. Addison Wesley, Reading (1998)
3. Leveson, N.G.: Analyzing software specifications for mode confusion potential. In: Proc. of the Workshop on Human Error and System Development (1997)
4. Beckert, B., Beuster, G.: A method for formalizing, analyzing, and verifying secure user interfaces. In: Liu, Z., He, J. (eds.) ICFEM 2006. LNCS, vol. 4260, pp. 55–73. Springer, Heidelberg (2006)
5. John, B.E., Kieras, D.E.: The GOMS family of user interface analysis techniques: comparison and contrast. ACM Transactions on Computer-Human Interaction 3(4), 320–351 (1996)
6. Swain, A., Guttman, H.E.: Handbook of Human Reliability Analysis with Emphasis on Nuclear Power Plant Applications. Sandia National Laboratories (1983)
7. Kirwan, B.: A Guide to Practical Human Reliability Assessment. Taylor and Francis, London (1994)
8. Hinton, A., Kwiatkowska, M.Z., Norman, G., Parker, D.: PRISM: A tool for automatic verification of probabilistic systems. In: Hermanns, H., Palsberg, J. (eds.) TACAS 2006. LNCS, vol. 3920, pp. 441–444. Springer, Heidelberg (2006)
9. Hansson, H., Jonsson, B.: A logic for reasoning about time and reliability. Formal Aspects of Computing 6(5), 512–535 (1994)
10. Kwiatkowska, M.Z., Norman, G., Parker, D.: PRISM: Probabilistic symbolic model checker. In: Field, T., Harrison, P.G., Bradley, J., Harder, U. (eds.) TOOLS 2002. LNCS, vol. 2324, pp. 200–204. Springer, Heidelberg (2002)

HieroMate: A Graphical Tool for Specification and Verification of Hierarchical Hybrid Automata

Ammar Mohammed and Christian Schwarz

Universität Koblenz-Landau, Computer Science Department, Koblenz, Germany,
{ammar,chrschwarz}@uni-koblenz.de

Abstract. In previous works, Hierarchical Hybrid Automata (HHA) have been proposed as a combination of UML state machine diagrams and hybrid automata to model complex and in particular multi-agent systems. This approach enables formal system specification on different levels of abstraction and expresses real-time system behavior. A prototype was implemented using constraint logic programming (CLP) as a framework to specify and verify these HHA. However, it still requires the user to write a CLP program in order to specify the HHA, which is a tedious and error-prone work. Therefore, this paper aims at simplifying the verification process by introducing a tool environment where a HHA model together with requirements are entered graphically and the process of verification is achieved automatically.

1 Motivation

Hybrid automata [12] are standard means for the specification and analysis of dynamical systems, where computational processes interact with physical processes. Basically, hybrid automata are state machines for describing discrete-event systems, augmented with differential equations for the treatment of continuous processes. Their formal semantics make them accessible to formal validation of systems in safety critical environments.

It is possible to prove desirable features as well as the absence of unwanted properties for the modeled systems automatically using hybrid automata verification tools. They are used for the specification of numerous applications, e.g. in the fields of robotics [18], logistics [15], and multi-agent systems [16].

However, because of the high complexity of the interactions among agents and the dynamics of the environments, the specification and verification of many of these systems is a highly demanding task. Therefore, it was proposed in [9] to combine the formal semantics of hybrid automata along with software engineering, precisely UML state machines [19]. These Hierarchical Hybrid Automata (HHA) allow complex systems to be modeled with different levels of abstraction and provide a formal way to analyze the behavior of the modeled systems. A concrete prototype implementation of the proposed combination is presented in [17]. In this framework, a hierarchical model is specified as a constraint logic program (CLP) and can be verified using an abstract state machine also written in CLP.

Usually, hierarchical specifications of hybrid systems are transformed into flat standard finite hybrid automata (see e.g. [4,20]). Composing of concurrent automata leads

B. Mertsching, M. Hund, and Z. Aziz (Eds.): KI 2009, LNAI 5803, pp. 695–702, 2009.

to the state explosion problem, because the number of states in the resulting flat automaton is the product of the number of states of all concurrent automata. Therefore, the main novelty of the proposed approach [9,17] is the direct verification of the hierarchical model without flattening the automata components first. However, the process of specifying HHA with CLP is a tedious task. The user needs to write hundreds of lines of CLP code to specify a certain problem. Moreover, the writing has to be done carefully in order to avoid side effects that may result in unwanted behavior.

To address this problem, the main contribution of this paper is to introduce a system which simplifies the verification procedure. The specification can be done graphically like with UML modeling tools and be verified directly. It is sufficient for the user to focus only on the specification process rather than focusing on both specification and the CLP implementation. This can reduce mistakes, which may occur due to the CLP implementation. The HieroMate tool accepts specifications and properties to be proven by visual interaction. Then it generates proper CLP code and passes it to a constraint solver which verifies it by means of reachability analysis.

In the following, we are discussing related works (Sect. 2), before we are describing the concept of HHA (Sect. 3). Then we will introduce the HieroMate tool (Sect. 4) and end up with conclusions (Sect. 5).

2 Related Works

In the last decade, there have been various tools developed based on hybrid automata to model and verify real time systems. These tools range from simple specifications (e.g. timed automata [1]) to more general ones like hybrid automata. Related to our work, Kronos [21] and UPPAAL [5] are tools to model and verify timed automata. In these tools, a system model is given graphically and verified automatically. However, these tools are restricted to timed automata and have no capability to add a hierarchical model.

In more general tools for hybrid automata, like HyTech [13] and PHAVer [8], the concurrent automata have to be specified in a proper plain text representation written in the tool's description language. Afterwards, the specification is given to a model checker in order to verify a given requirement. In contrast to our work, these tools neither have a graphical specification, nor they can handle hierarchical models directly. Instead, they require the user to manually flatten such models before it can be specified in a proper format.

Similar to the core of our verification framework, Urbina [20] and Jaffar et al. [14] employ CLP to implement concurrent hybrid automata, where constraint solvers together with logic programs are used as verification engine. But both systems require manual composition of concurrent automata into one single automaton. Then a CLP specification has to be created for this resulting automaton. This is of course a tiresome work, especially for specifying multi-agent systems.

Other authors engaged CLP for the implementation of timed automata [6,7,10], where a CLP implementation is given to each automaton participating in the mode, then a driver program has to take all the possible composition paths of the participated automata. In contrast to our CLP verification framework, these systems neither provide a graphical model nor hierarchy.

3 Hierarchical Hybrid Automata

Hybrid automata [12] typically consist of several components which operate concurrently and communicate with each other. Each automaton is represented graphically as a state transition diagram dialect like state charts, augmented with mathematical formalisms on both transitions and locations. HHA depend on both hybrid automata and state machines. We now give a definition of their syntax.

Definition 1 (Hierarchical Hybrid Automaton). *A Hierarchical Hybrid Automaton (HHA) is a tuple* $(X, S, s_0, \beta, \alpha, Init, Inv, Flow, E, Jump, Events)$. *The components are as follows:*

- *A finite set* $X = \{x_1, x_2, \ldots, x_n\}$ *of continuous, real valued variables.*
- *A finite set S of states which is partitioned into three disjunct sets* S_{simple}, S_{comp}, *and* S_{conc} *— called simple, composite and concurrent states.*
- *The root state* $s_0 \in S \setminus S_{simple}$.
- *A function* $\beta \subset (S \setminus \{s_0\}) \times (S \setminus S_{simple})$ *which assigns a parent to each state but the root. The parent of a concurrent state must not be a concurrent state as well.*
- *A function* $\alpha \subset S \times 2^S$ *which assigns a set of initial states to each state. A simple state has none, a composed state one and a concurrent state more than one initial states.* α *must respect the state hierarchy given by* β*: only direct descendants of a state may be in its initial state set.*
- *A labeling function Init which assigns an initial condition, a predicate over X, to each state.*
- *A labeling function Inv which assigns an invariant condition, a predicate over X, to each state.*
- *A labeling function Flow which assigns a flow condition, a predicate over* $X \cup \dot{X} = X \cup \{\dot{x}_1, \dot{x}_2, \ldots, \dot{x}_n\}$, *to each state.*
- *A set* $E \subseteq S \times S$ *of transitions among states. E must respect the state hierarchy given by* β*: only siblings may be connected by transitions.*
- *A labeling function Jump which assigns a jump condition, a predicate over* $X \cup X' = X \cup \{x'_1, x'_2, \ldots, x'_n\}$, *to each transition.*
- *A labeling function Events which assigns a set of event labels to each transition.*

The distinction between simple, composite, and concurrent states belong to the definition of state charts [11] and has become part of UML [19]. It is useful for expressing the overall system on several levels of abstraction. Inspired by the definition of a configuration in state charts [11], we define a configuration of a HHA as follows:

Definition 2 (Configuration). *A configuration c of a HHA is a tree.* s_0 *is the root node. For each state s in the tree must hold:*

- *s is a leaf node iff s is a simple state,*
- *s has exactly one child* s', $\beta(s') = s$ *iff s is a composed state.*
- *s has the children defined by* $\alpha(s)$ *iff s is a concurrent state.*

The current *situation* of the whole system can be characterized by a triple (c, v, t) where c is a configuration, v a valuation of the variables, and t the current time.

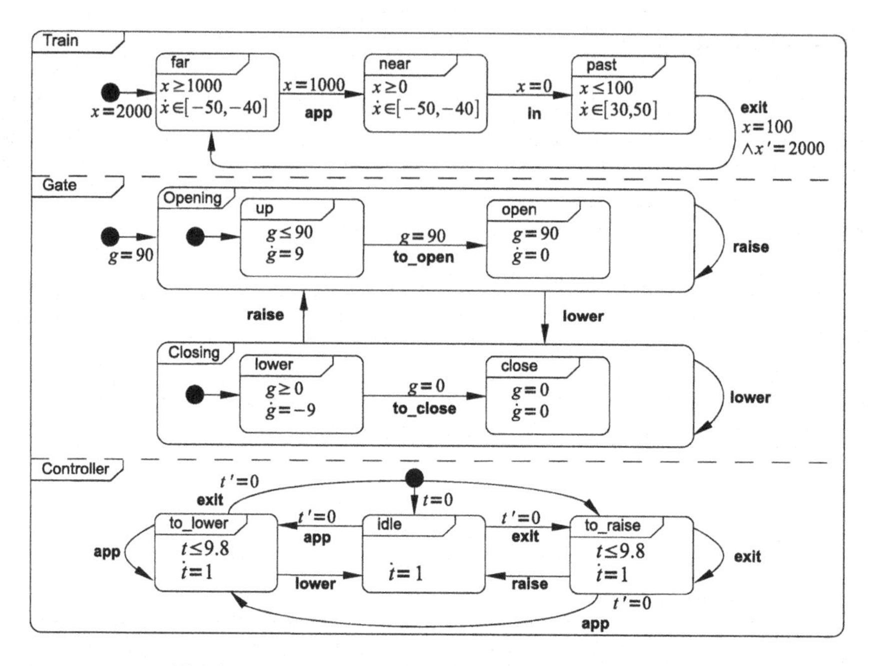

Fig. 1. Specification of the train gate controller automata

The evolution of the system can be defined using the initial, flow, invariant and jump conditions as well as the synchronization labels. Where the initial condition must hold when entering the regarding state, the invariant denotes the condition to stay there, the flow condition describes the timely evolution of the variables in that state and the jump condition describes conditions on the variables before and after following a transition. In flow conditions the derivative of the variables with respect to time can be referred by $\dot{X}$. In jump conditions X denotes the values of the variables before and X' the value of the variables after taking the transition. For a complete formal description of the semantic the reader is referred to [9].

Our verification framework can handle – like most of the model checking tools – linear automata only, according to plain linear hybrid automata [12] we define linear hierarchical hybrid automata as follows:

Definition 3 (Linearity). *A HHA is linear if and only if the initial, invariant, flow and jump conditions are boolean combinations of linear inequalities and the flow conditions contain variables from* $\dot{X}$ *only.*

4 The HieroMate Tool

In [17] an abstract state machine for model checking of hybrid automata by means of reachability analysis has been developed. This framework requires the user to specify the hybrid automaton as a CLP program. With the HieroMate tool we introduce a system which simplifies this process significantly. The user can specify the automaton graphically, which is the most natural way of describing hybrid automata. In addition, the tool enforces the syntax for linear hierarchical hybrid automata (Def. 1 and Def. 3) which helps to find mistakes early. The specification is then translated to CLP code and passed to a constraint solver which executes the model checking. The result is then shown to the user.

The tool environment of HieroMate is composed of different layers, namely the graphical user interface (GUI), an internal logic, and a CLP-solver as back-end.

The GUI consists of three major parts which are an interface to edit and view the specification of the automaton, an interface to view and edit the requirement and an additional interface to view the output of the CLP query.

Using the first interface, the user can edit the automaton by adding and removing states, transitions, and variables. Additionally, he can edit the jump condition and synchronization labels of transitions and the flow, invariant and initial conditions of states. Furthermore, the user can specify the initial states of composed states and change the appearance of the automaton by rearranging the states. Most of these actions are context sensitive so only legal options are shown and executed – for instance a context menu hides the "Add Concurrent State" option, if the actual state is a concurrent one, and transitions that break the state hierarchy are not allowed.

The interface for editing the requirement presents to the user a visualization of the state hierarchy tree, where he can mark the states he wants to have checked for reachability. If more than one child of a concurrent state is marked, that means, that all these states have to be reached in the same time. Checking more than one child state of a composed state means that the requirement is fulfilled if one of these states is reached. Later versions of HieroMate will enable the user to specify the reachability of certain variable values as well.

The answer interface is very basic at the moment and only shows the result of reachability analysis. But as the requirement interface will allow more complex requirements in future, here might be shown additional information as a trace which leads to the specified configuration of the automaton or the value of certain variables.

The internal logic uses a tree to represent the states' parent/child relationship and a directed graph for the relationship between siblings. The transitions can be labeled with synchronization labels (which are represented by a set of strings) and jump conditions (which are boolean combinations of atomic linear constraints like "$A \leq 10.0$" or "$A' \geq -1.0$" represented as a tree). States can be labeled with flow, invariant and initial conditions, which are represented in the same manner as jump conditions. This internal representation is then transformed into a CLP program.

For example a transition from state a to b both children of a state p involving a variable V is translated to `trans(T,a,[[],[V1]],b,[[],V2]):-V1$>20,V2$=0.`, if the jump condition of that transition was $V > 20 \wedge V' = 0$. In a similar way the flow and invariant conditions are translated to flow-clauses. The state hierarchy together with the initial condition and the initial states is transformed to init-clauses. Additionally a start-clause is added which specifies the topmost state in the hierarchy. Finally, the abstract state machine [17] and the translated requirement are attached to the program and it is passed to the constraint solver.

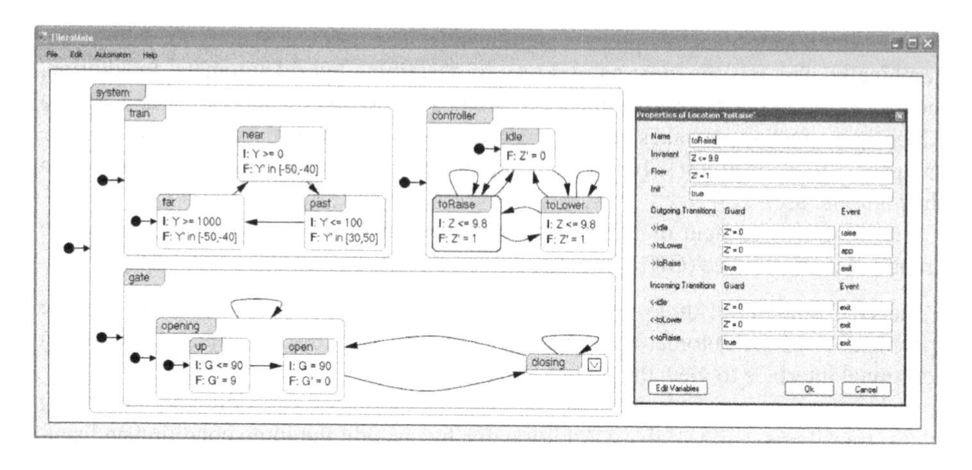

Fig. 2. A screenshot of HieroMate when modelling the train-gate example

The CLP solver used is eCLiPSe Prolog [2] which offers an interface for being embedded in other applications. It executes the model checking and produces some output. This output is passed back to the internal logic and presented to the user via the GUI.

4.1 Train-Gate Example

Fig. 1 shows the train-gate example. In this model, a road is crossing a train track, that is guarded by a gate, which must be lowered to stop the traffic when the train approaches, and raised after a train passed the road. The gate is supervised by a controller that has the task to receive signals from the train and to issue lower or raise signals to the gate.

The train starts at position $X = 2000\,\mathrm{m}$ and moves with some speed between $30\frac{\mathrm{m}}{\mathrm{s}}$ and $50\frac{\mathrm{m}}{\mathrm{s}}$ towards the railroad crossing which is located at $X = 0\,\mathrm{m}$. When it reaches the point $X = 1000\,\mathrm{m}$ the controller notices that a train is approaching, which is modeled by an event called “app”. As the train moves on, it will eventually reach the crossing after which it is called to be in the “past” state. After it has moved 100 m in this state, X will be reset to 2000 m in order to model a loop and the controller notices that the train has left, which is modeled by the “exit” event.

After the controller has received the “app” or “exit” event, it takes up to 9.8 s until it starts moving the gate by the events called “lower” or “raise”.

The gate starts at an initial position of $G = 90^\circ$. After receiving the “lower” event it will start moving down with a constant speed of $9^\circ/s$. After 10 seconds it will be fully closed. At any time the closing process can be interrupted by the controller by invoking the “raise” event just like the opening process can be interrupted by the “lower” event.

One possible security rule is now, that the gate must be closed while the train is at the crossing or up to 100 m behind. This can be modeled by the requirement that the automaton must not enter configurations where the gate is in another state than “closed” while the train is in the “past” state. This can be checked via reachability analysis.

The specification can be entered into the HieroMate tool as it is shown in Fig. 2. It is then automatically transformed into a CLP specification. Furthermore, the requirement can be entered through the requirement interface, it is transfered to CLP as well. The intermediate CLP representation has been compared with a former implementation [17] of a similar problem which had been created manually. No but small syntactic differences could be found. So it is not surprising that the reachabilty analysis gives the same results in both cases.

5 Conclusion

In this paper, we presented a tool environment that simplifies the specification process of hierarchical hybrid automata and supports the user in the task of creating correct automata. A graphical specification and a requirement can be given to the tool and it will convert them into a specification written in CLP. Then, the resulting CLP specification will be checked using an abstract state machine in terms of reachability analysis automatically.

Future work includes the possibility of generating input for related hybrid automata tools (like PHAVer or HyTech), which would allow us to automatically compare these tools in terms of verified requirements and efficiency. Furthermore, users will be allowed to automatically apply hybridization methods [3] for approximations of nonlinear behavior, which will allow us to examine more complex problems.

References

1. Alur, R., Dill, D.: A theory of timed automata. Theoretical Computer Science 126(2), 183–235 (1994)
2. Apt, K., Wallace, M.: Constraint logic programming using ECLiPSe. Cambridge University Press, Cambridge (2007)
3. Asarin, E., Dang, T., Girard, A.: Hybridization methods for the analysis of nonlinear systems. Acta Informatica, 451–476 (2007)
4. Banda, G., Gallagher, J.P.: Analysis of linear hybrid systems in CLP. In: Hanus, M. (ed.) Pre-Proceedings of LOPSTR 2008 – 18th International Symposium on Logic-Based Program Synthesis and Transformation, pp. 58–72. Technical University of Valencia, Spain (2008)
5. Behrmann, G., David, A., Larsen, K.G.: A tutorial on UPPAAL. In: Bernardo, M., Corradini, F. (eds.) SFM-RT 2004. LNCS, vol. 3185, pp. 200–236. Springer, Heidelberg (2004)
6. Ciarlini, A., Frühwirth, T.: Automatic derivation of meaningful experiments for hybrid systems. In: Proceeding of ACM SIGSIM Conf. on Artificial Intelligence, Simulation, and Planning, AIS 2000 (2000)
7. Delzanno, G., Podelski, A.: Model checking in CLP. In: Cleaveland, W.R. (ed.) TACAS 1999. LNCS, vol. 1579, pp. 223–239. Springer, Heidelberg (1999)
8. Frehse, G.: PHAVer: Algorithmic verification of hybrid systems past HyTech. In: Morari, M., Thiele, L. (eds.) HSCC 2005. LNCS, vol. 3414, pp. 258–273. Springer, Heidelberg (2005)
9. Furbach, U., Murray, J., Schmidsberger, F., Stolzenburg, F.: Hybrid multiagent systems with timed synchronization – specification and model checking. In: Dastani, M.M., El Fallah Seghrouchni, A., Ricci, A., Winikoff, M. (eds.) ProMAS 2007. LNCS (LNAI), vol. 4908, pp. 205–220. Springer, Heidelberg (2008)
10. Gupta, G., Pontelli, E.: A constraint-based approach for specification and verification of real-time systems. In: Proceedings of IEEE Real-time Symposium, pp. 230–239 (1997)

11. Harel, D., Naamad, A.: The STATEMATE semantics of statecharts. ACM Transactions on Software Engineering and Methodology 5(4), 293–333 (1996)
12. Henzinger, T.: The theory of hybrid automata. In: Proceedings of the 11th Annual Symposium on Logic in Computer Science, New Brunswick, NJ, pp. 278–292. IEEE Computer Society Press, Los Alamitos (1996)
13. Henzinger, T.A., Ho, P.-H., Wong-Toi, H.: HyTech: The next generation. In: IEEE Real-Time Systems Symposium, pp. 56–65 (1995)
14. Jaffar, J., Santosa, A., Voicu, R.: A CLP proof method for timed automata. In: IEEE International on Real-Time Systems Symposium, pp. 175–186 (2004)
15. Mohammed, A., Furbach, U.: Modeling multi-agent logistic process system using hybrid automata. In: Ultes-Nitsche, U., Moldt, D., Augusto, J.C. (eds.) Proceedings of the 6th International Workshop on Modelling, Simulation, Verification and Validation of Enterprise Information Systems, MSVVEIS 2008, pp. 141–149. Insticc Press (2008)
16. Mohammed, A., Furbach, U.: Multi-agent systems: Modeling and verification using hybrid automata. In: Proceedings of the 7th International Workshop on Programming Multi-Agent Systems (ProMAS 2009), Budapest, Hungary, May 10-15 (2009)
17. Mohammed, A., Stolzenburg, F.: Implementing hierarchical hybrid automata using constraint logic programming. In: Schwarz, S. (ed.) Proceedings of 22nd Workshop on (Constraint) Logic Programming, Dresden, pp. 60–71 (2008); University Halle Wittenberg, Institute of Computer Science. Technical Report 2008/08
18. Murray, J., Stolzenburg, F.: Hybrid state machines with timed synchronization for multi-robot system specification. In: Bento, C., Cardoso, A., Dias, G. (eds.) Proceedings of 12th Portuguese Conference on Artificial Intelligence, pp. 236–241. Institute of Electrical and Electronics Engineers (IEEE), Inc. (2005)
19. Object Management Group, Inc. UML Version 2.1.2 (Infrastructure and Superstructure) (November 2007)
20. Urbina, L.: Analysis of hybrid systems in CLP(R). In: Freuder, E.C. (ed.) CP 1996. LNCS, vol. 1118, pp. 451–467. Springer, Heidelberg (1996)
21. Yovine, S.: Kronos: A verification tool for real-time systems. International Journal on Software Tools for Technology Transfer (STTT) 1(1), 123–133 (1997)

Building Geospatial Data Collections with Location-Based Games

Sebastian Matyas, Peter Wullinger, and Christian Matyas

Chair of Computing in the Cultural Sciences
Laboratory for Semantic Information Processing
Otto-Friedrich-University Bamberg
{firstname.lastname}@uni-bamberg.de

Abstract. The traditional, expert-based process of knowledge acquisition is known to be both slow and costly. With the advent of the Web 2.0, community-based approaches have appeared. These promise a similar or even higher level of information quantity by using the collaborative work of voluntary contributors. Yet, the community-driven approach yields new problems on its own, most prominently contributor motivation and data quality. Our former work [1] has shown, that the issue of contributor motivation can be solved by embedding the data collection activity into a gaming scenario. Additionally, good games are designed to be replayable and thus well suited to generate redundant datasets. In this paper we propose semantic view area clustering as a novel approach to aggregate semantically tagged objects to achieve a higher overall data quality. We also introduce the concept of semantic barriers as a method to account for interaction betwen spatial and semantic data. We also successfully evaluate our algorithm against a traditional clustering method.

Keywords: Games and interactive entertainment, Knowledge acquisition.

1 Introduction

The traditional method of collecting geospatial data (geographic information combined with semantic annotations) is to send surveyors out into the field and subsequently postprocess the collected data, a process which is both expensive and time consuming. Goodchild [2] suggests replacing the experts by a community of voluntary contributors and thus to reduce survey costs significantly.

However, there are good arguments in favor of the traditional approach: Keeping the motivation of the voluntary contributors high is not an easy task. Additionally, ensuring acceptable data quality is also a problem as voluntary contributors are often inexperienced and community collected data is often of low quality.

Finding solutions to these problems is not trivial, as both are closely related to the well-known knowledge acquisition bottleneck (see e.g. [3]). Ahn and Dabbisch [4] argue that contributor motivation may be upheld by embedding the

B. Mertsching, M. Hund, and Z. Aziz (Eds.): KI 2009, LNAI 5803, pp. 703–710, 2009.

data collection process into an enjoyable game. Nonetheless, data quality assurance remains a tedious task and it is a challenge to integrate quality control into a game without spoiling the overall game experience.

One potential solution to the data quality problem is game replayability: Games, and in particular location based games are usually designed to be replayable, i.e. let the player pass through similar situations repeatedly in different game instances. If it is possible to adapt the game rules, such that each repeated situation produces a data point, this creates a large, inaccurate, but redundant dataset. To successfully exploit the gathered data, two new challenges have to be tackled: Designing replayable games and aggregating the redundant data points. In this paper we describe a method to aggregate geospatial data points.

The rest of the paper is structured as follows: After a review of related work (section 2), we proceed to the main contributions of our paper: (1) We illustrate how specific game rules can promote the collection of redundant data (section 3). (2) We use the data points (section 4) obtained in CityExplorer game rounds as input for a novel clustering algorithm that makes use of the unique structure of the collected data points (section 5). (3) We evaluate our algorithm against real game data to show the improvements in comparison to a traditional clustering algorithm (section 5.1). (4) We conclude the paper with the introduction of the concept of "*semantic barriers*" (section 5.2) as a method to account for semantic dissimilarity in spatial clustering and give an outlook onto further improvements (section 6).

2 Related Work

It has long been discussed in the games research community to gather geospatial data in a game [5]. CityExplorer [1], however, is the first working location-based game that specifically collects geospatial and semantic data of real world objects. Earlier work like [6] either focuses on only one kind of data (semantic or geospatial) and/or does not feature a quality control element. Lately, Bell et al. [7] presented the Eyespy game to identify digital photos suitable for navigation tasks, but their approach does not make proper use of data semantics, too.

Spatial clustering has so far relied mainly on geometric similarity measures between points of interest (POIs), consisting of a single spatial coordinate pair (e.g. [8] or [9]) to find assortative objects. Semantic information is most often used only for preliminary filtering if at all [10]. Snavely et al. [11] present an new approach to cluster images based on viewpoints and SIFT keypoints. Their similarity measurement is based solely on the keypoints and uses GPS data only for visualization purposes. They are able to propagate the semantic data associated to parts of the image to other images that feature the same part, but do not use this kind of data in their similarity computation process.

3 Game Rules Promoting Multiple Measurements

The goal of CityExplorer is to seize the majority of segments in an initially unexplored city-wide game area by placing virtual "followers" (*markers*) within

Fig. 1. Inaccessibility of an object and reconstructed view areas for two markers

them (see [1] for details). CityExplorer consists of an outdoor component where players use mobile phones to place markers and an online component in which they upload their markers and the game status is visualized[1]. Placing markers is only admissible at real-world locations of predefined categories of non-movable objects. Two basic rules make CityExplorer replayable:

Player-generated Content: In CityExplorer players can choose object categories freely. This makes it possible to reuse the same game area several times.

Freely Eligible Game-relevant Locations: Markers may be set anywhere in the game area. The only restriction is the list of available categories. The fact that players can transform any non-moveable object of interest (OI) into a game-relevant location greatly improves replayability.

Together, both rules make up an interesting property of CityExplorer: OIs are usually tagged not only once but multiple times, sometimes even in the same game round.

With only these basic rules, however, each marker would be represented only by a single spatial coordinate associated with a category. To improve clustering, we have requested the players to follow a certain procedure when *marking* an OI: Players have to take a photo of the object and subsequently move as near as possible towards the object (see figure 1) before selecting the object's category. We record the position (as GPS coordinates) of the player in both steps.

4 Aggregating Semantically Tagged Spatial Data Points

Community collected spatial data points are usually not accurate. Errors may occur in both the spatial coordinates as well as the semantic data of the dataset: The measuring errors of consumer-grade GPS devices are sometimes in the range of several meters and proper categorization of objects is often non-trivial even to experts.

[1] see `http://www.kinf.wiai.uni-bamberg.de/cityexplorer`

In general, we can view the collected data points as approximations of an abstract, errorless structure. Formally, we call the errorless structure an **object of interest** (OI). An OI is a pair $OI \equiv_{\mathrm{def}} \left\langle \overrightarrow{poi}, cat \right\rangle$, consisting of

poi The **point of interest**, a 2D-dimensional spatial coordinate that yields the geographic location of the POI.

cat A piece of semantic information. For simplicity, we assume that *cat* is one from a predefined set of categories. Our algorithm is extensible to more complex information sets.

As described above, players of CityExplorer place markers on the game fields to identify the approximate locations of OIs. These markers only represent vague information. Formally, a marker is a 3-tuple $mrk \equiv_{\mathrm{def}} \left\langle \overrightarrow{apl}, va, cat \right\rangle$ consisting of

APL The **assumed POI location**, a 2D-dimensional geographic coordinate vector, that gives an estimate of the actual POI.

VA **view area** (VA). While the APL is an estimate of the POI, the view area represents the geographic area, where the POI is assumed to be found. Shape and size of the view area are depend on the measuring error.

cat A **category identifier** (*cat*) that uniquely associates a marker with a user defined category.

Semantically tagged data points may contain errors in two dimensions: The location and the tag. Since all game-relevant locations can be freely chosen by the players not all of them may actually be accessible (see e.g. figure 1). In this case, the APL is really only an approximation of the actual POI. Given the above game rules, we make the following assumptions: (1) The viewing direction is from the POV in the direction of the APL. (2) The actual object is "behind" the APL (as viewed from the POV) (3) The cameras' aperture angle further limits the location of the object. We have therefore decided to restrict the potential locations of objects to viewing trapezoids. A schematic overview is given in figure 1. Miscategorization errors occur, when the tag assigned by the user does not match with the desired categorization. In particular, user assigned categories rarely extend to the desired level of intensional information about an object. We propose to use semantic similarity to tackle this problem.

5 Semantic View Area Clustering

Different players of the same or other game sessions can place a marker on the same object. It is even possible for a single player to set different markers on the same object using different categories if applicable. For example, a player could place a marker at *objectA* using "*building*" as a category and a second marker at *objectA* using "*restaurant*", if both categories are available in the game session. To expedite data quality, it is desirable to re-integrate those duplicate markers, a process commonly known as "*clustering*".

We propose *semantic view area clustering* (svac) as an algorithm that makes use of marker structure without increasing complexity. The algorithm consists of a preprocessing step followed by the actual clustering operation.

In the preprocessing step, we modify the view areas of the individual markers taking into account the semantic data associated with the markers. We measure the similarity between two marker categories as $sim_C(m1, m2)$. For each pair of markers, where the category-similarity is below a certain threshold, we set $m1.VA \leftarrow m1.VA - m2.VA$, where $-$ is the topological difference operation (see figure 2, line 4). This cropping step removes any overlap between markers, whose categories are too different with regard to $sim_C(m1, m2)$.

In the main clustering step, we make use of the cropped view areas. We first pick a marker *m1* from the list of unclustered markers and create a new cluster containing only *m1*. We store the view area of *m1* as the *currentArea*. The cluster is extended by searching for an additional unclustered marker *m2* whose view area overlaps with *currentArea*. If such a marker is found, it is added to the

```
CROPVIEWAREAS(Markers, threshold)
1  for ⟨m1, m2⟩ ∈ Markers × Markers
2        do if m1 ≠ m2 and If sim_C(m1, m2) < threshold
3             ▷ Crop marker view area, unless semantically similar
4               then m1.VA ← m1.VA − m2.VA
```

Fig. 2. View Area Cropping

```
SVAC(Markers)
 1  O ← Markers                                  ▷ Open/unprocessed marker list
 2  C ← ∅                                                          ▷ Cluster list
 3  while O ≠ ∅
         ▷ Create a new cluster with a marker from the open list
         ▷ Combine currentArea with the view area of the new marker
 4          do m1 ← PICKANY(O); O ← O \ {m1}
 5              currentArea ← GETVIEWAREA(m1)
 6              c ← NEWCLUSTER(m1)
 7              m2 ← FINDOVERLAPPING(currentArea, O, m1.cat)
                  ▷ Find all markers with overlap in the same category.
 8              while m2 ≠ NIL
 9                    do O ← O \ {m2}
10                        currentArea ← COMBINEVIEWAREAS(
11                          GETVIEWAREA(currentArea, m2))
12                        c ← c ∪ {m2}
13                        m2 ← FINDOVERLAPPING(currentArea, O, m1.cat)
14              C ← C ∪ {c}
15  return C
```

Fig. 3. Simple View Area Clustering

current cluster and its view area is combined with *currentArea* (see line 10 and 11 in figure 3). If no more overlapping markers are found, we finalize the current cluster and repeat the above steps until all markers are clustered. Pseudo-code for the algorithm is given in figure 3.

5.1 Evaluation

For evaluation, we compared svac against the $k - means$ algorithm found in [9]. We computed Ashbrooks $k - means$ algorithm with a cluster radius of 80m. We test two versions of svac that differ in the implementation of the COMBINEVIEWAREAS() function: (1) view area union and (2) view area intersection.

To compute the similarity of two markers our *svac* algorithm needs an external similarity function $sim_C(A, B) : Cat \times Cat \mapsto \mathcal{R}$. We use Resnik's terminological similarity [12] measure with a WordNET backend, using the existing tags as instances.

Precision and recall values of the algorithms were compared against a manual reference clusterinq of the around 700 data points from four different CityExplorer game rounds played in Bamberg (see [1] for details). Additionally, we computed the F_β score (see [13]), using the weighted (with $\beta = 3$) harmonic mean of the precision and recall values for a class of data points in the reference clustering.

The results in table 1 show an increase in precision with regard to k-means. The difference between the intersection and the union variant of svac algorithm is only marginal for our test data set. Nonetheless, we note, that the intersection variant is less prone to the *chaining effect*, known from single-linkage algorithms [14]. The lower recall is due to the fact that although more clusters with only one member were correctly found the percentage of correctly identified larger clusters went slightly down.

Table 1. Clustering Results

Algorithm	Precision	Recall	F_β
k-means	0.85	0.82	0.80
svac_catsim (intersect)	0.90	0.72	0.85
svac_catsim (union)	0.89	0.74	0.81

5.2 Semantic Barriers

Despite its simplicity, the algorithm exhibits some interesting properties, which we illustrate by the following example: Consider the scenario in figure 4. $mrk1$ and $mrk2$ are of the same, $mrk3$ of a different category. If we do not perform view area cropping, $mrk1$ and $mrk2$ would be incorrectly combined into a single cluster, albeit both are clearly separated by the view area for $mrk3$. We call this separative marker, respectively its corresponding object, a *semantic barrier* as

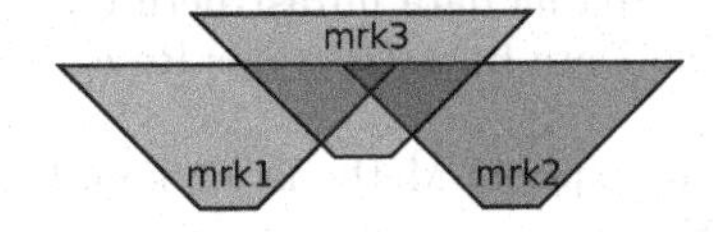

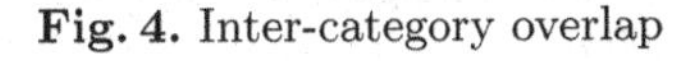

Fig. 4. Inter-category overlap

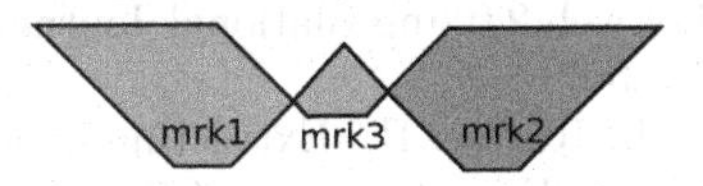

Fig. 5. Cropped overlap Splinter polygons non-adjacent to the APL have been removed

it forms a spatial barrier between semantically similar markers $mrk1$ and $mrk2$. Note that *semantic barriers* are not only found in this particular scenario, but play an important part in every aggregation scenario [15].

To evaluate the usefulness of this modification, the collection of more evaluation data is required, as the current evaluation data set does not contain semantic barrier situations.

6 Conclusion and Outlook

In this paper we illustrated that the addition of two simple game rules are sufficient for a location-based game to produce redundant, semantically tagged geospatial data points. In [1] we already showed that a game like CityExplorer that features these two rules is able to produce large amounts of such data. We presented *svac*, a novel clustering algorithm, that enables us to aggregate multiple, inaccurate measurements of a real-world object to obtain better quality data. We presented semantic barriers as an approach to use semantic information to separate spatially proximate data points, which we expect to generalize well to different scenarios (e.g. [11]).

We believe that the algorithm yields a good balance between complexity and performance and may serve as grounds for future improvements. Manual evaluation of the cropped view areas showed, that the binary, "all-or-nothing" decision to crop view areas is often too radical. A future version of the algorithm should take into account the relative area of overlap and support weighting of overlapping regions using semantic similarity. Additionally, The current algorithm to determine semantic similarity only makes use of the hyponym/hypernom relationship of concepts. We can improve clustering performance, if we take into account the part-of relationship. For example, the categories "*gargoyle*" and "*historical building*" share a high spatial coappearance.

References

1. Matyas, S., Matyas, C., Kiefer, P., Schlieder, C., Mitarai, H., Kamata, M.: Designing location-based mobile games with a purpose - collecting geospatial data with cityexplorer. In: ACM SIGCHI International Conference on Advances in Computer Entertainment Technology: ACE, pp. 244–247. ACM, New York (2008)

2. Goodchild, M.F.: Citizens as voluntary sensors: Spatial data infrastructure in the world of web 2.0. International Journal of Spatial Data Infrastructures Research 2, 24–32 (2007)
3. Olson, J., Rueter, H.: Extracting expertise from experts: Methods for knowledge acquisition. Expert systems 4(3), 152–168 (1987)
4. von Ahn, L., Dabbish, L.: General techniques for designing games with a purpose. Communications of the ACM, 58–67 (August 2008)
5. Capra, M., Radenkovic, M., Benford, S., Oppermann, L., Drozd, A., Flintham, M.: The multimedia challenges raised by pervasive games. In: ACM international conference on Multimedia, pp. 89–95. ACM, New York (2005)
6. Casey, S.K.B., Rowland, D.: The gopher game: a social, mobile, locative game with user generated content and peer review. In: International Conference on Advances in Computer Entertainment Technology: ACE, vol. 203, pp. 9–16 (2007)
7. Bell, M., Reeves, S., Brown, B., Sherwood, S., MacMillan, D., Ferguson, J., Chalmers, M.: Eyespy: supporting navigation through play. In: CHI: International Conference on Human Factors in Computing Systems, pp. 123–132. ACM, New York (2009)
8. Morris, S., Morris, A., Barnard, K.: Digital trail libraries. In: ACM/IEEE-CS Joint Conference on Digital Libraries, JCDL 2004, pp. 63–71. ACM, New York (2004)
9. Ashbrook, D., Starner, T.: Learning significant locations and predicting user movement with gps. In: ISWC: IEEE International Symposium on Wearable Computers, Washington, DC, USA, pp. 101–108. IEEE Computer Society, Los Alamitos (2002)
10. Gösseln, G.v., Sester, M.: Integration of geoscientific data sets and the german digital map using a matching approach. International Archives of Photogrammetry, Remote Sensing and Spatial Information Sciences 35 (2004)
11. Snavely, N., Seitz, S.M., Szeliski, R.: Photo tourism: exploring photo collections in 3d. In: ACM SIGGRAPH 2006 Papers, pp. 835–846. ACM, New York (2006)
12. Resnik, P.: Semantic Similarity in a Taxonomy: An Information-Based Measure and its Application to Problems of Ambiguity in Natural Language. Journal of Artificial Intelligence 11(11), 95–130 (1999)
13. Zhao, Y., Karypis, G.: Evaluation of hierarchical clustering algorithms for document datasets. In: CIKM 2002: International Conference on Information and Knowledge Management, pp. 515–524. ACM, New York (2002)
14. Eckes, T., Rossbach, H.: Clusteranalysen. Kohlhammer (1980)
15. Matyas, S.: Collaborative spatial data acquisition - a spatial and semantic data aggregation approach. In: AGILE International Conference on Geographic Information Science, Aalborg University (2007)

Design Principles for Embodied Interaction: The Case of Ubiquitous Computing*

Rainer Malaka and Robert Porzel

University of Bremen,
TZI, Digital Media Group
Bremen, Germany
{lastname}@tzi.de

Abstract. Designing user interfaces for ubiquitous computing applications is a challenging task. In this paper we discuss how to build intelligent interfaces. The foundations are usability principles that are valid on very general levels. We present a number of established methods for the design process that can help to meet these principle requirements. In particular participatory and iterative so-called human centered approaches are important for interfaces in ubiquitous computing. In particular the question how to make interactional interfaces more intelligent is not trivial and there are multiple approaches to enhance either the intelligence of the system or that of the user. Novel interface approaches, presented herein, follow the idea of embodied interaction and put particular emphasis on the situated use of a system and the mental models humans develop in context.

1 Introduction

It is well known that user interfaces for computational devices can constitute a challenging matter for both their users and their designers [1,15]. By definition an interface is at the boundary of two entities, which - in the case of user interfaces lies between humans and machines while other interfaces lie, for instance, between networks and computers. In ubiquitous and pervasive computing scenarios [33], we face the problem, that there might be no clear boundaries any more. Computers are truly no longer visible at the end and they can disappear from the user's conscious perception. We will, therefore, face the challenge to build an interface for something that is rather shapeless and invisible. Thereby traditional means of employing affordances [20] may no longer apply and the critical notions of context-awareness and situated interaction come into play.

Generally speaking, new forms of enabling meaningful and felicitous interactions in dynamically changing contexts have to be found. A specific proposal for an alternative form of interaction - motivated by the shortcomings of so-called *representational* approaches for including context - has been put forward under the heading of *interactional approaches* to context [7]. In short, this approach - situated itself in the embodied interaction framework [4] - seeks to employ the interface to enable the user to make intelligent decisions, rather than relying on the system to make a decision. This particular

* The research presented in this paper was supported by the German Research Foundation (DFG) as part of the Collaborative Research Centre 637 "Autonomous Cooperating Logistic Processes".

B. Mertsching, M. Hund, and Z. Aziz (Eds.): KI 2009, LNAI 5803, pp. 711–718, 2009.

way of fleshing out the principles of embodied interaction, however, is thwarted in the ubiquitous computing scenario introduced above where that option is no longer viable or - in a manner of speaking - has disappeared together with the interface itself.

In this work, we will propose an alternative solution that provides an extendable combination of representational and interactional instruments in an embodied interaction approach viable for everyday computing. While prototype implementations exists [18] together with a range of prior work and experimental studies, the focus of this work is not to describe a specific technical realization or evaluation thereof, but to flesh out the principles for designing computational blueprints for approaches that are based on the awareness of the shortcomings of prior approaches as well as on the contributions that come from ethnomethodological and psychological considerations [27,9].

In the following, we will go in more detail through these principles and will introduce some general approaches for designing user interfaces. We will see that we one learn from interface design for other - classical - scenarios, and still apply many of those user interface design principles for ubiquitous computing as well. A central aspect thereby is design process that helps to find the right sequence of steps in building a good user interface. After discussing these general aspects of user interface design we will focus on the specific needs for fleshing out embodied interaction in ubiquitous computing scenarios and finally on how to build intelligent user interfaces - or to put it the other way around - to circumvent interfaces that do not facilitate efficient and felicitous interaction.

2 Prior Art

The design of a good user interface is an art, which has been ignored for a long time in the information and communication business. Frequently developers implemented what they thought useful and assumed it to be beneficial for the respective users. However, most users are not software developers and their way of interacting with technology can differ radically. The result is technology that despite its functionality is intuitively usable only for a small group of people. For others it can be highly inefficient, frustrating or even unusable. Some of the highlights of this dilemma can be found in the communication with the user when something goes wrong. Error messages notifying the user often are of use for the developer, e.g. to help in his efforts in debugging the system, but frequently a user will not be able to understand what happened.

In contrast, today usability plays an important role and many systems are designed with more consideration for intuitive and (fail)safe usage. On the one hand this is due to obligations concerning accessibility, but also due to the fact that many systems do not differ greatly in functionality and technical detail and vendors have to diversify their products mainly in terms of their *look and feel*. There exists now a wealth of empirical methods, tools, and guidelines, which all help to develop good user interfaces [28,25,6]. However, there is not one single recipe whose application guarantees perfect results. One essence of usability engineering - in all cases - is to work iteratively in order to achieve the goal of better usability.

Consequently, much prior work on iterative and agile development still applies for designing intelligent user interfaces for ubiquitous computing, while they are also valid for other user interfaces such as *visible* web- or desktop interfaces. The general process of human-centered design could even be applied for products such as machines, appliances and other artifacts [19]. As a matter of fact, one can see this as a truly generic process. With this perspective good usability is a property that is generic featuring similar design processes for multiple domains which aligns perfectly with the way in which ubiquitous computing seeks to integrate everyday computing into the objects of our normal life. From this point of view, then, usability engineering and ubiquitous computing are both concerned with the usability of everyday things, whether they be soft (digital) or hard (physical) components of the intertwined ecosystem.

As a result, from its very beginning on ubiquitous computing had usability in its focus. Mark Weiser's idea of ubiquitous computing encompasses *invisible* interfaces that are so naturally usable that they literally become invisible for the users conscious perception [34]. This notion is expressed by the philosophers Martin Heidegger and hos student Georg Gadamer who called such an interaction with things that we use without conscious awareness things that are *ready-to-hand* or at our *horizon* [13]. In this phenomenologist view, the meaning of the things is actually derived from our interaction with them. Such a view on interactive artifacts can be adopted in ubiquitous computing and is closely related to the notion of embodiment [5,7,12]. This is a fundamental shift from the classical positivist approach that was taken in computational systems. Specifically, this shift goes from modeling the real world via simplifying abstractions to an embodied approach that puts the users and their context at the heart of the matter.

The shift from classical positivist approaches to interactional phenomenologistic approaches is pertinent for finding suitable new forms of interfaces for ubiquitous computing. This is case on the one hand if applications of ubiquitous computing are to be used in *rich* dynamic real-world settings - as opposed to *poorer* closed world settings which feature one or a specific number of contexts and all other contexts have been abstracted away, usually in an implicit manner [22] - then, as a consequence, the way in which *meaning* is afforded for the user will, in fact, evolve in the course of action. But also if applications of ubiquitous computing should become natural extensions of our physical abilities, they must be designed such that they do not need more conscious interference from their users than other natural embodied forms of interaction. Please note that this notion of being *invisible* does not necessarily imply *not there*, but rather present without requiring conscious interaction with the form, but rather natural employment of the form as one finds it in language-based interactions among human interlocutors.

The common examples for interacting with physical objects employ our body parts. When we grasp a coffee cup, we just take it and we do not think and plan how to make our hand grasp the handle and to make our arm to bring the hand there. In this sense, the arm and hand are invisible but also very present. Thus, speaking of interfaces for ubiquitous computing as being *invisible* or computers that are *disappearing* we actually speak of things that are present and *ready-to-hand*. As a result, ubiquitous computing artifacts that one interacts with might not be visible as computers or components. Humans have economical models how things works that are internalized to such a degree that we do not have to think about them, unless we do so professionally [14]. As stated above, these so-called *mental models* of how things work play an important role in designing good user interfaces as well as in designing other everyday things [19,20]. Don

Norman emphasizes that providing a good design comes down to providing the following mappings in an adequate manner: the *design model* must be mapped to the system image, the users must be able to map their mental model(s) to the system which must allow the user to map its image to the users' models.

The question is now, how can a system image support the appropriate user's mental model. The answer - with the notion of embodiment in mind - must bring the meaning of things into the things themselves - thereby enabling a user to derive the meaning of something from the interaction with it, for example, via its appearance that signals some properties indicating how to use it. This, of course, brings up the central notion of affordance again where the idea of affordances now is to bring knowledge into the world and interface instead of having it in the mind, e.g. remembering a sequence of commands or button presses or what to do with chairs. Many usable things in our environment let us know by their physical appearance how to use them. A chair, for instance, does not need a label, manual or instructions on how to sit on it. We just see and know that it can serve as a chair in this context because we know what we can do with it, enabled via embodied simulations that *run on* the respective mental models.

Consequently, this notion of affordances has been adopted in the form of *virtual* affordances for computer interfaces and numerous metaphors on our computer screens signal functionalities, e.g., mouse pointers and scrollbars. With the advent of ubiquitous computing, an affordance becomes again more literally a property attached to the physical properties of things. Not surprisingly, therefore, many ubiquitous computing objects include tactile interfaces or objects with both physical and virtual properties.

3 Principles of Embodied Interaction for Ubiquitous Computing

There are a number of consequences following from assuming this perspective and line of argumentation for fleshing out embodied interaction principles for ubiquitous computing [15]:

- Support mental models - humans employ mental models to understand and to predict how things react to their actions. The system image should support such mental models and make it therefore understandable and usable;
- Respect cognitive economy - humans re-use their mental models and if well-established mental models for similar things exist then they constitute a good basis for providing a means to understand what a new artifact could afford;
- Make things visible and transparent - in order to understand the state of an object it should be obvious what is going on at the appropriate level of granularity, e.g., both containers and folders can indicate how loaded they are;
- Design for errors - mappings between the users' models and the systems sometimes fail and many *human errors* are, in fact, mapping errors. Therefore, systems must assist users in finding a solution for their task even if something went wrong, for which there are a number of techniques, e.g., allowing undo-actions or checks;
- Internal and external consistency - things within an application should work consistently, for instance, pushing a certain type of button, e.g. a specifically colored one, always carries some consistent signal. As a common example the color red

has been conventionalized in our culture for denoting stops, which is also pertinent for external consistency where expectations users may have from usage of other applications, i.e. if we add some ubiquitous computing technology to an artifact, e.g. when turning a *normal* cup into a smart cup, a user will still expect the artifact to retain its normal function(s), or expect the cup to work also as a cup.

Additionally, for the case of our invisible interface forms, where some set of (inter)actions are afforded by the environment that has been instrumented by means of ubiquitous computing technologies. As specified above this instrumentation should only add new affordances or augment - or extend - already existing affordances without obscuring or disabling them. Situated interaction in such an environment becomes feasible by means of situational awareness and corresponding contextual computing approaches.

It has been pointed out that current implementations of context-dependency or context-awareness in computational systems follow an almost standardized path [7]. Firstly, a set of possible environmental states of contextually relevant parameters are defined; then, rules are implemented that try to match sensory inputs to one of the given states during runtime. Within these types of applications context-awareness is fundamentally provided by such matching processes and context itself is represented by the predefined and stored set of environmental settings.

Hereby it is not only difficult to determining the appropriate settings or states of the pertinent parameters, but also the fundamental problem of this approach to contextual computing hinges of the question of how one can pre-compile all the settings and parameters that may become pertinent in advance. When applications of ubiquitous computing are to be used in the aforementioned *rich* dynamic real-world settings - it becomes impossible to define these settings and parameters based solely on past research, surveys, testing, own experience, and on the purpose of the particular system alone.

Correspondingly, given versatile form instruments, as applied in multimodal systems [32,16] it becomes virtually impossible to predict all the possible input and the corresponding contextual dependencies on which their interpretation might hinge. But even in seemingly less murky waters human behavior can hardly be predicted as pointed out frequently by the example of cell phone use as watches, alarm clocks, flash-lights or short message terminals. These examples show that people may use and interact with technology in unexpected ways. This reveals a fundamental problem of implementing a predefined set of settings as such approaches will inevitably not scale to cover possible interactions and behavior that will occur or evolve in future. One reason for this problem is that context has been approached as a representational problem by assuming the following properties of context [8]:

- context is a form of information, i.e. context is seen as something that can be known, represented and encoded in software systems;
- context is delineable, i.e. it is thought to be possible to define what counts as context for a specific application in advance,
- context is stable, i.e. while context may vary from application to application, it does not vary from instance to instance of an interaction with an application;
- context and activity is separable, i.e. context is taken to describe features of the environment within which an activity takes place but the elements of the activity do not belong to context itself.

So far we presented a number of techniques for building good interfaces. We also saw how the view of embodied interaction can be used as a paradigm for ubiquitous computing. In general, a technical solution can be called *intelligent* for two reasons:

(i) there is some built-in intelligent computation that solves some otherwise unsolvable problem;
(ii) using the system, a user can solve an otherwise unsolvable problem, even though the system itself does not actually do anything intelligent.

Suppose that calculating the logarithm of a number is a hard problem for a human, then a calculator is a good example for case (i) and an abacus would be an example for (ii). The calculator solves the problem for the human and the abacus empowers users to solve the problem on their own. It has also been pointed out that the classical approach of artificial intelligence is a rationalist one and according to this approach a system should model the knowledge that human experts have and thus emulate human intelligence. In this sense, the *intelligence* moves from the user to the system. This approach is valid for many cases, e.g., if expert knowledge is rare and non-experts should also be able to work with a system. As discussed above, the interactional approach seeks to make the interaction more intelligent. This fits to many new trends in artificial intelligence where embodied intelligence is viewed as a property that emerges from the interaction of an intelligent agent with the environment.

In this view, even communities of simple and light-weight agents can cooperate to perform intelligent behavior and can produce emergent interaction forms without full reflective and conscious knowledge of the world [29,30]. Also from this perspective all of the above-mentioned material already describes how to build an intelligent ubiquitous interface. Because the processes for designing human-centered systems are just the right techniques for designing intelligent interactive systems, we already defined to an extend how to build intelligent user interfaces. Instead of leaving all intelligence to either the system, the user or the interaction, we can also try to get the best of all worlds and combine these techniques to a cooperative system where both the system and the user cooperate with their knowledge on solving some tasks supported by intelligent interaction techniques.

4 Concluding Remarks

In the sense discussed above, we can make the system more intelligent by enabling the system, the interaction and the user to find appropriate forms, mappings and functions in a cooperative context-adaptive manner. Intelligent user interfaces techniques exist for implementing such a blueprint adhering to the outlined principles. More principled techniques can also now be used to put more knowledge and reasoning into the system. Besides state of the art methods such as data bases, expert systems, heuristic search and planning, a number of more recent developments have caught a lot of interest by researchers and practitioners in the field. Especially the development of foundational infrastructures [17], dedicated and re-usable design patters [10] for modeling world knowledge in formal ontologies make it feasible to describe not only ground domain models, but also contextually reified descriptive models without resorting to higher logics [11]. Semantic technologies and formal models of world knowledge, therefore, had a great renaissance in the last couple of years. In context of the Semantic Web project [3], ontologies have been established as a standard method for capturing complex relations of objects and events in the world. Ontologies have also been successfully used in user interfaces in order to give the system a better understanding on the domain of an interaction and for understanding contextual information supplied by web services and *wrapped* sensors [21].

In terms of implementing a more interactional context-adaptivity that includes user and situation as well as prior interactions in addition to ground and pragmatic models of the domain at hand [24] several viable and more dynamic representational approaches have been forthcoming [23]. Hereby, among the costs of enabling interactional context-adaptivity is that the formal notion of correctness is replaced by one of *plausibility*. As we have discussed context plays a central in ubiquitous computing. Context-dependent user interfaces can greatly enhance the usability of these systems. However, context can also be challenging because it can depend on a huge number of parameters and it is still hard to acquire the needed formalization of the meaning of contexts and to learn the relations between descriptive and ground entities autonomously. To remedy this new approaches, e.g. in the form of human computation [2], can be applied [31], as well as so-called category games [26].

References

1. Adelstein, F., Gupta, S., Richard, G., Schwiebert, L.: Fundamentals of Mobile and Pervasive Computing. McGraw Hill, New York (2004)
2. Ahn, L.V.: Human computation. In: K-CAP 2007: Proceedings of the 4th international conference on Knowledge capture, pp. 5–6. ACM, New York (2007)
3. Berners-Lee, T., Hendler, J., Lassila, O.: The semantic web. Scientific American (May 2001)
4. Carroll, J.: Human-Computer Interaction in the New Millennium (ACM Press). Addison-Wesley Professional, Reading (2001)
5. Clark, A.: An embodied cognitive science? Trends in Cognitive Sciences 3(9), 345–351 (1999)
6. Dix, A.J., Finlay, J., Abowd, G.D.: Human-computer interaction, 3rd edn. Pearson Prentice-Hall, Harlow (2004)
7. Dourish, P.: Where the Action Is: The Foundations of Embodied Interaction. The MIT Press, Cambridge (2001)
8. Dourish, P.: What We Talk About When We Talk About Context. Personal and Ubiquitous Computing 8(1), 19–30 (2004)
9. Dourish, P.: Responsibilities and implications: Further thoughts on ethnography and design. In: Proc. ACM Conf. Designing for the User Experience DUX 2007, Chicago, IL (2007)
10. Gangemi, A.: Ontology design patterns for semantic web content. In: Gil, Y., Motta, E., Benjamins, V.R., Musen, M.A., et al. (eds.) ISWC 2005. LNCS, vol. 3729, pp. 262–276. Springer, Heidelberg (2005)
11. Gangemi, A., Mika, P.: Understanding the semantic web through descriptions and situations. In: Meersman, R., Tari, Z., Schmidt, D.C. (eds.) CoopIS 2003, DOA 2003, and ODBASE 2003. LNCS, vol. 2888, pp. 689–706. Springer, Heidelberg (2003)
12. Gibbs, R.: Embodiment and Cognitive Science. Cambridge University Press, Cambridge (2005)

13. Heidegger, M.: Einfuhrung in die Metaphysik. English. Yale University Press, New Haven (1959)
14. Lakoff, G., Johnson, M.: Metaphors We Live By. University of Chicago Press, Chicago (1980)
15. Malaka, R.: Intelligent user interfaces for ubiquitous computing. In: Max Mühlhäuser, D.I.G. (ed.) Handbook of Research: Ubiquitous Computing Technology for Real Time Enterprises. Springer, Heidelberg (2008)
16. Malaka, R., Haeussler, J., Aras, H., Merdes, M., Pfisterer, D., Joest, M., Porzel, R.: Intelligent interaction with a mobile system. In: Wahlster, W. (ed.) SmartKom - Foundations of Multimodal Dialogue Systems, pp. 505–522. Springer, Berlin (2006)
17. Masolo, C., Borgo, S., Gangemi, A., Guarino, N., Oltramari, A., Schneider, L.: Wonderweb deliverable d17: The wonderweb library of foundational ontologies (2003), `http://wonderweb.semanticweb.org/deliverables/documents/D18.pdf` (last accessed: 04/18/2009)
18. Mudersbach, G., Khan, A.M., Sharma, G., Sayah, S.A., Awad, M., Othman, A.: Intelligent interaction. In: Proceedings of the fifth Student Interaction Design Research Conference, Eindhoven, the Netherlands, April 2009, pp. 22–23 (2009)
19. Norman, D.: Psychology of Everyday Things. Basic Books, New York (1988)
20. Norman, D.: Affordances, conventions, and design. Interactions 6(3), 38–41 (1999)
21. Oberle, D., Lamparter, S., Eberhart, A., Staab, S.: Semantic management of web services. In: Benatallah, B., Casati, F., Traverso, P. (eds.) ICSOC 2005. LNCS, vol. 3826, pp. 514–519. Springer, Heidelberg (2005)
22. Porzel, R., Gurevych, I.: Towards context-adaptive utterance interpretation. In: Proceedings of the 3rd SIGdial Workshop, Philadelphia, USA, pp. 90–95 (2002)
23. Porzel, R., Gurevych, I., Malaka, R.: In context: Integrating domain- and situation-specific knowledge. In: Wahlster, W. (ed.) SmartKom - Foundations of Multimodal Dialogue Systems, pp. 269–284. Springer, Heidelberg (2006)
24. Porzel, R., Zorn, H.-P., Loos, B., Malaka, R.: Towards a seperation of pragmatic knowledge and contextual information. In: ECAI 2006 Workshop on Context and Ontology, Riva del Garda, August 28 (2006)
25. Preece, J., Rogers, Y., Sharp, H.: Interaction Design: beyond human - computer interaction. Wiley, New York (2002)
26. Puglisi, A., Baronchelli, A., Loreto, V.: Cultural route to the emergence of linguistic categories. Proceedings of the National Academy of Sciences (June 2008), 0802485105+
27. Sebanz, N., Bekkering, H., Knoblich, G.: Joint action: Bodies and minds moving together. Trends in Cognitive Sciences 10, 70–76 (2006)
28. Shneiderman, B.: Designing the User Interface. Addison Wesley, Reading (1997)
29. Steels, L.: The origins of ontologies and communication conventions in multi-agent systems. Autonomous Agents and Multi-Agent Systems 1, 169–194 (1998)
30. Steels, L.: Spontaneous lexicon change. In: Proceedings of COLING-ACL, ACL, pp. 1243–1249 (1998)
31. Takhtamysheva, A., Porzel, R., Krause, M.: Games for games: Manipulating context in human computation games. In: KDD 2009 Workshop on Human Computation, Paris, June 28 (2009)
32. Wahlster, W.: Smartkom: Symmetric multimodality in an adaptive and reusable dialog shell. In: Günter, A., Kruse, R., Neumann, B. (eds.) KI 2003. LNCS (LNAI), vol. 2821, Springer, Heidelberg (2003)
33. Weiser, M.: The computer for the 21st century. Scientific American 265(3), 66–75 (1991)
34. Weiser, M.: Creating the invisible interface. In: Szekely, P. (ed.) Proceedings of the 7th annual ACM symposium on User interface software and technology. ACM Press, New York (1994)

Multidisciplinary Design of Air-Launched Space Launch Vehicle Using Simulated Annealing

Amer Farhan Rafique, He LinShu, Qasim Zeeshan, and Ali Kamran

School of Astronautics, Beijing University of Aeronautics and Astronautics, Beijing, China
afrafique@yahoo.com, afrafique@sa.buaa.edu.cn

Abstract. The design of Space Launch vehicle is a complex problem that must balance competing objectives and constraints. In this paper we present a novel approach for the multidisciplinary design and optimization of an Air-launched Space Launch vehicle using Simulated Annealing as optimizer. The vehicle performance modeling requires that analysis from modules of mass, propulsion, aerodynamics and trajectory be integrated into the design optimization process. Simulated Annealing has been used as global optimizer to achieve minimum Gross Launch Mass while remaining within certain mission constraints and performance objectives. The mission of ASLV is to deliver a 200kg payload (satellite) to Low Earth Orbit.

Keywords: Multidisciplinary Design Optimization, Air-launched Space Launch Vehicle, Simulated Annealing.

1 Introduction

The design of Air-Launched Space Launch Vehicle (ASLV) carrying payload to Low Earth Orbit (LEO) is a complex problem that must balance competing objectives and constraints. The whole process requires close interaction between diverse disciplines like aerodynamics, propulsion, structure, trajectory etc. separately albeit coordinated through a system level set of design requirements. This type of segmented design process requires much iteration and invariably leads to design compromises. Therefore the need arises for Multidisciplinary Design and Optimization (MDO) and the use of an artificial intelligence learning tool that can control design process.

Significant research has been performed in rocket-based vehicle design and optimization using various optimization techniques [1,2,3,4,5,6,7,]. The above literature and many more gives us insight that Genetic Algorithm (GA) has been the most attractive choice of designers for MDO of Space Launch Vehicles (SLVs). However in this research effort we propose a new innovative strategy wherein we apply SA for design and optimization of ASLV.

2 Simulated Annealing

Simulated annealing was originally proposed by Metropolis in the early 1950s as a model of the crystallization process. It was only in 1980s, that independent research

B. Mertsching, M. Hund, and Z. Aziz (Eds.): KI 2009, LNAI 5803, pp. 719–726, 2009.

done by [8] and [9] noted similarities between physical process of annealing and some combinatorial optimization problems.

The main advantage of SA is that it can be applied to large problems regardless of the conditions of differentiability, continuity, and convexity that are normally required in conventional optimization methods. The algorithm starts with generation of initial design; new designs are randomly generated in the neighborhood of the current design. The change of objective function value between new and current design is calculated as a measure of the change of temperature of the system. At the end of search, when temperature is low the probability of accepting worse designs is very low. Thus, the search converges to an optimal solution [10,11]. SA has been shown to be a powerful stochastic search method applicable to a wide range of problems [6,12,13] but application of SA on MDO of SLV has not been attempted before. The optimization strategy and parameters applicable to MDO of ASLV are shown in Figure 1.

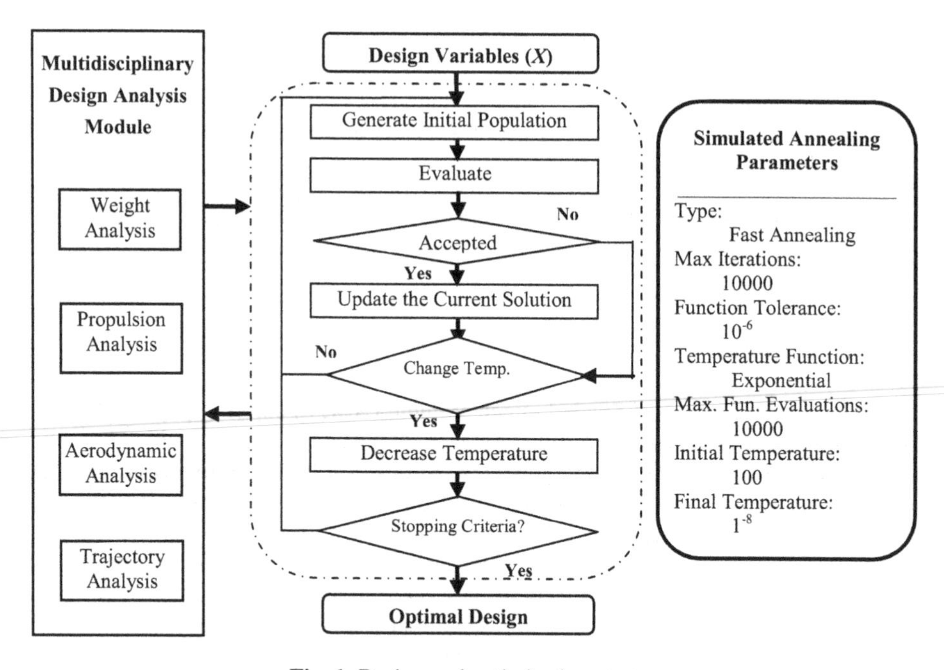

Fig. 1. Design and optimization strategy

In this case a set of design variables *(X)* with upper *(UB)* and lower bounds *(LB)* is passed to SA which sends candidate design vectors to aerodynamics, weight and sizing, propulsion and trajectory analyses modules and performs its further operations. The constraints are calculated and handled by external penalty function. The algorithm runs in a closed loop via optimizer until optimal solution is obtained. Total computation time of whole process is about 18 hours on Intel(R) Core(TM)2 Duo CPU @2.84 GHz with 8GB RAM.

3 Multidisciplinary Design Problem of ASLV

Multidisciplinary design of SLV is an iterative process, requiring a number of design iterations to achieve a balance of emphasis from the diverse inputs and outputs. The major tasks of multidisciplinary design are 1) mission definition; 2) payload requirements, trade studies, and sensitivity analysis; 3) physical integration of different disciplines; and 4) technology assessment and technology development roadmap. Flow chart of multidisciplinary design approach is given in Figure 2.

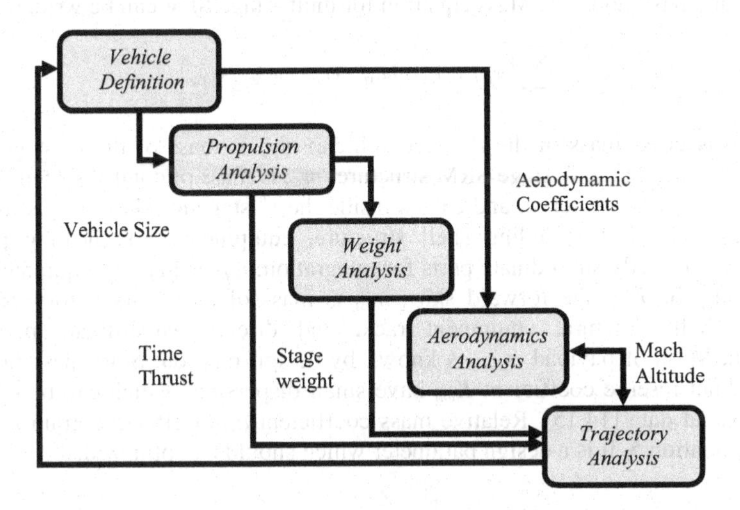

Fig. 2. Multidisciplinary Design Process

Vehicle Definition: ASLV has to be carried on mother aircraft to certain altitude and then launched in horizontal direction (Flight Path Angle = 0 deg) which is opposite to ground launched SLV. The baseline design is launched from approx 12 km at Mach number of 0.8. The propulsion system is solid fuelled Solid Rocket Motor (SRM) and number of stages is fixed as three.

Propulsion Analysis: describes important parameters like thrust, burn time, mass flow rate and nozzle parameters [14]. In this analysis, we have not restricted to particular shape of grain at the conceptual design level, rather a variable grain shape factor (k_{si}) is used to represent the burning surface area of grain (S_{ri}) as a function of grain length (L_i) and diameter (D_i).

Thrust, vacuum specific impulse and mass flow rate are used, as described by [15];

$$F_{1..N} = I_{sp}^{vac}\, \dot{m}_{gni} - p_a A_{ei} \tag{1}$$

$$I_{sp}^{vac} = I_{sp}^{a} + \left(\frac{p_e}{p_c} \right)^{\frac{\gamma-1}{r}} \frac{R_c T_c}{g_o^2 I_{sp}^{a}} \tag{2}$$

$$\dot{m}_{gni} = \rho_{gni} u_i S_{ri} \tag{3}$$

Where N is number of stages, p_a is atmospheric pressure, A_{ei} is nozzle exit area, p_e is exit pressure, p_c is chamber pressure, R_c is gas constant, T_c is temperature in combustion chamber, g_o acceleration due to gravity, (I_{sp}^{a}) average specific impulse, ρ_{gn} is density of grain and u_i is burning rate of grain.

Weight Analysis: Using a combination of physics-based methods and empirical data, the weight of the major components for solid stages are determined from [15]. Total mass of multistage ASLV includes masses of propellants and their tanks, related structures and payload mass. Mass equation for multi-stage SLV can be written as:

$$m_{01} = m_{PAY} + \sum_{i+1}^{n} \left(m_{gni} + m_{sti} + m_{svi} + m_{asi} + m_{fei} + m_{fsi} \right) \tag{4}$$

Where m_{01} is gross mass of the ith stage vehicle; m_{gni} is mass of the ith stage SRM grain; m_{sti} is mass of the ith stage SRM structure; m_{svi} is mass of control system, safety self-destruction system, servo, and cables inside the ith stage aft skirt; m_{asi} is mass of the ith stage aft skirt including shell structure, equipment rack, heating protect structure, and directly subordinate parts for integration; m_{fei} is mass of equipment and cables inside the ith stage forward skirt; m_{fsi} is mass of the ith stage forward skirt including shell structure, equipment rack, and directly subordinate parts for integration. Mass of payload m_{PAY} is known by design mission. Skirt mass ratio N_i, and propellant reserve coefficient K_{gni} have small dispersions which can be selected from statistical data [14,15]. Relative mass coefficient μ_{ki} of effective grain is given below in Equation 5. It is a design parameter which should be optimized.

$$u_{ki} = {m^{e}_{gni}} \Big/ {m_{oi}} \tag{5}$$

Structure mass fraction (α_{sti}) is the main parameter for designing a multistage SLV. It is dependent upon structural material, grain shape, as well as the parameters of internal ballistics of SRM. This structure mass fraction is ratio of the sum of the chamber case mass (m_{cc}), cementing layer mass (m_{cl}), nozzle mass (m_n) and insulation liner mass (m_{in}) to the grain mass (m_{gni}), as shown in Equation 6:

$$\alpha_{sti} = {(m_{cc} + m_{cl} + m_n + m_{in})} \Big/ {m_{gni}} \tag{6}$$

Aerodynamics Analysis: In conceptual design phase of SLV, rapid and economical estimations of aerodynamic coefficients are required. For this purpose U.S. Air Force Missile DATCOM 1997 (digital) [16] has been widely used. DATCOM is capable of quickly and economically estimating the aerodynamics of wide variety of design configurations and in the different flow field regions that SLV encounters.

Trajectory Analysis: Computational cost penalty and unavailability of detailed data during conceptual design phase compels use of 3 Degree-of-Freedom (DOF) trajectory analysis instead of 6DOF trajectory simulation [17]. The trajectory analysis depends on inputs from aerodynamic, mass and propulsion modules. The flight program and results obtained from the other disciplines are used to compute the flight trajectory. ASLV is treated as a point-mass, and flight in 2D over spherical non-rotating earth is assumed. Forces acting on ASLV are given in Figure 3.

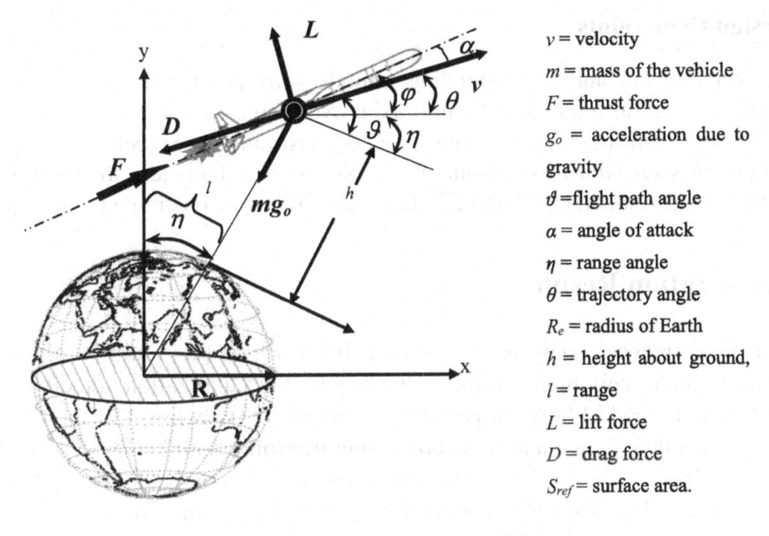

Fig. 3. Forces acting on ASLV

3.1 Design Objective

In aerospace vehicle design in general and in SLV design in particular minimum take-off weight concepts have traditionally been sought as weight (or mass) is strong driver on vehicle performance and cost. For the present research effort, design objective is to minimize Gross Launch Mass (GLM) of entire vehicle, Eq (7) and (8).

$$^{Min}\,GLM\ (X) \tag{7}$$

$$X = f(\mu_{\text{ki}}, D_i, p_{ci}, p_{ei}, k_{si}, u_i, \alpha_m, a) \tag{8}$$

3.2 Design Variables

Governing design variables of ASLV problem are given in Table 2.

Table 1. Discipline-wise design variables

Discipline	Design Variable	Symbol	Units
Structure + Propulsion	Mass Coefficient of Grain	μ_{ki}	ratio
Structure + Propulsion + Aerodynamics	Body Diameter	D_i	m
Structure + Propulsion	Chamber Pressure	p_{ci}	bar
Structure + Propulsion	Exit Pressure	p_{ei}	bar
Structure + Propulsion	Grain Shape Coefficient	K_{si}	
Propulsion	Grain Burning Rate	u_i	mm/s
Aerodynamics + Trajectory	Max Angle of Attack	α_m	deg
Aerodynamics + Trajectory	Launch Maneuver Variable	a	

3.3 Design Constraints

Mission velocity (v) and corresponding altitude (alt) are formulated as trajectory constraints. Structural requirements limit axial overload (O_x) to be lower than 12g for 1st and 2nd stage. During launch maneuver α_m is constrained to be below 22 deg and is ensured that it is zero during transonic phase. Nozzle exit diameters are constrained to be less than stage diameters. 1st and 2nd stage diameters are constrained to be equal.

4 Optimization Results

Detailed performance analysis of ASLV, being proposed, fulfills all mission requirements under certain constraints. The required circularization velocity (> 7600 m/s) and altitude (> 450 km) are perfectly achieved. O_x is below 12g for 1st and 2nd stage, α_m is less than 22deg and it is zero during transonic regime (Fig. 4). Optimized design variables (X^*) lie between their upper and lower bounds (Table 3). Minimum GLM achieved is 20.5 Mg and is comparable to existing system [18].

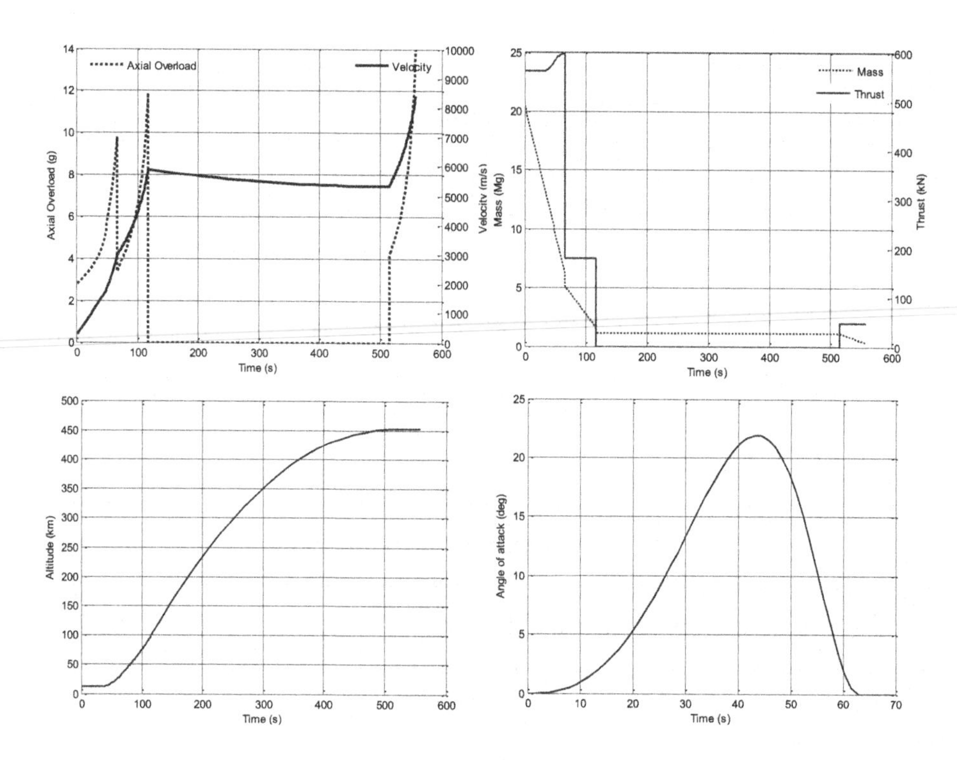

Fig. 4. Performance graphs of ASLV

Table 2. Optimum Values of Design Variables

	Symbol	Units	LB	UB	Optimum Value
Mass Coefficient of Grain 1st Stage	μ_{k1}	ratio	0.60	0.75	0.6994
Mass Coefficient of Grain 2nd Stage	μ_{k2} / μ_{k1}	ratio	1.00	1.10	1.0000
Mass Coefficient of Grain 3rd Stage	μ_{k3} / μ_{k2}	ratio	1.00	1.10	1.0002
Body Diameter 1st and 2nd Stage	D_1	m	1.20	1.40	1.3000
Body Diameter 3rd Stage	D_3	m	0.80	1.00	0.8504
Chamber Pressure 1st Stage	p_{c1}	bar	55.0	75.0	66.048
Chamber Pressure 2nd Stage	p_{c2}	bar	55.0	75.0	60.314
Chamber Pressure 3rd Stage	p_{c3}	bar	55.0	75.0	60.024
Exit Pressure 1st Stage	p_{ei}	bar	0.10	0.16	0.1400
Exit Pressure 2nd Stage	p_{e2}	bar	0.08	0.14	0.1278
Exit Pressure 3rd Stage	p_{e3}	bar	0.08	0.14	0.0999
Grain Burning Rate 1st Stage	u_1	mm/s	5.00	8.00	6.0976
Grain Burning Rate 2nd Stage	u_2	mm/s	5.00	8.00	6.9985
Grain Burning Rate 3rd Stage	u_3	mm/s	5.00	8.00	6.0996
Grain Shape Coefficient 1st Stage	K_{s1}		1.50	2.30	2.1814
Grain Shape Coefficient 2nd Stage	K_{s2}		1.50	2.30	2.2998
Grain Shape Coefficient 3rd Stage	K_{s3}		1.50	2.30	2.0167
Max Angle of Attack	α_m	deg	1.00	22.0	21.911
Launch Maneuver Variable	a		0.01	0.10	0.0166

5 Conclusion

Simulated Annealing as optimizer has been successfully implemented to solve the complex multidisciplinary design problem of ASLV which has never been attempted before. SA proved able for the multidisciplinary design of ASLV considering propulsion, mass features, aerodynamics and trajectory performance objectives and mission constraints. The results of this preliminary design can be used as a basis for detailed design. The optimization results and performance are to be considered as preliminary (proof-of-concept) only, but they can be compared to existing systems [18] and can be applied in conceptual design of similar space launch systems.

Such a design strategy allows vehicle designers to rapidly consider number of fully converged design alternatives in a very short time without sacrificing design detail, thus improving the quality of whole design process. Application of SA as optimizer for the MDO opens new avenues in design and optimization of aerospace systems.

References

[1] Bayley, D.J., Hartfield, R.J., Burkhalter, J.E., Jenkins, R.M.: Design Optimization of a Space Launch Vehicle using a Genetic Algorithm. Journal of Spacecrafts and Rockets 45(4), 733–740 (2008)

[2] Le Moyne, R.: Multidisciplinary Cost and Performance Optimization of Two-Stage Liquid Propulsion Based Launch Vehicle, AIAA Paper 2008-2642

[3] Jahangir, J., Masoud, E., Jafar, R.: Multidisciplinary design optimization of a small solid propellant launch vehicle using system sensitivity analysis. Structure and Multidisciplinary Optimization 38, 93–100 (2009)
[4] Bayley, D.J., Hartfield, R.J., Burkhalter, J.E., Jenkins, R.M.: Design Optimization of a Space Launch Vehicle Using a Genetic Algorithm. In: 3rd AIAA Multidisciplinary Design Optimization Specialist Conference, AIAA Paper 2007-1863
[5] Bayley, D.J., Hartfield, R.J.: Design Optimization of a Space Launch Vehicles for Minimum Cost Using a Genetic Algorithm. In: 43rd AIAA Joint Propulsion Conference and Exhibit, AIAA Paper 2007-5852
[6] Tekinalp, O., Saadet, U.: Simulated Annealing for Missile Trajectory Planning and Multidisciplinary Missile Design Optimization, AIAA Paper 00-0684
[7] Rafique, A.F., Zeeshan, Q., LinShu, H.: Conceptual Design of Small Satellite Launch Vehicle Using Hybrid Optimization. In: International Conference on Nonlinear Problems in Aviation and Aerospace. ICNPAA 2008 (2008)
[8] Kirkpatrick, S., Gelatt Jr., C.D., Vecchi, M.: Optimization by Simulated Annealing. Science 220(4598), 498–516 (1983)
[9] Cerny, V.: Thermodynamical approach to the Traveling Salesman Problem: An efficient simulation algorithm. Journal of Optimal Theory and Applications 45(1), 41–51 (1985)
[10] Chattopadhyay, A., Seeley, C.E.: Simulated Annealing Technique for Multiobjective Optimization of Intelligent Structures. Journal of Smart Materials & Structures 3, 98–106 (1994)
[11] Aarts, E.H.L., Korst, J.H.M., Van Laarhoven, P.J.M.: A Quantitative Analysis of The Simulated Annealing Algorithm: A case study for the traveling salesman problem. Journal of Statistical Physics 50(1–2), 187–206 (1988)
[12] Tekinalp, O., Bingol, M.: Simulated Annealing for missile optimization: developing method and formulation techniques. Journal of Guidance, Control, and Dynamics 27(4), 616–626 (2004)
[13] Osman, I.H., Laporte, G.: Metaheuristics: A bibliography. Annals of Operations Research 63, 513–623 (1996)
[14] Sutton, G.P., Biblarz, O.: Rocket propulsion elements, 7th edn. Wiley-Interscience, New York (2001)
[15] LinShu, H.: Launch vehicle design. BeiHang University Press, Beijing (2004)
[16] Blake, W.B.: Missile DATCOM: User's manual-1997 FORTRAN 90 revision. Wright-Patterson AFB, Oklahoma (1998)
[17] Zipfel, P.H.: Modelling and simulation of aerospace vehicle dynamics. AIAA, Reston (2007)
[18] Isakowitz, S.J.: International Reference Guide to Space Launch Systems, 3rd edn. American Institute of Aeronautics and Astronautics (1999)

Stochastic Feature Selection in Support Vector Machine Based Instrument Recognition

Oliver Kramer and Tobias Hein

Department of Computer Science
Technische Universität Dortmund
44227 Dortmund, Germany

Abstract. Automatic instrument recognition is an important task in musical applications. In this paper we concentrate on the recognition of electronic drum sounds from a large commercially available drum sound library. The recognition task can be formulated as classification problem. Each sample is described by one hundred temporal and spectral features. Support Vector Machines turn out to be an excellent choice for this classification task. Furthermore, we concentrate on the stochastic optimization of a feature subset using evolution strategies and compare the results to the classifier that has been trained on the complete feature set.

1 Introduction

Instrument recognition is an important task in music information retrieval. Onset and beat detection as well as instrument recognition belong to the first necessary steps from transforming subsymbolic music data to symbolic representations, e.g. transcription. In the taxonomy of problem classes in music by Kramer, Stein and Wall [8] instrument recognition belongs to the music analysis and subsymbolic branch. We concentrate on the recognition of digital drum sounds, as this genre is of high practical importance. For this purpose we formulate the instrument recognition task as classification problem and use a multi-class support vector machine (SVM) to assign correct labels. Classification with SVMs exhibits various degrees of freedom, e.g. the choice of kernel and regularization parameters or the selection of features. In various classification approaches with SVMs the application of stochastic search methods turned out to be an elegant solution. In this line of research we apply evolution strategies to the instrument recognition task, i.e. the selection of musical features that describe the samples.

In the next Section 2 we formulate the instrument recognition task as classification problem, describe the underlying classifier and give a short survey of related work. Section 3 describes the data base of drum samples and the features we generate to describe each sample. In the following Section 4 we describe the evolutionary feature selection approach and show an experimental analysis with regard to five feature set sizes.

B. Mertsching, M. Hund, and Z. Aziz (Eds.): KI 2009, LNAI 5803, pp. 727–734, 2009.

2 Large Margin Based Instrument Recognition

At first, we describe the evolutionary large margin based classification algorithm. We want to learn a strong classification algorithm that is capable of recognizing electronic drum sounds in real scenarios.

2.1 Instrument Recognition as Classification Task

Instrument recognition can be treated as classification task. We assume that each electronic drum sample[1] is represented by a feature vector $\boldsymbol{x}$. Section 3 describes the musical features we use. Given a training set $\mathcal{L} \subset \mathcal{S}$ consisting of features with class labels (i.e. drum sound names) we train a classifier to predict correct class labels on the training set that minimizes the classification error on the test data set $\mathcal{S}$. In our scenario of electronic drum sound recognition seven labels are given. We use the LIBSVM implementation from Chang *et al.* [3] for our multi-class instrument recognition task. The stochastic search algorithms are based on our own implementation.

2.2 Large Margin Based Instrument Classification

Support vector machines are the most famous kernel methods and are usually used for classification and regression tasks. Classical SVMs are based on three concepts: First, finding the hyperplane that separates the data of different classes based on the maximum margin principle, second, mapping the data samples to the Hilbert $\mathcal{H}$ space with a kernel function $K(\boldsymbol{x}_i, \boldsymbol{x}_j)$ where the data is linearly separable. The resulting decision function is

$$f(\boldsymbol{x}) = \text{sign}\left(\sum_{i=1}^{l} y_i \alpha_i^* K(\boldsymbol{x}_i, \boldsymbol{x}_j) + b\right) \tag{1}$$

Third, the problem can be transformed into the dual form and solved more easily by means of Lagrange multipliers.

$$W(\alpha) = \sum_{i=1}^{l} \alpha_i - \frac{1}{2} \sum_{i,j=1}^{l} \alpha_i \alpha_j y_i y_j K(\boldsymbol{x}_i, \boldsymbol{y}_j) \tag{2}$$

subject to $\sum_{i=1}^{l} \alpha_i y_i = 0$ and $0 \leq \alpha_i \leq C$ for $i = 1, \ldots, l$. The coefficients α_i are the solution to the quadratic optimization problem. The vectors $\boldsymbol{x}_i$ are the support vectors. For a detailed introduction to SVMs we refer to Bishop [2]. Parameter C is the regularization parameter that controls the trade-off between maximizing the margin and minimizing the function complexity, i.e. the L_1-norm of the margin slack vector.

[1] For the sake of a better understanding: In machine learning a sample is one data example, in music a sample is a digital sound fragment, in digital signal processing a sample is one scanned value.

2.3 Related Work

In this section we shortly present related work in the field of machine learning based instrument recognition, in particular with regard to percussive sounds, and previous attempts of optimizing kernel methods with evolutionary algorithms.

Music Instrument Recognition. Most of the approaches that focus on instrument classification try to recognize natural instruments like piano or strings. Herrera-Boyer *et al.* [7] gave an overview of instrument classification methods. Only few work concentrates on electronic sounds although this genre is closer to computer music production and software applications. Van Steelant *et al.* [13] concentrated on the classification of only two electronic percussive instruments, i.e. bass drums and snare drums, with SVMs. They achieved a fast learning time with linear kernels and concentrated on a set of about 5,000 drum sounds. Gillet *et al.* [5] address the problem of separation and transcription of percussion sounds from polyphonic music. Their system is based on a harmonic-noise decomposition, time-frequency masking, an improved Wiener filtering-based separation method and the automatic selection of relevant features. They make use of grid search to determine SVM parameters. The classification accuracy varies between 50 and 80 percent taking three classes (base drum, snare drum and hi-hat) into account.

Stochastic Search in Classification with Kernels. The application of evolutionary methods turn out to be an elegant solution for some machine learning optimization tasks. Ruxandra *et al.* directly solves the primal SVM optimization problem by means of an evolutionary algorithm for regression [11] and classification tasks [12]. Mierswa [9] balances the classification accuracy and complexity of SVMs by means of multi-objective evolutionary algorithms. Friedrichs *et al.* [4] have tuned the kernel parameters of support vector machines concentrating standard Gaussian and general Gaussian kernels that rotate and scale the features with a positive definite matrix. They performed a coarse tuning with grid search and a fine tuning using Covariance Matrix Adaptation. The experimental results reported achieved a slight improvement of about 1% for the classification error. We have also tried to evolve the parameters of a Gaussian kernel, but were only able to achieve an improvement that was not statistically relevant. For an introduction to the problem of feature selection we refer to Guyon *et al.* [6].

3 Data Description

3.1 Sample Data Base

Our library of electronic drum sounds consists of 3,300 samples from two commercial libraries[2]. We emphasize that the drum samples are real-world data – containing "dirty" samples that are pitched atypically or contain two or more superposed instruments. The data sets comprise seven instrument classes, i.e.

[2] One-shot drum samples of *Techno Trance Essential* and *House Essential*, Überschall.

base drums, snare drums, claps, open hi-hats, closed hi-hats, and cymbols (chrashes and rides), and tom-toms. Most samples are based on famous digital and analog drum synthesizes like Roland TR-909, Roland TR-808 or Yamaha RY30. The programs, algorithms or digital circuits with their specific parameterizations lead to *typical* sounds. The average length of all samples is about 456 ms or 20126 samples. Before the training of the SVM the sample descriptors, i.e. the features, are computed offline. For each sample the computation of all feature takes approximately 3 seconds on a 2.0 GHz AMD-64 processor.

3.2 Sample Descriptors

To describe the sound samples we use typical features that have been proposed in the set of audio features of the CUIDADO project [10]. We make use of 10 features which can be devided into the following sets of temporal features and spectral features:

1. Temporal features: rms, zero crossing rate, centroid, crest-factor
2. Spectral features: centroid, crest-factor, skewness, kurtosis, flatness, rolloff

These features are extracted from the entire sound signal and 9 different frequency ranges, thus resulting in a total set of 100 features characterizing each sound sample.

Due to the limited space we restrict ourselves to the description of two features and refer to [10] for the definition of all sample descriptors. The zero crossing rate x_{zcr} – as an example for a temporal feature – is a commonly used feature in music information retrieval and speech recognition systems. It counts the number of times, the signal s_t crosses the axis of abscissae within period T. Let $\delta(\cdot)$ be one if its argument is true. Then

$$x_{\mathbf{zcr}} = \frac{1}{T}\sum_{i=1}^{T} \delta(s_t s_{t-1} < 0) \tag{3}$$

is the zero crossing rate.

Another example for a useful audio feature we use is the temporal crest factor x_{cf}. It is the peak amplitude of a waveform divided by its rms-value (root mean square)

$$x_{\mathbf{cf}} = \frac{|x|_{peak}}{\sqrt{\frac{1}{T}\sum_{i=1}^{T} x_i}} \tag{4}$$

The crest-factor is an important measure because it is easy to calculate and provides a measure for how much impacting is occurring in a waveform. This information is lost in a fast fourier transformation that cannot differentiate between impacting and random noise.

4 Stochastic Feature Selection

Features have an essential influence on the classification results. The question arises: what is the optimal set of music descriptors $\mathcal{F}$ for our drum recognition task? In the following we use evolutionary optimization techniques for the automatic selection of relevant musical features. For this sake we use a $(\mu + \lambda)$-Evolution Strategy (ES). For a comprehensive introduction to ES we refer to Beyer *et al.* [1].

4.1 Evolutionary Model

We search for an optimal subset of features $\mathcal{F}' \subset \mathcal{F}$ of a given size $\phi = |\mathcal{F}'|$ that minimizes the classification error, i.e. the number of wrong predictions with regard to the whole tuning set. The evolutionary algorithms is modeled as follows.

1. Representation: each individual $(\boldsymbol{x})_i$, $1 \leq i \leq \lambda$ consists of a list of features $(\boldsymbol{x})_i = \{x_1, \ldots, x_{|\mathcal{F}'|}\}$, with $x_j \in 1, \ldots 3,300$ defining $\mathcal{F}'$.
2. Population model: We use a (5+25)-EA, i.e. $\mu = 5$ parental and $\lambda = 25$ offspring individuals.
3. Initialization: Initial training samples are added randomly with uniform distribution and may occur multiple times.
4. Recombination: We do not use recombination, each solution is selected with uniform distribution from the parents.
5. Mutation: Each component x_i of vector $\boldsymbol{x}$ is mutated with probability $p_m = 1/|\mathcal{F}'|$.
6. Selection: Plus-selection selects the μ best solutions from all $\mu + \lambda$ solutions of the previous generation.
7. Fitness: The classification error on the tuning set of samples $\mathcal{T}$ guides the search.
8. Termination condition: we stop the algorithm after 500 iterations.

As stated in Section 3 each sample is described by 100 features. We seek for the feature set $\mathcal{F}'$ of sizes $\phi = 15, 30, 45, 60$, i.e. the ϕ best features that result in the best instrument recognition performance. To train the SVM we divide the set of 3,300 samples in each run into three sets of same size, i.e. a learning set $\mathcal{L}$ the classifier is trained with, a tuning set $\mathcal{T}$ and an evaluation set $\mathcal{E}$. The SVM learns on the learning set $\mathcal{L}$ using feature set $\mathcal{F}'$ defined by $\boldsymbol{x}_t$ in generation t. To evaluate the classification error $\delta_{\mathcal{T}}$, we test the SVM on the tuning set $\mathcal{T}$ and use $\delta_{\mathcal{T}}$ as quality (fitness measure) for the ES. After the run of the ES we evaluate the quality of the evolved feature set $\mathcal{F}'$ on the evaluation set $\mathcal{E}$.

4.2 Experimental Results

Table 1 summarizes the results of our stochastic feature selection approach with four different feature set sizes and five runs each. It shows the classification error $\delta_{\mathcal{T}}$ on the tuning set $\mathcal{T}$ in the last generation $t = 500$ and the error $\delta_{\mathcal{E}}$ on the evaluation set $\mathcal{E}$. In each run we randomly select the three sets $\mathcal{L}$, $\mathcal{T}$, and $\mathcal{E}$. For

Table 1. Stochastic evolution of feature set $\mathcal{F}'$ after $t = 500$. Classification error $\delta_\mathcal{T}$ shows the error on the tuning set $\mathcal{T}$ while $\delta_\mathcal{E}$ counts the correct labels on the evaluation set $\mathcal{E}$.

	15		30		45		60		100	
	$\delta_\mathcal{T}$	$\delta_\mathcal{E}$	$\delta_\mathcal{T}$	$\delta_\mathcal{E}$	$\delta_\mathcal{T}$	$\delta_\mathcal{E}$	$\delta_\mathcal{T}$	$\delta_\mathcal{E}$	$\delta_\mathcal{T}$	$\delta_\mathcal{E}$
best	0.8173	0.8045	0.8309	0.8173	0.8373	0.8127	**0.8391**	0.8127	0.8218	0.8318
mean	0.8009	0.7840	0.8123	0.7923	0.8236	0.7945	**0.8274**	0.8072	0.8057	0.8087
worst	0.7900	0.7664	0.7982	0.7809	0.8118	0.7782	0.8136	0.7973	0.7873	0.7890
dev	0.0117	0.0153	0.0105	0.0132	0.0101	0.0120	0.0082	0.0059	0.0092	0.0111

the purpose of comparison the table also shows the results of the SVM that is trained with all features and the last column of the table shows the results of 20 runs of the SVM with all features, but stochastically selected training, test and evaluation set.

We can observe that in each case the achieved accuracy on the tuning set $\mathcal{T}$ is better than the accuracy on the evaluation set $\mathcal{E}$. This is easy to explain as the stochastic search procedure adapts to the training set and optimizes the feature selection in order to achieve a maximal classification quality on $\mathcal{T}$. The error $\delta_\mathcal{E}$ on the evaluation set is a measure for the ability of our model to generalize. As the figures are comparable, this ability is quite high. With $\delta^*_\mathcal{T} = 0.8391$ the best accuracy on the training set $\mathcal{T}$ has been achieved on the optimized feature set of size $\phi = 60$. The best accuracy on the evaluation set has been achieved for the size $\phi = 30$, but in average also for size $\phi = 60$. But we can summarize that almost all classification results achieve similar accuracies – after feature set optimization. Of course, the training considering all features reaches the high accuracies of over 80% at once, without further optimization. But the results confirm the importance of proper features, because a comparably small, but optimized feature set ($\phi = 15$) can achieve similar results like a set taking all features ($\phi = 100$) into account. In all runs of the small feature sets one third of the chosen features concerned the whole feature spectrum. Furthermore, it is remarkable that the RMS-feature in different frequency bands was chosen often.

Figure 1 shows the fitness development of the evolutionary feature selection process in terms of classification error on the tuning set $\delta_\mathcal{T}$ and the evaluation set $\delta_\mathcal{E}$. The feature set sizes $\phi = 15, 30, 45$, and 60 are tested. At the beginning the accuracy of the trained SVM with a subset of features is comparably small in all runs. But we can clearly observe the reduction of the classification error in the course of evolution both on the evaluation and on the tuning set. For $\phi = 15$ classification accuracies lie closely together and show almost the same development. For the other feature set sizes we can observe a gap between the errors on both sets. In contrast to the averaged figures of table 1 for $\phi = 30$ and $\phi = 45$ the accuracy on the evaluation set is even higher during the course of evolution. For $\phi = 60$ the accuracy on the tuning set is higher. We can observe that the accuracy on the evaluation set can deteriorate.

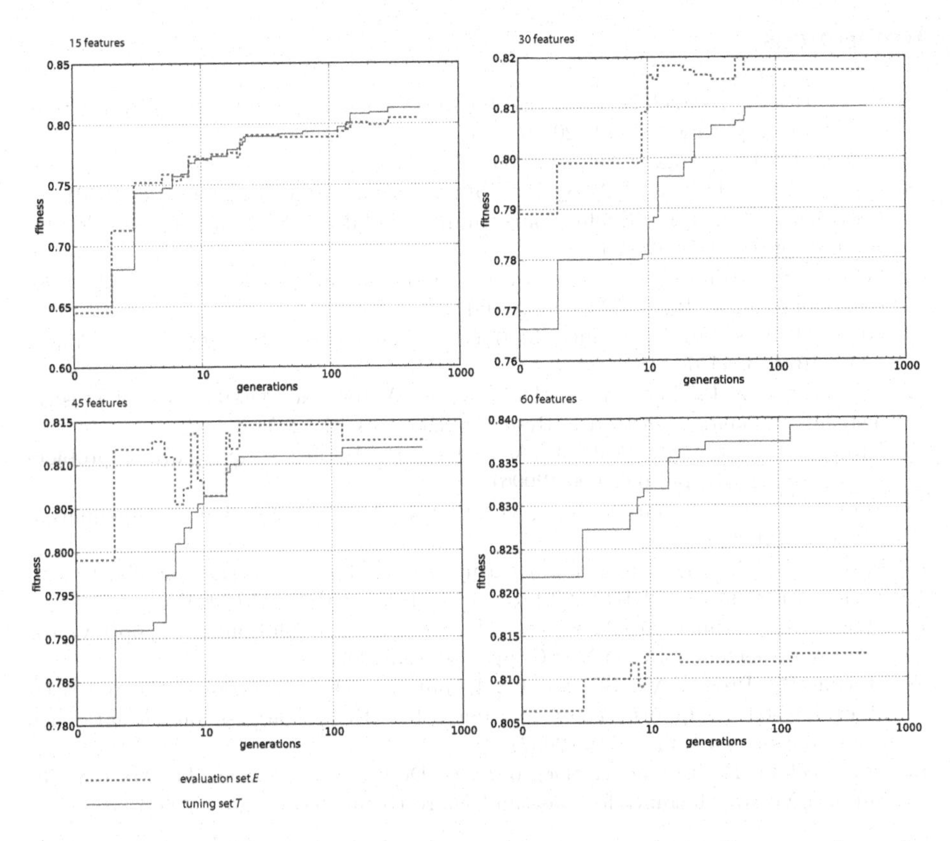

Fig. 1. Fitness development of the best runs with regard to the fitness on the evaluation set with various feature set sizes $\phi = 15, 30, 45$, and 60 on a logarithmic scale (generations).

5 Conclusion

Instrument recognition of digital drum sounds is an interesting classification problem of practical relevance. We took the seven most important instrument classes into account. Our SVM based classification approach showed a classification accuracy of over 80%. Taking the fact into account that many of the samples are disturbed and dirty – which are usual conditions in musical praxis – the classification error is excellent. The classifiers can directly be used to automatically classify huge sets of drum samples. In the end, all trained classifiers have shown similar classification errors. Our analysis has proven the importance of features: A small, but evolutionary optimized feature set can achieve the same accuracy like a large set of features. But it is questionable if the practitioner should invest the effort to optimize the feature set if the classifier itself is able to taking the relevant features into account. Our future work will concentrate on the recognition of drum samples within songs and audio mixes. We will try to answer the question whether other instruments disturb the classification results.

References

1. Beyer, H.-G., Schwefel, H.-P.: Evolution strategies - A comprehensive introduction. Natural Computing 1, 3–52 (2002)
2. Bishop, C.M.: Springer (August 2006)
3. Chang, C.C., Lin, C.J.: LIBSVM: a library for support vector machines (2001)
4. Friedrichs, F., Igel, C.: Evolutionary tuning of multiple SVM parameters. Neurocomputing 64, 107–117 (2005)
5. Gillet, O., Richard, G.: Transcription and separation of drum signals from polyphonic music 16(3), 529–540 (2008)
6. Guyon, I., Elisseeff, A.: An introduction to variable and feature selection. J. Mach. Learn. Res. 3, 1157–1182 (2003)
7. Herrera-Boyer, P., Peeters, G., Dubnov, S.: Automatic classification of musical instrument sounds. New Music Research 32(1), 13–21 (2003)
8. Kramer, O., Stein, B., Wall, J.: Ai and music: Toward a taxonomy of problem classes. In: ECAI, pp. 695–696 (2006)
9. Mierswa, I.: Evolutionary learning with kernels: A generic solution for large margin problems (2006)
10. Peeters, G.: A large set of audio features for sound description (similarity and classification) in the CUIDADO project. Tech. rep., IRCAM (2004)
11. Stoean, R., Dumitrescu, D., Preuss, M., Stoean, C.: Evolutionary support vector regression machines. In: SYNASC, pp. 330–335 (2006)
12. Stoean, R., Preuss, M., Stoean, C., Dumitrescu, D.: Concerning the potential of evolutionary support vector machines. In: IEEE Congress on Evolutionary Computation, pp. 1436–1443 (2007)
13. Van Steelant, D., Tanghe, K., Degroeve, S., De Baets, B., Leman, M., Martens, J.: Support Vector Machines for Bass and Snare Drum Recognition (2005)

Author Index

Abkar, Aliakbar 639
Abou Assali, Amjad 564
Adrian, Benjamin 249
Alimohammadi, Abbas 639
Althoff, Klaus-Dieter 556
André, Elisabeth 500
Arens, Michael 73
Artmann, Stefan 363, 387
Atzmueller, Martin 233
Autexier, Serge 444
Aziz, Muhammad Zaheer 315
Azizi, Amir 114

Bach, Kerstin 556
Barthelmes, André 169
Baumann, Michael 648
Beckert, Bernhard 687
Beckstein, Clemens 363
Bee, Nikolaus 500
Beierle, Christoph 273
Belli, Fevzi 427
Ben Amor, Heni 492
Benzmüller, Christoph 289
Bercher, Pascal 57
Berger, Erik 492
Bergmann, Kirsten 508
Betz, Christoph 9
Bidot, Julien 17
Biundo, Susanne 17
Breuss, Michael 419
Buschmeier, Hendrik 508
Butz, Martin V. 193, 460

Cesta, A. 49
Chumerin, Nikolay 339
Combaz, Adrien 339
Condell, Joan 265

Debes, Klaus 395
de Boer, Viktor 136
Debray, Bruno 564
de Melo, Gerard 281
Dengel, Andreas 249, 581, 664
Dietrich, Dominik 419, 444
Dietzfelbinger, Martin 33
Doktorski, Leo 73

Ebadi, Hamid 639
Eberling, Markus 548
Edelkamp, Stefan 1, 33
Eggert, Julian P. 597
Eilers, Dirk 411
Eminov, Mubariz 427

Finzi, A. 49
Flötteröd, Gunnar 532
Fratini, S. 49
Fürnkranz, Johannes 65

Gadzicki, Konrad 331
Gärtner, Thomas 153
Ganascia, Jean-Gabriel 371
Garbacz, Paweł 379
Geßler, Sascha 17
Goesele, Michael 161
Gokce, Nida 427
Gottfried, Björn 572
Gretton, Arthur 144
Gross, Horst-Michael 395, 484, 597
Guhe, Markus 323
Günther, Maik 185

Hanser, Eva 265
Hartanto, Ronny 41
Hees, Jörn 249
Hein, Tobias 727
Heinrich, Gregor 161
Hellbach, Sven 597
Helmert, Malte 9
Hertzberg, Joachim 41
Hilkenmeier, Frederic 680
Hoder, Kryštof 435
Hollink, Vera 136
Hudlet, Volker 664
Hund, Marcus 97

Iqbal, Taswar 347

Jegelka, Stefanie 144
Jung, Bernhard 492

Kalesse, Sören 484
Kamran, Ali 719
Kerber, Manfred 656
Kern-Isberner, Gabriele 273
Khaluf, Lial 648
Kienberger, Julian 500
Kipp, Michael 524
Kissmann, Peter 1
Klubertanz, Georg 631
Kluegl, Peter 233
Klügl, Franziska 631
Knieper, Tobias 648
Knorr, Rudi 411
Koch, Patrick 177
Kolarow, Alexander 597
Kolomiyets, Oleksandr 225
Kopp, Stefan 508
Körner, Edgar 597
Kramer, Oliver 169, 177, 727
Kruijff, Geert-Jan M. 241
Kruse, Michael 648
Kruse, Rudolf 476
Kuhn, Lukas 403
Kulicki, Piotr 379
Kurata, Yohei 452

Lämmel, Gregor 532
Lamotte-Schubert, Manuel 281
Langer, Falk 411
Le, Dinh Khoi 347
Lechniak, Marek 379
Lehmann, Nico 648
Lenne, Dominique 564
Lermontow, Alex 648
Lin, Zuoquan 615
LinShu, He 719
Lipinski, John 257
Lison, Pierre 241
Liwicki, Marcus 581
Lohweg, Volker 347
Lunney, Tom 265

Maier, Paul 403
Malaka, Rainer 711
Mandow, L. 25
Manyakov, Nikolay V. 339
Mattmüller, Robert 57
Matyas, Christian 703
Matyas, Sebastian 703
Maus, Heiko 664
Mc Kevitt, Paul 265
Merken, Patrick 339
Mertsching, Bärbel 89, 97, 315
Mesgari, Mohammad Saadi 639
Messinger, Christian 648
Michaelsen, Eckart 73
Miksatko, Jan 524
Missura, Olana 153
Moens, Marie-Francine 225
Mohammed, Ammar 695
Müller, Michael 128
Müller, Steffen 484

Neumann, Florentin 355
Neves, Herc P. 339
Nguyen, Nhung 516
Nissen, Volker 185
Nolte, Michael 347

Orlandini, A. 49

Panitzek, Kamill 65
Pardowitz, Michael 589, 607
Park, Sang-Hyeun 65
Pauli, Josef 128
Pease, Alison 323
Pedersen, Gerulf K.M. 193, 460
Pérez de la Cruz, J.L. 25
Petersen, Nils 106
Petruseva, Silvana 672
Pfeiffer-Leßmann, Nadine 540
Pommerening, Florian 468
Porzel, Robert 711
Pourreza, Hamid Reza 114
Prade, Henri 306
Priess, Heinz-Werner 123
Puppe, Frank 233

Rafique, Amer Farhan 719
Reichenberger, Andrea 355
Reichle, Meike 556
Reithinger, Norbert 209
Richard, Gilles 306
Richter, Simon 648
Rindsfüser, Guido 631
Ritter, Helge 589, 607
Roller, Roland 209
Roth-Berghofer, Thomas 664
Rudolph, Günter 169
Rühr, Thomas 403

Sachenbacher, Martin 403
Sadeghipour, Amir 508
Sandamirskaya, Yulia 257
Schaffernicht, Erik 395
Scharlau, Ingrid 81, 123, 680
Schattenberg, Bernd 17
Schäufler, Christian 363
Scheffler, Tatjana 209
Schiller, Marvin 289
Schmidt, Thomas 648
Schölkopf, Bernhard 144
Schöner, Gregor 257
Schwarz, Christian 695
Schweizer, Immanuel 65
Seipel, Christoph 648
Senftleben, Dennis 128
Shad, Rouzbeh 639
Shafik, M. Salah E.-N. 89
Shi, Hui 452
Smaill, Alan 323
Sriperumbudur, Bharath K. 144
Stalph, Patrick O. 193
Steffen, Jan 589, 607
Steinbrecher, Matthias 476
Stephan, Volker 395
Strass, Hannes 298
Stricker, Didier 106
Suda, Martin 281
Sutcliffe, Geoff 281
Suykens, Johan A.K. 339
Swars, Daniel 648

Thielscher, Michael 298
Tönnies, Hauke 201
Torfs, Tom 339
Tronci, E. 49
Trypuz, Robert 379
Tünnermann, Jan 680

Vafaeenezhad, Alireza 639
van Elst, Ludger 249
Van Hoof, Chris 339
Van Hulle, Marc M. 339
van Someren, Maarten 136
Vogt, David 492
Volkhardt, Michael 484
Volkov, Vladimir 623
von Luxburg, Ulrike 144
Voronkov, Andrei 435

Wachsmuth, Ipke 516, 540
Wagner, Markus 687
Wang, Hongqi 153
Wang, Rui 217
Weber, Markus 581, 664
Weiß, Katharina 81
Westphal, Matthias 468
Wischnewski, Patrick 281
Wissner, Michael 500
Wölfl, Stefan 468
Wrobel, Stefan 153
Wullinger, Peter 703

Yazicioglu, Refet Firat 339

Zeeshan, Qasim 719
Zhang, Xiaowang 615
Zhang, Yi 217
Ziegler, Martin 355